Introduction to Basics of Pharmacology and Toxicology

Mageshwaran Lakshmanan •
Deepak Gopal Shewade •
Gerard Marshall Raj
Editors

Introduction to Basics of Pharmacology and Toxicology

Volume 3 : Experimental Pharmacology : Research Methodology and Biostatistics

Springer

Editors
Mageshwaran Lakshmanan
Department of Pharmacology
Thanjavur Medical College
Thanjavur, Tamil Nadu, India

Deepak Gopal Shewade
Department of Pharmacology
Jawaharlal Institute of Postgraduate Medical Education and Research (JIPMER)
Puducherry, Pondicherry, India

Gerard Marshall Raj
Department of Pharmacology
All India Institute of Medical Sciences (AIIMS) Bibinagar
Hyderabad, Telangana, India

ISBN 978-981-19-5342-2 ISBN 978-981-19-5343-9 (eBook)
https://doi.org/10.1007/978-981-19-5343-9

This Springer imprint is published by the registered company Springer Nature Singapore Pte Ltd.
The registered company address is: 152 Beach Road, #21-01/04 Gateway East, Singapore 189721, Singapore

Dedicated to the "three institutes" that made me what I am now—Govt Tuticorin medical College, JIPMER, and Thanjavur Medical College.
Mageshwaran Lakshmanan
Thanjavur, Tamil Nadu, India
May 2022

Dedicated to all my mentors including my parents (for their love, care, and affection), my darling "chinnu" @ Dr. Rekha (for her unconditional love and exceptional support), and finally to my little wonder Gershin and cutie pie Riya (for lending their time).
And in loving memory of my dear Balabalajee Sir!
Gerard Marshall Raj
Hyderabad, India
May 2022

Foreword

I take pleasure in writing the Foreword for Volume III of *Introduction to Basics of Pharmacology and Toxicology* being edited by Drs. Mageshwaran Lakshmanan, Deepak Gopal Shewade, and Gerard Marshall Raj. This particular volume deals with experimental pharmacology, biochemical pharmacology (instrumentation in drug analysis), research methodology, and biostatistics.

At present, due to several reasons like advances in biotechnology, the internet, and an improvement in the economy, leading to great numbers of scientific researchers, biomedical research is advancing at an unprecedented pace and rate. This research also involves basic and clinical pharmacology and new drugs are coming out into the market at a rapid pace. In this light, the contents of this volume will be very useful for those in the business of studying about and bringing out new drugs. This is because this volume covers many important topics in experimental pharmacology, several important drug screening methods, and instrumentation in drug analysis. In addition, the book has two important sections, one on research methodology and the other on biostatistics.

This volume, as well as the other three volumes of this work, will be of great value to M.D. students in pharmacology as well as M.Pharm. students, M.Sc. pharmacology students, Ph.D. in pharmacology students, as well as all others engaged in teaching and researching in pharmacology and its allied sciences.

Department of Pharmacology
and Clinical Pharmacology,
Christian Medical College
Vellore, Tamil Nadu, India

Jacob Peedicayil

Preface to Volume 1: General and Molecular Pharmacology: Principles of Drug Action *(https://link.springer.com/book/10.1007/978-981-32-9779-1)*

It was in the mid-2016s, when I had been in the phase of transformation from a postgraduate student to a teaching faculty, this idea of writing a book especially for M.D. Pharmacology postgraduates came out from nowhere. Though it was a light-bulb thought, I had been ruminating on that since then.

And it was in the late 2017s and early 2018s, I had actually started working on this herculean task. The first thing I did was to browse through the various syllabi of Indian Universities imparting M.D. Pharmacology course. I could retrieve (online search) around 20 syllabi of Universities in India from the States of Bihar, Delhi, Karnataka, Kerala, Madhya Pradesh, Maharashtra, Odisha, Puducherry (UT), Punjab, Tamil Nadu, Uttar Pradesh, and West Bengal. Then, I sifted through the course contents and also the recommended reference books and textbooks to finalize the topics to be covered.

The ultimate goal of this book is to cater to a prospective M.D. Pharmacology postgraduate appearing for his/her formative or summative (final) assessment examinations. Hence, I had divided the book into exam paper-wise volumes.

Volume 1: General and Molecular Pharmacology: Principles of Drug action

- *deals about General Pharmacology*

Volume 2: Essentials of Systemic Pharmacology: From Principles to Practice

- *deals about Systemic Pharmacology*

Volume 3: Experimental Pharmacology: Research Methodology and Biostatistics

- *deals about Experimental Pharmacology*

Volume 4: Pharmacology and Therapeutics

- *deals about Clinical Pharmacology*

For this project, I had included five other contributors from the field of Pharmacology, namely, Dr. Mageshwaran, Dr. Abialbon, Dr. Avinash, Dr. Neel, and Dr. Nishanthi. They are the major contributors of this four-volume book and also serve as the primary editors of the other three volumes. I know them both personally and professionally for the past 8–10 years. They are a bunch of young-yet-proficient and enthusiastic academicians whom I suppose could satisfy the demands of this project. They are currently working in different medical institutes and pharmaceutical industries around the globe.

This book **(Volume 1: General and Molecular Pharmacology: Principles of Drug action)** is divided into the following five parts.

Part I on *Historical Aspects of Drug Discovery* comprises renowned contributions to the field of Pharmacology by personalities both from the Indian arena and from world over—including the Nobel Laureates.

Part II on *General Pharmacological Principles* discusses in detail about the various facets of general pharmacology ranging from sources of drugs; routes of drug administration; basic pharmacokinetics including drug transporters and pharmacodynamics including adverse effects; drug interactions to structure–activity relationships.

Special Topics in Pharmacology encompassing drug information; pharmacogenetics; chrono- and ethno-pharmacology; pharmaco-epidemiology and -economics; orphan drugs; fixed-dose combinations; translational and reverse pharmacology; gene and stem cell therapy and principles of prescription writing are compiled under **Part III**.

Part IV is about *Toxicology* including two chapters on environmental toxicology and basic principles of management of drug poisoning.

Part V deals about *Molecular Biology in Pharmacology* with chapters on PCR, blotting techniques, antisense oligonucleotides among others.

The M.D. Pharmacology postgraduates rely on multiple texts for their assessment examinations which could be highly time-consuming, especially during the preparation period. Moreover, by referring to different texts of varied patterns can at times be extremely distracting to a potential exam-going student. Hence, a review book of this sought can be in a way lessen the arduous task of referring to numerous books or materials with mixed patterns.

The book could also be of use to young pharmacologists working in different job portfolios ranging from teaching-research faculty in academia, pursuing medical advisory roles in pharmaceutical industry to drug safety physicians in pharmacovigilance sectors.

The key features of our review book are

- Point-wise listing of facts.
- Inclusion of around 145 figures, tables, and boxes.

These features make the book more concise and precise. Thereby, they make the reading all the more easier and reproducible.

Wishing you all a very happy reading!

Puducherry, India
April 2019

Gerard Marshall Raj

Preface to Volume 2

This book aims to present the most important information and concepts relating to the clinical use of drugs in a concise yet comprehensive way. The use of chemical substances to modify altered pathophysiology towards normalcy has been one of the major modes of clinical management of human diseases. A good understanding of pharmacodynamic mechanisms of actions would not only be helpful to understand a drug's use in a particular disease but also to theorize possible adverse drug reactions and possible off-label uses. Pharmacokinetic aspects of a drug help select an appropriate drug in relevant conditions and thereby personalize the treatment based on the disease, patient, and drug factors.

In the words of the late Dr. Steven A. Dkhar, my professor, getting to know a drug is akin to getting to know a person. A drug has its positives and its negatives. While the positives are what we desire and shun the negatives, the fact that you really cannot separate both will eventually dawn on you. Understanding the negatives helps better utilize drugs or, rather, avoid the unnecessary use of drugs. Hence a balanced approach to understanding the drugs would benefit the budding pharmacologist.

The properties of clinically useful drugs (both pharmacokinetic and pharmacodynamic) have been presented in a concise form rather than verbose text. Concepts related to drugs' use have been explained with diagrams whenever needed. Considerations that would help put the knowledge obtained to clinical practice are listed whenever possible. The drugs are dealt with based on organ systems whose pathology they are most likely to treat. Since most of them have actions on multiple organ systems, care has been taken to present the spectrum of action of a drug on multiple systems and how it contributes to the adverse drug profile of the drug.

This book is possible owing to the result of many brilliant minds working together as a team. This book along with three other volumes has been the result of the passion and hard work of Dr. Gerard Marshall Raj. He has been instrumental in bringing the project to fruition. The other main contributors are Dr. Magheshwaran, Dr. Nishanthi, Dr. Neel Shah, and Dr. Avinash. Various others have also contributed individual chapters. The accumulation of such knowledge and wisdom is not possible if not for the numerous great minds who have toiled to discover/study these pharmacological phenomena and record them diligently.

The key features of our review book are:

- Point-wise listing of facts.
- Inclusion of around 417 figures, tables, and boxes.

While this book is primarily targeted to postgraduates specializing in pharmacology, it can be a handy and useful reference to anyone who is involved in clinical medicine. We hope this book is of special use to postgraduates for their exam preparation.

Saint Kitts and Nevis, West Indies
January 2020

Abialbon Paul

Preface

Experimental pharmacology is an ever-evolving branch of pharmacology. From traditional animal experimentation to modern high-throughput screening techniques, the subject of experimental pharmacology has seen drastic progression in research medicine. In addition, various international ethical regulations and guidelines have been implemented over the past five decades to improve the quality of research. This book has up-to-date information regarding the recent developments in various fields of experimental pharmacology and standard guidelines for the same.

Besides experimental pharmacology, a budding pharmacologist should also acquire sufficient knowledge in analytical pharmacology, research methodology, and biostatistics to carry out research work in flawlessly scientifically valid methods. Thus, this book has been designed to integrate various research sector topics as a primary goal. All the topics have been discussed in-depth and written pointwise for easy recollection of postgraduates during exam preparation.

The chapters of various sectors in research in this book were written by experts in respective fields. Prof. Shewade, Dr. Gerard, and I have scrutinized all the chapters and made them into concise, apothegmatic, and absorbing texts.

The key features of our review book are:

- Point-wise listing of facts.
- Inclusion of around 270 figures, tables, and boxes.

I hope you enjoy the book and recollect the points efficiently.

Thanjavur, Tamil Nadu, India — Mageshwaran Lakshmanan
November 2021

Acknowledgments

I place my wholehearted gratitude to the publisher Springer Nature for having given the due shape to this proposal. Foremost, I wish to express my special thanks to Dr. Gaurav Singh (the former Acquisition Editor) who provided valuable suggestions during the initial stages of this project. I would also wish to extend my sincere regards to Ms. Raman Shukla, the Commissioning Editor, Mr. Jayesh Kalleri, the Production Editor and their team for their concerted efforts in preparing this book.

As the corresponding editor, I am greatly indebted to my friend and co-editor Dr. Mageshwaran who took the primary responsibility of editing this volume and to the editors of other volumes of this book, namely, Dr. Abi, Dr. Nishanthi, Dr. Avinash, and Dr. Neel—but for them, the whole proposal would not have seen the light of the day. I am equally indebted to all the 46 contributors of this book for extensively devoting their precious time and mind.

I also place my sincere regards and profound thankfulness to Retd. Senior Professor (Dr.) Deepak Gopal Shewade from the Department of Pharmacology, Jawaharlal Institute of Postgraduate Medical Education and Research (JIPMER) for his continuous guidance as a senior editor of this book.

I would also like to acknowledge Professor (Dr.) Jacob Peedicayil from the Department of Pharmacology and Clinical Pharmacology, Christian Medical College, Vellore, India, for readily accepting to write a Foreword for this book.

I also thank Dr. R. Raveendran, Senior Professor of the Department of Pharmacology, JIPMER, for his support in plagiarism detection.

I would like to extend my deep sense of gratitude to Prof. (Dr.) Vikas Bhatia, Executive Director, AIIMS Bibinagar, Hyderabad, for being a patron of par-excellence and constantly guiding us in all our endeavors including this book. I would also like to sincerely thank Prof. (Dr.) Rahul Narang, Dean, and Prof. (Dr.)

Neeraj Agarwal, Medical Superintendent, AIIMS Bibinagar, Hyderabad, for their continuous motivation in all our academic pursuits.

Nothing is possible without the Divine Grace ("GOD's love is so wonderful!"). Therefore, I thank the Almighty for showering good health and well-being to complete this book.

Gerard Marshall Raj
Hyderabad, Telangana, India
May 2022

Contents

Editors and Contributors

About the Editors

Mageshwaran Lakshmanan has completed his M.B.B.S. degree from Tuticorin Government Medical College (Affiliated to The Tamil Nadu Dr. M.G.R. Medical University) and M.D. Pharmacology from JIPMER, Puducherry, India. He is currently working as Senior Assistant Professor of Pharmacology at Thanjavur Medical College, Tamil Nadu, India, and has 8 years of teaching experience. He is involved in research activities related to preclinical toxicity studies, drug interactions, and therapeutic drug monitoring. He is a member of the Pharmacovigilance Unit, the Institutional Ethical Committee and the Medical Education Unit of Thanjavur Medical College. In addition to his Departmental activities, he is also the Academic Officer In-Charge of Thanjavur Medical College. He manages undergraduate, postgraduate medical, and paramedical student admission and examinations, coordinates IEC, and acts as nodal officer of scholarship programs and ECFMG. He is well experienced in using various statistical tools and orients the postgraduates of all departments with the research methodology and bio-statistical concepts. He has many publications in various national and international journals.

Deepak Gopal Shewade is a retired Senior Professor of Pharmacology from JIPMER, Puducherry. He had joined as Assistant Professor in the same Institute in 1986. Earlier, he worked at BITS, Pilani, for 7 years as faculty and completed his Ph. D. He did M. Pharm. (Pharmacology) from Nagpur University. His areas of interest are pharmacogenomics, rational drug therapy, analgesics, and GIT pharmacology. He has completed an Indo-French project under ICMR-INSERM on warfarin dose prediction based on genetic and environmental factors. He has more than 100 publications. He has guided 10 M.D. and 8 Ph.D. students. He has visited Thailand, Uganda, and France for academic and research purposes. In 2004, he was deputed to Uganda by the Ministry of External Affairs to start B. Pharm. course at MUST, Mbarara. As per Google Scholar, his h-index is 21 and i10-index is 35.

Gerard Marshall Raj holds an M.B.B.S. and M.D. in Pharmacology. Currently, he is working as an Assistant Professor in the Dept. of Pharmacology, All India Institute of Medical Sciences (AIIMS), Bibinagar, Hyderabad, Telangana, India. He has contributed and co-edited the books *Introduction to Basics of Pharmacology and Toxicology Volume 1: General and Molecular Pharmacology: Principles of Drug action* (Springer Nature; ISBN: 9789813297784) and *Introduction to Basics of Pharmacology and Toxicology Volume 2: Essentials of Systemic Pharmacology: From Principles to Practice* (Springer Nature; ISBN: 9789813360082). He has also contributed book chapters in other books. Currently, he is the Faculty In-Charge of Hospital Pharmacy and Hospital Management Information System (HMIS) at AIIMS, Bibinagar. He has received multiple prizes for academic and research activities—including the *Dr. Peter Sleight Memorial Award* for outstanding individual performance and *Dr. Salim Yusuf Team Award* for best team performance during a course on "Randomized Controlled Trials" organized by St. John's Medical College and Research Institute, recently. He has acted as a resource person in various workshops, training programs, CMEs, and delivered invited talks as a guest speaker. He was awarded the "Prof. M. N. Ghosh Gold Medal" for outstanding performance during the M.D. Pharmacology course at JIPMER, Puducherry, and has published many research articles, reviews, commentaries, case reports, and letters-to-the-editor in peer-reviewed and indexed journals. He has acted as a reviewer for several elite journals and also for the *National Institute for Health Research (NIHR), UK*—Global Health Research Unit and Groups Programme scrutinizing funding research proposals.

Contributors

Mangaiarkkarasi Adhimoolam Department of Pharmacology, Sri Venkateshwaraa Medical College Hospital and Research Centre, Puducherry, India

Nishanthi Anandabaskar Department of Pharmacology, Sri Manakula Vinayagar Medical College and Hospital, Puducherry, India

Vimala Ananthy Department of Pharmacology, MGMC & RI, Puducherry, India

Avinash Arivazhahan Sanofi India Ltd., Chennai, Tamil Nadu, India

Sivaranjani Arumugam Department of Micro-Nano Mechanical Science and Engineering, Graduate School of Engineering, Nagoya University, Nagoya, Japan

Jayapriya Balamurugan Department of Pharmacology, Virudunagar Medical College, Virudunagar, Tamil Nadu, India

Lakshmi Balasundaram Department of Pharmacology, Jawaharlal Institute of Postgraduate Medical Education & Research (JIPMER), Puducherry, India

Bertilla Benjamin Community Health Centre, Irikkur, Kannur, Kerala, India

Rammohan Bondha Department of Pharmacology, Anna Medical College, Montagne Blanche, Mauritius

Parayil Varghese Christapher Department of Pharmacology, AIMST University, Bedong, Kedah Darul Aman, Malaysia
Department of Pharmacology, Al Shifa College of Pharmacy, Perinthalmanna, Malappuram, Kerala, India

Madhavi Eerike Department of Pharmacology, All India Institute of Medical Sciences (AIIMS) Bibinagar, Hyderabad, Telangana, India

Dhivya Elango Department of Pharmacology, Jawaharlal Institute of Postgraduate Medical Education and Research (JIPMER), Karaikal, Puducherry, India

Selvalaxmi Gnanasegaran Department of Pharmacology, Vinayaka Mission's Medical College and Hospital, Karaikal, Puducherry, India

Lourdu Jafrin Department of Pharmacology, Indira Gandhi Medical College and Research Institute, Puducherry, India

Ilanthamizhan Jayakumar Department of Pharmacology, Jawaharlal Institute of Postgraduate Medical Education & Research (JIPMER), Puducherry, India

Manu Jose Department of Pharmacology, Government Medical College, Kozhikode, Kerala, India

Sushil Kiran Kunder Novo Nordisk India Ltd., Bengaluru, Karnataka, India

Mageshwaran Lakshmanan Department of Pharmacology, Thanjavur Medical College, Thanjavur, Tamil Nadu, India

B. Maharani Department of Pharmacology, Indira Gandhi Medical College and Research Institute, Puducherry, India

Aarthi Manoharan Multi-Disciplinary Center for Biomedical Research, Vinayaka Mission's Research Foundation, Aarupadai Veedu Medical College and Hospital, Puducherry, India

Rupadevi Muthaiah Department of Pharmacology & Toxicology, Jacobs School of Medicine and Biomedical Sciences, The New York State University at Buffalo, Buffalo, NY, USA

R. Padmapriya Department of Pharmacology, Jawaharlal Institute of Postgraduate Medical Education & Research (JIPMER), Puducherry, India

Subramani Parasuraman Department of Pharmacology, AIMST University, Bedong, Kedah Darul Aman, Malaysia

Abialbon Paul Department of Pharmacology & Clinical Skills, Medical University of the Americas, Charlestown, Nevis, West Indies

Rekha Priyadarshini Department of Pharmacology, All India Institute of Medical Sciences (AIIMS) Bibinagar, Hyderabad, Telangana, India

Gerard Marshall Raj Department of Pharmacology, All India Institute of Medical Sciences (AIIMS) Bibinagar, Hyderabad, Telangana, India

Charukesi Rajulu Department of Biotechnology & Bioinformatics, JSS Academy of Higher Education & Research, Bannimantap, Mysuru, Karnataka, India

Kirubakaran Ramakrishnan Department of Clinical Pharmacology, Jawaharlal Institute of Postgraduate Medical Education & Research (JIPMER), Puducherry, India

Muthu Meera Ramasubramanian Department of Pharmacology, Velammal Medical College Hospital and Research Institute, Madurai, Tamil Nadu, India

Gomathi Ramaswamy Department of Community Medicine & Family Medicine, All India Institute of Medical Sciences (AIIMS) Bibinagar, Hyderabad, Telangana, India

Ramasamy Raveendran Department of Pharmacology, Jawaharlal Institute of Postgraduate Medical Education & Research (JIPMER), Puducherry, India

Gandham Ravi Department of Pharmacology, All India Institute of Medical Sciences (AIIMS) Bibinagar, Hyderabad, Telangana, India

S. C. Santosh Kumar Department of Pharmacology, Jawaharlal Institute of Postgraduate Medical Education & Research (JIPMER), Puducherry, India

Jhananey Senthilvelan Department of Internal Medicine, Chengalpattu Govt. Medical College, Chengalpattu, Chennai, Tamil Nadu, India

Neel Jayesh Shah Quality & Compliance Services Inc., Mississauga, ON, Canada

Kiruthika Sivagourounadin Department of Pharmacology, Jawaharlal Institute of Postgraduate Medical Education & Research (JIPMER), Karaikal, Puducherry, India

Umamaheswari Subramanian Department of Pharmacology, Sri Venkateshwaraa Medical College Hospital & Research Centre, Puducherry, India

B. P. V. Sunanda Department of Pharmacology, Anna Medical College, Montagne Blanche, Mauritius

Arun Kumar Tharkeshwar Laboratory of Neurobiology, VIB-Center for Brain & Disease Research, Leuven, Belgium

Amod Tilak Pfizer India Ltd., Mumbai, Maharashtra, India

Amruta Tripathy Department of Pharmacology, All India Institute of Medical Sciences (AIIMS), Bhubaneswar, Odisha, India

Vinayagamoorthy Venugopal Department of Community and Family Medicine, All India Institute of Medical Sciences (AIIMS), Deoghar, Jharkhand, India
Department of Community Medicine, Sri Manakula Vinayagar Medical College and Hospital, Puducherry, India

Saranya Vilvanathan Department of Pharmacology & Clinical Skills, Medical University of the Americas, Charlestown, Nevis, West Indies

Mourouguessine Vimal Department of Pathology, Sri Manakula Vinayagar Medical College and Hospital, Puducherry, India

Suja Xaviar Department of Pharmacology, Jawaharlal Institute of Postgraduate Medical Education & Research (JIPMER), Puducherry, India

Alphienes Stanley Xavier Department of Pharmacology, Sri Ramachandra Medical College and Research Institute, SRIHER (Deemed University), Chennai, Tamil Nadu, India
Clinical Pharmacology, Sri Ramachandra Medical Centre, Porur, Chennai, Tamil Nadu, India

Part I

Experimental Pharmacology

Experimental Methodologies Involved in the Discovery of Drugs

1

Abialbon Paul

Abstract

The drug discovery process has evolved from a simpler process identifying active ingredients in traditional medications to searching complex and large chemical libraries for specific molecules with identified or predicted drug binding properties. Such progress has only been made possible due to the advancements in Pharmacological sciences including better understanding of structure activity relationships and quantification of molecular interactions. Drug discovery techniques can be broadly divided into 'in silico', 'in vitro' and 'in vivo' techniques. The *in silico* techniques have made good use of the data available from the human genome project and receptor structure data to predict and simulate molecular level binding and interactions. Hence, they help select potential candidates for preclinical studies from larger chemical libraries in shorter turnaround times than ever before. Various preclinical studies are now conducted in cell lines as miniature assays that further improve the candidate molecule selection process. Having better leads and candidates reduces the risk of failure further down in the drug discovery process saving money and time. The following chapter discusses some of these assays that are routinely used as a part of the drug discovery process.

Keywords

Drug discovery · Assays · Methods · In vivo · In silico · In vitro

A. Paul (✉)
Pharmacology & Clinical Skills, Medical University of the Americas, Nevis, Saint Kitts and Nevis
e-mail: a.paul@mua.edu

M. Lakshmanan et al. (eds.), *Introduction to Basics of Pharmacology and Toxicology*, https://doi.org/10.1007/978-981-19-5343-9_1

1.1 Basic Principles of Drug Discovery

Drug discovery is a complex and challenging process. In the earlier days, drug discovery was principally identifying the active principles present in various traditional remedies or from natural sources, often serendipitously. Our understanding of human biology has come a long way since then. The complete sequencing of the human genome after the Human Genome Project has made cloning various biological target molecules possible. Concurrent advancements in pharmacological sciences have led to a better understanding of structure-activity relationships and better quantification of molecular interactions.

Drug discovery is now a highly organized capital intensive process. Large libraries of chemical molecules are categorized and stored in chemical libraries that can be used to rapidly screen and identify leads for further development. Such chemical libraries enable high throughput screening. High throughput screening refers to the large number of chemical molecules that are screened in a shorter time than conventional methods. Since chemical libraries are the cornerstone of preclinical drug development, the branch of combinatorial chemistry has developed to a large extent and deals with developing and maintaining such large libraries. Chemical compounds are maintained as individual compounds or as mixtures and in differing concentrations. Various types of chemical synthetic methods are used to create a large number of products. One such is called divergent synthesis. It is a chemical synthetic strategy that tries to improve the efficacy of chemical synthesis. For example, in diversity-oriented synthesis compounds which can generate a large number of products on reaction with different reagents are stored as skeletons. Hence, a lesser number of molecules can be stored without sacrificing the diversity a chemical library needs. Various processes involved in such high throughput screening and lead optimization in preclinical drug development are discussed in depth in Volume 4.

The rest of the chapter discusses briefly some of the *in-silico* and *in-vitro* techniques used in drug discovery. The rest of the book contains descriptions of *in-vivo* techniques.

1.2 *In-silico* Techniques

The term *'in-silico'* refers to the mode of experimentation by computers. *In-silico* pharmacology is also called *computational therapeutics* or *computational pharmacology*. It is a rapidly growing area that uses data from historical records and biological experiments to run computational simulations and create models that can predict, hypothesize, and help in the discovery of newer drugs and therapeutics.

Advantages:

- Finding new agonists and antagonists by mining chemical libraries using the data collected by in silico techniques like QSAR.
- Reduces the number of actual *in-vitro* and *in-vivo* trials done by eliminating non-effective compounds.

***Disadvantages*:**

- Protein flexibility is not well accounted for in many computational models.
- The data used for computational models come from many *in-vitro* studies and *in -vivo* studies on lower animals rather than from humans. Hence, the model which is predicted may not always be well suitable for humans, thus reducing the applicability of the model.

Some of the *in-silico* methods are discussed below.

1.2.1 Quantitative Structure-Activity Relationship (QSAR)

- A mathematical model is constructed that can quantify the biological effect of a chemical substance based on the chemical nature of the substance.
- The model is created by running multiple simulations of the historical data we have based on various ligand-receptor interactions. If a good mathematical model can be constructed, then this model can be used to predict the effects of new chemical compounds before they are synthesized or tested on the biological medium.
- C-QSAR is a database containing more than 18,000 equations that predict the biological effects based on the physio-chemical properties of drugs.
- QSAR models are routinely employed to predict the toxic effects of a drug. For example, if a chemical has an aromatic amine or an epoxide in its chemical structure, it is more likely to be genotoxic. Such rules are learned by training the algorithm with a larger training set from databases.
- While QSAR tries to quantify the biochemical action with its physical and chemical action, it does not always provide reliable insight into the mechanism of action of these compounds. Various substances can act through various mechanisms to produce similar actions in a biological medium.

1.2.2 2D-QSAR and 3D-QSAR

- Descriptors are numerical representations of chemical structures. Various types of descriptors can be used to improve the model.
- The simple QSAR studies use one-dimensional descriptors like molecular weight, refractive index, partition coefficient, lipophilicity, etc. This model is simple but it may not always be predictive of biological phenomena.
- 2D-QSAR uses two-dimensional data. The topological representations of a chemical are used in the equation to create a 2D-QSAR model.
- 3D-QSAR brings more complexity to the model by using the 3D structure of the chemical to model its biological action. While the 3D-QSAR is more complex, it can have a better prediction of the action of various proteins and large drug molecules.
- Molecular alignment is very important in 3D-QSAR as chiral molecules can have different docking patterns, different affinity to the receptors, and differences in activating the receptor.

- Recent developments have resulted in the use of a lot of data and consequently higher dimensions of QSAR are now possible. 6D-QSAR is available.

1.2.3 Molecular Docking

- Molecular docking is the estimation of the binding affinity between a protein target (receptor) and a ligand. The free energy of the complex is estimated by running multiple simulations computationally.
- There are two types of molecular docking: receptor-based docking and ligand-based docking.
- In a receptor-based method, the structure of the target receptor is known. This information is used to screen compounds from a chemical library for an appropriate fit. The appropriateness is evaluated by a scoring function. The substances with higher scores are then further evaluated by other methods.
- In ligand-based methods, the target receptor is not known but there are known inhibitors available. Using the structure of the known receptor, the chemical library is screened for molecules that have a similar 3D structure and pharmacophore. These compounds are more likely to act on the target receptor where the inhibitor binds.

1.2.4 Virtual High Throughput Screening

- Virtual screening is similar to molecular docking in that a large number of chemical substances in the chemical libraries are screened for a potential fit for a target protein receptor/enzyme.
- This is relatively new and has now become an integral part of all new drug discoveries. It uses various techniques to automatically and quickly come up matches which can be further evaluated for leads.
- Virtual High Throughput Screening is less expensive than a conventional High Throughput Screening as the real assays need not be carried out. This also makes Virtual Screening faster and less expensive.

1.2.5 Pharmacophore Mapping

- A pharmacophore is a group of features in a chemical molecule along with their relative orientation in space that can interact with a target protein. Pharmacophore mapping is a process of identifying all the pharmacophores in a chemical entity and mapping them on a distance map.
- Pharmacophore mapping relies more on the atomic properties of the drug rather than on the chemical connectivity patterns.
- It has an advantage over QSAR that it takes into account the conformational flexibility of drugs that bind to receptors and takes into account all the pharmacophores in a drug.

- Pharmacophore mapping takes the two-dimensional data of the lower dimension QSAR to create a three-dimensional profile that can be used to search a large chemical database for matches.

1.3 *In-vitro* Techniques

In-vitro techniques literally mean tests conducted in glass containers. These tests are performed outside the biological environment in laboratory containers like test tubes, flasks, Petri dishes, and microtiter plates.

Difference Between *In-vitro* and *Ex-vivo*
It is important to note that *ex-vivo* also refers to outside the biological environment. However, *ex-vivo* refers to tests, procedures, and experiments performed on cells, tissue lines, or organs outside the body. *In-vitro*, on the other hand, refers to testing in test tubes with artificial media and not the original biological medium.

Some of the *in-vitro* techniques are described below.

1.3.1 Absorption Assays

- It is essential to confirm oral absorption before an oral drug can be formulated for clinical trials. Human colon carcinoma cells (Caco-2) cultured on a polycarbonate membrane as a monolayer can be used to measure absorptivity in the human intestine.
- The expression of CYP 3A4 on these cells can be enhanced to mimic the drug metabolism by the gut wall. The expression of P-glycoprotein can also be mimicked by this cell culture and can be useful to study potential P-gp substrates and inhibitors.

1.3.2 Metabolic Stability Assays

- Metabolic stability assays use microsomes or hepatocyte cultures to confirm the stability of the drugs after first-pass following oral administration.
- These assays are useful to estimate the half-life and clearance of the experimental drug. They also help in the estimation of the initial dose to experiment in dose-ranging studies.
- The effect of chemical and structural modification of the drugs can be confirmed by metabolic stability studies.

1.3.3 CYP450 Pathway Elucidation

- Pathway elucidation to identify the most common isoforms of the CYP enzymes that metabolize the drug is carried out later in the development of drugs. The most

common technique is the use of known substrates and inhibitors of these enzymes and their effect on the metabolism of the tested drug. Monoclonal antibody-based inhibition can also be carried out. Incubation with cDNA sequences is another technique for CYP450 pathway elucidation.
- The most important point to be considered in these assays is the genetic variability of these isoforms. The enzymes should be chosen from multiple donors to avoid the influence of one or a few genotypes on the results.

1.3.4 Inhibition Potential

- These are the assays that are designed to predict the drug-drug interaction due to a drug's potential to inhibit certain CYP450 isoforms.
- Recent techniques use a microplate with multiple enzymes embedded and all are tested and quantified at one by a high-throughput process. Various ways to measure drug metabolites are by the use of radionuclide isotopes, HPLC, and LCMS/MS.
- These studies are usually performed early in the drug development process so they can inform the safety of the drug in advance before in-depth clinical testing is done.

1.3.5 Induction Potential

- Similar to the inhibition potential, the induction potential a drug has on the various CYP isoform can be studied by induction assays.
- CYP3A4, CYP1A, CYP2C9, and CYP2E1 and known to be inducible in humans when exposed to certain substrates. Hence, these are the common enzymes studied for induction potential.
- Binding to the nuclear receptors PXR correlates well with CYP3A4 induction and hence can be used as a surrogate marker to determine the induction potential of the drug candidate.

1.3.6 Metabolic Profiling

- Metabolic profiling refers to the measurement of different types of metabolites produced in different organ systems in an organism and also among different species.
- Metabolic profiling is essential in early drug development to explain species-specific differences in toxicity. For example, if a drug does not produce a metabolite in a lower species but produces it in humans, the animal model cannot be used for toxicity studies.
- This is done by incubating the drug with a fraction of cell cultures from various organs and various species and measuring the end products with various analytical methods. Very precise measurements are made possible by the development of LCMS/MS.

- For example, in the development of efavirenz, nephrotoxicity was observed in rats. Nevertheless, metabolic profiling showed that the pathways of nephrotoxicity were absent in humans and hence the drug was not eliminated in early drug testing.

Limitations of Various In vitro Methods

- Do not replicate the physiologic conditions like body temperature, electrolyte levels, and extracellular composition.
- Do not mimic actual cell densities seen in organs
- Do not replicate cell-cell interaction as seen in complex organ systems. *In-vitro* studies often use only one cell line in an assay.
- Do not replicate an adequate supply of nutrients and other growth-inducing and inhibiting factors seen physiologically.
- Validation and acceptance by certain regulatory agencies of certain new assay types can be challenging.

1.3.7 Antisense Technology

- Antisense technology is a powerful technique for target validation. It uses chemically modified RNA which are oligonucleotides that are complementary to the target mRNA being studied.
- Binding of the antisense oligonucleotide to the target mRNA prevents the translation process and results in low or undetectable levels of the encoded protein.
- Example: Rats injected with P2X3 antisense oligonucleotides has no heightened pain responses confirming their role in chronic pain.
- Because the oligonucleotides need to be present to prevent proteins from encoding, their action is reversible when they are discontinued. This is an advantage compared to gene knockout models where the action is not easily reversible.
- Disadvantages include limited bioavailability of antisense oligonucleotides, the need for parenteral administration, toxicity, and possible non-specific unknown actions.

1.4 *In-vivo* Techniques

In-vivo techniques refer to tests and experiments done directly on the animal or living organism. While *in-vivo* techniques form the major focus of forthcoming chapters in this book, a few techniques are described below.

1.4.1 Gene Knockout Animals

- An animal with a gene or a gene fragment deleted from its genome is called a knockout animal. Mice are the most commonly preferred animals for knockout models as the entire mouse genome has been sequenced.

- Knockout animals give more information on the function of a normal gene and its role in various physiological and pathological pathways.
- For example, the P2X7 knockout mice helped in the confirmation of the role of this ion channel in neuronal inflammation and chronic pain. This led to target validation and helped in the search for newer drugs for chronic inflammatory diseases of the neurons.
- Knockout models tend to have a high mortality rate and most studies are feasible in the younger ones due to the shorter life span. The genes might have different functions and different expressions in various tissues than their human counterpart limiting the extrapolation of data to humans.
- The lethality of various knockout experiments is overcome by using inducible or tissue-specific gene knockouts. This is more technically challenging.

1.4.2 Transgenic Animals

- Transgenic animals have a gene or genes from another species artificially transferred into their genome. Mice are the predominantly used species. They help in understanding the gene function in terms of disease risk and progression.
- Transgenic animals are created either by the addition of genetic material by microinjection into the zygotes or by site-directed mutagenesis of the target gene by homologous recombination in the stem cells. When a new gene is injected into the zygote by microinjection, the organism is sometimes called a knock-in animal as opposed to a knockout animal with a gene deletion.
- Transgenic animals are advantageous as they allow phenotypic expression to be confirmed and the effect of the therapy on the phenotype.
- Transgenic mice are also used in the production of antibodies.
- The use of transgenic animals is time-consuming and very expensive.

1.4.3 Monoclonal Antibodies Mediated Gene Manipulation

- Monoclonal antibodies can be used to turn off the function of a receptor by blocking it. The specificity of the monoclonal antibodies is an advantage over the small molecule inhibitors as off-target side effects are less or non-existent with monoclonal antibodies.
- The disadvantages include immunological reactions, cost, and difficulty in administration.

1.5 Assays

An assay is a procedure that investigates the nature of a chemical substance and forms the backbone of drug characterization in the drug discovery process. There are exceedingly numerous assays that have been developed to identify various aspects of drugs. Some examples are given in the Table 1.1.

Table 1.1 Examples of assays used in drug discovery

Type of assay	Assay name	Notes
Toxicity assays	Brine shrimp lethality assay	Biologically active drugs are often toxic to the larvae of shrimp. Hence, potential drug candidates can be screened for biological activity by monitoring their toxicity to shrimp larvae. Artemia salina is commonly used for this purpose.
	Brine shrimp microwell cytotoxicity assays	This is a modified microplate assay using the eggs of Artemia salina.
Antimicrobial assays	Agar diffusion assay	The test organism is incubated in an agar plate and multiple wells are made in the agar to add known concentrations of the test substance. The zone of inhibition of the organism is measured to assay the antibacterial activity.
	Microtiter plate assay	Usually done on a 96-well microtiter plate containing two extracts of 12 microorganisms and multiple concentrations of test substances. This increases the throughput of the agar diffusion assay.
	Direct bioautography assay	This is a modification of the agar diffusion method in which the antibacterial compounds are transferred from the chromatographic plate to the agar and multiple zones of diffusions are compared and the best antibiotic is identified.
Antimitotic assays	Sea urchin egg assay	Antimitotic effects of various drugs like vinca alkaloids and podophyllotoxins can be demonstrated by observing the mitotic activity of sea urchin and starfish oocytes in the presence of the test sample.
Genotoxicity assays	SOS chromotest	This is a type of biochemical induction assay where the lysogenic induction of a lambda-lac z fusion phage by using Escherichia coli is tested. The SOS gene (sifA gene) is fused with the lac z gene and alterations in this gene can be identified by calorimetric methods.
	Ames test	The ability of drugs to cause mutations in strains of Salmonella typhimurium is tested. This is one of the best short term assays using bacteria. They are better at testing reverse mutations.
Hepatotoxicity assays	Carbon tetrachloride-induced cytotoxicity	The hepatoprotective effects of various plant extracts and chemical substances can be tested by their ability to reduce or prevent the hepatotoxicity caused by CCl_4 on hepatocyte cell lines or livers of mice.
	Galactosamine induced cytotoxicity	This method is an in vitro assay using D-galactosamine to cause hepatocyte toxicity.

Bibliography

Batool M, Ahmad B, Choi S. A structure-based drug discovery paradigm. Int J Mol Sci. 2019;20 (11):2783. https://doi.org/10.3390/ijms20112783.

Bylund DB, Enna S. Receptor binding assays and drug discovery. Adv Pharmacol. 2018;21–34 https://doi.org/10.1016/bs.apha.2017.08.007.

Dhuri K, Bechtold C, Quijano E, Pham H, Gupta A, Vikram A, Bahal R. Antisense oligonucleotides: an emerging area in drug discovery and development. J Clin Med. 2020;9 (6):2004. https://doi.org/10.3390/jcm9062004.

Eder J, Herrling PL. Trends in modern drug discovery. New Approaches to Drug Discovery. 2015;3–22 https://doi.org/10.1007/164_2015_20.

Lu X, Yang H, Chen Y, Li Q, He SY, Jiang X, Feng F, Qu W, Sun H. The development of pharmacophore modeling: generation and recent applications in drug discovery. Curr Pharm Des. 2018;24(29):3424–39. https://doi.org/10.2174/1381612824666180810162944.

Pinzi L, Rastelli G. Molecular docking: shifting paradigms in drug discovery. Int J Mol Sci. 2019;20(18):4331. https://doi.org/10.3390/ijms20184331.

Sacca R, Engle SJ, Qin W, Stock JL, McNeish JD. Genetically engineered mouse models in drug discovery research. Methods Mol Biol. 2009;37–54 https://doi.org/10.1007/978-1-60761-058-8_3.

Wang T, Wu MB, Lin JP, Yang LR. Quantitative structure–activity relationship: promising advances in drug discovery platforms. Expert Opin Drug Discovery. 2015;10(12):1283–300. https://doi.org/10.1517/17460441.2015.1083006.

Laboratory Animals 2

Mageshwaran Lakshmanan

Abstract

Though the current modern era has introduced equivalent alternatives for animal research like *in-silico*, *in-vitro* methods, and Computer-Assisted-Learning, animal experimentation cannot be overlooked entirely and put aside, especially in research. The 'hit molecule' should be administered to an animal before humans at some point in time for the concern of safety. Various mammals ranging from humble mice to large horses have been used in the experimental laboratory for drug research. Invertebrates, fishes like zebrafish, avians like pigeon and chicken have been used as an alternative to mammals. Every animal has its own merits and demerits. Moreover, the recent decade has seen the development of various genetic animal models that exhibit similar human pathological conditions. This chapter will review the salient feature, advantages, disadvantages, and genetic models of laboratory animals.

Keywords

Laboratory animal · Genetic model · Mouse · Rat · Guineapig

2.1 Introduction

For many decades, since modern science's evolution, animal experiments have been a fundamental part of research and medical teaching. Though the current modern era has introduced equivalent alternatives for animal research like *in-silico*, *in-vitro* methods, and Computer-Assisted-Learning, etc., animal experimentation cannot be overlooked entirely and put aside, especially in research. This is because the 'hit

M. Lakshmanan (✉)
Department of Pharmacology, Thanjavur Medical College, Thanjavur, Tamil Nadu, India

M. Lakshmanan et al. (eds.), *Introduction to Basics of Pharmacology and Toxicology*, https://doi.org/10.1007/978-981-19-5343-9_2

molecule' should be administered to an animal before humans at some point in time for safety concerns. Furthermore, the choice of an animal model with a close resemblance to humans for evaluating drugs is a vital step in drug development. Hence, the pharmacologist should be aware of the anatomical peculiarities, physiological similarities, biochemical resemblance, research suitability, and recent genetic modifications of various experimental animals.

The experimental animals are broadly divided into the following categories:

- Mammals—Rodents (Mouse, Rat, Guinea pig, Hamster, and Gerbil)
- Mammals—Non-rodents (Rabbit, Cat, Pig, Dog, Monkey, Sheep, and Horse)
- Avian (Pigeon, Chicken)
- Amphibian (Frog and toad)
- Pisces (Zebrafish)

2.2 Mouse

2.2.1 Scientific Name

- Genus: *Mus*
  - Species: *musculus* (Common), *famulus* (India), *domesticus* (Western Europe and Africa), *spretus* (Western Mediterranean), *fragalicauda* (Thailand), *macedonicus* (Eastern Mediterranean), and *spicilegus* (Central Europe)
- Genus: *Peromyscus*
  - Species: *maniculatus* (North American deer mouse), *leucopus* (American White-foot mouse)
- The genus *Mus* is commonly used, while *Peromyscus* is rarely used

2.2.2 Description

- Body Length: 6–9 cm
- Tail length: 6–10 cm
- Weight: 12–30 g
- Breeding season: Throughout the year
- Gestational age: 19–21 days
- Litter size: 3–12
- Sexual maturity: 5–7 weeks
- Average life span: 2 years

2.2.3 Salient Points of the Mouse as Experimental Models

- The mouse is one of the earliest animal model used in experimentation:
  - William Harvey in 1678 for studying circulation and reproduction
  - Joseph Priestley in the eighteenth century for studying respiration
  - Gregor Mendel in the nineteenth century for studying gene inheritance
- The mouse is the smallest experimental rodent that was used in the laboratory. Easy to keep, cheap, and requires only minimal space for housing
- Mice have the shortest gestational duration and hence, a good model for reproductive toxicity studies
- Commonly used in acute toxicity studies
- The mouse genome was sequenced in 2002, and it revealed 17,000 genes with identifiable human orthologs that can be manipulated to study more than 1000 human diseases. Mice have a similar genome to humans (>90% conservation) and hence an ideal animal of choice for genetic studies and inherited human diseases

2.2.4 Outbred Stocks of the Mouse

- *Swiss albino Mouse:*
  - It is created by the outbreeding of two albino males and seven females in Lausanne, Switzerland
  - As it is an outbred stock, it is extensively used in toxicological studies due to high variability between the animals
  - Not used as an animal model for specific genetic disease for which inbred or transgenic mice are a better model
- *J/NU Mouse:*
  - It is also called 'outbred-homozygous-athymic-nude mouse ($Foxn1^{nu}$/$Foxn1^{nu}$). It lacks thymus and hair due to a recessive mutation in the '*nu*' gene.
  - Tumor cells can be successfully transplanted in this strain and studied, as T cell development is absent due to lack of thymus.
  - It is a standard model *in vivo* model for anticancer drug testing.
- *CD-1 mouse:*
  - It is derived from the outbreeding of a group of Swiss albino mice in Lausanne, Switzerland.
  - It is commonly used for testing the reproductive toxicity of various chemicals, diphenylamine, atrazine, oxalates, carbaryl, etc., in chemical industries.
  - Despite albino status, the visual acuity of the CD-1 mouse is good.
- *ICR-mouse:*
  - The Institute of cancer research (ICR) mouse was also derived from Swiss albino outbreeding and selected as a separate mouse line for fertility.
  - It has a rapid growth rate with a high litter yield. The occurrence of spontaneous tumors is less in this outbred stock.

- Besides these, various outbred stock of mouse-like Carworth Farms-1 (CF-1), Swiss-Webster (SW), OF-1, Mouse-outbred-rock-only (MORE), and many more are available in the market but are uncommonly used.

2.2.5 Inbred Strains of the Mouse

- *C57BL/6J mouse:*
  - It is also called the B6/J mouse. C57BL/6J mouse is the most commonly used inbred strain. It is the first inbred mouse strain with successful complete genome sequencing.
  - They have low bone density when compared to other strains and develop loss of hearing upon aging. They are resistant to sound-induced-seizures and anthrax toxins but are susceptible to atherosclerosis, diet-induced-obesity, and diabetes mellitus-type 2.
  - B6/J mice are commonly used in transgenic mice production.
  - Variants like C57BLKS/J, C57BL/10J, C57BL/6NJ, C57BL/6J DIO, and C57BL/10SnJ are available in the market across the world and is used for specific research fields.
- *Non-obese diabetic mouse (NOD) mouse:*
  - It is also called as NOD/ShiLtJ strain. The NOD mouse is a polygenic model commonly used for type-1 diabetes (autoimmune-mediated).
  - It has a mutation in CTLA-4 gene-exon-2 that leads to the failure of suppression of T-cell immune response.
  - NOD mouse presents with hyperglycemia due to pancreatitis with the infiltration of leukocytes in the pancreatic islets. Defects in NK cell function, antigen presentation, and complement-5 malfunction lead to inflammation of the pancreas that mimics type 1 diabetes pathophysiology.
  - Female NOD mouse develop type-1 diabetes faster than males.
  - NOD mouse often presents with SCID and hearing impairment.
- *CB-17 Severe combined immunodeficiency (CB17-SCID) mouse:*
  - CB17-SCID is an albino mouse strain derived in Fox-Chase Cancer Centre in 2005.
  - Due to spontaneous mutation in $prkdc^{scid}$, both B cell and T cells are entirely underdeveloped in this strain.
  - Around 20% of CB17-SCID mouse exhibits significant low levels of IgG even at 12 weeks of age. Antibody response to a particular antigenic material is not mounted. This makes CB17-SCID an ideal model for anticancer drug evaluation.
- *Friend Leukemia Virus B (FVB) Mouse:*
  - FVB inbred strains are derived from the National Institute of health mice in 1970 by observing that some strains were susceptible to Friend-Leukemia-Virus (FVB).

  - FVB mice are blind. They have a homozygous mutation in the *PDE6B* gene resulting in loss of rods in the retina within 9 weeks of birth. Circadian rhythm is severely affected in this strain due to early blindness.
  - The oocyte of the FVB strain is large when compared to other strain, and hence they are an ideal strain for transgenic research.
  - Squamous cell carcinoma can be induced easily in this strain.
- *Murphy Roths Large (MRL) mouse:*
  - It is also called MRL/MpJ mouse. This strain is derived in the Wistar Institute of Philadelphia, the USA, in 1999.
  - They exhibit autoimmune disorders in their later stage of life.
  - MRL mouse strain is famous for its ability to regenerate the tissue at a faster rate without any scar on healing. In their skin, the hair follicles and the sebaceous gland complex also shows a higher degree of regeneration. Cardiac tissues can also be regenerated in this strain. Female MRL strains heal faster than males.
  - MRL mouse has an inbuilt capacity to resist muscular dystrophy. MRL strains are extraordinarily docile, and even the male MRL strains fight very rarely.
- *DBA-1 and DBA-2 mouse:*
  - DBA inbred strains were derived by Jackson Laboratory in 1909. DBA is one of the oldest inbred strains of mice. Subsequently, in 1929, two sub-strains, namely DBA-1 and DBA-2, were derived.
  - DBA-1 and DBA-2 are also called D1 and D2 mouse respectively. Both DBA-1 and DBA-2 are non-albino strains.
  - DBA-1 strains are commonly used as a model of evaluation of rheumatoid arthritis. After injection with collagen-type-II, the DBA-1 strains develop polyarthritis that is almost similar to humans. Synovitis, bone, and cartilage erosion are seen in the DBA-1 strain that mimics rheumatoid arthritis in humans.
  - DBA-2 strains are commonly used as a model of atherosclerosis, glaucoma, and cochlear pathology. DBA-2 strains mutation in the $Cdh23^{ahl}$ gene that results in progressive hearing loss by the third month of age. Upon aging, DBA-2 strains also develop glaucoma that closely mimics human hereditary glaucoma with the dispersion of iris pigmentation and atrophy, synechia, and elevated IOP.
  - DBA-2 strains show significant intolerance to morphine and alcohol. They are naturally CD94 deficient with the absence of expression of CD94/NKG2A receptors.
- *Ob/Ob mouse or obese mouse or B6-ob mouse:*
  - The B6-ob mouse strain was derived by a spontaneous mutation in V/Le strain with subsequent backcrossing for 45 generations in the Jackson Laboratory, Maine, in 1949.
  - The leptin gene (Lep^{ob}) is defective in this strain leading to hyperphagia, rapid weight gain, and obesity. The mouse gains weight three to four times higher than the regular mouse.

  - B6-Ob mouse exhibits hyperglycemia, hyper-insulinemia, altered hormonal secretion from the pituitary, and impaired wound healing, making this strain an ideal model of choice for evaluating metabolic syndrome, type-2 diabetes mellitus, and infertility.
  - This strain is hypothermic, hypo-metabolic, and sub-fertile.
  - Adipogenesis is increased in the bone marrow, and the bone mass of the hindlimbs are decreased in this strain.
- *Japanese Waltzing mouse:*
  - It is a 'Fancy' inbred mouse strain derived from the *Mus musculus* subspecies *molossinus* in Japan.
  - They have inherent cochlear defects, and hence they cannot move forward in a straight line; instead, they twirl in small circles (similar to the dance- waltz). Hence, they are called Waltzing mouse.

2.2.6 Genetically Modified Mouse

- A mouse model in which its genome is altered by biotechnology is called a genetically modified mouse. The first genetically modified mouse was produced in 1974 at the Massachusetts Institute of Technology. They can be produced via the following methods:
  - *Retroviral infection method:* Retrovirus is used as a vector to insert the 'small DNA inserts' into developing embryos of the mouse. The limitations of this technique are that only a small portion of DNA can be inserted, and random unpredicted insertion by retrovirus can lead to the expression/silencing of non-target phenotypes.
  - *DNA microinjection method:* The DNA construct of the desired transgene can be directly injected into the mouse oocyte's pronucleus to create a transgenic mouse. However, this method possesses similar limitations to the retrovirus infection method. To overcome this limitation, the technique has been upgraded to microinjection of fertilized mouse egg using 'endonuclease-based-reagent' that can create a transgenic mouse with minimal off-target gene modifications.
  - *Gene-targeted transgene method:* As a first step, the embryonic stem cells of the mouse are targeted for gene manipulation by introducing primarily 'loss-of-function' at specified loci. Later, the mouse's blastocysts were injected with this modified embryonic stem and transferred to a female mouse resulting in successful chimeric offspring. These chimeric offspring are intercrossed repeatedly to obtain the mutated homozygous transgenic mouse.

- *Types:*
  - *Transgenic mouse:* Gene modification by the methods mentioned earlier can result in the random insertion of the desired genome anywhere in the host genome and produces a transgenic mouse. The primary use of transgenic mice in research is in understanding the pathophysiology of cancer and non-communicable disease. The following are a few examples of a transgenic mouse.

    MEF2A mouse: Overexpression of muscle-specific MEF2A protein due to a mutation in *mef2a* gene resulting in diabetes phenotype

    GLUT4 series mouse: Alteration in expression of *Glut4* gene resulting in diabetes phenotype

    LKB1-DN/B mouse: Overexpression of skeletal-muscle specific LKB1 resulting in metabolic syndrome phenotype

    AMPK series mouse: Overexpression of AMPK protein due to alteration in *ampk* gene resulting in obesity and diabetes phenotype

    Coll-11a2-Smad6 mouse: Alteration in expression of chondrocyte-specific smad6 resulting in dwarfism and osteopenia phenotype

    Apart from these, more than 100 types of transgenic mice are commercially available in the market, and describing all is beyond this chapter's scope
  - *Constitutive knock-out mouse:* Gene modification method results in permanent loss of function of the gene in the entire cells of the mouse produces Constitutive knock-out mouse. This type of mouse is often employed to understand the change in anatomy, physiology, biochemical process, and behavior. Also, constitutive knock-out mice are used in the identification of novel cancer genes. The following are a few examples of constitutive knock-out mice used in research

    BKO mouse: Acid-beta-galactosidase enzyme is deficient due to *Glb1* gene loss resulting in GM1-gangliosidosis phenotype

    ATF4 mouse: Activating transcription factor-4 is absent due to the loss of the *Atf4* gene resulting in microphthalmia phenotype

    ADAMTS13 mouse:Loss of function of ADAMTS13 metalloproteinase due to knock-out of *Adamts13* gene resulting in Thrombotic thrombocytopenic purpura phenotype

    PEX11a mouse: Loss of function of Pex11 protein in kidney and liver lead to the development of hepatic steatosis with renal disorder phenotypes

    FAM16a-L53P-tg mouse: Used to study hypomyelinating leukodystrophies in the central nervous system
  - *Conditional knock-out mouse:* Gene modification method results in loss of gene function in a specified organ in a specified time produces conditional knock-out mouse. This model mimics better than the constitutive knock-out model in terms of the development of cancer in humans. The gene knock-out can be regulated spatially and temporally in this model Example:

    TAK1flox mouse and Ubc13 flox mouse: used to study immune abnormality

    NDRG1lfox mouse: used to study Charcot-Marie-Tooth disease type4D

    FIR flox mouse: used to study various cancers

Sik-3 flox mouse: Used to study abnormal metabolism of lipids and carbohydrates

BIG1 and arf1 mouse: Used to study cellular vesicular trafficking mechanisms

– *Knockin mouse:* In this mouse, the genome is inserted, resulting in overexpression or production of a new protein that mimics humans' disorders. Example

nNOS-Cre mouse: used to study various neurological disorders related to nitric oxide synthase systems

MBP- Cre mouse: Used to study myeline generative diseases

CD72-Y7F mouse: Used to study systemic lupus erythematous

2.2.7 Limitations of the Mouse as Experimental Models for Human Diseases

- Owing to their tiny size, the detection and dissection of lesions developed in the mice are very difficult.
- The atherosclerotic mouse model shows an extremely low incidence of rupture of plaque leading to thrombus/myocardial infarction in mice, while in humans, plaque rupture is the primary cause of myocardial infarction.
- An anticancer drug that was shown effectiveness in mouse was later found to be ineffective in humans due to underlying different regulatory mechanisms between mice and humans. Example: Endostatin.
- Chemical moiety that was proven carcinogenic in mice had been proven non-carcinogenic in humans and later was approved for use in humans. Example: Chloramphenicol.
- Majority of cancer models of mouse show development of mesothelial sarcomas while in humans epithelial carcinoma are the most common cancers.
- Spontaneous regressions of cancers are uncommon in adult humans, while it is common in the adult mouse.
- The basal metabolic rate per gram of body tissue in the mouse is seven-time higher than that of adult humans. Moreover, mice also have a higher mitochondrial density in the cell than humans. Thus the mouse is very prone to oxidative damages due to reactive oxygen species than humans that may confound the study of anti-oxidant drugs.

2.3 Rat

2.3.1 Scientific Name

- Genus: *Rattus*
- Species:
  - *novergicus* (Norway brown rat- most commonly used)
  - *rattus* (Black/Roof rat—less commonly used)
  - *xanthurus* (Sulawesian white-tailed rat—less commonly used with the longest tail of 25–35 cm)
  - *osgoodi* (Southern Vietnam Osgood's rat—less commonly used)

2.3.2 Physical Description

- Body Length: 12–14 cm
- Tail length: 12–15 cm
- Weight: Male: 300–500 g, Female: 250–350 g
- Breeding season: Throughout the year
- Gestational age: 21–22 days
- Duration of estrous cycle: 4–5 days
- Litter size:6–12
- Sexual maturity: 9–14 weeks
- Average life span: 2–4 years

2.3.3 Salient Points of the Rat as an Experimental Animal

- Rat is the most commonly used lab animal besides mouse and guinea pig in academic and pharmaceutical research sectors, especially in acute and chronic toxicity studies, carcinogenicity, and mutagenicity studies.
- A rat can be trained easily when compared to other animals. Hence rat is the lab animal of choice for studying behavior, conditional reflexes, and various neurological disorders.
- The rat has two parts of the stomach—the upper two-fifth is translucent and non-secretory (rumen), and the lower three-fifth is glandular (antrum with pylorus). The lower portion of the stomach of a rat is identical to the human stomach. Moreover, Rat lacks gall bladder, and hence constant bile presence in the duodenum causes the continuous secretion of HCl even under fasting state. Hence antipeptic ulcer drugs are evaluated better in rat models.
- The rat has the fastest liver regeneration period (less than 7 days) and is commonly used for liver physiology, effects of drugs on the liver, and outcome of various types of liver surgeries.

- Being small and easy administration of drugs (intraperitoneally), rats are commonly used to study analgesics' effects using a standardized tail-flick method.
- Various rats' organs express particular receptors in high concentration, and hence rat organs became standard tissue of choice for evaluation of agonist and antagonist of that receptor. For example:
  - Bioassay of adrenaline using rat uterus
  - Bioassay of acetylcholine using rat colon
- The MHC complex of rats has been thoroughly studied by genome sequencing, and hence rat is often used to study biological control over the functional activity of MHC complexes. Due to the same reason, the rat is also used for xenograft and allograft transplant experiments.
- The rat has a shorter gestation period, with high litter size and shorter sexual maturity duration. Because of these facts, the rat is commonly employed to test the reproductive toxicity of drugs.
- Rats have a high intrinsic healing capacity. Hence rats are most useful in assessing the various parameters of implant like residence time, rate of degradation, and safety of biomaterials that are implanted via the subcutaneous or intramuscular route.
- Seven days old rat pups are similar to human infants and are extensively studied for the hypoxic-ischemic model for stroke and evaluation of drugs for cerebral palsy.

2.3.4 Outbred Stocks of Rat

- *Wistar rat:*
  - Wistar is the first rat model that was standardized for research purposes. Wistar is an outbred stock rat developed in 1906 by Wistar Institute, USA.
  - Wistar rat is also an albino rat with a broad head, long ears, and average body length.
  - The tail length of this stock is always lesser than the body length.
  - The substrains Wistar-Unilever rat and Wistar-Hannover rats are outbred stock. On the other hand, Wistar-Furth and Wistar-Kyoto rats are inbred strains.
- *Sprague Dawley rat:*
  - It was developed in 1945 by Sprague Dawley Incorporation from the Wistar stock.
  - They are albino rats and relatively docile when compared to wild rats.
  - This stock is commonly used in all aspects of biomedical research.
  - It has a low incidence of spontaneous tumors and a high reproductive rate.
  - It has a narrow head with a longer body length when compared to other stocks. The tail is always longer than the body in Sprague Dawley rats.
- *Long Evan rat:*
  - This stock was developed in 1915 by crossing several wild grey male rats with female Wistar rats.

  - This stock is also called 'hooded rat' because of the typical fur coat color. The head and its upper part of the trunk have greyish to black color fur, and other parts of the body have varying proportions of albino fur. Occasionally, a white hood with brown body parts can also be seen.
  - This stock has a high resistance to respiratory problems when compared to other stocks. Thus Long Evan stock is preferred to study surgical procedures that require high inhalational anesthetic use.

2.3.5 Inbred Strains of Rat

- *Spontaneously hypertensive rat:*
  - This strain was developed in 1960 by inbreeding of Wistar-Kyoto strain with hypertension.
  - Hypertension develops spontaneously around the fifth week to the sixth week of age. The systolic blood pressure can reach up to 200 mmHg in adult rats.
  - This rat is the preferred strain for studying and understanding the pathophysiology of hypertension, stroke, and the evaluation of several anti-hypertensive drugs.
  - Further modification of this strain led to the development of 'stroke-prone SHR' in which the rat will die of stroke due to severe hypertension.
- *Zucker Fatty rat:*
  - Zucker fatty rat (ZFR) was developed by Zucker in 1961. This strain comprises another substrain called Zucker diabetic fatty rat (ZDF rat).
  - They have a mutation in the leptin receptor leading to increased food intake (hyperphagia). The rat can reach a weight of up to 1.5 kg. Obesity is apparent from the fourth week of age.
  - The ZFR is not overtly diabetic even though they show poor glucose tolerance. However, the substrain- ZDF rats also have significant hyperglycemia and dyslipidemia besides obesity and became overtly diabetic within the tenth week of their age.
  - Diabetic changes are characteristically seen in males than females in ZDF rats. Feeding females with high-calorie food convert them to overt diabetic like males.
  - This strain is extensively used in research concerned with obesity and diabetes.
- *OLETF rats:*
  - The Otsuka-Long-Evans-Tokushima fatty rats are inbred strains derived from the spontaneously diabetic Long Evans rats in 1994.
  - Late-onset hyperglycemia, mild obesity, and hyperinsulinemia are their characteristics features.
  - The diabetic changes are clinically visible by the 18th week of their age. The progressive degeneration of islets of the pancreas due to beta-cell apoptosis is the underlying pathology for the development of diabetes.

- *Goto-kakizaki rats:*
  - This inbred strain is derived from repeated inbreeding of Wistar rats that were showing glucose intolerance by Goto in 1976.
  - This strain is used as a non-obese model rat model for type II diabetes. Hyperglycemia occurs in this strain by insufficient insulin response rather than insulin resistance, as seen in obese models like ZDF rats and OLETF rats.
- *BB rat:*
  - This strain was called *biobreeding rat* and was developed from inbreeding of Wistar rats in 2005.
  - This strain shows features of diabetes in their eighth week of age. Lymphopenia is an additional feature seen in this strain.
  - This strain is used in the research area interlinking diabetes with autoimmunity.
- *Other inbred strains:*
  - Besides these strains, various inbred strains of rat-like *ZDSD rats*, *Brattleboro rats, Rowett nude rats, Fuzzy rats, Shorn rats, Royal-College of surgeon (RCS) rats*, and *Lewis rats* are also used in research.
  - Rowett nude rats lack thymus and are used in upper respiratory infection studies. Fuzzy rats are also a nude rat that develops progressive renal failure in their first year of age.
  - RCS rat has an inherent capacity to develop retinal degeneration due to mutation in the *MERTK* gene. Lewis rat has a tendency for a higher incidence of spontaneous tumor and leukemias.

2.3.6 Transgenic and Knock-out Rats

- Similar to the mouse, the rat genome has been extensively studied. Hence the gene of rats can be knocked-out or knocked-in, resulting in the production of transgenic rats.
- Transgenic rats have similar pathophysiology to human diseases and can provide better data quality than outbred stocks and inbred rat strains.
- Long Evans and Sprague Dawley rats are commonly used to produce transgenic rats. The following methods can produce them:
  - Plasmid transgene method
  - Bacterial artificial chromosome transgene method
  - Pronuclear microinjection of Zinc-finger-nuclease-mRNA method
  - CRISPR/Cas9 gene manipulation method
- Transgenic rats like Fisher 344 rats, Crl:LE rats, CrL:Wi rats, and Crl:SD rats have been commonly used in research.

2.3.7 Limitation of the Rat as an Experimental Animal

- The rat does not have a vomiting center, and hence anti-emetic drugs cannot be evaluated using it.
- The pancreas of rats is very diffuse, and hence complete and total pancreatectomy is extremely difficult. Hence, rats are a poor model for type I diabetes mellitus studies.
- The rat has a coprophagic habit (eat their stool). Thus the pharmacokinetic parameters of drugs with GI elimination can be significantly affected when the rat is used.
- The cartilages of rats are very thin, and the joints are small. Hence rats have a limited role in chondral-defect repair studies.
- The rat has less (4 times) platelet response to thrombin and prolonged (3 times) clotting time when compared to humans. Moreover, their coagulation systems have significant interspecies differences (e.g., Severe hemorrhage occurs after Vitamin K deficient diet only in Wistar but not in Sprague-Dawley rats). Hence, rats have a limited role in the evaluation of anticoagulant drugs.
- The rat has very 'loose skin' (without any adherence strength to structures under the elastic part of its skin) that is in contrast to human skin. Moreover, the primary method of wound healing in rats is 'wound contraction' in contrast to 're-epithelization' in humans. Hence, rats have a limited role in the evaluation of drugs for soft tissue trauma.

2.4 Guinea Pig

2.4.1 Scientific Name

- Genus: *Cavia*
- Species:
  - *porcellus* (domesticated and commonly used)
  - *aperea* (Brazilian guinea pig—nondomesticated)
  - *fulgida* (Shiny guinea pig—nondomesticated)
  - *tschudii* (Montane guinea pig—nondomesticated)
  - *magna* (Largest size, nondomesticated, called as greater guinea pig)

2.4.2 Description

- Body Length: 20–40 cm
- Tail length: Not visible externally
- Weight: Male: 500–1500 g
- Breeding season: Throughout the year

- Estrous cycle: 13–20 days
- Gestational age: 59–72 days
- Litter size: 3–6
- Sexual maturity: 3–5 weeks
- Average life span: 4–5 years

2.4.3 Salient Points of the Guinea Pig as an Experimental Animal

- Vitamin C was first identified by using guinea pig in 1907. Like humans, guinea pig lacks the enzyme L-gulonolactone-oxidase and hence unable to synthesize vitamin C. Thus the diet of guinea pig must contain Vitamin C supplementation at the rate of 200 mg–1 g/l in water.
- Guinea pigs are highly susceptible to Mycobacterium tuberculosis infection. In fact, Mycobacterium tuberculosis was first discovered using guinea pig in 1882 by Robert Koch.
- Besides tuberculosis, etiological agents for infectious diseases like amoebiasis, brucellosis, diphtheria, typhus fever, and yellow fever were identified using guinea pigs.
- Guinea pigs are highly sensitive to histamine and allergens. Hence guinea pigs are commonly employed in allergen testing, skin prick testing, studies related to asthma, and anaphylactic reactions.
- Various tissues of guinea pigs act as a standardized tool for assaying chemical compounds:
  - Guinea pig bronchus—screening for bronchodilators like beta-2 agonists and muscarinic antagonists.
  - Guinea pig trachea—screening of beta-blocker activities.
  - Guinea pig terminal ileum—screening of spasmodic compound, anti-histaminic drugs, etc.
  - Guinea pig uterus—bioassay of adrenaline
  - Guinea pig aorta—alpha-adrenergic blockers
- Guinea pig serum contains an asparaginase enzyme that shows anti-leukemic activity.
- A test for deafness called Preyer reflex (whistling sound moves the outer ear) can be easily tested in guinea pigs. Moreover, the ear structure of the guinea pig resembles closely to humans. Hence guinea pigs are commonly used in hearing research. Cochlear mechanical mechanisms were discovered by using guinea pig in 1961. Nowadays, guinea pigs are commonly employed in research related to the regeneration of inner ear hair cells.
- The complement system of immunology was first discovered in the blood by using a guinea pig.
- The placental structure of guinea pigs resembles humans, and their gestation period can also be divided into three trimesters like humans. Hence guinea pigs are preferred in studies related to pre-eclampsia and gestational hypertension.

- The immune system functions of the guinea pig are closely similar to humans. Thus, guinea pigs were used extensively for the development of vaccines. Vaccines for diphtheria and TB were discovered using guinea pigs.
- Due to their relatively larger arterial size than rat/mouse, guinea pigs are preferred in studying the aortic atherosclerosis models. Moreover, guinea pig carries cholesterol in LDL lipoprotein, unlike other rodents. Hence guinea pigs act as a good model for studying interlinking lipid metabolism and atherosclerosis.

2.4.4 Strains of Guinea Pig

- Unlike rats and mice, only limited numbers of outbred stock and inbred strains are available for the guinea pig.
- *Dunkin-Hartley* stock is the most commonly used outbred stock. It is an albino stock derived from a short-haired English Guinea pig.
- *Wright strain 2 (NIH 2) and Heston strain 13 (NIH 13)* are the two most commonly used inbred strains of guinea pig in research.

2.5 Hamster

2.5.1 Scientific Name

- Eighteen species of hamsters are currently identified and have been classified under seven genus
- Genus: *Mesocricetus*
  - Species:
    auratus (Syrian or Golden hamster or Teddy Bear Hamster—commonly used)
    brandti (Turkish hamster or Brandt's hamster)
    newtoni (Romanian Hamster)• Genus: *Cricetulus*

- Species:
  griseus (Chinese hamster—commonly used)
  barabensis (Chinese striped hamster or striped dwarf hamster)
  longicaudatus (long-tailed hamster)
- Other genus includes A*llocricetulus, Cansumys, Cricetus, Phodopus,* and *Tscherskia.* Only the Syrian golden hamster and Chinese hamster are commonly used in research

2.5.2 Description

- Body Length: 15–20 cm
- Tail: diminutive and fluffy
- Weight: 100–400 g
- Breeding season: Throughout the year
- Gestational age: 2–3 weeks
- Litter size: 5–6
- Sexual maturity: within 6 weeks
- Average life span: 1.5–3 years

2.5.3 Salient Points of Hamster as an Experimental Animal

- Hamsters have distinct cheek pouches, unlike other rodents. The cheek pouches are devoid of intact lymphatic drainage. Hence this site is used as the best experimental model for tissue transplantation research.
- Chinese hamsters have an inherent tendency to spontaneously develop diabetes mellitus due to defective beta cells of the pancreas.
- Unlike a mouse, hamsters can accept the human cytokines, IL-12, and GM-CSF to the full extent and shows the corresponding responses. Hence hamsters are a better model for immunogenicity testing than other rodents.
- Hamsters were extensively used to study the pathogenesis of various viruses like West Nile virus, Yellow fever virus, Nipah virus, Ebola, and Marburg Virus.
- Hamsters were also extensively used to study disease pathogenesis of leptospirosis, Leishmaniasis, Clostridium infections, Schistosomiasis, and amoebiasis.
- Hamster cheek pouch can also be used for evaluation of various chemicals with carcinogenic potential (e.g., DMBA can cause oral squamous cell carcinoma in hamsters).
- Hamster cheek pouch also act as excellent tissue of choice to:
  - Assay microcirculation using various prostaglandins
  - Study the reperfusion injury after ischemia
- Hamsters are also an excellent model for chronic smoke inhalation-induced carcinoma in the respiratory tract. European hamsters are more suited than Syrian hamsters for this purpose due to their large size.
- Hamsters exhibit a peculiar feature that cancer could act as a contagious agent amongst them. E.g., Reticulum cell sarcoma can be transmitted between hamsters via the *Aedes aegypti* mosquito.

2.6 Gerbil

- *Scientific name:* More than 180 species of gerbil has been identified and classified under 16 genera. The commonly used gerbil in the lab is *Meriones unguiculatus* (Mongolian gerbils).

- *Description:* It is also known as a sand rat or clawed jirds. The body length is between 10 and 15 cm, and the tail length is as same as the body length. An average adult gerbil weighs about 50–100 g. Gerbils often exhibit monogamy and attain sexual maturity by the 12th week of age. The gestational age is about 21–26 days, and the litter size ranges from 1 to 8 pups. Gerbils are expected to live for 2–5 years.
- *Salient points of gerbil as an experimental model:*
  - Gerbil exhibits similar hearing patterns and curves to humans and is extensively used in auditory research like guinea pigs.
  - Gerbil can act as a good model for studying pathological aspects of aging.
  - Gerbils spontaneously exhibit epilepsy when placed in a new lab environment or by repeated handling. Hence gerbil can act as a suitable animal model for analyzing seizures.
  - The gerbil is a well-known animal model for the study of gastric carcinogens.
  - Due to a high average life span than mouse/rat, gerbils are used to study the carcinogenic effects of various compounds with long latency periods.
  - While aging, gerbils exhibit the features of amyloidosis spontaneously. It can also occur when gerbils are experimentally infected with the filarial worm. Hence gerbils are suitable animal models for amyloidosis.
  - Gerbils also tend to form aural cholesteatoma, with a prevalence of about 50% at 24 months of age. Thus gerbils are used in the evaluation of the pathogenesis of aural cholesteatoma.

2.7 Rabbit

- *Scientific name:* 20 species of rabbits have been discovered and classified under 11 genera. The commonly used rabbit in the lab is *Oryctolagus cuniculus* (European rabbit) and their outbred stocks and inbred strains.
- *Description:* An average adult male rabbit weighs about 4–5 kg. Females are slightly heavier than males. Rabbits exhibit polygamy and attain sexual maturity by the 8th - 12th week of age. Their gestational age is about 28–35 days, and the litter size ranges from 1 to 14 pups (average 8). The average life span of the rabbit is around 1–2 years.
- *Strain and stocks:* New Zealand albino rabbit is a commonly used strain. Other strains like Dutch, Watanabe heritable hyperlipidemic (WHHL) rabbit, and Flemish Giant are also used.
- *Salient points of the rabbit as an experimental model:*
  - Rabbits are very docile and do not show aggression. Hence they are effortless to handle and make observations effectively.
  - Have shorter reproductive cycle duration, and hence rabbits are utilized in reproductive toxicity studies like rats and mice.

  - Rabbits are the excellent animal model of choice for pyrogen testing as rabbits' skin is susceptible to irritants.
  - The ears of rabbits are large with visible veins in them. Hence studies involving repeated blood collection can choose the rabbit as an animal model.
  - Due to the relatively large size of eyes compared to other small lab animals, rabbits are often used to study the action of drugs on the pupil, ophthalmic antifungal and antibiotic preparations, and study the pathogenesis of fungal, viral, and bacterial keratitis.
  - The anterior chamber of the rabbit's eye can accommodate the IOL designed for human purposes. Hence rabbit can be used for the evaluation of novel surgical techniques involving IOL.
  - Rabbits are an excellent animal model to evaluate nano-emulsion eye drops and other novel ophthalmic drug delivery systems.
  - Traditionally, rabbits were used for pregnancy identification in humans. Injection of serum from a pregnant human female induces ovulation in female rabbits.
- *Limitations:*
  - Wild rabbits with intact fur exhibit expression of atropinase enzyme leading to failure of the abolition of light reflex by atropine. New Zealand albino rabbit lacks this enzyme, and hence atropine can be tested in this strain.
  - Rabbit lacks adrenergic vasodilator nerves. Hence the phenomenon of vasomotor reversal of Dale could not be performed in rabbits.
  - Rabbit lack a vomiting center, and hence the evaluation of anti-emetic cannot be done in them.
  - The retina of the rabbit is poorly vascularized, with higher rod cell concentration when compared to humans. Thus effects of drugs on rabbit retina cannot be translated to retinal damages of glaucoma and other human retinal diseases.

2.8 Cats

- *Scientific name: Felis catus.*
- Cats are a good model of choice for neurological research, especially in balance, movement, hearing, and motor neuron disorders of spinal injury.
- Feline Leukemia Virus and Feline immunodeficiency virus (FIV) have been studied using domestic cats, and the results have been correlated with human leukemia and HIV, respectively. Thus FIV infected cats act as a good model for studying the pathogenesis of AIDS.
- Cats are highly susceptible to *Helicobacter felis* infection, similar to humans for *H.pylori.* Hence the feline helicobacter model helps in understanding the pathogenesis of ulcer by Helicobacter infection.
- Cat nictitating membrane is the good tissue of choice to bioassay adrenaline and histamine.

- Unlike humans, a paradoxical response by histamine is observed in cats. Histamine produces bronchodilation and vasoconstriction in cats.
- Traditionally, cats are used for studying the hemodynamic effects of various cardiovascular drugs.
- *Limitation:* Righting reflex cannot be studied in cats as it quickly regains this reflex even after falling from a high altitude.

2.9 Pig

- *Scientific name: Sus-scrofa domestica.*
- *Breeds:* Duroc pigs, Landrace pigs, Yorkshire pig, Large white pig, Mexican hairless pig, and Clawn miniature pigs.
- Pigs share anatomical and biochemical similarities like humans. Hence pigs are a better research model when compared with other non-rodent species.
- Coronary artery circulation, size of the heart, and blood vessels of the pig are identical to humans. Similar to the right-side dominant blood supply in the heart observed in 90% of the human population, pigs also share the same right-side dominant system. Moreover, collateral coronary circulation is absent in pigs (unlike dogs and monkeys where it is present) result in complete infarct on occlusion. Hence pigs are an excellent animal model to study heart circulation and myocardial infarction.
- Pig's pancreas is similar to humans in terms of histological, functional, and anatomical effects. A surgical pancreatotomy-induced diabetes model can be produced effectively in pigs.
- Pigs act as a standard model for reconstructive surgical methods and wound healing models as pig's skin has a similar dermal and epidermal composition to humans.
- Pigs were extensively used for studying intravascular coronary stents leading to the development of recent stents like bioresorbable stents and drug-eluting stents.
- Pigs are a popular model for studying aneurysms. Surgical induction of saccular aneurysms in pigs has been standardized.
- Pigs are preferred for analyzing ventricular assist devices and other novel transplant techniques.

2.10 Dog

- *Scientific name: Canis familiaris*
- *Breeds:* Mongrel and Beagles are commonly used breeds for experimentation
- Dogs are mainly used for studying the effects of drugs on the cardiovascular system
- Dogs are a good model for the surgical induction of diabetes. Diabetes and the role of the pancreas and insulin in diabetes were discovered by using experimentation on dogs

- The Vagus nerve and cervical sympathetic nerve run together in dogs, and hence stimulation can lead to complex variations in blood pressure
- Dogs are the animal of choice for studies on digestion and gastric secretion as they can be 'conditioned' easily
- Dogs can also develop hypertension spontaneously like humans

2.11 Monkeys and Other Non-human Primates

- *Scientific name:* Out of many non-human primates like lemurs, lorises, tarsiers, gibbon, gorilla, Orang-utan, Chimpanzees, the rhesus monkey (*Macaca mulatta*) is the commonly used species in the lab.
- Monkeys are used mainly when results from rodent species are not effectively translated to humans. Thus results from monkeys have closer resemblances to humans.
- Monkeys are the best animal model of choice for psychopharmacology due to their high neurological development and similar brain structure to humans.
- Monkeys have been used to develop polio vaccines, analyze the effects of the life-support system in infants, study novel techniques of dialysis, transplant rejection, and various stroke rehabilitation therapies.
- Research on Parkinsonism involving deep brain stimulation utilizes monkeys often.
- The pharmacokinetic profile, allergic sensitivity, and safety profile of the drug are very similar to the humans when estimated using monkeys.

2.12 Horse

- *Scientific name: Equus caballus.*
- Horse genome sequencing was completed in 2009. To date, more than 90 hereditary diseases of horses have been identified with similar patterns in humans.
- The insulin resistance pattern of the horse is similar to humans. Like humans, horses also develop 'Equine metabolic syndrome' with insulin resistance and abnormal fat distribution. Hence, horses are used to study insulin resistance and its response to overfeeding.
- Domestic horses are used to study the behavioral pattern in depression.
- Horses have been utilized in the lab to generate anti-serum for snake venom, diphtheria toxin, tetanus toxin, etc., in the past decades.
- *Limitations:.*
  - Due to its larger body size, feeding and housing are very difficult. Moreover, due to the high cost/animal in the horse, data generation from larger samples cannot be done.
  - Horses cannot vomit routinely but do in an extreme situation due to the presence of a powerful esophageal sphincter. Hence evaluation of anti-emetic drugs is limited in horses.

2.13 Sheep

- *Scientific name: Ovis aries.*
- Sheep has a similar anatomical structure and physiological function of the respiratory system to humans. Sheep can be used as an animal model for studying asthma, COPD, and interstitial lung diseases.
- The sheep model sensitized with a nematode (*Ascaris suum*) is a standard model to study the basic mechanism of lower respiratory tract inflammatory reactions and evaluate drugs affecting the lower respiratory tract's inflammation.
- The sheep model for asthma by dermatophagoides is a standard ovine model developed in Australia for studying aeroallergens in asthma.
- Premature lambs were used to study the pathogenesis of surfactants in the development of infant respiratory distress syndrome, leading to surfactant development as a drug for the same.
- Due to their long life span of 10–12 years, sheep are used to study neurodegenerative disorders like Alzheimer's disease, Huntington's disease, and Parkinsonism. The transgenic sheep model (created by microinjection of mutant HTT gene) is a standardized model for Huntington's disease.

2.14 Chicken

- *Scientific name*: *Gallus domesticus* and *Gallus gallus.*
- Chicken tissues are widely used in various in-vitro studies owing to their comfortable and surplus access.
- The fact that the virus can induce tumors was identified using chicken in 1900 by Peyton Rous.
- Chick embryos and it's amniotic/chorionic membrane have played a significant role in developing various vaccines like yellow fever vaccine, chickenpox vaccine, polio vaccine, rabies vaccine, and smallpox vaccine.
- Standard models of chicken are:
  - Chick comb method (an assay of androgenic substances)
  - Chick chorioallantoic membrane assay (to test angiogenesis)
  - Aversive discrimination in chicken (to test learning and memory)
  - UCD-200 chicken strain for obesity
  - Catalepsy antagonism using White Leghorn chicken
  - Scleroderma model in a chicken
  - Chicken embryo trachea model to measure the ciliary beat frequency
- Besides these, chicken is being utilized in broad research areas like growth, performance, embryology, fertility, toxicology, endocrinology, genetics, and neurobiology.

2.15 Pigeon

- *Scientific name: Columbia livia.*
- Pigeon belongs to the family Columbidae, which comprises 300 species.
- Pigeons are extensively used in the screening of anti-emetic activity. The pecking of the beak by a pigeon is considered as the vomiting response. Cisplatin, emetine, ipecac syrup, digitalis, and ditolyganidine are the agents used to induce emesis in pigeons.
- The following are a few standardized models developed using pigeon:
  - Intravenous injection of digitalis induced emesis in pigeon
  - Punished behavior model in pigeons for evaluation of chlorpromazine and amobarbital
  - Apomorphine-induced pecking behavior in pigeon to study the effects of neuroleptics
  - Auto-shaping procedure model in pigeons for learning and memory
  - Spontaneous arteriosclerosis model in pigeons
  - Pigeon crop method to assay prolactin hormone

2.16 Frog

- *Scientific name:* The amphibian species frogs are classified under order Anura that includes more than 7100 species. The commonly used lab frogs are *Xenopus laevis* (native African frog), Rana *tigrina* (Indian bullfrog) and *Rana hexadactyla* (Indian green frog), and *Bufo bufo* (Common toad).
- Frogs have longer legs than toads and have thinner skin than toads. Frogs lay eggs in the cluster while toads lay eggs in strands. Frogs are usually moist, and toads are dry. Toads can be found in dry areas, while frogs are usually seen in aquatic conditions.
- Frogs breathe primarily through their skin which is highly permeable to oxygen. Certain species of frog skin secretes a wide variety of substances like alkaloid (epibatidine), toxin (tetrodotoxin, zetekitoxin, chiriquitoxin), irritants, peptides (bombesin), antimicrobial-peptides (brevinin, ranlexin, nigrocins, dermaseptin), hallucinogens, and convulsants.
- The concept of bioelectricity (involvement of electrical transmitters in the generation of muscle activity) was discovered by Luigi Galvani in the eighteenth century using frog sciatic nerve preparation.
- Otto Loewi discovered Vagusstoff (later as acetylcholine) and awarded Nobel Prize for the same in 1936 using frog isolated heart experiments.
- Frogs have been extensively used in research to the extreme extent that nowadays, frogs for academic and research purposes are totally banned in several countries.

- A frog's heart contains three chambers in contrast to other lab animals (4 chambers in the heart). The frog's heart model is a standard method to study the chronotropic, inotropic, dromotropic, and bathmotropic effects of various drugs in the heart.
- In frogs, adrenaline is the neurotransmitter for the sympathetic system in contrast to humans, where non-adrenaline is the neurotransmitter for the same.
- Traditionally, frogs were used for the diagnosis of pregnancy.
- Showing light on the frog's eye bleaches rhodopsin in retinal layers and takes nearly 1 h for resynthesizing rhodopsin. Thus frogs were used for evaluating the retinal toxicity of compounds.
- The frog was used for the biological standardization of cardiac glycosides like digoxin. Digoxin produces ventricular systolic arrest and a widely dilated atrium.
- Ciliary motility of the frog esophagus was a standard method to assess the activity of spasmodic and antispasmodic drugs.

2.17 Zebra Fish

- *Scientific name: Danio rerio.*
- Zebra fishes are freshwater fish with blue horizontal stripes on each side of their body. Almost 70% of the human genome is present in the zebrafish genome, making the zebrafish a potential animal experiment model.
- Zebrafish's breeding time is 10 days, which is shorter than all the rodents and hence used in reproductive toxicity studies.
- The entire period of development of zebrafish from fertilized to young fish can be observed under a microscope.
- Transgenic zebrafish are relatively easier to create than rodents. This is because fertilized eggs of zebrafish develop externally while rodents it happens inside the body.
- The following pathological conditions have been successfully modeled using zebrafish:
  - Duchenne muscular dystrophy using *dystrophin* gene knock-out zebrafish model.
  - Human melanoma model using the *BRAF* gene knock-in zebrafish model.

Bibliography

Burggren WW, Warburton S. Amphibians as animal models for laboratory research in physiology. Inst lab animal Res J. 2007;48:260–9.

Guénet JL, Bonhomme F. Wild mice: An ever-increasing contribution to a popular mammalian model. Trends Genet. 2003;19:24–31.

Hedrich HJ, Bullock G. The laboratory mouse. London: Elsevier Academic Press; 2004.

Hoffman RA. The golden hamster. Ames, Iowa: Iowa State University Press; 1968.

Jackson F, Scott PP. Lab Anim. 1970;4:135–7.

McGrath P, Li CQ. Zebrafish: a predictive model for assessing drug-induced toxicity. Drug Discov Today. 2008;13(9–10):394–401.

Schwentker V. The gerbil. A new laboratory animal. Ill Vet. 1963;6:5–9.

Simon GA. The pig as an experimental animal in biomedical research. Israel J Vet Med. 1993;48: 161–7.

Thomas J, Ill G, Smith GJ, Robbery W. The rat as an experimental animal. Science. 1989;245 (4915):269–76.

Care and Handling of Laboratory Animals 3

Subramani Parasuraman

Abstract

Care of laboratory animals is one of the essential parts of biomedical research. The laboratory animals such as a mouse, rat, guinea pig, rabbit, hamster, gerbils, ferret, dog, and cat should be maintained in a hygienic environment provided with good ventilation, day/night cycle, and humidity condition. The living environment of the experimental animals should be maintained uniformly throughout their lifespan to minimize stress and anxiety. Handling is an important part of the maintenance of laboratory animals. Inappropriate handling of laboratory animals makes them become stressed and anxious. Importantly, the researcher or animal handler should have appropriate training and may attend the courses related to animal welfare to eliminate or minimize stress. In this chapter, care and handling of laboratory animals are discussed.

Keywords

Environment · Rodent · Non-rodents · Stress

3.1 Introduction

Experimentation on the animal is an essential component of biomedical research, to understand the biochemical mechanism(s), pharmacology of the drugs and molecular mechanism(s). In the history of biomedical research, animals have been used repeatedly. Early Greek physician-scientists, Aristotle (384–322 BC) and Erasistratus (304–258 BC) have performed experiments on living animals. From antiquity to the nineteenth century, large numbers of animals were used to study

S. Parasuraman (✉)
Department of Pharmacology, AIMST University, Bedong, Kedah Darul Aman, Malaysia
e-mail: parasuraman@aimst.edu.my

M. Lakshmanan et al. (eds.), *Introduction to Basics of Pharmacology and Toxicology*, https://doi.org/10.1007/978-981-19-5343-9_3

physiology. At the beginning of the twentieth century, Paul Ehrlich (1854–1915) introduced a modern research system based on chemical compounds for pharmacological screening in animal models, which in turn led to the birth of chemotherapy. The care of laboratory animals is one of the essential parts of biomedical research. The laboratory animals should be maintained in a hygienic environment provided with good ventilation, day/night cycle, and humidity condition. The living environment of the experimental animals should be maintained uniformly throughout their lifespan to minimize stress and anxiety. Also, animals should be allowed to have a routine social life to enhance their quality of life. Handling is an important part of the maintenance of laboratory animals. Inappropriate handling of laboratory animals makes them stressed and anxious. Importantly, the researcher or animal handler should have appropriate training and may attend the courses related to animal welfare to eliminate or minimize stress. In this chapter, animal ethics - which is a pre-requisite for the animal experiments, followed by care and handling of laboratory animals are discussed.

3.2 Animal Ethics

Prior approval should be obtained from the Institute Animal Ethics Committee for use of animals in biomedical research. Various laws and acts have been in place to control the use of animals in an unethical manner and ensure the use of a minimum number of animals in the experimentation without pain or with only a minimum pain. Organizations like Committee for Purpose of Control and Supervision on Experiments on Animal (CPCSEA), the National Institute of Health (NIH), American Psychological Association (APA) and the Organization for Economic Co-operation and Development (OECD) provide the guidelines for animal use in scientific experiments and their maintenance.

3.3 Housing and Environment

In biomedical research, drug screenings are performed using live animals or animal tissue preparations. The results obtained from animal experiment depend on animals' health and their living environment. The unhygienic condition and noisy environment may affect the study outcome. The European Convention for the Protection of Vertebrate Animals used for Experimental and other Scientific Purposes provides that:

> *Any animal used or intended for use in a procedure shall be provided with accommodation, an environment, at least a minimum of freedom of movement, food, water, and care, appropriate to its health and well-being. Any restriction on the extent to which an animal can satisfy its physiological and ethological needs shall be limited as far as practicable. In the implementation of this provision, regard should be paid to the guidelines for accommodation and care of animals set out in Appendix A to this Convention.*

Appendix A of the European Convention (Council of Europe, 1986) briefly describes general care and accommodation of vertebrate animals used for experimental and other scientific purposes.

An animal house should be designed to provide a suitable living environment including a specific requirement for social contact and exercise for the species to be housed. The animal house should be maintained hygienically and allow the animals to leave in the appropriate day/night cycle at specific room temperature and humidity. Maintenance of normal body temperature is essential for animal well-being. Animals should be housed within temperature and humidity ranges appropriate for the species, to which they can adapt with minimal physiologic alteration and stress. The recommended dry-bulb macro environmental temperature for mouse, rat, hamster, gerbil, guinea pig is 20–26 °C; the rabbit is 16–22 °C; nonhuman primate is 18–29 °C and for farm animals 16–27 °C. The animal house should be provided with ventilation systems to maintain air quality and a stable environment.

3.4 Handling of Commonly Used Animals in Biomedical Research

Around 50–100 million animals are used for experimentation every year. The total number of animals used for experimentation in the USA in 2010 was almost 1.37 million excluding rats and mice. The most commonly used animals in the experimentation are mice, rats, guinea pigs, rabbits, and hamsters. Broadly the experimental animals are classified into three categories, *viz.*, rodents (mouse, rat, hamster, guinea pig), non-rodents (rabbit, monkey, dog), and miscellaneous (frog, zebrafish, pigeon). The difference in handling methods can influence not only the welfare of animals but also the observation and results of the research conducted on the aforementioned animals. Handling methods also significantly alter the immune status either reducing or enhancing and have important methodological implications. The handling of animals itself is a stressful procedure and animals may display increased body temperature, heart rate, and respiration rate.

3.4.1 Handling of Mice

According to the paleontological data, the three main commensal species *Mus musculus domesticus* (Western European house mouse), *Rattus rattus* (house rat/black rat), and *Passer domesticus* (house sparrow) evolved during the Natufian from local wild progenitors. Mice are the commonly used animal species in biomedical research and are the smallest rodents used in the laboratory. Mice are highly adaptable animals and generally easy to restrain. The mice should be grasped or lifted by the tail from the cage and placed on a surface or cage top. The scruff can be grasped between the thumb and forefinger whilst maintaining a grip on the tail. The laboratory animals should be handled calmly and confidently and exaggerate or sudden movements should be avoided. The researcher or caretaker should use a soft tone when talking to animals, which may alert them to a non- threatening approach and reinforce a caring attitude in the handler.

3.4.2 Handling of Rat

The rat (*Rattus norvegicus domestica*) is one of the major experimental animals in biomedical research. This strain is derived from Norway rat which originated in the area between the Caspian sea and Tobolsk. Rats are typically docile animals. The rat should be handled gently by grasping it around its shoulders. Restraint should be very gentle and not too tight as this will obstruct the rat's respiration.

Rats are commonly handled using the following techniques and which are routinely being used in laboratories:

- *Rat transport restraint:* The rat is taken out from the cage by gently gripping the center or base of its tail with index finger and thumb and placed on a rough surface or in a new cage. In this method, suspending the rat by tail or by its upper body for a long-time should be avoided.
- *Shoulder or Two fingers restraint:* The rat is handled dorsally. The rat thorax is held using thumb and forefinger firmly and thorax of the rat can be supported by remaining fingers.
- *Encircling or Three fingers restraint:* In this technique, the rat is used to grasp around the thorax with thumb and index finger.
- *Grip or Four fingers restraint:* Head and jawbones of the rat are seized between the knuckles of the index and middle finger of the handler. Then thorax is grasped by the thumb and remaining fingers under both forelimbs of the rat.
- *Neck hold or Scruff restraint:* The rat tail is grabbed with the thumb and index finger of the non-dominant hand and the palm of the other dominant hand is placed on the rat's back. Loose skin on the dorsal neck is being gathered by using thumb and fingers. The thumb and index finger are used to stabilize the head of the rat and the remaining three fingers are used to stabilize the trunk.
- *Restraint tubes or Boxes:* This method is designed for easy access to the rat tail. The rat is grasped around the thorax and the rat's head is placed into the opening of the restraint tube. Rat is then released and grip is maintained on its tail.
- *Cone restraint:* The rat is handled through the base of the tail firmly with the thumb and index finger and the rat is placed in the cone with the head first.
- *Cloth or Towel restraint:* Use a piece of cloth or towel to wrap the rat to restrain it. The cloth is placed over the palm of handler take them in the hold of choice and tuck.

3.4.3 Handling of Guinea Pig

Guinea pigs should be handled calmly with care to prevent distress and injury to the animal and handler. Guinea pigs are prey animals and become tamer over time.

Guinea pigs can become easily stressed with handling for extended periods and do not like being handled too often. To restrain a guinea pig, the handler must be rapid and smooth, to avoid frightening the animal. The handler's thumb is positioned beneath the jaw of the guinea pig. The back of the guinea pig is supported by the handler's other hand.

3.4.4 Handling of Rabbit

Rabbits are very docile, non-aggressive, and easy to handle. Daily human contact may reduce handling induced stress. Rabbit can become aggressive or nervous and inflict severe bites from their powerful hind legs if they are not handled or restrained properly. The rabbit should be held by grasping a large fold of loose skin over the shoulders with one hand and either supporting or grasping the rear feet with the other hand. The bones of rabbits are fragile. Failing to support on to the back feet result in the animals kicking and try to escape, which may cause a bone fracture or severe spinal injury. Importantly, rabbits should not be held by their ears because of a high probability of causing cervical luxation and death.

3.4.5 Handling of Hamster

Hamsters can be restrained by placed beneath the palm of one hand. Then be restrained by the scruff (grasped between the thumb and forefinger) starting with the skin near the front of the shoulders.

3.4.6 Handling of Gerbils

Several species of gerbil have been used as experimental animals in biomedical research; however, Mongolian gerbils are most commonly used. Gerbils can be handled quite freely and should be cupped using one or two hands, or can be scruffed. Gerbils should be never restrained by the tail as the skin of the tail is delicate and tends to tear easily.

3.4.7 Handling of Ferret

The adult ferret can be held by grasping around the neck and shoulders and the other hand supporting the animal's hindquarters.

3.4.8 Handling of Dog

A dog should be handled with proper support. The dog can be restrained in a sitting position or lateral recumbency for injections and minor procedures. For venepuncture, a dog can be restrained on a table with one arm around its neck and another hand is free to restrain the body, if necessary. Muzzles are used when a snappy or potentially aggressive dog must be handled.

3.4.9 Handling of Cat

Cats are naturally nervous, cautious animals, suspicious of strange surroundings and strange people. Cat is often cooperative adequately to be restrained on a table by the loose skin at the back of the neck and hips. Cats can also be restrained by 'stretching' - by holding the scruff with one hand then gently holding and extending both hind limbs.

3.5 Animal Welfare

In 1959, Russell and Burch proposed the replacement of animals with non-sentient alternatives and reduce to a minimum number of animals used in experiments. The concept of "Alternatives" was restated by Smyth (1978) as "All procedures which can completely replace the need for animal experiments, reduce the number of animals required, or diminish the amount of pain or distress suffered by animals in meeting the essential needs of man and other animals."

Adapting Reduce, Refine, Replace (3Rs) principles in the animal research is advisable to minimize the number of animals in the biomedical research (reduce), decrease in the incidence or severity of 'inhumane' procedures applied to those animals (refine) and replaced by alternative methods, whenever possible (replace).

The 3Rs principles assist in improving the welfare of experimental animals in numerous ways including concerns about scientific animal use, focus on individual animals, adapts and responds to new information, balances the needs of science and the needs of the animals and assists in improving animal welfare through uniting different groups who have an interest in the welfare of animals used in science including scientists, veterinary professionals, policymakers, humane organizations, and the general public. Adopting the 3Rs in biomedical animal research will empower researchers in justifying their research (robustness of the experimental design), and results (validity of the scientific outputs) in higher-quality science.

Bibliography

Badyal DK, Desai C. Animal use in pharmacology education and research: the changing scenario. Indian J Pharmacol. 2014;46(3):257–65.

Balcombe JP, Barnard ND, Sandusky C. Laboratory routines cause animal stress. Contemp Top Lab Anim Sci. 2004;43(6):42–51.

Belmaker M, Brown AB. A new look at "on Mice and Men": should commensal species be used as a universal indicator of early sedentism? In: Marom N, Yeshuran R, Weissbrod L, Bar-Oz G, editors. Bones and identity: zooarchaeological approaches to reconstructing social and cultural landscapes in Southwest Asia. Oxford: Oxbow Books; 2016.

Bosch F, Rosich L. The contributions of Paul Ehrlich to pharmacology: a tribute on the occasion of the centenary of his Nobel Prize. Pharmacology. 2008;82(3):171–9.

Doke SK, Dhawale SC. Alternatives to animal testing: a review. Saudi Pharm J. 2015;23(3):223–9.

Fenwick N, Griffin G, Gauthier C. The welfare of animals used in science: how the "Three Rs" ethic guides improvements. Can Vet J. 2009;50(5):523–30.

Festing S, Wilkinson R. The ethics of animal research. EMBO Rep. 2007;8(6):526–30.

Flecknell P. Replacement, reduction, and refinement. ALTEX. 2002;19(2):73–8.

Franco NH. Animal experiments in biomedical research: a historical perspective. Animals (Basel). 2013;3(1):238–73.

Gill TJ, Smith GJ, Wissler RW, Kunz HW. The rat as an experimental animal. Science. 1989;245 (4915):269–76.

Guide for the Care and Use of Laboratory Animals. National research council (US) committee for the update of the guide for the care and use of laboratory animals. 8th ed. Washington, DC: National Academies Press (US); 2011.

Guidelines on the Care of Laboratory Animals and Their Use for Scientific Purposes. I. Housing and Care. The universities federation for animal welfare. 1987. Available from: https://royalsociety.org/-/media/Royal_Society_Content/policy/publications/1987/10705.pdf, Last assessed 14 July 2019.

Handling Methods of Laboratory Mice and Rats. Student report 657, Uppsala 2016. Available from: https://stud.epsilon.slu.se/9461/1/Lyden_F_160708.pdf. Last assessed 05 July 2019.

Manzoor M, Raza S, Chaudhry B. Proficient handling and restraint of the laboratory animal rat (*Rattus norvegicus*) facilitate essential biochemical and molecular level studies in biomedical science. IOSR J Pharmacy Biol Sci. 2013;6(2):21–33.

Mapara M, Thomas BS, Bhat KM. Rabbit as an animal model for experimental research. Dent Res J (Isfahan). 2012;9(1):111–8.

Smith R. Animal research: the need for a middle ground. BMJ. 2001;322(7281):248–9.

Sneddon LU, Halsey LG, Bury NR. Considering aspects of the 3Rs principles within experimental animal biology. J Exp Biol. 2017;220(17):3007–16.

[illegible]

Biological Sample Collection from Experimental Animals

4

Subramani Parasuraman and Ramasamy Raveendran

Abstract

Biological samples such as blood, urine and other body fluids are collected from laboratory animals for specific purposes. The methods used for the collection of biological samples should be the least stressful and painful. Prior permission should be obtained from the animal ethics committee for collection of biological samples from the animals. In the laboratory, rats, mice, guinea pigs and rabbits are most commonly used for pharmacological experiments. The frequency of blood collection depends on animal species and nonterminal blood collection is limited up to 10% of total circulating blood volume in healthy adult animals. Apart from blood samples, urine and saliva samples are collected from animals for investigation. In this chapter, blood, urine and saliva collection methods from laboratory animals are discussed.

Keywords

Blood sample · Animal ethics · Rodents · Non-rodents

S. Parasuraman (✉)
Department of Pharmacology, AIMST University, Bedong, Kedah Darul Aman, Malaysia
e-mail: parasuraman@aimst.edu.my

R. Raveendran
Department of Pharmacology, Jawaharlal Institute of Postgraduate Medical Education & Research (JIPMER), Puducherry, India

M. Lakshmanan et al. (eds.), *Introduction to Basics of Pharmacology and Toxicology*, https://doi.org/10.1007/978-981-19-5343-9_4

4.1 Introduction

Biological samples such as blood, urine, and other body fluids are collected from laboratory animals for specific purposes. Commonly the biological samples are collected from the laboratory/experimental animals for the estimation of haematological, biochemical, and immunological parameters and also to produce antibodies. The methods used for the collection of biological samples should be the least stressful and painful. Prior permission should be obtained from the animal ethics committee for the collection of biological samples from the animals. Blood, urine, and saliva sample collection methods from the laboratory animals are described here.

4.2 Collection of Blood

Blood is collected from laboratory animals for pathological, biochemical, haematological, pharmacokinetic, and immunological investigations. The method of collection, frequency of sampling and volume of the blood sample to be drawn depends on the animal species and experimental needs.

4.2.1 General Principles

- The method(s) used for blood sample collection from experimental animals should be approved by the Animal Ethics Committee.
- The method used for blood sample collection should be less stressful and less painful.
- The blood sample may be collected without or with anesthesia (Tables 4.1 and 4.2).
- The frequency of blood collection depends on animal species (for rodents, once in 2 weeks is preferred). If the study requires multiple blood samples, lagomorphs (e.g. hares, rabbits) can be used.
- Nonterminal blood collection in healthy adult animals is limited to 10% of total circulating blood volume on a single occasion, and blood collection can be repeated every three or four weeks at intervals. A maximum of 1% of the animal's total blood volume, or 0.6 ml/kg/day, can be collected every 24 h when repeated blood samples are required at short intervals.
- The estimated blood volume in adult animals is summarized in Table 4.3. The animals will require fluid replacement if the sample volume exceeds >10% of their total blood volume. The National Institutes of Health (NIH) recommends Lactated Ringer's solution (LRS) for fluid replacement. An adequate measure should be provided to prevent hypovolemia if the sampling volume exceeds >30% of the total circulatory blood volume of the experimental animals.

Table 4.1 Non-terminal blood sample collection methods during experimentation

Species	Blood collection route(s)	Sedation recommended	Anesthesia not required	Anesthesia required (local/general)
Mice	Saphenous vein		✓	
	Dorsal pedal vein		✓	
	Tail vein			✓
	Tail snip			✓
	Orbital sinus			✓
	Jugular vein			✓
Rat	Saphenous vein		✓	
	Sublingual vein	✓		
	Dorsal pedal vein		✓	
	Tail vein			✓
	Orbital sinus			✓
	Jugular vein			✓
	Temporary cannula			✓
	Blood vessel cannulation			✓
Guinea pig	Saphenous vein		✓	
	Blood vessel cannulation			✓
	Tarsal vein			✓
Rabbit	Marginal ear vein/artery			✓
	Central auricular artery		✓	
	Saphenous vein	✓		
	Jugular vein	✓		
Ferret	Cephalic vein		✓	
	Saphenous vein		✓	
	Jugular vein	✓		
	Cranial vena cava	✓		
	Blood vessel cannulation			✓
Hamster	Saphenous vein		✓	
	Cephalic vein		✓	
	Jugular vein	✓		
	Cranial vena cava	✓		
Gerbils	Lateral saphenous vein		✓	

(continued)

Table 4.1 (continued)

Species	Blood collection route(s)	Sedation recommended	Anesthesia not required	Anesthesia required (local/general)
	Cranial vena cava	✓		
Dog, Cat	Cephalic vein		✓	
	Saphenous vein		✓	
	Jugular vein		✓	
Pig	Ear vein	✓		
	Cranial vena cava	✓		
	Jugular vein	✓		

Table 4.2 Terminal blood sample collection methods

Species	Blood collection route(s)	Anesthesia required
Rat	Cardiac puncture	✓
	Posterior vena cava	✓
	Orbital sinus	✓
Mice	Cardiac puncture	✓
	Posterior vena cava	✓
	Orbital sinus	✓
Rabbit	Cardiac puncture	✓
Guinea pig	Cardiac puncture	✓
Ferret	Cardiac puncture	✓
Hamster	Cardiac puncture	✓
Gerbil	Cardiac puncture	✓
Dog, Cat	Cardiac puncture	✓
Pig	Cardiac puncture	✓

4.2.2 Blood Sample Collection Through Saphenous Vein

Requirements include animal, cotton, alcohol cotton ball, towel, electric trimmer, capillary tube, sample collection tubes, 20G needle and rodent handling gloves. From mice, rats, guinea pigs, rabbits, ferrets, hamsters, gerbils, dogs, and cats the blood sample is collected from the lateral saphenous vein with aseptic precautions. The animal is restrained using an animal restrainer or manually. The back of the animal's hind leg is shaved using an electric trimmer until the saphenous vein is visible (Fig. 4.1a). The collection site is cleaned with an alcohol cotton ball. The hind leg is immobilized and slight pressure can be applied gently above the knee joint. Using a 20G needle, the vein is punctured, and an adequate volume of blood is collected in a syringe with a needle or a capillary tube. The puncture site is cleaned with sterile cotton and compressed to stop the bleeding (Fig. 4.1b).

Table 4.3 Estimated blood volume in adult laboratory animals

Species	Reference weight (g)	Blood volume (ml/kg)[a]	Total blood volume (TBV), normal adult (ml)	Safe volume for single bleed (ml)[b]	Bleed out volume (ml)
Mouse	18–40	58.5	Male 1.5–2.4 Female 1.0–2.4	0.1–0.2	Male 0.8–1.4 Female 0.6–1.4
Rat	250–500	54–70	Male 29–33 Female 16–19	Male 2.9–3.3 Female 1.6–1.9	Male 13–15 Female 7.5–9
Hamster	85–150	78	Male 6.3–9.7 Female 7.1–11.2	Male 0.6–0.9 Female 0.7–1.1	Male 2.9–4.5 Female 3.3–5.2
Gerbil	55–100	66–78	Male 4.5–7 Female 3.8–6	Male 0.4–0.7 Female 0.4–0.6	Male 2.2–3.5 Female 1.9–2.9
Guinea pig	700–1200	69–75	Male 59–84 Female 48–63	Male 6–8 Female 5–6	Male 29–42 Female 24–31
Rabbit	1000–6000	57–65	58.5–585	5–50	31–310
Ferret	600–2000	70	42–140	4–14	21–70
Dog	–	70–110[c]	900–1170[d]	90–110	–
Cat	–	47–65	140–200	14–20	–
Pig(large white)	–	56–69	13,200–15,000	1320–1500	–
Pig (yucatan)	–	56–69	4200–4800	420–480	–
Sheep	–	58–64	4060–4480	400–450	–
Goat	–	57–90	3990–6300	400–630	–
Cattle	–	60	27,000–36,000[e]	2700–3600	–
Horse	–	75	33,750–45,000[e]	3375–4500	–
Marmoset	–	60–70	21–24.5	2.1–2.4	–
Rhesus macaque	–	55–80	Male 420–770 Female 280–630	Male 42–77 Female 28–63	–
Long-tailed macaque	–	50–96	Male 280–560 Female 140–420	Male 28–56 Female 14–42	–

Reproduced with permission from the NC3Rs: https://www.nc3rs.org.uk/blood-sample-volumes. Table adapted from Wolfensohn & Lloyd (2003), Handbook of Laboratory Animal Management and Welfare, third Edition

[a]A blood volume estimate for a single species may not reflect differences among individual breeds or variations due to age, size, or illness

[b]Single bleed of 10% total blood volume

[c]Much breed variation

[d]Beagle

[e]Assumes adult weight 450–600 kg

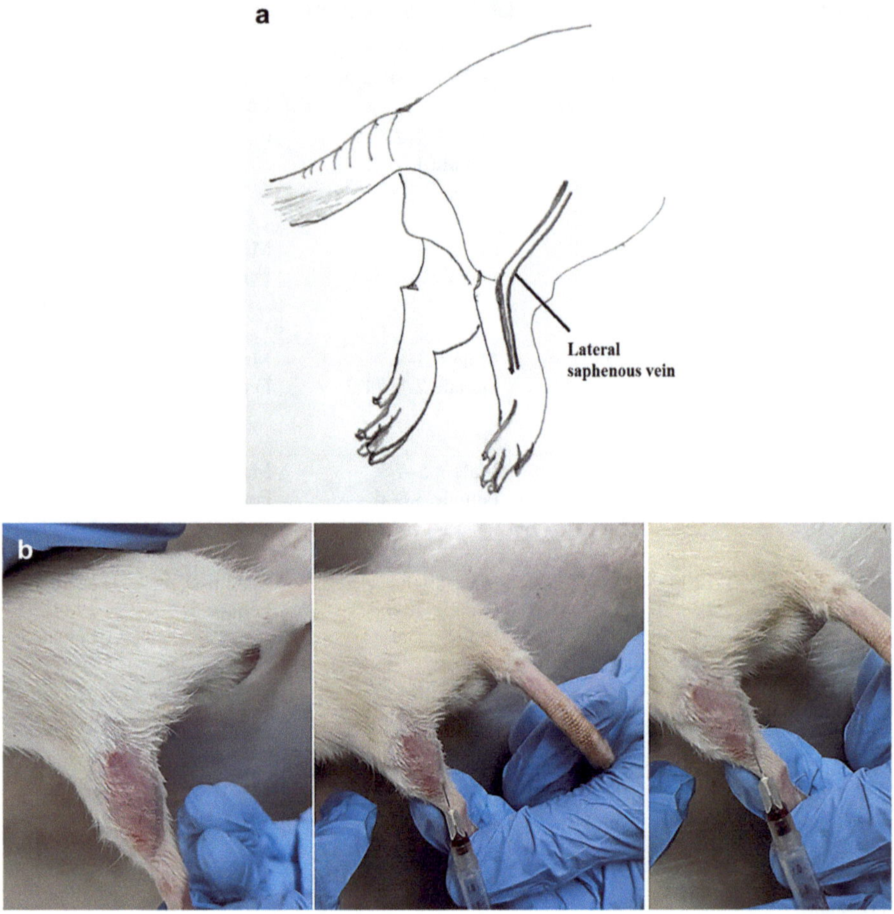

Fig. 4.1 (**a**) Identification of rat lateral saphenous vein. (**b**) Blood sample collection from rat lateral saphenous vein

Caution:

- On the collection site, a topical anesthetic agent may be applied to reduce pain
- Avoid multiple attempts
- Collecting more than four samples per day is not recommended

4.2.3 Blood Sample Collection Through Dorsal Pedal Vein

Requirements include rodent handling gloves, animal, cotton, alcohol cotton ball, 23G/27G needle, capillary tube, and blood-sample-collection tubes. From mice and rats, the blood sample is collected from the dorsal pedal vein with aseptic precautions. The animal is restrained using an animal restrainer or manually. The medial dorsal pedal vessel is positioned on top of the hindfoot around the ankle. The foot is cleaned using alcohol and the dorsal pedal vein is punctured with a 23G/27G needle. The blood drops that would appear on the surface of the skin are collected in a capillary tube. The puncture site is cleaned with sterile cotton and compressed to stop the bleeding (Fig. 4.2).

Caution:

- On the collection site, a topical anesthetic agent may be applied to reduce pain
- Avoid multiple attempts

4.2.4 Blood Sample Collection Through Tail Vein

Requirements include rodent handling gloves, animal, animal warming chamber, LRS, towel, alcohol, antiseptic solution, alcohol cotton ball, cotton and sample collection tube. From mice and rats, the blood sample is collected from the tail vein with aseptic precautions. The tail vein blood sample collection method is recommended for the collection of the large volume of the blood sample in rodents. The animal is restrained using an animal restrainer and the temperature is maintained at around 24–27 °C. The blood sample collection site is cleaned with alcohol/ antiseptic solution. If the tail vein is not visible, the tail may be dipped in warm water (approx. 40 °C). Before 30 min of sample collection, a topical anesthetic agent should be applied on the tail to minimize the pain. Immobilize the a tail in non-dominant hand rotate and align the needle (beveled edge should be facing up)

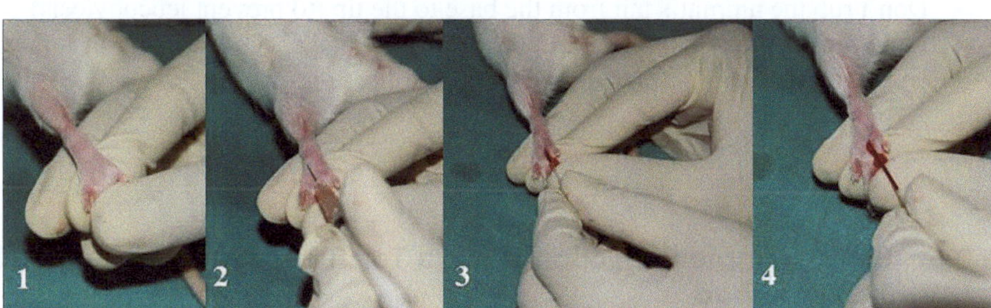

Fig. 4.2 Blood sample collection from rat dorsal pedal vein (Reproduced with permission from Phcog.Net, Bangalore, India)

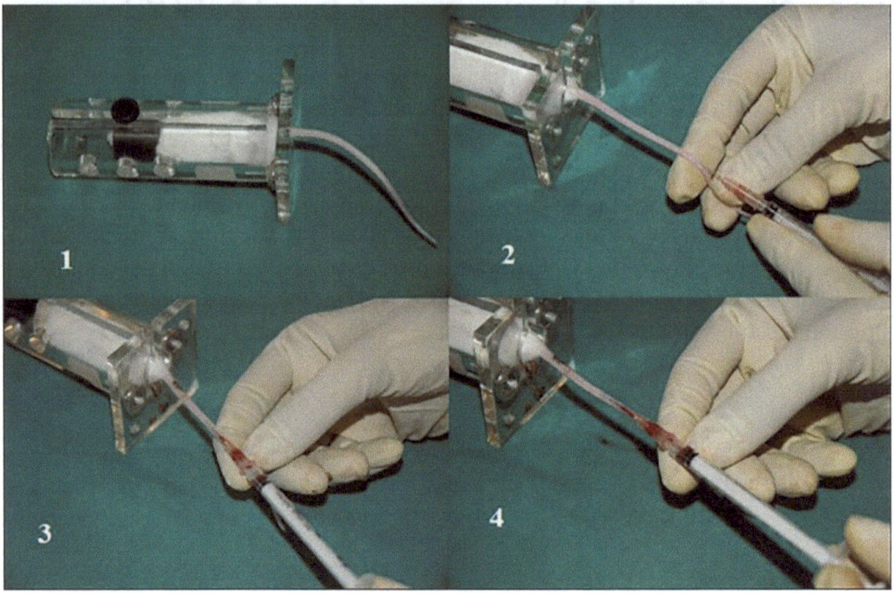

Fig. 4.3 Blood sample collection from mice tail vein (Reproduced with permission from Phcog. Net, Bangalore, India)

parallel to the tail. Genteelly a 23G needle is inserted into the tail vein starting at the tip of the tail. In case of difficulties, cut 0.5–1 cm of the skin surface, puncture the vein with a needle or blood collection needle, and collect blood with a syringe with a needle or a capillary. After completion of tail vein blood collection, pressure/silver nitrate solution or ointment is applied to stop the bleeding. A temporary surgical cannula can be used if multiple samples are needed for the experiment. The animal (rodent) restrainer is cleaned frequently to prevent pheromonally-induced cross-infection or stress (Fig. 4.3).

Caution:
- Don't rub the animal's tail from the base to the tip (to prevent leucocytosis)
- Fluid replacement may be required when accidental excess bleeding occurs
- Use surgical cannula for multiple sample collection
- Avoid multiple attempts

4.2.5 Blood Sample Collection Using a Tail Bleed (Tail Snip and Tail Nick) Method

Requirements include animal, surgical/scalpel blade (for both tail snip and tail nick method), cotton, anesthetic agent and blood sample collection tubes. The tail bleed blood sample collection method is recommended only for mice. The mouse is restrained using a suitable mice restrainer. The mouse tail is wiped with warm water (approx. 40 °C) to clean and cause slight vasodilation. Blood sample collection from the tail snip and tail nick methods is carried out under general anesthesia.

For the tail snip blood sample collection method, the tail is extended, and the end of the tail (0.5–1 mm for mice) is cut with the surgical/scalpel blade (preferably a number 11 blade).

In the tail nick blood sample method, the tail is extended, and a cut is made with the surgical blade about two-thirds the distance from the backend, directly over the lateral tail vein.

Caution:

- This method should be avoided because it can cause potential permanent tail damage
- It should be done under terminal anesthesia only
- Do NOT use hot water to clean the tail
- Scissors should not be used in tail nick method
- Cut made using scissors is crushing and promotes blood clotting and decreases blood flow
- Avoid multiple attempts

4.2.6 Blood Sample Collection Through Orbital Sinus

Requirements include animal, anesthetic agent, capillary tube, LRS, blood sample collection tubes and cotton. The blood sample collection from the orbital sinus is recommended for rats and mice. Orbital sinus bleeding is performed in the terminal procedure. This method is also recommended to use with recovery in experimental circumstances with scientific justification (where the researcher needs a large volume of a blood sample). Blood sample collection from the orbital sinus is carried out under general anesthesia. The non-dominant hand's thumb and forefingers are used to scruff the mouse/rat, and the skin surrounding the eye is pulled taut. A capillary tube is inserted into the medial canthus of the eye (30-degree angle to the nose). Slight thumb pressure is sufficient to puncture the tissue and enter the plexus/sinus. Blood will flow through the capillary tube once the sinus/plexus is punctured. After collection of the required volume of the blood from plexus, the capillary tube is gently removed, and the eye is wiped with sterile cotton. Gentle finger pressure is applied to the eyes to stop bleeding (Figs. 4.4 and 4.5). After 30 min of blood sample collection, the animal is examined for postoperative and periorbital lesions. The recommended volume of sample collection through orbital sinus bleeding is 0.2 mL with recovery/ 0.5 mL with non-recovery in 20 g adult mouse and 2.0 mL in 200 g adult rat.

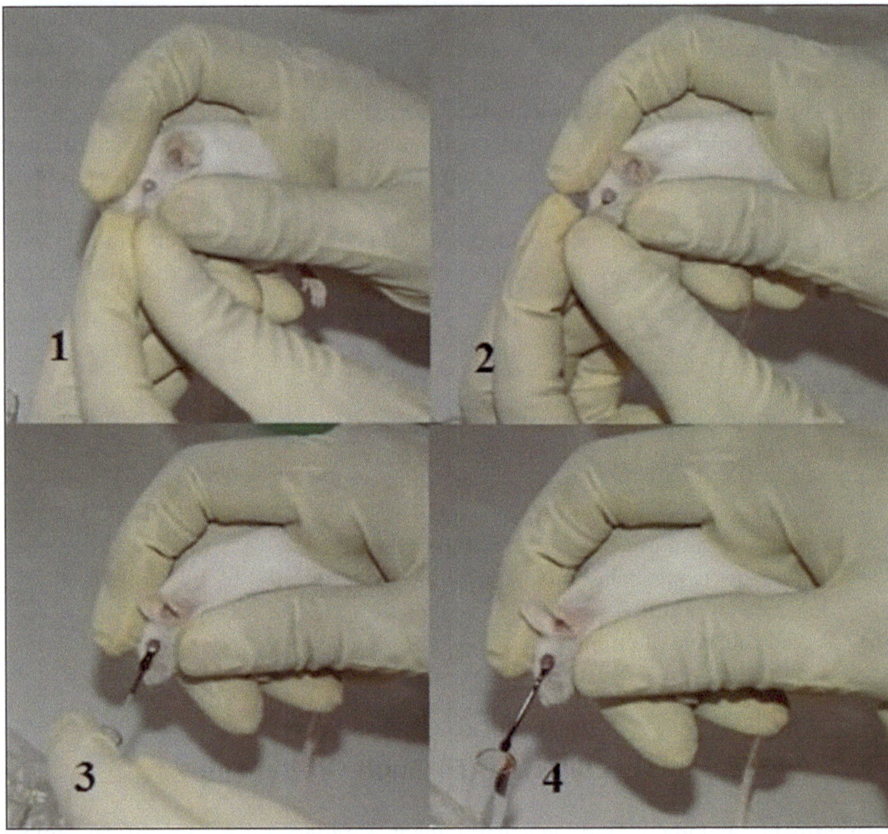

Fig. 4.4 Blood sample collection from mouse orbital sinus (Reproduced with permission from Phcog.Net, Bangalore, India)

Caution:

- Adequate training is required
- Repeated blood sample collection is not recommended
- Avoid multiple attempts
- A 2-week interval between two bleedings is recommended

4.2.7 Blood Sample Collection Through Jugular Vein

Requirements include animals, cotton, anesthetic agent, alcohol cotton ball, 25G needle, and blood-sample-collection tubes. The jugular vein blood sample collection is recommended for blood sample collection in mice, rats, rabbits, ferrets, hamsters, dogs, cats, and pigs. This method is used to collect micro volumes to a milliliter of a

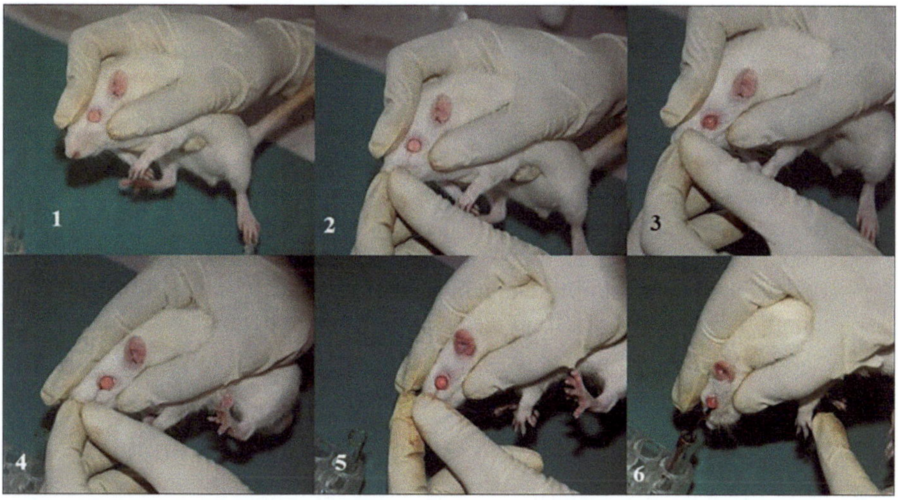

Fig. 4.5 Blood sample collection from rat orbital sinus (Reproduced with permission from Phcog. Net, Bangalore, India)

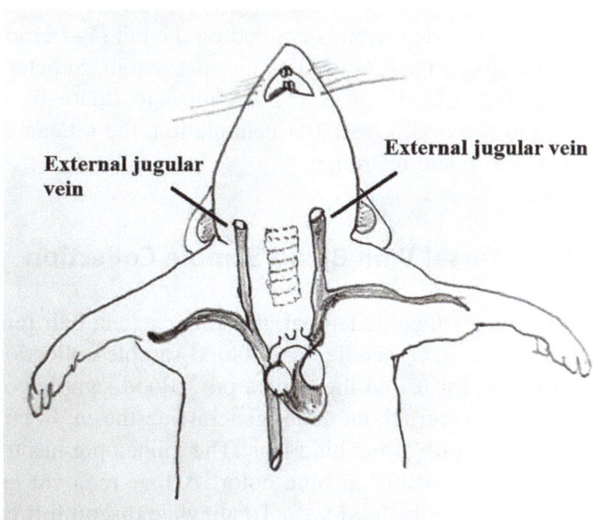

Fig. 4.6 Identification of rat external jugular vein

blood sample. Blood sample collection from the jugular vein is carried out under sedation or inhalation anesthesia/general anesthesia. The animal is restrained manually. The neck region of the animal is shaved and kept in a hyperextended posture. The jugular veins appear blue (in the neck region) and are found 2–4 mm lateral to the sternoclavicular junction (Fig. 4.6). A 25G needle is inserted in the caudocephalic direction (back to front) and blood is withdrawn gently to prevent collapsing these small blood vessels. The animal is handled gently and not more than

3–4 mm of the needle is to be inserted into the blood vessel. After the withdrawal of blood, the needle is slowly removed, and gentle finger pressure is used to stop bleeding.

Caution:
- Adequate training is required
- Repeated blood sample collection is not recommended
- Avoid multiple attempts; it may collapse the vein
- If the blood collection attempt fails, the needle is gently removed and monitored for bleeding (If no bleeding, one more attempt can be made)

4.2.8 Blood Sample Collection Using a Temporary Cannula

Requirements include animal, anesthetic agent, alcohol cotton ball, cotton, 22G intravenous catheter, animal warming chamber, and blood-sample-collection tubes. This technique is appropriate for use in the rat. Normally, temporary cannulation is made in the rat tail vein and used for a few hours. The rat is restrained and local/topical anesthetic cream is applied on the tail (1–2 cm above the tail tip). The tail is cannulated with a 22G (0.90) mm intravenous catheter. Rat is restrained at a warm temperature (37–39 °C for 5–15 min) to dilate the blood vessels and to have continuous blood flow. After cannulation, the rat can be restrained either manually or using a rodent restrainer.

4.2.9 Tarsal Vein Blood Sample Collection

Requirements include animal, anesthetic agent, hair removal cream, cotton, alcohol cotton ball, 22G needle, and blood-sample-collection tubes. This technique is appropriate for use in the guinea pig. Blood sample collection from the guinea pig tarsal vein is carried out under general anesthesia. In large animals, the tarsal vein is identified in one of the hind legs. The guinea pig has to be restrained properly. The tarsal vein is visible in blue color. A hair removal cream is used to remove the surface hairs in the hind legs. To alleviate discomfort/pain, topical local anaesthetic medication is administered to the collection site. After 20 min, the blood sample is collected slowly using a 22G needle. After sample collection, the needle is removed gently, and mild thumb pressure is applied with a finger for 2 min to stop bleeding (Fig. 4.7).

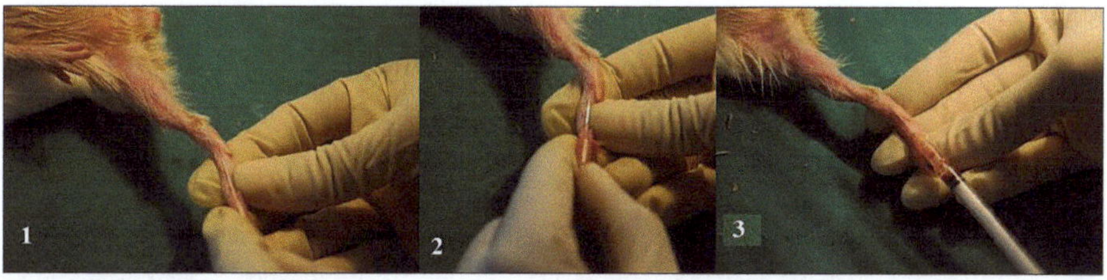

Fig. 4.7 Blood sample collection from guinea pig tarsal vein (Reproduced with permission from Phcog.Net, Bangalore, India)

Caution:
- Adequate training is required
- Repeated blood sample collection is not recommended
- Avoid multiple attempts, it may collapse the vein
- A maximum of three samples can be collected per leg
- Collection of more than six blood samples in any 24 h is not recommended
- Each sampling, 0.1–0.3 ml of blood can be collected

4.2.10 Marginal Ear Vein/Artery Blood Sample Collection

Requirements include animal, anesthetic agent (topical), cotton, 26G needle, alcohol cotton ball, 95% v/v alcohol, ortho-xylene, topical vasodilator, surgical blade, and blood sample collection tube. This method is appropriate for use in rabbits. The animal is restrained using a suitable animal restrainer. The rabbit ear is cleaned with alcohol and a topical anesthetic agent is applied on the collection site 10 min before sampling. A topical vasodilator (orthoxylene) can be applied on the collection site to dilate blood vessels, if required. A 26G needle is inserted on the marginal vein/artery to collect blood. After sampling, the needle is gently removed from the marginal vein/artery and cleaned with sterile cotton. In the collection site, clean sterile cotton is kept, and finger pressure is used to stop the bleeding (Fig. 4.8). Otherwise, a small incision is made with a size 11 surgical blade on the marginal ear vein and blood sample is collected in a sample collection tube. Later, the marginal vein is cleaned with sterile cotton, and finger pressure is used to stop the bleeding (Fig. 4.9).

4.2.11 Blood Sample Collection Through Cephalic Vein

Requirements include animal, cotton, 25–27G needle, alcohol cotton ball, cotton, and blood sample collection tube. This technique is appropriate for use in the ferret, hamster, dog, and cat. The animal is restrained manually or using a suitable animal restrainer. The cephalic vein is situated on the dorsal surface of the front leg. This

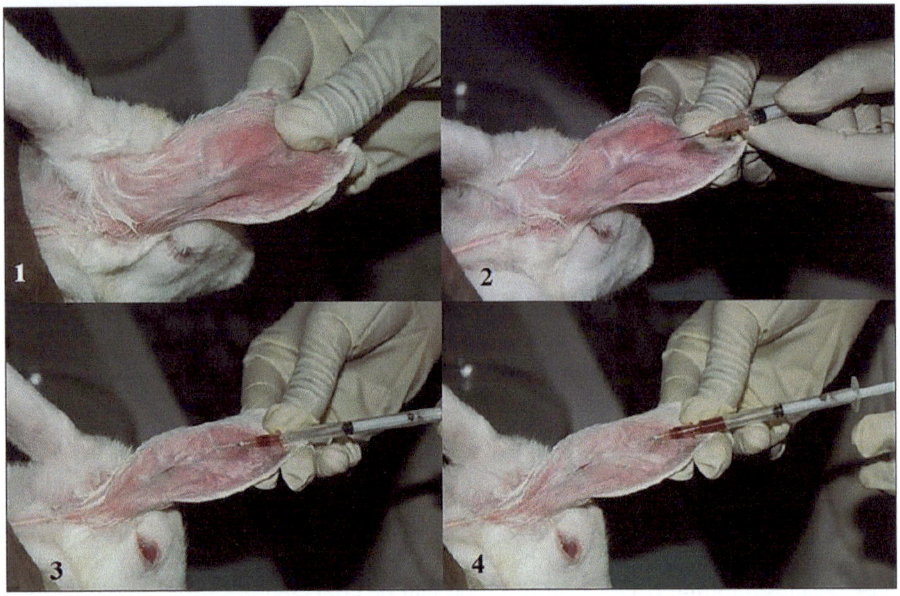

Fig. 4.8 Blood sample collection from rabbit central artery using 26G needle (Reproduced with permission from Phcog.Net, Bangalore, India)

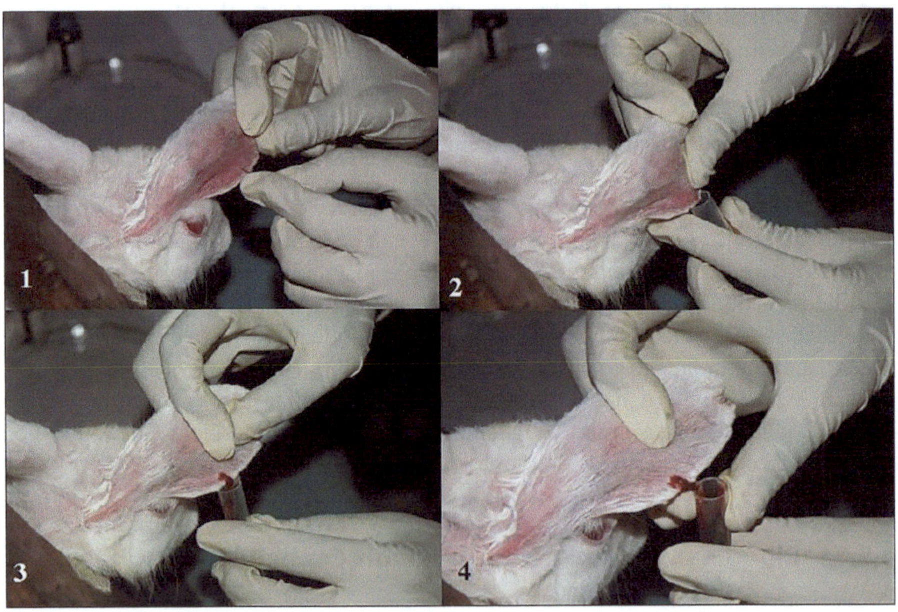

Fig. 4.9 Blood sample collection from rabbit marginal ear vein using incision method. (Reproduced with permission from Phcog.Net, Bangalore, India)

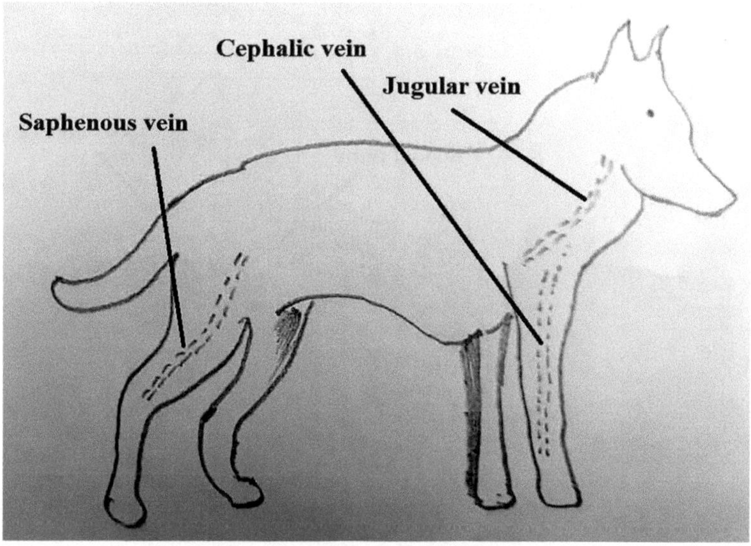

Fig. 4.10 Identification of saphenous vein, cephalic vein and the jugular vein in dog

method is quick and appropriate to collect a small volume of blood (2–5 ml) using a 25–27G needle. Blood is collected from the peripheral vein in the forelimb, starting distally (near the paw) and working proximally (away from the paw). This vein should be preheated and occluded with a tourniquet. A needle is inserted into the blood vessel and blood is collected using a syringe. A temporary cannula can be used to collect multiple samples in a short period. After sample collection, the needle is removed gently and finger pressure is applied on the collection site for hemostasis, and the tourniquet is removed (Fig. 4.10).

4.2.12 Blood Sample Collection Through Cranial Vena Cava

Requirements include animal, cotton, anesthetic agent, 23–25G needle with 1–2 ml syringe, and blood sample collection tube. Venipuncture of the cranial vena cava is suitable for collecting large blood volumes in rats, ferrets and large animals. Blood from cranial vena cava is collected under short-term inhalation anesthesia. The rat is restrained at a warm temperature. It is placed in dorsal recumbency with forelimbs pointing caudally and parallel to the body. The blood sample collection site is 0.3–0.8 cm lateral to the manubrium, located immediately cranial to the first rib. The needle is inserted 0.2–1 cm in the same direction until blood starts to flow at a 30-degree angle to the thermal pad in the direction opposite to the femoral head. After collecting the blood sample, the needle is carefully removed and light thumb pressure is applied at the site of the insertion for 30 s to stop bleeding. Collection of >20% of the total blood volume of experimental animal is not recommended for any 2-week period (Fig. 4.11).

Fig. 4.11 Identification of cranial vena cava in rat

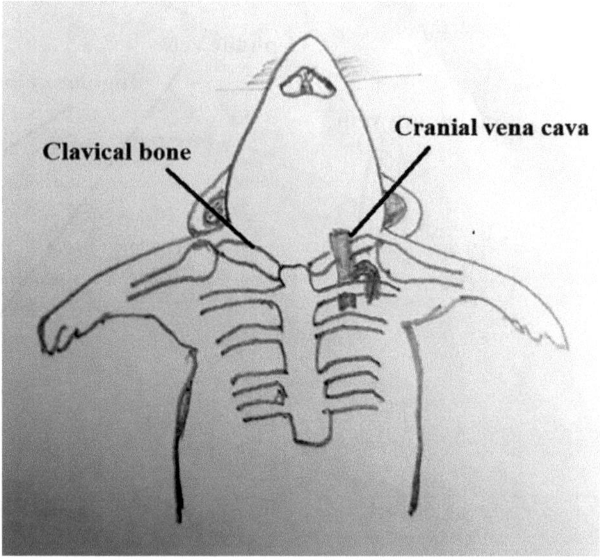

4.2.13 Blood Sample Collection Through Posterior Vena Cava

Requirements include an animal, cotton, anesthetic agent, small glass rods, surgical scissor, scalpels and surgical blade, 1–5 ml syringe with 21–25G needle and blood-sample-collection tube. This method is appropriate for collecting the large volume of blood samples in rodents at the end of the study. The animal is anesthetized, 'V' or 'Y'-shaped cut is made in the abdomen, and the intestinal tissues, small intestine and large intestine are gently removed. The liver is pushed forward and the posterior vena cava is identified (between the kidneys). A 21–25G needle is inserted on posterior vena cava to collect blood. This method is repeated 3 or 4 times to collect a large volume of a blood sample.

4.2.14 Blood Sample Collection Through Cardiac Puncture

Requirements include animal, anesthetic agent, surgical blade, 1–5 ml syringe with 19–25G needle, towel, cotton, tube for thoracotomy (internal diameter of 0.1–0.3 mm), and blood-sample-collection tubes. In general, blood sample collection through cardiac puncture is recommended to collect good quality and a large volume of blood samples from the experimental animals at the end of the study. In this method, the appropriate needle (19–25G) is used for collecting blood samples with or without thoracotomy. To avoid collapse, the blood sample will be collected carefully from the heart, preferably from the ventricle (Fig. 4.12).

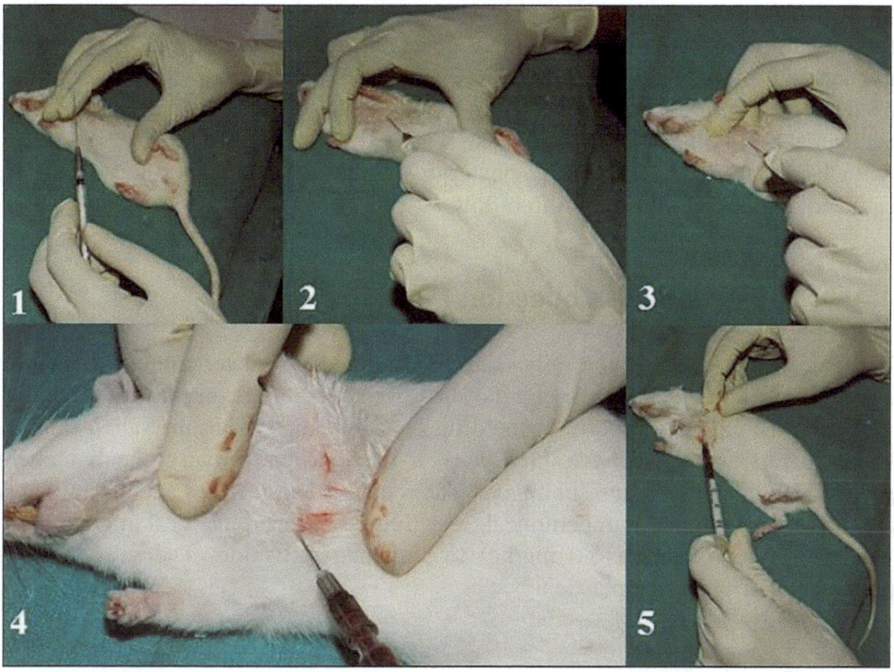

Fig. 4.12 Blood sample collection through cardiac puncture in rat (Reproduced with permission from Phcog.Net, Bangalore, India)

4.2.15 Adverse Effects of Blood Sample Collections

Infections (due to problems with aseptic technique), infection at the site of sample collection, skin abrasions, seroma formation, haemorrhage, hypovolaemia, stress, bruising (may be caused by venipuncture), phlebitis, thrombosis, nerve damage, swelling and scarring are adverse effects of blood sample collection. These effects can be avoided or minimized by using appropriate experimental care.

Hematoma, corneal ulceration, keratitis, pannus formation, rupture of the globe, damage to the optic nerve, and necrotic dacryoadenitis of the Harderian gland are uncommon adverse effects observed in orbital sinus bleeding.

4.3 Urine Sample Collecting Methods

Several behavioral, biochemical, metabolic, nutritional, physiological, toxicological, and urological studies required urine sample collection from experimental animals. Urine sample is collected from experimental animals and used for qualitative (clinical veterinary purpose) or quantitative (experimental scientific purpose) analysis.

In small animals such as rats and mice, the urine sample is collected by housing the animals in metabolic cages. From mice and rats, the urine sample is also collected by manual urinary bladder compression, cystocentesis (a common clinical technique used to collect urine samples directly from the urinary bladder), and urinary bladder catheterization (only in female animals).

In large animals (e.g., dogs), the urine sample is collected by cystocentesis, catheterization, and mid-stream free-flow.

4.4 Saliva Sample Collection Method

In the preclinical studies, the saliva sample is collected to determine the metabolic enzyme levels and to study the effect of investigational drugs on salivary secretion. The pharmacokinetic property of the drug is also studied using the saliva sample. Rodents (rats and mice) have three pairs of macroscopic salivary glands namely parotid, submandibular, and sublingual. The rodent is anesthetized and the animal is administered with an intraperitoneal injection of pilocarpine (1.0 mg/kg body weight) or isoproterenol (5.0 mg/kg) to stimulate salivation. The saliva that is produced is dripped in a plastic tube.

4.5 Conclusion

Biological sample collections from the experimental animals are one of the important procedures in pharmacokinetic studies, pharmacological screening of drugs, and biomedical research. Researchers should ensure the method used to collect biological samples from the experimental animals is ethically approved and induces minimum/no pain.

Acknowledgments The authors are grateful to Phcog.Net, Bangalore, India, and NC3Rs, London, for giving permission to reproduce Figs. 4.2, 4.3, 4.4, 4.5, 4.7, 4.8, 4.9 and 4.12 from 'Parasuraman S, Raveendran R, Kesavan R. Blood sample collection in small laboratory animals. J Pharmacol Pharmacother. 2010;1 (2):87-93' and Table 4.3 from 'https://www.nc3rs.org.uk/blood-sample-volumes', respectively.

Bibliography

Amano O, Mizobe K, Bando Y, Sakiyama K. Anatomy and histology of rodent and human major salivary glands: -overview of the Japan salivary gland society-sponsored workshop. Acta Histochem Cytochem. 2012;45(5):241–50.

Bagavant H, Trzeciak M, Papinska J, Biswas I, Dunkleberger ML, Sosnowska A, Deshmukh US. A method for the measurement of salivary gland function in mice. J Vis Exp. 2018;(131):57203. https://doi.org/10.3791/57203.

Birck MM, Tveden-Nyborg P, Lindblad MM, Lykkesfeldt J. Non-terminal blood sampling techniques in guinea pigs. J Vis Exp. 2014;92:e51982. https://doi.org/10.3791/51982.

Blood Collection (Guideline). Available from: https://animal.research.uiowa.edu/iacuc-guidelines-blood-collection, Last assessed 30 June 2019.

Blood Sample Volumes., Available from: https://www.nc3rs.org.uk/blood-sample-volumes, Last assessed 31 July 2019.

Head V, Eshar D, Nau MR. Techniques for nonterminal blood sampling in black-tailed prairie dogs (Cynomys ludovicianus). J Am Assoc Lab Anim Sci. 2017;56(2):210–3.

Hem A, Smith AJ, Solberg P. Saphenous vein puncture for blood sampling of the mouse, rat, hamster, gerbil, guinea pig, ferret and mink. Lab Anim. 1998;32:364–8.

Hoff J, Rlagt LV. Methods of blood collection in the mouse. Lab Anim. 2000;29:47–53.

Jekl V, Hauptman K, Jeklová E, Knotek Z. Blood sampling from the cranial vena cava in the Norway rat (Rattus norvegicus). Lab Anim. 2005;39(2):236–9.

JoVE Science Education Database. Lab animal research. Cambridge, MA: Blood Withdrawal I. JoVE; 2019.

Kurien BT, Everds NE, Scofield RH. Experimental animal urine collection: a review. Lab Anim. 2004;38(4):333–61.

Lu C, Anderson LC, Fenske RA. Determination of atrazine levels in whole saliva and plasma in rats: potential of salivary monitoring for occupational exposure. J Toxicol Environ Health. 1997;50(2):101–11.

Parasuraman S, Raveendran R, Kesavan R. Blood sample collection in small laboratory animals. J Pharmacol Pharmacother. 2010;1(2):87–93.

Paulose CS, Dakshinamurti K. Chronic catheterization using vascular-access-port in rats: blood sampling with minimal stress for plasma catecholamine determination. J Neurosci Methods. 1987;22:141–6.

Romero AC, Bergamaschi CT, de Souza DN, Nogueira FN. Salivary alterations in rats with experimental chronic kidney disease. PLoS One. 2016;11(2):e0148742. https://doi.org/10.1371/journal.pone.0148742.

Sheldon CC, Sonsthagen T, Topel J. Animal restraint for veterinary professionals. Animal restraint for veterinary professionals. 2nd ed. Amsterdam: Elsevier; 2016. p. 91.

Talhada D, Andrade A, Faustino-Rocha AI, Teixeira-Guedes CI, Teixeira J, Arantes-Rodrigues R, Vasconcelos-Nóbrega C, Gil da Costa R, Oliveira PA. Recommendations for urine and urinary bladder collection in chemical carcinogenesis assays with rodent. Scand J Lab Anim Sci. 2015;41:1–8.

Vachon P, Moreau JP. Serum corticosterone and blood glucose in rats after two jugular vein blood sampling methods: comparison of the stress response. J Am Assoc Lab Anim Sci. 2001;40(5): 22–4.

Van Herck H, Baumans V, Brandt CJ, Boere HA, Hesp AP, Van Lith HA, Schurink M, Beynen AC. Blood sampling from the retro-orbital plexus, the saphenous vein and the tail vein in rats: comparative effects on selected behavioural and blood variables. Lab Anim. 2001;35(2):131–9.

Vogel HG, editor. Drug discovery and evaluation: pharmacological assays. 3rd ed. Berlin, Heidelberg, New York: Springer; 2008.

Wolfensohn & Lloyd. Handbook of laboratory animal management and welfare. 3rd ed. USA: Blackwell publishing; 2003.

Yoburn BC, Morales R, Inturrisi CE. Chronic vascular catheterization in the rat: comparison of three techniques. Physiol Behav. 1984;33:89–94.

Anesthesia and Euthanasia of Experimental Animals

5

Subramani Parasuraman and Parayil Varghese Christapher

Abstract

Appropriate anaesthetic regimen is an essential component of laboratory animal experimental design to ensure analgesia and reversible loss of consciousness to facilitate surgical procedures. Pre-anesthetic medication and postoperative analgesic agents are playing vital role in animal experimentation to reduce inhalational or injectable anaesthetic dose and to reduce postoperative pain, respectively. This chapter describes the role of anesthesia in experimental animals, different stages of anesthesia with its characteristic features and role of pre-anesthetic medication with different classes of pre-anesthetic agents. This chapter also details about the types of anesthesia and, how to monitor its depth and the role of postoperative analgesia. Euthanasia, an important ethical aspect of animal experimentation with different methods to perform it and guidelines are also highlighted in this chapter.

Keywords

Anesthesia · Euthanasia · Pre-anesthetic medication · Postoperative analgesia

S. Parasuraman (✉)
Department of Pharmacology, AIMST University, Bedong, Kedah Darul Aman, Malaysia
e-mail: parasuraman@aimst.edu.my

P. V. Christapher
Department of Pharmacology, AIMST University, Bedong, Kedah Darul Aman, Malaysia

Department of Pharmacology, Al Shifa College of Pharmacy, Perinthalmanna, Malappuram, Kerala, India
e-mail: christapher@alshifacollegeofpharmacy.ac.in

M. Lakshmanan et al. (eds.), *Introduction to Basics of Pharmacology and Toxicology*, https://doi.org/10.1007/978-981-19-5343-9_5

5.1 Anesthesia of Experimental Animals

5.1.1 Anesthesia

Providing an appropriate anesthetic regimen is an essential part of animal experimental design. Anesthesia (from Greek "without sensation") has profound effects on the physiological functions of animals, and this can have a noticeable effect on experimental data. Not only anesthetic agents, effective use of analgesia following the surgical procedures is also important and the agent used to induce analgesia and anesthesia is without or with minimum adverse effects. In biomedical research, the experiments which either require recovery or terminal experimental procedures are performed under anesthesia.

5.1.2 Stages of Anesthesia

General anesthesia is a reversible drug-induced experimental coma with loss of protective reflexes, resulting from the administration of one or more general anesthetic agents. There are four stages of anesthesia as described in Table 5.1. These four stages are evident in animals with slow anesthesia-inducing agents such as ether, whereas these stages are not recognizable with the use of faster anesthesia-inducing agents such as thiopentone.

Table 5.1 Stages of Anesthesia

Stages	Physiological effect
I. Stage of analgesia	Heart and respiratory rate increase Dilation of pupils
II. Stage of excitation (delirium)	Irregular respiration Dilated pupils Increased motor reflexes Nystagmus (rapid involuntary eye movements) Opisthotonus (spasm of the muscles of the back)
III. Stage of tolerance. This stage is divided into four steps	
Plane 1: light anesthesia	Regular respiration, narrow pupils, presence of most reflexes
Plane 2: surgical anesthesia	Narrow pupils, no eyelid reflex, presence of corneal reflex, relaxed skeletal muscles, flat respiration, good analgesia
Plane 3: deep anesthesia	Only corneal reflex present, very flat respiration, dilated pupils
Plane 4: no reflexes	The animal loses all respiratory efforts
IV. Stage of asphyxia (medullary paralysis)	Paralysis of intercostal muscles leads to breathing cessation

Table 5.2 Commonly used pre-anesthetic medications for laboratory animal experiments

Pre-anesthetic medications	Use
Sedatives or tranquilizers: • minor tranquilizers without autonomic effects: meprobamate, diazepam • major tranquilizers with autonomic side effects: propionyl-promazine, acetylpromazine, azaperone, dehydrobenzperidol, detomidine, xylazine	• To reduce aggression and apprehension or fear • Stress-free induction of anesthesia • To reduce the dose of anesthetic agent needed to produce the desired level of anesthesia
Analgesics: • methadone, meperidine, fentanyl	• To reduce the dose of anesthetic agent needed to produce the desired level of anesthesia
Anticholinergics: • atropine 0.05–0.1 mg/kg body weight • glycopyrrolate 0.01–0.5 mg/kg body weight	• To reduce bronchial and salivary secretions • To protect the heart from vagal inhibition • To avoid cardiopulmonary problems

5.1.3 Pre-Anesthetic Medication

Most of the laboratory animals are docile. They are nervous or aggressive to have them well sedated before the induction of anesthesia. To reduce nervousness or aggressions, laboratory animals can be treated with pre-anesthetic medications such as analgesics, anticholinergics, sedatives, and tranquilizers (e.g., chlorpromazine). Though the pre-anesthetic medications help to reduce the dose of general anesthetic agent, they are not commonly used in rodents due to handling-induced stress. The commonly used pre-anesthetic medications and their uses are summarized in Table 5.2.

5.1.4 Types of Anesthesia

5.1.4.1 Local Anesthesia

Local anesthetic techniques are adjunct to anesthesia and used to block transmission of impulses in nerve fibers and to reduce or eliminate sensation. Local anesthetics help to desensitize the surgical site and reduce general anesthetic requirements that lead to animal comfort. In animals, local anesthesia is produced by topical or surface anesthesia, infiltration anesthesia, extravascular infiltration techniques, and intravascular infiltration techniques. Commonly recommended local anesthetic agents on laboratory animal experiments are summarized in Table 5.3.

5.1.4.2 General Anesthesia

Inhalational Anesthetics: Inhalational anesthetic agents are delivered to an animal along with oxygen (up to 5%) for the induction of anesthesia using a precision vaporizer. Usually, anesthesia is maintained with 21–33% oxygen in the inhalation mixture.

Table 5.3 Commonly recommended local anesthetic agents on laboratory animal experiments

Local Anesthetic	Onset time	Duration	Use
2 or 4% lignocaine HCl	5 min	30–45 min	Desensitization of affected area
4% lignocaine HCl as spray			Surface analgesia of mucous membranes
4% cocaine or lignocaine			Surface analgesia of eyes
Lignocaine or amethocaine jelly			Desensitization of abrasions, eczematous areas
Lidocaine (4 mg/kg)	5–15 min	1–2 h	Desensitizing the surgical site
Bupivacaine (1–2 mg/kg)	15–30 min	4–6 h	Desensitizing the surgical site
Mepivacaine (4 mg/kg)	5-15 min	1.5–3 h	Desensitizing the surgical site

An inhalational agent has its own advantages and disadvantages. They offer rapid onset of action, quick recovery time and, easy control of the depth of anesthesia. Inhalational agents do not have an analgesic effect and the procedure requires specialized equipment (induction chambers, facemasks, and endotracheal tubes) to induce anesthesia.

Methoxyflurane, enflurane, sevoflurane, and halothane are used as inhalational anesthetic agents. Ether is no longer recommended as it irritates the respiratory tract and due to its environmental hazards. Inhalational anesthesia can be stopped by removing the supply of evaporated anesthetic agent or increase the concentration of oxygen in the system for 5 min.

Isoflurane: Isoflurane is commonly used as an inhalational anesthetic agent in rodents for the past few years because of its quick on- and offset of action (induction time: 3–5 min at 5%) and low metabolism rate. In rodents, repeated administration of isoflurane is aversive than a single administration.

Sevoflurane: Sevoflurane is non-flammable, sweet-smelling, highly fluorinated methyl isopropyl ether that is used as an inhalational anesthetic agent. It has lower blood solubility and more rapid induction compared to isoflurane or halothane.

Others: Halothane, methoxyflurane, desflurane, and nitrous oxide are less commonly used as inhalational anesthetics in animals. Overall, all inhalational anesthetics depress the cardiopulmonary functions in a dose-dependent manner; a higher degree of myocardial depression and reduction in cardiac output are observed with halothane than nitrous oxide (halothane $\geq$ methoxyflurane $>$ isoflurane $=$ sevoflurane $=$ desflurane $>$ nitrous oxide).

The potency of Inhalational Anesthetics: The minimal alveolar concentration (MAC) of an anesthetic agent used to compare the inhalational anesthetic potency is affected by various factors. The factors affecting the anesthetic potency is summarized in Table 5.4.

Injectable Anesthetics: Injectable anesthetics are administered through intravenous, intramuscular, subcutaneous, or intraperitoneal route. Commonly used

Table 5.4 Factors affecting anesthetic potency

Factors that decrease MAC	Factors that increase MAC	Factors that NOT affect MAC
• Hypotension • Anemia (PCV < 13%). • Hypothermia • Metabolic acidosis • Extreme hypoxia (PaO_2 < 38 mmHg) • Age: older animal require less anesthetic • Premedication (opioids, sedatives, tranquilizers) • Local anesthetics • Pregnancy • Hypothyroidism	• Hyperthyroidism • Hypernatremia • Increasing body temperature (increases the cerebral metabolic rate of the brain)	• Type of stimulation • Duration of anesthesia • Species -MAC varies by only 10–20% from species to species • Sex • $PaCO_2$ between the range of 14 and 95 mmHg • Metabolic alkalosis • PaO_2 between the range of 38 and 500 mmHg • Hypertension • Potassium

MAC: Minimal alveolar concentration, $PaCO_2$: partial pressure of carbon dioxide, PaO_2: Partial pressure of oxygen, PCV: Packed-cell volume

Table 5.5 Commonly recommended anesthetic agents for laboratory animal experiments

Animal species	Short-duration anesthesia	Medium-duration anesthesia	Long-duration anesthesia
Mice	Isoflurane (inhalation)	Xylazine + ketamine (5 mg + 100 mg i.m.)	Xylazine + ketamine (16 mg + 60 mg i.m./i.p.)
	Halothane (inhalation)	Xylazine + ketamine (5 mg + 100 mg i.m.)	Xylazine + ketamine (16 mg +60 mg i.m./i.p.) or Urethane (1200 mg/kg i.p.)
Guinea pig	Isoflurane (inhalation)	Xylazine + ketamine (2 mg + 80 mg i.m.)	Xylazine + ketamine (4 mg + 100 mg i.m.)
Rabbits	Isoflurane (inhalation)	Xylazine + ketamine (5 mg + 15–30 mg i.m.)	Xylazine + ketamine (5 mg + 100 mg i.m.)

Atropine (0.02 mg/kg s.c./i.m.) is used as a pre-anesthetic medication for all these species to reduce salivation, bronchial secretion and protect the heart from vagal inhibition (Reproduced with permission from Parasuraman et al., 2010)

injectable anesthetic agents are barbiturates, chloralhydrate, ketamine, hypnotic agents, xylazine, and urethane. Rodents vary significantly in their sensitivity to various anesthetics. Strain, age, sex, body composition, health status, and genetic manipulation are few of the factors that contribute to anesthetic sensitivity. Commonly recommended anesthetic agents for laboratory animal experiments are presented in Table 5.5.

Barbiturates: Barbiturates are gamma-aminobutyric acid ($GABA_1$)-mimetic agents that inhibit the release of norepinephrine and glutamate. There are three classes of barbiturates, namely, long, short, and ultra-short acting. For laboratory animals, short and ultra-short acting barbiturates are used predominantly (sodium pentobarbitone at 50–90 mg/kg i.p.; sleep time: 60–120 min and thiopental at 30–40 mg/kg i.p.; sleep time: 10–20 min). Thiopental is administered through either intravenous or intraperitoneal route of administration and not through the intramuscular route of administration, because of its high histotoxicity.

Chloral hydrate: Chloral hydrate is a relatively old soporific compound. By using it for anesthesia cardiovascular side effects are often observed. The range of dosing is very narrow. Its use for laboratory animals is therefore limited. Intraperitoneal injections in rats can lead to paralytic ileus.

Ketamine: Ketamine is considered as a dissociative anesthetic agent, which is characterized by catalepsy, analgesia, amnesia, and immobilization. Ketamine has a wide margin of safety and a marked analgesic effect that prevents spinal sensitization by inhibiting NMDA (*N*-methyl-*D*-aspartate) receptors. Norketamine is the main metabolite of ketamine and it is less potent than ketamine as an anesthetic agent. Ketamine induces mild respiratory depression while preserving cardiovascular function by stimulating the cardiovascular system resulting in increases in heart rate, blood pressure, and cardiac output which are mediated through the sympathetic nervous system. Ketamine increases the tone of skeletal muscles and this can be prevented by simultaneous administration of diazepam or xylazine.

Hypnotic agents: Hypnotics produce a very deep sleep without analgesia (metomidate, a non-barbiturate imidazole). The combination of metomidate with neuroleptic agents (e.g., azaperone) used to induce anesthesia in pigs. As a single compound, metomidate can only be used to anesthetize birds.

Xylazine: Xylazine is an α_2 adrenergic receptor agonist and it produces mild to moderate analgesia in most animals. In general, xylazine is combined with ketamine to produce surgical anesthesia. It causes cardiovascular and respiratory depression and these effects can be reversed by the administration of specific antagonists such as atipamezole or yohimbine. Xylazine as a single compound can only be used to anesthetize cattle.

Urethane: Urethane (ethyl carbamate) was formerly used as a hypnotic agent. It can be used at the appropriate dose to produce a long-acting (about 10 h) anesthesia in rodents. It has a wide safety margin and has little effect on normal blood pressure and respiration. Urethane acts non-selectively on multiple neurotransmitter receptors including acetylcholine, glycine, NMDA, AMPA (α-amino-3-hydroxy-5-methyl-4-isoxazole propionic acid), and GABA-A (gamma-aminobutyric acid type A) receptors. Its use in animals is very limited due to liver toxicity and carcinogenic properties.

Tribromoethanol: Tribromoethanol (500 mg/kg, i.p.) is a non-pharmaceutical grade anesthetic agent, has been used in laboratory mice due to its ready availability, rapid anesthetic induction, and recovery times. Tribromoethanol is a white crystalline powder. When reconstituted and administered, it produces generalized CNS depression, including depression of cardiovascular and respiratory centers.

5.1.5 Monitoring of Depth of Anesthesia

It is essential to make sure that the animal is in complete anesthesia before the commencement of surgical procedures. After the administration of the anaesthetic agent, the animal will become quiet, still, and minimally responsive to handling. Non-aversive stimulation such as pedal and righting reflex and muscle tone are used to check the animal response.

Pedal reflex is the response observed if the limb is retracted towards the body and this response will disappear when the animal is deeply anaesthetized.

Righting reflex/labyrinthine righting reflex corrects the orientation of the body when it turns back to its normal upright position, but after anesthesia, this reflex should be absent. Muscle tone is the continuous and passive partial contraction of the muscles and, it will diminish after anesthesia. The animal will become still, posture is low to the ground and the body becomes flaccid and soft.

5.1.6 Postoperative Analgesia

The use of animals in biomedical research is increasing every year and most of these studies involve surgical procedures. Postoperative management of experimental animals is an essential part of the study, where measures are taken to reduce pain. The rodent Grimace Scale is used to assess pain in non-humans. Grimace Scale consists of "facial action units" (FAU) (orbital tightening, nose/cheek appearance, ear and whisker positions) which are scored using still images as depicted in Fig. 5.1.

Opioids such as buprenorphine, butorphanol, and non-opioids such as ketoprofen, flunixin, and paracetamol are commonly used to manage postoperative pain.

5.2 Euthanasia of Experimental Animals

Animal euthanasia (euthanasia from Greek: εὐθανασία; good death) is allowing the animal to die by withholding extreme medical measures. According to the *American Veterinary Medical Association (AVMA) Guidelines on Euthanasia*, an ideal euthanasia results in rapid loss of consciousness followed by respiratory or cardiac distress and loss of brain function thereby minimizes the anxiety and distress experienced by the animal. All animal euthanasia must be performed by trained personnel using a method approved by the animal ethical committee or institutional regulatory body. The method of euthanasia should be appropriate for the species and the age of the animals. In general, the experimental animals are euthanatized by physical or inhalation methods. A combination of inhalation (primary) followed by a physical

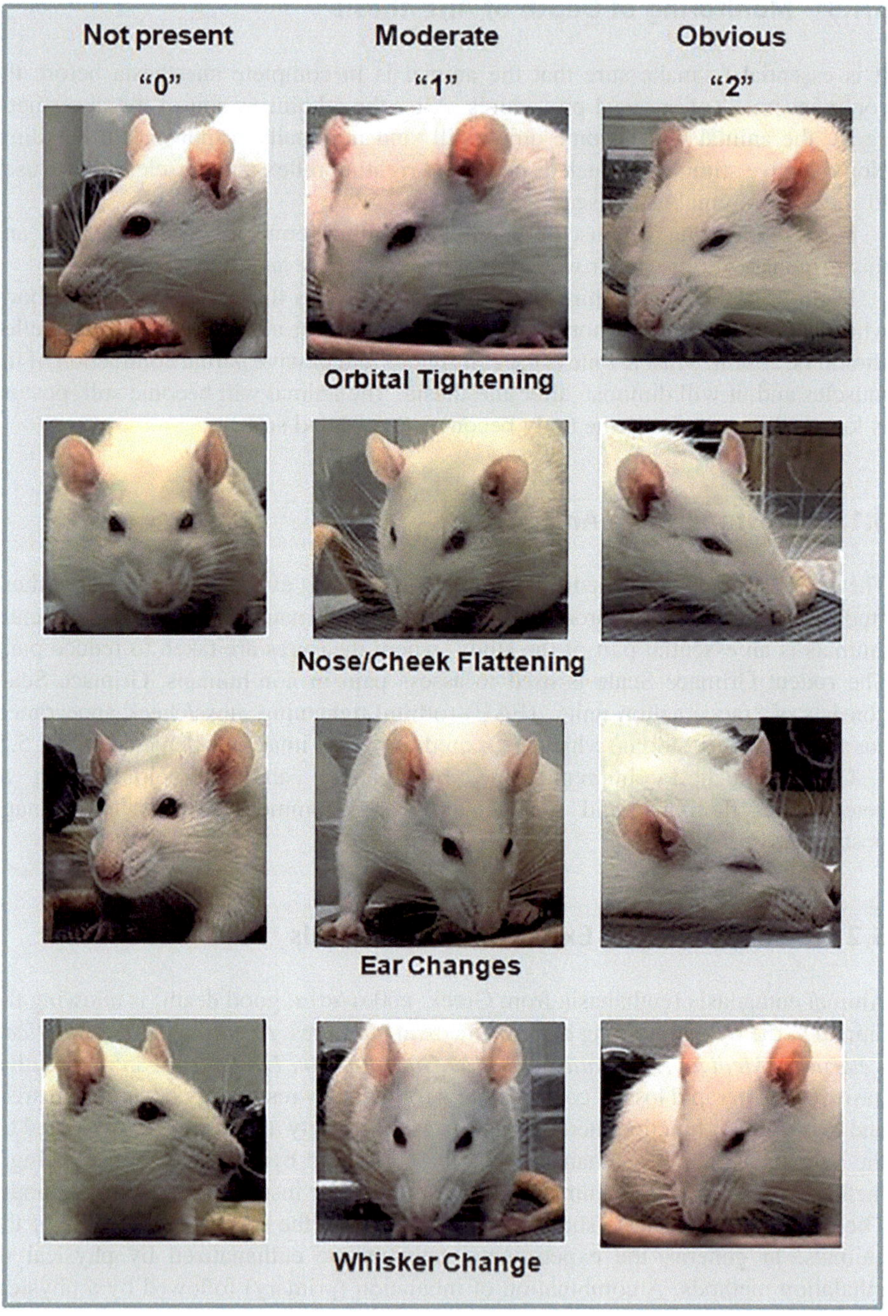

Fig. 5.1 The Rat Grimace Scale (Reproduced with permission from 'Sotocinal SG, Sorge RE, Zaloum A, Tuttle AH, Martin LJ, Wieskopf JS, et al., The Rat Grimace Scale: a partially automated method for quantifying pain in the laboratory rat via facial expressions. Mol Pain. 2011;7:55')

Table 5.6 Euthanasia methods in experimental animals

Euthanasia method		Animal species (Dose)
Inhalation method	Inhaled anesthetic (halothane, isoflurane, sevoflurane or desflurane)	Mouse, rat, hamster, guinea pig
	Carbon dioxide (80%)	Mouse, rat, hamster, guinea pig,
Injectable anesthetic overdose	Sodium pentobarbitone	Mouse (150 mg/kg i.p.), rat (dose: 100 mg/kg i.v. or 150 mg/kg i.p.), hamster (300 mg/kg i.p.), guinea pig (150 mg/kg i.p.), rabbit (120 mg/kg i.v.), cat (1 00 mg/kg i.v. or 200 mg/kg i.p.), dog (100 mg/kg i.v.), ferret (120 mg/kg i.p.), primate (100 mg/kg i.v.)
	[a]T-61 Euthanasia Solution (via a catheter)	Rabbit (0.3 ml/kg i.v.), cat (0.3 ml/kg i.v.), dog (0.3 ml/kg i.v.), primate (0.3 ml/kg i.v.)
Physical methods	Cervical dislocation	Mouse, rat
	Decapitation	Mouse, rat, hamster, guinea pig,
	Concussion (with subsequent exsanguination)	Rat, guinea pig, rabbit,
	Stunning with a captive bolt	Rabbit
	Microwave irradiation	Rat
Exsanguination	Cardiac Perfusion	injectable anesthetic agents are recommended to anesthetize the animal before perfusion
Unacceptable methods		Nitrous oxide alone, nitrogen or argon asphyxiation

[a]Prior to administration of T-61 euthanasia solution, the animal should be anesthetized with xylazine/ketamine

(secondary) method is recommended for all rodents as they, especially neonates, resist inhaled agents such as CO_2. The animal species-specific euthanization methods are summarized in Table 5.6.

After euthanization, in all species and for all methods of euthanasia, death of animals must be confirmed by the trained person after assessing the vital signs such as heart function, pupillary response to light and respiratory pattern as the lack of movement or observable absence of breathing are not accurate signs to confirm the death. Carcasses of euthanized animals must be disposed of as per the institutional bio-waste disposal guidelines or as per local regulations.

In conclusion, millions of animals of various species are used for animal experimentation, and the number seems to be increasing every year. Thus, the ethical conduct of animal experimentation is considered as one of the primary norms of these experimentations. Anesthetics in animal experimentation significantly reduce pain and stress to the animals. Physical, chemical or gaseous agents are employed to render the animal to safe and effective anesthesia. Pre- and post-anesthetic medication also play a vital role in the significant reduction of pain and stress in higher animals. Euthanasia is another major ethical consideration for the terminal completion of animal experiments. It reduces the pain and stress experienced by animals during the terminal methods. Framing new ethical guidelines, and development of safe and highly effective anesthetic regimen are still going on to render safe and efficient anesthesia and euthanasia.

Acknowledgments The authors are grateful to SAGE journals for granting permission to reproduce Fig. 5.1 from 'Sotocinal SG, Sorge RE, Zaloum A, Tuttle AH, Martin LJ, Wieskopf JS, et al., The Rat Grimace Scale: a partially automated method for quantifying pain in the laboratory rat via facial expressions. Mol Pain. 2011;7:55.'.

Bibliography

Anesthesia Guidelines.: Rats Available from: https://www.researchservices.umn.edu/services-name/research-animal-resources/research-support/guidelines/anesthesia-rats. Last assessed 7 July 2019.

Animal Care and Use Standard. Available from: https://staff.unimelb.edu.au/__data/assets/pdf_file/0008/2046986/STANDARD-Humane-killing-of-reptiles-FINAL.pdf. Last assessed 7 July 2019.

Azimaraghi O, Sistani MN, Abdollahifar MA, Movafegh A, Maleki A, Soltani E, Shahbazkhani A, Atef-Yekta R. Effects of repeated exposure to different concentrations of sevoflurane on the neonatal mouse hippocampus. Braz J Anesthesiol (English Edition). 2019;69(1):58–63.

Cressey D. Best way to kill lab animals sought. Nature News. 2013;500(7461):130.

Faller KM, McAndrew DJ, Schneider JE, Lygate CA. Refinement of analgesia following thoracotomy and experimental myocardial infarction using the mouse grimace scale. Exp Physiol. 2015;100(2):164–72.

Fish R, Danneman PJ, Brown M, Karas A, editors. Anesthesia and analgesia in laboratory animals. Academic; 2011.

Flecknell P. Laboratory animal anaesthesia. 3rd ed., Academic, 2009. pp. xxiii-xxiv, 21, 23.

Gargiulo S, Greco A, Gramanzini M, Esposito S, Affuso A, Brunetti A, Vesce G. Mice anesthesia, analgesia, and care, Part I: anesthetic considerations in preclinical research. ILAR J. 2012;53(1): E55–69.

Gonfalone AA. Sleep and gravity. Med Hypotheses. 2018;113:81–4.

Hickman DL, Johnson SW. Evaluation of the aesthetics of physical methods of euthanasia of anesthetized rats. J Am Assoc Lab Anim Sci. 2011;50(5):695–701.

Hill WA, Tubbs JT, Carter CL, Czarra JA, Newkirk KM, Sparer TE, Rohrbach B, Egger CM. Repeated administration of tribromoethanol in C57BL/6NHsd mice. J Am Assoc Lab Anim Sci. 2013;52(2):176–9.

Hohlbaum K, Bert B, Dietze S, Palme R, Fink H, Thöne-Reineke C. Severity classification of repeated isoflurane anesthesia in C57BL/6JRj mice-assessing the degree of distress. PLoS One. 2017;12(6):e0179588. https://doi.org/10.1371/journal.pone.0179588.

Holland AJ. Laboratory animal anaesthesia. Can J Anesth. 1973;20(5):693–705.

Ilkiw J. Local anesthesia and local anesthesic techniques. World small animal veterinary association world congress proceedings, 2001. Available from: https://www.vin.com/apputil/content/defaultadv1.aspx?pId=8708&meta=generic&catId=18041&id=3843658&ind=12&objTypeID=17. Last assessed 7 July 2019.

Keating S. Small animal local and regional anesthesia. Available from: https://vetmed.illinois.edu/wp-content/uploads/2016/09/75.-Keating-Local-and-Regional-Anesthesia-in-Small-Animals.pdf. Last assessed 7July 2019.

Kurdi MS, Theerth KA, Deva RS. Ketamine: current applications in anesthesia, pain, and critical care. Anesth Essays Res. 2014;8(3):283–90.

Lee C, Jones TA. Effects of ketamine compared with urethane anesthesia on vestibular sensory evoked potentials and systemic physiology in mice. J Am Assoc Lab Anim Sci. 2018;57(3): 268–77.

Leung V, Zhang E, Pang DS. Real-time application of the rat grimace scale as a welfare refinement in laboratory rats. Sci Rep. 2016;6:31667.

Morgan CJ, Curran HV. Independent scientific committee on drugs (ISCD) ketamine use: a review. Addiction. 2012;107(1):27–38.

Parasuraman S, Raveendran R, Kesavan R. Blood sample collection in small laboratory animals. J Pharmacol Pharmacother. 2010;1(2):87–93.

Rat and Mouse Anesthesia and Analgesia (2016). https://animalcare.ubc.ca/sites/default/files/documents/Guideline%20-%20Rodent%20Anesthesia%20Analgesia%20Formulary%20%282016%29.pdf. Last assessed 7July 2019.

Richardson CA, Flecknell PA. Anaesthesia and post-operative analgesia following experimental surgery in laboratory rodents: are we making progress? Altern Lab Anim. 2005;33(2):119–27.

Sotocinal SG, Sorge RE, Zaloum A, Tuttle AH, Martin LJ, Wieskopf JS, Mapplebeck JC, Wei P, Zhan S, Zhang S, McDougall JJ, King OD, Mogil JS. The rat grimace scale: a partially automated method for quantifying pain in the laboratory rat via facial expressions. Mol Pain. 2011;7:55.

Tremoleda JL, Kerton A, Gsell W. Anaesthesia and physiological monitoring during in vivo imaging of laboratory rodents: considerations on experimental outcomes and animal welfare. EJNMMI Res. 2012;2(1):44.

Underwood W, Anthony R, Cartner S, Corey D, Grandin T, Greenacre CB, Gwaltney-Bran S, McCrackin MA, Meyer R, Miller D. AVMA guidelines for the euthanasia of animals. Schaumburg, IL: American Veterinary Medical Association; 2013.

Vogel HG, Vogel WH, editors. Drug discovery and evaluation: pharmacological assays. 3rd ed. Springer Science & Business Media; 2008. p. 2029–37.

Equipments in Experimental Pharmacology 6

R. Kirubakaran

Abstract

Experimental Pharmacology as a science has been delivering high-quality evidence and advancements in Pharmacology for decades. The experiments in the field of Pharmacology contribute right from the basic step of molecule identification to marketing a drug. Experiments in biomedical research involve subject, intervention (experiment) and the results. Though experiments in Pharmacology date back to many centuries but only with the invention of various tools (equipment) and new techniques the discipline of experimental pharmacology attained a new paradigm. The invention of various equipments facilitated the experiments done in-vitro, animal or in humans. In this section, we will see some of the commonly used and basic instruments used in experimental Pharmacology.

Keywords

Pole climbing · Bio-assay · Organ bath

6.1 Organ Bath

- An organ bath is a classical experimental unit used in physiological and pharmacological experiments on tissue preparations.
- It has its own range of applications in research, development, and teaching. It consists of various small units assembled in a chamber made up of transparent thermoplastic material commonly known as **Perspex™**glass.
- The main purpose of the organ bath in any experimental set up is to sustain and maintain the tissue for several hours in an external environment.

R. Kirubakaran (✉)
Department of Clinical Pharmacology, JIPMER, Puducherry, India

M. Lakshmanan et al. (eds.), *Introduction to Basics of Pharmacology and Toxicology*, https://doi.org/10.1007/978-981-19-5343-9_6

6.1.1 History

- The conventional single unit student organ bath was designed by Rudolf Magnus and later came the double unit organ bath by Gaddum.
- The organ baths as a part of an experimental tool are being used in pharmaceutical research and academics for more than 100 years.
- In the drug development field, the traditional organ baths are replaced by multiple-unit automatic organ baths, which offers high throughput screening and rapid assay results (Fig. 6.1).

6.1.2 Contents

The organ bath holds most of the vital things needed for bio-assay. (Fig. 4.1)

- *Water bath:*
  - It is the outer chamber of the organ bath made up of transparent acrylic glass aka Poly(methyl methacrylate—PMMA).
  - This chamber is filled with water for the maintenance of external temperature and also holds other major parts.
  - It has two openings one for emptying the water from the bath itself and the other one directly from the tissue chamber bottom for draining the Physiological Salt Solution (PSS).
- *Electric heater and Thermostat:*
  - The heating coil helps to increase the water temperature in the organ bath and the thermostat ensures that the set temperature is maintained throughout the procedure.
  - It is important to maintain the temperature constant throughout the experiment to reduce errors and variability.

Fig. 6.1 Organ bath and the contents; (**a**) Water bath; (**b**) Electric heater and Thermostat; (**c**) Electric stirrer (stirrer is hidden behind the heating coil); (**d**) Organ tube or tissue chamber (**e**) Glass coil

  - The working temperature for various tissues ranges from 34–39 °C. The allowed variability in temperature during the experiment is ±1 °C.
- *Electric stirrer:* The purpose of the electric stirrer is to dissipate the temperature uniform throughout the setup.
- *Organ tube or tissue chamber:*
  - The organ tube is a cylindrical glass tube in the organ bath where the tissue is suspended, maintained thriving with a physiological salt solution, and where test substances are administered.
  - It has an inlet at the bottom sidewards to allow PSS from the reservoir and an outlet along the axis of the organ tube for drainage of PSS.
  - The PSS is not filled from the top rather from the bottom through the side inlet.
  - The tube is also marked with variable volumes and should be kept constant throughout the experiment.
  - The test substance is administered directly into the glass tube from the top. *The volume of the glass tube varies from 5–100 ml.Remember the volume measured and marked is from the clamp level, so it will be wise to mark and maintain the position of PSS outlet clamp as constant.*
  - The glass tube houses the tissue holder where one end of the tissue is tied and it also serves as an aerator.
- *Glass coil:*
  - The coiled glass structure between the tissue bath and PSS reservoir remains submerged in the water bath.
  - The function of this coiled structure is to maintain the PSS temperature close to the water bath.
  - The volume of this coiled structure should be adequate to act as an intermittent reservoir of PSS at a temperature as same as that of a water bath.

6.2 Kymograph

- Kymograph is a device used to record changes in spatial position over time. In simple words, it is a device used to graphically measure the changes in tissue length or pressure in a system.
- It consists of a rotating drum with paper attached on which the stylus records the tracings.

6.2.1 History

- The Kymograph as a research tool was first invented by German physiologist Carl Ludwig in 1847.
- It was one of the early equipment used to measure blood pressure in animals during experimentation.

Fig. 6.2 Kymograph with lever for speed adjustments

- The kymograph which is being used now commonly known as the smoked drum was invented by Sherrington and Starling.
- The drums were designed to rotate clockwise to get readings from left to right direction.
- Initially designed as a mechanical brass drum, later underwent modifications with steel and electrical motors for rotation.
- The speed of rotation was also made adjustable (eight different speeds) for use in multiple physiological and pharmacological experiments (Fig. 6.2, 6.3).

6.2.2 Procedure

- Kymogram: A standard paper is available commercially to be pasted on a rotating drum and smoked before use.
- The paper has a glossy surface on one side which is to be smoked with the help of benzene or kerosene-soaked burning cotton, the other surface is rough which is pasted on the drum.
- The soot deposited should be of a smooth thin layer to avoid resistance to moving stylus.

Fig. 6.3 Digital kymograph

- The smoked drum is adjusted and fixed on the shaft at an appropriate height and brought in close to the writing point of the lever.
- The speed of the drum can be set from 0.12 mm/s to 640 mm/s depending on the experiment and tissue.
- Generally, for fast contracting tissue, the speed of the drum is kept fast and vice versa (Fig. 6.4).
- An alternative to smoked drums, plain paper with ink writing portion is also used. Further modifications for better quality included continuous ink writing kymograph, capillary ink pen, and ink- vibrator pen.
- The disadvantage with ink-based kymograph is ink spilling and blotting of paper, whereas the disadvantage with the smoked drum is the smoking process and the recordings need fixing with the use of shellac or varnishing agents.

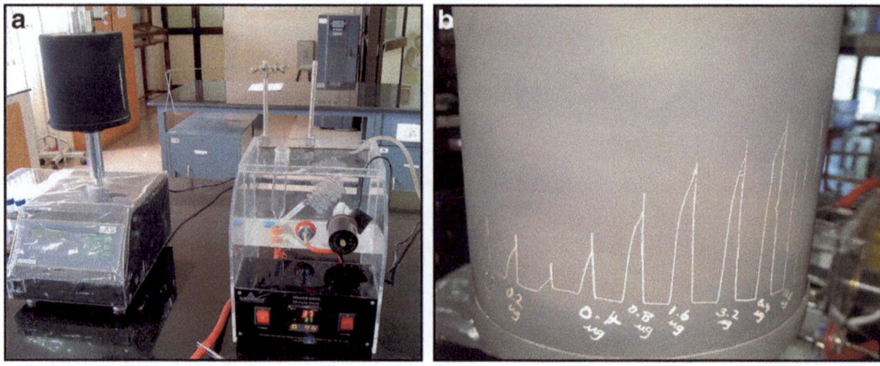

Fig. 6.4 (**a**) Bio-assay setup; (**b**) tracing of hisamine assay on guinea pig ileum

"Fast tissue" and "Fast drum"

Concept: If the drum is kept in high speed for slow contracting tissue then, you may end up using full length of the kymogram (paper) for a single response.

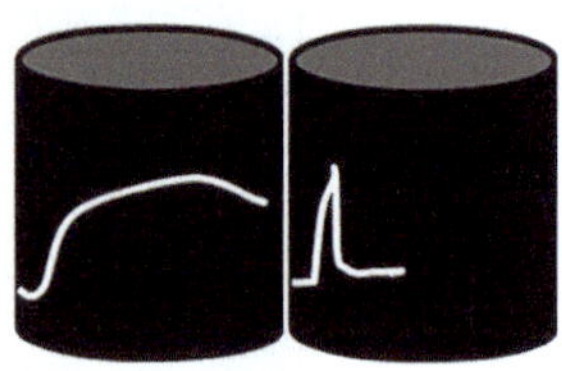

6.3 Tissue Levers

- The writing lever helps in the magnification and tracing of the change in tissue length. Ideally, a lever should be weightless to allow smooth movement over the smoked drum and also should be rigid enough to avoid bending.
- Aluminum, stainless steel, balsa wood, and drinking straw are used as levers. The lever used in isolated tissue preparation belongs to the "**class type I**" lever (Fig. 6.5).

Starling heart lever: The stylus is attached perpendicular to the drum. The fulcrum is at one end (Fig. 6.6).

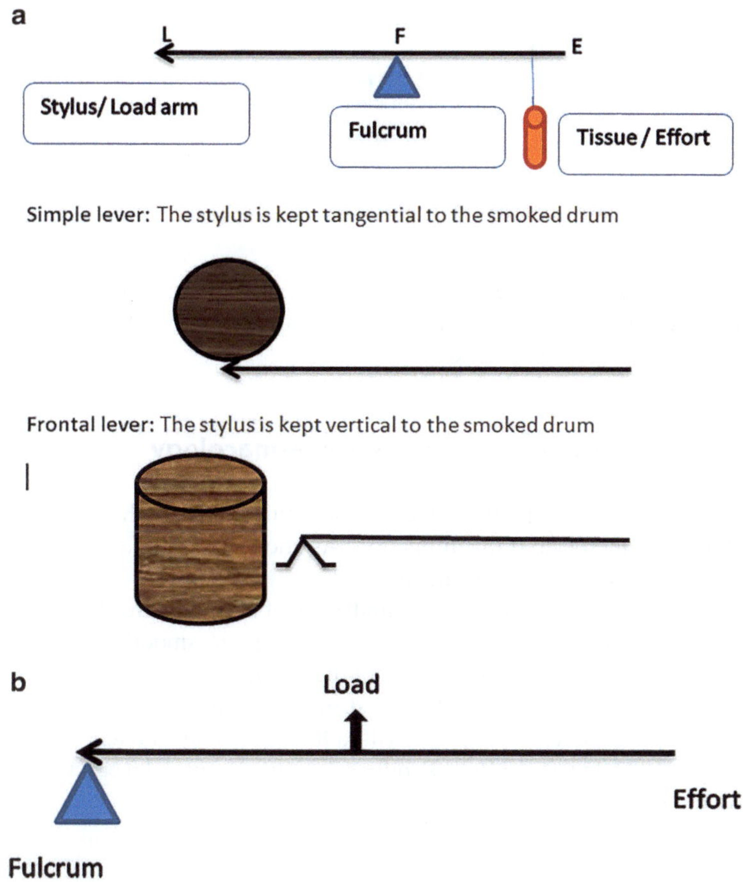

Fig. 6.5 Schematic working concept

6.3.1 Magnification

The distance between the fulcrum and writing point (X) is kept more than the distance between the fulcrum and attached tissue (Y), the magnification increases.

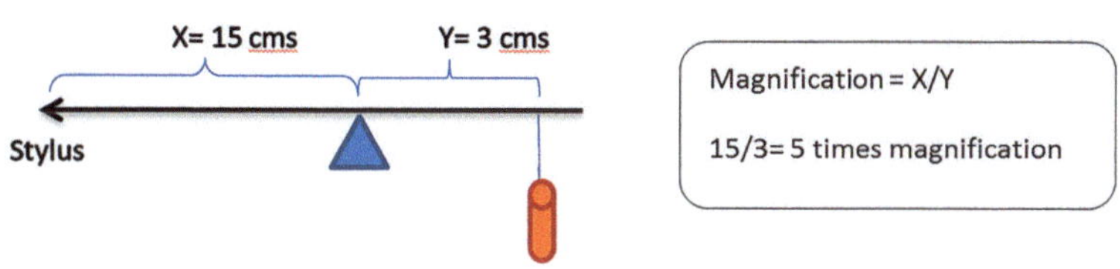

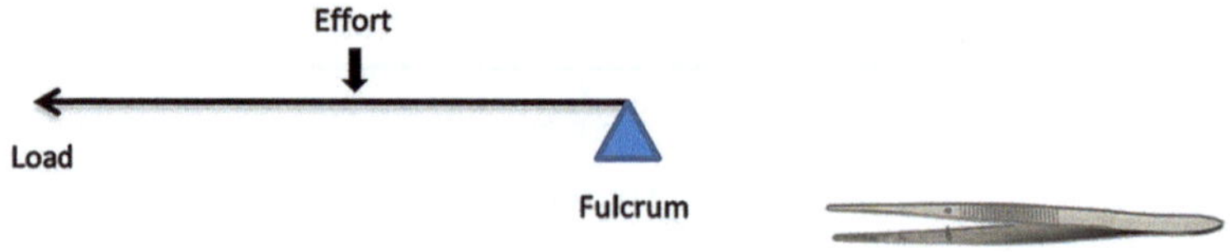

Fig. 6.6 Schematic working concept of type III lever

Magnification is fixed depending on the tissue being experimented. Ileum preparations usually need 5–10 times the magnification and for uterus bioassay 1–2 times magnification is sufficient.

6.3.2 Levers Used in Experimental Pharmacology

- *Simple lever:* Made of stainless steel or aluminum. The length can be adjusted at fulcrum to get the desired magnification. As shown in Fig. 4.5 (a)in type I lever the stylus is kept tangential to the drum.
- *Frontal lever:* It is also made up of stainless steel or aluminum. The writing point is attached to a freely movable hinge which helps in smooth movement of the lever. It is kept perpendicular to the moving drum.
- *Starling's heart lever:* Consists of a strip of aluminum or steel frame witch multiple holes, notches, and spring with a hook for adjustment.
- Other important levers include Brodie's lever (universal lever), Gimbal lever, Auxotonic lever.

6.4 Analgesiometer

- Evaluation of the analgesic activity of a compound can be done in mice or rats by inducing various painful stimuli and comparing the threshold before and after test substance.
- The painful stimuli can be in the form of physical, thermal, chemical, or electrical.
- Careful selection of animals for the experiment is very essential to reduce errors (e.g. animals with motor problems or disability may contribute to the false-positive increase in pain threshold, because of their inability to escape from the painful stimulus).

6.4.1 Basic Principle

- The inherent nature of animals to get rid of the noxious stimuli is exploited to measure the pain threshold.
- The time duration taken by an animal from the time of administration of noxious stimuli to the first attempt to get rid of the stimuli is considered to be the pain threshold.
- A compound with analgesic activity tends to increase the pain threshold. (i.e. more effort or time is needed to induce pain compared to baseline or control group).
- The pain threshold parameter can be either compared between the test compound and positive control or before and after administration of the test compound.
- The response time is increased only by a central analgesic in a hot plate or tail-flick test.

6.4.2 Hot Plate Analgesiometer

- A hot plate as a screening tool was first described by G. Woolfe and D. Macdonald in 1943 and later modified and validated by Nathan B. Eddy and D. Leimbach in 1950.
- The hot plate test evaluates the pain threshold, based on the thermal pain reflex duration of the animal.
- The equipment consists of a base (the control unit) and a hot plate chamber (cylindrical/box) (Fig. 6.7).

6.4.3 Procedure

- The animals can be screened for locomotor problems in an activity cage and (or)in analgesiometer, the animals which don't show any thermal reflex response within the cut of time (i.e. 30 s) can be avoided for further experiments.
- The temperature knob is used to set plate temperature at 55 ± 1 °C.
- A foot-operated or ordinary stopwatch can be used to measure the reaction time.
- The reaction time can be measured before drug administration and after drug administration at specific pre-set time points like 20, 60, and 90 min.
- The cut off time for the rat is 30 s and for mice it is 20 s.
  (i.e. Even if the animal doesn't show any reaction in 30 s, immediately the animal should be taken out of the hot plate to avoid injury)
- Endpoints:
  - Paw licking
  - Jumping response

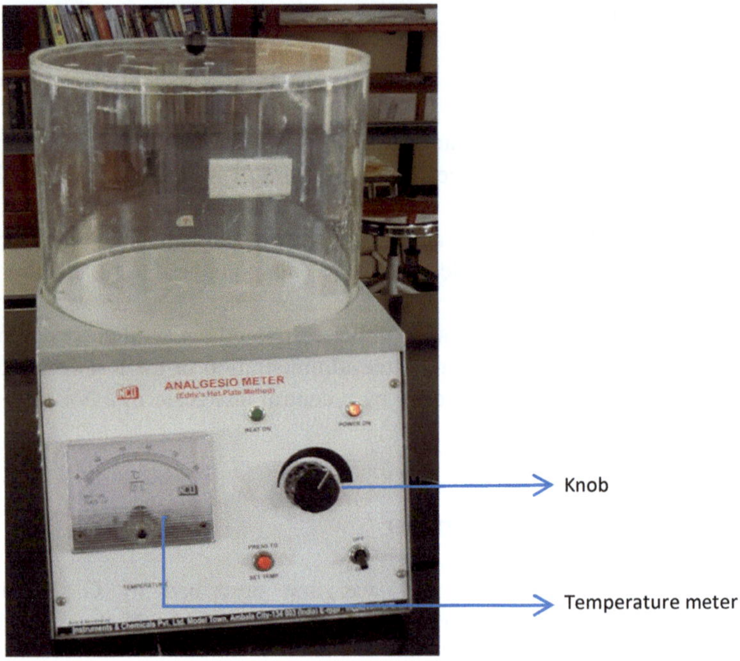

Fig. 6.7 Eddy's hot plate analgesiometer

- The stopwatch should be stated at the time of dropping the animal on a hot plate and stopped immediately at the first episode of the above-mentioned endpoints.
- The time duration taken by the animal to respond to pain can be compared before and after drug administration. The experiment can also be done in two different groups of animals and comparing the mean/median reaction time.
  – Positive control: Morphine 5 mg/kg i.p.

6.5 Digital Analgesiometer:Tail-Flick Method

- The basic principle involved in this test is the animal response to the radiant heat (Avoidance of noxious stimuli by tail-flick).
- The time taken by the rat to flick the tail away from radiant heat is measured and compared before and after the administration of the drug.
- The drug with analgesic activity is expected to prolong the tail-flick (Fig. 6.8).
- The equipment consists of a temperature adjustment knob, an inbuilt timer, and a heating coil at the top
- The metal platform where the tail is kept above the heating coil has an inlet and outlet for continuous circulation of water. (To prevent direct heat)

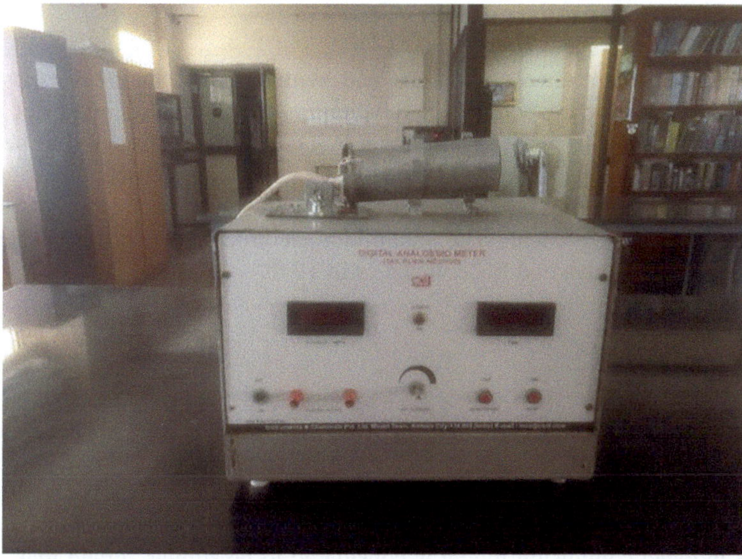

Fig. 6.8 Digital analgesiometer using radiant heat

- The heating coil kept few centimeters away is the source of radiant heat. (fixed at the top of the machine next to restrainer holder)

6.5.1 Procedure

- Animals are selected by a preliminary screening with the same equipment. The animals which show more than a one-second variation in response duration of response at 15 min interval or more than 3 s from group mean duration of response are avoided for experiments.
- The rat is kept in a restraining container with tail protruding.
- The tail is cleaned with cotton soaked in water or ethanol and allowed to dry.
- Fix the restrainer on the holder provided in analgesiometer with the tail pointing towards the heating coil side.
- Keep the distal one-thirds portion of the rat tail at a fixed distance above the heating coil and start the stopwatch simultaneously. The proximal portion of the tail is to be avoided because of thick keratinized skin.
- The duration taken for escape reflex, i.e. the characteristic tail flick is noted.
- The time duration taken by the animal to respond to pain can be compared before and after drug administration at regular intervals of 15 min.
- The experiment can also be done in two different groups of animals and comparing the mean/median reaction time.

  (*Note: Ambient temperature can also influence the response time in animals, so it is ideal to maintain a uniform ambient temperature throughout the experiment*)

6.6 Electroconvulsiometer

- A seizure is characterized by signs and symptoms of abnormal excessive or synchronous neuronal activity in the brain. Animal models may not exactly match the seizure or epilepsy in human beings, but are considered to be one of the important models in the evaluation of the antiepileptic activity of a compound.
- A seizure can be induced in an animal by a stimulus like chemical (Pentylenetetrazol -PTZ) or electrical. The maximum electroconvulsive seizure test was developed by Toman and his colleagues in 1946.

6.6.1 Principle

- Maximum electroconvulsive seizure (MES): The rats and mice are given a maximum current of 150 and 50 mA for 0.2 s respectively.
- The characteristic course of seizure-like activity is seen in animals which include tonic flexion, extension, clonic convulsion, stupor, and recovery/death.
- MES induced convulsion model is considered to be similar "grand mal" type in humans. Abolishing or reducing the duration of the extensor phase by a test drug is considered to have anti-epileptic action (Fig. 6.9).
- The equipment front panel consists of knobs to select the duration and strength of the current.
- Output electrodes are connected with ear clips or corneal tips.
- It is also fitted with a display or analog meter showing the intensity of the current.

Fig. 6.9 Electro convulsiometer

- A single press of the start button will deliver the current (prefixed strength and duration)

6.6.2 Procedure

- *Screening of animals:* Rats weighing 150–250 or mice weighing 20–40 g are first screened with preliminary maximum current (150 mA for rats and 50 mA for mice) delivered via corneal or ear electrodes.
- The animals which show characteristic hind limb extension phase of MES in three different trials on different days are only selected for the procedure.
- The same animal should not be used for subsequent MES on the same day. Ideally, screening should be completed 1 week before the actual experiment.
- On the day of the experiment, the animals are divided into three groups with 8–10 animals in each group.
- The test substance can be administered in increasing doses (e.g. one group 30 mg/kg, group II 100 mg/kg, group III control).
- Specific time points should be predetermined and fixed to see the effect of the test substance.
- Either corneal electrodes or ear electrodes in fixed to deliver the current. If the corneal electrode is used a topical anesthetic (0.5% tetracaine HCl) should be instilled before electrode placement.
- Ear electrodes are preferable in terms of the part of the brain stimulated as well as avoid the risk of eye injury. Saline soaked cotton can be kept between the ears and clip to avoid injury as well as to facilitate conduction.
- Endpoints: can be all or none (e.g. presence or absence of Hind limb extension) or the total duration of hind limb extension.

6.6.3 Threshold for Electroconvulsions

- Mice usual threshold for electric induced seizure will be 6–9 mA or 90–140 V. A drug with antiepileptic activity is expected to increase the threshold for electric seizure(I.e. more energy is needed to produce seizure)
- *The MES or threshold test can only screen compounds against generalized tonic clonic seizures.*

6.7 Elevated Plus-Maze

- Handley and Mithani in 1984 reported elevated plus-maze as an experimental tool in an anxiety-related animal study. Later it is used studying the anxiety behavior of rodents, screen compounds for anxiogenic and anxiolytic properties (Fig. 6.10).

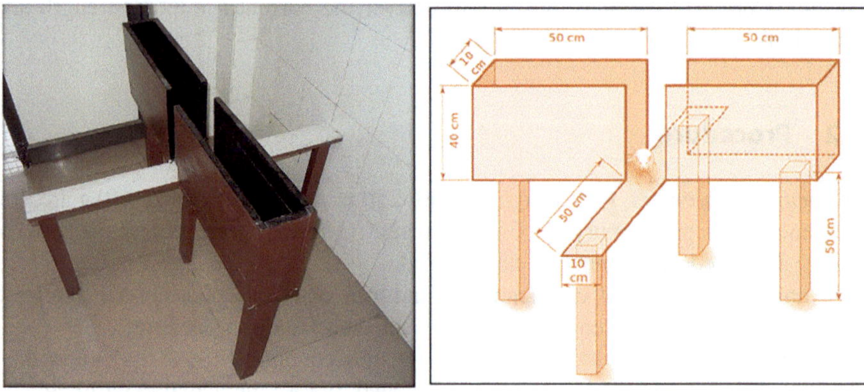

Fig. 6.10 Elevated plus maze; schematic representation (right side)

6.7.1 Equipment

- The elevated plus-maze consists of two closed arms and two open arms
- Dimensions for Rat: Closed arm: (L × B × H = 50 x 10 × 40 cm); Open arm: (L × B) = (50 × 10 cm)
- Dimensions for mice: Closed arm: (L × B × H = 16 × 5 × 12 cm); Open arm: (L × B) = (16 × 5 cm)

6.7.2 Principle

- Rodents have innate fear to open spaces and also have the desire to explore new environments.
- In anxious states, they tend to prefer closed dark spaces over open spaces. This behavior is exploited in screening drugs for anxiolytic activity.
- In the elevated plus-maze, the closed arms offer dark enclosed space which the rodents prefer, the exploration of open arms is considered to be an anxiolytic effect.

6.7.3 Procedure

- Screening: Rats/mice should be screened in an activity cage or actophotometer for locomotor problems
- Habituation: Animals should be allowed to acclimatize to the testing room for 45–60 min before testing
- Animals are divided into two groups individually weighed and the respective drug is administered (Group I - Test; drug Group II - Control)

- The maze should be cleaned with 70% ethanol before the experiment (Smell can induce stress in animals)
- At a predetermined time point, the rat/mouse is kept in the middle of the elevated plus-maze with the head of the animal facing the open arm. The animal is allowed to explore the maze for 5 min
- Video recording the activity helps in the retrieval of undisturbed data, the presence of an experimenter close to the maze will force the animal to stay in a closed space thus affects the results
- After 5 mins of exploration, the mouse is removed and again cleaned with 70% ethanol and allowed to dry. (There should be no fecal matter or urine smell in the maze)
- Parameters noted: First arm entry, number of entries to the closed and open arms, time spent in both arms are measured
- The various automatic video analyses software is available for interpretation. One disadvantage with elevated plus maze is that the time spent in the middle of both arms is not included in any analysis. Moreover, since it is the starting point for the animal, the animal may prefer the same location considering safe. It has been observed that animals may spend up to 30% of the total time of exploration in the middle area where the animal is released
- Other mazes with subsequent modifications are elevated zero maze and elevated T maze but elevated plus-maze is still the preferred tool my majority of the researchers

6.8 Actophotometer

- The determination of spontaneous motor activity in an animal is very much useful in selecting animals for other experiments where the motor activity is expected to be normal.
- For example—Elevated plus maze test, rotarod test, test on memory and spatial orientation, hot plate analgesiometer test—all the mentioned tests require a normal baseline motor activity to avoid errors.
- Measurement of spontaneous activity in itself can be used as a screening procedure for various CNS stimulants and depressants.
- The Actophotometer also is known as an activity cage is made up of steel or transparent perspex glass.
- The cage where the animal has placed consists of 4 infrared beams on each adjacent wall and 4 sensors on opposite walls.
- When the animal cuts 2 or more beams by the movement it is automatically picked by the equipment and displays the number of activity on the external monitor.
- The cage is also provided with a lid to prevent the impact of external factors on motor activity.
- It is also provided with the bottom pan for the collection of waste and cleaning.

Fig. 6.11 Actophotometer

- The transparent activity cage is fitted with vertical beamers and sensors, so it can also measure vertical along with horizontal movements (Fig. 6.11).

6.8.1 Principle

- Excitation of the Central Nervous System reflects an increase in spontaneous motor activity.
- Similarly, a compound which excites or depress the CNS activity will also produce an increase or decrease the spontaneous motor activity.
- Actophotometer is a useful tool in the quantitative measurement of motor activity.
- When the photocells are interrupted inside the cage, the machine converts the signals into activity counts.

6.8.2 Procedure

- Habituate the animal to lab condition 45-60mins before the experiment.
- Switch on the instrument and verify that there is no activity recorded from the empty cage.
- Warm the floor of the activity cage. (2–3 min)
- Place the animal in the center of the activity cage and close the lid.

- Start the timer and reset the activity counter simultaneously.
- Record the readings for 5 min and remove the animal from the chamber.
- Administer the test drug to animals and repeat the same procedure at the time points at which the drug action is expected.
- Percentage change in motor activity can be calculated after the drug administration to know the CNS stimulant or depressant action of test substances.
- Clean the activity cage with 70% ethanol and allow the cage to dry.

6.9 Rota-Rod Apparatus

- Sedatives and anti-anxiety drugs can also produce motor incoordination and muscle relaxation. Rotarod test can be used to determine whether the drug produces motor incoordination or muscle relaxation in animals
- It can also be used to measure endurance and motor deficits following traumatic brain injury (Fig. 6.12)

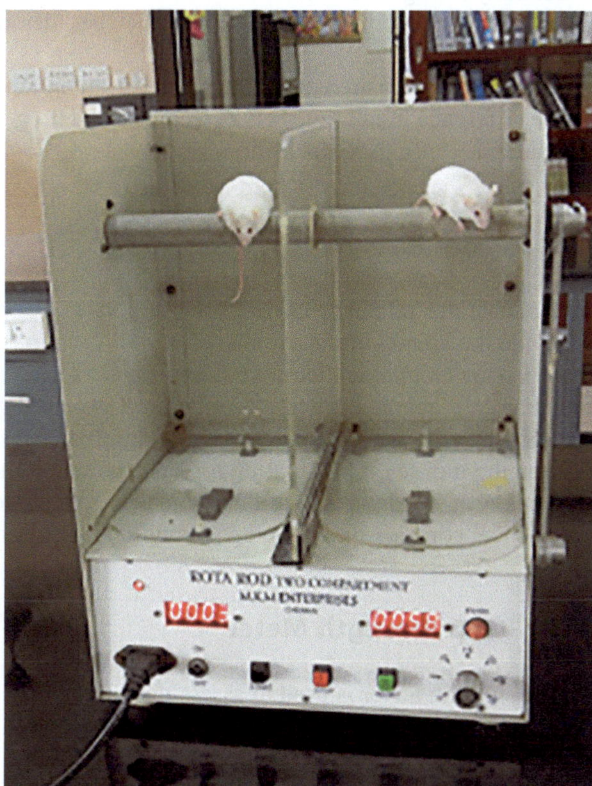

Fig. 6.12 Rota rod apparatus

6.9.1 Equipment

- The rotarod apparatus consists of a rotating rod of 3 cm diameter (for rats it is replaced with 6 cm diameter rod), the speed adjustable knob, and two or more compartments where the animal is tested.
- Between the compartments, a disc barrier or plastic barrier is placed.
- The front panel is fitted with a timer display which will be cut off automatically when the animal falls from the rod to the surface fitted with the stop button.
- Rotarod apparatus with 5 or 6 compartments are also available for time conservation.
- A multiple rod rotation speed menu is available to choose from, but usually, most of the experiments are conducted at a uniform speed of 3–5 rotation per minute.
- The speed of the rotation can be adjusted to make the normal animal hold on to the rod for 3–5 min.

6.9.2 Principle

- Loss of muscle grip indicates either weakness or relaxation.
- Centrally acting muscle relaxants or sedatives can cause loss of muscle grip. This loss of grip or incoordination can be demonstrated with a rotarod performance test.
- Animals with impaired coordination fall from the rod early than normal animals.
- The difference in fall time between control and drug-treated animals can be taken as an index of muscle relaxation.

6.9.3 Procedure

- Habituate the animal to lab condition
- Divide the animal into 3 groups
- Screen the animal on the rotarod for 2 min at a speed of 3 revolutions per minute and discard animals which fall off from the rotarod within 120 s.
- Inject saline to one group, test drug in different dose to the other two groups
- Keep the mouse on the rotating rod and switch on the timer
- The time at which the mouse falls from the rod is noted for each group and a direct statistical comparison of average duration can be done

6.10 Grip Strength Meter

- Grip strength meter is used to assess the neuromuscular function in animals. The equipment can be utilized in the evaluation of toxins, muscle relaxants, or disease affecting muscular strength.

6.10.1 Principle

- The effect of drugs on CNS activity is reflected externally as increased motor activity in case of stimulants or decreased activity, loss of tone, or muscle grip if the drug is a CNS depressant. With the use of grip strength meter, the drugs causing muscle relaxation can be identified by a percentage decrease in grip strength.
- The apparatus consists of a rectangular steel grip called a base plate and force sensor which is connected to a transducer that transmits the signal to the output display.separate steel grids are available for rats and mice (Fig. 6.13).

6.10.2 Procedure

- Habituate and allow the animals to acclimatize to laboratory condition for 40–60 min
- Divide the animal into three groups (Group I- Saline; Group II- Diazepam; Group III- test substance)
- The mouse is lifted over the baseplate by the tail so that the forepaw is allowed to grasp the steel grip
- Reset the output reading and gently pull the animal away from the steel grid by the tail until the grip is released
- The machine automatically records the maximum strength
- Time points: Baseline (i.e. 0 min) and 15, 30 min after drug administration
- Percentage change at 15 and 30 min can be calculated and used for analysis

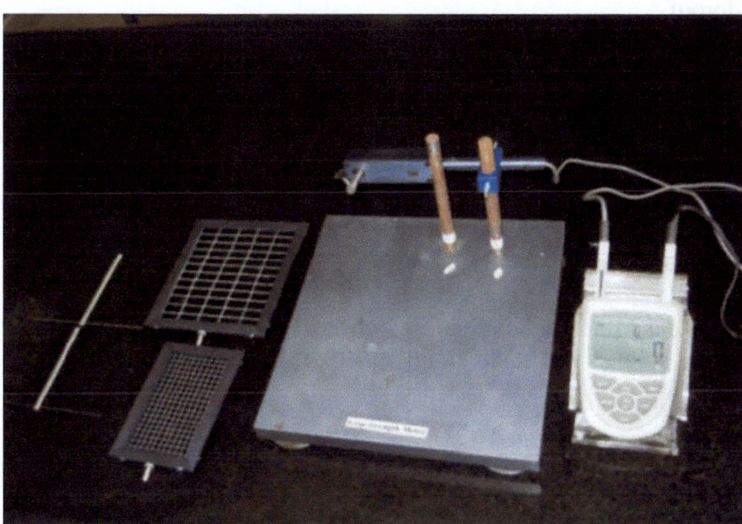

Fig. 6.13 Grip strength meter

6.11 Cook's Pole-Climbing Apparatus

- Cook's pole climbing apparatus is used in the screening of a compound for antipsychotic activity and the evaluation of cognitive behavior especially related to learning and retention.

6.11.1 Principle

- In a conditioning method, animals are trained to respond for a cue in a particular way (e.g. climbing of pole, moving to a safe chamber) to avoid noxious stimuli.
- In this apparatus the cue is sound (buzzer) and a noxious stimulus is an electric shock.
- If the animal climbs to pole after a buzzer, it is the conditioned response (remember animals are trained already!).
- If an animal moves to pole after an electric shock, it is an unconditioned response.
- Antipsychotic affects the conditioned response but do not affect unconditioned response (*i.e. a trained animal when administered with antipsychotic may not climb the pole after sound but will move to safety after an electric shock*) (Table 6.1).

6.11.2 Equipment

- Experimental chamber (25 × 25 × 25 cm or 40 × 40 × 40 cm) with transparent front panel for visualization of the animal reaction to stimulus and it is also soundproof
  - (Other animals in the lab should not perceive the buzzer frequently)
- Floor grid which is electrified for applying of noxious stimuli—(2–2.5 mA)
- A buzzer is fitted to provide sound cues
- A pole is attached to hang from the top of the chamber (but it does not touch the grid floor) (Fig. 6.14)

6.11.3 Procedure

- Habituate animal to lab conditions
- Divide the animals into three groups (Saline, Test drug, CPZ)

Table 6.1 Drug class identification using pole climbing test

	Conditioned response	Unconditioned response
Antipsychotics—therapeutic dose/ less dose	Absent	Present
CNS depressants	Absent	Absent
Antipsychotics—high dose	Absent	Absent

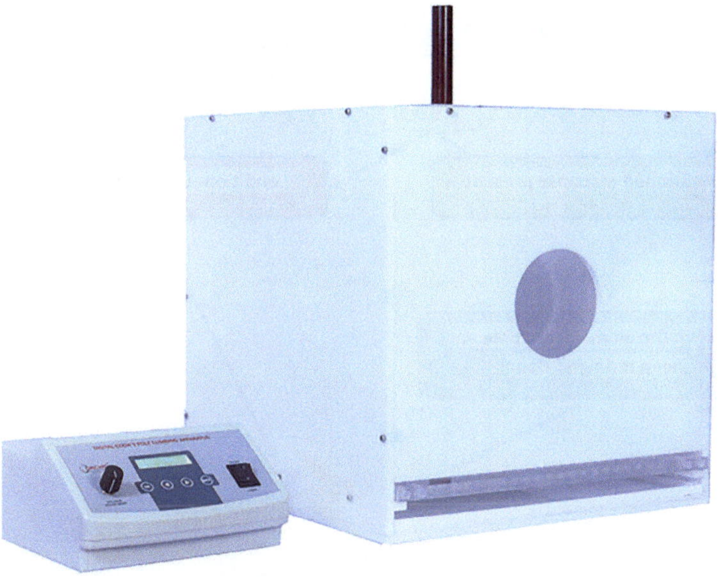

Fig. 6.14 Cook's pole-climbing apparatus

- *Training:*
- The animal is kept in the chamber and allowed to explore for 1 min
- A buzzer sound followed by a series of shock is delivered. This is done concurrently for 30 s or till the animal climbs on to the poll (whichever is earlier)
- The training is repeated for 5–10 trials a day for 8 consecutive days
- At least 30 s gap should be there between the trials
- Only rats trained for a response to conditioned stimuli should be taken for further study (cut off: 8/10 time should respond to conditioned stimuli- either buzzer/ light)

6.11.4 Experiment

- The animals are administered the respective drug as per group
- The animal is kept in the chamber and conditioned stimulus is given. (Sound/ light)
- Further steps are explained in Fig. 6.15

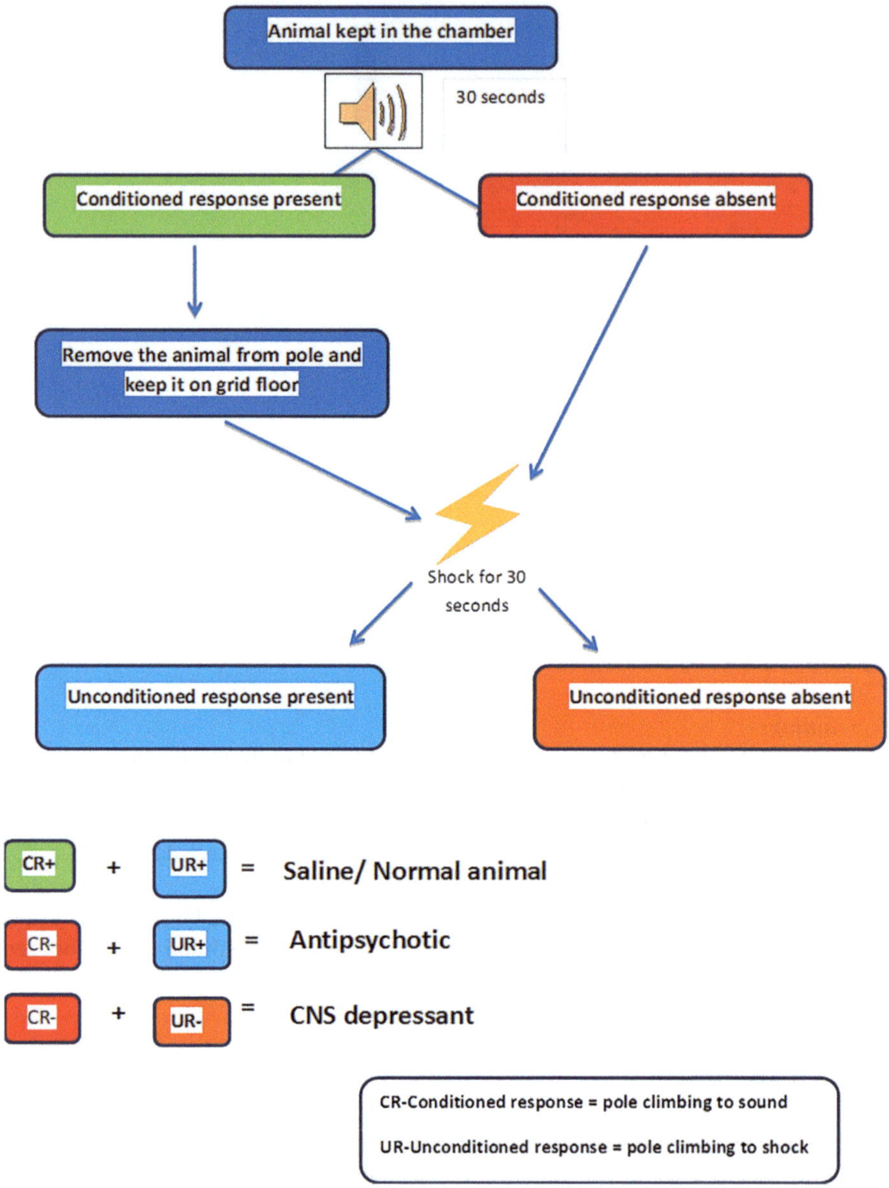

Fig. 6.15 Schematic workflow for screening compounds using pole climbing apparatus

Bibliography

Albuquerque AA, Carvalho MT, Evora PM, De Nadai TR, Celotto AC, Evora PR. In vitro reactivity ("organ chamber") of guinea pig tracheal rings—methodology considerations. Ann Transl Med. 2016;4(11):216–20.

Berge OG, Garcia-Cabrera I, Hole K. Response latencies in the tail-flick test depend on tail skin temperature. Neurosci Lett. 1988;86(3):284–8.

Bhosale UA, Yegnanarayan R, Pophale PD, Zambare MR, Somani RS. Study of central nervous system depressant and behavioral activity of an ethanol extract of Achyranthesaspera (Agadha) in different animal models. Int J Appl Basic Med. 2011;1(2):104.

Bogo V, Hill TA, Young RW. Comparison of accelerod and rotarod sensitivity in detecting ethanol- and acrylamide-induced performance decrement in rats: a review of experimental considerations of rotating rod systems. Neurotoxicology. 1981;2(4):765–87.

Castel-Branco MM, Alves GL, Figueiredo IV, Falcão A, Caramona MM. The maximal electroshock seizure (MES) model in the preclinical assessment of potential new antiepileptic drugs.

Cook L, Weidley E. Behavioral effects of some psychopharmacological agents. Ann N Y Acad Sci. 1957;66(3):740–52.

De Luca A, Tinsley J, Aartsma-Rus A, van Putten M, Nagaraju K, de La Porte S, Dubach-Powell J, Carlson G. Use of grip strength meter to assess the limb strength of MDX mice. SOP DMD_M. 2008;2(001):5–11.

Eddy NB, Leimbach D, Synthetic analgesics. II. Dithienylbutenyl-and di-thienyl butylamines. J Pharmacol Exp Ther. 1953;107(3):385–93.

Fry CH. Experimental models to study the physiology, pathophysiology, and pharmacology of the lower urinary tract. J Pharmacol Toxicol Methods. 2004;49(3):201–10.

Ghosh MN, Vedasiromoni JR. Fundamentals of experimental pharmacology. 6th ed. Kolkata; 2015. isbn:8190296507.

Hamm RJ, Pike BR, O'DELL DM, Lyeth BG, Jenkins LW. The rotarod test: an evaluation of its effectiveness in assessing motor deficits following traumatic brain injury. J Neurotrauma. 1994;11(2):187–96.

Handley SL, Mithani S. Effects of alpha-adrenoceptor agonists and antagonists in a maze-exploration model of 'fear'-motivated behavior. Naunyn Schmiedeberg's Arch Pharmacol. 1984;327(1, 1):–5.

Isca.co.in. 2019. [cited 31 July 2019]. Available from: http://isca.co.in/ANI_VAT_AND_FISHRY/lab_manual/Laboratoy%20Manual%20for%20neuropharmacology.pdf.

Jones BJ, Roberts DJ. The quantitative measurement of motor incoordination in naive mice using an accelerating rotarod. J Pharm Pharmacol. 1968;20(4):302–4.

Kamila S, Madhav NS, Sarkar CN. Pharmacological evaluation of transcranially applied lignocaine gel as a novel amnesic agent on rodents. Int J Pharm Sci Invent. 2014;3(7):13–21.

Kwan A. " Do not kill guinea pig before setting up apparatus:": the kymograph's lost educational context. Teor Vedy/Theory Sci. 2016;38(3):301–35.

Kymograph Apparatus - Sherrington Recording Drum [Internet]. Orchidscientific.com. 2019. [cited 31 July 2019]. Available from: https://orchidscientific.com/product/kymograph-apparatus-sherrington-recording-drum.

Leo LM, Pamplona FA. Elevated plus maze test to assess anxiety-like behavior in the mouse. PLoS One. 2014;e1211

Löscher W. Critical review of current animal models of seizures and epilepsy used in the discovery and development of new antiepileptic drugs. Seizure. 2011;20(5):359–68.

PANAChE [Internet]. Panache.ninds.nih.gov. 2019 [cited 31 July 2019]. Available from: https://panache.ninds.nih.gov/Model1B.aspx.
Sreemantula S, Nammi S, Kolanukonda R, Koppula S, Boini KM. Adaptogenic and nootropic activities of aqueous extract of *Vitis vinifera* (grape seed): an experimental study in the rat model. BMC Complement Altern Med. 2005;5(1):1.
Toman JE, Swinyard EA, Goodman LS. Properties of maximal seizures, and their alteration by anticonvulsant drugs and other agents. J Neurophysiol. 1946;9(3):231–9.
Woolfe G, MacDonald AD. The evaluation of the analgesic action of pethidine hydrochloride (Demerol). J Pharmacol Exp Ther. 1944;80(3):300–7.

Bioassay and Experiments on Isolated Muscle Preparations

7

Mageshwaran Lakshmanan

Abstract

Bioassay is detecting/measuring the potency or concentration of active principle in preparation using living cells by observing the specific pharmacological effect. It can be done in-vitro, in-vivo, or ex-vivo technique. Bioassay is commonly used to measure or isolate active principles whose chemical composition is unknown. Various methods like direct endpoint, matching, bracketing, and multiple point assays are available for assaying agonists. For bioassaying various endogenous compounds like histamine, NA, Ach, and serotonin, the muscles of various living organisms are being used. The dorsal muscles of leech have nicotinic receptors and are highly sensitive to acetylcholine. Guinea pig ileum is highly sensitive to histamine as it is rich in histamine receptors. Similarly, jejunum of rabbit is used to assay sympatholytics. Thus, various muscle tissues are selected for assaying specific chemical compounds. The choice of PSS, type of aeration, amount of tension provided, and temperature setting are specific for each preparation.

Keywords

Matching assay · Bioassay · Multiple point assay · PSS · Muscle preparation

7.1 Definition

Bioassay is defined as detecting and measuring the concentration of the substance in preparation or estimating the active principle's potency in a unit quantity of preparation using biological methods by observing specific pharmacological effects on living cells, micro-organisms, or animals.

M. Lakshmanan (✉)
Department of Pharmacology, Thanjavur Medical College, Thanjavur, Tamil Nadu, India

M. Lakshmanan et al. (eds.), *Introduction to Basics of Pharmacology and Toxicology*, https://doi.org/10.1007/978-981-19-5343-9_7

7.2 Classification

- In-vitro bioassay techniques: It utilizes cell culture techniques. The condition provided to the cells are not similar to the living environment, and cells survive by nutrition support from media. The effects of the test compound can be assayed with certain standard conditions that be extrapolated to living conditions to a certain extent. E.g., Use of stem cells, Microbes (bacteria), Cell culture, etc.
- *In-vivo bioassay methods:* A living animal or group of animals is used for this assay wherein the effects of test compounds in terms of biological effect or response of biological effect are studied directly. E.g., Studying the effect of skeletal muscle relaxant on the rabbit.
- *Ex-vivo bioassay methods:* In this technique, a part of the tissue from the living animal is taken outside the body and utilized for the assay. It is different from the in-vitro technique because, in ex-vivo, an organ or part of the animal's tissue is taken as whole and studied for a shorter period (till the tissue is viable). The physiological salt solution provides the nutrition for the survival of the tissue for that particular short duration. E.g., Estimation of histamine concentration using guinea pig ileum.

7.3 Indications

- The active principle of drugs is unknown.
- Unable to isolate active principle, e.g., insulin, posterior pituitary extract, etc.
- The chemical method is either not available, and if available, it is too complex, or insensitivity is observed with low doses.
- Unknown chemical composition, e.g., long-acting thyroid stimulator.
- Variable chemical composition but has same pharmacological action, e.g., cardiac glycosides isolated from diff sources, catecholamines, etc.

7.4 Principles

- The same response should be measured for the active principle to be assayed in all animal species
- Under identical conditions, the pharmacological response observed should be reproducible. [e.g., Histamine should produce ileum contraction in the same species under identical weight, age, sex, strain/breed, etc.]
- The sample and standard should have the same principle and activity
- Any variation observed between the individual unit must be accounted for or should be minimized
- A different aspect of the same substance can be measured with bioassay when compared to chemical assay (e.g., testosterone & metabolites)

7.5 Purpose

- To estimate the potency of a drug quantitatively
- To standardize vaccines, toxins, protein based drugs, and antiseptics that contain uniform pharmacological activity
- Determination of specificity of a chemical compound
- Assaying compounds with complex structure (e.g., Vitamine B12)

7.6 Methods of Bioassay for Agonist

- An agonist may produce a quantal response or graded response.
- In graded response, the response is proportional to the dose. The response may lie between no response and the maximum response. On the other hand, quantal response means that the response is all or none.
- *Dose-response curve:* The effect on a biological system due to various levels of exposure (dose) is described by the 'dose-response relationship' curve. Dose levels are usually expressed in mg/m^3 or parts per million for aerosols/vapors, and for solids, it is expressed as mg/kg body weight of the test animal. These levels can be plotted on a graph against the response (Fig. 7.1). The dose-response curve helps comprehend various concepts like median effective dose, median lethal dose, and therapeutic index.
- *Quantal-dose response:* When a response to a drug in a quantal (population) fell under two categories (observed—not observed, present-absent) and plotted against log-dose, the curve obtained is called the quantal-dose response curve. The shape and parameters of the quantal-dose response curve are almost similar to the graded dose-response curve. Information regarding the safety margin of the drug (e.g., therapeutic index) can be obtained with a Quantal-dose response curve.

Fig. 7.1 Dose-response curve (DRC)

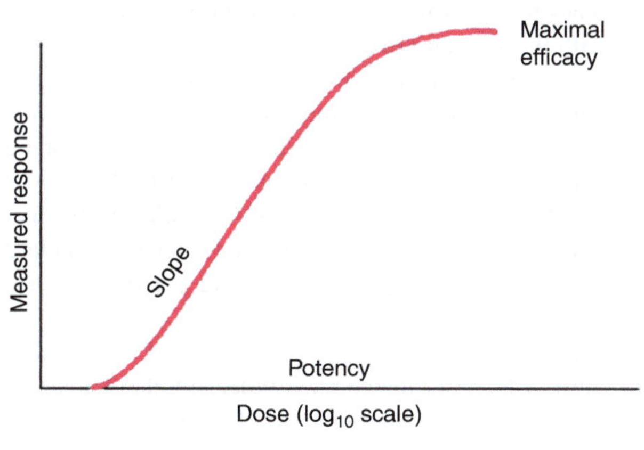

- *Graded responses:* A single animal (or patient) gives a graded response to the graded doses. As the dose is increased, the response also increases. With graded responses, one can obtain a complete dose-response curve in a single animal. The drugs producing graded responses can be bioassayed by matching (bracketing method) or graphical method.

7.6.1 Direct-End-Point Method

- In this method, the threshold dose of the agonist that produces the desired effect is measured on each animal. Subsequently, it is then compared to the average results of group of animals. Fixed doses are administered to groups of animals when it is impossible to measure individual effective doses or if animals are not available, and the percentage of mortality is determined at each dose level.
- The concentration of the given unknown agonist can be calculated by using the formula:

$$Conc.of\ unknown = \frac{(Threshold\ dose\ ofstandard)}{(Threshold\ dose\ of\ the\ test)} \times Conc.of\ standard$$

- Examples:
  - Digitalis induced cardiac arrest in guinea pigs
  - Hypoglycemic convulsions in mice
  - Digitalis induced head drop in rabbits
  - Calculation of LD_{50} in mice or rats

7.6.2 Matching Method

- The matching method is the simplest and crude method to calculate the concentration of the test sample. In the matching assay, a constant dose of the standard is bracketed by varying test doses until an exact match is obtained for the test dose with the standard dose.
- The reason for selecting a constant dose of standard but not the test while matching is that the tissue sensitivity can be monitored over the time when the standard dose is kept constant. The dose of the standard that is to be selected for matching can be done by constructing DRC with the standard. Any response of standard from 20 to 70% of DRC is selected for matching.
- Now the test response is repeated after the standard response. By adjusting the dose of the test solution, matching between the standard and test is attempted (Fig. 7.2)
- Once matching is obtained, a response of half the dose of the standard and half the dose of test that match each other are also taken. This is followed by taking the response for the double dose of standard and test that matched previously. All

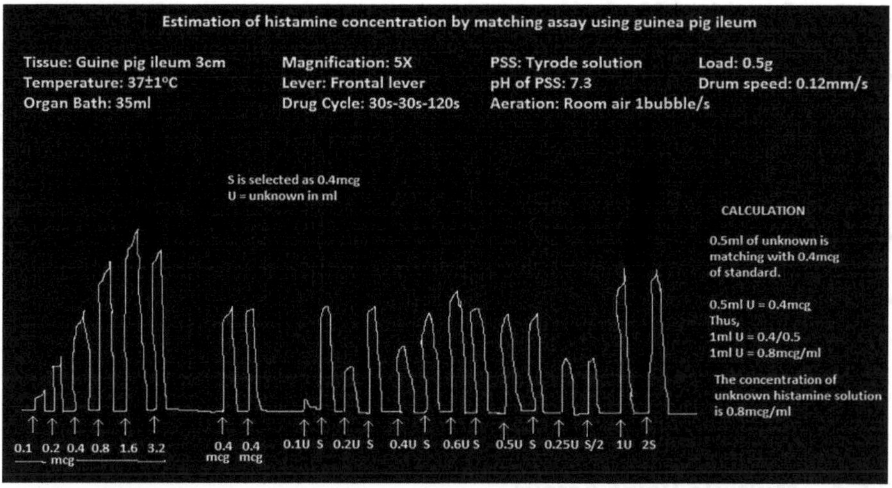

Fig. 7.2 Matching assay of an unknown solution of histamine on guinea pig ileum

these responses should match between test and standard (i.e., S = T, 2S = 2T, and S/2 = T/2). This confirms that the particular volume of the test is producing the response that matches with the particular volume of known concentration of the standard solution.

- Finally, the concentration of the test sample can be determined by the following formula:

$$Conc.of\ unknown = \frac{(Dose\ ofstandard)}{(Dose\ of\ the\ test)} * Conc.of\ standard$$

- Limitations:
  - Many times, the kymogram will run out of space for recording in a matching assay.
  - High rate of subjective error.
  - A dose-response relationship could not be established.
- Nevertheless, a matching assay will be helpful when the sensitivity of test drug preparation is unstable.
- *Bracketing method:* A slight variation of the matching method where two standards are selected (S1 and S2), and the response of the test solution is attempted to 'get bracketed' between the responses of S1 and S2. The concentration of the test drug is then calculated by interpolation through the dose-response curve.

7.6.3 Graphical Method

- The graphical method requires an assumption of the dose-response relationship. Initially, the log-dose-response curve for the standard is plotted. Then the response produced by the test sample is noted. Finally, the concentration of the test drug is derived from the graph directly by interpolation (explained in Fig. 7.3a, b)
- The characteristic of log- dose-response curve is that it is linear in the middle (20–80%). Thus, the comparison should be made within the range only
- Advantage:
  - it is a simple method
  - chances of errors are less (provided the sensitivity of preparation is stable)

7.6.4 Multiple-Point Assays

- Other methods based on the dose-response relationship include 3-points, 4-points, 6-points, and 8-points methods.
- The responses are repeated many times, and the mean value on each is taken in a multiple-point assay. Hence, the chances of errors are minimized in multiple-point assay methods.
- 3-*point assay method:*
  - In this method, two doses of the standard and one dose of the test are used. Log dose-response [LDR] curve plotted with varying concentrations of standard drug solution and given test solution
  - Select two standard doses, s1 & s2 [in 2:3 dose ratio], from the linear part of LDR [let the corresponding response be S1, S2]
  - Choose a test dose 't' with a response T between S1 and S2.
  - Record 4 sets of responses in the following order

    S1→S2 →T
    T→S1→S2
    S2→ T →S1
    S1 →S2→T*
  - The mean value of responses is calculated and plotted against log dose. The amount of standard that produces the same response as produced by the test is determined graphically or mathematically by following three steps
  - Step-1: Calculation of dose ratio $(d) = s2/s1$
  - Step-2: Calculation of log-relative potency $(M) = \frac{T-S1}{S2-S1}\log(d)$
  - Step-3: Calculation of unknown concentration $= \left(\frac{s1}{t}\right) \times \text{antilog}(M)$
- *4-point assay method:*
  - In this method, two doses of the standard and two doses of the test are used. Log dose-response [LDR] curve plotted with varying concentration of standard drug solution and given test solution

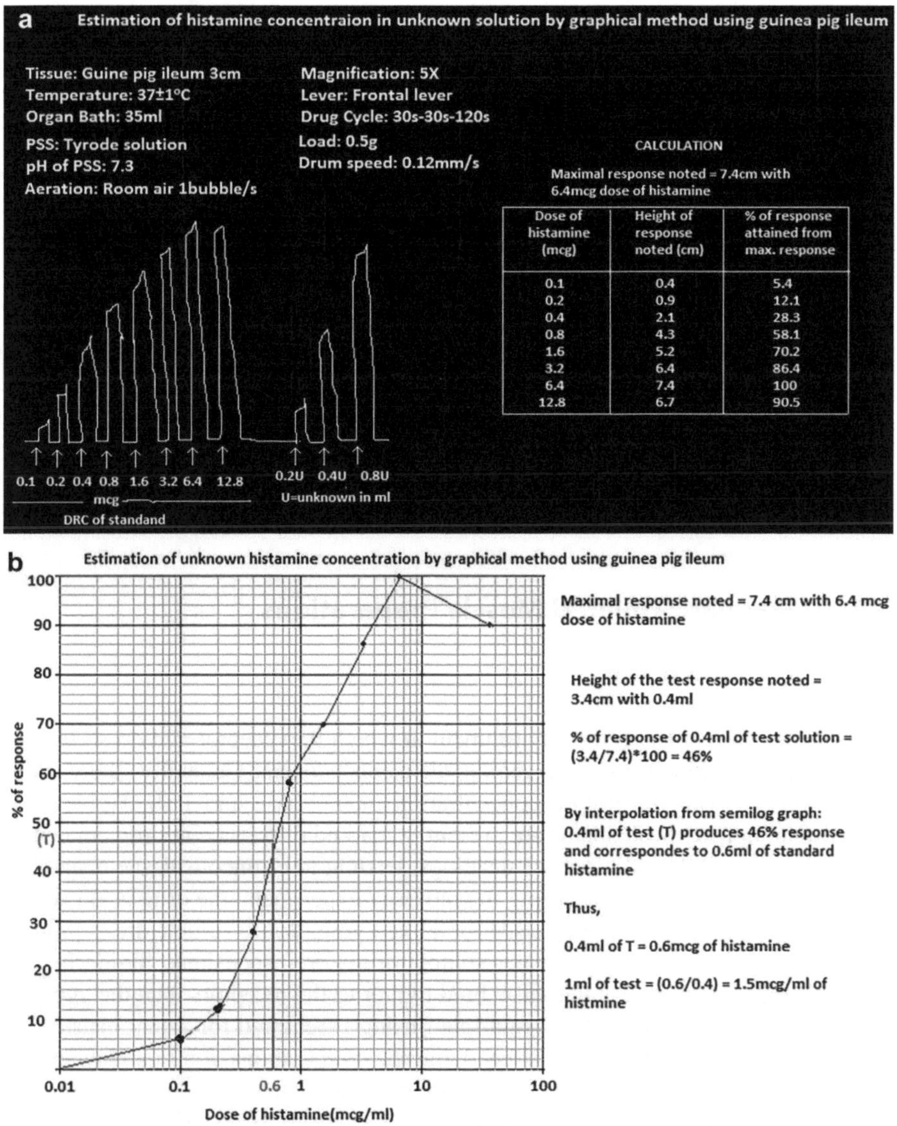

Fig. 7.3 (**a**) Graphical method of bioassay—Constructing DRC of standard and measuring the response of test solution (**b**) Graphical method of bioassay—Estimating histamine concentration in test solution by graphical method

- Select two standard doses, s1 & s2, from linear part DCR [let the corresponding response be S1 and S2]*
- Choose two test doses t1 & t2, with response T1 & T2 between S1 and S2.
- Record 4sets data follows (By Latin square design)

S1→S2→T1→T2
S2→T1→T2→S1
T1→T2→S1→S2
T2→S1→S2→T1

- The mean values of the responses are calculated and plotted against log-dose. The amount of standard the produces the same response as produced by the test is determined graphically or mathematically by the following three steps
- Step-1: Calculation of dose ratio (d) = s2/s1
- Step-2: Calculation of log-relative potency $(M) = \frac{(S2-T2)+(S1-T1)}{(S2-S1)+(T2-T1)} \log(d)$
- Step-3: Calculation of unknown concentration $= \left(\frac{s1}{t1}\right) \times \text{antilog}(M)$

- *Six points and eight points assay:* These methods have high accuracy and precision than other methods in estimating the actual value of unknown concentration with reduced error rates. However, these methods are uncommonly used due to limitations like long procedural durations, loss of tissue sensitivity to the drug after a particular time, and problems in recording all the responses in single kymograph tracings, etc.

7.7 Standard Isolated Muscle Preparations

7.7.1 Terminologies

- *Biophase:* It represents the environment in which the receptor and drugs interaction occurs without any intervening diffusion barriers. Generally, there exists equilibrium with the drug's concentration in biophase and PSS. However, if neuronal uptake mechanisms are present, a marked difference in drug concentration between PSS and biophase develops, leading to erroneous tissue response.
- *Drug tissue contact time:* During the bioassay experiment, the drug is allowed to interact with receptors in the tissue for a fixed time. Water-soluble drugs produce a response faster than poor-water soluble drugs or suspensions. Similarly, in a fixed contact time, the antagonists bind tighter to the receptors than the agonists. Thus, after fixed contact time, multiple rinsing is required for antagonists.
- *Inherent muscle tone:* Few isolated muscle preparations for bioassay exhibit inherent background tone (sustained submaximal contraction) due to the production of prostaglandins intramurally. In the stomach strips of rat, duodenum/jejunum of rabbit, and ascending colon of guinea pig, an inherent muscle tone is present. Lack of glucose or oxygen in PSS or applying relaxants can abolish this inherent muscle tone.
- *Isotonic muscle contraction:* It represents the shortening of the muscle during contraction. It can be recorded under steady load, isotonically. A simple frontal lever can measure isotonic muscle contraction.
- *Isometric muscle contraction:* It represents the development of tension inside the muscle without shortening during contraction. It can be measured using an electronic force transducer effectively.

- *Auxotonic muscle contraction*: It represents the combination of isotonic and isometric muscle contraction in variable proportions simultaneously. It is commonly seen in many isolated muscle experiments. It can be measured using a lever like a spring lever or auxotonic pendulum lever.

7.7.2 Muscle Preparations Used in Bioassay

- *Guinea pig ileum:* The ideal PSS for guinea pig ileum is a Tyrode solution maintained between 35 and 37 °C with room air. The longitudinal muscle in guinea pig ileum shows no difference between the development of tension and the percentage of shortening upon agonist administration. The ratio of EC_{50} of isometric to isotonic for guinea pig ileum is 1. Instead of Tyrode PSS, the Ringer solution can also be used. The usual tension fixed for guinea pig ileal tissue is 0.5–1 g. Guinea pig ileum is used to assay agonists and antagonists of histamine, muscarinic, serotonergic receptors. Guinea pig ileum contracts to both bradykinin and angiotensin II and is used to assay ACE inhibiting activity. The spasmogenic activity of vasopressin can also be measured using guinea pig ileum.
- *Guinea pig tracheal chain:* The suitable PSS for guinea pig tracheal chain is Krebs solution maintained at 37 °C with carbogen. Isotonic contraction is measured using a frontal lever with 0.5–1 g tension and 5–10 magnification. The Guinea pig tracheal chain contains alpha and beta-adrenergic receptors, muscarinic and histamine receptors. It is used to assay various anti-asthmatic drugs.
- *Guinea pig vas deferens:* The ideal PSS for guinea pig vas deferens is Tyrode. Ringer, McEwen, and Krebs PSS can also be used as an alternative. The temperature should be maintained at 29, 31, 32–34, and 37 °C for McEwen, Kreb's, Tyrode, and Ringer, respectively. Isotonic contractions are measured using a frontal lever with 0.5–1 g tension and 4–8 magnification. Guinea pig vas deferens contain prejunctional-alpha-2 adrenergic and IP_3 receptors. It is commonly used to assay alpha-adrenergic agonists and antagonists. For nerve stimulation studies, guinea pig vas deferens are isolated along with hypogastric nerves.
- *Guinea pig seminal vesicles:* The annular and longitudinal muscles of guinea pig seminal vesicles are innervated by the sympathetic system. It contains alpha-1 adrenergic receptors and is used to study sympatholytic activity. Ringer or McEwen solution can be used as PSS for guinea pig seminal vesicles. It should be maintained at 32 °C with carbogen. Both isotonic and isometric contractions can be measured using frontal lever and force transducer, respectively. The tension provided for measuring isotonic contraction is 0.2–0.5 g with eight-folds of magnification.
- *Guinea pig atria:* The suitable PSS for guinea pig vas deferens is McEwen, Ringer, or Krebs. The temperature should be maintained at 29–30, 31, 32–37, and 30 °C for Ringer, McEwen, and Kreb's, respectively. Carbogen should be used if Kreb's is used as PSS, and oxygen should be used for Ringer and McEwen PSS. Isotonic contractions are measured using a Starling heart lever with 0.5–1 g

tension and seven–ten-folds of magnification. Guinea pig atria are used to study beta-1 sympatholytic activity, calcium antagonism, and histamine-2 antagonism.

- *Guinea pig ureter:* Tyrode or Kreb's solution is used as PSS for guinea pig ureter. For tyrode PSS, the ureter should be maintained at 37 °C with a carbogen. For Kreb's PSS, the ureter should be maintained at 34 °C with oxygen. After the addition of KCl to the organ bath, the guinea pig ureter shows phasic-rhythmic contractions. Hence, guinea pig ureter is used to study potassium channel antagonists. Moreover, the guinea pig ureter is used as a model to study the peristaltic activities of the ureter. Guinea pig ureter is also used to study the interaction between yohimbine and NA efflux.
- *Guinea pig papillary muscle:* Ringer solution is suitable PSS for guinea pig papillary tissue. It should be maintained at 36 °C with carbogen. Guinea pig papillary muscle is used to study calcium antagonism, action potential and refractory period, and antiarrhythmic activities.
- *Guinea pig stomach:* It is used to study gastric motility and the intracellular mechanism regulating the stomach's motility. The circular muscles from the corpus region of guinea pigs' stomach are commonly used. The stomach strip should be maintained in Krebs-Henseleit solution with carbogen at 37 °C. Isometric contractions can be measured using force-displacement transducers like Grass-FT03. Isotonic contractions can be measured using a simple frontal lever under 1 g tension.
- *Guinea pig gall bladder:* It is used to study the activity of drugs on the motility of the gall bladder. The longitudinal strips of the gall bladder should be maintained in the Krebs solution at 37 °C with carbogen. At least 1 h of stabilization is required for the tissue before testing the activity. Guinea pig gall bladder contains M_3 receptors, and hence it is used to evaluate antimuscarinic agents.
- *Guinea pig taenia caeca:* It contains muscarinic receptors and is used to evaluate antimuscarinic and muscarinic agonists. The suitable PSS for this tissue is McEwen which is maintained at 37 °C with carbogen. Isotonic contractions are measured using a simple frontal lever with a tension of 1 g and six-fold magnification.
- *Rabbit duodenum:* McEwen PSS is used for this tissue and maintained at 37 °C with carbogen. The relaxation effect is measured using a frontal lever with 4 g tension and nine-fold magnification. The rabbit duodenum contains H_2, alpha, and beta-adrenergic, and serotonin receptors. However, the rabbit duodenum is standardly used to evaluate the relaxing effect of adrenaline and nor-adrenaline. The critical element in studying the relaxing effect of sympathetic drugs on the rabbit duodenum is 'tension provided' and 'tone produced.' Only after obtaining the ideal tone, the relaxation effect can be observed from 'fall from the baseline' in the kymograph.
- *Rabbit jejunum and ileum:* Ringer PSS is used for this tissue and maintained at 34–38 °C with room air. The isotonic contraction is measured using a frontal lever with 1–3 g tension and five-fold magnification. Rabbit jejunum is used to assay postsynaptic alpha-1, presynaptic alpha-2, and beta-1 adrenergic receptor

modulators. Rabbit ileum has muscarinic receptors and is used to study the spasmolytic activity of antimuscarinic agents.

- *Rabbit colon:* The rabbit colon can be used to study cholinergic agents as it contains muscarinic receptors. However, it is not commonly preferred as other tissues like guinea pig ileum have good standardization for cholinomimetic bioassay than rabbit colon. Rabbit colon can be used for studying the inhibition of chloride secretion activity.
- *Rabbit aorta:* The rabbit aorta contains alpha-adrenergic, histaminergic, and muscarinic receptors and responds to agonists and antagonists of the above receptors. Adrenaline, nor-adrenaline, histamine, and acetylcholine contract rabbit aorta. The rabbit aorta is usually maintained in the Krebs-half bicarbonate PSS at 37 °C with oxygen. Isotonic contraction can be measured using a frontal lever with 2–4 g tension and seven–ten-fold magnification. Rabbit aorta is also used to study calcium channel antagonism activities and angiotensin II-induced contraction studies.
- *Rabbit Heart:* It is used to assay beta-adrenergic agents. The suitable PSS is Langendorff perfusate and maintained at 37 °C with oxygen. Staling heart lever is used to measure isotonic contraction with 4 g tension and four–nie-fold magnification.
- *Rabbit atria:* It is used to assay beta-adrenergic agents. The suitable PSS is Krebs (double glucose) or McEwen. Rabbit atria should be maintained at 37 °C with oxygen for Krebs and at 29 °C with carbogen for McEwen PSS.
- *Rabbit gastrocnemius with sciatic nerve:* This tissue is used to study the neuromuscular blocking agents. Unlike other tissues of rabbits (ex vivo), this is an in-vivo method. Both sciatic nerve and gastrocnemius are not entirely removed from the animal. The test drugs are administered in the rabbit's marginal vein, and the effect of drugs on muscle contraction is noted by electrical stimulation.
- *Rat Colon:* The initial part of descending colon is used for bioassay, and it is distinguished by the presence of diagonal strips of muscle on the surface. Adrenergic agonists, substance P, prostaglandin-F_{2a}, and angiotensin II can be assayed using rat colon. DeJalon PSS is suitable for rat colon, and it is maintained at 25 °C with carbogen or oxygen. Isotonic contractions can be measured using a simple frontal lever with 2 g tension and four-fold magnification.
- *Rat duodenum:* It is the best tissue to assay bradykinin. Substance P contracts the rat duodenum while bradykinin relaxes it. DeJalon PSS is suitable for rat duodenum, and it is maintained at 30–31 °C with carbogen or oxygen. Isotonic relaxation/contraction can be measured using a simple frontal lever with 0.5 g tension and 15-fold magnification.
- *Rat stomach:* Longitudinal muscles of the rat stomach are used to assay acetylcholine and serotonin. Histamine and prostaglandin-E2 are also assayed using rat stomach. Tyrode PSS is suitable for rat stomach, and it is maintained at 37 °C with carbogen or oxygen. Dejalon or Krebs PSS are alternative to tyrode for rat stomach. Isotonic contractions can be measured using a simple frontal lever with 1 g tension and five–16-fold magnification.

- *Rat uterus:* Oxytocin, serotonin, histamine, and adrenaline can be assayed using rat uterus. Prior sensitization of virgin rat uterus with estrogen is required for assaying serotonin. Bradykinin, oxytocin, adenosine, and substance P stimulate the rat uterus while histamine (indirectly by releasing NA) and noradrenaline (directly) relaxes rat uterus by acting on beta-adrenoceptors (predominant receptor). The excitatory alpha adrenoceptors are temporarily seen in the rat uterus during late pregnancy, after estrogen therapy, and five days after parturition. Dejalon (25–36 °C with carbogen), Krebs (37 °C with carbogen), McEwen (34 °C with oxygen), and Locke (37 °C with carbogen) are suitable PSS for rat uterus. Relaxation or contraction can be measured using a simple frontal lever with 0.5–4 g tension and four–ten-fold magnification.
- *Rat anococcygeus muscle:* The two anococcygeus muscle along with ventral nerve band in the rat is used for assaying muscarinic, serotonergic, and alpha-adrenergic agents (all produces stimulation). The rat anococcygeus muscle preparation is free from ganglion supply and innervated only long adrenergic neurons. Krebs PSS is suitable for rat anococcygeus muscle, and it is maintained at 36–37 °C with carbogen or oxygen. Isotonic contraction can be measured using a simple frontal lever with 0.2–1 g tension and ten-fold magnification.
- *Rat phrenic nerve-diaphragm:* Rat phrenic nerve-diaphragm preparation is used to assay skeletal muscle relaxants and various toxins affecting sodium and potassium channels. Tyrode (double glucose) PSS is suitable for rat phrenic nerve-diaphragm, and it is maintained at 31–37 °C with carbogen or oxygen. Isotonic contraction can be measured using a simple frontal or spring lever with 0.5 g tension and eight–18-fold magnification.
- *Rat vas deferens:* It is used to assay alpha-sympatholytic activity. The alpha-agonist causes rapid contraction in rat vas deferens. Upon washing the agonist, the rat vas deferens goes for quick relaxation. Tyrode PSS is suitable for rat vas deferens, and it is maintained at32 °C with carbogen. Isotonic contraction can be measured using a simple frontal lever with 0.5 g tension and six–eight-fold magnification. Ringer or Krebs PSS can also be used as an alternative PSS to tyrode.
- *Hamster stomach:* It is used to assay Prostaglandin-E and PG-F. Other stimulators like serotonin and histamine are ineffective in this preparation and thus making hamster stomach ideal for assaying PG-E and PG-F. Amongst prostaglandin, the hamster stomach is most sensitive to PG-E. Krebs PSS is suitable for hamster strips, and it is maintained at 37 °C with air or oxygen. Isotonic contraction can be measured using a simple frontal lever with 0.5–2 g tension and 0.5–ten-fold magnification.
- *Cat splenic strip:* It is used to assay noradrenaline, adrenaline, isoprenaline, and histamine. Splenic strips of 25X5mm are maintained in Kreb's solution at 37 °C with carbogen. Isotonic contraction can be measured using a simple frontal lever with 1–1.5 g tension and six–ten-fold magnification.
- *Cat papillary muscle:* It is used to study restoration of mechanical force by cardiac glycosides after creating reduced performance by prolonged electrical stimulation. Ringer PSS is suitable for cat papillary muscle and maintained at

36 °C with oxygen. Contraction by electrical stimulation can be recorded using the polygraph.

- *Frog rectus abdominis:* It is used to assay acetylcholine and skeletal muscle relaxants. It is more sensitive than mammalian tissue to cholinomimetic drugs. As experiments on frogs are banned in India since 1999, frog rectus abdominis is not used nowadays for bioassay.
- *Leech dorsal muscle:* It is one of the oldest tissues used to assay acetylcholine and nicotine (both cause contraction). Anticholinesterase like physostigmine potentiates the contraction of Ach in leech dorsal muscle. The two dorsal muscles in the leech are isolated and mounted in the organ bath of 10 ml volume. Ringer (bicarbonate) solution is suitable for leech dorsal muscle and should be maintained at 20–25 °C with 97% oxygen and 3% CO_2.
- *Chicken ileum:* Lately, chicken ileum has been proven as an alternative to mammalian ileum. Chicken ileum contracts to histamine, acetylcholine, PGs, and serotonin and is used for assaying the same. Tyrode is the ideal PSS for chicken ileum. Ringer PSS is not suitable as it produces spontaneous activity. Chicken ileum should be maintained at 37 °C with 1 g tension and five-fold magnification.

Bibliography

Ghosh MN. Fundamentals of experimental pharmacology. 1st ed. Calcutta: Scientific book agency; 1984. p. 93–7.

Kurtel H, Yegen BC, Dedeoglu A, Ulusoy NB, Oktay S. Muscarinic receptor subtypes of guinea-pig gallbladder smooth muscle. Archives internationales de pharmacodynamie et de therapie [Internet]. 1990. [cited 2021 Aug 6];308

Nirmala P, Elandevan K, Chidambaram N, Santhakumari AS. Isolated chick ileum for bioassay of acetylcholine. Indian J Pharmacol. 2013;45(3):312–3.

Spencer N, Walsh M, Smith TK. Does the guinea-pig ileum obey the "law of the intestine"? J Physiol. 1999;517(Pt 3):889–98.

Vane JR. A sensitive method for the assay of 5-hydroxytryptamine. Br J Pharmacol. 1997;120 (Suppl 1):142–7.

Experiments on Anaesthetized Intact Animals

8

Amruta Tripathy

Abstract

Whole animal experiments have a historical background, and are still used in a few isolated screening experiments. Dogs, cats and rats are arguably the most commonly used animals for intact whole animal experiments today. Measurement of blood pressure and spleen volume are commonly performed on dogs, while the nictitating membrane model of cat is a classical example for whole animal experiments. Spinal cat is another classical model used to study dose responses and drug interactions. A major merit of these intact whole animal models is that we can obtain a real time effect of various drugs on the body as a whole. This chapter details the most popular and commonly used models in this domain.

Keywords

Whole animal · Spinal cat · Decerebrate cat · Intact animal models

8.1 Introduction

Screening of various substances in animals can be done in two possible ways:

- Tissue assay
- Whole animal assay

Whole animals like dogs, cats, rats have been used by many scientists in experimental pharmacology for drug assay purposes, especially in autonomic, cardiovascular,

A. Tripathy (✉)
Department of Pharmacology, All India Institute of Medical Sciences (AIIMS), Bhubaneswar, Odisha, India

M. Lakshmanan et al. (eds.), *Introduction to Basics of Pharmacology and Toxicology*, https://doi.org/10.1007/978-981-19-5343-9_8

and central nervous systems. Albeit whole animal experiments are not commonly performed anymore, knowledge about them is quite imperative in the field of Pharmacology.

8.2 Intact Animal Models n Dog

8.2.1 Blood Pressure Recording in the Dog

- The various intact animal models that have been used to study the cardiovascular system involve the studying of a large number of inter-related factors that affect the functioning of the system.
- Blood pressure changes are an index of comparison very commonly (if not routinely) chosen in a pharmacology laboratory. This parameter is regulated by an interplay of various factors like autonomic transmission, central nervous system, electrolyte levels, etc.
- Various drugs have been found to modulate blood pressure. Hence, while screening a new compound in pre-clinical models, it becomes imperative to know if the new compound affects blood pressure. Blood pressure recording is one of the most standardized methods used in preclinical studies.
- There are two methods to measure blood pressure in animals:
  (i) Non-invasive methods like Doppler and oscillometry.
  (ii) The invasive or direct method, which is considered the gold standard in measuring blood pressure.
- The method for the direct method of blood pressure measurement is as follows:
  - The entire procedure is performed in strict aseptic conditions.
  - The animal is anesthetized using thiopentone and then laid down in the supine position.
  - A suitable peripheral artery is chosen for catheter placement among the following arteries - dorsopedal (most common), femoral, plantar metacarpal, metatarsal, radial, coccygeal, auricular.
  - Once a suitable peripheral artery is selected, a small nick, just enough to let the catheter pass through is made in the artery.
  - Once the catheter is placed, it is connected to a pressure transducer via non-compliant tubing filled with heparinized saline.
  - The pressure transducer, can, in turn, be connected to an appropriate monitor, which displays the blood pressure as well as the arterial waveforms
  - Potential complications that can arise from catheter placement include hemorrhage, inflammation, infection, risk of thrombosis, and tissue ischemia distal to the catheter insertion site.
  - In case the central or carotid artery is supposed to be catheterized, a midline incision is given on the skin of the neck between lower larynx & upper thorax following which trachea is exposed after retracting the pre-tracheal muscles. A tracheal cannula is introduced into a gap between two tracheal rings. The cannula is held in position by a ligature and it allows free-breathing without

any obstructions due to respiratory secretions. One of the carotid arteries is identified and a small nick is made in the artery wall to allow an arterial catheter filled with an anticoagulant to be inserted into it. The catheter is connected on the exterior to a transducer and a monitor.

8.2.2 Spleen Volume Measurement

- Recording of spleen volume in a closed constant- pressure system gives an idea regarding the contraction of the splenic capsule. The splenic capsule contains α receptors; thus any α adrenergic drug will increase the splenic capsule contractions. This contraction is recorded with Marey's tambour, an instrument that graphically records the contractions.
- The steps for measuring splenic volume are as follows:
  - The entire procedure is performed in strict aseptic conditions.
  - The carotid blood pressure recording is set up as described above.
  - The abdomen is opened by an upper half midline incision.
  - The spleen is lifted very delicately from its position and is very carefully fitted into a spleen oncometer.
  - The oncometer is carefully placed in the abdominal cavity after carefully sealing off the edges with paraffin wax.
  - The outlet tube is connected using rubber tubing to a Marey's tambour.
  - Response from a control strip is taken first and subsequently, the effect of drugs on contractility and splenic volume is assessed.

8.2.3 Diuretic Study in Dogs

- Whole animal experiments can be performed in dogs for evaluating the diuretic or anti-diuretic effect of a drug. This preparation is employed for clearance studies as well. The process is depicted in Fig. 8.1.

8.3 Intact Animal Models in Cat

8.3.1 Nictitating Membrane

- The nictitating membrane in cats (most common) and dogs has been used to study sympathetic transmission and drugs affecting it. The muscle of the nictitating membrane has a very low resting tone; hence it is incapable of relaxing any further and can only be used to demonstrate contractions.

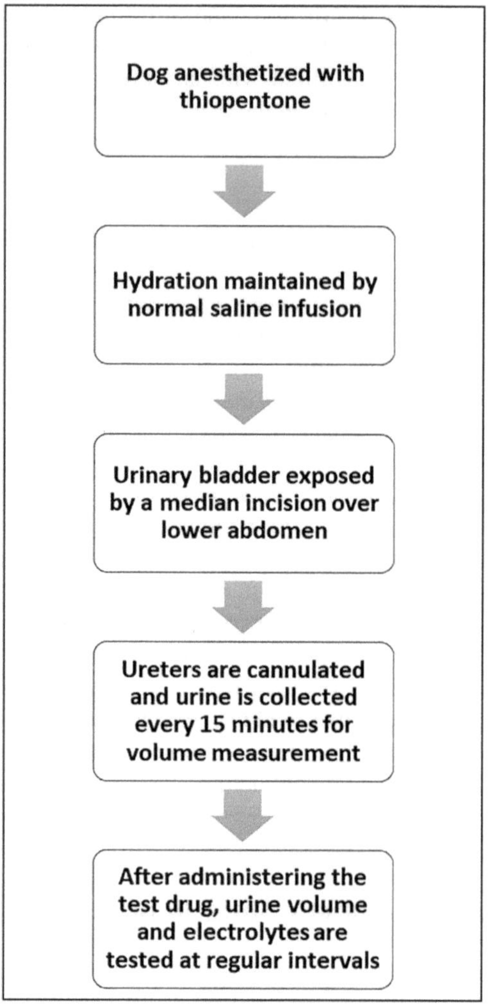

Fig. 8.1 Methods for diuretics study in dogs

- Cats are more suitable as they have a more evolved nictitating membrane that responds better to the stimulation of the cervical sympathetic trunk. Additionally in dogs, as the vagal nerves and cervical sympathetic trunk run together, stimulation of the sympathetic trunk causes central vagal stimulation as well.
- The entire method for recording nictitating membrane contractions has been described in Fig. 8.2. Chloralose is the anesthetic of choice here as it causes minimum inhibition of autonomic functions and produces stable anesthesia thereby causing minimal fluctuations in the tone of the nictitating membrane or the blood pressure.

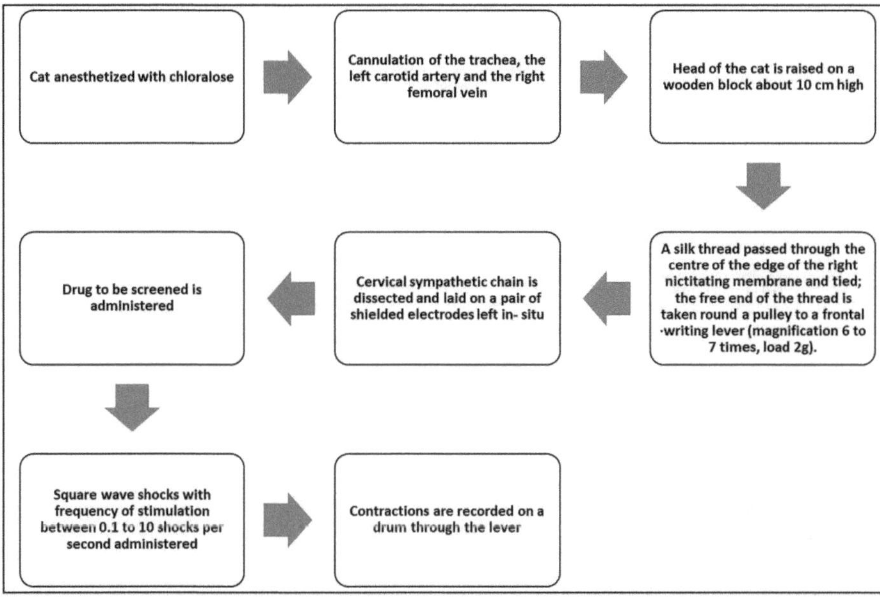

Fig. 8.2 Method to record nictitating membrane contractions

8.3.2 Spinalisation of the Cat

- Numerous experiments have been performed in animals after resecting out the spinal cord and evaluating the effect on the locomotion of the animal. These experiments shed light on the effect exerted by various ascending and descending pathways of the spinal cord on various functions. Such knowledge is important for improving the design of various types of therapeutic approaches in spinal-cord-injured patients.
- Resection of the spinal cord can be done by two methods:
  (a) Posterior approach
  (b) Anterior approach
- The steps of spinalisation of a cat by the posterior approach are depicted in Fig. 8.3.
- In the anterior approach, the spinal cord is resected out anteriorly after the tracheal cannulation to prevent turning the cat. The neck is hyperextended in this approach. The trachea and esophagus are divided, the anterior spinal muscles are retracted which exposes the first cervical vertebrae.
- The spinal cat model is predominantly used for observing the effects of various drugs acting on the autonomic nervous system. Blood pressure is the most common parameter measured in a spinal cat as it is an excellent surrogate marker for the functioning of the autonomic nervous system.
- *Uses of the model include:*

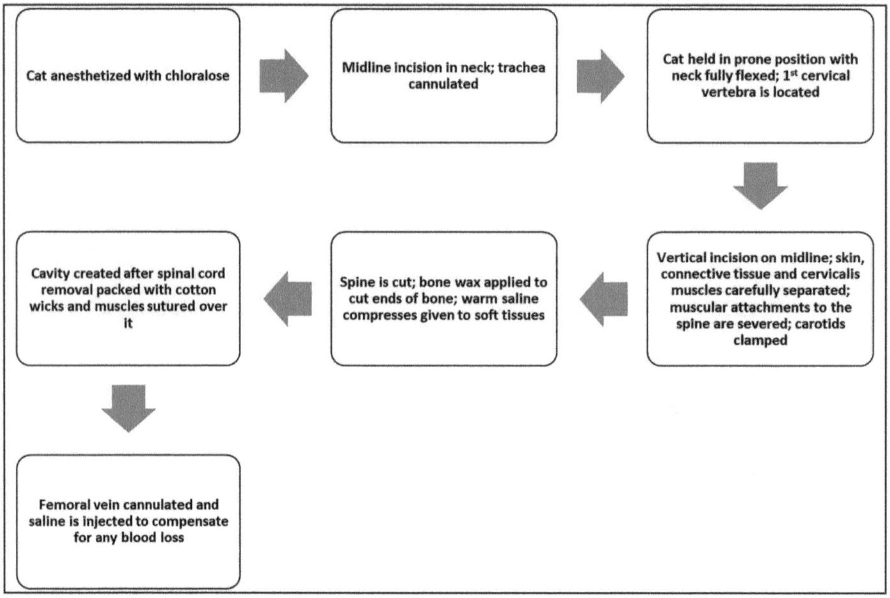

Fig. 8.3 Spinalisation of the cat (posterior approach)

- Study of drug dosage response of a drug by eliciting the effect of the unknown drug on blood pressure in a spinal cat
- Study of various drug interactions like synergism and potentiation by measuring the effect of the drugs on blood pressure in a spinal cat

8.3.3 Open Chest Preparation

- This method is used to measure blood pressure in a cat by directly attaching a pressure transducer to the heart of the animal. The open chest preparation of a cat not only helps in determining the response to drugs acting on the sympathetic nervous system but also helps in knowing pressure changes after vagal stimulation or inhibition (parasympathetic nervous system).
- The steps included in this preparation are as follows:
  - The cat is anesthetized with chloralosed and placed in the supine position.
  - An incision is made over the sternum; the sternum and the sternocostal junctions are divided with a bone cutter.
  - The cut edges of the chest wall are retracted; pericardium is slowly opened to expose the heart.
  - A suture is taken through the wall of the left atrium and the thread is either tied to a Starling lever recording on a kymograph or a sensitive pressure transducer connected to a physiograph.

8.3.4 Muscle Nerve Preparations

- These are used for evaluating the effects of drugs acting on the neuromuscular junction. The twitches obtained by stimulation of the nerve would provide a good index of neuromuscular transmission and at the same time, any direct muscle stimulation could help to pinpoint the exact site of action of the drug.
- Tibialis anterior and gastrocnemius are the most commonly used muscles for this experiment. The cat is first anesthetized. Depending on the muscle being used, the preparation of either muscle is depicted in Fig. 8.4.
- Once the muscle has been isolated, a bipolar electrode is placed on the nerve to the respective muscle and stimuli delivered to the muscle nerve preparation from a square wave pulse generator.
- The twitches generated are recorded on a smoked drum. Analysis of these twitches helps in evaluating the nature of the drug.

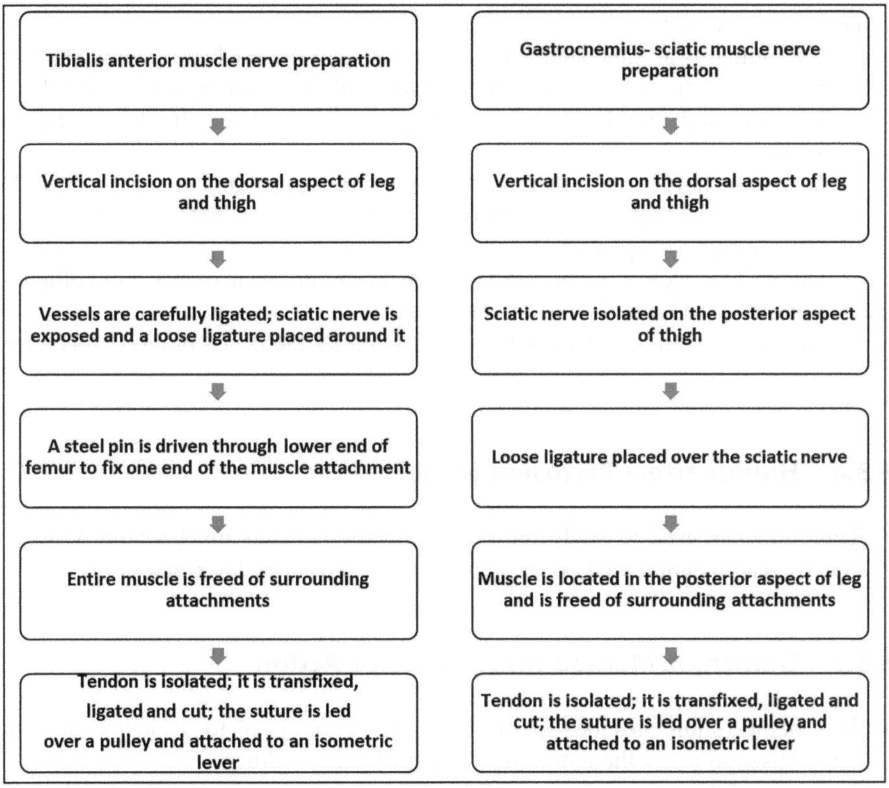

Fig. 8.4 Different techniques of muscle nerve preparations

8.3.5 Models to Study the Action of Centrally Acting Skeletal Muscle Relaxants

- Muscle tone is also influenced by drugs that act centrally by disrupting supraspinal facilitatory or inhibitory influences (via polysynaptic pathways) on the basic monosynaptic reflex arc.
- Various phenomena like linguomandibular reflex, clonus, decerebrate rigidity represent the disinhibition of spinal neurons from supraspinal influences. These reflexes will be ameliorated by centrally acting muscle relaxants.
- *Monosynaptic reflexes in the cat:*
  - The animal is anesthetized using chloralose and the trachea is cannulated. The hip and the knee are adjusted at 90° flexion.
  - A suture is passed through the tendoachilles and connected to an isometric lever and a kymograph. A patellar hammer is fixed over the knee joint such that it lightly taps the quadriceps femoris tendon to elicit the knee jerk.
  - Any reduction in the reaction to the knee jerk proves the central action of the administered muscle relaxant.
- *Decerebrate cat:*
  - The animal is anesthetized using chloralose and the trachea is cannulated.
  - The cat is then placed in a prone position with the head flexed completely.
  - The skin and muscles overlying the parietal region are divided to expose the bone.
  - A trephine hole is made 0.65 cm from the midline, on the line joining both the ears. The dura mater is carefully opened and a 5.0 cm pliable strip of plastic is inserted into the opening.
  - The strip is slowly maneuvered inside the brain such that a mid-collicular section of the brain stem is affected.
  - The patellar reflex of the cat is elicited 1 h after the decerebration procedure.

8.3.6 Hindquarters Perfusion Model

- This model is used to study the effect of various drugs on the cardiovascular system. The detailed method of preparation of the cat is described in Fig. 8.5.

8.3.7 Domenjoz Method for Antitussive Action

- This method stimulates the superior laryngeal nerve in cats to produce cough and can be used to study the antitussive action of various drugs.
- The steps of this procedure are as follows:
  - The cat is anesthetized using thiopentone. The femoral vein is cannulated for administering the drug.

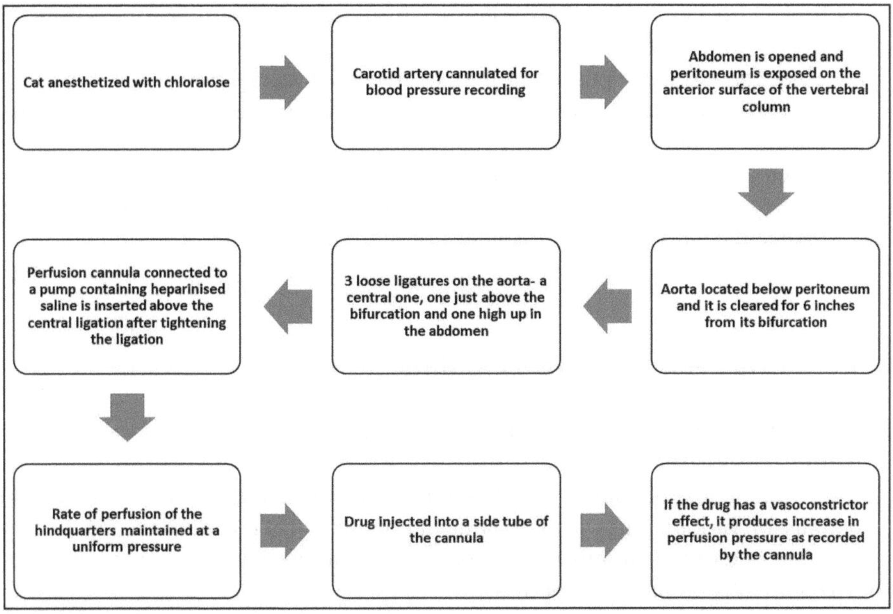

Fig. 8.5 Hindquarter preparation in cat

- The trachea is cannulated and connected to a Marey's Tambour which records the respiratory movements of the animal.
- The superior laryngeal nerve is located near the carotid artery and isolated. The laryngeal end of the nerve is ligated. As superior laryngeal nerve carries the afferent impulses for cough reflex, any stimulation of the nerve would produce a cough reflex.
- The unknown drug is first injected through the femoral vein cannula and the nerve is then stimulated. In case the unknown drug has any antitussive action, it will suppress the cough reflex.

8.4 Intact Animal Models in Guinea Pigs

8.4.1 Konzett–Rossler Method for Bronchodilator Activity

- Guinea pigs are highly sensitive to histamine; hence they are the models of choice for testing any drug with a suspected bronchodilator effect.
- The steps of this procedure as follows:
  - Guinea pig is anesthetized with urethane. A midline incision is made on the anterior aspect of the animal exposing the trachea.

- The internal jugular vein is located and cannulated with a 3- arm glass cannula, two of which are connected to a respirator.
- Since the respirator pushes air into the tracheobronchial tree against a particular resistance that is recorded, any change in the resistance will alter the tracings produced.
- Histamine will cause bronchoconstriction and increase the height and frequency of the tracings whereas any drug with bronchodilator property will reduce the amplitude and frequency of these tracings.

8.5 Intact Animal Models in Rats

8.5.1 Blood Pressure Recording in Rats

- Studying the fluctuations in blood pressure is an important measure to assess the activity of various drugs acting via the autonomic or the cardiovascular system. Rats are one of the various animal models used to measure blood pressure.
- The animal is prepared as follows:
  - Rat is anesthetized using thiopentone. A midline incision is made on the neck keeping the animal in the supine position.
  - The trachea is cannulated and the common carotid artery of one side is ligated at the superior end and clamped at the other end.
  - A cannula is inserted into the artery and connected via a three-way stopcock to a Condon mercury manometer.
  - The entire system is flushed with heparin saline solution.
  - Blood pressure is recorded by releasing the arterial clamp.
  - Drugs are pushed into the system by cannulating either the internal jugular vein or peripheral veins like a femoral vein and injecting the drug through the vein.

8.6 Conclusion

Experiments on intact anesthetized animals have often considered the gold standard for the screening of various drugs acting on various organ systems. Unlike isolated tissue preparations, intact animals show the overall effect of different regulatory pathways on a particular parameter e.g. blood pressure. These experiments thus provide a real-time effect of the drugs on the body as a whole. However, as these procedures are invasive, highly sensitive, require expertise, are quite cumbersome to perform and considered unethical, experiments on intact anesthetized animals have been substituted by various non- invasive methods wherein the model is replicated as a 3- dimensional model and the effects of various drugs can be studied without actually performing the experiments.

Bibliography

Claire Roberts RVN. An overview of direct (invasive) blood pressure monitoring. Vet Nurs J. 2016;31:218–21.

Ghosh MN. Fundamentals of experimental pharmacology. 6th ed. Kolkata: Hilton and Company; 2015.

Rossignol S, Belanger M, Chau C, Giroux N, Brustein E, Bouyer L, et al. The spinal cat. In: Kalb RG, Strittmatter SM, editors. Neurobiology of spinal cord injury. Contemporary Neuroscience. Totowa, NJ: Humana Press; 2000.

Sheth UK, Dadkar NK, Kamat UG. Selected topics in experimental pharmacology. 4th ed. Bombay: The Kothari Book Depot; 1972.

Vogel HG, editor. Drug discovery and evaluation pharmacological assays. 2nd ed Berlin: Springer-Verlag Berlin Heidelberg; 2002.

Bioassay of Standard Compounds

9

B. P. V. Sunanda and Rammohan Bondha

Abstract

Bioassay is defined as the estimation of the potency of an active principle in a unit quantity of preparation or detection and measurement of the concentration of the substance in a preparation using biological methods (i.e. observation of pharmacological effects on living tissues, microorganisms or immune cells or animals). The biological indicators are blood pressure, blood glucose, muscle contraction and inhibition of growth of microorganisms, etc. Bioassay is more sensitive than chemical assay. A bioassay or biological assay is a biological testing procedure for estimating the concentration of a pharmaceutically active substance in a formulated product or bulk material. In contrast to common physical or chemical methods, a bioassay results in detailed information on the biological activity of a substance. Over the last decade bioassays and bio statistical analysis have become more important in effectively controlling the quality of biopharmaceutical development and manufacturing. The purpose of bioassay is to ascertain the potency of a drug and hence it serves as the qualitative identification and quantitative measurement of the amount of active principle in pharmaceutical preparation or biological material. Measurement of concentration of a drug from magnitude of its biological effect. The principle of bioassay is to compare the test substance with the international Standard preparation of the same and to find out how much test substance is required to produce the same biological effect, as produced by the standard. The standards are internationally accepted samples of drugs maintained and recommended by the expert committee of the biological standardization of W.H.O.

B. P. V. Sunanda (✉) · R. Bondha
Department of Pharmacology, Anna Medical College, Montagne Blanche, Mauritius

M. Lakshmanan et al. (eds.), *Introduction to Basics of Pharmacology and Toxicology*, https://doi.org/10.1007/978-981-19-5343-9_9

Keywords

Bioassay · Standard preparations · Biological activity of a substance · Receptors · Potency

9.1 Bioassay on Acetylcholine

- The dorsal leech muscle and the frog rectus abdomen's muscle are most tactful tissues to Ach. For the bioassay of Acetylcholine, the colon is not the acceptable tissue, consequently it may give uncertain responses. This preparation are commonly used in isolated tissue experiments on account of contented isolation and more handling resistance than other tissues with variable spontaneous responses.
- *Methods used:*
  - Acetylcholine (ACh) utilize of rat ascending colon tissues.
  - Acetylcholine (ACh) utilize of rat/descending colon.
  - Acetylcholine (ACh) utilize of rat uterus.
- To begin with, one or three centimeters (cm) of rat colon are generally taken for the bioassay of noradrenaline (NA) and adrenaline (Adr) like preparation. NA is highly sensitive to the colon. The above methods also performed on the ascending or descending colon. By keeping the colon at 4 °C for 24 h the sensitivity likely increases. It must consider that after taking out any tissue from the fridge, it should intend-to attune the room temperature. It is also sensitive to the substance P, modest to prostaglandins and angiotensin are also sensitive to the above preparations. It is found that the extracellular calcium (Ca^{2+}) stimulate the calcium-sensing receptor (CaSR) and mediates an rise in inositol 1, 4, 5-trisphosphate, which is involved in the contraction and relaxation of the colon.
- *Acetylcholine (ACh) assay using ascending colon of the rat:*
  - Keep animal (rat) for fasting at least for 24 h.
  - By the procedure stunning (a strong blow) on the head, sacrifice the rat. After that keep the rat on the dissecting board (DB).
  - The rat should be fix on the DB by tying its legs with a thread.
  - To expose the abdominal viscera by making a vertical midline incision on rat abdomen.
  - Through the cecum must find out the ascending colon. Identify the cecum and go forward towards the rectum at least 5.2–7.5 cm.
  - Cut a little piece (1.4–3.6 cm) accordingly to inner organ bath volume and clean it with moderately warm water or PSS by using a syringe. later bind the two ends with a thread finally bind it to the balanced horizontal lever. Left the tissue for layoff not more than 30 min.
  - By serial adding of dilution which are prepared solutions like standard and test drugs and then record the response. choose the experimental method and

record the responses of the prepared drug. Quantify the recorded responses appropriately to the selected experimental methods.

- *Acetylcholine (ACh) assay using descending colon of the rat:*
  - Keep animal (rat) for fasting at least for 24 h. Same like above strong blow on head procedure sacrifice the animal. Following the procedure, keep the animal on the dissecting board (DB).
  - Just follow the above proceedings to bind the rat on DB. To expose the abdominal viscera by making a vertical midline incision on rat abdomen.
  - Find out the rectum and then back at least 5.2–7.3 cm to collect the descending colon.
  - Cut a little piece (1.4–3.6 cm) accordingly to inner organ bath volume and clean it with moderately warm water or PSS by using a syringe. later bind the two ends with a thread finally bind it to the balanced horizontal lever. Left the tissue for layoff not more than 30 min.
  - By serial adding of dilution which are prepared solutions like standard and test drugs and then record the response. choose the experimental method and record the responses of the prepared drug. Quantify the recorded responses appropriately to the selected experimental methods.
  - Quantify and graph mainly on the method of experimentation taken on, either 3-point or matching, etc.
- *Acetylcholine (ACh) assay using rat uterus:*
  - The response of a Drug towards the uterus tisuess in the presence of various estrous cycles differs due to the expression of the different receptors. The α-adrenergic receptor is prominent, and it causes relaxation. which is blocked by α-receptor antagonists. Since the uterus has no inherent tone but relaxation is observed by physiological antagonism of the contractile in response of above preparation or other drugs. In the estrous stage (the size and uterus vascularity are increased), suddenly contractions are increased. It is found that after priming with estradiol, there is an excessive expression of α-excitatory adrenoreceptors. Other agents like serotonin, norepinephrine, etc., also be assayed with the uterus. Contraction of uterine muscles in non-pregnant animals, whereas relaxation with PGE_2.
  - Depend upon the age of the animals the preparation response mainly various, cause of difference in the estrus cycle. It is fast contracting tissue and shows a spontaneous response. It responds to the minimal dose of Ach at 5.3×10^{-5} M and carbachol at 10^{-4} M. The other drugs like adrenaline (3×10^{-6} M), noradrenaline (3×10^{-5} M), Isoprenaline (10¯6 M), ephedrine (10^{-3} M), and tyramine (10^{-3} M) manifest the accurate to the uterus preparation.
  - Something vital to promote the estrus cycle into the normal female animal before the experiment. Artificially induced estrus cycle has the differs the sensitivity of a tested drugs such as for bioassay of 'oxytocin' or Ach.

- Take stilbestrol 0.1 mg/kg (20 μg/ml), subcutaneous 24 h before the experiment starts, whereas for assay of serotonin on the uterus is primed for three days at the dose of 0.3 mg/100 ml. Other drugs like histamine (mediated by H_1), adrenaline, and noradrenaline (mediated by β-receptor) are as well sensitive to the uterus.
- Female rat (FR) weighing around 150–300 g is taken, and 24 h before the experiment, the rat is primed with 0.1 mg/kg stilbestrol, IM. By the procedure stunning (a strong blow) on the head, sacrifice the FR. After that keep the FR on the dissecting board (DB).
- Just follow the above proceedings to bind the FR on DB. To expose the abdominal viscera by making a vertical midline incision on rat abdomen (also see the identification and collection of tissue/muscle section).
- Identify the vaginal orifice. Trace the vaginal orifice interiorly to find the two horns of the uterus. Tailor the tissue in respective to the inner organ bath (handle the tissue minimally at the middle part, which can reduce the sensitivity of the tissue).
- Choose the experimental method (matching bracketing, 3-point, or 4-point assay). The evaluate and graph depend upon the method of experimentation taken, either 3-point, 4-point, matching, etc.

9.2 Bioassay of Adrenaline

- Adrenaline (Adr) or epinephrine is a hormone released from the adrenal gland. It is one of the neurotransmitters in the body.
- *Principle:* The estimation of potency of the test drug by comparing with the rise in blood pressure of the test sample with the standard preparation of adrenaline.
- *Preparations:*
  - Cat spinal blood pressure
  - rabbit duodenum (Isolated)
  - fowl caecum (Isolated)
  - muscle tone Inhibition
  - rat uterus (Isolated)
- *Potion of standard solution:* According to the adrenaline monograph in the Indian Pharmacopoeia should fulfill all the purity tests. In certain rotation of 4% w/v solution of the standard Adr in 0.1 N HCl should in between −5.5 and 5.5 °C.
- The dilutions of adrenaline are prepared by adding saline in an appropriate concentration is 1:10,000. The test solution is also prepared in saline.
- *Models using Anaesthetized Dog:*
  - An averaged dog (11–13 kg) is anesthetized by pentobarbitone sodium with a dose of 25 mg/kg, I.V.

- If necessary, artificial ventilation is given through a tracheal tube. Blood Pressure is recorded on a kymograph, carotid artery is cannulating with an arterial cannula and same with a mercury manometer.
- The dog should be given 0.001–0.002 mg of atropine sulfate to paralyze the vagal nerve. This can be confirmed by stimulating the vagal nerve by electrical stimulation. after electrical stimulation there will be no fall in Blood Pressure by atropinized.
- Through the venous cannula inject Adr into the femoral vein. Two succeeding responses of the similar dose of adrenaline are noted. The same responses with the same doses of Adr implies that there is no change in Blood Pressure, indicates the dog is now ready for the assay.
- Inject both the test and the standard samples are given in alternate until both produce the same rise in Blood Pressure. At the intervals of five minutes Injections are given. There by comparing the potency of the test sample with that of the standard.

• *On Spinal Cat:* The spine of the cat is exposed. The spine is demolished by passing a stiff wire down the spinal canal. This causes a precipitous fall in BP. Then adrenaline is given in the I.V route and observed for the rise in BP.
• *Method Use on Rabbit's Duodenum:*
  - An averaged rabbit weighing 2–3 kg is killed by the head blow procedure and bled to death. The rabbit is dissected. The abdominal parts are exposed to isolate the duodenum. In the inner bath of mammalian organ bath duodenum is suspended.
  - Bioassay of Adr Inner bath is fill up with Tyrode solution, which is continued at 37.5 °C. The muscle is acclimatized for 30 min. Regular contractions are recorded applies to the isotonic frontal writing lever on the kymograph.
  - The duodenum is relaxing by Adr, by the way same property had been taken up to consideration to finding out the potency of the test sample.
• *On Rat's Uterus by using De Jalon's Method:*
  - De Jalon initially demonstrated the method. An average weight of 100–150 g virgin female rat is used in this experiment. Follow the above procedure to scarify the rat. The uterine horns are isolated by opening of abdomen and positioned in a petri dish which have de Jalon solution.
  - Suspend the ligated uterine horns in a mammalian organ bath containing modified Ringer solution having the following composition is used. KCl 0.46 g; NaCl 9.01 g; Dextrose 0.5 g; $CaCl_2$ 0.06 g; $NaHCO_3$ 0.5 g and distilled water 1000 ml. This composition prevents the rhythmic contraction of the uterus. It is preserved at 37 °C and bubbled continuously with air. Subsequently carbachol (0.75 mcg/ml) has given to record the two indistinguishable contractions for 30 s.

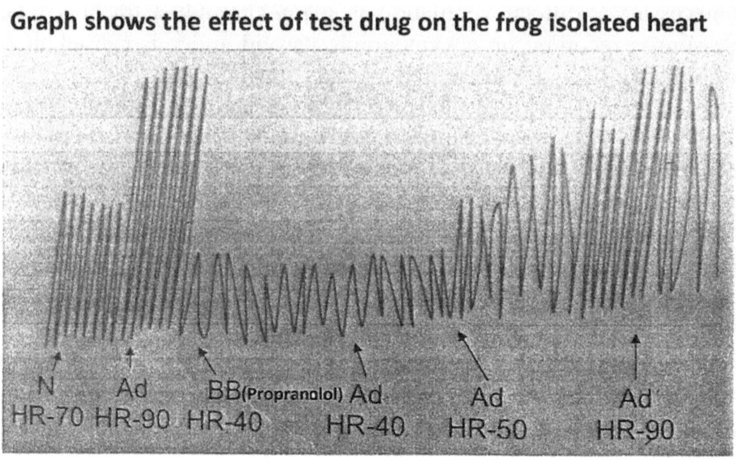

Fig. 9.1 Graph on the isolated frog heart

- The decrease percentage of carbachol-induced contractions is recorded by choosing three various doses of standard Adr and one intermediate dose of test adrenaline. Response are recorded by 3 various doses of standard Adr and one intermediate dose test Adr. identify percentage of reduction in carbachol-induced contractions have been recorded.

- *Method of Straub's:* In this procedure, through the aorta and valves a glass cannula is passed into the ventricle of the heart in such away the solution is pumped up and down in the cannula with each beat. Behind the cannula attach a graph paper to evaluate the force of contraction of the heart's systole at each beat. By attaching the edge of the ventricle to lever by a thread, to record each beat. Adr increases the force of contraction of the ventricle (Fig. 9.1). The increased force produced by the test sample is contrast with the standard preparation.

9.3 Bioassay of Histamine

- Sir Henry Dale first discovered histamine. Its synthesis occurs in the mast cella, CNS, Parietal cell of gastric mucosa, and periphery. Histamine functions as an autocoid and one of the mediators cause the inflammatory allergic responses. It has a vital role in gastric acid secretion.
- In the part of an immune response by basophils and mast cells found in nearby connective tissues releases the histamine. Through Histamine increases the permeability of the capillaries to white blood cells and some proteins to permit them to engulf pathogens in the infected tissues. Histamine present in all mammalian tissue derivative fromβ- imidazolylethylamine.

- *Methods used:*
  - Cat or dog Blood Pressure
  - Guinea pig ileum
  - Guinea pig uterus
- *Assay of Histamine by using guinea pig ileum:*
  - Guinea pig ileum is highly sensitive to histamine. Histamine contractile response towards the ileum is mediated by the H_1 receptor. H1 receptors predominantly found in the ileum, bronchi, and capillaries. The ileum is preferred for the experiment because of less mesentery innervation to the ileum, and, most of the receptors are present. Nevertheless, cut small portion (10 cm) of ileum attached to the cecum contains more of the α-excitatory receptor, which should be excluded.
  - Keep guinea pig for fasting at least for 1–2 days. By the procedure stunning (a strong blow) on the head, sacrifice the rat. After that keep the guinea pig on the dissecting board (DB).
  - The guinea pig should be fix on the DB by tying its legs with a thread.
  - To expose the abdominal viscera by making a vertical midline incision on rat abdomen.
  - There two different ways to Identify the ileum, (1) find out the cecum then come back at least 10 cm, or (2) find out the stomach, then go forward to identify the ileum 10 cm before cecum (also see in identification and collection of tissue section). Take out the desired length of ileum approximately 2–3 cm or as the, and the remaining should be kept in the refrigerator for future use. The ileum is trimmed away from the mesentery or attached tissues. Any waste present in the ileum is cleaned by a gentle push of PSS with the help of a syringe (preferably warm PSS at 30–35 °C).
  - Then, the ileum has two edges, one edge is attached with a hook into the tissue holding arm, and another edge is tied to the lever (attachment should be in the way of which the intestine is vivo as far as concern).
  - The experimental design is selected 4-point, 3- point, or any other design mentioned. Then, make the DRC of standard and test drug. Then estimation and graph depend on the experimental method/design adopted neither 3-point, 4-point nor matching, etc.
- *Anesthetized dog or cat:*
  - Dog or cat is anesthetized by using chloralose or barbiturate and prepared for the measuring of BP.
  - By injecting a standard solution of 0.05, 0.1, 0.15 μg of histamine based on per kg body weight is given for 5 min interval to determine the sensitivity. An a standard dose producing a fall in BP of about 20 mmHg by injected with changing doses of the test at regular intervals, and a matching assay is done.

9.4 Bioassay of 5HT (Serotonin)

- Serotonin are derived from naturally occurring amine from the tryptophan & regularly found in plants, animal tissues, some fruits and insect venoms. In humans, it is found in GIT & CNS of enterochromaffin cells. The limbic system, raphe nucleus, cortex, hypothalamus, amygdala, caudal nucleus, mid-brain, vomiting center & spinal cord are predominantly with serotonergic system. Serotonergic neurons regulate sleep, body temperature & mood. The melatonin is derived from serotonin.
- *Methods used:*
  - Assay using rat uterus
  - Assay using Rat stomach
- *Assay of serotonin by stomach fundus of the rat:*
  - The bioassay of 5-HT is carrying out by using stomach fundus, by contemplate fundus is the more sensitive tissue among all the parts of stomach. The fundus can be identified by their color gray and located above the pink think pyloric region. Drugs like histamine are less or insensitive to the rat fundus because of the absence or lack of H_1 receptors in the stomach muscle.
  - Other than serotonin some more drugs which are sensitive against the fundus are ACh, bradykinin, and prostaglandin (PGE2). In this assay the very important step is Muscle preparation using stomach fundus. Perhaps, both types of muscle used in the experiment, transverse cut made to prepare the tissue.
  - At least for 24 h Keep the animal (rat) under fasting and water is given *freely*. By the procedure stunning (a strong blow) on the head, sacrifice the rat. After that keep the rat on the dissecting board (DB).
  - The rat should be fix on the DB by tying its legs with a thread
  - To expose the abdominal viscera by making a vertical midline incision on rat abdomen.
  - Identify the stomach, cut and separate the stomach attachment at the closing part of the esophagus and down at the duodenum.
  - Identify the fundus (grey part), which discriminated from the pyloric region that is pink in appearance. Through wash of stomach contents with the distill water and then through lesser curvature open the fundus. Then, cut the fundus into two equal parts through a midline incision on rat abdomen.
  - For preserving the longitudinal muscle, take turn to opposite vertical cut should be given, and then tie the thread on both end of the muscle and stretch out the muscle. Then, tie into the organ bath and leave for relaxation at least 30–45 min. Select the experimental design and get the response with standard and test drugs.
  - The evaluation and graph depend on the experimental method/design of adopted, either 3- point, 4- point or matching, etc.

9.5 Angiotensin II Bioassay

- Contraction in rabbit aorta -induced by Angiotensin II:
  - Adult male albino rabbits are sacrificed by exsanguination. The thoracic aorta is identified and removed without any damage. By gentle rubbing, the vascular endothelium is removed.
  - Spiral aortic strips around 2.5–3.5 mm wide are prepared. The length of the strip should be at least 3 cm.
  - Mount the aortic strips are in the 5 ml organ bath that contains Krebs-Hensseleit preparations at 37 °C and aerated with carbogen. The strips are kept undisturbed for one hour for equilibration with 1 g tension. Baseline contractions are monitored. The strips are washed every 15 min for 1 h.
  - After one hour, angiotensin II (0.1–300 mM) is add on, and the record the contractions. Angiotensin is added cumulatively, and contraction curves are recorded. After completion of each curve, wash the aortic strips four times and allowed to return to the normal baseline tension.
  - By using this method on new compound can also be tested, to find out the angiotensin II antagonistic activity.

9.6 Bioassay of Vasopressin

- Vasopressin (antidiuretic hormone) is secreted from the posterior pituitary gland. It has a potent vasoconstriction and antidiuretic activity. It is used for upper esophageal bleeding (20 units/15 min). It is also used for diabetes insipidus (5–20 units).
- *Principle:* Under the conditions of suitable methods of assay are used to evaluate the activity of vasopressor by contrast the activity of the preparation with that of the standard preparation of arginine vasopressin.
- *Standard Preparation:* The 1st International Standard preparation of Arginine vasopressin established in 1978, contains human albumin and citric acid with freeze-dried synthetic arginine vasopressin peptide acetate, which are supplied in ampoules containing 8.20 Units.
- other appropriate preparation and their potency has been determined in relation to that of the International Standard.
- *Methods of Bioassay:*
  - Rat blood pressure method
  - Antidiuretic activity in ethanol rats
- *Method Using Blood Pressure of Anesthetized Rats:*
  - Average weighing rats of 300 g are anesthetized with 1.75 ml/kg of 25% Urethane administrated intra peritoneally. It is treated with a ganglion blocker like hexamethonium (1 mg) to produce a steady baseline. Cannulate the external jugular vein and internal carotid artery to inject the drug and record

the blood pressure respectively. Doses of references standard + test drug are injected in a randomized 2 × 2 assay.
  - Inject the volume of 0.1–0.5 ml.
  - The initial dose of reference standard may be tried in the range of 6–10 micro-units. Relative potency is calculated. This method can also be used for the assay of Epinephrine and Norepinephrine.
- *Method using Diuretic activity highly anesthetized rats:*
  - Rat weighting 250–300 g are hydrated and anesthetized with an oral dose of 5 ml/100 g f 2.5% ethanol. To collect the urine, cannulate the urinary bladder. The tail vein is cannulated for injection of drugs. Urine output is measured after the oral administration of 0.5% ethanol. After injecting both standard and test drugs, the decrease of urine output is estimated, and relative potency is calculated.

9.7 Bioassay of ACTH

- ACTH (Adrenocorticotropic hormone, corticotropin) is a tropic polypeptide hormone (39 amino acids) secreted by the anterior pituitary gland. ACTH stimulates the production of cortisol, a steroid hormone essential for regulating glucose, protein, and lipid metabolism, suppressing the immune system response, and support to maintain blood pressure.
- Regulation of cortisol synthesis: When cortisol level falls, the hypothalamus synthesis corticotropin-releasing hormone (CRH). CRH stimulates the hypothalamus to release ACTH; this will stimulate the adrenal gland.
- *Preparations:*
  - Corticotropin injection: Is obtained from mammals' pituitary gland which is a polypeptide of sterile solution, in an appropriate diluent. The range of potency must be in 80.0–120.0% of USP corticotropin units. This preparation has been taken as the reference standard for the bioassay.
- *Method of Assay:*
  - By giving pituitary ACTH decreases the ascorbic acid in the adrenals. The main function of ACTH administration is to deplete the adrenal ascorbic acid. This has been used for a quantitative assay of ACTH.
  - An average weighted Male Wistar rat (100–200 g), has removed pituitary gland by surgery (hypophysectomies) one day before the test. Constant number at least 35 (selectively 60) of hypophysectomies rats are necessary to perform one test with three different doses of test preparation and standard Number of hypophysectomies rats required.
  - Solution: the 5 units of test or standard preparations are dissolved in 0.25 ml of 0.5% phenol solution and finally dilute with 8.1 ml of 15% gelatin solution. Now every 0.5 ml contain 300 mU ACTH solution.

- Solution A- 3 ml of solution A have been diluted with 6 ml gelatin solution. Now the concentration is reduced to 100 mU ACTH/0.5 ml.
- Solution B again 3 ml of solution B have been diluted with 6 ml of gelatin solution, now at end solution contains 33 mU ACTH/0.5 ml the Bioassay of ACTH Procedure Continued.
- Distribute 6 groups of the hypophysectomized rats through randomly. Administered 0.5 ml of various concentration of test and stranded preparations to Each rat subcutaneous 0.5 ml. After the injection of 3 h, anesthetized the animals, and remove the both adrenals, weight the both adrenals without ant extraneous tissue.
- The rats are sacrificed, and the skull is opened to verify the completeness of hypophysectomy. The adrenals are integrated into a glass tubes which contains 200 mg pure sand and 8.0 ml of 4% trichloroacetic acid, finally ascorbic acid can determine. The ratio of potency, including confidence limits, is calculate with the 3 + 3-point assay.

9.8 Bioassay of Estrogen

- Estrogen compounds are important in the menstrual and estrous reproductive cycles. They are the chief female sex hormones. Female sex hormones are steroid hormones, where as some synthetic ones are not steroidal. This have role in some oral contraceptives, in postmenopausal women as hormonal replacement therapy, and hormone replacement therapy. In women estrone (E1), estradiol (E2), and estriol (E3) are the 3 primaries naturally occurring estrogen. Estrogen receptors like αand β which can be act as activates and inhibits the action of estrogen by the transcriptional activity of the receptors. They may have high potential to produce the adverse health effects, which includes developmental and sexual effects. Consequently, is required to assess and evaluate a drug as a possible estrogen agonist or antagonist.
- *Principle of the test:* The sensitivity on an animal test system by an Uterotrophic Bioassay relies. The assay conducted in Hypothalamic-pituitary-ovarian axis is not functional, leading to low endogenous levels of circulating estrogen. This will indicate a low uterine weight and a high range of response with the estrogen's administration.
- *2 estrogen-sensitive conditions in the female rodents to meet this essential:*
  - Immature females after weaning and before puberty.
  - young adult females after ovariectomy.
- *Selection of animal species*: Wister strains of rats and Sprague-Dawley.
- *Preparation of animals:* The animals should be identified uniquely. Preferably, immature animals should be caged with dams or foster dams until weaning during acclimatization. Before starting the study, the acclimatization period should be about five days for young adult animals and immature animals. Each treated and

control group should include at least six analyzable animals (for both immature female and ovariectomized female protocols).

- *Procedure for ovariectomy:* ovariectomy must occur between 45 and 62 days of age, for this procedure under take in a female rat of both test and control groups. At least not less than of 14 days must pass between ovariectomy and the day one of administration. allow the uterus to lapse to a minimum of constant baseline.
- *Bodyweight:* In the ovariectomy model, body weight and uterine weight are not corresponded due to the uterine weight is affected by female sex hormones like estrogens but do not have any role on the growth factors and the body size. On the opposite, in a weanling animal the body weight is related to uterine weight while they are maturing in 35 days. Thus, at the early days of the study, the variations in weight of animals used in the immature model should be minimal and not more than ±20% of the average weight.
- *Dosage:* To find out the estrogen antagonists, the test chemical and estrogen antagonist is administered to immature or ovariectomized female rats together, e.g., 17α-Ethinylestradiol. With the reference of estrogen agonist should administrated minimum of two different test chemical doses in two test groups. The control group and a test group will receive the reference agonist at the same dose.
- *Administration of doses:* administered the test compound by oral or subcutaneous injection.
- *Observations:* observed the weight and food consumption. On daily bases should measure the weight of All animals that close to 0.1 g, that should started likely before beginning of treatment. The food intake results should be measured in grams per rat per day.
- *Measurement and Dissection of uterus weight:* The objective of this assay is to measure both the uterus weights of wet and blotted.in a wet weight uterus incorporate the uterus and the luminal fluid. The blotted weight uterus is measured after the luminal contents have been expressed and excluded.
- *Alternative investigations:* After hematoxylin and eosin (HE)-staining, must examined histopathological findings. after the weighing the uterus buffered with 10% neutral buffered formalin. According to the finding may be investigated then vigina.

9.9 Bioassay of Oxytocin

- Oxytocin is secreted by the posterior pituitary along with vasopressin (ADH). Oxytocin is a nonapeptide.
- In 1909, first used pituitary extract to induce labor. In 1953, eventually du Vigneaud resolved the controversy about the ADH and uterine stimulants. Those activity due to one substance or two of them and also two separate principles, when he isolates both Oxytocin and Vasopressin. They are nonapeptides which vary in their positions 3 and 8.

- *Principle:* By analyzing the activities like decrease of BP, contraction of the uterus and milk ejection, the Potency of standard preparation of oxytocin/ vasopressor.
- *Standard preparation of oxytocin solution:* the potency of standard preparation has been determined by corelating with the international standard. The oxytocin preparation contains of free dried synthetic oxytocin peptide with human albumin citric acid of 12.5 units or any other suitable preparation.
- *Preparation of test sample solution:* follow the above procedure to prepare the sample solution.
- *procedure A- By depression of blood pressure in chicken*
  - Animals: cockerel weighing 1.2–2.3 kg.
  - Organ: Anaesthetised, prolonged, and constant high BP should be maintained.
  - Expose gluteus primus muscles (thigh) and remove popliteal artery and crural vein. Cannulate the popliteal artery and record BP response. Cannulate the both popliteal artery and crural or brachial vein to record the BP and to inject the prepared two standard doses solution with saline of dose 0.1–0.5 ml respectively. The dose must cause a decrease in BP (required dose between 20 and 100 mUnits). The interval between two injections is usually between 3 and 10 min and depends on how BP returns to the normal range.
  - Diluted test preparation with saline to get the same response as standard. The ratio between standard and test need to be equal. Another solution must be used if the animal is insensitive to repeated injections.
  - Standard bioassay calculations are applied, and results are inferred.
- *Procedure B - By contraction of the rat uterus*
  - Test animals: female rat 120–200 g.
  - Inject 100 ug of oestradiol benzoate IM into female rats before the experiment starts. Before the assay, confirm the rats are in oestrus or pre oestrus by checking their vaginal smear.
  - Scarify the rat and tie one horn of the uterus in an organ bath fill of a PSS. Maintain the bath at the temperature of 32 °C and bath liquid required dose between 10 and 50 units/ml. Oxygenate the PSS with a carbogen and record the contraction of muscles.
  - Record contraction produced by adding two doses of standard preparation (required dose 10–50 m-units/ml). When maximum contraction is reached, replace the bath liquid with a new PSS solution. The dose should be added at the regular interval of 4–6 min. Record the contraction of test preparation similar to standard.
  - The ratio between two doses of test & 2 doses of the standard should be equal. Must maintained the constant ratio throughout the assay.by using a standard statistical method quantify the all responses & calculate the result of the assay.
- *Method C - By evaluate the milk-ejection pressure in a lactating rat:*
  - Test animals: lactating rats, 3–21 days after parturition, 300 g.
  - Separate the animals from litter & 30–60 min later, anesthetize them using pentobarbitone sodium (i.p). Attach rat to an operating table by its hind legs and quit the front legs free.

- Cannulate trachea with a short Pressure Equalizing tube of a width 2.6 mm in such that to make sure a free airway; if necessary, supply artificial respiration only.
- Cannulate an external jugular or femoral vein with a Pressure Equalizing tube width of 0.4 mm filled with stranded saline & closed. Shave the surrounding the under inguinal and abdominal and excise the tip of one part among them, ideally the lower inguinal. Insert a Pressure Equalizing tube of width 0.3 & 0.6 mm, to a depth sufficient to gain an exact assessment of pressure (3.5–10.5 mm depth) into the chief duct, which can open into the cut surface and tie strongly with a ligature. To an appropriate strain gauge transducer is Connect to this cannula, such that is used for recording arterial BP in rats.
- Fill the cannula with a 3.8% w/v of sodium citrate/normal saline contain of 50 Units of heparin sodium/ml to avoid clotting of milk. Next to cannulation, to clear milk from the tip of the cannula must inject 0.05–0.2 ml of solution into the teat duct by a transducer.
- If any obstruction arises from milk ejected into the cannula the procedure can be repeated during the assay.
- Maintain slight tension by clamping the strain gauge is applied on the teat and its alignment is maintained and fix the gauge to the potentiometric recorder adjusted to give full-scale deflection for an to increase the milk-ejection pressure up to 5.3 kPa should adjust the given full scale deflection. Inject all solutions of 1 ml syringe graduated in 0.01 ml through the venous cannula thorough wash them with 0.2 ml of normal saline.
- The Prepare of standard and test in a normal saline solution, injected volume in between 0.1 and 0.4 ml. select 2 doses of standard Ppn so that lower dose have the milk-ejection pressure is about 1.35 kPa and for the high doses as about 2.7 kPa.
- As an initial estimation, a low dose in between 0.1 and 0.4 mU and a high dose in between 1.55 and 2.5 times, perhaps used. select 2 doses of the test preparation with the similar ration of inter-dose and match the effects of the standard preparation doses as near as propable.
- Choose two different doses of the test solution with the similar inter-dose ratio with compare the effects of doses of the standard solution as accuracy as possible. Two standard and two test doses should be given according to irregular block or a Latin square method & minimum four responses to every -recorded. Measure all responses & evaluate the responses of the assay by standard statistical methods.
- The methods used for assaying oxytocin are summarized in Table 9.1.

Table 9.1 Different bioassay methods of oxytocin

Method A		Method B	Method C
Depression of blood pressure		Contraction of uterus	Milk ejection preparing in lactating rat
Animal	Cockerel, 1.2–2.3 kg	Female rat 120–200 g	Lactating rat, 300 g, 3–21 days parturitiction
Organ	Anesthetized, prolonged & constant high BP Gluteus Primus muscle (thigh) & remove both the popliteal artery & crural vein. Cannulate Popliteal artery & record BP response.	Inject 100ug oestradiol by IM. Confirm rat in oestrus/pre oestrus by vaginal smear. Kill rat & suspend one horn of uterus in organ bath contain Na, K, Ca, Mg chloride $NaHCO_3$, $NaHPO_4$ and dextrose-bath 32 °C	Anesthetize (Pentobarbitone Na IP). leave the front legs free and tie hind legs at 37 °C. if required cannulate trachea-respirate artificially. Shave skin around inguinal abdominal teats (Pressure 3–10 mm) Connect cannula with gauge transducer & fill with 3.8% Na citrate/saline which containing 50 Units heparin/ml which prevent milk clotting. Next clean the milk from the tip of the cannula by inject 0.05–0.2 ml of this solution by using transducer into the teat duct. so that slight tension is applied to teat by clamp strain gauge & its natural alignment is preferred & connect the gauge to a potentiometric recorder adjusted to -for increase milk ejection pressure of 5.3 kPa
Standard solution	With saline 0.1–0.4 ml Inject as two doses of stranded solution into cannulate vein evaluate BP decrease of BP	10–50 U/ml Oxygenate the solution with carbogen and record the contraction by addition of two doses of standard.	0.1–0.4 ml Two test doses preparation with the similar inter-dose ratio.
Interval	3–10 min between two injections depend on the rate of BP returns	3–5 min	3–5 min
Test	Diluted test preparation with saline get the same response		Two standard and two test doses—irregular block/ Latin square method & minimum four responses to every -recorded

Bibliography

Adam HM, Hardwick DC, Spencer KEV. Assay of histamine on the isolated guinea-pig intestine by the method of superfusion. Br J Pharmacol Chemother. 1954;9(3):360–6.

Musabayane CT, Brimble MJ, Balment RJ. Renal sodium retention and vasopressin induced kaliuresis in ethanol anaesthetised rats. European Journal of Endocrinology. 1985;110(2): 214–20.

Robinson ICAF, Walker JM. Bioassay of oxytocin on the superfused mammary gland of the mouse, using an automatic apparatus. Br J Pharmacol. 1974;50(3):409–17.

Twarog BM. Notes on the bioassay of serotonin. Methods Med Res. 1961;9:183–5.

Vizi ES, Hársing LG Jr, Duncalf D, Nagashima H, Potter P, Foldes FF. A simple and sensitive method of acetylcholine identification and assay. Bioassay combined with minicolumn gel filtration or high-performance liquid chromatography. J Pharmacol Methods. 1985;13(3): 201–11.

Principles of EC_{50}, ED_{50}, pD_2 and pA_2 Values of Drugs

10

Avinash Arivazhahan

Abstract

The effects of most drugs depend on the dose that is provided to the subject. In line with this, we have a concept called "dose descriptors", which takes into account the relationship between a particular dose of a drug and the corresponding effect produced by it. The commonly used "dose descriptors" include EC_{50} (half maximal effective concentration), LC_{50} (median lethal concentration), ED_{50} (median effective dose), TD_{50} (median toxic dose), LD_{50} (median lethal dose), etc. This chapter attempts to elucidate the basics of these determinants. In addition, we will also look at the drug toxicity studies that are commonly performed for any new drug.

Keywords

Dose descriptors · Lethal dose · Toxic dose · Acute toxicity · Limit test · Main test

10.1 The Concept of "Dose Descriptors"

As is popularly said, the dose is what turns a drug into poison. The importance of drug dose has been understood right from caveman days. Every drug has an optimal dose which is bordered by a sub-optimal and a supra-optimal dose, as depicted in Fig. 10.1.

This relationship between the dose of a drug and the effect that the drug produces is determined by a few factors that are collectively called "dose descriptors". The most commonly used dose descriptors are explained below.

A. Arivazhahan (✉)
Sanofi India Ltd, Chennai, Tamil Nadu, India

M. Lakshmanan et al. (eds.), *Introduction to Basics of Pharmacology and Toxicology*, https://doi.org/10.1007/978-981-19-5343-9_10

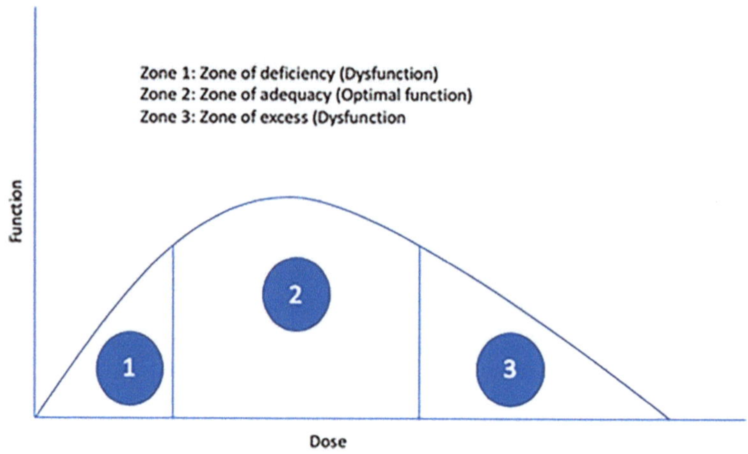

Fig. 10.1 A representative figure showing the importance of "optimal dose" of a drug

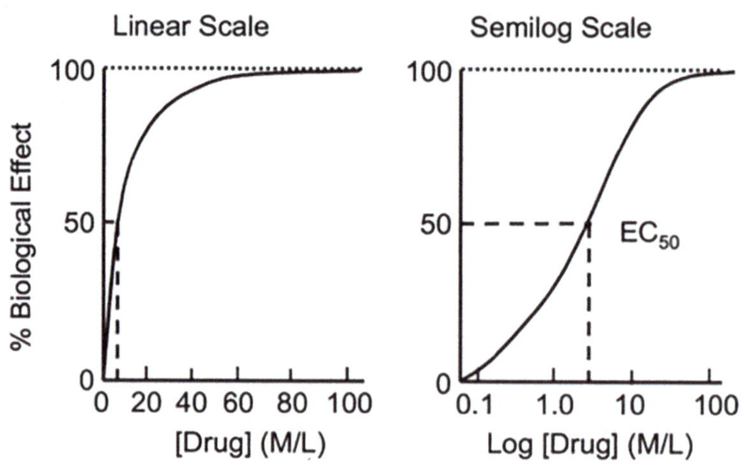

Fig. 10.2 Dose-response relationship when plotted on a linear scale (on the left) and on a semi-logarithmic scale (on the right), denoting the EC_{50}

- *EC_{50} (Half maximal effective concentration):*
  - EC_{50} represents the effective concentration (EC) of any drug at which the half-maximal response is obtained (50% of the maximal response). A graphical representation is depicted in Fig. 10.2.
  - The EC_{50} is a very important parameter (ED_{50} is another) in comparing the potency of different drugs. In short, the potency of a drug can be considered to be inversely proportional to the EC_{50} value. An example is shown in Fig. 10.3
- *LC_{50} (Half lethal concentration/Median lethal concentration):*
  - LC_{50} denotes the lethal concentration (LC) of the drug at which half the population (of rodents or other species) is observed to die (50% mortality).

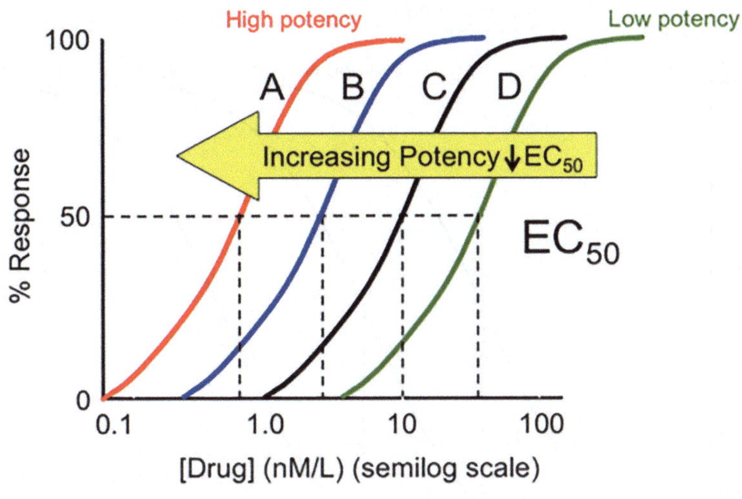

Fig. 10.3 Four drugs (A, B, C, D) are compared to check for potency. As observed, higher the EC_{50} value, lower is the potency (Drug A has the highest potency, while D has the least)

- *ED_{50} (Median effective dose/Median therapeutic dose):*
  - This is very similar to the concept of EC_{50}, with the only difference being the replacement of concentration with dose. ED_{50} is defined as the dose of a drug that can produce a desired or expected response in 50% of the test population.
- *TD_{50} (Median toxic dose):*
  - TD_{50} is the dose of a drug that can produce a prespecified toxic effect in 50% of the test population. Although TD_{50} is placed between ED_{50} and LD_{50} theoretically, it is not always the case in practicality. This is because sometimes, toxic effects may not be dose-dependent. Also, many of the toxic effects seen clinically are time-dependent (as in the case of carcinogenesis or teratogenesis, which cannot be determined based on short-term studies). Further, if the toxic effects are unpredictable, they cannot be prespecified, thus negating the utility of TD_{50}.
  - The ratio of TD_{50} to ED_{50} is defined as the therapeutic index (TI). Sometimes, it is defined as the ratio of LD_{50} to ED_{50}, particularly when preclinical rodent studies form the context.
- *LD_{50} (Median lethal dose):*
  - LD_{50} is conceptually analogous to TD_{50} since it also deals with toxicity. However, a major difference here is that toxicity being tested for is death or mortality of the experimental animal. LD_{50} is defined as the dose of the drug that can produce mortality in 50% of the test population. While LD_{50} estimation is considered as an important parameter in animal toxicity studies, it cannot be taken as an absolute reflector of mortality. This is due to the reason that pharmacokinetics of a drug might alter the toxicity of a drug. For example, several drugs with systemic toxicity may induce mortality based on the speed

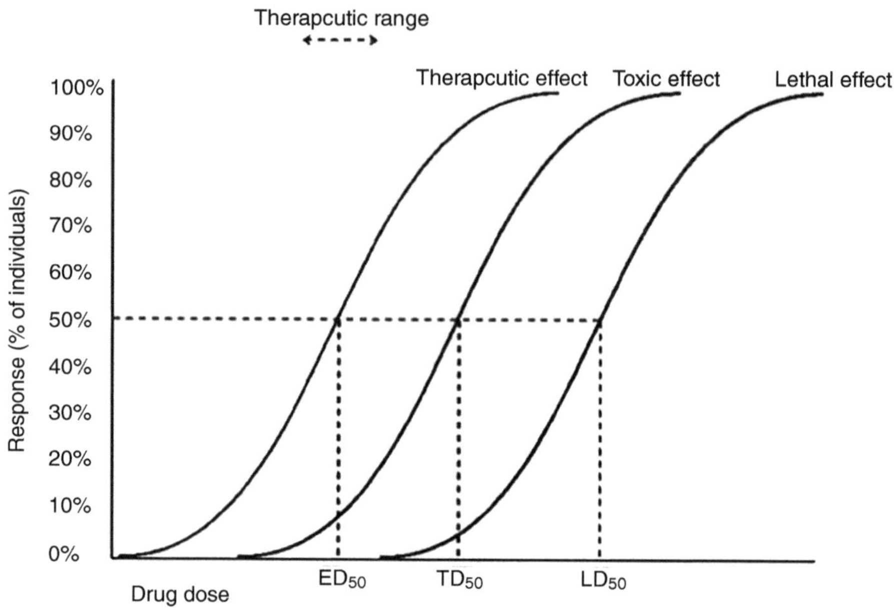

Fig. 10.4 A simple pictorial depiction of ED_{50}, TD_{50} and LD_{50} shown as a continuum; however, this continuum may not be always seen in practicality

at which they are injected intravenously (thereby suggesting that concentration and dose are different parameters, and both may be equally important) (Fig. 10.4).

- *NOAEL (No Observed Adverse Effect Level):*
  - NOAEL denotes the highest tested dose or concentration of a drug that has been observed to produce no adverse effects on the experimental animal (in simple terms, it signifies the maximal safest dose of a drug). Effects may be produced, but adverse effects are not seen. Higher the NOAEL value, lower is the systemic or chronic toxicity of the drug.
- *LOAEL (Lowest Observed Adverse Effect Level):*
  - LOAEL signifies the lowest tested dose or concentration of a drug that is known to cause an adverse effect. Sometimes, the first effect produced by a drug might be adverse in nature. This adverse effect may be any modification of morphology or physiology of the animal species.

  NOAEL and LOAEL are determined by comparing the effects of the drug with a control group of the same animal species that are exposed to a similar environment (Fig. 10.5).
- *pD_2:*
  - pD_2 essentially denotes the potency of the agonist. It is expressed as the negative logarithmic value of EC_{50}.

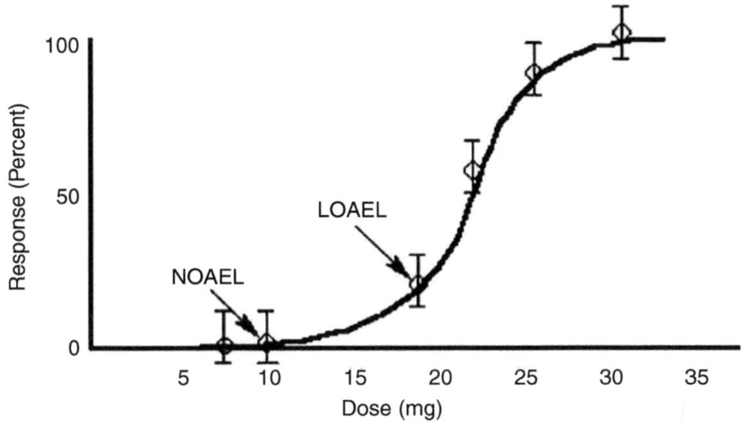

Fig. 10.5 Graph showing NOAEL (at 10 mg dose) and LOAEL (at around 18 mg dose)

- *pA_2:*
  - pA_2 denotes the strength or potency of an antagonist. pA_x is the (negative logarithmic value of) concentration of an antagonistic drug that can produce 'x'-fold shift in the agonist dose-response curve or graph. Hence, pA_2 is the concentration required to induce a two-fold shift in the curve (expressed as a negative logarithmic value).
  - The pA_2 value is derived from the Schild's plot (developed by Sir Heinz Otto Schild), which helps to differentiate between competitive and non-competitive drugs.

10.2 Preclinical Toxicity Studies

Preclinical toxicity studies are a mandatory step in evaluating a drug for its safety. These toxicity studies can be grouped into acute, subacute, and chronic toxicity studies. The acute toxicity studies are explained in brief here.

Acute toxicity studies are limited to those adverse or toxic effects that occur within 24 h of dosing (single or multiple doses) of the drug when given orally or via dermal route, and within 4 h when given via inhalational route. Rats are preferred for oral and inhalational toxicity studies, while rats and rabbits are used for acute dermal toxicity studies. The Organisation for Economic Cooperation and Development (OECD) has prescribed guidelines for chemical drug testing in animals. The '4' series under these guidelines (401–493) relate to toxicity studies of drugs. As an example, acute oral toxicity estimation (by Up-and-Down procedure) is detailed below.

There are three commonly used procedures for acute oral toxicity: The up-and-Down method, fixed-dose procedure, and conventional LD_{50} method. The Up-and-Down method for acute oral toxicity determination was first described by Dixon and Mood and later modified by Bruce. This test has a mandatory component called the

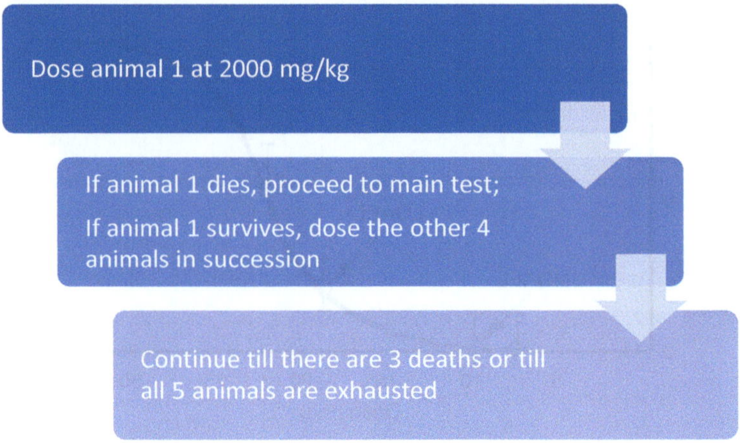

Fig. 10.6 Limit test procedure for acute oral toxicity

'main test' and an optional pre-requisite called the 'limit test', which is mainly used for drugs with low toxic potential.

- *The limit test:*
  - This test uses a total of five animals (preferably female rats) in a sequential manner. It can be done with doses of 2000 or 5000 mg/kg. The procedure is simplified in Fig. 10.6.
- *The interpretations of the limit test are as follows:*
  - If the first animal dies, then the limit test is stopped prematurely and the main test is commenced.
  - If there are 3 or more deaths, then the LD_{50} of the drug is less than 2000 mg/kg (or 5000 mg/kg). Once the third death happens, the limit test is stopped, and the main test is commenced.
  - If there are less than 3 deaths, the LD_{50} is more than 2000 mg/kg (or 5000 mg/kg).
- *The main test:*
  - Here again, the animals are dosed sequentially at rough intervals of 48 h. The dose step used here is 3.2. If the animal survives, then the next animal receives a higher dose (stepped up by the progression factor of 3.2). If the animal dies, then the next animal receives a lower dose (stepped down by the progression factor of 3.2). The initial dose is decided based on the available data on the drug. If limited data exists, then a standard dose of 175 mg/kg is used as the initial dose.
  - Observation of the test animals is critical during the first 24 h of dosing with focus needed in the initial 4 h. Each animal is observed for a total of 2 weeks, unless dead. The LD_{50} values and the confidence limits are computed, and results are drawn.

– Based on the LD_{50}/LC_{50} values obtained from these toxicity tests, the drugs can be classified into 5 classical categories.

Bibliography

Bruce RD. An up-and-down procedure for acute toxicity testing. Fundam Appl Tox. 1985;5:151–7.

Dimmit S, Stampfer H, Martin JH. When less is more:efficacy with less toxicity at the ED50. Br J Clin Pharmacol. 2017;83(7):1365–8.

Dixon WJ. The up-and-down method for small samples. J Am Stat Assoc. 1965;60:967–78.

Sarker K, Sarker A, Ghosh A, Sen DJ. Posological dose of therapeutics over prophylaxis in toxicological serendipity. Int Educ Sci Res J. 2016;2(11):61–9.

Physiological Salt Solutions

11

Saranya Vilvanathan

Abstract

During pharmacological experiments on isolated tissues, in order to maintain the normal physiological conditions, artificial salt solutions are freshly prepared and used to suspend the tissue in an organ bath. It helps keep the tissue alive, provides nutrients to the tissue as well as acts as a buffer that mimics the tissue fluid in ionic composition. Alterations in the composition, pH, temperature, aeration of the solution can affect the experimental outcomes. The choice of salt solution used for each experiment is specifically decided on the species of origin of the animal tissue that is experimentally tested. Ringer solution is a polyionic non-alkalinising isotonic crystalloid solution used for amphibian tissues. The variant of Ringer solution called Ringer-Locke is preferred for mammalian isolated heart tissue experiments. Tyrode solution is the isotonic solution preferred for mammalian smooth muscle tissue and in tissue culture. De-Jalon is similar to Ringer-Locke with one-fourth of $CaCl_2$ and half the amount of glucose, and is commonly used for isolated rat uterus preparation. Krebs-Henseleit (Krebs) and its variants are commonly used for mammalian isolated organ, especially for nerve responses.

Keywords

Isolated tissue · Organ bath · Ringer · Krebs · Salt solution

S. Vilvanathan (✉)
Department of Pharmacology & Clinical Skills, Medical University of the Americas, Nevis, Saint Kitts and Nevis
e-mail: s.vilvanathan@mua.edu

M. Lakshmanan et al. (eds.), *Introduction to Basics of Pharmacology and Toxicology*, https://doi.org/10.1007/978-981-19-5343-9_11

11.1 Introduction

Physiological salt solutions are the artificially prepared solutions that are used to keep animal tissues alive under controlled conditions while performing a pharmacological experiment on isolated tissues.

The roles of these solutions are:

1. to keep the isolated tissues alive,
2. to provide nutrients to the tissues,
3. act as a buffer and
4. mimic the body tissue fluid (in ionic composition) providing a homeostatic environment outside the body for them to elicit a physiological response in a controlled setting.

11.2 General Principles in Preparation of Physiological Salt Solution

Physiological salt solutions are substitutes to tissue fluid and the first and foremost requirement to meet the criteria is isotonicity, that is, the prepared solution should be isotonic to tissue fluid. Alterations in this principle may lead to shrinkage of cells, blotting, or loss of physiological function. Various precautions should be taken during the preparation of salt solutions to maintain isotonicity including:

- *Double distilled deionized water:* Salt solutions are ideally prepared from double-distilled deionized water and the reagents should be of analytical grade. Water acts as a vehicle to dissolve various ingredients. Anhydrous salts are commonly preferred and when salts with water of crystallization are used, its weight should be suitably adjusted.
- *Balance of cations and anions:* The salt solutions contain a mixture of cations, anions, and glucose in distilled water. The role of each of these ions is very significant and as follows:
  - Sodium is responsible for maintaining the excitability, contractility, and rhythmicity of muscle, and nerves
  - Sodium chloride also helps to adjust the isotonicity of the solution
  - Potassium ions are responsible for the increase in relaxation of heart, increased neuromuscular transmission, and excitability of nerves
  - Calcium increases the force of contraction and tone of the heart and decreases the excitability of nerves
  - Magnesium is responsible for contraction of smooth muscle
  - Sodium bicarbonate is used to adjust the pH of the solution
  - Sodium dihydrogen phosphate can also be used as a buffer to compensate for any change in pH
- *Stock solution:* Ideally, the physiological salt solutions should be prepared on the day of the experiment. However, for the sake of convenience, a 20-fold stock

solution can be prepared the night before and can be diluted later, to the ideal composition, at the time of experiment according to the need. While preparing a stock solution, sodium bicarbonate and glucose is not added and should be added freshly at the time of the experiment. This is because sodium bicarbonate when stored with calcium for a long time can precipitate calcium carbonate. Moreover, glucose can encourage the growth of microorganisms when stocked in the solution overnight. Therefore, both sodium and glucose are freshly weighed and added to the stock solution, just before the experiment.

- *The pH of solution:* pH of various salt solutions varies between 7.3 and 7.8 depending on the organ used in the experiment. When the pH is acidic (low pH), the tone of the preparations tend to decrease and the effect of the drug will be altered. Conversely, when the pH is high (alkaline environment), it can increase the activity of the tissue. Tissue activity during the experiment can lead to the accumulation of metabolites and alteration of pH that necessitates the change of solution frequently or adding up of buffering agents such as bicarbonate or phosphate in the solution.
- *Temperature:* for obtaining a consistent effect in the mammalian tissues, ideally the temperature should be maintained constant. The low temperature in the bath can increase the tone of the tissue causing smaller contractions with increased contraction-relaxation time. Mammalian tissue must be bathed in a warm solution with temperature adjusted to 37 °C (e.g. isolated rabbit intestine). The temperature should be decreased in some experiments to reduce the myogenic contraction, the warmed solution adjusted to 32 °C (e.g. Guinea pig ileum). Amphibian tissues (e.g. frog rectus abdominis) survive at room temperature (25 °C).
- *Aeration:* Air, pure O_2, or carbogen (95% O_2 with 5% Co_2) is required for the proper functioning of the tissues. Bubbling of gases inside the bath-tube can also help with the stirring of the solution and facilities the diffusion of drugs in the solution. Commonly, pure oxygen is used for heart tissues, normal air for intestines, and carbogen for uterus preparations.
- *Glucose:* Glucose supplies energy for the isolated tissue, acts as a nutrient, and helps with the contractility of the tissue. It should be freshly weighed and added to the salt solution, just before the experiment.

11.3 Commonly Used Physiological Salt Solution

- *Frog-ringer solution*: It is named after Sydney Ringer, who determined that a solution perfusing the frog's heart must contain sodium, potassium and calcium salts in definite proportions if the heart is to be kept beating for long. It is a balanced polyionic non-alkalinizing isotonic crystalloid solution that contains physiologic concentrations of sodium, potassium, calcium, and chloride. Ringer established the relative importance of the concentration of each of these ions as well as the long-term necessity to prevent or reverse the acidification associated with contraction.

Composition: NaCl–65 g, KCl–1.4 g, $NaH_2PO_4 \cdot H_2O$–0.1 g, Glucose–20 g, $NaHCO_3$–2 g, and $CaCl_2$–1.2 g. The measurements given are the amounts to be dissolved in 10 l of distilled water. Frog-Ringer solutions are aerated with O_2 or air is used for frog heart and tissues.

The addition of sodium lactate in place of sodium bicarbonate makes it a Ringer-lactate solution (Hartmann solution), commonly used for replacing fluid and electrolytes in humans. Products of lactate metabolism can counteract acidosis and maintain a stable pH.

Alterations in the concentrations of these ions in the Ringer solution have been proven to cause progressive depression in cardiac activity (in the frog's heart), especially the heart rate, which can be due to toxicity or abnormality in ionic concentrations of the fluid.

- *Ringer-Locke solution:*
  Mammalian Ringer solution or Ringer-Locke or Locke's solution is used for mammalian isolated heart or other tissues. It contains an increased concentration of sodium and less glucose compared to the Frog-Ringer solution. Frog-Ringer solution can be prepared by adding 400 ml distilled water to 1 l of the Ringer (Locke) solution.

  Composition: NaCl–91.5 g, KCl–4.2 g, Glucose–10 g, $NaHCO_3$–1.5 g, $CaCl_2$–2.4 g, Aerating gas–Pure O_2. Ringer solutions are aerated with air or O_2 and are used for mammalian isolated heart or other tissues.

***Variants of Ringer*:**

- Calcium-free Ringer solution is identical to Ringer, except that $CaCl_2$ is omitted.
- Calcium-free depolarizing Ringer solution: contains NaCl–94 mM, KCl–60 mM, $NaHCO_3$–6 mM, dextrose–11 mM
- Zero-calcium depolarizing Ringer solution: it is similar in composition to calcium-free Ringer solution except that it contains EGTA (Ethylene glycol-bis-β-aminoethyl ethyl-N-N-tetra acetic acid) 0.01 mM.
  Note: Ringer, Frog-Ringer, and De Jalon solutions do not contain magnesium or phosphate ions.
- *Tyrode solution:*
  Tyrode solution is isotonic with interstitial fluid and is used for physiological experiments involving mammalian smooth muscles and in tissue culture.

  Composition: NaCl–80 g, KCl–2 g, $MgCl_2 \cdot 6H_2O$–1 g, $NaH_2PO_4 \cdot H_2O$–0.5 g, Glucose–10 g, $NaHCO_3$–10 g, $CaCl_2$–2 g, aerating gas–95% O_2, 5% CO_2, air or pure CO_2. The measurements given are the amounts to be dissolved in 10 l of distilled water and the final pH of the solution to be adjusted to 7.3–7.4.

  It resembles Ringer-lactate solution in its composition except that it contains, bicarbonate or phosphate as a buffer (not lactate); glucose acts as an energy source; in addition to normal cations, it also contains magnesium which helps with smooth muscle contraction.
- *Krebs-Henseleit (Krebs) solution:* It is commonly used for mammalian isolated organ, specially for nerve responses.

Composition: NaCl–69 g, KCl–3.5 g, $MgCl_2 \cdot 6H_2O$–1.1 g, $NaH_2PO_4 \cdot H_2O$–1.4 g, Glucose–20 g, $NaHCO_3$–21 g, $CaCl_2$–2.8 g, aerating gas–95% O_2 + 5% CO_2. The measurements given are the amounts to be dissolved in 10 l of distilled water.

Variants of Kreb's Solution:

- Calcium-free Krebs solution (zero-calcium EDTA solution): to deplete calcium in the isolated tissue, sodium-EDTA (around 0.3 mM) is added to calcium-free Kreb's solution.
- Potassium-rich Kreb's solution: is made by increasing KCl concentration up to 47.3 mM at the expense of an equivalent amount of NaCl.
- Low-Sodium Krebs (glucose 5.55 mM) solution (63 mEq/L): is prepared by reducing NaCl to 38 mM and adding sucrose 150 mM to maintain the tonicity of the solution.
- Low-sodium Krebs solution (25 mEq/L): is prepared by omitting NaCl completely and adding sucrose 220 mM.
- K^+ free Krebs solution: is prepared by omitting KCl and KH_2PO_4 and adding NaH_2PO_4 1.2 mM and sucrose 9.5 mM.
- Glucose-free Krebs solution: is prepared by omitting glucose.
- Kreb's solution may contain 2 x 10_{-5} M choline to feed the cholinergic nerves and preserve their acetylcholine stores in coaxially stimulated isolated Guinea pig ileum.
- Modified Kreb's-Ringer's bicarbonate: it contains an increased concentration of sodium bicarbonate and glucose and decreased concentration of KCl and $MgSO_4$. It is usually aerated with 5% CO_2 in O_2 and used for mammalian isolated organ, especially for nerve responses.
- *De Jalon solution*

  De Jalon is similar to Ringer-Locke, except that it contains one-fourth of $CaCl_2$ and half the amount of glucose.

  Composition: NaCl–90 g, KCl–4.2 g, Glucose–5 g, $NaHCO_3$–5 g, $CaCl_2$–0.6 g, Aerating gas–95% O_2 + 5% CO_2. It is commonly used for testing isolated rat uterus preparation.
- *McEwen solution:*

  Composition: NaCl–76 g, KCl–4.2 g, NaH_2PO_4. H_2O– 1.4 g, Glucose–20 g, $NaHCO_3$–21 g, $CaCl_2$–2.4 g, Aerating gas–95% O_2 + 5% CO_2.

  The summary of composition of above explained salt solution for preparation of 10 L quantity are given in Table 11.1. After the final preparation the pH should be in the range of 7.3–7.4. For all the solutions, calcium chloride should be added at last. Due ot hygroscopic nature, $CaCl_2$ and $MgCl_2$ containers should be always kept airtight. In case of improper storage, water from atmospheric moisture will be added to the salt and may cause error in achieving final molarity of calcium or magnesium ions.

Table 11.1 Composition of common physiological salt solution for the quantity of 10 L

Salts	Frog-Ringer solution	Ringer-Locke solution	Tyrode solution	Kreb's solution	De Jalon solution	McEwen solution
NaCl	65 g	91.5 g	80 g	69 g	90 g	76 g
KCl	1.4 g	4.2 g	2 g	3.5 g	4.2 g	4.2 g
$MgCl_2 \cdot 6H_2O$	–	–	1 g	1.1 g	–	–
$NaH_2PO_4 \cdot H_2O$	0.1 g	–	0.5 g	1.4 g	–	1.4 g
$NaHCO_3$	2 g	1.5 g	10 g	21 g	5 g	21 g
$CaCl_2$	1.2 g	2.4 g	2 g	2.8 g	0.6 g	2.4 g
Glucose	20 g	10 g	10 g	20 g	5 g	20 g
Aerating gas	air	Pure O_2	O_2 or air	O_2 + 5% CO_2	O_2 + 5% CO_2	O_2 + 5% CO_2

11.4 Choice of Physiological Salt Solution for Specific Tissues

- For amphibian tissue use frog–Ringer solution
- For mammalian or avian skeletal muscle use Kreb's solution
- For intestine use Tyrode's solution
- For heart muscle use Ringer-Locke solution
- For the rat-uterus preparation use De Jalon's solution
- Mammalian tissues sometimes require aeration with O_2 + 5% CO_2

Bibliography

Ghosh MN. Fundamentals of experimental pharmacology. 6th ed. Kolkata: Hilton and Company; 2015.

Mathews FS. X. physiological salt solution. Ann Surg. 1904;40(2):243–54.

Glaxo Wellcome Pharmacology Guide, In vitro pharmacology: physiological salt solutions, 2019. http://www.pdg.cnb.uam.es/cursos/Barcelona2002/pages/Farmac/Comput_Lab/Guia_Glaxo/chap2b.html Accessed 12 July 2019

Validation of Animal Models 12

Umamaheswari Subramanian

Abstract

Animal research provides a major contribution to the discovery of new compounds and its mechanism of action. It also deals with the pharmacokinetics profile and determination of safe dose of a compound which is to be tested in humans. There is a necessity to choose an appropriate animal model for preclinical research in order to carry out a clinical trial. Research can be performed on already existing validated animal model or by validating a newer model. Validation criteria of an animal model changes from one to another based on the purpose of the model (fit-for-purpose). Face validity, predictive validity and construct validity ensures the closeness of the animal model to humans. In addition to these validity, few more criteria have been added to assess and optimise the animal model, i.e. epidemiology, symptomatology, natural history, end points, genetics, and biochemical parameters, pharmacological and histological features. There is no single animal model which can satisfy all types of validity for any disease. Even though shortcomings are inevitable, these models pave way for the safer research study in humans. One can choose an animal model closer to an ideal one. Thus validation plays a crucial role in translation of animal research to humans.

Keywords

Predictive · Face and construct validity · Animal model validation

U. Subramanian (✉)
Department of Pharmacology, Sri Venkateshwaraa Medical College Hospital & Research Centre, Puducherry, India

M. Lakshmanan et al. (eds.), *Introduction to Basics of Pharmacology and Toxicology*, https://doi.org/10.1007/978-981-19-5343-9_12

12.1 Introduction

Animal research provides a major contribution to the discovery of new compounds and its mechanism of action. It also deals with the pharmacokinetics profile and determination of safe dose of a compound which is to be tested in humans. Various animal species have also been used to study the disease pathogenesis. Animal welfare organisations have made the regulatory bodies in the field of animal research to follow stringent ethical guidelines while handling them. 3R has been followed as a guideline (Reduce, Replace and Refine).

12.2 Animal Model

One of the definition for animal model has been given by Held based on Wessler's original definition: "a living organism in which normative biology or behaviour can be studied, or in which a spontaneous or induced pathological process can be investigated, and in which the phenomenon in one or more respects resembles the same phenomenon in humans or other species of animal."

12.2.1 History -Animals in Research

The use of animals for purposes like understanding the human physiology date backs to ancient period. Till date it is being used judiciously for the betterment of human life (Fig. 12.1).

12.2.2 Classification of Animal Model

Animal models are classified based on various factors (disease -course, symptoms) (Tables 12.1, 12.2, and 12.3).

12.2.3 Utilization of Animal Models in Scientific Field

The ultimate use of animal models is to deal with the translation of the data obtained from the same to humans for the better health care. Extrapolation of the results from one species to the other is based on the evolution and the morphological and physiological similarity. The toxicity data obtained through animal models are used to determine the safer dose in humans. Certain information from these models do not get translated because of lack of relevance in humans. Significant contribution have been achieved by animal models to research and development:

- Type I diabetes treated with insulin - first demonstrated in the dog (Banting and McLeod)

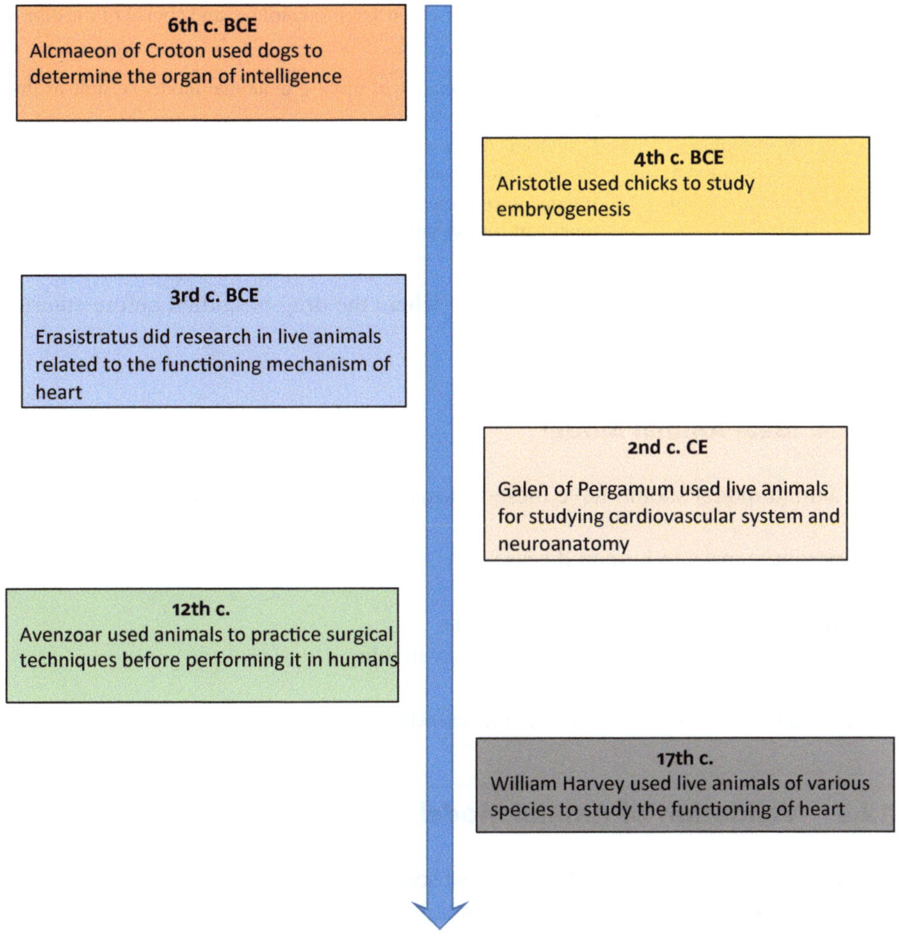

Fig. 12.1 Timeline of research activities using animals

Table 12.1 Types of animal models based on research design

S. No.	Model	Aim
1	Exploratory	Animal model are used to explore the fundamental mechanism(cellular, organ level) and identify it as normal or abnormal
2	Explanatory	Animal models are used to break down and understand the complex functional mechanism
3	Predictive	Animal models used to quantify the effect of treatment(curative) and also the toxic nature if any

- Evaluation of ex vivo, in vitro, and *in silico* models to attain better validation of extrapolated data from animal models
- Experimental design and methodology—crucial role to get valid extrapolation

Table 12.2 Different types of animal models based on symptomatology and course of the disease

S. No.	Model	Description
1	Homologous animal model	Symptoms and course of disease in animal model is same as that of humans
2	Isomorphic animal model	Symptoms of the disease in animal model is same as that of humans but causative factor of disease is different
3	Partial animal model	It helps to reproduce only a certain part of human disease and treatment for the same

Preclinical research is a preliminary level where the drug is studied before entering the clinical phase.

12.2.4 Ideal Animal Model

Any animal model should have these following characteristic features:

1. Appropriateness to human disease
2. Translation to humans
3. Genetic uniformity or closer to humans
4. Cost and availability- cheap & easily available
5. Generalizability of the results
6. Ethical consideration should be addressed

12.3 Validation of Animal Model

Validation means the relation between test score and its quality to measure. It should cover internal validity (replicability), generalizability, predictive and construct validity. There is a necessity to choose a proper and appropriate animal model for preclinical research in order to carry out a clinical trial which will eventually yield results through which humankind can be benefitted. Hence selection of the same requires validation. Research can be performed on already existing validated animal model or by validating a newer animal model.

Animal model is being called so only when it has been validated whereas it is called as putative animal model if it is yet to undergo validation. Validation criteria of an animal model changes from one to another based on the purpose of the model (fit-for-purpose). It is impossible for a single animal model to achieve validity with all criteria. In order to obtain the maximum number of validity criteria, optimal combination of models should be considered.

Criteria for valid animal model includes:

Table 12.3 Animal models- Advantages & Disadvantages

S. No.	Model	Example	Advantage	Limitations
1	*Induced (experimental) models:* Healthy animals induced to produce a specific disease	Diabetes mellitus induced with encephalomyocarditis virus.	Choice in selecting the species	Majority of these models are either isomorphic or partial
2	*Spontaneous (mutant) models*: The animal model with natural genetic variation/ mutation	Athymic nude mouse	1. Phenotypic similarity with respect to disease between humans and animal (face validity) 2. This model plays significant role in the development of therapeutic regimen in humans	Genetic impairment will lead to various compensatory responses which will differ between human and animals
3	*Genetically modified models:* Genetic modification in the animal-transgenic animals	Use of mutagen (ethylnitrosourea) to induce genetic mutation in animals	1. Able to identify the cause and genetic involvement of certain diseases 2. Able to study about gene-gene interaction, gene-environment interaction and the effects produced due to alteration in genetic pathway	Phenotyping of the animals found to be difficult
4	*Negative models:* Animal models which remain unresponsive towards a pathogen which causes infection in human	Gonococcal infection in rabbit (where rabbit do not have susceptibility)	Provides physiological basis for resistance and its mechanism	
5	*Orphan models:* Disease occurs in animals but not yet discovered in humans (but can be identified later)	1. Bovine spongiform encephalopathy 2. Feline leukemia virus.	Useful information provided by the orphan models in case humans get infected by the same infection.	

1. Face validity
2. Construct validity
3. Predictive validity

The above mentioned validity criteria comes under the broader component called external validity.

12.3.1 Face Validity

The symptoms induced in animal model should match the symptoms of humans, i.e. phenotypic similarity. Example Type I Diabetes mellitus in humans and insulin requiring animal (BB rats). These two conditions respond well to the insulin therapy. Most of the spontaneous animal models exhibit face validity.

12.3.2 Construct Validity

Presence of homology between the human and animal model on genomic level is called construct validity. Transgenic disease models comes under this category. Animal model with this validity will give inputs about the occurrence of a disease condition with the effects of genetic change and interaction of gene with the environment.

12.3.3 Predictive Validity

It shows how much an animal model can predict an unknown aspect of the human disease or its therapy. It major contribution is by assessing the cure of a disease. It do not go in for specific cause of a disease and its treatment.

All these three criteria were proposed by Wilner. These three criteria falls under external validity and provides basis for generalisation of the results obtained from animal studies to humans. Belzung and Lemoine proposed a nine validity criteria in case of models related to psychiatric disorders. Tricklebank and Garner suggested other criterias in addition to Wilner's—internal validity (third variable influencers), external validity (results can be generalized), convergent validity, discriminant validity. It is mostly not possible to have an animal model with all these above criteria/validity, but researchers insist the presence of predictive validity which is the most important one.

Any animal model is said to be valid if it resembles humans with respect to the above said aspects as depicted in the figure (Fig. 12.2).

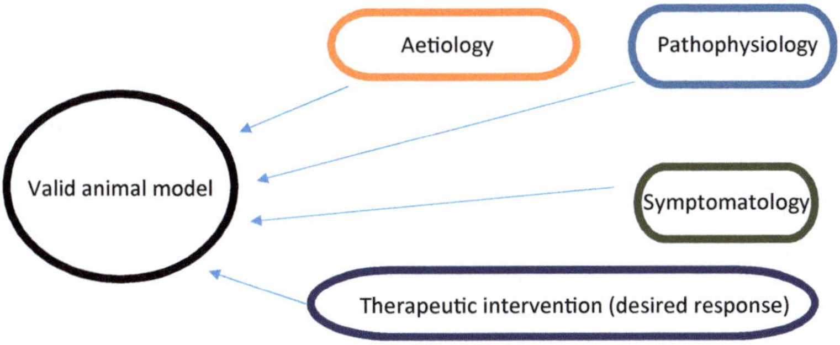

Fig. 12.2 Significant components—Valid animal model

12.3.4 Internal Validity

Internal validity means the results or outcomes among animals (different treatment groups) varies only with respect to the intervention. If so those animal models are considered to be with adequate internal validity. Internal validity may be reduced if there is any flaw in the design of animal model and conduct of the study. Certain bias (selection bias, performance bias, attrition bias and detection bias) will affect the internal validity.

Reliability—consistent results (similar experimental condition).
Replicability—reproducibility.
High replicability & reliability—good internal validity.

12.3.5 Factors Affecting Internal Validity

1. Rearing conditions
2. Housing conditions
3. Social hierarchy
4. Gender
5. Age
6. Time of testing
7. Day-night cycle
8. Health condition of research animal
9. Calibration of research equipment

These factors need to be addressed to increase the internal validity (Table 12.4).

Animal models can either be holistic or reductionist. Holistic model- the model as a whole (symptoms, behaviour, underlying mechanism, etc.) is comparable to human targets. Reductionist model- any specific feature of the animal model is alone comparable to the human targets.

Table 12.4 Different validity of animal models

S. No.	Validity of animal model	Description
1	Discriminant validity	The model proposed is based on the divergent property where the measured parameter do not relate as expected. i.e. The concept and its correlating factors diverge
2	Mechanistic validity	As such, the behaviour resemblance or similarity between the animal model and human targets is not enough but the underlying mechanism for those behaviour should be identical
3	Convergent validity	The model proposed is based on the convergent property where the measured parameter do relate as expected. i.e. The concept and the variables measured are closely related
4	Biomarker validity	The animal model and the human targets may have variations in the biomarkers due to the difference in the species, but if both of them produce the same symptoms and also matches the underlying mechanism, then it is considered as valid animal model
5	Target validity	Target being studied in the animal model should be comparable to that of human in terms of mechanism/function

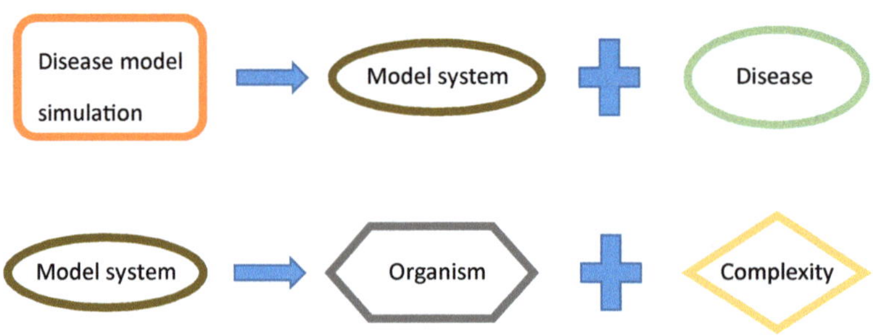

Fig. 12.3 Disease model simulation and model system

12.4 Process of Developing an Animal Model

Developing an animal model should be done by a series of steps in a meticulous manner. Researcher should have the knowledge about certain terminologies before going into this process (Figs. 12.3, 12.4, and 12.5).

12.5 Validation Process for Predictive Validity

Among face validity, predictive validity and construct validity, main focus of validation process is on predictive validity. The following steps are involved in the validation process:

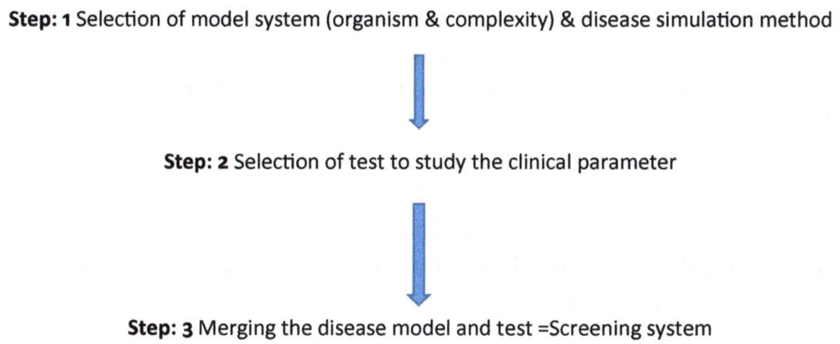

Fig. 12.4 Valid screening system reflects predictive validity which is a crucial component for an appropriate animal model. It is obtained by stepwise approach as depicted in the figure

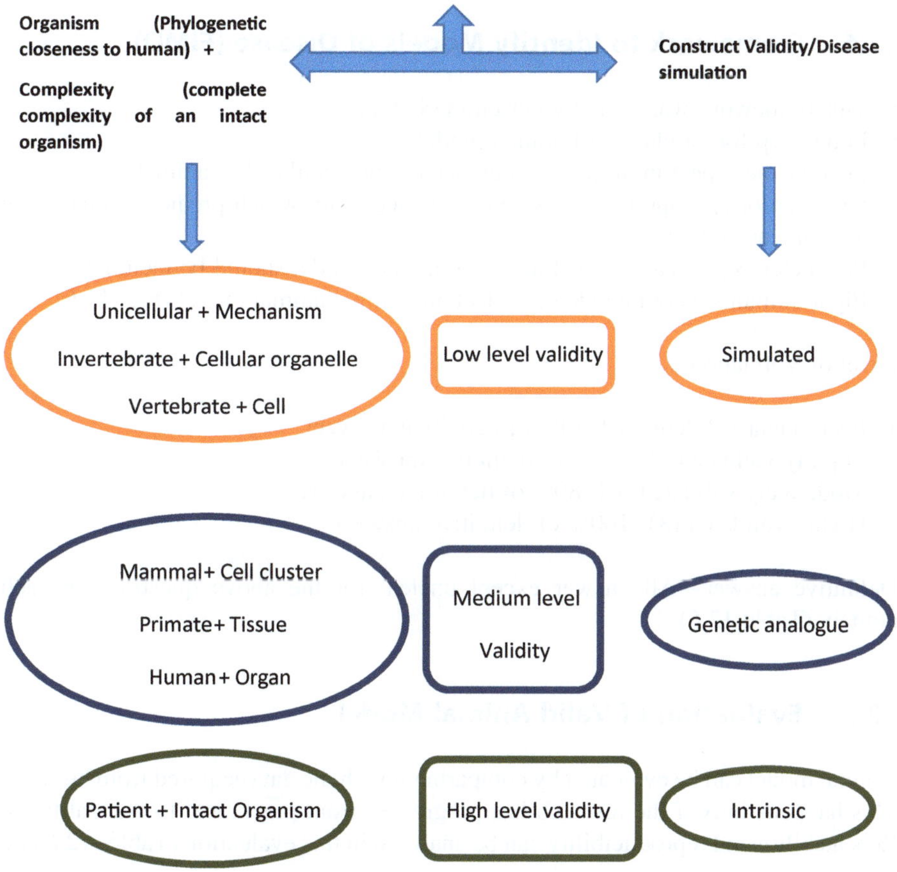

Fig. 12.5 Disease model validity (organism with complexity + construct validity)

1. Test development: Development of diagnostic/therapeutic intervention to predict the outcome of the same in a specific disease
2. Pre-validation: Test is performed in two or more laboratory (relevance & reliability)
3. Validation: Test must be relevant and also reliable for more than one purpose. Data analysis should be done followed by further evaluation.
4. Independent assessment: The study data and conclusion must be published after peer reviewing the same and should be done by the independent panel.
5. Regulatory body acceptance

This validation process can also be performed in a retrospective manner, not necessary to be unidirectional all the time. But retrospective method found to show less reliability.

12.6 Framework to Identify Models of Disease (FIMD)

- This framework was done by Ferreira GS et al.
- Initial step for validation of animal models
- Identify the aspect in humans which need to be simulated in animals
- Selection of appropriate disease model is necessary which predicts (outcome in humans) accurately
- Parameters significant for validation of animal model should be identified
- Eight domains were included to select an optimal animal model (Fig. 12.6)

Level of Validation:

1. Insufficient validation (0–40% of definitive answer)
2. Slightly validated (41–60% of definitive answer)
3. Moderately validated (61–80% of definitive answer)
4. Highly validated (81–100% of definitive answer)

Definitive answer—All answer except unclear for the above questions in each domain (Table 12.5).

12.7 Evaluation of Valid Animal Model

Animal model can be evaluated by comparing it with the data acquired from humans. This happens only if the research has progressed from preclinical to clinical trials. Repeatability and reproducibility can be analysed in this evaluation (Tables 12.6 and 12.7).

Discrepancy to clinical state is 3 if all three components (organism, complexity & disease simulation) shows discrepancy when compared with human disease

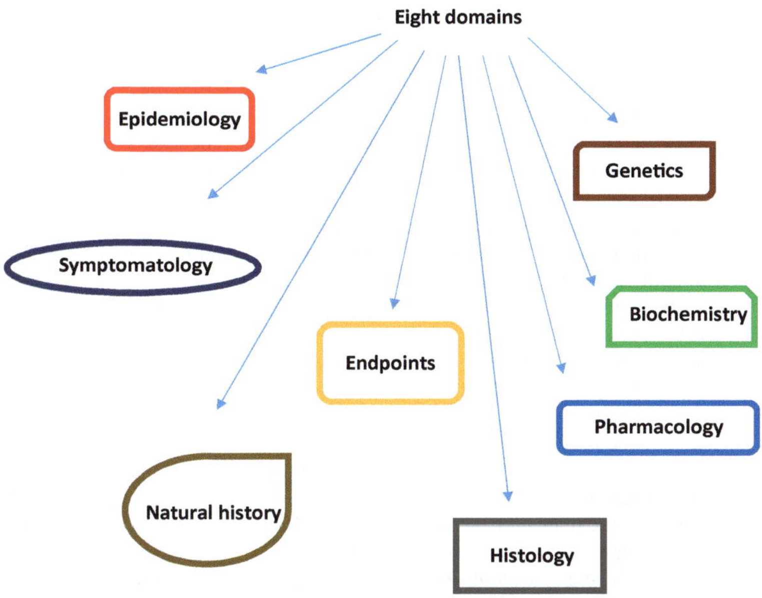

Fig. 12.6 Domains for framework to identify models of disease (FIMD)

condition, similarly discrepancy to clinical state is two if any of the two components did not match the human disease.

12.8 Limitations of Animal Model

- Certain clinical endpoints (biomarkers) are not reproduced with animal models hence following research process (preclinical study) is not achieved in few disease conditions.
- Quality of life cannot be assessed with the help of animal model.
- Ideal animal model do not exist.
- Experimental group with less number of samples exhibit insufficient power.
- Duration of follow up in animal models varies with the human.
- Disparity between animal and humans in the drug metabolic pathways.

12.9 Conclusion

Animal model need to be selected wisely based on the human disease condition and the predictive outcome. There are many animal models available for a single disease condition, appropriate model (proper validation) is selected based on the criteria of concern. Even though shortcomings are inevitable, these models pave way for the

Table 12.5 Weightage and questions to be raised under each domain for optimising animal model (FerreiraI et al. 2019)

S. No.	Question	Weightage
1	*Epidemiological validation:* 1. Is the model able to simulate the disease in the relevant sexes? 2. Is the model able to simulate the disease in the relevant age groups (e.g., juvenile, adult or ageing)?	**12.5** 6.25 6.25
2	*Symptomatology and natural history validation:* 1. Is the model able to replicate the symptoms and co-morbidities commonly present in this disease? If so, which ones? 2. Is the natural history of the disease similar to human's regarding 2.1 Time to onset 2.2 Disease progression 2.3 Duration of symptoms 2.4 Severity	**12.5** 2.5 2.5 2.5 2.5 2.5
3	*Genetic validation:* 1. Does this species also have orthologous genes and/or proteins involved in the human disease? 2. If so, are the relevant genetic mutations or alterations also present in the orthologous genes/proteins? 3. If so, is the expression of such orthologous genes and/or proteins similar to the human condition?	**12.5** 4.17 4.17 4.16
4	*Biochemical validation:* 1. If there are known pharmacodynamic (PD) biomarkers related to the pathophysiology of the disease, are they also present in the model? 2. Do these PD biomarkers behave similarly to humans? 3. If there are known prognostic biomarkers related to the pathophysiology of the disease, are they also present in the model? 4. Do these prognostic biomarkers behave similarly to humans?	**12.5** 3.125 3.125 3.125 3.125
5	*Aetiological validation:* Is the aetiology of the disease similar to humans?	**12.5** 12.5
6	*Histological validation:* Do the histopathological structures in relevant tissues resemble the ones found in humans?	**12.5** 12.5
7	*Pharmacological validation:* 1. Are effective drugs in humans also effective in this model? 2. Are ineffective drugs in humans also ineffective in this model? 3. Have drugs with different mechanisms of action and acting on different pathways been tested in this model? If so, which?	**12.5** 4.17 4.17 4.16
8	*Endpoint validation:* Are the endpoints used in preclinical studies the same or translatable to the clinical endpoints? Are the methods used to assess preclinical endpoints comparable to the ones used to assess related clinical endpoints?	**12.5** 6.25 6.25

safer research study in humans. One can choose an animal model closer to an ideal one. Thus validation plays a crucial role in translation of animal research to humans.

Table 12.6 Scoring system—validation of animal model (Sams-Dodd et al., 2006)

S. No.	Criterion		Score
1	Species	Human	4
		Non-human primate	3
		Non-human mammal	2
		Non-mammal	1
2	Simulation of disease	True	4
		Complex	3
		Pharmacological	2
		No	1
3	Face validity	More than 1 core symptom	4
		One core symptom	3
		One symptom	2
		No	1
4	Complexity	In vivo	4
		Tissue	3
		Cellular	2
		Sub-cellular/molecular	1
5	Predictivity	Graded for all pharmacology principles	4
		Graded for certain pharmacology principles	3
		All or none for certain pharmacology principles	2
		No or not shown	1

Table 12.7 Animal model validity based on drug screening system (Sams-Dodd et al., 2006)

S. No.	Drug screening method	Organism	Complexity	Disease simulation	Discrepancies to clinical state
1	Receptor binding assay	Mammal	Mechanism	None	3
2	Animal tissue	Mammal	Tissue	Artificial	3
3	Isolated organ	Mammal	Organ	Artificial	3
4	Animal disease model	Mammal	Intact organism	Artificial	2
5	Patient tissue	Patient	Tissue	True disease	1
6	Transgenic animal	Mammal	Intact organism	True disease	1(familial) 2(sporadic)
7	Human patient	Patient	Intact organism	True disease	0

Bibliography

Andersen ML, Winter LMF. Animal models in biological and biomedical research – experimental and ethical concerns. An Acad Bras Cienc. 2017;91(Suppl 1):1–13.

Bailey J, Balls M. Recent efforts to elucidate the scientific validity of animal-based drug tests by the pharmaceutical industry, pro-testing lobby groups, and animal welfare organisations. BMC Med Ethics. 2019;20(16):1–7.

Denayer T, Stohr T, Van Roy M. Animal models in translational medicine: validation and prediction. New Horiz Transl Med. 2014;2:5–11.

Ericsson AC, Crim MJ, Franklin CL. A brief history of animal modeling. Mo Med. 2013;110(3): 201–5.

Ferreira GS, Veening-Griffioen DH, Boon WPC, Moors EHM, Gispen-de Wied CC, Schellekens H, et al. A standardised framework to identify optimal animal models for efficacy assessment in drug development. PLoS One. 2019;14(6):1–17.

Ferreira GS, Veening-Griffioen DH, Boon WPC, Moors EHM, Meer PJK. Levelling the translational gap for animal to human efficacy data. Animals. 2020;10(1199):1–13.

Hau J. Animal models. In: Hau J, Hoosier GLV, editors. Handbook of laboratory animal science. 2nd ed. Boca Raton: CRC Press; 2003. p. 1–11.

Hau J. Animal models for human diseases. In: Conn PM, editor. Sourcebook of models for biomedical research. Totowa, NJ: Humana Press Inc; 2008. p. 3–8.

McGonigle P, Ruggeri B. Animal models of human disease: challenges in enabling translation. Biochem Pharmacol. 2013;87(1):1–39.

Pound P, Ebrahim S, Sandercock P, Bracken MB, Roberts I. Where is the evidence that animal research benefits humans? BMJ. 2014;328:514–7.

Robinson NB, Krieger K, Khan F, Huffman W, Chang M, Naik A, Yongle R, Hameed I, Krieger K, Girardi LN, Gaudino M. The current state of animal models in Research: a review. Int J Surg. 2019;10:1–21.

Sam-Dodd F. Strategies to optimize the validity of disease models in the drug discovery process. Drug Discov Today. 2006;11(7):355–63.

Sjoberg EA. Logical fallacies in animal model research. Sjoberg Behav Brain Funct. 2017;13(3): 1–13.

Varga OE, Hansen AK, Sandoe P, Olsson IAS. Validating animal models for preclinical research: a scientific and ethical discussion. ATLA. 2010;38:245–8.

Worp HB, Howells DW, Emily Sena ES, Porritt MJ, Rewell S, Collins VO, et al. Can animal models of disease reliably inform human studies? PLoS Med. 2010;7(3):e1000245.

Screening Methods for the Evaluation of Antidepressant Drugs

13

Alphienes Stanley Xavier

Abstract

Depression is a complex, heterogeneous, multifactorial disorder. It can be chronic and recurrent. Another challenge is the emergence of treatment-resistant depression. Identifying and validating novel treatment targets are the need of the hour to overcome these challenges. Even though the monoamine hypothesis is the widely accepted hypothesis to explain the pathophysiology of depression, it will not be enough to explain all the intricacies of the disorder. Experimental animal models are essential tools in studying pathophysiology and finding novel treatments for depression. An ideal animal model should possess face, construct, and predictive validities. It is an uphill task to develop a homologous model using rodents similar to all the depression manifestations experienced by humans. Apart from the classical stress-induced depression models, the researchers are developing and validating novel models involving neuro-inflammation, genetic, epigenetic dysregulation, and models using optogenetic tools.

Keywords

Depression · Animal models · Antidepressants · Validity · Neurobiological model · Stress-induced depression

A. S. Xavier (✉)
Department of Pharmacology, Sri Ramachandra Medical College and Research Institute, SRIHER (Deemed University), Chennai, Tamil Nadu, India

Clinical Pharmacology, Sri Ramachandra Medical Centre, Chennai, Tamil Nadu, India
e-mail: alphclinpharm@sriramachandra.edu.in

M. Lakshmanan et al. (eds.), *Introduction to Basics of Pharmacology and Toxicology*, https://doi.org/10.1007/978-981-19-5343-9_13

13.1 Introduction

Major Depressive Disorder (MDD) is the second leading cause of global disability. MDD is a complex, heterogeneous, multifactorial, and often chronic mood disorder affecting around 120 million people worldwide. The disorder has a life lime prevalence of almost 20% and a significant reason for suicide. The various symptoms of depression include depressed mood, mood swings, anhedonia, sleep disturbances, inability to concentrate, lack of energy, disturbed appetite, change in body weight, excessive guilt, feeling of worthlessness, suicidal thoughts, etc.

A disorder of such a considerable prevalence still has many unanswered questions in its pathophysiology. The most widely discussed hypothesis for the pathogenesis of depression is the monoamine hypothesis. In this theory, depression is associated with a functional deficit of nor-epinephrine (NE) and/or serotonin (5-HT) in certain areas of the brain. This is also strengthened by the role of drugs that facilitate monoamine transmission as effective antidepressant drugs. These drugs either inhibit the reuptake of NE, 5-HT, or the degradation by the enzyme Mono Amine Oxidase. There are many discrepancies also noted in this hypothesis, and many researchers have suggested alternative mechanisms underlying the pathophysiology of depression.

13.2 Animal Models of Depression

The two major uses of animal models in depression are to study the underlying complexity of the disease and to identify as well as test the molecules for a cure. There are three kinds of animal models for a mental disorder like depression. They are:

- Homologous models—Have identical causes, symptoms, and treatment characteristics of humans
- Isomorphic models—Have identical symptoms and treatment characteristics
- Predictive models—Exhibit only the treatment characteristics of a disease, helpful when the exact cause of the disease

13.3 Criteria for Valid Animal Models of Depression

An ideal model should have three major criteria of validity:

- Face validity—the resemblance between the behaviors modeled in the animals and the symptoms of depression in humans (Phenomenological similarity)
- Predictive validity—capacity of the model to successfully predict treatment response/performance in humans, a model with predictive validity should be able to identify true positive and true negative treatment responses

- Construct validity (theoretical rationale)—analogous to the disorder in its etiology, biochemistry, symptomatology, and treatment characteristics. It is considered as a vital criterion for a valid animal model. Etiological validity—analogous to etiology can be considered as a separate validity

13.4 Genetic Models

13.4.1 Traditional Genetic Mouse Models

Candidate genes involved in the pathophysiology are either silenced or overexpressed in genetically engineered mice to study the functional consequences. There are two main approaches in identifying candidate genes, they are forward genetics and reverse genetics. In the forward genetics approach, using simple mutagenic techniques, large random mutations are induced in animals, followed by selective breeding and development of animal lines with the desired phenotype (depression-like behavior). Then the responsible gene/genes will be identified. The reverse genetic approach involves genetic manipulations to develop either loss or gain of function mutants. Knock-out mice models are commonly used where a particular gene is disrupted to produce a loss of function mutants. Since the introduction of more advanced techniques such as RNA interference technology and virus-mediated genetic manipulations, refined control of spatial and temporal gene expression can be achieved.

Earlier models target genes in the monoamine pathway, and models such as the 5-HT1A receptor knock-out mice model, nor-epinephrine transporter knock-out model are used. Later models target glucocorticoid receptor, corticotrophin-releasing hormone receptor, Brain-Derived Neurotrophic Factor (BDNF), etc. Recently epigenetic mechanisms like histone acetylation, DNA methylation are being evaluated for identifying candidate genes. Traditional genetic models have only construct validity and presumed etiological validity with limited or no face and predictive validities. They target only specific proteins, which on manipulation affect the entire body, not only the brain. Most of the genetic manipulations are done in mice, but the available behavioral characteristics of depression phenotype have been done with rats, which further complicates the interpretations.

13.4.2 Models Using Optogenetic Tools

Optogenetic tools are a rapidly emerging approach in the field of neuroscience. They help us to stimulate or block a particular neuronal activity with high anatomical, temporal, and genetic precision. Through viral vectors, gene encoding opsin (light-sensitive ion channel) along with its promoter is injected into the animal's brain. Exposure to a particular wavelength of light can activate opsin, resulting in neuronal activation or inhibition. This approach has been tried to induce behavioral patterns like hedonic, anhedonic-like, reward-seeking, active or passive coping. The optogenetic approach is still in its early phase, and it could be technically challenging and expensive.

13.4.3 Selective Breeding

Rodents with specific physiological or behavioral abnormalities can be selected and bred over several generations to yield in-bred strains. These in-bred line animals are similar to each other genetically, which helps the researchers study the underlying genetic background and molecular mechanisms responsible for the abnormal behavior. These inbred animals are also used in the research to screen for antidepressant efficacy. Flinders Sensitive rat line (FSL) is a selectively bred rat line based on its sensitivity to an anticholinesterase agent, diisopropyl fluorophosphate. FSL rats show reduced appetite, psychomotor function, altered circadian rhythm, along with sleep and immune abnormalities. Wistar-Kyoto (WKY) rat strain was originally developed as normotensive controls for spontaneously hypertensive rats. WKY rats are hyper-reactive to stress, dysregulation of the HPA axis, and exhibit depression-like behavior in various behavioral tests. Two sub-strains of WKY rats are also developed based on the behavioral response in the forced swim test. Other inbred strains used as depression models are high anxiety/low anxiety behavior lines, high/low ultrasonic vocalization (USV) lines, and high/low yawning rats. These inbred animal strains show good face and predictive validities, but their construct and etiological validities are questionable.

13.5 Stress Exposure

13.5.1 Models of Acute Stress

13.5.1.1 Water Wheel Model

This is a model of "Behavioral despair." The animals will be placed in a desperate situation of no escape. After a period of struggle, the animals will stop trying to escape from the situation, which is the point of despair. The drugs with antidepressant activity can delay this despair, which can be used as an outcome measure to test the drug's efficacy.

The apparatus used for this model includes a plexiglass tank of size $20 \times 8 \times 18$ cm. The tank can be filled with water up to 9 cm in height. A wheel consists of a plexiglass shaft with six paddles (each 0.5 cm width) will be placed in the tank. The paddle will move when more than 5 g of weight is placed over that so that that wheel will rotate. Once a mouse is placed in the tank, initially, it will swim vigorously and try to escape from the tank. On finding the water wheel, the animal will try to cling to the wheel paddle.

Nevertheless, the paddle will move, and the continuous effort of the mouse will rotate the wheel. After a period of struggle, the animal will stop trying and stay afloat, motionless keeping its head outside the water surface. The number of wheel rotations and duration of the effort to cling on to the wheel can be noted as outcome measures. The drugs with antidepressant activity can increase the rotations and effort, thus delaying behavioral despair.

13.5.1.2 Forced Swim Test

This is also a model of "Behavioral despair." The adult animal will be placed in a cylinder with water (40 × 18 cm, rats) with no escape provisions. After a period of struggle, the rat will become immobile. Again it will try to escape and then followed by the phase of immobility. The total duration of immobility within a stipulated time limit can be taken as an outcome measure. This immobility is considered an act of despair, and the drugs with antidepressant activity can reduce the duration of immobility.

13.5.1.3 Tail Suspension Test

To induce behavioral despair, in this test, the mice will be suspended from a wire upside down with their nostril touching the water surface. The mouse will agitate and struggle to correct its posture, which is followed by a phase of immobility. Both the phases of agitation and immobility alternate. Over time the phase of immobility will be prolonged because of despair. Duration of immobility during the observation period will be noted as the outcome measure. A computerized real-time observation of an animal's mobility in the suspended position will give an accurate measurement to study the antidepressant activity of the study intervention. The duration of immobility will be reduced by the drugs which possess antidepressant property.

13.5.1.4 Learned Helplessness Test

This model consists of two phases. In the first phase, chronic mild stress is introduced without a chance to escape. The classic model is inescapable shock treatment, where repeated mild shock (0.8 mA every minute each of 15-s duration) is given through floor grid after placing the rat in a compartment. There will not be any escape route to avoid foot shock, and the exposure duration is up to 1 h. In the second phase, "conditioned avoidance training" will be provided. During this phase, a cue will be given before the shock; the cue can be a sound (buzzer) or light signal. Simultaneously, an escape route will be opened through which the animal can move to an un-electrified compartment to avoid shock. Failure to express escape response to the noxious stimulus after the training phase is seen with depressed animals. The drugs with antidepressant properties can reduce this 'escape failure.'

13.5.2 Chronic Stress Models

13.5.2.1 Chronic Mild Stress Model (CMS)

The chronic Mild Stress model is one of the extensively validated and more realistic models for depression. This model focuses on anhedonic behavior in animals after exposure to chronic mild stress. Anhedonia is an inability to feel pleasure in previously pleasurable activities. CMS model involves exposing the animals to series of mild stressors in an unpredictable manner. The stressors can be isolation, crowded housing, food or water deprivation, disruption of the dark-light cycle, tilting of home cages, dampened bedding, etc. The duration of exposure should be from a minimum of 2 weeks to even months. Initial models of chronic stress, as explained by Katz and Hersh, included harsh stressors such as electric shock, cold swimming, and shaker stress, etc. Willner modifies the model using mild stressors, which closely resembles real-world factors. Sucrose consumption test or sucrose-water preference test is used to assess the anhedonic behavior. The decrease in sucrose consumption is being observed as an outcome measure of a decrease in reward sensitivity. Chronic administration of agents with antidepressant action can improve anhedonic behavior by improving sucrose consumption.

The rats will be matched according to their sucrose intake behavior before the experiment. After chronic exposure to mild stress, the majority of the rats will reduce their sucrose consumption. Some animals can be more resilient (less 10% reduction in sucrose intake), and some can be more vulnerable (more than 30% reduction). Animals will recover from an anhedonic state with chronic administration of antidepressants. Almost 50% of animals can be refractory to treatment, similar to treatment-refractory depression seen in a real-life situation.

13.5.2.2 Chronic Restraint Stress Model

Restraining the animal can be used as a chronic predictable stressor to induce depression. After 2 h of restraining continuously for 14–21 days, the animals have shown depressive behavior, aggression, and increased corticosteroid levels. This kind of predictable chronic stress tries to simulate stresses of humans, such as daily repetition of a stressful job, social or financial stress, and familial stresses, which are predictable and repetitive.

13.5.2.3 Chronic Social Defeat Stress

Exposure to unfamiliar adult animals can be a stressful situation for adolescent animals. The aggression of adult animals towards adolescents will lead to the experience of chronic social defeat, stress, and depression. Experimental cages with two equal compartments and a partition in between are used for this procedure. Adult male rodents will be placed in a compartment for 5 days. On the sixth day, the adolescent animal (4 weeks old, male rodent) will be placed in the other compartment. The partitions can be removed, and both the animals will be allowed to interact. Adult animal shows aggression towards the younger ones, and the adolescent one expresses defensive behavior. Depending on the severity of aggression shown by the adult animal, the duration of interaction can be 5 min or shorter. This kind of repeated exposure for 2 weeks can induce depression in the adolescent animal because of stress-induced by chronic social defeat to the adults.

13.5.3 Isolation Induced Hyperactivity

Socially deprived animals have shown depressive behavior, characterized by increased spontaneous locomotor activity, decreased response to external stimuli, and stereotype behavior. To induce isolation, rats are housed separately in cages without any auditory or visual contact with other rats for 10–15 days. Molecules with antidepressant action block this isolation-induced depressive behavior.

13.5.4 Early Life Stress Model

Unfavorable events and experiences during early life (prenatal, postnatal) play a substantial role in the risk of developing the affective disorder later in life. Maternal separation is the extensively studied model for early life stress, but researchers found that the parameters studied in this model to be inconsistent. A recently developed chronic early life stress model, the dam-pup interaction was disrupted by limiting the nesting and bedding material in cages. Repeated stress-provoking exposures in early life have shown long-lasting effects in learning, memory in adult life.

13.6 Secondary or Iatrogenic Depression Models

Depression due to secondary causes such as disease conditions (stroke, pancreatic cancer, and hypothyroidism), medications are difficult to be differentiated from endogenous or primary depression. A detailed insight into iatrogenic depression and primary or stress-induced depression will help us identify the underlying pathophysiology and targets for pharmacological intervention.

13.6.1 Hormones of HPA Axis

The hypothalamus-Pituitary-Adrenal axis is activated by stressful stimuli, which increases serum glucocorticoids. According to the 'cortisol hypothesis,' persistent activation of the HPA axis is responsible for certain manifestations of depression. HPA axis dysregulation can be either way in depressed individuals. Individuals with elevated glucocorticoid concentration are benefitted from the pharmacological antagonism of glucocorticoid receptors. In atypical depression seen with post-traumatic stress disorder, fibromyalgia, and chronic fatigue syndrome, there will be increased sleep, appetite, as well as reduced circulating glucocorticoid levels. Depression is also observed as an adverse effect of chronic glucocorticoid administration. Animals exposed to chronic steroid administration also display depressive symptoms. Elevated serum cortisol due to excess Corticotrophin Releasing Factor (CRF) is also discussed as an etiological factor in depression, and there are some models developed based on this observation.

13.6.2 Retinoic Acid Derivatives

Isotretinoin, a retinoic acid derivative used in acne treatment, has been associated with the development of depression and suicidal behavior. Mice chronically treated with isotretinoin have displayed increased immobility phase in forced swim test and tail suspension test. These changes are associated with decreased hippocampal metabolism and neuronal proliferation.

13.6.3 Cytokines and Immune System Dysregulation

Pro-inflammatory cytokines such as interferon-alpha and recombinantly derived proteins have induced depression as an adverse effect. Immune function and depression possess a bidirectional relationship. Certain inflammatory cytokines induce a depression-like state, as well as depression also produces significant changes in immunity. Model of neuro-inflammation is tried by injecting lipopolysaccharide bacterial endotoxin. The induced inflammation model has shown depression-like behavior (decreased sucrose preference, increased despair behavior) and characterized by increased expression of pro-inflammatory cytokines (IL-1β, TNF-α) in the brain. These changes could be reversed by antidepressant drugs.

13.7 Pharmacological Models

Administration of some pharmacological agents can disrupt monoamine balance and induce a depression-like behavior. These models possess etiological and face validities. Nevertheless, their predictive and constructive validities are questionable.

13.7.1 Reserpine Induced Hypothermia

Reserpine is an indole alkaloid derived from the roots of Rauwolfia serpentine (Indian climbing shrub). It blocks the Vesicular Monoamine Transporter-2 (VMT-2) irreversibly and inhibits the uptake of catecholamines (dopamine and nor-epinephrine) into the storage vesicles. Reserpine was approved for use in the treatment of hypertension because of its sympatholytic activity. Reserpine-induced depression was correlated with its monoamine-depleting activity, and this observation helped to frame the neurochemical model of depression.

Administration of reserpine at the dose of 2.5–5 mg per kg subcutaneously to the mice induces ptosis, hypothermia, and catalepsy. The rectal body temperature of mice can be measured every 30 min for 2–3 h. The ability of a compound to inhibit reserpine-induced hypothermia can be considered due to its mechanism of improving monoamine levels at the synapses. This model is used as an early screening tool for assessing antidepressant activity.

13.7.2 Amphetamine Potentiation

Amphetamine is an indirectly acting sympathomimetic and a CNS stimulant. It increases the release of nor-epinephrine, serotonin, and dopamine from nerve terminals. It also causes a redistribution of monoamines from the storage vesicles to the cytoplasm. The action of amphetamine will be potentiated in animals repeatedly administered with an antidepressant because of monoamine excess. This potentiation leads to increased locomotor activity, hyperthermia, and stereotypical effects such as sniffing, licking, gnawing, biting.

Rats can be pretreated with the drug of interest for 2 weeks. Ninety minutes after the last dose, D-amphetamine 5–10 mg per kg is administered intra-peritoneally. After 30 min, the animal can be placed in the cage with photocells to observe its' locomotor activity for 1 h. Potentiation of amphetamine's action is seen with drugs increasing monoamine levels similar to tricyclic antidepressants and monoamine oxidase inhibitors. This method also helps to differentiate antidepressants and neuroleptics.

13.7.3 Apomorphine Antagonism

Apomorphine, a dopamine agonist, will induce hypothermia, stereotype, and climbing behavior in mice. Drugs possessing antidepressant activity will antagonize apomorphine-induced hypothermia. Neuroleptic agents block the only stereotype, climbing behavior without affecting apomorphine-induced hypothermia.

13.7.4 Psychostimulant Withdrawal Paradigm

The withdrawal effect of animals treated with CNS stimulants like amphetamine, cocaine can display changes similar to depression. These animals show reward deficits and an increase in the immobility phase in tail suspension, forced swim tests. This model can be used as a tool to screen for antidepressant activity.

Tetrabenazine antagonism in mice, potentiation of nor-epinephrine toxicity, yohimbine toxicity enhancement, and 5-hydroxytryptophan potentiation are the other pharmacological methods to screen antidepressant drugs.

13.8 Miscellaneous

13.8.1 Resident Intruder Paradigm in Rats

The social behavior of the resident rat to the unfamiliar intruder rat is observed as a result of depression. The resident rat displays aggressive behavior to the intruder rat. Antidepressant drugs reduce this aggressive behavior. Resident and intruder rats for the experiments will be procured from completely different sources to avoid prior contact between the groups. Social, non-social, agonistic behavior between resident and intruder rats will be observed during each social encounter.

13.8.2 Muricidal Behavior in Rats

Female rats of the Holtzman strain show compulsive mouse killing behavior irrespective of their satiety status, which has been due to a depressive state. Prior treatment with pilocarpine can convert the non-muricidal rat to a muricidal one. Molecules with the antidepressant property will reduce this muricidal behavior. Only rats showing muricidal behavior within 5 min of the presentation of mice can be selected for this experimental model.

13.8.3 Olfactory Bulbectomy Model

Olfactory function is extremely important in rodents. The olfactory system, along with the amygdala, and hippocampus form the part of the limbic region, contributing to the emotional and memory components of behavior. Bilateral olfactory bulbectomy (OBX) can result in endocrinal, immune, neurochemical, and behavioral changes in animals resembling major depression disorder in humans. Hyperactivity, changes in social behavior, enhanced nocturnal activity, deficits in learning and memory, and changes in taste-aversion behavior are the changes noted after bilateral OBX. Chronic administration of drugs with antidepressant activity can correct most of the changes that occur after olfactory bulbectomy, and this can be used as a model to evaluate molecules for antidepressant action. Despite possessing face and predictive validities, this model lacks etiological and construct validities.

13.8.4 Disruption of Circadian Rhythm

A significant disruption in the circadian rhythm has been noted in patients with depression. Stabilizing the rhythm can have a positive impact on depression symptoms. Disruption of circadian rhythm in animals can induce depression in animals, which can be used as a screening tool. This model has a unique etiological validity and specific face validity. Further research works are being undertaken to refine this model.

13.9 Behavioral Endpoints

The major symptoms of Major Depressive Disorder are anhedonia, depressed mood, sleep disturbance, change in appetite/sleep, psychomotor changes, and social withdrawal. These manifestations can be reproduced in the behavioral animal models of depression. Nevertheless, some of the symptoms seen in humans, such as a feeling of worthlessness, excessive guilt, suicidal ideation, or suicidal attempt, cannot be seen with rodent models. The behavioral endpoints in experimental models of depression are:

- Anhedonia—Sucrose preference test, cookie test
- Despair—Forced swim test, tail suspension test
- Hopelessness—Learned hopelessness test
- Apathy—deficit in goal-directed behavior (impaired nest building, reduced maternal care, reduced social interest, and reduced self-grooming)
- Anxiety (exposure to new/unfamiliar environment)
- Abnormalities in eating behavior
- Sleep disturbance
- Psychomotor agitation or retardation (Agitation–Isolation induced hyperactivity, Retardation-immobility in forced swim test)
- Cognitive impairment (Object recognition, subject recognition tests)

13.10 Conclusion

MDD is a complex, heterogeneous disorder. Many factors, either alone or in combination, contribute to the pathogenesis of depression. Available rodent models may not be able to capture all the symptom spectrum of depression, but they are important tools in investigating molecular, genetic, epigenetic, and environmental risk factors in understanding the pathogenesis of depression. They are also vital tools in the identification of novel targets for antidepressants and assessing treatment responses. Using a combination of animal models will help to study different aspects of antidepressant efficacy. The understanding of depression now not only relies on the monoamine hypothesis but also includes inflammation, stress signaling pathways, apoptosis, growth factors, environment, nutrition, comorbidities, genetic and epigenetic pathways of regulation. A detailed understanding of these underlying mechanisms will guide us to identify targets and refine newer molecules in overcoming challenges in the pharmacological management of depression.

Bibliography

Czéh B, Fuchs E, Wiborg O, Simon M. Animal models of major depression and their clinical implications. Prog Neuro-Psychopharmacol Biol Psychiatry. 2016;4(64):293–310. https://doi.org/10.1016/j.pnpbp.2015.04.004.

Krishnan V, Nestler EJ. Animal models of depression: molecular perspectives. Curr Top Behav Neurosci. 2011;7:121–47. https://doi.org/10.1007/7854_2010_108.

Planchez B, Surget A, Belzung C. Animal models of major depression: drawbacks and challenges. J Neural Transm (Vienna). 2019;126(11):1383–408. https://doi.org/10.1007/s00702-019-02084-y.

Wang Q, Timberlake MA 2nd, Prall K, Dwivedi Y. The recent progress in animal models of depression. Prog Neuro-Psychopharmacol Biol Psychiatry. 2017;3(77):99–109. https://doi.org/10.1016/j.pnpbp.2017.04.008.

Screening Methods for the Evaluation of Antiepileptic Drugs

14

Jayapriya Balamurugam and Mageshwaran Lakshmanan

Abstract

Seizure is sudden disorganized uncontrolled electrical activity in the brain, leading to alteration of behaviors, movement, or level of consciousness. Due to the varying etiology and presence of various types of seizures, there is a need for specific drugs for each type of seizure. In preclinical trials, drugs are tested on healthy rodents, often male, from uniform living conditions, with uniform etiology and pathology of seizures. The age of onset and duration of disease will be similar. Drugs will be given before the onset of seizures, no other drug will be co-administered, and there will be no previous history of antiepileptic drug intake or comorbidities. Therefore, the drugs which are found to be effective in preclinical studies may not be effective in clinical trials. The currently available screening methods include in-vitro techniques like $GABA_A$ and $GABA_B$ receptor binding measurement in rat cortex, GABA uptake in rat's cerebral cortex, modulation of NMDA, glutamate, and glycine receptors by test drugs, and neurotoxicity assays. Screening methods also include various in-vivo techniques like electrically-induced seizure models, chemically induced seizure models, and surgically induced seizure models. Apart from these models, various genetic models of epilepsy have been developed for studying rare types of seizures. This chapter will summarize the basic techniques, calculations, advantages, and disadvantages of screening methods for anti-epileptic activity.

J. Balamurugam
Department of Pharmacology, Virudunagar Medical College, Virudunagar, Tamil Nadu, India

M. Lakshmanan (✉)
Department of Pharmacology, Thanjavur Medical College, Thanjavur, Tamil Nadu, India

M. Lakshmanan et al. (eds.), *Introduction to Basics of Pharmacology and Toxicology*, https://doi.org/10.1007/978-981-19-5343-9_14

Keywords

Antiepileptic · Screening · Maximal electroshock seizure · Picrotoxin · PTZ · Bicuculline · GTCS · Absence seizure

14.1 Introduction

India has 12 million people with epilepsy, which accounts for one-sixth of the global burden of 70 million. Nearly 8–10% of the population gets seizures once in their lifetime. Seizure is sudden disorganized uncontrolled electrical activity in the brain, leading to alteration of behaviors, movement, or level of consciousness. Seizure becomes epilepsy when a minimum of two unprovoked seizures occur more than 24 h apart.

Based on the etiology, seizures can be unprovoked seizures or acute symptomatic seizures. In unprovoked type, seizures occur due to unknown etiology or pre-existing brain lesion or progressive CNS disorder. The risk of recurrence is more in this type. In acute symptomatic type, seizures occur due to any systemic or brain insult. Systemic causes could be drug withdrawal, alcohol withdrawal, metabolic disturbances like hypo or hypernatremia, hypomagnesemia, hypocalcemia, hypoglycemia, non-ketotic hyperglycemia, uremia, hyperthyroidism, hypoxia, dialysis disequilibrium syndrome, or porphyria. Brain insults may be a stroke, trauma, encephalitis, subdural hematoma, etc. The risk of future epilepsy is low with this type of acute symptomatic seizure.

Based on the onset of electrical activity in the brain, and hence the type of presentation, seizures can be focal or generalized.

•

Focal seizures with retained awareness were previously called simple partial seizures. The symptoms depend on the part of the cortex involved:

Occipita—flashing lights
Motor cortex—rhythmic jerking movements (jacksonian seizure)
Parietal cortex—distortion of spatial perception
Frontal lobe—sudden speech difficulties

Focal seizures with impaired awareness (complex partial seizures) are the most common type seen in adults with epilepsy. It usually lasts less than 3 min. Patients may seem to be awake but do not respond to external stimuli. Automatisms may be present. Among the generalized seizures, the most common type is Generalized tonic-clonic seizures. It starts with loss of consciousness with or without a scream or choking sound. The tonic phase lasts for 10–20 s, the clonic phase for 30–90 s, and the postictal phase for minutes to hours. Absence seizures usually last 5–10 s and may occur even hundreds of times in a day. Other types of generalized seizures include myoclonic, atonic, clonic, and tonic seizures.

Due to this varying etiology and types of seizures, there is a need for specific drugs for each condition. In preclinical trials, drugs are tested on healthy rodents, often male, from uniform living conditions, with uniform etiology and pathology of seizures. The age of onset and duration of disease will be similar. Drugs will be given before the onset of seizures, no other drug will be co-administered, and there will be no previous history of antiepileptic drug intake or comorbidities. So drugs which are found to be effective in preclinical studies may not be effective in clinical trials. Even though we have got many drugs for the symptomatic treatment of epilepsy now, nearly one-third of seizures remain uncontrolled. Hence, there is still a search for drugs for resistant seizures, age or gender-specific drugs, and drugs that can be disease-modifying with better efficacy and tolerability.

The currently available screening methods are:

In vitro methods
In vivo models
Genetic models

14.2 In-vitro Methods

14.2.1 $GABA_A$ Receptor Binding Measurement in Rat Brain

Principle: Radiolabelled [^{3}H]-Muscimol and [^{3}H]-GABA is used as a ligand to measure $GABA_A$ receptor binding activity. Unlabeled GABA or bicuculline methiodide is used to measure the non-displaceable binding (blank). The amount of specific radioligand binding to the $GABA_A$ receptor is estimated by subtracting blank from the total binding. Fixed concentration (10 or 50 μM) of the test compound is challenged with [^{3}H]-Muscimol binding. The competitive binding nature of the test compound with $GABA_A$ receptor is obtained by deriving the K_i value of the test compound that competes with [^{3}H]-Muscimol. For procedure and calculation, refer to Chap. 17.

14.2.2 $GABA_B$ Receptor Binding Measurement in Rat Brain

- Recent studies have discovered the role of $GABA_B$ receptors in the maintenance of the sleep cycle. Baclofen and 2-OH-saclofen are well-established $GABA_B$ receptor agonists and antagonists, respectively.
- The principle and calculation are the same as described in Sect. 17.2.1. Different buffers (Triton X-100 + Tris citrate) and agonists ([^{3}H]-Baclofen) are used here. The final protein concentrate achieved is 1 mg/ml.

14.2.3 GABA Uptake in the Cerebral Cortex of the Rat

- *Principle:* Uptake of GABA by GABA transporter can be inhibited by using nipecotic acid, guvacine, tetrahydro-isoxazolo-pyridol(THPO), and hydroxyl-nipecotic acid. Radiolabelled GABA is used to measure the uptaking by the transporter in the presence of the test compound.
- *Procedure:*
  - The cerebral cortex of rats is obtained after decapitation and dissection. It is then centrifuged and homogenized to obtain tissue pellet and suspended in Ringer solution.
  - For assaying, triplicate sets of microcentrifuge tubes are used. To each test tube, 60 μl of Ringer, 100 μl of test compound/vehicle, and 800 μl of tissue suspension are added.
  - The totals are incubated at 25 °C and the controls at 0 °C for 10 min. To this, 40 μl of radiolabelled GABA is added and reincubated for another 10 min.
  - After 10 min, the tubes are centrifuged at 13,000 × *g* for a minute, and supernatant fluid is aspirated. The samples are again subjected to centrifugation at 13,000 × *g* for 15 min after adding and incubating with solubilizer at 90 °C for 3 min.
  - After 15 min, 40 μl of supernatant fluid is aspirated and subjected to the calculation of CPM value in a scintillation counter.
- *Calculation:* The difference between the CPM at 0 and 25 °C gives the 'active uptake.' For each test compound concentration, the percentage inhibition is calculated, and IC_{50} values are obtained using log-probit analysis

14.2.4 GABA Release in Hippocampal Slices of Rats

- *Principle:* $GABA_B$ receptors located pre-synaptically act as auto-receptor and inhibit the GABA release. Therefore, theoretically, antagonism on $GABA_B$ autoreceptors can increase GABA release and cause hyperpolarization leading to reduced epilepsy. Radiolabelled GABA is used to measure the release of GABA in the presence of the test compound.
- *Procedure:*
  - The hippocampal tissue slices of rats are obtained after decapitation and dissection. The slices are incubated with radiolabelled GABA (0.05 μM) for 15 min at 37 °C using HEPES-Kreb's Ringer solution as a buffer.
  - The slices are then washed twice, incubated again for 15 min, and finally resuspended in the medium.
  - Now the tissue is incubated in the tubes with and without test compound at 37 °C for 15 min. This duration is taken as the 'GABA release period.' At the end of the release period, tubes are centrifuged (500 g) for 1 min, and the medium is separated from the tissue.
  - Using 0.13 N perchloric acid, the tissue is homogenized and subjected to scintillation spectroscopy to calculate CPM values.

- *Calculation:* The amount of radioactivity obtained in the presence and absence of the test compound is estimated and expressed as a percentage of the total radioactivity. The difference in radioactivity percentage between samples with and without test compound provides the effect of the test compound on the 'GABA release.'

14.2.5 [3^H]-CPP Binding Assay for Glutamate Receptors in Rat Cerebral ortex

- *Principle:* The 3-carboxy-piperazin-propyl-phosphonic acid (3-CPP) is a selective antagonist of NMDA receptors. The direct interaction between the test compound and the NMDA receptor is evaluated by estimating the inhibition of binding of the radiolabelled 3-CPP to the NMDA receptor.
- *Procedure:*
  - Reagents used:
    Buffer: Tris HCl (0.5 M and 50 mM)
    0.0001 M L-glutamic acid
    50 μl of 10 mM of [3^H]-CPP
    Various concentrations of test compound
    0.05% v/v of Triton-X100 for lysing tissue
  - The tissue pellets of the rat cerebral cortex are obtained using standard homogenization technique and suspended in 50 mM of Tris buffer. The final tissue pellet is prepared after reincubation, washing, and centrifugation with Triton-X100.
  - To each assay test tubes, 380 μl of distilled water, 50 μl of Tris Buffer, 20 μl of 0.0001 M L-glutamic acid, 50 μl of [3^H]-CPP, and 500 μl of tissue homogenate are added and incubate with agitation at 25 °C for 20 min.
  - The binding is terminated using the standard technique, supernatant fluid is discarded, and the pellet is washed and suspended in scintillation fluid. Specific binding is measured using a scintillation counter.
- *Calculation:*
  - Total binding is estimated by measuring binding of ^{3}[H]-CPP.
  - Non-specific binding is estimated by measuring the binding of L-glutamic acid. Specific binding is estimated by the difference between total binding and non-specific binding.
  - Percent inhibition: 100–specific binding as a percentage of the control value. Using log-probit analysis, the IC_{50} calculations are performed.

14.2.6 [3^H]-TCP Binding Assay for NMDA Receptor Complex in Rat Cerebral Cortex

- *Principle:* The NMDA receptor has a phencyclidine (PCP) binding site, which is near to its ion channel. PCP is a noncompetitive antagonist of the NMDA receptor. The thienyl derivative of PCP is called thienyl-cyclohexyl-piperidine (TCP). The direct interaction between the test compound and the PCP binding site in the NMDA receptor complex is evaluated by estimating the inhibition of binding of the radiolabelled 3-TCP to the PCP binding site in the NMDA receptor.
- *Procedure:*
  - Reagents used:
    Buffer: HEPES (0.1 M and 10 mM)
    0.0001 M L-glutamic acid
    0.00001 M L-glycine
    20 μl of 0.0001 M of PCP
    50 μl of 2.5 nM of [3^H]-TCP
    Various concentrations of test compound
  - The tissue pellets of the rat cerebral cortex are obtained using standard homogenization technique and suspended in 10 mM of HEPES buffer. The final tissue pellet is prepared after reincubation, washing, and centrifugation with HEPES buffer.
  - The assay is performed in the presence (stimulated) and absence (basal) of L-glutamic acid and glycine. To each assay test tubes, 380 μl of distilled water, 50 μl of HEPES Buffer, 20 μl of 0.0001 M L-glutamic acid 20 μl of 0.0001 M L-glutamic acid, 20 μl of 0.00001 M L-glycine, 50 μl of [3^H]-TCP, and 500 μl of tissue homogenate are added and incubated with agitation at 25 °C for 120 min. For basal assays, glutamic acid and glycine are not added in the tubes.
  - The binding is terminated using the rapid filtration technique. Specific binding is measured using a scintillation counter.
- *Calculation:*
  - The difference between the binding in the presence and absence of PCP provides the specific binding. In basal conditions, the specific binding observed will usually be 50% of total binding, and in stimulated conditions, the specific binding observed will be 90% of total binding. Alteration of these values in the presence of various concentrations of test compounds is noted, and PCP site-NMDA receptor antagonism is noted.

14.2.7 [H^3]-Glycine Binding Assay for NMDA Receptor Complex in Rat Cerebral Cortex

- The principle and calculation are similar, as explained in Sect. 14.2.6.
- The direct interaction between the test compound and the glycine binding site in the NMDA receptor complex is evaluated by estimating the inhibition of binding of the radiolabelled 3-glycine to the glycine binding site in the NMDA receptor.

14.2.8 [H^3]-Strychnine Binding Assay for the Strychnine-Sensitive-Glycine Receptor in Rat Cerebral Cortex

- The principle and calculation are similar, as explained in Sect. 14.2.6.
- The direct interaction between the test compound and the strychnine binding site in strychnine-sensitive-glycine-receptor is evaluated by estimating the inhibition of binding of the radiolabelled 3-strychnine to the strychnine binding site in strychnine-sensitive-glycine-receptor.

14.2.9 [^{35}S]-TBPS Binding Assay for Picrotoxin Site in GABA Receptor Complex in Rat Cerebral Cortex

- *Principle:* The GABA receptor complex has a picrotoxin binding site which blocks the opening of the chloride channel in the GABA receptor. Thus, picrotoxin acts a convulsants. The t-butyl-bicyclo-phosphorothionate (S-TBPS) is a potent convulsant that acts similar to picrotoxin. Radiolabelled TBPS [^{35}S]-TBPS is used to assay the interaction of test compound between the picrotoxin binding site in GABA receptor complex.
- *Procedure:*
  - Reagents used:
    Buffer: Tris (0.05 M and 5 mM)
    10 μl of 0.0001 M picrotoxin
    25 μl of 2 nM of [^{35}S]-TBPS
    Various concentrations of test compound
  - The tissue pellets of rat cerebral cortex are obtained using standard homogenization technique, suspended in Tris buffer, and stored overnight at −70 °C. On the day of assay, the tissue pellets are thawed and suspended in the buffer by standard technique.
  - To each assay test tubes, 190 μl of distilled water, 25 μl of HEPES Buffer, 10 μl of 0.00001 M picrotoxin/25 μl of 2 nM of [^{35}S]-TBPS, and 250 μl of tissue homogenate are added and incubated with agitation at 25 °C for 150 min.

  - The binding is terminated using rapid filtration and vacuum filtration technique. Specific binding is measured using a scintillation counter.
- *Calculation:*
  - The difference between the binding in the presence and absence of picrotoxin provides the specific binding. The specific binding observed will usually be 85–90% of total binding.
  - Alteration of specific binding in the presence of various concentrations of test compounds is noted, and picrotoxin-site-GABA/BZD/Cl receptor complex antagonism is noted.

14.2.10 Other In Vitro Techniques

- Besides estimating the interaction of test compounds with various receptors like GABA, NMDA, and strychnine-sensitive glycine receptors, many in-vitro techniques like
  - electrical recording from guinea pig hippocampal slices,
  - electrical recording of isolated brain cells using patch-clamp techniques,
  - monosynaptic reflex and single motorneuron activity recording using the spinal cord of neonate rats, and
  - neurotoxicity assays using cell culture of neurons are also implemented to determine the test compound's activity as anti-epileptic.

14.3 In Vivo Techniques

In vivo methods can be classified based on the mechanism of induction of seizures or the type of seizure for which the animal model is being used.

Based on the induction of seizures, the in vivo models can be classified as:

1. Electrically induced seizure models
2. Chemically induced seizure models
3. Seizure induction using surgery: Seizures induced by focal lesions
   - Seizures induced by kindling.
   - Post hypoxic myoclonus.

14.3.1 Electrically Induced Seizure Models

Model	Procedure	Effect of stimuli	Effect of drugs	Calculation	Model for	Effective drugs	Principle
Maximal electroshock seizures	Test drug in varying doses given orally or Intraperitoneally. High-intensity shock given to the rodent's head (2–5 times the threshold current) 30 min after IP injection or 60 min after oral administration of the drug. Observed for 2 min	Tonic flexion—**tonic extension** tonic convulsions	Abolishment of tonic hind limb extensor phase	Percentage of animals with hindlimb extension abolishment in test drug group/ Percentage of animals with hindlimb extension abolishment in control group × 100 gives the percentage protection by the test drug ED50 (dose required to protect 50% of the animals) also calculated	GTCS, CPS	Phenytoin, Carbamazepine, lamotrigine	Electroshock mediate the induction of synchronous neural discharges in the brain leads to development of seizures.
The threshold for maximal electroconvulsions	Electrical stimulation of linearly increasing intensity of 5–30 mA for 0.2 s is given through corneal or ear electrodes to mice.	Clonic seizure and tonic hindlimb extension occurs	Elevation of the current threshold by 20%	The current at which clonic seizure and tonic hindlimb extension occurs is noted as minimal and maximal seizure threshold, respectively	Screening of drugs for GTCS		The ability of AED to alter the seizure threshold

MES and PTZ models do not take into account the individual variation of seizure susceptibility. So threshold model should be used to determine the effect of drugs on the seizure threshold. It also allows the detection of the pro-convulsant property of the chemical, if any.

14.3.2 Chemically Induced Seizure Models

Chemical agents can induce seizures after both systemic and central (CNS) administration.

The commonly used systemic agents to induce generalized seizures are Pentylenetetrazole, picrotoxin, bicuculline, 4-aminopyridine, 3-Nitropropionic Acid, isoniazid, penicillin, gamma hydroxyl butyrate, strychnine, NMDA, homocysteine, allylglycine, methionine, sulfoximine, flurothyl inhalation, pilocarpine, thiosemicarbazide, kainic acid, DDT, beta CCM, DMCM.

The chemicals used centrally to induce focal seizures include penicillin, kainic acid, pentylenetetrazol, and quinolinic acid.

Model	Procedure	Effect of stimuli	Effect of drugs	Calculation	Model for	Drugs	Principle
Pentylene tetrazol(PTZ)/ Metrazol in mice	Standard or test drug is given SC or IP or orally. PTZ injected 15 min after **sc** injection, 30 min after IP, or 60 min after oral administration of standard/test drug and observed for an hour	Clonic seizures (spike-wave complexes in EEG) In higher doses – tonic seizure (sharp hypersynchronized poly-spikes in EEG).	Delay in the onset of seizures and no, of animals that did not develop seizures in the test group compared to the control group	Protection percentage by test drug = Percentage protection in test group/ Percentage protection in the control group. ED50 calculated. The time interval between PTZ and seizure onset was measured and compared to the control group	Absence seizures and myoclonic seizures Muscle relaxant and anxiolytic property	Ethosuximide, Valproate Chlordiazepoxide, Phenytoin, Phenobarbitone	PTZ antagonize postsynaptic GABA receptors
Picrotoxin induced convulsions in mice	Test or standard drug is given orally or IP. Picrotoxin is given sc after 60 min of oral or 30 min of IP administration of standard/test drug. Observed for 30 min	Clonic seizures, tonic seizures, and death	Time of onset of seizures and time to death are noted Inhibition of seizures compared to vehicle control	Percentage inhibition in test group/percentage protection in the control group ED 50 Time response curve with test drug 30,60,120 min before picrotoxin	Used to characterize drugs with CNS activity	Similar to PTZ Model. No added advantage over PTZ for drug screening	Picrotoxin is a GABA antagonist
Bicuculline induced seizures in rats	Standard drug or test drug is given orally. One or 2 hours later, Bicuculline injected IV	Tonic convulsions within 30 s (Myoclonic seizures in mice)	Control of seizures	No. of animals protected/ total no. of animals in that group × 100 = Percentage of protected animals ED50 also calculated	Specific for antiepileptic activity	Benzodiazepines	Bicuculline is a GABA antagonist

(continued)

(continued)

Model	Procedure	Effect of stimuli	Effect of drugs	Calculation	Model for	Drugs	Principle
4-aminopyridine induced seizures in mice	Test drugs n varying doses given IP. After 15 min, 4 aminopyridine given sc.	Hyperactivity, trembling, intermittent clonus followed by hindlimb extension, tonic seizures, opisthotonus, and death	Abolishment of the behavioral signs	ED50 calculated using the percentage of protected animals calculated for each dose	To differentiate the MOA of anticonvulsants	Phenytoin, CBZ, PB, Valproate. Not abolished by BZD, NMDA antagonist, CCB.	4-aminopyridine, a K channel antagonist, enhance spontaneous and evoked neurotransmitter release. This leads to stimulation of non NMDA type EAA receptors leading to seizures.
3 Nitropropionic acid-induced seizures in mice	Test drug given, followed by 3-nitropropionic acid IP	Occurrence of seizures	Latency of seizure occurrence increased	ED50, LD50 estimated. Percentage of animals with seizures, latency, mortality in 2 h are evaluated			3-nitropropionic acid, a natural toxin, inhibits succinate dehydrogenase of mitochondrial complex II irreversibly, leading to impaired energy metabolism.
Isoniazid induced convulsions in mice	Isoniazid injected sc, 60 min after oral or 30 min after IP administration of standard or test drug and observed for 120 min	Tonic seizures, clonic seizures, and death	Suppression of seizures or death	Percentage suppression of seizures or death in the test group relative to the control group is calculated ED50 calculated.	model for status epilepticus	BZD with pyridoxine	In patients with a seizure disorder, isoniazid can precipitate seizures due to its GABA synthesis inhibition property

Systemic penicillin test in cats/rats	Penicillin G is given IM to control and antiepileptic test drug treated group	An hour after injection, recurrent intermittent episodes of arrested activity, staring, myoclonus, facio-oral twitchings may progress to GTCS. Lasts for 6–8 h	Abolishment of seizures	Percentage suppression of seizures or death in the test group relative to the control group is calculated. ED50 is also calculated	Screening of drugs for absence and myoclonic seizures	Ethosuximide, valproate	Penicillin selectively blocks GABA mediated inhibitory postsynaptic potentials
Gamma hydroxyl butyric acid (GHB) in rats	The test drug is given IP. GBH is given IP after 60 min. EEG recorded an hour before to 4 h after GBH administration.	Behavioral arrest, stupor, and spontaneous and auditory evoked myoclonic jerks. Baseline EEG progress to brief bursts of spikes, continuous spikes, hypersynchronous spikes, and waves in EEG	Shortened duration or abolishment of seizures	Duration of seizures & Time interval between seizure onset and EEG normalization is measured	Petit mal epilepsy	Valproate, ethosuximide, trimethadione Phenytoin is not effective.	GHB, a minor metabolite of GABA in the brain, is an endogenous epileptogen. Its synthesis is blocked by valproate and enhanced by 3-mercaptopropionic acid and kainite
Strychnine induced convulsions	Mice given test/ standard drug orally followed an hour later by IP injection of strychnine. Observed for an hour	Tonic extensor convulsions and death.	Timing of the convulsion and death noted	ED50 amount of drug which protects 50% of the test group animals	Used to find the character of CNS active drugs, e.g., Drugs for Anxiety	Not controlled by standard antiepileptic drugs	Glycine-induced postsynaptic inhibition of motor and interneurons in the spinal cord is antagonized by strychnine.

Among the chemical models, strychnine, pilocarpine, NMDA and kainite are used to study the mechanism of drug resistance as the seizure induced by these agents are not controlled by the standard antiepileptic drugs.

The effective drugs against 4-aminopyridine induced seizures were similar to those effective against the Maximum Electroshock (MES) test. So this model can be used to differentiate the mechanism of action of anticonvulsant drugs.

14.3.3 Seizures Induced by Surgery

14.3.3.1 Seizures Induced by Focal Lesions

- *Principle:* Seizures can be induced by topical application or intracerebral injection of many chemicals and metals and by some cerebral lesions
- *Procedure:* Anaesthetized rats placed in a stereotactic apparatus are injected with the chemicals through a burr hole in the cranium. Chemicals, e.g., Kainic acid, are given in various doses. Electrodes are placed in the skull, and an EEG polygraph is recorded
- *Calculation:* Seizures and EEG readings are studied during the acute phase and also for up to 2 months in both control and test group of animals.

Various chemicals used for inducing focal lesions include:

- Kainic acid—intrahippocampal injection - stimulation of glutaminergic receptors
- Alumina (aluminum hydroxide) cream application
- Cobalt powder implantation—inhibition of GABAergic transmission at focal sites
- Tungstic acid colloidal gel injection—resistant to standard antiepileptic drugs
- Penicillin topical application
- Saturated FeCl3—subpial injection—recurrent seizures
- Zinc sulfate—intracerebral injection
- Antibodies to brain gangliosides—intracerebral injection
- Cholinergic agonist microinjections,
- Atropine—topical application,
- Tetanus toxin—intrahippocampal injection,
- Strychnine—visual or somatosensory cortex injection
- Bicuculline electrophoretic application
- Liquid nitrogen probe—Local freezing of cortex

14.3.3.2 Seizures Induced by Kindling

- *Principle:* Repeated sub-convulsive electrical stimulation of amygdala leads to after discharge initially. Later, due to the self-perpetuating effect, more neurons get recruited, leading to spontaneous seizures, which last for nearly 7 months after termination of stimulation.
  - Alternative methods that can be used for kindling include Electroshocks through Corneal electrodes, stimulation of neocortex or hippocampus, repeated sub-convulsive doses of PTZ
- *Procedure:* Electrodes are implanted in the amygdala of rats and allowed to recover for a week. After a week, repeated weak electrical impulses are applied briefly to the amygdala. The animal goes through five stages of seizures, class 1–5, during the daily electrical stimulation period. When stage 5 seizures occur, the animal is said to be fully kindled. This takes nearly 10–15 days, i.e., for the animals to develop spontaneous seizures.
- Racine's seizure severity scale

Stage/Class	Behavior
1	Immobility, mouth and facial movements like sniffing, ear twitching
2	Head nodding
3	Unilateral forelimb clonus
4	Rearing with bilateral forelimb clonus
5	Rearing and falling, loss of righting reflex, GTCS

- Now the test drug is given IP or orally. Animals are tested for seizure latency, severity, duration, and after discharge duration (prolonged spike activity which appears in neurons after an initial brief response to the short stimulus) a day before and after the drug.
- *Calculation:*
- Seizure latency, severity, duration, and after discharge period were compared between the test and control groups of animals. Drug efficacy can also be assessed by the increase in current intensity needed to elicit after discharge.
- ED50 can be calculated for the abolishment of generalized seizures, i.e., stages 4 & 5, or focal seizures, i.e., stage 1 3 or amygdala after discharges.

Thus this model can be used to evaluate drugs for their anti-epileptogenesis property and secondarily generalized seizures and complex partial seizures.

14.3.3.3 Post Hypoxic Myoclonus in Rats

Principle: This model was developed to mimic the post-hypoxic myoclonus in humans. Rats show susceptibility to audiogenic stimulus for 3–14 days post-cardiac arrest.

Procedure: After 12–24 h of fasting, rats are anesthetized, intubated, and ventilated with 30% O_2 in N_2O. ECG and BP are recorded. Then the animals are paralyzed, and cardiac arrest is induced by intracardiac KCl and stopping ventilation.

After 10 min, the animals are resuscitated by turning on the ventilator with 100% O_2 and over the next 2–4 h, weaned from the ventilator and extubated.

- Myoclonus induced in these animals by auditory induction of 45 clicks from a metronome. The animals are scored as follows:
- No response – 0.
- Ear twitch – 1
- Ear and head jerk – 2
- Ear head and shoulder jerk – 3
- Whole-body jerk – 4
- Severe whole body jerk leading to jump – 5

The score summed for all the 45 clicks for each rat.
Thirty minutes later, a test/standard drug is given intraperitoneally. After 60 min, again audiogenic stimulus was given, and myoclonus was scored.

Clonazepam and valproate are found to be effective in this model and not phenytoin.

Calculation: Pre and post-drug myoclonus scores are analyzed to find the efficacy of the drug.

14.4 Classification of Animal Models Based on the Type of Seizure Disorder

14.4.1 Models Used for Status Epilepticus

(i) *Rat Kainite Model*
Principle: Kainate induces complex partial seizures and secondary generalization and has been used as a model for complex partial seizures, drug-resistant epilepsy, status epilepticus, and hippocampal lesions in the rat.

Procedure: Rats are implanted with electrodes extradurally over the frontal and parietal cortex under anesthesia. A week later, a control or test drug is given intraperitoneally. After 15 min, kainite is given intraperitoneally. EEG and behavioral changes are observed for 4 h.

A sustained ictal pattern in EEG for more than 20 min without >1-min interruptions is considered status epilepticus.

Calculation: Latency to develop and the duration of status epilepticus, percentage of animals protected from status epilepticus are calculated.

(ii) *Pilocarpine/Lithium-pilocarpine induced status epilepticus*
Principle: Pilocarpine, a cholinomimetic agent, induces status epilepticus. Animals pretreated with lithium need a lesser dose(1/tenth dose) of pilocarpine.

Procedure: Rats are anesthetized and implanted with two single contact recording electrodes on the skull corresponding to the parietal cortex and one bipolar deep recording electrode in the right hippocampus. A week later, lithium chloride is given IP. The next day methylscopolamine is given SC to prevent peripheral effects of the convulsant. Half an hour later, pilocarpine is given to induce status epilepticus. Test drug or standard drug (diazepam) is given 1 h after the onset of status.

Behavioral changes and EEG changes are noted down. EEG shows continuous spiking 30–60 min after pilocarpine administration. Bilateral EEG cortical activity and unilateral hippocampal activity are recorded for the duration of status epilepticus.

After 14 days, animals are sacrificed under anesthesia, and the cell density of the hippocampus is quantified.

Calculation: Protection percentage from status epilepticus and neuronal damage between the different groups are analyzed.

(iii) *Self-Sustained Status Epilepticus (Electrical stimulation of hippocampal perforant pathway):*

Principle:

Electrical stimulation of the perforant pathway leads to self-sustained status epilepticus in rats. This model can also be used to study drug-resistant epilepsy.

Procedure:

Under anesthesia, bipolar stimulation electrodes are placed into the angular bundle of the perforant path and a bipolar recording electrode in the ipsilateral dentate gyrus. Stimulation of the perforant pathway electrodes induces SSSE.

The test drug is injected IV through the tail vein either 20 min before or 10–40 min after this.

Calculation:

Duration of SSSE, number of seizure episodes, the average duration of each episode, cumulative seizure time, and number of spikes per hour are analyzed. EEG and electrographic activity are analyzed manually and using software, respectively.

Other models for status epilepticus include:

- Lithium-methomyl induced seizures in rats: Methomyl, a carbamate anticholinesterase given after pre-treatment with lithium
- Homocysteine thiolactone administration in rats with actively epileptogenic cortical cobalt lesions
- Prolonged electrical stimulation of the basolateral amygdala in rats

14.4.2 Model for Infantile Spasm

14.4.2.1 Genetic Models of Chronic Infantile Spasms:

(i) *Arx Knock-In Mouse Model:*

Arx protein is a transcription factor important in the maturation and migration of developing GABAergic interneurons in humans. The most common mutation is the expansion of the polyalanine tract in the Arx gene. Mice generated with seven additional alanine codons in the polyalanine tract develop behavioral spasms on days 7–11 of postnatal life. These animals do not respond to ACTH or vigabatrin but respond to 17β-estradiol

(ii) *Arx Conditional Knock-Out (cKO) Mouse Model:*
Arx knocks out mice pups of age 14–17 days show behavioral seizures and abnormal EEG activity.

(iii) *APC cKO Mouse Model:*
Mutations in proteins that modulate β-catenin signaling lead to infantile spasms. Removal of Adenomatous polyposis coli, a negative regulator of β-catenin/Wnt signaling, leads to behavioral spasms and EEG abnormalities in neonatal and adult mice. These mice respond to standard treatment of Infantile spasms.

Acquired models:

(iv) *NMDA induced seizures:*
Principle: NMDA induces seizures in pups which were sensitized by betamethasone in prenatal life.

Procedure: Betamethasone given to pregnant rats on day 15 of gestational age at 8 and 6 pm. This is supposed to sensitize the brain of pups to NMDA-induced seizures. After delivery, pups are injected with NMDA on day 15 of postnatal life. This induces recurrent episodes of behavioral changes - twisting of tail and arching, followed by loss of righting reflex and flexion spasms. The drug for infantile spasm, ACTH, delays the onset of NMDA-induced seizures.

Calculation: Duration of delay in onset of seizures by ACTH is measured.

This method requires validation

(v) *TTX rat model:*
Tetrodotoxin, a sodium channel blocker, is infused into the dorsal hippocampus of rats. After a week, the animals show behavioral changes with or without spasms and focal seizures of head posturing and forelimb clonus. EEG shows hypsarrrhythmia in the contralateral side of TTX infusion. This model responds to treatment with vigabatrin

(vi) *Multiple-hit rat model:*
Doxorubicin and lipopolysaccharide are infused into the lateral ventricle of rat pups on day 3 of postnatal life. P-chlorophenyl alanine is given 2 days later intraperitoneally. Pups develop spasms at day 4 of postnatal life and last till day 13.

Multiple hit in adult animals induces clonic and tonic-clonic seizures.

This model is usually refractory to standard treatment, and vigabatrin is found to suppress the spasms transiently.

(vii) *Chronic Early Stress Rat Model:*
Stress is induced in early life by mixing rat pups from different litters and providing cages with restricted bedding and nesting materials. Nearly half of the animals develop behavioral spasms and EEG spikes, and nearly 10% develop seizures later in life.

14.4.3 Models for GTCS

- MES
- Chemical convulsants—PTZ, Picrotoxin, Bicuculline, Penicillin, Methionine sulfoximide
- Metabolic derangements—hypoxia, hypoglycemia, hypercarbia, uremia, hyperthermia, drug withdrawal
- Genetic models—Photosensitive baboons, audiogenic seizures, totterer and El mice, genetically epilepsy prone rats, Mongolian gerbil, drosophila shakers

14.4.4 Models for Complex Partial Seizures

- Kindling
- Chemicals—kainic acid, tetanus toxin
- In vitro models

14.4.5 Models for Absence Seizures

Chemical convulsants—Systemic penicillin, gamma hydroxyl butyrate, thalamic stimulation.

14.4.6 Models for Simple Partial Seizures

Acute: Topical convulsants eg. topical penicillin, bicuculline, pricortoxin, strychnine, pilocarpine, atropine.

Chronic: Cortical implantation of cobalt, tungsten, zinc, iron, cryoinjury, ganglioside antibody injection.

14.5 Genetic Models of Epilepsy

14.5.1 Genetic Animals Models with Spontaneous Seizures

- *Epileptic dogs:* These dogs develop both focal seizures and GTCS. Nearly 30% of the animals with GTCS and more no. of dogs with focal seizures were resistant to conventional antiepileptic drugs, and hence can be used for testing drugs useful in resistant seizures. However, as this is a model for chronic epilepsy, it cannot be used for screening antiepileptic drugs; also, there is the problem of getting the required number of epileptic dogs.
- *Rats with spontaneously occurring petit mal epilepsy:* These tremor rats (tm/tm) have high seizure frequency and are a better model for absence seizures and acute and chronic drug efficacy studies than the PTZ test. EEG of these animals shows

7–11/s spontaneous spike-wave discharges and typically occurs after the age of 14–18 weeks and lasts for its lifetime. Antiepileptic drug dose, which decreases the spike-wave discharge duration by 50% in 1 h after drug administration, is noted as the ED50.

- *Tottering mice:* These mice present with either
  - focal motor seizures - controlled by diazepam and not by ethosuximide and valproate or
  - Spike waves petit mal seizures (homozygous tottering mice) blocked by ethosuximide, diazepam, and phenobarbitone and not by phenytoin.
  - The dose which reduces the no. of spikes is noted as the effective dose. The problem with this model is getting homozygous animal models.
- *AE mice:* It is an inbred mouse strain with too infrequent spontaneous seizures. So it is not routinely used to test the antiepileptic drug efficacy but is more responsive to electrical induction of seizure.
- *Quaking mice:* These mice have an autosomal recessive genetic disorder that leads to hypo-myelination. Homozygous mice develop spontaneous and induced myoclonic and GTCS. Seizures are induced by the handling of the animals, which was blocked by phenytoin, phenobarbitone, valproate, and carbamazepine but not by ethosuximide and diazepam. This model can be used for screening drugs effective against focal seizures in humans.
- *Syrian golden hamsters:* These animals develop drug-resistant spontaneous myoclonic and GTCS induced by mild stress. Nevertheless, the clinical picture does not correspond to any of the human seizure types.
- Out of the above six genetic models, 'rats with spontaneous petit mal epilepsy' is used for acute drug efficacy studies, and 'Epileptic dogs and rats' are used for chronic drug efficacy studies. Other genetic animal models for epilepsy include Genetic epileptic WAG/RiJ rat, Lethargic lh/lh mouse, GAERS rat (Genetic Absence Epilepsy Rat from Strasbourg Models for absence seizures), Stargazer mutant mouse (Generalized non-convulsive spike-wave seizures with behavioral arrest resembling the human absence seizure), Learner mutant mice (Severe ataxia and atrophic cerebellum), and spontaneous epileptic rats (Mating of tremor rat (tm/+) with zitter homozygous rat zi/zi)

14.5.2 Genetic Animal Models with Absence Like Seizures and Tonic Convulsions

- NER—Noda epileptic rat: Inbreeding of rats with spontaneous tonic-clonic seizures in Crj:Wistar
- Epileptic rat mutant with spontaneous limbic like seizure: Bred from successive mating and selection from an inherited cataract rat
- Kindling prone and kindling resistant rats: Breeding of rats with fast and slow amygdala kindling rats
- FH—Flathead rat: Genetic model in the early postnatal period

- WER—Wakayama epileptic rat: Wistar rat mutant with both tonic-clonic seizures and absence seizures

14.5.3 Genetic Animal Models with Reflex Seizures

- These are genetically predisposed animal models in which seizures do not occur spontaneously but can be induced by sensory stimuli like sound or light. Nevertheless, the occurrence of seizures in response to sensory stimuli is less in humans (5%). The advantage of this model is the easy induction of seizures without chemical or electrical means. The animal models include
  - Photosensitive baboons (Papio papio),
  - Photosensitive fowls,
  - Fayoumi strain of chickens (Fepi),
  - EL mouse,
  - Audiogenic seizure susceptible mice,
    Audiogenic seizure susceptible rats (Wistar Audiogenic Rats WAR),
  - Mechanically stimulated mice,
  - Genetically epilepsy prone rat (GEPR),
  - Mongolian gerbil with reflex seizures,
  - Szmutant hamster (also a model for paroxysmal dystonia)
  - Cats with generalized photosensitive epilepsy—caused by long term intramuscular low dose penicillin injections
- Sound-induced seizures in mice are sensitive to all antiepileptic drugs and hence can be used as a screening model for potential anticonvulsant drugs, whereas the sound-induced seizures in GEPR are blocked by drugs that are effective in MES test. So GEPR is used as an alternative to the MES test. GEPR-3 strain develops moderate or clonic convulsions, and GEPR-9 strain develops more severe tonic extensor convulsions.
- Reflex seizures in Mongolian gerbils occur in response to bright light, audiogenic stimuli, vigorous shaking of the cage, different handling techniques, new environment, air blast at the back of the animal for 15 s, etc. Young animals (7–10 weeks) present with facial myoclonic 'minor' seizures, whereas older animals (7 months) present with generalized myoclonic or tonic-clonic 'major' seizures. Minor seizures were more susceptible to valproate, ethosuximide, and BZD, and major seizures to phenytoin, phenobarbitone, and carbamazepine. So young animals are used as models for petit mal epilepsy and older animals for GTCS. These animals can also be used as a model to study the mechanism of generalization of epileptic seizures, as the severity of seizures increases up to 7 months of age in these animals. The drawback of this model is that the animals go for long postictal refractoriness after seizures and can be tested only once a week. So care must be taken not to induce seizures and postictal refractoriness while administering drugs.
- EL mice develop seizures in response to vestibular stimulation, like altering the equilibrium of the mice. This can be used as a model for complex partial seizures

with secondary generalization. Nevertheless, this model is used mainly for the study of seizure mechanisms rather than for drug evaluation.

- It is not clear if the genetic models with reflex seizures are superior to the traditional PTZ or MES tests for evaluating the drug efficacy or screening of anticonvulsant properties. Nevertheless, once the drug efficacy is tested in small animal species, higher models of dogs and baboons can be used to understand the behavioral changes caused by the drugs in animal models.

14.6 Conclusion

As spontaneous seizures in animals are rare and sporadic, various in vitro and in vivo models have been developed. Selection of proper model is mandatory to get relevant results. Rather than convenience and familiarity, the selection of model should depend on the research question, type of seizure to be modeled. For example, in vitro models are chosen if the effect of drugs on membrane ion channels is studied, and in vivo mammals are chosen if the effect of surgical intervention in controlling seizures is studied. No one model can answer all the research questions. There are several types of epilepsies, and none of the models typically imitates clinical epilepsy. Thus, it is always better to validate drugs using several animal models.

Bibliography

Guillemain I, Kahane P, Depaulis A. Animal models to study aetiopathology of epilepsy: what are the features to model? Epileptic Disord. 2012;14(3):217–25.

Kupferberg H. Animal models used in the screening of antiepileptic drugs. Epilepsia. 2001;42 (Suppl 4):7–12.

Löscher W. Genetic animal models of epilepsy as a unique resource for the evaluation of anticonvulsant drugs. A review. Methods Find Exp Clin Pharmacol. 1984;6(9):531–47.

Löscher W. Animal models of epilepsy for the development of antiepileptogenic and disease-modifying drugs. A comparison of the pharmacology of kindling and post-status epilepticus models of temporal lobe epilepsy. Epilepsy Res. 2002;50(1–2):105–23.

Vogel H. Psychotropic and neurotropic activity. In: Vogel HG, editor. Drug discovery and evaluation. Springer, Berlin: Heidelberg; 2007. p. 459–93.

Screening Methods for the Evaluation of General Anaesthetics

15

Kiruthika Sivagourounadin

Abstract

General anaesthetics are drugs which induce a reversible state of loss of consciousness with reduced motility, attenuated response to painful stimuli along with abolition of somatic and autonomic reflexes. The general anaesthetics are classified into groups as inhalational and intravenous anaesthetics. Small laboratory animals like mice and rats are commonly used for the screening of general anaesthetics. Electroencephalographic (EEG) burst suppression threshold test is used for evaluating the sensitivity of central nervous system to intravenous anaesthetics in rats. New Zealand white rabbits are used for assessing safety profile of intravenous anaesthetics. While screening for volatile anaesthetic agents, rats are used for determination of minimum alveolar concentration of test agent. This chapter mainly provides an overview of the screening methods involved in the evaluation of inhalational and intravenous anaesthetics.

Keywords

Screening · Anaesthetic drugs · Inhalational · Intravenous

15.1 Introduction

General anaesthetics are drugs which induce a reversible state of loss of consciousness with reduced motility, attenuated response to painful stimuli along with abolition of somatic and autonomic reflexes.

The general anaesthetics are classified into two broad groups namely:

K. Sivagourounadin (✉)
Department of Pharmacology, Jawaharlal Institute of Postgraduate Medical Education & Research (JIPMER), Karaikal, Puducherry, India
e-mail: kiruthikas.kkl@jipmer.edu.in

M. Lakshmanan et al. (eds.), *Introduction to Basics of Pharmacology and Toxicology*, https://doi.org/10.1007/978-981-19-5343-9_15

- Inhalational anaesthetics: e.g. nitrous oxide, ether, halothane, isoflurane and other fluorinated anaesthetics
- Intravenous anaesthetics: e.g. thiopentone, propofol, diazepam, midazolam, ketamine, fentanyl, etc.

During the evaluation of general anaesthetics, it is important to determine the endpoints which are to be assessed as a measure of effects of general anaesthesia. Further for assessing the safety profile of the test drug, its effect on cardiovascular and respiratory parameters also have to be determined.

15.2 Screening of Intravenous Anaesthetic Agents

15.2.1 Evaluation of Intravenous Anaesthetic Agents in Mice

- Small laboratory animals like mice and rats are commonly used for the screening of intravenous anaesthetics.
- The primary endpoint is the estimation of dose which produces loss of righting reflex.
- The time taken for the onset of action and the total duration of action is the secondary parameters that are determined.
- The lethal dose is also estimated for acute toxicity testing of these drugs.
- Procedure:
  - The test agent is injected intravenously through the tail vein of male mice weighing about 18–22 g. When the mouse injected with the test agent is placed on its back and if it is not able to move its head or body and recover from this position within 1 min, then it indicates the loss of righting reflex. The number of animals and the dose at which loss of righting reflex occurs are noted.
  - For acute toxicity testing, the animals are observed for 24 h and the lethality produced is noted.
  - For estimating the onset and duration of anaesthetic action, groups of 20 mice are required. Each animal is placed in an observation cage which is maintained at room temperature of 24 ± 1 °C, devoid of any kind of stimulation during the study period.
  - When the loss of righting reflex occurs in the mouse injected with the test agent, it indicates the onset of anaesthetic action.
  - When the mouse is placed on its back and if it can return to normal position (recovery of righting reflex) within 15 s, then it indicates the recovery of the animal from anaesthesia.
  - The duration of anaesthesia is the time interval between the loss and recovery of righting reflex.
  - From the data obtained, based on the method elaborated by Litchfield and Wilcoxon (1949), the median anaesthetic dose (AD_{50}) and the median lethal dose (LD_{50}) will be determined from the dose-response curves.

15.2.2 Electroencephalographic (EEG) Threshold Test in Rats

- The EEG threshold test was first described by Bolander *et al* for evaluating and comparing the potency of various CNS depressant agents. Korkmaz and Wahlstrom had devised a protocol of EEG burst suppression threshold test for evaluating the sensitivity of the central nervous system (CNS) to intravenous anaesthetics in rats.
- The CNS effects mediated by the anaesthetic agent are analysed by observing EEG changes induced by the test drug.
- Procedure:
  - Adult Sprague–Dawley rats are housed in a controlled environment with a temperature of 23 ± 1 °C and a 12:12 reversed light/dark cycle. The rats are shifted to a tube restrainer, about 24 h before the start of the EEG threshold test. Using appropriate measures and surgical care, electrodes used for EEG recording are sewed to animal's scalp above frontal cortex.
  - The rat is placed on a warm cloth and the test agent is infused from an infusion pump through a needle inserted to the lateral tail vein. The scalp electrodes are connected to the EEG recorder using crocodile clips. The normal EEG recordings of an awake rat, usually are of low amplitude with a frequency of 30 cycles per second.
  - Initially, after infusing the test agent, the amplitude increases and the frequency decreases slightly. During this phase of infusion, jerks or convulsive episodes are observed in the animal based on the test drug. By continuing the infusion, the frequency gets reduced further and is followed by periods of burst suppression along with loss of righting reflex.
  - The threshold criterion is considered to be attained if the burst suppression lasts for 1 s. This threshold period of burst suppression lasting for 1 s is called the '***silent second***'. The time taken to reach the silent second is recorded and the rats are shifted to the recovery room. For calculating the threshold dose of the test agent, the time taken to reach the threshold criterion is multiplied with the dose administration rate of anaesthetic.
  - The optimal dose administration rate of an anaesthetic is obtained by determining the threshold doses at different dose administration rates of the test agent.

15.2.3 Evaluation of Safety of Intravenous Anaesthetics in Rabbits

- Borkowski et al. devised a protocol to test the effect of intravenous anaesthetics on cardiovascular and respiratory parameters in rabbits.
- Procedure:

Table 15.1 The various parameters which are measured in rabbits for evaluating the safety of intravenous anaesthetics

Parameters	The technique of evaluating it
1.Degree of muscle tension	
Jaw tone	Using the index finger, lower law of rabbit is pulled to open
Leg muscle tone	By flexing and extending the right rear leg.
Corneal reflex	A moist cotton swab is placed on the cornea.
Palpebral reflex	A dry cotton swab is used to touch medial canthus.
Ear pinch reflex	A compressive force is applied on the ear using an alligator clip
Pedal withdrawal reflex	A compressive force is applied on the right rear fifth digit at distal phalanx using an alligator clip
2. Cardiopulmonary parameters	
Heart rate	From the tracings of the physiological recorder.
Mean arterial blood pressure	
Respiratory rate	
Respiratory pattern	
Arterial blood pH, the partial pressure of oxygen and partial pressure of carbon dioxide	Through samples for arterial blood gas analysis
3.Rectal body temperature	

– Under halothane anaesthesia, an 18-gauge catheter is inserted into the left carotid artery of adult New Zealand white rabbits to facilitate direct blood pressure recording and to collect the sample for arterial blood gas analysis.
– After a recovery period of 24 h, the rabbit is placed in a sling and to monitor respiratory rate and pattern, a pneumograph is fitted around the animal's caudal thorax at the level of tenth to 12th ribs.
– By determining the cardiovascular parameters and responses to noxious stimuli before and after the infusion of the test drug, each animal will be serving as its control.
– After withdrawing sample for arterial blood analysis, the 18-gauge catheter is connected to a transducer for recording blood pressure.
– Before recording the baseline or control values, an adaptation period of 10 min is given. A 22-gauge catheter is inserted into the right marginal ear vein for administering the test agent. About one-third of the test drug's dose is injected for 1 min. After the rabbit gets relaxed, it is removed from the sling and is placed on a heating blanket in left lateral position.
– The animal's response to noxious stimuli and its degree of muscle tension are measured. It is done by assessing the following parameters, while the rabbit is at the sling and again after induction of anaesthesia at 15 min interval when the rabbit is in left lateral position. (Table 15.1)

15.3 Screening of Inhalational Anaesthetic Agents

- In 1946, for evaluating the efficacy and safety of inhalational anaesthetic agents, Robbins has defined the AD_{50} and LD_{50}.
- When mice are placed in a rotating jar in the presence of a known concentration of anaesthetic, the concentration which causes loss of righting reflex for 15 s in 50% of the animals is considered as AD_{50}.
- The concentration of the test drug which led to apnoea in 50% of the mice within 10 min of administration is considered as LD_{50}. The ratio LD_{50}/AD_{50} is considered as the safety index of the anaesthetic.

15.3.1 Screening of Volatile Anaesthetics in Mice or Rats

- For evaluating the anaesthetic effect of the test agent, rats or mice are exposed to a measured amount of the test agent present inside in a sealed glass container of known volume. Then the volume% concentration of the test agent can be determined based on the density and molecular weight of the volatile liquid.
- Procedure:
  - Male NMRI mice of 20–25 g weight or male Wister rats of 250–300 g weight are used. Oxygen is flushed for 1 min into a wide-mouthed, screw-capped glass jar of 3 l volume. Then through a syringe placed at the bottom, a measured amount of volatile test liquid is calculated and administered into the jar, to yield vapours with 1.25 vol% concentration or in logarithmic multiples of 1.25% (0.63, 2.5, 5,10). The jar is closed and rotated gently for facilitating the evaporation of the test agent.
  - Then either one rat or five mice are quickly placed into the jar and it is closed immediately. The jar is gently rotated for every 15 s and the time taken for loss of righting reflex (induction time) in each animal is noted. This step is continued until all animals inside the jar are anaesthetized.
  - It should be noted that the induction time for each animal must be between $\geq$30 s and $\leq$ 5 min.
  - If the induction time is less than 30 s or more than 5 min, the concentration of test agent is adjusted (either decreased or increased) until the adequate anaesthetic concentration is established.
  - Thereafter the animals are retained inside the jar for 10 min and then they are shifted out to room air for recovery.
  - The time at which each animal can walk or move in an upright position is noted. This time (observed in minutes) indicates the recovery time from anaesthesia for each animal.
  - Post-anaesthetic analgesic effect of test drug is assessed by gently pressing the base of the tail for every minute until the animal has recovered from anaesthesia.

  - The mean induction time and the mean recovery time are calculated and 24 h survival rate is observed for determining the latent toxic effects of the test agent.

15.3.2 Determination of Minimal Alveolar Anaesthetic Concentration (MAC) in Rats

- In 1963, Merkel and Eger coined the term "minimum alveolar anaesthetic concentration" (MAC) and used it as an index for comparing different anaesthetic compounds.
- In 1964, Saidman and Eger defined MAC for humans "as the concentration at which 50% of the patients did not move in response to surgical incision".
- In 1973, Waizer et al. described a method for determining the minimum alveolar concentration of anaesthetic in rats.
- Procedure:
  - Adult Sprague Dawley rats are used for determining MAC of an anaesthetic agent. A gas-tight plastic cylinder with rubber stoppers for closure at both ends is taken and each rat is placed individually in a single cylinder. Two holes are made in the rubber stopper at the distal end of the cylinder. The rat's tail and rectal temperature probe are inserted individually in each of this hole.
  - The anaesthetics are administered using conventional vaporizers and are delivered at an average inflow rate of 1 l/min to each rat through ports present at the proximal end of the cylinder.
  - For determining MAC, the test agent is administered at an initial concentration which allows movement of the rat as a response to noxious stimulation. A tail clamp is applied for 30 s or until the animal moves its head or limbs and then the partial pressure of the test agent is measured using gas chromatography.
  - If the animal moves then the partial pressure of the test drug is increased by 0.2 or 0.3% atmospheres. After equilibrating for 30 min, the tail clamp is applied again and the partial pressure of the test agent is again measured using gas chromatography. This procedure is continued until the partial pressures causing movement-nonmovement of the animal are determined for each rat.
  - MAC is determined as the average of the partial pressures of the test agent which prevented the animal's movement in response to noxious stimuli (tail clamp)
- Other methods to determine MAC
  - In 1975, Davis at al described protocol for determining MAC of halothane in New Zealand white rabbits.
  - In 1988, Eger et al. described protocol for determining MAC of fluorinated anaesthetics in pigs.

15.3.3 Evaluation of Safety of Inhalational Anaesthetics in Rats

- Apart from determining the median effective dose (ED_{50}) of a test agent, it is important to evaluate its effect in cardiovascular and respiratory parameters. Further, it is also important to determine the maximal effective dose (ED_{95}) and minimally lethal dose (LD_5).
- Procedure:
  - Adult male Sprague Dawley rats are used. Each rat is placed in a transparent chamber and through an opening in the chamber, the rat's tail is protruded out.
  - The test agent is administered using Draeger vaporizer and a gas analyser standardised with a mass spectrometer is used to monitor the level of the anaesthetic agent inside the chamber. Rectal temperature is maintained at 37 °C using a heating pad.
  - Each rat is exposed individually to one particular prefixed concentration of test agent for 30 min and then the presence or absence of endpoints of anaesthesia is observed. (Table 15.2)
- After assessing the anaesthetic endpoints, rats are sacrificed and the whole brain is dissected out. The brain tissue concentration of the test agent is determined by gas chromatography.
- The therapeutic safety index (LD_{50}/ED_{50}) and the standard safety margin (SSM) are calculated by the formula Standard safety margin $= [(LD_5 - ED_{95})/ED_{95}] \times 100$
- The percentage by which ED_{95} has to be increased before LD_5 is reached is represented by a standard safety margin.

Table 15.2 The various anaesthetic endpoints which are assessed in rats for evaluating the safety of inhalational anaesthetics

Parameters	The technique of evaluating it
Righting reflex	The animal is placed on its side and if it fails to right itself with all its 4 ft on the floor within 15 s, the loss of righting reflex has occurred.
Purposeful movement in response to noxious stimuli	1 kg weight is placed on the middle of the rat's tail for 1 min and in response to the noxious stimuli, the animal tries to move its head or legs. The presence or absence of this purposeful movement is observed.
Heart rate response	In response to noxious stimuli, an increase in heart rate of >1% occurs. The presence or absence of this increase in heart rate is observed.
Lethal effect	Tracheostomy is done and rats are ventilated at 60 strokes/min using an endotracheal catheter. On mechanical ventilation, the static pressure of 7 mmHg in the femoral artery is considered as the lethal endpoint.

Bibliography

Arendt-Nielsen L. Safety of intravenous and inhalation anaesthetics. In: Vogel HG, Maas J, Hock FJ, Mayer D, editors. Drug discovery and evaluation: safety and pharmacokinetic assays [Internet]. Berlin, Heidelberg: Springer; 2013. [cited 2020 Aug 25]. p. 277–85. https://doi.org/10.1007/978-3-642-25240-2_9.

Bolander HG, Wahlström G, Norberg L. Re-evaluation of potency and pharmacokinetic properties of some lipid-soluble barbiturates with an EEG-threshold method. Acta Pharmacol Toxicol (Copenh). 2009;54(1):33–40.

Kallman MJ. General anaesthetics. In: Hock FJ, editor. Drug discovery and evaluation: pharmacological assays [Internet]. Cham: Springer International Publishing; 2016. [cited 2020 Aug 25]. p. 1593–607. https://doi.org/10.1007/978-3-319-05392-9.

Kissin I, Morgan PL, Smith LR. Comparison of isoflurane and halothane safety margins in rats. Anesthesiology. 1983;58:556–61.

Saidman LJ, Eger EI II. Effect of nitrous oxide and narcotic premedication on the alveolar concentration of halothane required for anaesthesia. Anesthesiology. 1964;25:302–6.

Turner RA. Depressants of the central nervous system. In: Screening methods in pharmacology [Internet]. Elsevier; 1965 [cited 2020 Aug 25]. p. 69–86. Available from: https://linkinghub.elsevier.com/retrieve/pii/B9781483232669500104

Waizer PR, Baez S, Orkin LR. A method for determining the minimum alveolar concentration of anaesthetic in the rat. Anesthesiology. 1973;39:394–7.

Zalucki O, Van Swinderen B. What is unconsciousness in a fly or a worm? A review of general anaesthesia in different animal models. Conscious Cogn. 2016;44:72–88.

Screening Methods for the Evaluation of Antipsychotic Drugs

16

Manu Jose and Mageshwaran Lakshmanan

Abstract

Psychotic disorder represents a spectrum of disorders ranging from delusional and brief psychotic disorders to schizoaffective disorder and chronic schizophrenia. Due to the discovery of various classes of antipsychotic drugs, modern medicine has seen a significant change in the management of various psychotic disorders in the past seven decades. In vitro techniques to study the interaction of the test compound with D_1, D_2, 5-HT, and adrenergic receptors in brain tissue will provide insight into the 'selection of lead molecule.' Similarly, in-vivo behavioral techniques like the golden hamster test, cotton rat test, pole climbing avoidance test, etc., can provide information regarding the antipsychotic efficacy of test molecules. Genetic models of psychosis like DISC-1 transgenic mice, Sandy mice, NR-1H mice, and hooded Wistar rats provide knowledge about the mechanism of action of antipsychotic test molecules under development. This chapter will review the various in-vitro, in-vivo behavioral methods and genetic models used to screen drugs with potential antipsychotic activity.

Keywords

Antipsychotic · Screening · Behavioral test · Golden hamster test · Pole climbing avoidance · Transgenic mice

M. Jose
Department of Pharmacology, Government Medical College, Kozhikode, Kerala, India

M. Lakshmanan (✉)
Department of Pharmacology, Thanjavur Medical College, Thanjavur, Tamil Nadu, India

M. Lakshmanan et al. (eds.), *Introduction to Basics of Pharmacology and Toxicology*, https://doi.org/10.1007/978-981-19-5343-9_16

16.1 Introduction

Psychotic disorder represents a spectrum of disorders ranging from delusional disorders and brief psychotic disorders to schizoaffective disorder and chronic schizophrenia. In the past seven decades, modern medicine has seen a significant change in the management of various psychotic disorders due to discovering and refining various classes of antipsychotic drugs. Though the development of an ideal animal model for evaluating all psychotic disorders is highly impossible, recent research has focused on the molecular level alteration in psychotic disorders using various genetically altered animal models. Thus, the development of anti-psychotic to restore the deranged molecular pathway requires the use of various in-vitro techniques (to mark the lead molecule), behavioral animal models (to identify the efficacy), and various genetic models of psychosis (to establish the mechanism of action). This review summarizes the various techniques and models involved in the screening of drugs with potential antipsychotic activity.

16.2 In-Vitro Studies

16.2.1 D_1 Receptor Assay Using [H]-SCH-23390 in Rat Brain

- The D_1 receptor in various locations of the CNS plays a vital role in the pathophysiology of psychosis. They are as follows:
  - The negative symptoms seen in schizophrenia, namely, lack of motivation, anhedonia, and poverty of speech, are due to decreased activation of the D_1 receptor.
  - Interaction of the D_1 receptor with the NMDA receptor at nucleus accumbens is essential for information-processing, and the same is lost in schizophrenia.
  - Imbalance of D_1 and D_2 receptor due to predominant loss of D_1 receptor concentration in prefrontal cortex causes a cognitive deficit in schizophrenia
- *Principle:* SCH-23390 and SCH-39166 are selective D_1 receptor antagonists. The direct interaction between the test compound and the D_1 receptor is evaluated by estimating the inhibition of binding of the radioactive ligand of SCH-23390/SCH-39166 to the D_1 receptor.
- *Procedure:*
  - Rat brain tissue is homogenized using standard techniques, and the final pellet is suspended in the Tris Buffer (0.05). The ^{3}[H]-SCH23390 is diluted to a 10 nM concentration. For determination of non-specific binding, d-Butaclamol-20 μL is taken. The test compounds are serially diluted (10^{-5}–10^{-8} M) and kept ready.
  - To each test tubes, 50 μl of Tris buffer, 380 μl of water, 20 μl of vehicle/test compound/d-butaclamol, 50 μl of ^{3}[H]-SCH23390, and 500 μl of tissue suspension are added and incubated at 37 °C for 30 min.
  - The assay is stopped using the rapid filtration technique, and the radioactivity is measured using a scintillation counter.

- *Evaluation:*
  - Total binding is estimated by measuring binding of ^{3}[H]-SCH23390.
  - Non-specific binding is estimated by measuring the binding of d-butaclamol.
  - Specific binding is estimated by the difference between total binding and non-specific binding.
  - Percent inhibition: 100–specific binding as a percentage of the control value.
  - Using log-probit analysis, the IC_{50} calculations are performed. At least 3–4 different concentrations of the test compounds in triplicate should be used to determine the IC_{50}.

16.2.2 D_2 Receptor Assay Using ^{3}[H]-Spiroperidol in Rat Brain

- The D_2 receptor in various locations of the CNS plays a vital role in the pathophysiology of psychosis. They are as follows:
  - Blockade of D_2 receptor causes increased glutamate release in striatum (mainly in the right side), leading to reward-motivation and improved cognition in psychosis.
  - D_2 receptor concentration is decreased in the anterior cingulate gyrus and thalamus in psychosis leading to abnormal dopamine transmission from the thalamus to the prefrontal cortex.
  - Increased mRNA concentration of D_2 receptor in the frontal cortex is seen in schizophrenia leading to increased D_2 receptor concentration and supersensitivity to dopamine and can cause the failure of antipsychotic therapy.
  - Blockade of more than 80% of D_2 receptors causes the extrapyramidal syndrome.
- *Principle:* Spiroperidol is a selective D_2 receptor antagonist. The direct interaction between the test compound and the D_2 receptor is evaluated by estimating the inhibition of binding of the radioactive ligand of ^{3}[H]-Spiroperidol to the D_2 receptor.
- *Procedure:*
  - Rat brain tissue is homogenized using standard techniques, and the final pellet is suspended in the M-Tris Buffer (0.05). The ^{3}[H]-Spiroperidol is diluted to 0.25 nM concentration. For determination of non-specific binding, d-Butaclamol-10 mM is prepared and kept ready. The test compounds are serially diluted (10^{-5}–10^{-8} M) and kept ready.
  - The membrane preparations are incubated at 37 °C for 20 min with K/Na phosphate buffer (50 mM), various concentrations of test compounds, and 10 mM of d-butaclamol.
  - The assay is stopped using the rapid filtration technique, and the radioactivity is measured using a scintillation counter.

- *Evaluation:*
  - Total binding is estimated by measuring binding of ^{3}[H]-Spiroperidol.
  - Non-specific binding is estimated by measuring the binding of d-butaclamol.
  - Specific binding is estimated by the difference between total binding and non-specific binding.
  - Percent inhibition: 100–specific binding as a percentage of the control value.
  - Using log-probit analysis, the IC_{50} calculations are performed. At least 3–4 different concentrations of the test compounds in triplicate should be used to determine the IC_{50}.
- Instead of d-butaclamol, other selective D_2 receptor antagonists like Nemonapride (YM-09151-2) and eticlopride can also be used as an alternative.

16.2.3 D_3 Receptor Assay Using ^{3}[H]-Nafadotride in Rat Brain

- The D_3 receptors are mainly autoreceptors. Along with D_2 receptors, they modulate the pre-synaptic autoregulation of dopamine release. The D_3 receptors are present in the ventromedial area of the nucleus accumbens, especially in neurotensin neurons. The D_3 receptors exert a tonic stimulation of expression of neurotensin gene on these neurons, which is essential for the feedback-loop controlling cortical activity involving mood, behavior, and cognition. Alterations in D_3 receptors concentration and gene expression are noted in patients with schizophrenia.
- The 7-OH-DPAT(7-hydroxy-dipropyl-aminotetralin) is the selective agonist for D_3 receptors. Nafadotride and compound-3a (SB277011A) are selective antagonists for D_3 receptors.
- Using the same principle explained in Sects. 16.2.1 and 16.2.2, the interaction of the test compound with D_3 receptors can be estimated.

16.2.4 D_4 Receptor Assay Using ^{3}[H]-L-745-870

- The D_4 receptors are GPCR located mainly in the prefrontal cortex and hippocampus and play a significant role in regulating exploratory behavior, inhibitory avoidance, attention, and cognitive tasks.
- The D_4 receptors concentrations are increased in the striatum of schizophrenic patients. Similarly, the D_4 receptors gene polymorphism has also been documented in patients with various psychotic disorders.
- The compounds WAY-100635 and L-745870 are the selective agonist and antagonist of the D_4 receptors, respectively. On a side note, clozapine (atypical antipsychotic) has ten times higher affinity for D_4 receptors than the D_2 or D_3 receptors.
- Using the same principle explained in Sects. 16.2.1 and 16.2.2, the interaction of the test compound with D_4 receptors can be estimated.

16.2.5 In-Vitro Assay for Other Receptors Involved in the Evaluation of Antipsychotic Action

- Using the principles explained in Sects. 16.2.1 and 16.2.2, the interaction between the test compound with respective receptors can be estimated using an appropriate selective antagonist to that particular receptor. The following Table 16.1 summarizes the various radioligand employed.

16.3 Invivo: Behavioral Tests

16.3.1 Golden Hamster Test

- *Principle:* Golden Hamster develops characteristics fighting behavior when more than ten animals are crowded in the cages. This innate aggressive behavior can be suppressed by the neuroleptics (care should be taken that the dose administered should not impair thc motor function)
- *Procedure:*
  - Ten to twenty male golden hamsters are placed in the cages for 2 weeks duration and observed for the development of fighting behavior. After 2 weeks, when the animal is placed in the 2 l jar, it assumes a squatting/resting position during daytime. At this moment, when the hamster is held by blunted forceps, it utters an angry shriek, throws itself on its back, bites, and pushes the forceps away with its legs. The responses like 'turning,' 'shrieking,' and 'biting' are considered characteristic fighting behavior. Fighting behavior should be screened for all the hamsters before including in the experiment.
  - On the day of the experiment, hamsters are randomly allotted to different groups containing six animals each. Baseline fighting behavior count is noted and recorded.
  - The test drugs are administered via the appropriate route (oral/s.c/intraperitoneally). The fighting behavior is noted by applying the stimuli every 20 min for 3 h. The number of stimuli provided and the number of fighting responses

Table 16.1 In-vitro assay for various receptors involved in the evaluation of anti-psychotic activity

S. No.	Receptor subtype	Radioligand	Reference standard
1	α_1-adrenoceptors	^{3}[H]-Prazosin	Prazosin Phentolamine
2	α_2-adrenoceptors	^{3}[H]-Clonidine	Clonidine
3	β-adrenoceptors	^{3}[H]-CGP12177	Propranolol
4	5-HT_{1A} receptors	^{3}[H]-8-OH-DPAT	8-OH-DPAT
5	5-HT_2 receptors	^{3}[H]-Ketanserin	Ketanserin
6	μ opioid receptors	^{3}[H]-Dihydromorphine	Morphine
7	GABA	^{3}[H]-Muscimol	Muscimol

obtained is noted for each animal. Completing taming (no defense response from hamster) even after punching with the forceps can also occur and should be noted.
  - The animal that underwent complete taming should be screened for 'impairment of motor function' due to the drug. This is done by placing the animal on the 20-degree inclined board and noted for any sliding down. Impairment of motor function, if present, should also be noted.
- *Calculation:* The number of tamed hamsters and hamsters with motor incoordination are recorded for each dose. Using the values, ED_{50} for taming-dose and motor-incoordination are derived, and the ratio between the two provides the 'neuroleptic width.' The drugs with potential neuroleptic activity will have the 'neuroleptic width' of 1.3–1.5.
- *Advantages:*
  - Golden Hamster test can differentiate the sedatives and neuroleptics easily
  - Animal training is not required, and hence experiments can be conducted in a relatively shorter duration.
  - No expensive apparatus is required
- *Disadvantage:* Golden Hamster test shows no difference between taming and impaired motor coordination when tested for anxiolytics with pronounced muscle-relaxant properties.

16.3.2 Cotton Rat Test

- *Principle:* The cotton rat (*Sigmodon hispidus*) has an innate tendency to conceal itself at all times (innate flight reflex). The drugs with neuroleptic activity decrease this reflex. Moreover, side-by-side evaluation of motor coordination allows differentiating neuroleptics from sedatives.
- *Procedure:*
  - Young male cotton rats weighing 35–45 g are placed in a cage (25 × 30 × 20 cm) with a wire lid. An artificial tunnel made of a metal sheet (20 × 7 cm) is placed inside the cage. The tunnel should be shaped like a half-cylinder so that it can be easily detached and moved to other sites inside the cage. As soon as the tunnel is placed, the cotton rat hides immediately into the tunnel. Now the tunnel is lifted and placed in the other cage sites, and the cotton rat should hide again inside the tunnel immediately.
  - The cotton rats are divided into different groups (6 rats/group). The drugs of respective groups are administered via the intended route. After 15 min of drug administration, observation is recorded every 15 min for 3 h.
  - The presence or absence of immediate hiding nature is noted. If the animal did not hide, a stream of air-jet is blown inside the cage to startle the cotton rat, and the response is noted. If the animal is still not hiding even after air-jet, the rat is placed in the 20-degree inclined board and tested for motor coordination.
- *Calculation:* The number of animals that exhibited at least a single suppression of flight response is noted down in each group. The number of animals with motor

incoordination is also noted down for each group. The ED_{50} for suppressing flight response and motor incoordination is then derived, and the neuroleptic width is calculated.

- *Advantages:* The cotton rat test allows differentiation of neuroleptics from other centrally active drugs. Like the 'Golden hamster' test, animal training and expensive apparatus are not required to experiment.

16.3.3 Artificial Hibernation in Rats

- *Principle:* Rats can survive in a poikilothermic state (cold environment with reduced oxygen tension leading to anesthetized/immobilized state) for more than 20 h, and after warming up, a complete recovery is observed. Neuroleptic and opioids can augment the artificial hibernation state in rats.
- *Procedure:*
  - Male Wistar rats weighing 100–150 g are kept under overnight fasting, and the test compounds are administered the next morning 15 min before the start of the experiment.
  - Following drug administration, the rats are placed in ice-cold water for 2 min. The surfactant should be added to the ice-cold water to remove air accumulation inside the fur. The rats are then placed inside the 750 ml glass container with a hermetic seal and placed inside the refrigerator at 2 °C.
  - The container is opened every 10 min for 10 s for 1 h to allow gas exchange. At each time, the rat is examined for signs of artificial hibernation. Artificial hibernation is considered positive when the animal lies on its back with extremities stretched out. Minimal movement may be present due to rigor. Heart rate, respiratory rate, and rectal temperature are reduced.
  - The rat recovers completely when placed at room temperature within few hours.
- *Calculation:* The time taken for the rat to go for artificial hibernation between the control and test group is noted. The number of animals that underwent artificial hibernation with various doses of test drug group and control is also noted. The ED_{50} for artificial hibernation for the test drug is then calculated and compared with standard neuroleptics and opioids.

16.3.4 Catalepsy in Rodents

- *Principle:* Neuroleptics inhibits the nigrostriatal dopamine system leading to catalepsy in rodents. When the rodents fail to correct an unusual posture that was applied externally for a prolonged duration are considered 'presence of catalepsy.'
- *Procedure:*
  - Six male Sprague Dawley rats weighing 120–250 g are selected for each group, and the respective drugs are administered via the intended route.

  - The rat is then placed in the translucent plastic box with 2 cm bedding material. A wooden dowel (10 cm in height) is mounted inside the box. After 15 min of drug administration, the rat is observed to develop cataleptic signs for every 30 min up to 6 h.
  - The shoulder and one forepaw of the rat are lifted and placed on the wooden dowel. When the animal removes the paw, the time duration spent in that unusual posture is noted down, and the rat is repositioned again.
- *Calculation:* When the rat assumes this unusual posture for more than 60 s, it is taken as 'presence of catalepsy,' and the time taken to develop this state is also noted. The number of animals that develop a cataleptic state in each dose of test drug and control group is noted. The ED_{50} of the cataleptic state for the test drug can be evaluated and compared with standard drugs.

16.3.5 Pole-Climb Avoidance in Rats

- *Principle:* When a rodent is provided an unpleasant stimulus, it responds with avoidance and escape at the same time. Neuroleptics at lower doses suppress the avoidance response and at relatively higher doses suppress the escape response. Sedatives, on the other hand, suppress both escape and avoidance responses at the same doses.
- *Apparatus*: Cook's Pole-Climb avoidance apparatus consists of a chamber (25 × 25 × 40 cm) with a sound-attenuating facility. A grid floor is present in the apparatus that can deliver an electric shock. A speaker (2.8 kHz) and light (28 V) are situated at the top of the chamber. A stainless steel pole of 2.5 cm diameter is suspended through a hole in the center of the chamber's roof. The pole is suspended using a counterweight balance, and it is linked to the microswitch. When the pole is pulled down to 3 mm by weight of 200 g (rat's weight), the microswitch is activated, and the response is recorded. The sound from the speaker is used as a conditioned stimulus, and electric shock is used as an unconditioned stimulus.
- *Training of animals:* All the rats subjected to the experiment should be trained for 10 days. The rats can adapt for 1 min inside the chamber and get trained on avoiding the electric shock by responding to the conditioned stimuli (sound). Initially, the conditioned stimuli (sound) are given for 4 s, followed by sound +electric shock (1.5 mA) for 26 s. The trial should be terminated if the rat responds (by jumping on the pole) or after 30 s, whichever is earlier.
- *Animals and Groups:* Long Evans male rats are used in this experiment. The rat should weigh more than 200 g so that it can activate the micro-switch in the apparatus while jumping to the pole. Six rats per group are allotted.
- *Parameters Observed:*
  - The jumping of a rat on the pole as soon as it hears a sound (conditioned stimuli in the first 4 s) is called conditioned response (avoidance response). This indicates that the rat has learnt that electric shock follows the conditioned

stimuli (sound), and it tries to avoid it (avoidance response) by jumping to the pole.
  - When the rat jumps to the pole during the sound+electric shock period, then the response is called the unconditioned response or escape response.
  - The latency time for both the conditioned (avoidance response) and unconditioned (escape response) is noted for all rats in different groups for 10 days.
  - The number of avoidance and escape failures is noted in each group for 10 days.
- *Calculation:* The ED_{50} for avoidance response and escape response for the test drug is calculated and compared with standard neuroleptics and sedatives. In case of neuroleptics, the avoidance response is affected at lower doses, and the escape response is affected at the higher dose. In case of sedatives, both the avoidance and escape responses are reduced at the same dose.

16.3.6 Brain Self-Stimulation Test in Rats

- *Principle:* Electrical stimulation of loci in rat's brain produces pleasurable effects, and these effects produced by minute external electric pulse are positively reinforced by observing operant behaviors like 'pressing a lever' in rats. Neuroleptic drugs reduce/block self-stimulation.
- Male Wistar rats are selected, and an electrode is placed in the median forebrain bundle of the hypothalamus using stereotactic surgery. After ten of recovery, the rats are trained for pressing the lever for obtaining the pleasurable effects. The experiment should be started only after obtaining a consistent baseline response (which indicates that the animal has learnt to press the bar for pleasurable effect).
- The drugs are administered to respective groups, and the number of attempts of pressing the lever is noted for each group. The values are compared with the baseline value. Using various doses of the test drug, ED_{50} for the reduction in self-stimulation is calculated and compared with standard neuroleptics.

16.3.7 Pre-Pulse Inhibition of Startle Response in Rats

- *Principle:* Sensory filtering/gating is a reduced response by an organism to the non-threatening stimuli from the environment. Pre-pulse inhibition is one of the operational measures of 'sensory gating' and was developed for neuropsychiatric research in the 1980s. In Schizophrenia, impairment in the ability to filter out irrelevant sensory stimuli is present; hence schizophrenic patients have reduced 'pre-pulse inhibition' than normal subjects. This experiment applies the same observation in rats to identity the neuroleptic activity of the test compound. 'Pulse' represents 'startling response' administered as sound (120 dB). Pre-pulse represents a weak, non-startling sound (3 or 5 or 10 dB above background noise). When a pre-pulse is administered 30–500 ms before pulse, the magnitude of startling response is low compared to the 'pulse' alone. Dopamine agonists (psychomimetic) facilitate pre-pulse inhibition, and neuroleptics antagonize the pre-pulse inhibition.

- *Procedure:* Male Sprague Dawley rats are administered with respective group drugs, and after 10 min of drug administration, the rat is placed in the startling chamber for 5 min acclimation. The startling chamber consists of a closed box with a speaker mounted 24 cm above the floor. A preprogrammed piezo-electric-accelerometer in the chamber measures the magnitude of startling response as 'degree of motion.' Each rat is tested for four separate occasions with a non-test duration of 7 days. The startle amplitude for 'prepulse-pulse pair' and 'pulse only' is measured and recorded.
- *Evaluation:*
- Percentage of Pre-pulse inhibition = (Startle amplitude after prepulse-pulse pair)/(startle amplitude of pulse only) × 100
- Using different doses of the test drug, the ED_{50} of pre-pulse inhibition is calculated for the test drug and compared with standard neuroleptics.
- A direct comparison of the pre-pulse inhibition value of different groups can be made using ANOVA.

16.3.8 Other Behavioral Tests Used

- Tests like N40 Sensory gating, latent inhibition test in rats (described in detail in Chap. 20), foot shock-induced aggression are also employed to determine the neuroleptic activity of test compounds.

16.4 In vivo Test Based on Mechanism of Action

- Various in-vivo methods for evaluating the neuroleptic activity of test compounds have been developed over the years and are used successfully in research. Table 16.2. summarizes the *in-vivo* tests based on the mechanism of action.

16.5 Genetic Models of Psychosis

16.5.1 Dominant-Negative DISC-1 Transgenic Mice

- Disruption of the DISC-1 gene produces increased susceptibility to schizophrenia.
- DISC-1 gene-altered transgenic mice exhibits characteristic behavior of psychosis-like disturbance in sensory-motor gating, hyperactivity, anhedonia, and olfactory-associated behavior disturbances.
- Enlarged lateral ventricles of the brain are noted in DN-DISC1 mice, similar to anatomical changes in the brain of schizophrenic patients.

Table 16.2 In-vivo methods to evaluate the neuroleptic activity based on mechanism of action

S. No.	Test name	Principle	Procedure	Calculation
1	Amphetamine group toxicity in mice	Amphetamine being an indirect sympathomimetic, increases catecholamine. Grouping of mice also increases catecholamine and causes the death of mice within 24 hours. Neuroleptics reduce the death rate in mice groups.	• 10 mice/group (male sex) • Group: Vehicle, amphetamine, amphetamine +test drug, amphetamine+Neuroleptic, • After drug administration, all mice of the same group are placed in a glass jar (18 cm diameter) • Mortality is assessed at 1, 4, and 24 h of dosing	• 80% mortality will be seen usually in amphetamine treated group • Using different doses of the test drug, the ED_{50} for protection effect is calculated and compared with neuroleptics
2	Inhibition of amphetamine stereotype in rats	Amphetamine being an indirect sympathomimetic increases catecholamine and induces typical stereotactic behavior in rats like continuous sniffing, chewing or licking, and gnawing that can be prevented by neuroleptics.	• 6 rat/group (male sex) • Group: Vehicle, amphetamine, amphetamine + test drug, amphetamine+Neuroleptic • After drug administration, each rat is placed in a separate cage and observed for 60 min	• The number of rats that developed behavior in each group is noted, and the percentage of protection is calculated. • Using different doses of the test drug, the ED_{50} for protection effect is calculated and compared with neuroleptics
3	Inhibition of apomorphine climbing in mice	Apomorphine (dopamine agonist) induces peculiar climbing behavior (initial rearing followed by full-climbing activity) due to mesolimbic dopamine stimulation. Neuroleptics antagonize this behavior in mice.	• 10 mice/group (male sex) • Group: Vehicle apomorphine, apomorphine +test drug, apomorphine + Neuroleptic • After drug administration, each mice is observed for climbing behavior • Scoring: 0–All 4 paws on floor; 1-forepaws holding vertical bar; 2–all paws are holding the vertical bar	• The number of mice that developed climbing behavior in each group is noted, and the percentage of protection is calculated • Using different doses of the test drug, the ED_{50} for protection effect is calculated and compared with neuroleptics

(continued)

Table 16.2 (continued)

S. No.	Test name	Principle	Procedure	Calculation
4	Inhibition of apomorphine stereotype in rats	Apomorphine induces typical stereotactic behavior in rats like continuous sniffing, chewing or licking, and gnawing due to the striatal dopaminergic system. Neuroleptics inhibits this behavior	• 6 Wistar rat/group (male sex) • Group: Vehicle, apomorphine, apomorphine +test drug, apomorphine+Neuroleptic, • After drug administration, each rat is placed in a separate cage and observed for 60 min for every 10 min	• The number of rats that developed behavior in each group is noted, and percentage of protection is calculated. • Using different doses of the test drug, the ED_{50} for protection effect is calculated and compared with neuroleptics
5	Yawing/penile erection syndrome in rats	Dopamine agonists like apomorphine induce penile erection-yawning syndrome in rats. This syndrome is antagonized by neuroleptics	• 6 rat/group (male sex) • Group: Vehicle, apomorphine, apomorphine +test drug, apomorphine + Neuroleptic • Neuroleptics/test drug/vehicle is administered first to the respective group. After 30 min, apomorphine is injected. • Upright position after repeated pelvic thrust, engorged penis indicates penile erection. Rat may lick penis and eat the ejaculate. • The slow, wide opening of the mouth indicates yawning	• Number of yawning and penile erection occurred in each rat in all group is recorded • The number of rats in each group that did not develop the yawing/penile erection behavior is recorded. • Using different doses of the test drug, the ED_{50} for protection effect for yawing/penile erection behavior is calculated and compared with neuroleptics
6	Inhibition of mouse jumping	L-DOPA causes dopaminergic overstimulation in amphetamine pretreated mouse leading to a jumping response. Neuroleptics abolish this behavior.	• 6–10 CD-1 mice/group (male sex) • Test compound/Vehicle/ Neuroleptic are injected into their respective group • After 60 min, the mouse is injected with amphetamine, and after 15 min, L-DOPA is injected	• The number of mice that developed jumping behavior in each group is noted, and percentage of protection is calculated. • Using different doses of the test drug, the ED_{50} for protection effect is calculated and compared with neuroleptics

			intraperitoneally • The mouse is observed for the next 60 min	
7	Inhibition of Dizocipline (MK-801) induced behavior in mice	Dizocipline, an NMDA antagonist, produces locomotion and falling behavior in mice via dopamine-dependent and independent pathways. Neuroleptics inhibit this stereotypic behavior.	• 6–10 CD-1mice/group (male sex) • Group: Vehicle, Dizocipline, test compound + dizocipline, Neuroleptics+dizocipline. • Test compound/Vehicle/ Neuroleptic are injected into their respective group. • After 30 min, the mouse is injected with dizocipline. • After 15 min, the mouse is observed for falling and locomotion behavior for 1 h	• The number of mice that developed characteristic behavior in each group is noted, and percentage of protection is calculated. • Using different doses of the test drug, the ED_{50} for protection effect is calculated and compared with neuroleptics
8	Phencyclidine model of psychosis in rats	Phencyclidine, an NMDA antagonist, induces psychotic behavior in rats, evident by hyperlocomotion (correlates with positive symptoms of schizophrenia) and social isolation (correlates with negative symptoms of schizophrenia. Neuroleptics attenuate these behaviors.	• 6 Sprague Dawley rat/group (male sex) • Group: Vehicle, Phencyclidine, Phencyclidine +test drug, Phencyclidine+Neuroleptic, • Rats are allowed to acclimatize in an actophotometer. • Test compound/Vehicle/ Neuroleptic are injected into their respective group. After 30 min, the mouse is injected with Phencyclidine. • After 15 min, the mouse is observed for stereotypic behavior for 1 h	• The number of ambulations, fine movements, and rearing are counted for each rat in all group • Parameters are compared between the group using ANOVA.

(continued)

Table 16.2 (continued)

S. No.	Test name	Principle	Procedure	Calculation
9	Purposeless chewing in rats	Chronic administration of neuroleptics can induce purposeless chewing behavior in rats via dopaminergic-nicotinic mechanisms. Cholinergic agonist also induces the same behavior	• 6 rats/group (male sex) • Groups: Vehicle, Test compound, Sulpiride(neuroleptic) • Respective group drugs are administered and observed for chewing for 30 min	• The number of chewing is recorded and compared between the groups using ANOVA
10	Inhibition of apomorphine-induced emesis in dogs	Apomorphine, a dopamine agonist, produces emesis in dogs. Neuroleptics antagonize the emesis induced by apomorphine (by central dopamine blockade)	• 3–9 Beagle dogs/group (adult male sex) • Group: Vehicle, apomorphine, Test compounds + apomorphine, Neuroleptic + apomorphine. • Test compound/Vehicle/ Neuroleptic are injected into their respective group. • After 30 min, apomorphine is administered and observed for emesis (retching movement with opening of the mouth) or successful ejection of stomach content	• The number of dogs that developed characteristic emesis behavior in each group is noted, and percentage of protection is calculated. • Using different doses of the test drug, the ED_{50} for protection effect is calculated and compared with neuroleptics • This test fails to differentiate anti-emetic and neuroleptics effectively
11	Single unit recording of substantia nigra and ventral tegmental area in rats	An increase in the number of spontaneously firing dopaminergic neurons of substantia nigra(SN) and ventral tegmental area(VTA) occurs after acute treatment with neuroleptics. Repeated treatment with neuroleptics for 21 days decreases the same.	• 6 Wistar rat/group (male sex) • Group: Acute-Vehicle, Acute- test compound, Acute-neuroleptic, Chronic-vehicle, Chronic-Test compound, Chronic-neuroleptic. • Acute study observation starts after 2 h of drug administration, and chronic study observation starts after 2 h on the 21st day.	• A neuron is considered dopaminergic if ‘triphasic’ (positive-negative-positive) spike (amplitude = 0.4–1.5 mv; duration = 2.5 ms) occurs • Neurons firing an irregular pattern (3–9 Hz with occasional burst) with progressive increase of spike duration and reduction of spike amplitude is also considered

			• After 2 h, using stereotactic surgical guidance, by lowering electrode via the 12 tracks in SN and VTA, spontaneously firing dopamine neurons are estimated. • Extracellular neuronal sampling is also performed to count dopaminergic neurons. • Light microscopic findings after rat sacrifice are also obtained	dopaminergic • Parameters between the groups are compared using ANOVA
12	In Vivo Voltammetry	Monoamine metabolism and release can be monitored by an electrochemical technique called '*in-vivo voltammetry*' using stereotactically implanted carbon fiber micro-electrodes and a mini-optoelectronic system.	• 6 Sprague Dawley rats/group (male sex) • Group: Vehicle, test compound, Neuroleptic • Under anesthesia, using stereotactic surgery, micro-electrodes are placed under the left/right nucleus accumbens and contralateral anterior striatum. Auxillary and reference electrodes are placed on the dura • Recordings from the micro-electrodes are made via an optoelectronic system • Drugs of respective groups are administered, and voltammeter recordings are made again	• Absolute peak heights from the various electrodes are recorded before and after drug administration. • Absolute peak heights between the before and after drug are compared using paired 't' test and between the groups using the unpaired 't' test. • Percentage changes from the baseline peak height value are also compared.
13	Model for tardive dyskinesia (to identify ADR of neuroleptic)	Chronic administration of neuroleptic produces characteristic vacuous chewing movement (robust, purposeless chewing) and	• 6 Male Wistar/group (male sex) • Group: Vehicle, Neuroleptic, test compound • Drugs of the respective group	• Oral dyskinesia are quantified for 15 min • Comparison between the groups is made using ANOVA.

(continued)

Table 16.2 (continued)

S. No.	Test name	Principle	Procedure	Calculation
		tongue protrusion in rodents – an indicator of ADR of neuroleptics	are administered for 4 weeks. If haloperidol is used as a standard neuroleptic, it is administered once a week for 4 weeks. • Characteristics vacuous chewing movement or tongue protrusion are observed.	

16.5.2 $G_s\alpha$-Transgenic Mice

- The $G_s\alpha$-Transgenic mice have point mutation that prevents hydrolysis of bound GTP in GPCR. This transgene expression is mainly seen in the hippocampus, striatum, and cortex but not seen in the brainstem, thalamus, and cerebellum.
- The $G_s\alpha$-Transgenic mice show the reduced amplitude of cortically generated N40 that is similar to the schizophrenia phenotype.

16.5.3 Sandy Mice or Dysbindin-1 Mutant (DTNBP-1) Knockout Mice

- The DTNBP-1 gene is essential for coding dystobrevin-binding protein-1 that regulates the dopaminergic and glutamatergic transmission in the hippocampus.
- Reduction in DTNBP-1 expression is seen in schizophrenic patients.
- Knocking out of the DTNBP-1 gene produces disturbances in pre-pulse inhibition, social interaction, and altered memory processing in mice similar to schizophrenic patients.

16.5.4 NR-1 Hypomorphic (NR-1H) Mice

- NR-1 is an essential subunit of NMDA receptors, and reduced expression of this protein severely affects the functioning of NMDA receptors that mediates glutamate transmission.
- The NR-1 hypomorphic mouse expresses only $<10\%$ of normal NR-1 protein, leading to abnormal glutamate transmission in the brain. The resultant NMDA receptor hypofunction leads to social and cognitive abnormalities similar to schizophrenia.

16.5.5 Heterozygous Reeler Mouse

- Reelin protein is essential for neuronal development, dendritic spine development, and synaptic plasticity of GABAergic neurons in the hippocampus.
- Reeler mouse exhibits a 50% reduction in expression of reelin protein leading to accumulation of NADPH-phosphate in subcortical white matter.
- Neophobic behavior on elevated plus maze and decrease in pre-pulse inhibition is noted in reeler mouse suggesting the psychosis vulnerability.

16.5.6 Other Genetic Animal Models of Psychosis

- *NCAM-180 knockout mice:* It has increased lateral ventricle size in the brain and reduced pre-pulse inhibition similar to schizophrenic patients

- *Hooded Wistar rats:* It shows deficient pre-pulse inhibition and is employed in psychosis research
- *COMT transgenic mice:* Loos of COMT activity lead to increased dopamine action and produces psychosis-like features.
- *GAT1 knockout mice:* It exhibits hyperactivity, impaired novel object recognition, and latent inhibition due to loss of GABA transporter-1 protein (GAT-1)

Bibliography

Brisch R, Saniotis A, Bielau H, Bernstein H-G, Steiner J, Bogerts B, et al. The role of dopamine in schizophrenia from a neurobiological and evolutionary perspective: old fashioned, but still in vogue. Front Psychiatry. 2014;5:47.

Forrest AD, Coto CA, Siegel SJ. Animal models of psychosis: current state and future directions. Curr Behav Neurosci Rep. 2014;1(2):100–16.

Koszła O, Targowska-Duda KM, Kędzierska E, Kaczor AA. In vitro and in vivo models for the investigation of potential drugs against schizophrenia. Biomolecules [Internet]. 2020;10(1):160.

Maramai S, Gemma S, Brogi S, Campiani G, Butini S, Stark H, et al. Dopamine D3 receptor antagonists as potential therapeutics for the treatment of neurological diseases. Front Neurosci. 2016;10:451. https://doi.org/10.3389/fnins.2016.00451.

McArthur SG. Dopamine receptor subtypes, physiology and pharmacology: new ligands and concepts in schizophrenia. Front Pharmacol. 2020;11:1003. https://doi.org/10.3389/fphar.2020.01003.

Vogel H. Psychotropic and neurotropic activity. In: Vogel HG, editor. Drug discovery and evaluation. Berlin, Heidelberg: Springer; 2007. https://doi.org/10.1007/978-3-540-70995-4_6.

17 Screening Methods for the Evaluation of Sedative-Hypnotics

Manu Jose and Mageshwaran Lakshmanan

Abstract

Sedatives cause calmness, drowsiness, and reduced motor coordination, while hypnotics induce/maintain sleep. Due to the discovery of various classes of sedatives and hypnotics, modern medicine has seen a significant change in the management of insomnia, epilepsy, and anxiety. In vitro techniques to study the interaction of the test compound with $GABA_A$, $GABA_B$, Histaminergic, and adrenergic receptors in brain tissue will provide insight into the 'selection of lead molecule.' Various in-vivo techniques to study motor activity like the open field test, hole-board test, and method of intermittent observation provide information regarding the sedative efficacy of test molecules. Similarly, in-vivo tests like the incline phase test, chimney test, and Rotarod test delivers data for the effect of the test compound on muscle coordination. EEG registration in conscious cats, automated sleep system analysis in rats and experimental insomnia in rats provide data for the efficacy of test compounds on hypnotic activity. This chapter will review the various in-vitro, in-vivo behavioral methods and genetic models used to screen drugs with potential sedative and hypnotic activity.

Keywords

Sedative · Hypnotic · Screening · Rotarod method · Inclined phase test · Open field test

M. Jose
Department of Pharmacology, Government Medical College, Kozhikode, Kerala, India

M. Lakshmanan (✉)
Department of Pharmacology, Thanjavur Medical College, Thanjavur, Tamil Nadu, India

M. Lakshmanan et al. (eds.), *Introduction to Basics of Pharmacology and Toxicology*, https://doi.org/10.1007/978-981-19-5343-9_17

17.1 Introduction

The classic observations noted with sedative drugs are subduing of excitement, calmness, mild drowsiness, and decreased motor coordination. On the other hand, the hypnotics are the drug that induces/maintains sleep. It is now well-known that modulation of GABA receptors, histamine receptors, and serotonergic receptors can induce/promote sleep. Thus various in-vitro methods have been developed to assess the activity of testing drugs on these receptors before proceeding to the in-vivo testing. The screening tests for sedative/hypnotic activity can be classified as follows:

- In-vitro studies:
  - $GABA_A$ receptor binding measurement in rat brain
  - $GABA_B$ receptor binding measurement in rat brain
  - [^{3}H]-Mesulergine binding assay for $5HT_{2C}$ receptors
  - [^{3}H]-Mepyramine binding assay for H_1 receptors
- In-vivo studies -Tests for sedative activity:
  - Test for motor activity and behavior:
    Method of intermittent observation
    Open field test
    Hole-Board test
    Combined open field test
  - Test for muscle coordination:
    Inclined phase test
    Chimney test
    Grip strength test
    Rotarod method
- In-vivo studies -Tests for hypnotic activity:
  - Potentiation of hexobarbital sleep time in rats
  - Experimental insomnia in rats
  - EEG registration in conscious cats
  - Automated rat sleep system analysis

17.2 In-vitro Studies

17.2.1 $GABA_A$ Receptor Binding Measurement in Rat Brain

- *Principle:* Radiolabelled [^{3}H]-Muscimol and [^{3}H]-GABA are used as a ligand for measuring the $GABA_A$ receptor binding activity. Unlabeled GABA or bicuculline methiodide is used to measure the non-displaceable binding (blank). The amount of specific radioligand binding to the $GABA_A$ receptor is estimated by subtracting blank from the total binding. Fixed concentration (10 or 50 μM) of the test compound is challenged with [^{3}H]-Muscimol binding. The competitive binding nature of the test compound with $GABA_A$ receptor is obtained by deriving the K_i value of the test compound that competes with [^{3}H]-Muscimol.

- *Procedure:*
  - The frozen rat brain is treated with 50 mM Tris citrate buffer and homogenized using a tissue homogenizer. The protein concentrate (0.5 mg/ml) is obtained after multiple centrifuge-washing cycles.
  - Three borosilicate glass culture tubes on ice are taken. To the first tube, 4 nM of [^{3}H]-Muscimol is added with buffer making the volume of 1 ml. This is to measure the total radioligand binding. To the second test tube, 4 nM of [^{3}H]-Muscimol and 200 μM of bicuculline methiodide is added with buffer (1 ml volume) to measure the non-displaceable binding. In the third test tube, 4 nM of [^{3}H]-Muscimol and various concentrations of unlabeled test drug is added with buffer (1 ml volume).
  - It is best practice to perform the assay as duplicate or triplicate. Tissue suspension of 1 ml volume is added to each test tube, gently mixed using a vortex, and incubated at 4 °C for 30 min. After this, the binding reaction is terminated, and radioactivity is quantified using a Scintillation counter.
- *Calculation:*
  - The count per minute (cpm) data obtained from the Scintillation counter is converted to [^{3}H]-Muscimol-bound and plotted against added [^{3}H]-Muscimol.
  - Using the Scatchard plot of bound ligand Vs bound/free ligand, the dissociation constant (K_d) and maximum binding values (B_{max}) can be estimated.

17.2.2 $GABA_B$ Receptor Binding Measurement in Rat Brain

- Recent studies have discovered the role of $GABA_B$ receptors in the maintenance of the sleep cycle. Baclofen and 2-OH-saclofen are well-established $GABA_B$ receptor agonists and antagonists, respectively.
- The principle and calculation are the same as described in Sect. 17.2.1. Different buffers (Triton X-100 + Tris citrate) and agonists ([^{3}H]-Baclofen) are used here. The final protein concentrate achieved is 1 mg/ml.

17.2.3 [^{3}H]-Mesulergine Binding Assay for $5HT_{2C}$ Receptors

- The 5-HT_{2C} receptor plays a vital role in sleep induction and regulation in humans. The 5-HT_{2C} receptor binding assay can be performed using the human 5-HT_{2C} receptor membrane.
- Principle and calculation are similar to Sect. 17.2.1.

- Procedure: 50 nM Tris-Hcl buffer is used to dilute the human 5-HT_{2C} receptor membrane. 180 μL of membrane solution is added to the test solution and 10 μl of 1 nM [^{3}H]-Mesulergine (standard agonist for a 5-HT_{2C} receptor) and incubated for 60 min. Similar to Sect. 17.2.1. the binding reaction is terminated, and count per minute data is obtained by the Scintillation counter.

17.2.4 [^{3}H]-Mepyramine Binding Assay for H_1 Receptors

- Histamine is one of the vital neurotransmitters that have a physiological role in sleep besides serotonin and GABA.
- Like other receptors, the H1 receptor binding capacity can be estimated for the test drug by using radiolabelled [^{3}H]-Mepyramine and radiolabelled histamine in rat astrocytes cells or rat brain tissue.

17.3 In-Vivo Tests for Sedative Activity

17.3.1 Effects on Motility (Sedative or Stimulatory Activity)

- *General Considerations:*
  - Spontaneous motor activity in mice or rats can provide insight into the sedative or stimulatory nature of the test drugs.
  - Locomotion, grooming, eating, sniffing, rearing, and drinking are some of the spontaneous motor activities noted in rodents.
  - Thigmotaxis (the tendency of the rodents to remain close to the walls of the cage) and curiosity are also the recognized features of animal locomotor activity. Currently, automated equipment is being employed to measure the locomotor activity of rodents.

17.3.1.1 Method of Intermittent Observations

- This method was designed by Ther in 1953 using mice of either sex.
- Twelve mice are taken and divided into four groups of three mice each. One group serves as a control, and different doses of investigational drugs are administered to the other three groups.
- The influence of circadian rhythm can be avoided by conducting the experiment between 08:00 and 12:00 p.m. Test drugs are injected intraperitoneally. After 10 min, groups of mice are injected with 0.5 mg/kg methamphetamine (CNS stimulant). If no CNS stimulant action is seen, then the test drug has potential sedative activity.
- Observation should be made every minute for 1 h for any characteristic changes in grooming, sniffing, locomotion, and rearing. The control group (treated only with methamphetamine) has counts between 120 and 150. Reduction in the count in the test group indicates the potential sedative activity of the drug.

- *Advantage:* It is a reliable test for sedative drugs and compounds with central depressant activity, such as antihistaminics, neuroleptics, and hypnotics.
- *Disadvantage:* A skilled and trained observer is needed to get reproducible results. Therefore, various automated systems have been developed.

17.3.1.2 Open Field Test

- *Principle:* Interruption of light beams by rats or mice in a cage (open field) is used to assess general motor activity, locomotion, rearing, and locomotion speed.
- A square open field arena is constructed with two rows of eight photocells (infrared light-sensitive) placed 4 and 12.5 cm above the floor. Each photocell is spaced 9 cm apart, and the last photocell is located 2.5 cm from the wall. The animal's movement is recorded automatically by interruption of the photocell beams, and the following parameters can be evaluated.
  - In lower rows, the 6 interruption of photo beams indicates motor activity.
  - In upper rows, the interruption of photo beams indicates rearing.
  - In lower rows, 6 the interruption of photocells in the same direction indicates animal locomotion.
  - The time interval between successive photo beam interruptions indicates the speed of locomotion.
  - In lower rows, activation of photo beams along with the last photocell (spaced 2.5 cm from wall) indicates peripheral motor activity
  - In upper rows, interruption of photo beams and the last photocell (spaced 2.5 cm from wall) indicate peripheral rearing.
- Generally, an adult male rat (Sprague Dawley) is used for this experiment. The test drugs are administered subcutaneously 10–40 min before the test.
- Instead of infrared light beams, several automated systems are now using video analysis of constant camera capturing method, magnetic field techniques, and using radiant body heat detection by the sensors.

17.3.1.3 Hole-Board Test

- Open field test does not evaluate the 'curiosity' behavior of the anima. Thus, Boisser and Simon, in 1964, modified the open field setup by providing holes in the board. The rodents poke their nose into the holes that reflect the 'curiosity' 7 behavior.
- Currently, automated nose poking calculation can be done using the infrared light photocell technique.
- Usually, six animals are used for each test drug dose in test groups and six animals in control groups. After 30 min from the administration of test drugs, observations are made for5 min.
- Suppression of nose-poking activity is reduced in benzodiazepine-treated animals.

17.3.1.4 Combined Open Field Test

- The open field setup is combined with a hole board technique to measure the locomotion and curiosity at the same time. Wiescher first developed it in 1976. Currently, several automated setups are available to meet the purpose.
- Interruption of photocell beams inside the cage (count for motility) and interruption of photocell beams outside the cage (curiosity – nose poking) can be determined simultaneously.
- Certain stimulants can reduce curiosity but increases locomotion. Conversely, tranquilizers at a low dose can reduce curiosity without affecting locomotion. Hence, a combined open field test can identify the dissociation of effect between the locomotion and curiosity by test drug.

17.3.1.5 EEG Analysis from Rat Brain by Telemetry

- Using the radio-electro-encephalography technique, field-potential analysis from different brain parts of freely moving rodents can be performed. This radio-EEG has been accepted as a sensitive method for locomotion.
- Four stainless steel implants (which carry micro plug) are fixed in the base plate, frontal cortex, reticular formation, and thalamus of the adult male rat. After 2 weeks of surgery, using a four-channel radio transmitter, the field potentials are analyzed in real-time.
- The effect of test drugs can be compared with baseline and known stimulants or sedatives very quickly in real-time.

17.3.2 Tests for Muscle Coordination

17.3.2.1 Inclined Plane

- This method was developed by Allmark and Bachinski in 1949. This method is used to test drugs with muscle relaxant activity and differentiate neuroleptics from other centrally active drugs.
- Two rectangular plywood boards are connected using a hinge. One board acts as a base while the other act as an inclined plane. A mat made of rubber with 2 mm thickness is fixed to the inclined plane (set at 65°).
- Male Charles River mice are used in this experiment. After 30, 60, and 90 min of drug administration, mice are kept in the upper part of the inclined plane. The ability of the mouse to hang for 30 s or fall off is noted.
- The time at which maximum performance deficit is noted is taken as a peak time of the test drug.
- Advantage: Simple assay for muscle relaxant activity assessment
- Disadvantage: Not sensitive to differentiate anxiolytics from neuroleptics and neurotoxins.

17.3.2.2 Chimney Test

- Boissier introduced the 'chimney test' in 1960 for testing the effects of tranquilizers and muscle relaxants on motor coordination.

- A 30 cm glass cylinder is taken, and a mark is made at 20 cm from the base. Keeping the glass cylinder in the horizontal position, a mouse is allowed to enter near the mark at the end of the tube. When the mouse reaches the mark, the tube is tilted to the vertical position.
- The mouse now tries to climb out backward with the help of muscle coordination. This movement is similar to 'mountain and rock climbers' who pass 'chimneys' in the mountain.
- When the test drug has a potential muscle relaxation effect, the mouse could not climb out of the cylinder.
- The dose in which 50% of animals fail to climb out backward within 30 s provides the ED_{50}.

17.3.2.3 Grip Strength Test

- Grip strength test (described first by Boisssier and Simon in 1960) is used to evaluate muscle strength that can be influenced by sedatives, muscle relaxants, and toxins.
- Mice are used in this experiment. The setup consists of a thin wire frame supported with a metallic rod and connected to transducers. The mouse is suspended above the wire mesh by the tail and slowly dragged over the wire mesh. With their forepaws, the mouse will now grab the wire mesh and exerts a pulling force.
- The pulling force exerted by the mouse is reflected in the transducer, and the values are noted down. The mouse can be maintained in the same posture for not more than 5 s.
- The animals are tests for every 15 min after drug administration for 2 h. After 2 h, the animals are observed in their cages for their behavior. The changes in the grasping reflex are valid only when the animal shows normal behavior and motility in the cages after the test.
- Using control and test groups (10 animals in each group); the pull-force can be compared between the groups. Similarly, the percentage of loss of grasp reflex can also be compared between the groups.
- A minor modification of this method is that the mouse is allowed to grasp a metallic grid with its forepaw while being suspended by its tail. Gradually weights are added to the metallic grid, and the maximum weight that causes the animal to lose its grip can be noted and compared.

17.3.2.4 Rotarod Method: (Refer to Chap. 42)

17.4 In-vivo Tests for Hypnotic Activity

17.4.1 General Considerations

- The definition for 'hypnotic' differs between humans and animals. In man, hypnotics are the drugs that induce/maintains normal sleep (can be aroused) without any hangover effects. However, in animals, hypnotics translates into drugs that cause 'deeper central depression' with loss of righting reflex and muscle tone due to unconsciousness.
- Thus results/conclusions obtained from animal models for hypnotics have lesser extrapolation value for human use. As biochemical events occurring during sleep are multiple and complex in-vitro methods are not available. Most of the sedatives in higher doses produce hypnosis, and most of the hypnotics in lower doses produce sedation

17.4.2 Potentiation of Hexobarbital Sleeping Time

- Tranquilizers, sedatives, hypnotics, and high-dose anti-depressants are proven to prolong single-dose hexobarbital induced sleep. Loss of righting reflex is the significant criterion measured in this method.
- Since mice have rapid metabolic elimination of hexobarbital, they are used in this experiment.
- Control and test groups have ten mice each. After administration (30 min in case of i.v and 60 min in oral), righting reflex is monitored.
- Duration of loss of righting reflex is estimated and compared with the control group.
- Disadvantages: Compounds that inhibit liver hexobarbital metabolism can also prolong the time of hexobarbital anesthesia. This can be avoided by using barbital instead of hexobarbital (as liver microsomal enzymes do not metabolize barbital)

17.4.3 Experimental Insomnia in Rats

- A proposed suitable animal model for insomnia in patients is 'foot-shock induced insomnia in rats. This method was first described by James and Piper in 1978.
- After preparing rats for EEG and EMG, the rats are implanted with two disc nuchal and four silver epidural electrodes and allowed to recover for 10 days. EEG and EMG are recorded via polygraph.
- On the first day of the experiment, a vehicle is administered, and non-stress recordings are obtained. On the second day, the vehicle is administered again, and foot shock (0.5 mA pulse of 15 ms width for the 30 s at 1 Hz) is delivered for

8 h. Thirty minutes interval is provided between each shock period. The test drug is administered on the third day, and the same recordings are obtained for 8 h.
- Alteration in the sleep-wake cycle is reflected as an increase in slow-wave sleep I and a decrease in slow-wave sleep II and paradoxical sleep.
- Barbiturates and BZDs have been shown to oppose the changes in this model.
- Disadvantage: Expensive and time-consuming

17.4.4 EEG Registration in Conscious Cats

- Similar to the above technique explained in Sect. 17.4.3. electrodes are placed in multiple areas of the cat's brain. After recovery from surgery, freely moving cats are studied for the effects of hypnotics on the sleep pattern using EEG tracings.
- Placement of electrodes and wires:
  - Bipolar subcortical electrodes in dorsal hippocampus, amygdala, caudate nucleus, and reticular formation.
  - Cortical screw electrodes in medial suprasylvian, lateral suprasylvian, ectosylvian gyri, and anterior suprasylvian gyri.
  - Two steel wires (Teflon coated) in cervical neck muscles.
- After recovery from surgery, cats are placed in an experimental chamber, and all wires are connected to the socket. Recordings are made for 96 h (4 days). To avoid the 'first-night' effect in the sleep-wake cycle, experimental drugs are given on the third or fourth day.
- EEG, cervical muscle tone, and multiple-unit activity in reticular formation are recorded and analyzed for slow-wave sleep, wakefulness, and REM sleep. Changes made by the test drugs in the above parameters can be easily compared.

17.4.5 Automated Rat Sleep Analysis System

- This method was first described by Ruigt in 1988. This is an automated rat-sleep classification system that helps to classify psychotropic drugs into antidepressants, stimulants, and antipsychotics.
- Animals are subjected to 'electrode placement' surgeries in three places. Thus, three signals from parieto-occipital EEG, movement indicator signal, and nuchal EMG are obtained over an extended period from several rats.
- From this system, six sleep-wake stages are distinguished in rats. Amongst six stages, two are waking stages, and four are sleeping stages.
  - 'Active waking'—movement, high EMG, and theta activity
  - Quiet waking—no movement.
  - Quiet sleep—EEG spindles present
  - Deep slow-wave sleep—Prominent delta activity
  - Pre-REM sleep—low EMG, spindles against background theta activity
  - REM sleep—low EMG with theta activity.

- Usually, 32 rats are included and divided into eight rats per group. Drugs should be administered at the beginning of the 'light cycle' of rats. Two to threeweeks of wash-out time should be allowed after each experiment.
- Using several parameters from the hypnogram, the effect of the test drug on the sleep-wake cycle can be analyzed and compared with a placebo or standard drug.

Bibliography

DEWS PB. The measurement of the influence of drugs on voluntary activity in mice. Br J Pharmacol Chemother. 1953;8(1):46–8.

Enna SJ, McCarson KE. Characterization of GABA receptors. Curr Protoc Pharmacol. 2013;2(63): 1.7.1–1.7.20.

Hishinuma S, Sato Y, Kobayashi Y, Komazaki H, Saito M. Intact cell binding for in vitro prediction of sedative and non-sedative histamine H1-receptor antagonists based on receptor internalization. J Pharmacol Sci. 2008;107(1):66–79.

Joung H-Y, Kang YM, Lee B-J, Chung SY, Kim K-S, Shim I. Sedative-hypnotic and receptor binding studies of fermented marine organisms. Biomol Ther (Seoul). 2015 Sep;23(5):479–85.

Kallman MJ. Effects on behavior and muscle coordination. In: Hock F, editor. Drug discovery and evaluation: pharmacological Assays. Cham: Springer; 2016a. https://doi.org/10.1007/978-3-319-05392-9_26.

Kallman MJ. Hypnotic activity. In: Hock F, editor. Drug discovery and evaluation: pharmacological assays. Cham: Springer; 2016b. https://doi.org/10.1007/978-3-319-05392-9_29.

Screening Methods for the Evaluation of Anxiolytic Drugs

18

Gandham Ravi and Madhavi Eerike

Abstract

Once a new drug target is found and validated, it is then followed by the identification and optimization of the lead compound. Animal models using rats and mice will help us better understand both the pharmacological and toxicological properties of the lead compound. This is particularly important in the areas like neuroscience, which is facing many challenges to come up with the new drug molecules to treat various central nervous system disorders including anxiety, before testing them in human beings. Anxiety disorder is one of the most common health issues across the globe, affecting the quality of life significantly. Animal models for screening of the anxiolytic drugs are mainly categorized as models for conditional and unconditional responses, which will help us understand the etiology, neurobiology and there by the treatment of human anxiety disorders. There are many animal models available as a part of pre clinical studies for anxiety disorders using the rodents, with a good predictive, face and construct validity. Though these screening methods for anxiolytic drugs, using rodents act as very useful tools, new models are under development which will provide a better alternative.

Keywords

Animal model · Anxiolytic · Anxiety disorder

G. Ravi (✉) · M. Eerike
Department of Pharmacology, All India Institute of Medical Sciences (AIIMS) Bibinagar, Hyderabad, Telangana, India

M. Lakshmanan et al. (eds.), *Introduction to Basics of Pharmacology and Toxicology*, https://doi.org/10.1007/978-981-19-5343-9_18

18.1 Introduction

Pre-clinical studies in the form of animal experiments are a well-known way of understanding biological concepts of anxiety disorders. They play an essential role in understanding the spectrum of causes leading to psychological disorders and provide the basis for the early phases of drug development. Mice are most commonly used among the rodent family due to the following reasons:

- Easy to housing and can be accommodated in a given space
- Relatively easier to breed
- Mice genome is well studied

The preclinical studies on animals will help us understand the etiological basis, symptoms and management of anxiety in human beings, there by providing information on how the drugs used to treat anxiety in hums work and to come up with newer anti-anxiety medications. At the same time, a suitable animal model for anxiety should have the following features:

- Predictive validity—the model would decrease the anxiety behavior when treated with an anxiolytic drug
- Face validity—behavioral changes produced in the animal in response to the stressful stimulus would be similar to the changes observed in human beings
- Construct validity—mechanisms causing these behavioral changes in the form of anxiety should be similar to the human beings

Currently, there are various pre-clinical methods available to evaluate the anxiolytic compounds in the animals and are explained as follows.

18.2 Light/Dark Exploration Test

- This is a test for the unconditioned response, which means the animal is not exposed to nociceptive but just to a stressful stimulus. Routinely used rodent test to study the unconditioned anxiety-like behavior. Carried on the principle of approach-avoidance conflict between the drive to explore novel areas and an aversion to anxiety triggering conditions.
- The rodents show natural aversion to mild triggers like highly lit areas, open spaces, and a new ambiance. In this method, the animal is placed in a box that has a small dark portion and a large brightly illuminated portion (Fig. 18.1). Usually the animal prefers to stay in the dark area than in the over illuminated area, and the anti-anxiety drug will increase the exploratory behavior in the animal.
- This test can be used for studying both the anxiety-relieving and anxiety-provoking properties of the drugs in the animal. Behavioral measures commonly used to assess anxiety-like behavior are time spent in the light compartment, rearing, and the latency to emerge into the light. At the same time, total distance

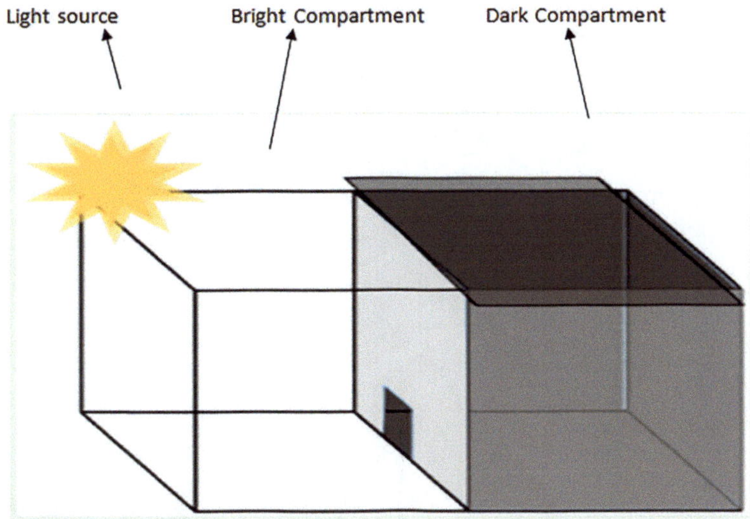

Fig. 18.1 Bright/Dark compartment box (Image source: https://www.creative-biolabs.com/drug-discovery/therapeutics/light-dark-box-test.htm)

travelled, and distance travelled in the dark compartment are used to assess the locomotor activity, and differences in baseline Light/Dark behavior and the response to anxiogenic drugs are measured.

18.3 Elevated Plus Maze Test

- It is a well-known and most commonly used model for testing the anti-anxiety drugs in the animal and has good predictive validity. This test also evaluates an unconditioned response.
- The system contains (Fig. 18.2) open arms and closed arms, which are elevated 50 cm above the floor. The animal is given access to all of the arms and is allowed to move freely between them. The behavior testing room should be soundproof and should not be over/under illuminated.
- This test is based on the principle of approach-avoidance conflict, where the animal is considered "less anxious" when it prefers to stay and explore the open arm and more anxious if it prefers to stay and explore the closed arm of the apparatus.
- When the animal is separated from its habituated habitat and kept in the apparatus, it will prefer to stay in the closed arm avoiding the open arm entry because it will perceive the former as safer than the latter portion. The anti-anxiety drugs will reverse this behavior. Parameters like total distance covered, the total number of entries into each arm, time spent by the animal in each arm, and the percent of entries into the open arms are calculated before and after administration of the

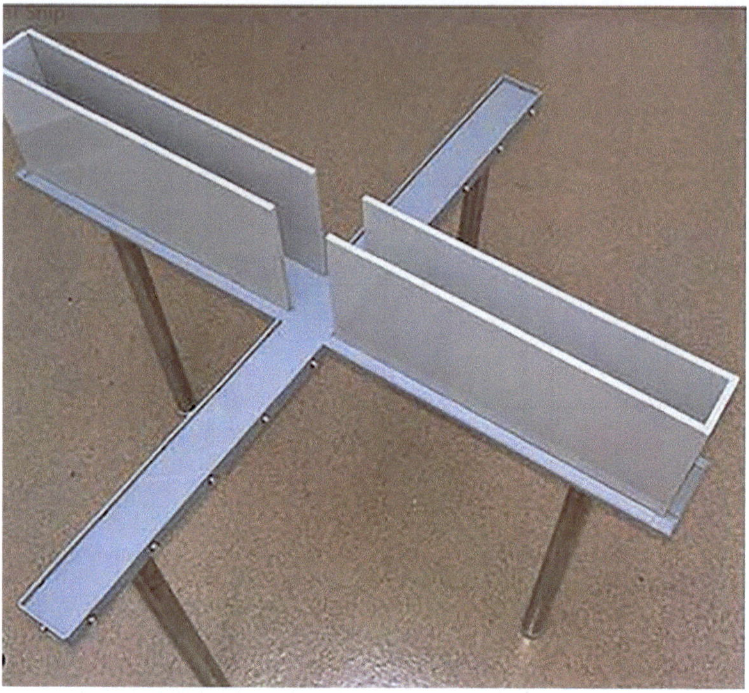

Fig. 18.2 Elevated plus maze (Image source: https://www.frontiersin.org/articles/10.3389/fnbeh.2017.00013/full)

testing drug. This test also can be used for studying both the anxiety-relieving and anxiety-provoking behavior in the animal.

18.4 Four-Plate-Test

- This is a test that evaluates conditioned response, in which the animal is exposed to a stressful and nociceptive stimulus. In this method, the animal is placed in an instrument that contains metal plates of similar size and shape forming the floor and is connected to a shock generator (Fig. 18.3).
- Usually, the animal will try to explore the new environment by crossing from one plate to the other. Whenever the animal crosses the plate, a mild electric shock will be applied, which is called the 'punished crossing.' The anti-anxiety drugs will increase the number of punished crossings.

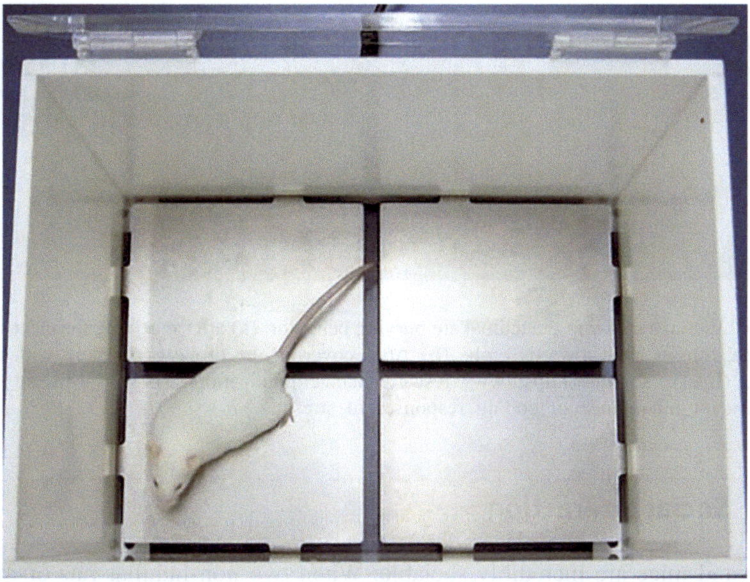

Fig. 18.3 Four Plate test (Image source: https://www.bioseb.com/en/anxiety-depression-disorder/588-aron-test-or-four-plates-test.html)

18.5 Fear Potentiated Startle Test

- It is a test of conditioned response. Based on the principle that when suddenly given a stimulus with sound, rats will respond by startle reflex in the form of closing the eyes and stiffening of the neck, which is potentiated if the acoustic stimulus is combined with an electric shock.
- The anti-anxiety drugs will inhibit the potentiation of the startle reflex.

18.6 Defensive Burying Test

- This test elicits an animal's response through two primary defensive strategies, flight or fight to a threatening or noxious object, and it is a test of conditioned response. Burying is a species-specific behavior of the animal in which it throws substances like sand in response to a threat stimulus.
- This response can be induced experimentally by placing the animal in a glass chamber with sawdust on the floor (Fig. 18.4). An electric probe is placed into the chamber through a hole which is used to deliver the shock. When the animal investigates the object, it receives a brief electric shock from the probe. Following the first shock with the probe, the animal will try to bury it by throwing the bedding material onto it. The anti-anxiety drugs will reduce the burying response of the animal or increase the number of contacts with the probe.

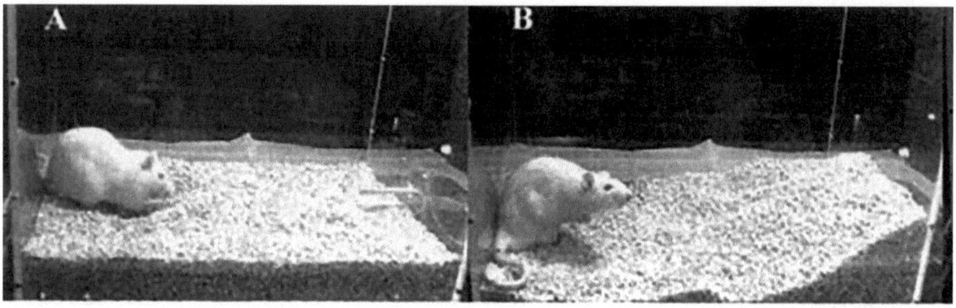

Fig. 18.4 Pictures showing the animal are burying behavior: (**a**) after receiving the electric shock, the animal throws sand towards probe (**b**) After covering the probe completely with the sand material. (Image source: https://www.researchgate.net/publication/8919857_X-linked_and_lineage-dependent_inheritance_of_coping_responses_to_stress)

18.7 Social Interaction

- Rats and mice are incredibly sociable. When two non-familiar rats or mice are placed in an open area, they show some behavioral activities like pouncing, chasing, social grooming, crawling over/under, charging, boxing, wrestling, pinning, anogenital sniffing, and biting.
- This social interaction is inhibited when the animals are placed in a new ambiance with bright illumination. The anti-anxiety drugs will increase the duration of social interaction between the animals in a novel environment to which they are not used to. This test is of an unconditioned response.

18.8 Separation-induced Ultrasonic "Distress" Vocalization

- This test elicits an unconditioned response. It is well known that the rodents produce ultrasonic vocalizations (USVs) following any stress-inducing stimulus that affects the emotional status of the animal. The frequency of these USVs varies from pups to adult animals. Recording of these USVs is commonly used to test drugs like anxiolytics.
- Three kinds of USVs are identified based on the age of the rodent. Rat pups produce USVs of 40-kHz when isolated from their parent rats, and anxiolytic drugs will decrease the generation of these stress-induced vocalizations by the pups. The same method is applied to other age group animals too.

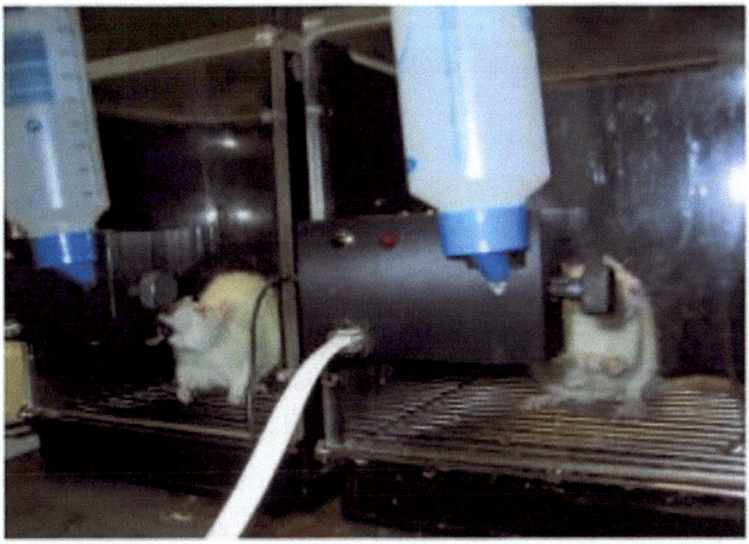

Fig. 18.5 Vogel apparatus (Image source: https://www.slideshare.net/utkarshalok/anxiolytic-screening-modelsutkarsh)

18.9 Vogel Conflict Test

- This is a conditioned test based on approach-avoidance conflict. The animal is placed in a plexy glass box to which attached one small compartment to arrange a water bottle containing a metallic tube extending into the large compartment through which the animal can drink water by licking (Fig. 18.5). The entire instrument has a grid floor, and electric shock will be applied through the drinking tube and the grid floor.
- The animal is allowed to drink through the tube, and shock is applied after every 20th lick, over a period of 3 min. This procedure is repeated before and after administering the anti-anxiety drug, and the total number of shocks animals received over 3 min were compared. The anxiolytic agent will increase the number of punishing shocks received by the animal.

18.10 Staircase Test

- **Rearing is an exploratory behavior of the animal wherein it** puts its weight on the hind legs, raises the forelimbs from the ground, and extends the head upwards.
- In the staircase test, the animal is placed in a closed environment with a small staircase inside (Fig. 18.6). The animal's rearing and stair climbing behavior is observed over a period of 3 min before and after giving the test drug. The anti-anxiety drugs will reduce the rearing behavior while increasing stair climbing.

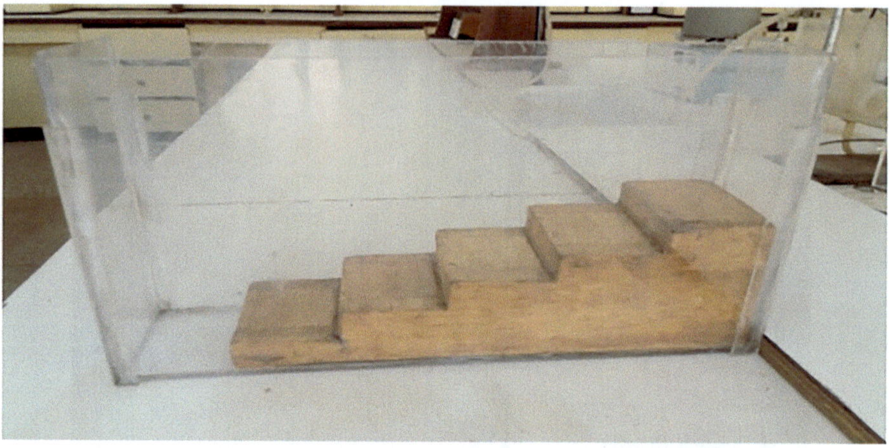

Fig. 18.6 Staircase test for anxiolytic activity (Image source: Abubakar AR, Haque M. Anxiolytic Testing of Medicinal Plants in Nigeria: Frequently Used Experimental Models. Medeniyet Medical Journal.2019;34(1):83–98 doi:10.5222/MMJ.2019.83604)

18.11 Hole-Board Test

- It is a test of an unconditioned response. The apparatus consists of a wooden board with holes placed at the same distance from each other in all directions.
- Animal's behavior like rearing, head-dipping into the holes, time taken for the first dip and the pace at which the animal dips head is observed. An agent with anti-anxiety properties will reduce the time taken for the first dip and increase the number and duration of the dips to the holes while decreasing the rearing behavior.

18.12 Open Field Test

- It is a test to evaluate unconditioned response. The animals are deprived of food for 20–24 h, after which they are placed in the center of the chamber with an open field, and food is provided in a small Petri dish.
- The total number of animals taking food and time taken for the first feed are noted before and after administering the test drug. The drug with anti-anxiety properties will decrease the time taken for first food intake and increase the number of animals taking the food.

18.13 Conclusion and Future Perspectives

- Even though the animal screening methods for anxiolytic drugs have a reasonable validity, the pace at which these drugs are being developed is very slow due to the fact that evaluation of these agents in clinical trials is very difficult.
- New models like Zebra Fish have been emerging as an important alternative to animal models studying central nervous system disorders like anxiety.

Bibliography

Boer SF, Koolhaas JM. Defensive burying in rodents: ethology, neurobiology, and psychopharmacology. Eur J Pharmacol. 2003;463(1–3):145–61. https://doi.org/10.1016/s0014-2999(03)01278-0.

Bourin M. Animal models for screening anxiolytic-like drugs: a perspective. Dialogues Clin Neurosci. 2015;17(3):295–303.

Bourin M, Hascoët M. The mouse light/dark box test. Eur J Pharmacol. 2003;463(1–3):55–65. https://doi.org/10.1016/s0014-2999(03)01274-3.

Hascoët M, Bourin M, Colombel MC, Fiocco AJ, Baker GB. Anxiolytic-like effects of antidepressants after acute administration in a four-plate test in mice. Pharmacol Biochem Behav. 2000;65(2):339–44. https://doi.org/10.1016/s0091-3057(99)00191-4.

Komada M, Takao K, Miyakawa T. Elevated plus Maze for Mice. J Vis Exp. 2008;(22):1088. https://doi.org/10.3791/1088.

Ohl F. Animal models of anxiety. Handb Exp Pharmacol. 2005;169:35–69. https://doi.org/10.1007/3-540-28082-0_2.

Rex A, Voigt JP, Voits M, Fink H. Pharmacological evaluation of a modified open-field test sensitive to anxiolytic Drugs. Pharmacol Biochem Behav. 1998;59(3):677–83.

Shekhar AI, McCann UD, Meaney MJ, et al. Summary of a National Institute of Mental Health workshop: developing animal models of anxiety disorders. Psychopharmacology. 2001;157:327–39.

Simiand J, Keane PE, Morre M. The staircase test in mice: a simple and efficient procedure for primary screening of anxiolytic agents, Psychopharmacology. 1984;84:48–53.

Simola N. Rat ultrasonic vocalizations and behavioral neuropharmacology: from the screening of drugs to the Study of Disease. Curr Neuropharmacol. 2015;13(2):164–79. https://doi.org/10.2174/1570159X13999150318113800.

Takeda H, Tsuji M, Matsumiya T. Changes in head-dipping behavior in the hole-board test reflect the anxiogenic and/or anxiolytic state in mice. Eur J Pharmacol. 1998;350:21–9.

Treit D, Engin E, McEown K. Animal models of anxiety and anxiolytic drug action. https://doi.org/10.1007/7854_2009_17.

Vanderschuren LJMJ, Niesink RJM, Ree JMV. The neurobiology of social play behavior in rats. Neurosci Biobehav Rev. 1997;21(3):309–26.

Vogel JR, Beer B, Clody DE. A simple and reliable conflict procedure for testing anti-anxiety agents. Psychopharmacologia (Berl). 1971;21:1–7.

Winslow JT, Insel TR. Serotonergic and catecholaminergic reuptake inhibitors have opposite effects on the ultrasonic isolation calls of rat pups. Neuropsychopharmacology. 1990;3(1):51–9.

Screening Methods for the Evaluation of Antiparkinsonian Drugs

19

Mageshwaran Lakshmanan

Abstract

Damage to dopaminergic neurons in the substantia nigra is well-established pathophysiology of Parkinson's Disease (PD); hence, drugs targeting (directly or indirectly) dopaminergic pathways have been investigated and used nowadays. In vitro techniques to study the interaction of the test compound with D_1, D_2, cholinergic, and serotonergic receptors in brain tissue will provide insight into the 'selection of lead molecule.' Various drug-induced in-vivo models for PD have been established using reserpine, methamphetamine, and neuroleptic and provide information regarding the anti-parkinsonism efficacy of test molecules. Similarly, toxin-induced in-vivo PD models using MPTP, rotenone, paraquat, and 6-OHDA delivers data for the effect of the test compound on neuroprotectivity in PD. Genetic models of PD like Parkin mouse, DJ-1 transgenic mice, LRRK2 knock-in mice, and mitopark mice provide information regarding the pathophysiology of PD and modulation of the deranged pathway by the test compound. This chapter will review the various in-vitro, in-vivo behavioral methods and genetic models used to screen drugs with potential anti-parkinsonism activity.

Keywords

Antiparkinsonism drugs · Toxin models · Reserpine · 6-OHDA · MPTP · Paraquat · Mitopark · Parkin mice

M. Lakshmanan (✉)
Department of Pharmacology, Thanjavur Medical College, Thanjavur, Tamil Nadu, India

M. Lakshmanan et al. (eds.), *Introduction to Basics of Pharmacology and Toxicology*, https://doi.org/10.1007/978-981-19-5343-9_19

19.1 Introduction

Millions of elderly worldwide are being affected by Parkinson's disease (PD) every year. Damage to dopaminergic neurons in the substantia nigra is well-established pathophysiology of PD; hence, drugs targeting dopaminergic pathways directly or indirectly have been investigated and successfully used nowadays. Besides classical motor deficits, many autonomic and psychological disturbances are being well recognized in PD nowadays. Multiple newer targets and genes for PD, like ARRK2, Dj-1, etc., have been studied in-depth, and newer drug targets that modulate these pathways are being evaluated. Thus selecting an ideal animal model for PD plays a crucial role in drug screening. This chapter reviews the various in-vitro and in-vivo techniques involved in screening drugs with potential anti-PD activity.

19.2 In vitro Techniques

19.2.1 Primary Microglial Cultures

- Microglia (brain-resident macrophages) is involved in the plasticity and maintenance of CNS neurons. Inflammation, an essential component of PD's pathophysiology, converts ramified, healthy microglia to 'round-amoeboid', 'hyper-ramified-stress primed,' and 'dark-dystrophic' microglia. When PD progresses, reduced phagocytosis, surveillance nature, and increased oxidative stress markers are noted in microglia.
- The cerebral cortex of rat pups (two days old) is used to produce primary microglial cultures. After cell separation using standard methods, the cell suspension is cultured in Minimum Essential Medium (MEM) for two to three days. Penicillin and streptomycin are added for the prevention of contamination. By gentle shaking, microglia cells are detached and collected. Astroglia (about 15%) are unavoidable contaminant with 85% of microglia during collection. Using poly-lysine coated well containing GM-CSF, microglia multiplication is enriched, and purity is achieved.
- These microglia cells are then assessed for neurodegenerative processes and activation of the inflammatory process in the presence of test drugs and various inflammatory stimuli like LPS, INF-γ, TNF-α, and ILs.
- Prevention of morphological changes after applying inflammatory stimuli in the presence of test drugs in culture can also be assessed.

19.2.2 Animal Striatal Slices

- Release of dopamine and ACh can be assessed in the animal striatal slices as the striatum is the primary structure affected in PD.

- After decapitation and opening of the skull, right and left striata are removed from the male Sprague Dawley rats and maintained in Ice-Cold Kreb's solution with 95% O_2, 5% CO_2 at 37 °C.
- The slices are then incubated with radio-labeled dopamine and choline. After standard washing and stabilization techniques, the slices are subjected to field stimulation (10–15 mA/cm^2 current strength; 2 ms pulse duration and 3 Hz frequency) for 5 min.
- The testing drug is added in the medium, and information about the inhibition, reuptake, or stimulation of release can be assessed.

19.2.3 Dopamine Receptors Assessment by Radioligand Binding

- For assessing antagonistic activity at D_1 receptors, [^{11}C]-halobenzapine (also called [^{11}C]-SCH23390), [^{11}C]-NNC112, [^{11}C]-NNC756, and [^{11}C]-NNC687 are used. [^{11}C]-SKF75670 and [^{11}C]-SKF82957 are the radioligand used for assessing the agonistic activity at D_1 receptors.
- For assessing antagonistic activity at D_2 receptors, [^{3}H]-raclopride, [^{3}H]-Spiperone, and [^{3}H]-nemonapride are commonly used. [^{11}C]-MNPA is the radioligand used for assessing the agonistic activity at D_2 receptors.
- Striatal tissues from the Sprague Dawley rats are used in this method. After tissue homogenization and centrifuge of the tissues, the radioligand was for the respective receptor study was incubated in the presence and absence of another standard antagonist and investigational drug at room temperature for 60 min.
- Specific binding is assessed by the difference between radioligand binding in the absence and presence of standard antagonist and test drug.

19.2.4 Dopamine Transporter Assessment by Radioligand Binding

- As explained in Sect. 19.2.3, the effect of the investigational drug on the dopamine transporter can also be assessed using radioligand.
- [^{3}H]-WIN35428 is the common radioligand for the dopamine transporter. Striatal tissues (mainly nucleus acumbens and putamen) from Sprague Dawley rats are homogenized and centrifuged.
- After stabilization with standard protocol, the radioligand is added and incubated at 4 °C for 120 min. Specific binding is assessed by the difference between radioligand binding in the absence and presence of standard antagonist and test drug.

19.2.5 Dopamine Release from Synaptosomes

- Animal brain tissue, after homogenization and fractionation based on their density or size, a subcellular fraction called 'synaptosomes' are obtained. In

synaptosomes, presynaptic portions of the synapse are fused with a large fraction of 'resealed postsynaptic entities.' Thus, synaptosomes can provide information about neurotransmitters, ion entry and exit, and membrane polarization.
- The release of dopamine and Acetylcholine in the presence of an investigational agent can be assessed using the synaptosomes for potential benefit in PD.
- Slices of the frontal cortex, striatum, and olfactory tubule from the mouse brain are obtained after standard dissection technique with forceps. Using the standard and convenient 'sucrose density method,' the synaptosomes are prepared.
- Synaptosomes are incubated with different concentrations of test drugs in the presence and absence of standard agonists and antagonists for 5 min at 37 °C. After this, 0.5 μCi of radiolabelled [3H]-dopamine is added and incubated for another 5 min at 37 °C.
- The degree of binding and details about the release or inhibition of dopamine from synaptosomes can be assessed by measuring radioactivity using scintillation counting.

19.2.6 Assessment of Neuroprotective Efficacy

- Neuroprotection is one of the critical components in the prevention of the progression of PD. The various existing antiparkinson drugs like selegiline, rasagiline offer neuroprotection.
- Human neuroblastoma cells are selected for assessing the neuroprotective effect of the investigational drug. After standard culturing techniques, the neuroblastoma cells are incubated with various concentrations of investigational drugs for 30 min.
- All the neuroblastoma cells in the culture plates are now subjected to stress by incubating with 100–300 μM hydrogen peroxide for 6 h.
- By performing a colorimetric MTT assay for cell viability, the neuroprotective efficacy of the investigational drugs can be assessed.

19.3 In vivo Models

- The in vivo animal models of PD:
  - provides insight into understanding the pathophysiology,
  - correlate the symptoms of PD with lesions,
  - evaluate the possible molecular mechanisms involved in the evolution of PD and test the potential drugs for PD.
- However, it should be noted that no animal model of PD has been able to reproduce all the pathophysiology of PD yet.
- The animal models of PD are classified into the following types:
  - Drug-induced PD models
  - Toxins induced PD models
  - Genetic PD models

19.4 Drug-Induced Animal Models of PD

19.4.1 Reserpine Model of PD

- *Principle:* The central catecholamine stores are depleted by reserpine via blocking vesicular reuptake and produce signs and symptoms similar to PD. Motor disorders will be apparent after 20 min of administration. Sedation, marked hypokinesia, hind limb rigidity, ptosis, and fixed facial expressions are seen within 1–2 h of administration.
- The reserpine induces signs and symptoms are drastically improved by administration of L-DOPA, leading to the discovery of 'dopamine depletion' as one of the primary mechanisms of PD.
- *Assessment:* Using a wooden box with the floor divided into 16 equal squares, the hypokinesia can be assessed in rats by counting the squares entered every 2 min for 2 h. By using, grasping test apparatus, muscular rigidity can be assessed. Grooming episodes and rearings are also observed and documented. The number of incidences of tongue protrusion is taken to assess the orofacial dyskinesia (equivalent to tardive dyskinesia in humans). The degree of catatonia is also assessed using an arbitrary scale by challenging all four paws of rats at various heights.
- *Advantage:*
  - An effective method in simulation of symptoms of PD.
  - Better prediction of in vivo pharmacological activity of investigational drugs.
- *Disadvantages:*
  - Reserpine produces only amine depletion without any neuronal damage. Hence, the neuroprotective effect of the test drug could not be assessed.
  - Depletion of neurochemical occurs only temporarily, and nigral dopamine levels are not affected effectively.
  - Depletion of other monoamines leads to 'confounding' in the observation.

19.4.2 Methamphetamine Model of PD

- *Principle:* Similar to reserpine, methamphetamine also causes dopamine depletion by acting on the dopamine transporter.
- Following the administration of methamphetamine, the motor symptoms and signs of PD occur similar to reserpine. Similar to reserpine, methamphetamine does not cause loss of nigral dopamine neurons.
- The methamphetamine model is an acute model for PD and does not reproduce the chronic pathological changes in PD.

19.4.3 Neuroleptics Induced PD Models

- *Principle:* Trifluoperazine, chlorpromazine, and thioproperazine (phenothiazine with piperazine side chain) are administered to produce dopamine antagonism in the nigrostriatal pathway and produce PD symptoms.
- Marked catatonia, muscular rigidity, and bradykinesia are well noted with neuroleptic administration in rodents.
- *Limitation:* Only the motor component of PD can be assessed effectively. Sensory and autonomic disturbances in PD cannot be assessed with this model.

19.4.4 Cholinomimetics Induced PD Models

- *Principle:* Cholinergic hyperactivity due to loss of inhibitory-dopaminergic nigrostriatal neurons is a well-known fact. Thus, administration of cholinomimetic drugs like tremorine, oxotremorine, arecoline, carbachol, and physostigmine into a rat's brain can be used to produce signs like ataxia, spasticity, lacrimation, salivation, and tremors.
- The test drug should be administered 1 h before the administration of cholinomimetics, and the efficacy of the test drug is noted by its ability to prevent the PD signs mentioned above.
- The following are the observational period for each sign:
  - Tremors → 10 s period every 15 min for 1 h
  - Salivation and lacrimation → at 15th and 30th minute
  - Hypothermia → variation from the basal temperature at the first, second and third hour
- The observational values are compared with the control (Normal saline), and standard Anti-PD drug (Benzatropine) administered animals.
- Oxotremorine gives reliable results than the other cholinomimetics drugs. The pH of the cholinomimetics should be balanced to avoid brain necrosis. Preparation of solution in CSF buffer is advised to avoid this issue. Minimal volume usage (0.5–2 ml) and slow administration should be followed to avoid brain tissue displacement.

19.5 Toxin Induced PD Models

19.5.1 MPTP Models in Monkeys

- *Principle:* The environmental toxin N-methyl-4-phenyl-tetrahydropyridine (N-MPTP) is converted to 1-methyl-4-phenylpyridine (MPP*) by the brain MOA-B activity resulting in the formation of reactive free oxygen radicals. Being lipophilic, MPTP readily crosses the blood-brain barrier, is metabolized to MPP* in astrocytes, and is stored in dopaminergic neurons via dopamine transporters. MPP* interferes with the mitochondrial electron transport chain

Table 19.1 Evaluation of PD symptoms in MPTP monkey models

		Score				
S. No.	Parameter	0	1	2	3	4
1	Movement	Normal	Reduced	Sleepy	–	–
2	Checking movement	Present	Reduced	Absent	–	–
3	Attention and blinking	Normal	Abnormal	–	–	–
4	Posture	Normal	Abnormal trunk	Abnormal trunk and tail	Abnormal trunk tail and limbs	Flexed posture
5	Balance and coordination	Normal	Impaired	Unstable	–	Falls
6	Reactions	Normal	Reduced	Slow	Absent	–
7	Vocalizations	Normal	Reduced	absent	–	–

(specifically complex I) in the dopaminergic neurons resulting in neurotoxicity due to energy depletion and oxidative damage.

- *Relevant history:* In 1983, Langston et al. reported the Parkinsonism-like features observed in the cluster of intravenous opioid drug abusers. Later it was identified that the opioid (methyl-phenyl-propinoxy-piperidine) was contaminated with the MPTP (accidental side product), resulting in Parkinsonism.
- The MPTP model is the most widely used model for PD, and Monkeys and mice are commonly used. Rats are not used and are an inferior animal model for PG because rats are naturally resistant to MPTP. However, MPP* can be administered directly into the rat's brain stereotaxically to induce PD.
- Usually, eight adult rhesus monkeys weighing 5–8 kg are administered with MPTP (cumulative dose of 18 mg/kg) for 5–8 days. PD features like rigidity, akinesia, postural tremors, eyelid closure, drooling, and flexed posture will be observed.
- *Evaluation:* Minimum score = 0 and maximum score = 17. The scoring distribution is as follows in Table 19.1.
- *Modification of standard method:*
  - MPTP is infused unilaterally via the Right coronary artery to produce hemiparkinsonism in Macaca monkeys
  - Unilateral Intranigral injection of MPTP in old rats can produce circling behavior in rodents
  - MPTP can be delivered via nasal route repeatedly to produce characteristic pathological changes of PD (degeneration of nigral dopaminergic neurons)
  - Marmosets can also be used instead of rhesus monkeys

19.5.2 Rotenone Induced PD Models

- Rotenone (also called nicouline) is an isoflavone compound that naturally occurs in various plants like *Derris elliptica, Derris involuta, Tephrosiavirginiana, Lonchocarpusurucu, etc.,* Traditionally rotenone is used as a pesticide, herbicide, and insecticide.
- *Principle:* Like MPTP, rotenone is also a lipophilic compound and mitochondrial complex I inhibitor that induces PD features upon administering in rats. Rotenone produces systemic mitochondrial complex I inhibition, while MPTP produces selective nigrostriatal dopaminergic neuronal inhibition. In addition to this, rotenone also increases iron accumulation in the substantia nigra via transferring receptor-2 leading to neuronal damage.
- However, rotenone produces selective dopaminergic neuronal damage upon chronic low dose administration, but systemic administration of a single dose (18 mg/kg) produces non-specific brain lesions with peripheral side effects.
- Achieving selective dopaminergic neuronal damage is difficult with rotenone as it produces noradrenergic and cholinergic neuronal damage at conventional dose administration.
- The preferred animal for the rotenone model is rodents, usually Lweis rats (as this strain is more sensitive to rotenone compared to other rat strains).
- The Rotenone model is the model which provides the first 'proof of concept' that nigrostriatal dopaminergic neurons have selective sensitivity to the mitochondrial complex I inhibitors. This model is also a prototype model to demonstrate that systemic mitochondrial damage can cause PD.
- The major limitation of the rotenone PD model is significant variability (30–50%) between batches of animals in developing lesions following systemic administration. The recent introduction of osmotic pump delivery of rotenone and intraperitoneal rotenone administration has been shown to reduce this variability.

19.5.3 Paraquat Model for PD

- *Principle:* Paraquat, a known pulmonary-toxicant, has structural similarity to MPP* (a metabolite of MPTP). However, paraquat does not cause mitochondrial complex I inhibition like MPP*. Instead, paraquat enters the brain via neutral amino acid carrier undergoes extensive redox-cycling resulting in the generation of reactive oxygen species that causes neuronal damage and produces PD features upon administration in animals.
- Paraquat has been used in pesticides, and hence it has been linked to one of the environmental causes of PD in humans. Upon administration in the mouse brain, paraquat has a half-life of 28 days and produces sustained lipid peroxidation. Thus, the paraquat model is useful in evaluating functional deficits and chronic development of lesions in PD.
- The paraquat model is ideal for replicating the preclinical stages of PD. This is because, after four weeks of administration, paraquat produces an insignificant

dopaminergic neuronal loss in the nigrostriatal pathway with increased dopamine neurotransmission. This is evident by the fact that tyrosine hydroxylase activity is increased after four weeks of administration.

19.5.4 Maneb-Paraquat Model

- Maneb, an ethylene-bisdithiocarbamate-manganese polymer, is a pesticide and fungicide
- It is commonly used along with paraquat as the 'mixed model' of PD.
- Coadministration of maneb and paraquat for the development of the PD animal model increases the success rate.
- This mixed model is useful in evaluating the role of environmental toxins in the development of PD in humans.

19.5.5 6-OHDA (6-Hydroxy Dopamine or Oxdopamine) Model of PD

- 6-OHDA PD model is one of the oldest and commonly used models. Important information regarding the physiological effect of dopamine depletion in PD, behavioral effect of nigral dopaminergic neuronal loss, and biochemical changes in that area were obtained by this model.
- *Principle:* 6-OHDA is structurally similar to dopamine. 6-OHDA cannot cross the blood-brain barrier, and therefore it should be injected into the brain locally using stereotactic techniques. After local administration, similar to dopamine, it enters the dopaminergic neurons via dopamine transporter and produces neuronal damage/death via the following mechanisms:
  - Generation of hydrogen peroxide, hydroxyl, and superoxide radicals by undergoing auto-oxidation and by the action of the MAO enzyme
  - Direct inhibition of mitochondrial complex I
  - The Resulting ROS generation increases cytoplasmic free calcium (due to mitochondrial membrane permeability), leading to neuronal death.
- Rat is the ideal animal for modeling 6-OHDA induced PD. Mice are also used in this model but less frequently than rats. This is because of difficulty during stereotactic surgery and less quantity of tissue availability for testing from mice.
- In rats, ventrolateral striatal lesions by 6-OHDA in rats causes alterations in sensorimotor orientations, the onset of movements, and fine motor behavior that are typically seen in PDs. On the other hand, the dorsomedial striatal lesions by 6-OHDA cause locomotion defect and 'drugs-induced spinning' effect. When apomorphine (direct dopamine agonist) is given, the direction of spinning (circling behavior) is seen towards the opposite side (contralateral side of the lesion). On the other hand, administration of amphetamine (indirectly acting DA agonist) produces spinning towards the lesion side (ipsilateral rotation). Thus asymmetrical motor behavior is observed in the 6-OHDA treated hemiparkinsonian rat models.

- Circling behavior can be monitored via digitalized 16-channel automated rotometer system. Clockwise or counterclockwise rotational behavior can be evaluated accurately using the digital rotometer.
- Alternatively, the asymmetrical motor behavior in the 6-OHDA treated hemiparkinsonian rats can be evaluated by 'elevated body swing test.' After seven days from 6-OHDA administration, the rats are accustomed to the Plexiglas box and kept in a neutral position with all four limbs on the ground. Now, the rat is elevated 2.5 cm from the ground by holding its tail for 60 seconds. When the rat moves the head out of the vertical axis to either side, it is recorded as 'one swing' and counted for 60 s. The proportion of left and right swings can be derived by dividing the number of swings of each side by the total number of swings made in 60 s. Rats are allowed to return to the neutral position before making the next swing test.
- The advantages of the 6-OHDA model are:
  - It produces progressive lesion development, and hence neuroprotective agents can be tested using this model
  - Non-motor behavioral changes like cognitive alterations, emotional disturbances can be successfully reproduced using this model
  - Depression, a key feature of PD, is successfully modeled using 6-OHDA treated rats.
- The major limitation of the 6-OHDA model is the failure of the formation of the Lewy body.
- 6-OHDA model is extensively used in the development of 'Neurotrophic Factor therapy' for PD. The role of glial-cell-derived NF, BDNF, and cerebral-dopamine-NF are evaluated as 'targeted gene delivery' methods using viral/non-viral/nanoparticles using 6-OHDA rat models nowadays.

19.5.6 Manganese

- *Principle:* Intrastriatal administration of manganese causes impaired energy metabolism by inhibiting complex I along with excitotoxic lesions and increases ROS formation leading to neuronal damage. At low concentration, Mn increases DA and its metabolite level. However, in higher concentrations, Mn decreases the level of DA in the neurons. Moreover, the valence of Mn plays a significant role in producing toxicity. Trivalent Manganese produces higher neuronal toxicity than the divalent manganese (as trivalent Mn is more potent than divalent Mn in inhibiting Complex I in neuronal mitochondria).
- After administering 10 mg of manganese (as $MnCl_2$), intraperitoneally in rats, locomotor function deterioration will be observed in 5 weeks.

19.6 Genetic Models of PD

19.6.1 α-Synuclein Transgenic Mouse Models

- The α-synuclein transgenic mouse models have over-expression of human α-synuclein (wither wild type or mutated).
- The first model in this series was developed by Masilah et al. that overexpresses wild-type human α-synuclein using Thy-1 promotor. A progressive accumulation of inclusions in substantia nigra, neocortex, and hippocampus were noted and stained positive for ubiquitin and α-synuclein. However, unlike in human PD, these inclusions had non-fibrillar composition. Moreover, no apparent DA neuronal loss was also observed. Thus Masilahα-synuclein transgenic mouse model was a less robust model for PD.
- Subsequently, thirteen different α-synuclein transgenic mouse model were developed using various promotors besides Thy-1 like prion, PDGF-β, and CamKIIα, resulting in overexpression of wild or mutant α-synuclein.
- Out of these models, protein inclusions were noted in 30% of models (4/13), and motor deficits were noted in 70% of models (9/13). The characteristic PD lesion—nigral cell loss was noted only in 20% of the models (2/13). Thus α-synuclein transgenic mice models do not appear to reproduce the human PD pathology.

19.6.2 α-Synuclein Transgenic Drosophila Model

- The first drosophila α-synuclein PD model was developed by Feany and Bender, which overexpresses wild or mutant types of α-synuclein in the fly's nervous system.
- In contrast to the α-synuclein mouse models, the drosophila models exhibit robust signs of PD. These adult transgenic drosophila flies exhibit progressive, age-dependent loss of DA neurons resulting in progressive loss of climbing ability. The formation of fibrillar α-synuclein was also noted in α-synuclein transgenic drosophila models, whereas non-fibrillar inclusion only was noted in rat models. The inherent resistance to neural loss in rodents probably explains the observed differences between the mouse and drosophila models.
- The major drawback of the Drosophila PD model is that DA neuronal loss documented was not consistently reproducible as few researchers observed only 50% of neuronal loss while others observed >90% neuronal loss. The discrepancy between the observations could be related to the type of tissue fixative used in the fly brain section. It has been proven that paraffin sections of the fly brain result in inconsistent preservation of DA epitopes. The alternate methods for better DA neuronal staining proposed are:
  - Cooled microwave infiltration with weak fixative
  - Agarose embedding with thick sectioning on vibratome
  - Using buffers with anti-oxidant protection of DA epitomes.

19.6.3 Parkin Transgenic Mouse Models

- Parkin functions as a ubiquitin ligase, causing proteosomal degradation and mutation of the parkin gene, leading to the development of AR-juvenile Parkinsonism.
- Deletion of exon-3, exon-2, and exon-7 can produce parkin transgenic mice. Nevertheless, these mice did not produce nigral cell loss, no protein inclusions were observed, and no characteristic PG motor impairment was noted. However, increased extracellular dopamine levels reduced DAT and VMAT levels, increased ROS-mediated damage, cognitive deficits, and reduced acoustic startle have been noted in these models.
- Mouse with Parkin 7 exon deletion has additional loss of non-adrenergic neurons in locus coeruleus area.
- Parkin mouse models are helpful in studying the pathophysiology of AR-juvenile Parkinsonism and possible drug targets.

19.6.4 Pink-1 Deleted Mutant Drosophila & Mice Model

- Another gene besides parkin that is related to AR-Parkinsonism is pink-1. The pink-1 protein, along with parkin protein, maintains the mitochondrial hemostasis in the DA neurons.
- In a healthy mitochondrial environment, Pink-1 is generally degraded by the mitochondrial processing peptidase (MPP) and transmembrane serine protease-PARL. This results in the failure of phosphorylation of parkin by the pink-1. However, in a damaged mitochondrial environment, pink-1 is not destroyed, resulting in phosphorylation of parkin and ubiquitin proteins leading to ubiquitination of substrate protein in the neurons.
- Loss of the pink-1 gene leads to excessive mitochondrial fusion, and the models with pink-1 mutation exhibits reduced evoked DA response with impaired mitochondrial functions in the striatum. Nevertheless, loss of nigral neurons and protein inclusions were not observed. The models also did not produce the characteristic motor impairment in PD.

19.6.5 DJ-1 Deficient Transgenic Mouse/Drosophila Models

- The Dj-1 protein functions as an anti-oxidant and oxidative stress sensor inside the DA neurons and prevents neurodegeneration. The expression of mitochondrial uncoupling proteins (UCP-4 & UCP-5) is increased by Dj-1, leading to reduced mitochondrial ROS production. Multiple stress adaptive pathways like Nrf2, PI3K/PKB, etc., are activated by Dj-1. Also, Dj-1 increases DA synthesis by upregulation of Dopamine decarboxylase and tyrosine hydroxylase activity.
- Mutations of Dj-1 cause AR-Parkinsonism. Deletions of exon-2, exon 1–5, exon 2–3, and exon-7 can produce various transgenic Dj-1 deficient mouse models.

- However, nigral cell loss was not observed in any of the Dj-1 deficient mouse models. Motor function impairment was noted in the exon-2 and exon 1–5 deleted Dj-1 transgenic mouse models.
- Biochemically, increased DA reuptake, increased mitochondrial hydrogen peroxide production, increased DAT levels in presynaptic membranes were observed in these models.
- These models are suited to evaluate the mechanisms by which Dj-1 functional loss occurs while aging and potential drug targets to reverse them.

19.6.6 LRRK2 Gene Knock-in Transgenic Mouse

- The pathology of Leucine-rich-repeat kinase-2 (LRRK2 or dardarin) mutation-induced PD is very similar to idiopathic PD. Mutation in LRRK2 will lead to the development of autosomal dominant PD. At present, the mutation of LRRK2 is the most common observation in the familial PDs (5–6%) and sporadic PD (1–3%).
- LRRK2, a protein with 2527 amino acids, exhibits GTP-dependent phosphorylation activity. Mutations in LRRK2 lead to over-activity of kinase resulting in neuronal toxicity.
- At the age of 18 months, after amphetamine administration, the LRRK2 transgenic PD mouse exhibit increased locomotor response.
- This model is helpful in the evaluation of LRRK2 role in the development of familial PD and possible drug targets.

19.6.7 Mitopark Mice

- Mitochondrial dysfunction is one of the putative causes for the development of sporadic PD. This is because midbrain DA neurons with respiratory-chain-deficits are found to higher in PD patients than non-PD patients.
- Thus, a conditional knockout Tfam (Mitochondrial transcription factor A) was created, and upon crossing them with DAT-Cre (Characterization of dopamine transporter), it produces mitopark mice models.
- The mitopark mice have decreased mtDNA expression and midbrain DA neurons with respiratory chain deficits leading to the development of adult-onset, slowly progressing, and motor function impairment. In addition to this, intraneuronal inclusions are also observed in this model.
- Upon administration of L-Dopa, the PD symptoms disappear in this model. Thus MitoPark mouse can act as a potential tool for evaluating drugs targeting mitochondrial damage and preventing neuronal damage in PD.

Bibliography

Bové J, Prou D, Perier C, Przedborski S. Toxin-induced models of Parkinson's disease. NeuroRx. 2005;2(3):484–94.

Cannon JR, Greenamyre JT. Neurotoxic in vivo models of Parkinson's disease recent advances. Prog Brain Res. 2010;184:17–33.

Kessler RM, Ansari MS, de Paulis T, Schmidt DE, Clanton JA, Smith HE, et al. High affinity dopamine D2 receptor radioligands. 1. Regional rat brain distribution of iodinated benzamides. J Nucl Med. 1991;32(8):1593–600.

Lim K-L, Ng C-H. Genetic models of Parkinson disease. Biochim Biophys Acta. 2009;1792(7): 604–15.

Martinez TN, Greenamyre JT. Toxin models of mitochondrial dysfunction in Parkinson's disease. Antioxid Redox Signal. 2012;16(9):920–34.

Meredith GE, Sonsalla PK, Chesselet M-F. Animal models of Parkinson's disease progression. Acta Neuropathol. 2008;115(4):385–98.

Salari S, Bagheri M. In vivo, in vitro and pharmacologic models of Parkinson's disease. Physiol Res. 2019;68(1):17–24.

Screening Methods for the Evaluation of Drugs for Learning and Memory

20

Manu Jose and Mageshwaran Lakshmanan

Abstract

Various diseases like Alzheimer's disease (AD), frontotemporal dementia, dementia with Lewy bodies, HIV-induced dementia, etc., affect humans' memory and learning, leading to loss of productivity and quality of life. Novel drugs targeting the newly identified pathological pathways of AD are being tested. In vitro techniques to study the interaction of the test compound with cholinergic receptor and acetylcholinesterase enzyme in brain tissue will provide insight into the 'selection of lead molecule.' Various in-vivo behavioral models of AD, like passive avoidance tests, active avoidance tests, and discrimination learning tests, provide information regarding the anti-AD efficacy of test molecules. Genetic models of AD like APP transgenic mice, bigenic mice, triple transgenic mice, MAPT-GR mouse, and APOE transgenic mouse provide information regarding the pathophysiology of AD and modulation of the deranged pathway by the test compound. This chapter will review the various in-vitro, in-vivo behavioral methods and genetic models used to screen drugs for learning and memory.

Keywords

Alzheimer's disease · Learning · Memory · Active avoidance test · Passive avoidance test · Discrimination test

M. Jose
Department of Pharmacology, Government Medical College, Kozhikode, Kerala, India

M. Lakshmanan (✉)
Department of Pharmacology, Thanjavur Medical College, Thanjavur, Tamil Nadu, India

M. Lakshmanan et al. (eds.), *Introduction to Basics of Pharmacology and Toxicology*, https://doi.org/10.1007/978-981-19-5343-9_20

20.1 Introduction

Various diseases like Alzheimer's disease (AD), frontotemporal dementia, dementia with Lewy bodies, HIV-induced dementia, etc., affect the memory and learning in humans leading to loss of productivity and quality of life. In older age, AD is the most common cause of dementia, with a prevalence of 44 million worldwide. Due to contemporary research, various pathological factors for the development of AD have been identified. Novel drugs targeting the newly identified pathological pathways of AD are being tested. There is no ideal model to test all the aspects of AD. Hence, batteries of tests involving various in-vitro and in-vivo behavioral methods are commonly employed to screen the drugs with the potential capacity to modulate learning and memory in animals.

20.2 In vitro Methods

20.2.1 In vitro Inhibition of Acetylcholine-Esterase (AChE) Activity in Rat Striatum

- *Principle:* Loss of cholinergic neurotransmission due to deterioration of cholinergic neurons is one of the reasons for decline for cognitive function in AD. Therefore, drugs with AChE blocking activity can increase the brain's acetylcholine level and prevents glutamate-induced excitotoxicity leading to improved cognition in AD. Using standard Ellman's procedure, the activity of the test drug in inhibiting the AChE can be estimated. Ellman's procedure uses acetylthiocholine instead of ACh as acetylthiocholine has similar qualitative kinetic behavior to ACh. Acetylthiocholine is acted upon by AChE and produces thiocholine that reacts with another reagent, dithiobisnitobenzoic acid (DTNB), to form a yellow product (5-mercapto-2-nitrobenzoic acid). Reaction intensity can be measured using a spectrophotometer.
- *Procedure:*
  - The corpora striata of the male Wistar rat is dissected and using the standard technique; the tissue homogenization is prepared.
  - The lyophilized enzyme AChE is diluted in 0.05 M phosphate buffer (pH 8), and this stock solution is stored at −20 °C. On the day of the experiment, the actual value of AChE preparation activity in units (U) should be determined, and the same volume should be used that particular day. One unit of AchE activity represents the amount of AChE stock solution required to convert 1 μmol of Acetylthiocholine in 1 min. For each day, the volume of AChE stock solution required to achieve one unit activity should be determined and used.
  - To the blank cuvettes following reagents are added:

    0.8 ml of 0.5 mM of DTNB mixed in 0.05 M phosphate buffer

    0.8 ml of 10 mM of acetylthiocholine mixed in 0.05 M phosphate buffer
  - To the Control cuvettes following reagents are added:

0.8 ml of 0.5 mM of DTNB mixed in 0.05 M phosphate buffer
0.8 ml of 10 mM of acetylthiocholine mixed in 0.05 M phosphate buffer
Estimated volume of AChE that provides 3.5 U of AChE-activity

- To the drug cuvettes following reagents are added:
  1 ml of test drug of various concentrations containing 25 μl of tissue suspension
  0.8 ml of 0.5 mM of DTNB mixed in 0.05 M phosphate buffer
  0.8 ml of 10 mM of acetylthiocholine mixed in 0.05 M phosphate buffer
  estimated volume of AChE that provides 3.5 U of AChE-activity are added
- After 10 min of incubation, the enzymatic activity is measured using a spectrophotometer (maximum absorption coefficient seen at 412 nm)

• *Calculation:*
  - To estimate the non-enzymatic hydrolysis of substrate, the blank values are determined for each run to the control, and this non-enzymatic hydrolysis should be subtracted from total hydrolysis.
  - The percentage of inhibition can be calculated using the formula:

$$\%of\ inhibition = \frac{slope\ of\ control - slope\ of\ drug}{slope\ of\ control} \times 100$$

  - IC_{50} value can be calculated using log-probit analysis.

20.2.2 In vitro Inhibition of Butyrylcholine-Esterase (BChE) Activity in Human Serum

• Human serum is the richest source of BChE. Instead of rat striatum, human serum can be used as an alternative.
• The principle and estimation are similar as explained above (refer Sect. 20.2.1).
• The lyophilized human serum is reconstituted to 3 ml distilled water, and 25 μl of the aliquot is added to the 1 ml of test drug solutions.
• Instead of acetylthiocholine, butyrylthiocholine (225.8 mg in 0.05 M phosphate buffer) is used in this experiment.

20.2.3 In vitro Estimation of Presynaptic Release of [^{3}H]- ACh and Other Transmitters from Rat Brain

• *Principle:* The presynaptic muscarinic autoreceptors control the release of ACh from the presynaptic neurons in various brain areas. Using radiolabelled ACh ([^{3}H]- ACh), the amount of ACh released by electrical stimulation from the presynaptic neurons in the presence of standard muscarinic agonist, AChE inhibitors, muscarinic antagonists, and test compounds can be compared.
• *Procedure:*

- *Reagents:*
  Krebs-Henseleit bicarbonate buffer (pH 7.4)
  100 nM of Methyl - [^{3}H]- Choline (with radioactivity of 80–90 Ci/mmol)
  20 µM of hemicholinium
  Serially diluted test compound and vehicle
- Rat brain slices from the cortex, striatum, and hippocampus are prepared using standard technique and incubated with oxygen and buffer for 30 min. After this, the buffer is emptied, and 100 nM concentration of Methyl - [^{3}H]-Choline is added and incubated with 2.5 ml buffer for 60 min. After incubation, the buffer is poured out, and slices are placed in stimulation chambers.
- Adding hemicholinium with buffer in the stimulation chamber prevents the uptake of [^{3}H]- Choline formed due to hydrolysis of released [^{3}H]- ACh.
- The addition of 2 µM of sulpiride and 5 µM methysergide will prevent the release of ACh due to Dopamine and 5-HT_2 receptor stimulation, respectively.
- An evoked release can be determined in the presence of various known muscarinic agonists, muscarinic antagonists, and AChE inhibitors.

• *Calculation:*
  - The percent fractional release (PFR) is the ratio of the amount of radiolabelled compound release to the amount present in the tissue. The spontaneous release (SP) is the average of two fractions preceding and the first fraction in that range after the stimulation period. Stimulated value (S) is the summed difference between PFR and SP
  - The stimulated value for the first period of electrical stimulation (S1) and stimulated value for the second period of electrical stimulation (S2) are determined for the control and test drug slices
  - The ratio of S2/S1 is calculated for all slices and compared with the standard agonists and antagonists

20.2.4 In-vitro [^{3}H]-Oxotremorine-M Binding to a Muscarinic Cholinergic Receptor in Rat Forebrain

• *Principle:* Oxotremorine is a partial central and peripheral muscarinic agonist, while Oxotremorine-M is a full agonist of the same. Using radiolabelled oxotremorine-M, the potential muscarinic agonistic activity of the test drug can be compared.
• *Procedure:*
  - Reagent:
    0.05 M Tris buffer (pH 7.4)
    0.5% (w/v) polyethylene-imine (used to soak the filters for 3 h to prevent binding of ligand to filter strip)
    100 nM of Methyl-[^{3}H]-oxotremorine acetate with 83–85 Ci/mmol

20 µl of 2 mM atropine sulfate (for determination of non-specific binding)
Serially diluted test compound and vehicle

- The forebrain of the male Wistar Rat is dissected, and the tissue pellet is made as per standard techniques after homogenization. The tissue pellets are suspended in the Tris buffer.
- Reagents are added in appropriate volume (50 µl of buffer, 380 µl of distilled water, 20 µl of atropine/drug/vehicle, 50 µl of radiolabelled oxotremorine-M, and 500 µl of tissue suspension and incubated for 45 min at 30 °C.
- The assay is stopped using the rapid filtration technique, and the radioactivity is measured using a scintillation counter.

• *Evaluation:*
- Total binding is estimated by measuring binding of ^{3}[H]-Oxotremorine-M.
- Non-specific binding is estimated by measuring the binding of atropine.
- Specific binding is estimated by the difference between total binding and non-specific binding.
- Percent inhibition: 100–specific binding as a percentage of the control value.
- Using log-probit analysis, the IC_{50} calculations are performed. At least 3–4 different concentrations of the test compounds in triplicate should be used to determine the IC_{50}.

20.2.5 In-vitro [^{3}H]-N-Methylscopolamine Binding in Presence/Absence of Guanylyimidophosphate

• *Principle:* Binding of muscarinic agonists to muscarinic receptors converts the high-affinity binding state of guanine nucleotide to a low-affinity state. On the other hand, the binding of muscarinic antagonists to muscarinic receptors has no effect. Therefore, based on the selective effect of guanine nucleotide on the affinity, the [^{3}H]-N-Methylscopolamine assay can distinguish the interaction of muscarinic agonist and muscarinic antagonist.

• *Procedure:*
- Reagent:

  0.05 M Tris buffer (pH 7.4)
  Distilled water
  1 mM atropine
  4 nM [^{3}H]-N-Methylscopolamine
  50 µM 5′-Guanylyimidophosphate

- Using the standard technique, the rat cerebella is dissected, homogenized, and tissue pellets are prepared and suspended in Tris buffer.
- Reagents are added in appropriate volume (1000 µl of 0.05 M Tris buffer, 780 µl of distilled water, 20 µl of 1mMatropine/drug/vehicle, 50 µl of [^{3}H]-N-Methylscopolamine, 50 µl of 5′-Guanylyimidophosphate, and 100 µl of tissue suspension and incubated for 90 min at 20 °C.

- The assay is stopped using the rapid filtration technique, and the radioactivity is measured using a scintillation counter.
- *Evaluation:*
  - Total binding is estimated by measuring the binding of [^{3}H]-N-Methylscopolamine.
  - Non-specific binding is estimated by measuring the binding of atropine.
  - Specific binding is estimated by the difference between total binding and non-specific binding.
  - Percent inhibition: 100–specific binding as a percentage of the control value.
  - Using log-probit analysis, the IC_{50} calculations are performed. At least 3–4 different concentrations of the test compounds in triplicate should be used to determine the IC_{50}.

20.2.6 Transfected Cell Lines

- The amyloid-β (Aβ) hypothesis of AD is well known, and increased levels of Aβ in the brain play a central role in the pathogenesis of AD. The Aβ is derived from the improper splicing of amyloid precursor protein (APP). The APP_{695} is the predominant splice variant observed in the AD.
- Therefore, the cell lines can be transfected with the APP_{695}-cDNA. This technique is used to evaluate the drugs with Aβ modulation effect.
- Under appropriate conditions, the N2a cells or HEK293 are transfected with APP_{695}-cDNA and grown in a culture medium (preferred is Dulbecco's modified Eagle medium).
- The cell lines are incubated with either control or various concentrations of test drug or vehicle for desired time. After successful incubation, by using siRNA directed against proteasome subunit- β5, the siRNA-directed silencing of proteasomal degradation of Aβ is studied.
- Cells lines like human retina pigment epithelial cell line, human fibroblast cells from Friedreich Ataxia patient, Leber's hereditary optic neuropathy, Pick's disease, and human-derived neuroblastoma cell line (SH-SY5Y) can be used to evaluate the Aβ toxicity assay.

20.2.7 Evaluation of β and γ Secretase Inhibition

- The β and γ-secretase enzymes are responsible for the production of Aβ from APP by proteolytic cleavage. Modulation of these enzymes can have a beneficial role in AD.
- Unlike γ secretase, the β-secretase does not carry any functional problems. Complete abolishment of β-secretase gene activity results in reduced levels of Aβ.

- Applying gene silencing techniques, peptide inhibitors for β and γ-secretase genes can be designed, applied in cell lines, and the reduction in the Aβ level can be quantified.

20.2.8 Cultures of Rat Hippocampal and Cerebral Cortex Neurons

- Various intracellular mechanisms like neuronal development, synaptic plasticity, learning, and memory are affected in AD. Thus cultures of rat hippocampal and cerebral cortex neurons are prepared and used to evaluate for testing the lead molecule in AD research.
- The pregnant Sprague-Dawley rats are nurtured using standard care. Once the pups are born, their cerebral cortex and hippocampus are removed and cultured in Dulbecco's modified Eagle medium for twelve days.
- The cells that express the glutamate receptors are used to study the intracellular mechanism for memory in the presence of various agonists and antagonists.

20.2.9 Study of Field Excitatory Postsynaptic Potential

- The primary neural basis for learning and memory is 'long term potentiation', a form of synaptic plasticity. The field excitatory postsynaptic potentials (fEPSPs) are evoked in the hippocampal regions and are studied for long-term potentiation.
- Rat hippocampal slices are prepared using standard technique and perfused with artificial CSF, 95% oxygen, and 5% CO_2. Using M-NaCl filled glass electrode, with a resistance of 2–20 mohm, the extracellular fEPSPs recordings are obtained from the stratum radiatum of the C1 region.
- Collateral-commissural fibers are stimulated using a bipolar stimulating electrode for every 30 s. The stimulus strength is adjusted so that it is close to half the maximum amplitude of fEPSPs.
- Before adding any drug to the tissue, a stable baseline for 15–20 min should be obtained. The mean percentage increase in the amplitude of spike from baseline after test drug application is compared with the controls.

20.2.10 Other in-vitro Techniques Used for Evaluating Activity in AD

- *Inhibition of respiratory burst in microglial cells/macrophages*: to determine the protective effect of test drug in oxidative damage to the neuronal cells.
- *Primary chicken neurons for evaluating Aβ toxicity:* Endogenous wildtype Aβ peptides are secreted in primary chicken neurons and can be used to evaluate the protective effect of the test drug. Both the human and chicken Aβ42 have identical sequences, and hence it provides reliable results.
- *Aggregated Aβ assay:* Aggregation of Aβ peptides is one of the main causes of neurodegeneration in AD. The lead molecule with potential anti-aggregatory

property can be evaluated using a commercial aggregated Aβ assay (Amorfix aggregated Aβ assay) in a fast and reproducible manner.

- *Evaluation of tau hyper-phosphorylation:* Hyper-phosphorylated tau protein dissociated from the microtubules leading to axonal flow damage and impaired neuronal viability. Thus, hyper-phosphorylation of tau protein causes neurodegeneration in AD, progressive supranuclear palsy, Pick's disease, and corticobasal degeneration- collectively called 'tauopathies. The ability of test compounds to prevent hyper-phosphorylation of tau protein can be evaluated in-vitro using the SH-SY5Y cell lines that over-express human tau441 isoform. Using immune blot analysis or immunosorbent assay, the reduction in the level of tau-phosphorylation in the SH-SY5Y by test compound can be estimated.

20.3 In-vivo Behavioral Models:Passive (Inhibitory) Avoidance Tests

20.3.1 Step Down Model

- *Principle:* By nature, rats and other rodents spend most of their time in the corners and close to the walls. The rat step-downs immediately to the floor when it is placed in the center of the elevated rectangular compartment and proceeds to explore for any enclosures. This behavior is reduced when an electric shock is provided after stepping down (passive avoidance) and indicates learning.
- *Procedure:*
  - A rectangular box (50 × 50 cm) with an electrifiable grid floor with a raised wooden rectangular platform at the center is used. The rat/mouse is placed at the center of the rectangular platform. The room of the experiment should be soundproof.
  - *Phase I (Habituation/Familiarization):* After placing the rat on the platform, the time taken to step-down is noted. If the rat did not step down in 10 s, it is returned to the cage.
  - *Phase II (Conditioning/Learning):* An unavoidable shock (1.5 mA; 50 Hz for 1 s) is applied when the rat steps down to the floor and returned to the cage.
  - *Phase III (retention test):* After 24 h of phase II, the rat is placed again on the platform, and the step-down latency (SDL) time is calculated. The experiment is complete when the rat steps down or spends 60 s on the platform.
- *Evaluation:* The SDL for the different groups is compared using the ANOVA.
- *Disadvantages:* Results might have high variability, and hence higher number of animals per group is required.

20.3.2 Step Through Model

- *Principle:* When placed in a bright environment, the rodents tend to isolate themselves to the darker one. When an unpleasant stimulus like the electric shock is applied as soon as the rodent enters the darker chamber, the behavior is reduced (passive avoidance) in subsequent testing and indicates learning.

- *Procedure:*
  - *Apparatus:* A large dark chamber is connected to a smaller, brighter chamber via a guillotine door. An electric bulb (12 V/7 W) is placed in a smaller chamber. The larger chamber is equipped with an electric grid floor to deliver an electric shock (1 mA for 1 s for mice and 1.5 mA for 2 s for rats). The guillotine door shuts automatically when the rodent enters the larger room.
  - *Phase I (Habituation/Familiarization):* After placing the rat in the brighter room at the maximum far distance from the guillotine door, the time taken to step-through the darker chamber (STL) is noted. The cut-off time is 90 and 180 s for mice and rats, respectively. Animals reaching the cut-off time should not be included.
  - *Phase II (Conditioning/Learning):* An unavoidable shock is applied when the rat steps through the darker room, removed quickly within 10 s and returned to the cage.
  - *Phase III (retention test):* After 24 h of phase II, the rat is placed again in the brighter room, and the step-through latency (STL) time is calculated. The experiment is complete when the rat steps through or spends maximum cutoff time in the brighter chamber. (Cut-off time in phase III: 300 s for mouse and 600 s for rat)
- *Evaluation:* STL values in Phase I and III are compared, and an increase of STL is considered as learning. The STL for the different groups is compared using the ANOVA.

20.3.3 Uphill Avoidance Model

- *Principle:* When placed on a slanted surface with the nose down posture, the rodent orient itself and tend to move towards the top (negative geotaxis). When an unpleasant stimulus like the electric shock is applied as soon as the rodent moves towards the top, the behavior is reduced (passive avoidance) in subsequent testing and indicates learning.
- *Procedure:*
  - *Apparatus:* The apparatus consists of a box (50 × 50 × 35 cm) with opaque plastic walls that can be inclined at any angle. The box floor consists of stainless steel grid bars(10 mm diameter) separated 1.3 cm apart. A tail electrode is present to deliver the electric shock.
  - *Phase I (Habituation/Familiarization):* The rat is fitted with a tail electrode and placed nose-down position in the slanting box. The latency period of turning 180 degrees and initiating to climb upwards is noted.
  - *Phase II (Conditioning/Learning):* An unavoidable shock is applied when the rat starts to climb, removed quickly within 10 s, and returned to the cage.

  - *Phase III (retention test):* After 24 h of phase II, the rat is placed again on the inclined box with a nose-down position. The latency time spent in the nose down position to 180-degree turn is noted down.
- *Evaluation:* The latency period values in Phase I and III are compared, and an increase in the latency period is considered learning. The STL for the different groups are compared using the ANOVA

20.3.4 Two-Compartment Test

- The two-compartment test is similar to the step-through test
- The only difference is that in a two-compartment test, the electric shock will not be applied. The time taken for the animal to enter from a large brighter compartment to a small darker compartment is measured and compared directly between the groups.

20.3.5 Trial to Criteria Inhibitory Avoidance

- Instead of measuring the latency period, the number of trials required for the animals to learn passive avoidance is the main criterion in this experiment.
- The procedure is similar to the step-through method with few minor modifications. After foot shock, the animal is not returned to the cage but placed again in the larger bright room. When the animal enters a dark room, foot shock is given, and the same is repeated. The experiment is concluded when the animal spends the 60 s in the bright room. The number of trials required to attend the passive avoidance is counted and indicates the speed of acquisition.
- After 24–48 h, the same procedure is repeated with a cut-off time of 180 or 300 s. The number of trails required to attend the criteria is noted (indicates retrieval process) and compared between the groups.
- The number of trials to reach the criteria is significantly increased with amnestic drugs. The test compounds can be administered before the test session, and the drug's effect on the retrieval process can be determined.

20.4 In-vivo Behavioral Models: Active Avoidance Tests

20.4.1 Runaway Avoidance Test

- *Principle:* In the active avoidance tests, a conditioned stimulus is followed by a noxious stimulus. The animals learn to control the administration of unconditioned stimulus by exhibiting apt reactions to the conditioned stimulus that is followed by noxious stimulus resulting in avoidance of noxious stimuli.
- *Procedure:*

  - *Apparatus:* A rectangular box with uniform illumination that contains a loudspeaker mounted 50 cm above the floor and a small guillotine door to the safe area.
  - The rat is allowed to explore the whole setup for 5 min. Now the guillotine door is closed, and the rat is placed in the light starting area. After 10 s, an acoustic buzz is provided, and the guillotine door is opened simultaneously. The electric shock is turned on after 5 s. The acoustic sound (CS) should be continued till the rat reaches the safe area. The rat should be left undisturbed for 60 s (intertrial interval). The exact process is repeated.
- *Evaluation:* The training is continued till the animal shows nine continuous avoidance out of 10 consecutive trials. After 24 h, the same process is repeated and the time taken to reach the same learning criteria is calculated and compared between the groups.

20.4.2 Shuttle Box (Two-way) Avoidance Test

- Unlike runaway avoidance, the rat is not handled between the trials in the shuttle box avoidance test. It is relatively difficult than the runaway avoidance test.
- The setup consists of a rectangular box divided into two compartments by a small guillotine door. Both the compartments have light (20 W bulb) and a grid to provide electric foot shock. A loudspeaker is mounted above and provides a buzz (CS).
- The rat is allowed to explore the setup for 5 min with all lights switched off and connecting door in the open state. Now the door is shut, and after 20 s, the compartment light in which the rat is present is turned on. The CS is now presented, and after 5 s, the foot electric shock is applied continuously (repeated cycles of 0.5 s on and 1.5 s off) till the rat enters the dark compartment.
- Once the rat enters the dark compartment, the door is shut, and shock and light are stopped. An interval of 90 s is allowed, and the light is switched on the compartment in which the rat is now present, and the same procedure is repeated. Thus the rat is shuttling between the compartments to avoid the noxious stimulus.
- Training is achieved when the rat successfully met the criteria for nine times out of ten consecutive trials. When the rat is tested after one or two days, the retention effect can be measured.
- The error count (number of times the rat not reaching the safe area) and the time taken for the animal to reach the safe area between various days and groups are noted and compared.

20.4.3 Jumping (One-way Shuttle) Avoidance Test

- *Principle:* The major setback of 20.3.6 and 20.3.7 tests is the inability to avoid animal handling or the requirement of high degree automation, both of which, if failed, will produce high variability in the results. The Jumping (one-way shuttle) test avoids these setbacks as the animal is returned to the start area either by force or spontaneously. In setup 20.3.6 and 20.3.7, the animal may reach the safe area by learning or by random wandering. However, the Jumping setup enhances the start-goal distinction by making the animal perform a distinct all or none response like jumping on the platform.

- *Procedure:*
  - *Apparatus:* A rectangular box (40 × 25 × 40 cm) with an electrifiable grid floor is fitted with a plastic pedestal (12 × 12 × 25 cm) in one of the narrow walls of the box. This provides a safe (goal) area. Another 'flush' is fitted along with the plastic pedestal, which can be retracted back to the wall (the goal area is now exposed) or pushed forward completely (the goal area is closed).
  - The rat is now placed into setup to explore for 5 min, with the goal area exposed for 2 s followed by a closed goal area for 2 s. Now, the trial starts with acoustic CS and exposed goal area. After 5 s of CS, electric shock is applied for every 2 s with continuous CS till the rat reaches the goal area.
  - After 30 s, the flush pushes the rat into the grid floor, and the same process is repeated. After 24–48 h, the same procedure is repeated.
- *Evaluation:* The time taken for the rat to reach the safe area by jumping on the different days between the groups is recorded and compared. The number of errors made by the rat between the groups is also compared.

20.5 In-vivo Discrimination Learning Tests

In the active and passive avoidance tests, only a single CS is applied, and the animal has no choice between the CS. However, in discrimination learning tests, multiple CS will be applied, and the animal has a choice to react between multiple CS and discriminated them.

20.5.1 Spatial Habituation Learning

- *Principle:* When a rodent is exposed to an unfamiliar environment, it exhibits exploratory behavior (out of curiosity) that is limited by precaution to avoid potential dangers. On repeated exposure, the rodent familiarizes with the environment leading to a decrease in exploratory time (spatial habituation learning). This test can evaluate both short-term (reduction in exploratory time within the trials) and long-term memory (reduction in exploratory time in trials conducted after 24, 48, or 96 h).
- *Procedure:* The rat is placed in a wooden, grey-painted rectangular box (60 × 60 × 40 cm) with a 25 W green/red bulb attached above to provide uniform illumination. The room of the experiment should be soundproof, and masking noise of 60 dB should be provided continuously. The rat is placed in the center or corner of the box and observed for exploratory behaviors like rearing, corner time, wall time, center time, and freezing are recorded for 10 min. The number of

boli deposited (defecation) is also counted. The procedure is repeated several times, and the decrease in exploratory behavior time due to spatial habituation can be appreciated. For long-term memory, the trials are performed after 24/48/96 h.
- *Evaluation:* The parameters measured are compared between the groups using ANOVA.

20.5.2 Spatial Habituation Learning

- *Principle:* The animals learn to discriminate the reward/punishment stimuli by selecting the correct pathway by remembering spatial orientation. Generally, left-right discrimination is commonly tested.
- *Procedure*: The apparatus consists of a 'Y' or 'T' shaped arm with an electrifiable grid floor. The common arm act as a starting point and the right arm is first kept as the goal area (devoid of electric shock or presence of food cup). The last 10 cm of both arms are disconnected from the rest of the setup by the swing door. The rat is placed in the far end of the common arm (starting point), electric shock (1 mA, 50 Hz, 0.5 s duration for every 3 s) is then applied after 5 s. The rat is guided and trained to reach the goal (kept in the right arm). Initially, the rat will make many errors, and those errors are not due to lack of remembrance but due to its exploratory behavior. After several trials, the rat is trained in such a way that criterion is attained (8 correct out of 10 trials) and returned to the home cage. The same procedure is repeated on the next day. Time taken to reach the criterion and the number of errors made are noted down. After 1 h, the goal area is shifted to the left side, and the same process is repeated.
- *Evaluation:* The number of trials the rat takes to attend the criterion on the same day and after 24–48 h between the groups is compared. The number of errors made is also compared.

20.5.3 Spatial Learning in a Radial Arm

- *Principle:* This model was developed by Olton et al. in 1976 and has been extensively used in learning and memory research. The animal locates the baited arm using the distal cues provided in the setup. Working memory and spatial reference memory can be studied in the rat using a radial arm setup. Information learnt by the rat is useful for many sessions and needed for an entire procedure in spatial reference memory procedures. On the other hand, information learnt by a rat is useful for one session only in working memory procedures. During the delay interval, the rat has to remember the information. The rat will be rewarded when it makes a correct choice.
- *Procedure:* An elevated wooden maze made of eight arms (50 × 5 × 2 cm) extending from a central platform (26 cm in diameter) is constructed. Good illumination is provided for the entire maze and loaded with numerous cues. At the end of the selected arm, food pellets are placed, and the rats are trained daily to

collect the food pellets by identifying the cues. The rat should obtain maximum rewards with minimum errors in eight consecutive trials.

- *Evaluation:* The number of errors made is noted down for every session and compared between the groups.
- *Modification of test:*
  - The rat is trained well, and the errors are noted down. Now, after a surgically induced hippocampal lesion, the same procedure is repeated. The time taken to achieve the criterion and the number of errors before and after surgery is compared.
  - Working memory procedure: The rat is forced to obtain a reward in a specific arm (4 arms). After a time delay, the rat should seek reward in the same arm (win-stay method). Alternatively, the arms are changed (reward to punishment and vice versa), and the rat should seek reward in new arms with reward (win-shift method).
- *Disadvantage:* This setup works mainly based on the appetitive nature of the maze. Thus drugs like anorectics or rats with hypothalamic lesions cannot be evaluated using this method.

20.5.4 Spatial Learning in the Water Maze (Morris Test)

- *Principle:* An escape platform is well hidden in the water maze, and no proximal cues to the platform are provided. The rat has to identify the platform by configuring cues located outside the tank. Decrease in path length to locate the platform and a decrease in latent time to escape indicates 'learning.'
- *Procedure:*
  - A circular water tank (water filled up to 20 cm depth) is divided into four equal quadrants using marks in the perimeter. A platform (22 cm height) is placed in the center of any quadrant so that 2 cm of the platform is visible above the water level. The position of the platform is unaltered during training days (reference memory procedure). The water should always be maintained at 25–26 °C. Now the rat is allowed to swim to reach the platform (cutoff time 60 s). If it fails to reach it by 60 s, it is gently guided to locate the platform and rest on it for 15 s, and the same process is repeated with a different starting position. The maximum number of trials should be 2–4 trials for 4–5 days. When the rat reaches the platform within 10 s, then it is trained.
  - Now the water is mixed with non-toxic opaque material, and the water level is raised so that platform is hidden. When the trained rat is placed again, using reference memory by cues, it will reach the platform. The latency time is noted down.
  - For testing the working memory, the rat can be trained to locate a new platform position. After 2–4 h of delay, the performance of the rat is evaluated.
- *Evaluation:* The latency time to reach the platform during training and after delay is compared between the groups.
- *Advantages:*

- Shorter duration of time for training animal.
- Water eliminated 'odor' based cues.
- Motivational or motor coordination problems, if any, in the rat can be identified.
- Animals need to be kept under food deprivation.

20.5.5 Visual Discrimination Test

- *Principle:* Visual information is processed from the retina to the cerebral cortex. The patterns in the visual information are compared with the images stored in the memory area when the animal is exposed repeatedly. Thus visual discrimination test is one of the crucial tests in experimental psychology and modern neurophysiology.
- *Procedure:*
  - *Prerequisites:* The animal should be devoid of pathological eye conditions, and adequate visual acuity should be ensured. The 'discriminant material' in the apparatus should be kept at an optimum distance for viewing for the animal for adequate time.
  - *Apparatus setup:* It consists of the start area, choice area, and goal area. The sliding door separates the start area from the choice area. The choice area is connected to the goal area via two separate doors that can be locked, and a plastic card with black-white patterns can be inserted in it. The floor of the start and choice area is electrifiable.
  - The animal is placed in the start area with all doors open and allowed to explore the setup for 5 min. Now the animal is placed in the start area again, and after 5 s, the sliding door is opened. Once the animal enters the choice area, after 5 s, the electric shock (50 Hz, 1 mA for 0.5 s for every 3 s) is applied until the animal reaches the goal area and is allowed to rest for 30s. The process is repeated till the animal masters it.
  - Now, two plastic cards with different patterns in black-white are inserted in the doors, which are located between choice and goal area. Out of two goal areas, the room to which the investigator wishes that the animal should not enter is called a 'negative room.' The door of this negative room is locked. The animal is kept in the start area, and the same process is repeated till the animal masters it.
  - After 24 or 48 h, the procedure is repeated to test for the retention of memory.
- *Evaluation*: The number of errors and the number of trials required in different drug groups is compared.

20.5.6 Olfactory Discrimination Test

- *Principle:* When compared to lights or sound, odor provides more discriminative control in animals significantly. In rodents, solving discriminative problems based on olfaction is equivalent to the acquisition rules of higher primates.
- *Procedure:*
  - Rat (Sprague Dawley) is used in this experiment. The rat should be deprived of water for at least 48 h. The apparatus consists of a rectangular box mounted with multiple photosensitive detector cells and light beam provisions.
  - The positive area consists of a positive odor (e.g., an odor of banana) with a water spout. The negative area consists of a negative odor (e.g., an odor of orange) with no water spout. The rat is placed in the rectangular box and trained to reach the positive area with a break-in light beam.
  - When the rat reaches the positive area successfully, it is presented with water (0.05 ml). When the rat reaches the negative area, it is presented with only light flashes without water. The inter-trial interval should be 15 s, and the whole session should not exceed 30 min per day. When the rat makes 90% correct choice or after reaching 400 trails, the session should be terminated.
  - The same procedure is repeated after 24 or 48 h to test the retention of memory.
- *Evaluation:* The number of trials required to reach the criteria and the number of errors made in different groups is compared using ANOVA.
- This experiment is used to evaluate NMDA receptor antagonists, centrally acting anticholinergics, and ACTH analogs.

20.6 In-vivo Experiments in Other Animals

20.6.1 Aversive Discrimination in Chickens

- One day old White-leghorn chickens are used in this experiment. The chicken is trained to peck the intended beads (e.g., red color) while avoiding the aversive beads (e.g., beads coated with methyl anthranilate).
- Three different stages have been identified in this method involving memory processing in chickens:
  - Short-term memory stage (10 min after learning)
  - Intermediate memory stage (20–50 min after learning)
  - Long term memory stage (60 min after learning)
- This method is helpful in evaluating drugs influencing memory formation.

20.6.2 Discrimination Studies in Aged Monkeys

- Since AD is associated with aging, the aged monkeys can be used to evaluate the drugs for AD. The setup usually consists of an automated reward-punishment

system with a camera and an automated score recording. The aged monkeys are trained to get rewards by pressing the correct lever, and the retention of memory is tested after 24–48 h.
- The main disadvantages of using aged monkeys in the experiment are the alteration of PK-PD parameters of testing drugs and high cost.

20.7 Genetic Models for Testing Memory and Learning

- Various genetic models are being developed to test the drugs influencing memory and learning in pathological conditions like AD, ALS, etc. The following is the list of genetic models currently available.
- *APP transgenic mouse:* It expresses familial autosomal dominant mutation of APP leading to APP accumulation within 8–12 months of age. All the features of AD are seen, but degeneration of cholinergic neurons is not present.
- *APP751 transgenic mouse:* It overexpresses human APP-751 leading to the development of AD features in an earlier stage of life. Altered behavior in Y – maze and water maze tasks are noted
- *K595N/M596L – APP transgenic mouse:* It overexpresses the Swedish mutation of human APP and develops amyloid plaques similar to AD in humans.
- *Bigenic mouse:* It develops both neurofibrillary tangles and amyloid plaques in the brain at the earlier stage of life. The main disadvantage is that the mouse becomes paralyzed soon, and difficult to test the behavioral methods.
- *Triple transgenic (3x-TG) mouse model:* This mouse has a mutation in presenilin-1 (PS-1), tau, and APP. Thus an age-dependent increase in soluble Aβ protein, neuronal accumulation, and extracellular deposition is seen by the seventh month of its age. Behavioral changes are noted from the third month onwards.
- *5x-FAD Mouse:* It overexpresses presenilin-1 (PS-1) and four mutant APP proteins, namely APP-695, Swedish APP (K595N/M596L), Florida APP (I716V), and London APP (V717I). It is the best model for studying the amyloid formation and intra-neuronal Aβ-42 induced neurodegeneration. Microscopic changes are evident from 1.5 months of age onwards. It has poor novel object recognition, reduced anxiety, poor conditioned behavior, and impaired performance in the Y-maze test.
- *APOE transgenic mouse models:* Mouse with knock-in mutation of APOE2˙or APOE3 or APOE4 genes will produce abnormal apolipoprotein-E and develop AD features in an earlier stage of life. The double mutant strain called APOE4/Trem*2R47H has abnormal apolipoprotein-E production with CRISPR/Cas-9 protein mutation and is used for studying late-onset AD.
- *MAPT (H2,1)-GR mouse:* It expresses human-microtubule-associated protein tau (MAPT), leading to AD features.
- *Other models:* Due to development in understanding the pathophysiology of AD, numerous genetic models focusing on particular pathways like ALZ7 mouse, ALZ17mouse, JNPL3 mouse, pR5 mouse, Abca7 Knockout mouse, Ceacam-1

knockout mouse, Plcg2 knockout mouse, etc., have been developed and being utilized.

20.8 Invertebrate Models

- Invertebrate models are easy to maintain and cheap to procure. The main advantage is that owing to its short life span, and the degenerative features are quickly produced in invertebrate models.
- *Caenorhabditis elegans* and *Drosophila melanogaster* are commonly used as a model for AD as their genome has been completely sequenced, and genetic manipulation can be done quickly.

Bibliography

Drummond E, Wisniewski T. Alzheimer's disease: experimental models and reality. Acta Neuropathol. 2017;133(2):155–75.

Esquerda-Canals G, Montoliu-Gaya L, Güell-Bosch J, Villegas S. Mouse models of Alzheimer's disease. J Alzheimers Dis. 2017;57(4):1171–83.

Götz J, Bodea L-G, Goedert M. Rodent models for Alzheimer disease. Nat Rev Neurosci. 2018;19(10):583–98.

LaFerla FM, Green KN. Animal models of Alzheimer's disease. Cold Spring Harb Perspect Med. 2012;2(11):a006320.

Myers A, McGonigle P. Overview of transgenic mouse models for Alzheimer's disease. Curr Protoc Neurosci. 2019;89(1):e81. https://doi.org/10.1002/cpns.81.

Vogel H. Drug effects on learning and memory. In: Vogel HG, editor. Drug discovery and evaluation. Berlin, Heidelberg: Springer; 2007. https://doi.org/10.1007/978-3-540-70995-4_7.

Screening Methods for the Evaluation of Antianginal Agents

21

Saranya Vilvanathan

Abstract

In humans, angina is a multifactorial disease influenced by multiple pathogenic factors having a variable course and outcome in each patient. The major limitation for an experimental model for angina is the inability to replicate the variable pathogenic features and hence cannot be used to assess the complete hemodynamic dysfunction. They are potentially used for testing various drugs in different settings to elucidate the pathologic sequence of events after angina. The different *In-vitro* models including Langendorff isolated heart preparation can be used to assess the heart rate, contractile force, cardiac output, coronary blood flow and left ventricular pressure. The different variants of this preparations and their advantages and applications in different settings can help with narrowing down of a particular high potency anti-anginal agent. *In-vivo* models, on the other hand, are done on animals such as dogs and pigs, where an angina model is experimentally induced by occlusion of coronary artery by surgical, electrical or chemical agents.

Keywords

Angina · Antianginals · Lagendorff · Ischemic heart disease

S. Vilvanathan (✉)
Department of Pharmacology & Clinical Skills, Medical University of the Americas, Nevis, West Indies
e-mail: s.vilvanathan@mua.edu

M. Lakshmanan et al. (eds.), *Introduction to Basics of Pharmacology and Toxicology*, https://doi.org/10.1007/978-981-19-5343-9_21

Table 21.1 List of animal models used for screening anti-anginal agents

	I. In-vitro models:	II. In-vivo models:
A.	Isolated heart (Langendorff) preparation	Occlusion of coronary artery
B.	Isolated Rabbit Aorta preparation	Microspheres-induced acute ischemia
C.	Calcium antagonism in the pithed rat	Isoproterenol-induced myocardial necrosis
D.	Relaxation of Bovine coronary artery	Stenosis-induced coronary thrombosis model
E.	Coronary artery ligation in isolated rat heart	Electrical stimulation-induced coronary thrombosis
F.	Isolated Heart-Lung preparation	Myocardial-ischemic preconditioning model
G.	Plastic casts technique in dogs	Models of coronary flow measurement

21.1 Introduction

Ischemic heart disease is characterized by the decreased blood supply to the heart's musculature, primarily due to the pathology in coronary arteries, which manifests in the patient as angina pectoris. The cardinal symptom of ischemic heart disease is angina pectoris, caused by periodic episodes of ischemia in the myocardial tissue due to an imbalance in the myocardial oxygen supply-demand relationship. The imbalance may be caused by the increase in myocardial oxygen demand or decreased myocardial oxygen supply or both.

The various animal models used for screening anti-anginal agents are listed in the Table 21.1 below:

21.2 In-vitro Models

21.2.1 Isolated Heart (Langendorff) Preparation

- Oscar Langendorff established this preparation in 1897.
- Principle: In an isolated heart preparation, retrograde perfusion of *in-situ* heart with oxygenated saline solution, through the aorta under constant pressure or constant flow closes the aortic valves during diastole. The solution will then be displaced into the coronary arteries flowing off into the right atrium and the coronary sinus. This preparation of heart under standard conditions can be used to study the positive inotropy, negative inotropy, coronary vasodilating effect of potential compounds.
- Procedure:

  Initially, to obtain heart for the preparation, Guinea pigs of either sex weighing around 500 g are assessed and are sacrificed by stunning. Anesthesia with barbiturate and maintaining the animal on artificial respiration is important and highly recommended to analyze and study the biochemical parameters in tissue and perfusate. Once after the animal is sacrificed, the chest cavity is opened quickly and the heart is dissected and removed. A petri dish with ringer's solution maintained at 37 °C is used to receive the organ. Once the heart is dissected, locate the major vessels including the aorta. Just below the point where the aorta divides, make a small nick to introduce a glass or plastic cannula. To position the cannula, tie it with two threads before initiating perfusion. Oxygenated ringer's solution or Krebs-Henseleit buffer solution is used as perfusate. Perfusion is initiated in a retrograde fashion from the aorta through the cannula, under constant pressure of around 40 mm Hg or flow. Transfer the heart to double-wall plexiform perfusion apparatus filled with water maintained at a constant temperature of 37 °C, using thermostat. Constant volume of perfusate is introduced from the reservoir maintained at temperature of 37 °C. The apex of the heart is attached through a string connected to a small steel hook. Potential compounds can be injected into the perfusion medium just above the aortic cannula to assess their effects on various cardiovascular parameters.

- Evaluation:
  To assess the heart rate, a chronograph attached to a polygraph can be used. A force transducer with a preload of 2.5 g can be used for isometric assessment of contractile force which can be recorded on a polygraph. The Statham pressure transducer P 23DB is used to measure the left ventricular pressure (LVP). The measured pressure on differentiation yields LV dp/dtmax. A drop counter can be used to measure the coronary flow. A mechanic electronic flowmeter containing a vertical pipe and magnetic valve can also be used for flow measurements. Electromagnetic flow probes attached to the outflow system can be used for cardiac output assessments.
- Applications of Isolated heart (Langendorff) preparation:
  - Positive inotropy: To assess the positive inotropic effect of a test compound, an experimental heart failure is induced by using an overdose of thiopental sodium (barbiturates) or calcium antagonists which in turn can reduce the cardiac force. The β-sympathomimetic activity of potential test compounds can be assessed by comparing its effect to a standard such as isoproterenol (β-agonist). From the change in left ventricular pressure (LVP), differentiation of LVP [dp/dt_{max}] and coronary flow measurements, the positive inotropic effect of the test compound can be assessed and compared to the test compound.
  - Negative inotropy:
    To assess the negative inotropic effects of test compounds, the effect of a β1-sympathethic stimulation is recorded first using a standard β-agonist drug, such as isoproterenol. The drug is given at dosage of 0.05–0.2 μg and the change in heart rate, increase in cardiac force and contractility are measured and recorded. Following this, the β-receptors in the heart are to be blocked using a standard β-blocker such as propranolol at dose of 0.1 mg. The attenuation in effects of isoproterenol after injection of β-blocker is measured and recorded. Similarly, the negative inotropic effect of the test compound can be tested by comparing its effects on isoproterenol attenuation with that of standard (propranolol).

- The coronary flow assessments done after injecting the potential test compounds can be used to study the vasodilating effects of these compounds on the coronary blood vessels.
- Antagonism of calcium channels: To assess the antagonistic effects of potential test compounds and the consequent vasodilating effect in the coronary arteries, a standard compound such as $BaCl_2$ is used to induce intense coronary vasospasm. The potential test compound with coronary vasodilating effect is to be injected, five minutes after $BaCl_2$ injection. The increase in coronary flow, following the injection of test compound is measured and recorded. The relaxing effect of the active test compounds due to calcium antagonism is correlated with the increase in coronary flow measurements.
- Potassium content in coronary outflow: In 1968, Lindner and Hajdu, while studying the effects of cardiac glycosides on Langendorff preparation described a method to assess the contractile force, coronary blood flow and the potassium content in the coronary outflow using flame photometry. The increase in potassium content in the coronary outflow following administration of cardiac glycosides correlates with the positive inotropic effect. This in turn can be used to assess the positive inotropic effects of test compounds.
- The effects of hypoxia and the potential damage to the myocardial tissue caused by myocardial ischemia and the subsequent hypoxic damage can be evaluated using the Langendorff model.
- Various metabolic studies, arrhythmogenic, antiarrhythmic, and anti-fibrillatory effects can also be assessed using the Langendorff method.
- This model can also be used to study EDRF (Endothelium derived relaxing factor) release from the coronary vasculature and electrophysiological evaluation of cardiovascular agents.

21.2.2 Gottlieb and Magnus Method (Balloon Method)

- Gottlieb and Magnus, in 1904, introduced the 'balloon method' which is regarded as one of the modified Langendorff technique. This preparation can be used to assess the preload and cardiac force in pressure units as well as record the left ventricular contraction curves with more accuracy.

- Procedure: Once after the Langendorff heart model is set up, a small balloon filled with water is attached to a tip of a catheter and inserted via the pulmonary veins into the left atrium, mitral valve and then the left ventricle. The catheter is fixed in placed by attaching and tying it to the stem of pulmonary veins. The balloon is inflated to a size that matches the left ventricular volume and fits in the ventricular cavity and hence the size of the balloon is to be determined based on the species of the animal used and its body weight. Using a three-way valve, the balloon can be extended to a given preload. When the heart is allowed to beat, the rhythmic force exerted on the balloon and its effect on the pressure transducer membrane can be assessed to record the ventricular contraction curves.
- Applications: The modified Langendorff – Balloon method can be used to assess the following additional parameters:
  - The contractile force development in the left ventricle and preload can be reproduced reliably in pressure units (mmHg)
  - The tracings of left ventricular contraction curves can be used for further calculations
  - Using a rate meter, continuous heart rate recordings can be assessed.

21.2.3 Doring and Dehnert (1998) Method

- Another modified method of Langendorff was described by Doring and Dehnert to assess the cardiac diameters in an isolated perfused guinea-pig heart using continuous ultrasonic receivers.
- In this method, ultrasonic recording of longitudinal and transverse cardiac diameter is done in a continuous and simultaneous manner using the ultrasonic receiver inserted in the ventricular chamber. For assessing the LV transverse diameter, insert the ultrasonic receiver into the right ventricular cavity via the right atrium to approximately the same height as that of the transmitter. For measurement of longitudinal diameter of the ventricle, additional transducers can be placed at the base and apex of the heart.

21.2.4 Isolated Rabbit Aorta Preparation

- Principle: Aortic rings are used to evaluate the smooth muscle contractile or relaxant activity. To induce contraction of the smooth muscle in aortic ring structure attached to the organ bath, potassium chloride or norepinephrine is added to modified Kreb's bicarbonate buffer solution used in the bath.
- Procedure:
  - Rabbits of either sex, weighing around 4000 g, are initially sacrificed with pentobarbital sodium pentobarbital given at a higher dose. A bilateral incision

is made to open the thorax. Dissect the heart and large blood blood vessels to remove the descending thoracic aorta. The dissected aortic vessel is quickly placed in Kreb's bicarbonate buffer solution maintained at 37 °C. Cut off eight rings each of 4 to 5 mm width and mount each in a 20 ml organ bath filled with Kreb's solution, under 1 g tension.
  - After 2 h stabilization period, add KCl to induce sustained contraction in the mounted tissue. Twenty minutes later, the test drug in added and the percent tissue relaxation induced by test drug can be assessed after 30 min. Different test drugs can be assessed for their relaxant effect with a time interval of at least 30 min in between addition of different drugs.
- Evaluation: ID_{50} and percentage relaxation caused by the test drug from the precontracted level are calculated. Those potential test compounds with calcium channel blocking effect can be assessed for their relaxing effect on the vascular smooth muscle tone.

21.2.5 Calcium Antagonism in the Pitched Rat

- Principle: This model can differentiate calcium entry blockers from other agents that do not directly block calcium entry.
- Procedure:
  - Sprague Dawley rats, weighing around 300 g are anesthetized by intraperitoneal injection of methohexitone sodium at a dose of 50 mg/kg. The rats are pithed through one orbit and are immediately put on artificial respiration after cannulating the trachea. The pithing rod can also serve as stimulating electrode to deliver square wave impulses at a supramaximal voltage of 0.5 Hz frequency. Continuous electrical impulses for a duration of 0.5 ms can be delivered to stimulate the thoracic spinal cord and initiate a cardio accelerator response. Only those rats which exhibits increase in heart rate (tachycardia) to about 100 beats/min are to be included in the study.
  - Dissect the neck to cannulate the jugular vein for test drug administration and insert a pressure transducer in the carotid artery to record the blood pressure.
  - When the cardio accelerator response is established for 3–5 min, calcium-channel blockers or β-blockers are administered.
- Before drug administration, the level of tachycardia is considered as 100%, and the change in response after drug administration is considered as a percentage of pre-dose tachycardia. ID_{50} is calculated and compared.

21.2.6 Relaxation of Bovine Coronary Artery

- Principle: Spiral strips of Bovine coronary artery can be used to assay the relaxation produced by different test compounds. Eicosanoids maintain the tonus of the coronary artery; prostacyclins induce relaxation, and thromboxane A_2 causes contraction.

- Procedure: The left descending coronary artery is dissected from the bovine heart and is preserved in cold oxygenated Kreb's solution. Small arterial spiral strips are made, one of which is tied and suspended in the organ bath of 4 ml volume at a tension of 2 g. Oxygenated Kerb's solution is added to the organ bath and the temperature of the organ bath is maintained at 37 °C. A mixture of antagonists is added to Kreb's solution to inhibit the action of endogenous acetylcholine, histamine, 5-hydroxy-tryptamine or catecholamines. Prostaglandin analogues, PGE_2 and PGI_2, each at dose of 100 ng/ml can be used as standards to induce contraction and pronounced relaxation respectively. The maximal response of the test compound can be expressed as percentage variation.

21.2.7 Coronary Artery Ligation in Isolated Rat Heart

- Principle: Langendorff technique can also be used to produce regional ischemia by applying a clamping the left coronary artery at its origin. Changes in reperfusion period can be observed after the removal of the clip. The efficacy of the test compounds with coronary vasodilating effects can be assessed by its ability to prevent the changes induced.
- Procedure: Wistar rats of either sex, weighing around 300 g are initially sacrificed by decapitation. Through a midline incision, dissect around and remove the heart from the thoracic cavity. Cannulate the aorta and perfuse the vessel with a non-circulated perfusion medium, just as in the Langendorff procedure. A silicone balloon in inserted into the left ventricle, inflated just in the size to fit in the cavity. The balloon is then attached to an artificial systemic circulation. A flowmeter probe and integrator connected in series is utilized to assess the stroke volume of the heart by measuring the fluid volume pressed from the balloon during each heartbeat. The isolated working model of heart is perfused with Kreb's buffer at 65 mm Hg for a period of around 20 min. The artery is clamped at its origin for around 15 min to induce acute myocardial ischemia. Remove the clamp to monitor and observe the changes during reperfusion for a period of 30 min. Hemodynamic parameters such as heart rate, left ventricular pressure, cardiac output and rate of coronary flow are measured. Samples are obtained from the coronary effluent to assess LDH (lactate dehydrogenase), CK (creatinine kinase), glycogen, ATP and lactate level determination. The potential test compound can be added to the perfusate medium just before occlusion or 5 min before reperfusion to record the changes in the hemodynamic parameters. The incidence and duration of ventricular fibrillation after treatment with test drugs are compared with controls.

21.2.8 Isolated Heart-Lung Preparation

- Principle: Isolated heart-lung preparation used to study various physiological and pharmacological processes has been established in rats.

- Procedure: Wistar rats, weighing around 400 g are anesthetized with intraperitoneal pentobarbitone sodium at a dosage of around 50 mg/kg. Cannulate the trachea and maintain the animal on artificial respiration. The chest cavity is opened, and ice-cold saline is injected to arrest the heart. The aorta, superior and inferior vena cava are cannulated. The heart-lung preparation is perfused with Kreb's-Ringer bicarbonate buffer (pH 7.4) containing rat RBC (hematocrit 25%). The perfusate is pumped from the aorta and is passed through the pneumatic resistance and collected in a reservoir maintained at 37 °C. It is then returned into the inferior vena cava, thus perfusing only the heart and lung. The test drug is administered 5 min after start of the experiment, and various parameters are recorded in the test drug and control group.

21.2.9 Plastic Cast Technique in Dogs

- Principle: Sudden or gradual occlusion of any one of the major coronary branches may stimulate development of collaterals. When the arterial coronary bed is filled with plastic, the collaterals can be visibly quantified.
- Procedure: Dogs (10–15 kg) are intravenously anesthetized with pentobarbital sodium (30 mg/kg). Cannulate the trachea and maintain the animals on artificial respiration. Open the chest cavity to expose the heart. After removing the pericardium, Ameroid cuffs are placed around the major coronary branches. The plastic materials gradually swell and occlude the lumen in 3–4 weeks. The animals are given a test drug or placebo for six weeks. Sacrifice the animals, remove their hearts and flush the coronary beds with saline. Liquid Araldite is filled in bulbus aortae, the coronary and venous tree. Care is taken to maintain uniformity of filling pressure, filling time, and viscosity of the filling material. After polymerization is complete, the tissue is digested with 35% KOH. Plastic casts obtained from test animals treated with drugs are compared with the controls, and the ability of test drug to induce the collaterals is assessed by checking the number and size of collaterals.

21.3 In-vivo Methods

21.3.1 Occlusion of Coronary Artery in Anesthetized Dogs and Pigs

- Principle:
  - In anesthetized open-chest dogs or pigs, occlusion of the left anterior descending artery can induce myocardial ischemia and infarction. The size of the infarct can be studied before and after administration of potential test compounds.
  - Gelatin mass of $BaSO_4$ injection given in the left coronary ostium can serve to delineate the postmortem area at risk of infarction in coronary arteriograms. Nitro-blue tetrazolium chloride stains done on myocardial tissue can be used to visualize the infarct size.

- Preparation for the procedure:
  Dogs of either sex weighing approximately 30 kg are initially anesthetized with pentobarbital sodium injection given intravenously as IV bolus of 35 mg/kg followed by continuous infusion of 4 mg/kg/h. Place the animal in right lateral position and maintain them on artificial respiration using a positive pressure respirator. Collect arterial blood to check for arterial blood gases. The ventilatory rate and the oxygen flow rate are adjusted to maintain physiological blood values: pH 7.47, PO_2: 100–140 mmHg, PCO_2: 32–40 mmHg. The saphenous vein is cannulated to inject potential test compounds. ECG is continuously monitored in lead II. Hemodynamic parameters including peripheral systolic and diastolic blood pressure, heart rate, left ventricular pressure, left ventricular end-diastolic pressure are measured. Cannulate the femoral vein and attach the cannula to a pressure transducer to monitor the systolic and diastolic blood pressure. To determine the left ventricular pressure (LVP), catheterize the left coronary artery using a Miller microtip catheter [PC 350]. A high-sensitivity scale is used to measure the left ventricular end-diastolic pressure [LVEDP]. The pressure curves are recorded and can be used to assess the heart rate and differentiated dp/dt max values.
- Experimental procedure:
  - A left thoracotomy incision is made between the fourth and fifth intercostal space and the heart is exposed. Open the pericardium of the heart and dissect around the left anterior descending artery [LAD] to expose it. Hemodynamic parameters are monitored throughout for a period of around 45 min to achieve steady-state conditions. Once after steady-state conditions are reached, the first diagonal branch of LAD is identified and ligated for a period of around 360 min. The arrhythmic activity, if any, developed after ligation is left untreated. In different schedules of administration, the control vehicle and potential test compounds are given through intravenous bolus or continuous infusion and the hemodynamic parameters are monitored and recorded through the entire procedure. Sacrifice the animal at the end of the procedure with overdose of pentobarbital sodium. The heart is removed and is used to determine the infarct area at risk.
- Post-mortem analysis of the heart tissue:
  - Coronary arteriograms are to delineate the anatomic post-mortem area at risk. A purse-string suture is placed around the left coronary ostium in the sinus of Valsalva; After inserting a cannula in the ostium, the suture is tightened and close. Through the cannula placed in the ostium, a 12% gelatin solution with suspended micronized $BaSO_4$ (maintained at temperature of 37 °C) is given under increasing pressure as follows: 2 min at 100 mmHg, 2 min at 150 mmHg, and 2 min at 200 mmHg. The heart is placed in crushed ice to gel the injectate. Remove the right ventricle. The left ventricle with the septum is then transversely cut into thin sections of approximately 1 cm from the apex to the base, till the site of occlusion. Assess each slice individually with

coronary angiograms made with an X-ray tube at 40 kV. The defect in opacity is indirectly determines by decrease in $BaSO_4$ filled vessels in infarct tissue, which can be recorded as a measure to determine the post-mortem area at risk of infarction.
  - To determine the infarct size, the slices are incubated in p-nitro-blue tetrazolium solution and the infarct tissue is visualized; blue/violet-stain indicates healthy tissue, unstained areas indicate necrotic tissue. For determination of the infarct area, the slices are placed on color transparency film and photographed. The left ventricle, infarct area and the area at risk of infarction are measured by planimetry from projections of all sizes; excluding the slice of apex and that containing the ligature.
- Evaluation:
  - The hemodynamic parameters are measured and evaluated. The change in parameters in drug-treated animals and controls can be compared to assess the efficacy of the test compound. Each parameter or characteristic can be assessed individually. The size of infarct area and area at risk are expressed in mean values $\pm$ SEM. Student's t-test, regression and correlation analyses are done for statistical assessments.
- Modifications of the method:
  - In 1986, Raberger et al. elaborated a modified version of the model in conscious dogs where in only transient myocardial dysfunction is induced.
  - In 1990, Hartman and Warltier described multivessel coronary artery disease model in conscious, chronically instrumented dogs
  - In 1998, Symons et al. experimented the attenuation of regional dysfunction in response to 25 cycles of ischemia (2 min) and reperfusion (8 min) of the left circumflex coronary artery in conscious swine after administration of a Na^+/H^+ exchange inhibitor.
  - In 1995, Klien et al. used intact pigs and found myocardial protection by Na^+/H^+ exchange inhibition in ischemic reperfused hearts.

21.3.2 Microspheres Induced Acute Ischemia

- Purpose and rationale:
  - The effect of potential test compounds on myocardial performance during acute ischemic left ventricular failure can be evaluated with this model. Repeated injection of microspheres (50 μm) into the left main coronary artery of anesthetized dogs can induce acute ischemia and left ventricular failure. Once the hemodynamic parameters are measured and recorded, test drugs can be given and the improvement in myocardial performance can be evaluated.
- Procedure:
  - Dogs (weighing 30 kg) of either sex are initially anesthetized with intravenous injection of pentobarbitone sodium (40 mg/kg) and maintained at a supplementary dose of 4 mg/kg/h. The trachea is cannulated, and the animal is

maintained on artificial ventilation, and the ECG is recorded continuously. Cannulate the brachial vein for administration of analgesics; the saphenous vein is cannulated for administering test compounds, and the femoral artery is cannulated and connected to a pressure transducer for measurement of systolic and diastolic pressure.
  - Through a left thoracotomy incision, the heart is exposed, and microspheres are injected through the angiogram catheter; initially as 10 ml and later as 5 ml bolus about 5 min apart
  - Embolization is terminated when left ventricular end-diastolic pressure [LVEDP] increases to 20 mmHg and heart rate increases to 200 beats/min. Test compound and the control agents are administered intravenously, and the changes produced in hemodynamic parameters are recorded.
- Evaluation:
  - Hemodynamic parameters such as LVEDP (measured using Miller microtip catheter), heart rate (measured using the pressure curve), mean pulmonary capillary pressure, mean pulmonary artery pressure (PAP), cardiac output is measured (using cardiac index computer and a balloon tip triple lumen catheter with the thermistor positioned in the pulmonary artery via the jugular vein).
  - Other parameters such as stroke volume, tension index, coronary vascular resistance, total peripheral resistance, pulmonary artery resistance can also be measured
  - Changes in the parameters for the drug-treated animals are compared to the vehicle controls.
  - Mean embolization time, the dose of microspheres, and number of microspheres are evaluated

21.3.3 Isoproterenol-Induced Myocardial Necrosis in Rats

- Rationale:
  - Administration of natural and synthetic sympathomimetics at high dosage can induce cardiac necrosis. The induction of these lesions can be partially or totally prevented by pre-treatment with drugs such as sympatholytics and calcium antagonists.

Table 21.2 Histopathological grading of myocardial necrosis induced by high-dose isoproterenol

Grading	Histological feature
Grade 0	No histological change observed.
Grade 1	Mild focal interstitial response observed.
Grade 2	Focal lesions observed in many sections; mottled staining and fragmentation of muscle fiber present.
Grade 3	Confluent retrogressive lesions with hyaline necrosis and fragmentation of muscle fiber is observed.
Grade 4	Massive infarct observed with occasional acute aneurysm and mural thrombi.

  - Rona et al, in 1959 described infarct like myocardial lesion induced by high dose isoproterenol in rats.
- Procedure:
  - Male Wistar rats, weighing around 200 g are placed in groups of 10 and are pre-treated with test drug or standard administered either orally or by subcutaneous route for one week. Both the groups are then administered high dose isoproterenol; 5.25 and 8.25 mg/kg on two continuous days. The symptoms in each group are carefully monitored and recorded and the mortality rate is noted and compared. Around 2 days (48 h) after administration of high dose isoproterenol, the rats are sacrificed. Autopsy is done where hearts are removed, weighed and frontal sections are sliced and embedded for histopathological examination.
- Evaluation:
  - The histological grading of the tissue is done as per the grades given in Table 21.2 below:

For each group, the main grade is calculated with the standard deviation to reveal significant differences.
Critical assessment of the method: Many authors have used the test to evaluate coronary active drugs, such as calcium antagonists and other cardioprotective drugs such as nitroglycerine and molsidomine (Vertesy et al. 1991; Classen et al.1993).

21.3.4 Stenosis-Induced Coronary Thrombosis Model

- Principle: When the coronary arteries are transiently clamped, the alteration and stagnation of coronary blood flow that results in the constriction site can cause transient platelet aggregation and thrombus formation. The cyclic flow variations after administration of test compounds can be recorded and compared with pre-treatment values.
- Procedure: Dogs are anesthetized using pentobarbitone sodium (40 mg/kg) intraperitoneally. Dissect and expose the left coronary artery (LCA). An electromagnetic flow probe placed in the proximal part of LCA can be used to measure blood flow in the vasculature. The artery is then transiently squeezed with a hemostatic clamp for 5 s and a plastic constrictor is placed around the artery at the site of damage. It is adjusted and changed several times till the desired constriction is achieved. The potential test compound is administered IV, and the cyclic flow variations are registered for 2–5 h and compared with pretreatment values. Dogs with repeated cyclic flow variations of the same intensity are used for experimental purpose

21.3.5 Electrical Stimulation-Induced Coronary Thrombosis

- Principle: Electrical stimulation induces thrombosis in the coronary artery. An alteration in coronary blood flow with transient platelet aggregation at the coronary constriction site is assessed using this model.
- Procedure:
  - German Landrace pigs (40 kg) are anesthetized with ketamine (2 mg/kg, i.m), metomidate (10 mg/kg, ip), xyaline (2 mg/kg, i.m.).
  - Tracheal intubation is done, and they are maintained on artificial respiration through a tracheal tube using a positive pressure respirator.
  - An electromagnetic flowmeter is placed on the proximal part of the left coronary artery (LCA) to measure coronary blood flow.
  - Vanadium steel electrode is placed in the vessel with the intimal lining and connected with the Teflon coated wire of 9-volt battery, potentiometer, and amperemeter. Tunica intima is stimulated with 150 μA for 6 h that causes occluding thrombosis
  - The test drug is given by subcutaneous route either during the electrical stimulation or 30 min following the stimulation
  - Hemodynamic parameters such as systolic, diastolic, and mean blood pressure, heart rate are measured by cannulating the femoral artery and connecting it to a pressure transducer.
  - Left ventricular pressure and left ventricular end-diastolic pressure are measured by inserting a micro dip catheter via the carotid artery retrogradely.
  - Time interval until thrombotic occlusion, thrombus size is determined.

21.3.6 Myocardial Ischemic Preconditioning Model

- Rationale:
  - Preconditioning the heart by a brief duration of ischemia and reperfusion, can reduce myocardial damage produced by prolonged ischemia and reperfusion.
  - Preliminary preconditioning of the myocardium reduces the Infarct size, leakage of cellular proteins indicative of myocyte death, improves post-ischemic ventricular function, and attenuates cardiac arrhythmia associated with frequent ischemia/reperfusion.
- Procedure:
  - Rabbits are anesthetized with ketamine (50 mg/ml)
  - The trachea is cannulated, and the animal is maintained on artificial respiration (30 respirations/min).
  - Catheterize the right femoral artery and vein for measurement of arterial pressure using pressure transducer and for the administration of drugs, respectively
  - Hemodynamic parameters are measured.

  - A 4–0 suture is looped around the marginal branch of LCA to facilitate coronary occlusion.
  - Ischemic preconditioning is induced by tightening the loop around the coronary artery for 5 min and loosening to reperfuse the myocardium for 10 min before subsequent 30 min occlusion.
  - After 30 mins of ischemia, the ligature is released for 120 min of reperfusion.
  - Prior to 30 min of occlusion, the rabbits are selected to receive ischemic preconditioning, no preconditioning or preconditioning, along with administration of test compound.
  - Animals are then sacrificed after the reperfusion duration.
- Evaluation:
  - Comparisons between systemic hemodynamic data and infarct size studies are analyzed by ANOVA using statistical software

21.3.7 Models of Coronary Flow Measurement

The anti-anginal properties of various compounds can be screened based on their coronary artery dilating property and measuring the coronary outflow in open and closed chest animal preparations.

Bibliography

Gupta SK. Drug screening methods. 3rd ed. New Delhi: Jaypee Brothers Medical Publisher (P) Ltd: Jaypee Brothers Medical Publishers; 2016.

Verdouw PD, van den Doel MA, de Zeeuw S, Duncker DJ. Animal models in the study of myocardial ischemia and ischemic syndromes. Cardiovasc Res. 1998;39(1):121–35.

Vogel H, editor. Drug discovery and evaluation: pharmacological assays. 3rd completely rev., updated and enlarged ed. Berlin, NY: Springer; 2008. p. 2007.

Screening Methods for the Evaluation of Antihypertensive Drugs

22

Saranya Vilvanathan

Abstract

Hypertension is the most common cardiovascular disease, the prevalence of which is increasing with advancing age. The applications of experimental hypertensive animal models include not only the study of various cardiovascular pathogenic changes during the course of development of hypertension but also to assess the effect of different interventions developed for the prevention and management of the disorder. Accurate measurement of blood pressure in animals can be done by non-invasive, tail cuff or Doppler, or invasive, indwelling catheter method. Reno vascular hypertension can be surgically induced by clamping the renal arteries which in turn can activate the renin-angiotensin-aldosterone system. Non-invasive rat models include hypertension induced by high-salt diet, fructose diet, mineralocorticoid injections causing salt and water retention or by psychogenic air-jet stimulation. The role of genetics in the etiology and pathogenesis of hypertension can also be studied in certain strains of rats that are salt-sensitive or spontaneously hypertensive. Dog and monkey models of hypertension were also developed to test the effects of various test drugs and to assess the cardiovascular parameters.

Keywords

Hypertensive models · Renin · Angiotensin · Cardiovascular parameters · Indwelling catheter

S. Vilvanathan (✉)
Department of Pharmacology & Clinical Skills, Medical University of the Americas, Nevis, West Indies
e-mail: s.vilvanathan@mua.edu

M. Lakshmanan et al. (eds.), *Introduction to Basics of Pharmacology and Toxicology*, https://doi.org/10.1007/978-981-19-5343-9_22

22.1 Introduction

Hypertension is a chronic cardiovascular disorder that occurs due to sustained elevation of systemic arterial pressure leading to progressive pathological changes in the vasculature and gradual hypertrophy of the left ventricle. The incidence and prevalence of hypertension increases with advancing age as well the risk of pathophysiological events including stroke, coronary artery disease, aortic dissection, renal insufficiency and so on. The etiology of hypertension is multifactorial with various genetic and environmenatal factors playing a significant role increasing the risk of hypertension.

Several animal models were eloborated to study the various cardiovascular changes that happen with the course of development of hypertension and to understand the pathogenesis of the different cardiovascular events. They give insight into the etiology and pathogenesis of the disease process. They also act as valuable tools to assess the effect of various interventions developed to prevent and manage this disorder.

22.2 Measurement of Blood Pressure in Animals

Accurate measurement of blood pressure is critical to the interpretation of various studies on the animal models of hypertension. The techniques or methods used to measure the blood pressure in rats and dogs (commonly used animals to study hypertension) are discussed below.

- *Measurement of blood pressure in rats*
  - *Non-invasive technique (Tail-cuff method):* This is the indirect non-invasive method of recording blood pressure in rats and is analogous to sphygmomanometry in humans. The influence of potential test compounds with antihypertensive effect can be studied and evaluated in spontaneous and experimentally induced hypertensive rats. In this method, a tubular shaped inflatable blood pressure cuff is attached to the base of rat's tail. Just distal to the cuff, a piezoelectric pulse detector is placed and then the cuff is gradually inflated well above the systolic blood pressure till the pulse is completely obliterated. The cuff is then slowly deflated till the pulse reappears and the changes are recorded in a polygraph.
  - *Invasive technique (In-dwelling catheter in conscious rats):*

    The most important advantage of this technique is that the blood pressure can be directly measured in conscious rats while eliminating the effect of anesthesia on cardiovascular system. Male Sprague Dawley rats are initially anesthetized with intraperitoneal injection of pentobarbitone sodium at dosage of 45 mg/kg. Following this, a midline abdominal incision is made, content of the abdomen removed and a trocar is passed through the psoas muscle just adjacent to aortic segment. A cannula is inserted through the trocar and the trocar is withdrawn from the body.

The other end of the cannula that comes out through the neck is then anchored to the neck tissue with silk suture. And the lower end of cannula is sutured to the psoas muscle. After blocking the neck end of the cannula with a tight-fitting stainless-steel needle, fill the cannula with heparin. Implant the other end to aorta. Puncture the aorta with a 27-gauge hypodermic needle and advance the tip of the PE catheter into the aorta. Replace the intestines and close the abdominal incision. Allow the rats to recover for a week and then remove the remove the stainless steel needle from the neck end of the cannula to flush it with heparin. The cannula can be attached to a pressure transducer (Statham P23Db) and the blood pressure recordings can be obtained in polygraph.

- *Blood pressure measurement in dogs:*

- Non-invasive method: Doppler technique
- Invasive catheter method

22.3 Animal Models of Hypertension

Most of the animal models of hypertension were developed based on the presumed etiological factors implicated as one of the etiology for development of hypertension in humans. Hence most of these animal models tend to share features that are also seen in humans. Some of these etiological factors associated with human hypertension that forms the basis for development of these animal models include genetic mutations, excessive salt intake and hyperactive RAAS (renin-angiotensin-aldosterone) axis. The list of experimental models for hypertension is tabulated in Table 22.1.

For an hypertensive animal model to be considered as an ideal model, the following criteria need to be fulfilled:

- The model should be feasible in small animals
- Should be simple in procedure and uniformly reproducible
- Should be able to assess and predict the anti-hypertensive potential of test compounds
- Should be comparable to some form of hypertension seen in humans

Table 22.1 Classification of experimental models of hypertension

Experimental models of hypertension				
I.In-Vitro models:	II. Animal models:			
	IIA. Rat models	IIB. Dog models	IIC. Monkey models	IID.Transgenic models
A. Endothelin-receptor antagonism in Porcine-isolated hearts	1.Renovascular hypertension: A. Goldblatt model or two-kidney one clip model [2K1C] B. One Kidney one clip [1K1C] model of chronic hypertension C. Two-kidney-two clip [2K2C] model for chronic hypertension	1.Chronic renal hypertension	1.Renin-inhibition in monkeys	1.Transgenic rats overexpressing mouse Ren-2 gene [TGR(mRen 2)
B. Monocrotaline induced pulmonary hypertension	2.Dietary hypertension:	2. Neurogenic hypertension		
C. ACE inhibition in isolated Guinea pig Ileum	3.Neurogenic hypertension: Pithed-rat models			
	4.Endocrine hypertension: DOCA-salt rats			
D. Isolated Guinea pig atria	5.Psychogenic hypertension: Air-jet stimulation-induced hypertension			
	6.Genetic hypertension: A. Salt-sensitive Dahl rats B. Spontaneously hypertensive rats			

22.4 In-Vitro Models of Hypertension

22.4.1 Endothelin Receptor Antagonism in Porcine Isolated Hearts

- *Principle:* Endothelin is a potent vasoconstrictor that can induce potent, sustained lasting contractions in isolated strips of blood vasculature, especially the porcine coronary artery, which is considered to possess ET_A receptors.

- *Procedure:* Dissect and isolate the left anterior descending coronary artery from the porcine heart and the strip off the vascular endothelium gently. The tissue strip is then suspended in an organ bath filled with Krebs-Henseleit solution bubbled with 95% O_2 and 5% CO_2 at 37 °C. Reference isometric contraction is obtained with 50 mM KCl, and later concentration-response curves are obtained with ET-1 and other agonists/test drugs added to the organ bath. Analysis of Schild plots will help derive the slopes and pA_2 values.

22.4.2 Monocrotaline Induced Pulmonary Hypertension

- *Principle:* Monocrotaline is a pyrrolizidine alkaloid, hepatotoxic and pneumotoxic agent that has the potential to cause progressive pulmonary hypertension leading to right ventricular hypertrophy and failure. Microscopic pathological changes in the structural architecture of blood vessels including degeneration and fragmentation of endothelial cells, edema in the perivascular tissue, red blood cells extravasation, increased muscularization of pulmonary arterioles and arteries are observed on exposure to the agent.
- *Procedure*: Sprague Dawley rats weighing around 250 g are given test drugs 1 week before the administration of 100 mg/kg monocrotaline injection. The animals are later sacrificed on days 4, 7, or 14 after which their major thorcic organs including the lungs and heart are removed. The major vessels around the heart are dissected carefully to remove the segments of pulmonary artery including the main pulmonary artery, right extrapulmonary and intrapulmonary artery. Once the segments are dissected and excised, the isolated tissue is suspended in organ bath fileld with Kreb's—Hensleit buffer. The bath is aerated with 95% O_2 and 5% CO_2 at 37 °C. After 1 h, KCl is added to the bath and the arteries are made to contract, and contractions are recorded with a lever transducer. The generated contractions and the maximum force produced are recorded and plotted as a function of applied force and recorded on a polygraph. The agonist drugs can be added to the organ bath to record the contrationa and relaxation of the tissue. The cumulative concentration response curve with recordings of response to KCl, norepinephrine and angiotensin II are plotted. The responses to each of the agent are plotted as negative log of agonist concentration and the mean responses are compared with Student t-test.

22.4.3 ACE inhibition in Isolated Guinea Pig Ileum

- *Principle:* Guinea pig ileum responds powerfully to Angiotensin II and Bradykinin. Angiotensin-converting enzyme (ACE) is a powerful enzyme that converts inactive angiotensin I into active angiotensin II; and is also involved in conversion of active peptide bradykinin into inactive products. Inhibition of ACE can decrease the activity of Angiotensin II and increase the activity of bradykinin. The degree of ACE inhibition of test drugs can be graded by its inhibitory action on the smooth muscle contraction (induced by Angiotensin II or Bradykinin) of Guinea pig ileum.

- *Procedure*: Guinea pig of either sex, weighinhg around 400gm, is selected and sacrificed by stunning. After tying a chord near the duodenum, the intestine is removed and separated from the mesentery. Tyrode solution is used to clean the intestine, and one of the distal pieces (which is more sensitive) is fixed to the organ bath. Angiotensin I is added (10 ng/ml) after 30 min equilibrium period, and the contraction is recorded. The test drug (ACE inhibitor) is then added, and the diminished contraction is recorded 5 min after the addition of the drug. A reverse response can be observed with the addition of bradykinin.

22.4.4 Isolated Guinea Pig Atria

- *Principle:* The β1 receptor blocking action can be evaluated using Guinea pig atria. Isoprenaline (β agonist) has positive ionotropic activity and can increase the frequency and intensity (force) of atrial contraction. It can also potentiate the contraction of electrically stimulated left atria. The addition of β1 sympatholytic drugs can decrease the isoprenaline-induced ionotropic action.
- *Procedure*: Guinea pig of either sex, weighing around 300 g, is used. The animal is sacrificed by stunning and exsanguination. The heart is removed, the right and left atrium are dissected out and suspended in a 50 ml organ bath filled with Krebs-Henseleit buffer with aeration (95% O_2 and 5% CO_2) at 37 °C. A lever transducer is used to record the contractions.
- Right atrium: After around 30 min, isoprenaline is added to the organ bath to induce inotropy. Cumulative dose is maintained starting from 0.5 μg/ml and consecutive doses at 3 min intervals. The bath is flushed for 3–5 min after the stable max plateau is achieved, and the test drug is then added. After 5 min, again, isoprenaline is added in the same concentration, and the response is observed. If the test drug has β1 receptor blocking activity, a higher concentration of isoprenaline will be necessary to attain the same response (potentiation and inotropy). The increase in the frequency and force of contraction is reduced when the same isoprenaline dose is added.
- Left atrium: A square wave stimulator of 92 impulses at 15 V for 1 min is used to stimulate the left atrium. After equilibrium is achieved, repeat the same as above with isoprenaline given at a dose of 0.050.15 μg/ml. The bath is flushed 3–5 times after the stable maximum plateau is reached, and the test drug is added to the organ bath. After 3 min, add isoprenaline in the same doses. If the test drug has β1 receptor blocking activity, the isoprenaline-induced activity is inhibited. The individual dose-responses are used to determine the IC50 values.

22.5 Rat models of Hypertension

22.5.1 Reno-Vascular Hypertension Models in Rats

This is the commonly used hypertensive animal model where in hypertension is experimentally induced in the animal models by constriction of renal artery and activation of peripheral renin-angiotensin-aldosterone [RAAS] axis and sympathetic nervous system.

In 1930, Goldblatt et al. described that when the renal arteries are partially constricted, it can induce hypertension in dog models and that this can be reproduced in other animal models including rats, rabbits and monkeys. Rats of around 150 g are chosen and anesthetized with hexobarbitone sodium given at a dose of 40 mg/kg. With a flank incision, the left renal artery is dissected and a silver clip of 0.2 mm diameter is used to clip the vasculature close to the aorta. The artery can also be ligated with a 4–0 silk suture to induce a constrictriction of more than 50%. The incision is closed and the animal is carefully monitored. The blood pressure recordings are noted. Approximately after 4 weeks of the ligation/clip procedure, when the recorded systolic blood pressure is more than 160 mmHg for two consecutive days, the animals are considered hypertensive. Three different variants of Glodblatt hypertensive model was described;

- Two kidney one clip (2K1C) hypertension
- One kidney one clip (1K1C) hypertension
- Two kidney two clip (2K2C) hypertension

22.5.1.1 Two Kidney One Clip (2K1C) Hypertension [Goldblatt Hypertension Model]

- *Principle:* In this model, one of the renal artery is constricted while the other renal artery on the contralateral side is left intact. In the rat, clamping the renal artery for a period of around 4 h can induce acute renal hypertension by activating the renin-angiotensin system. When the constriction is released and the vessel is reopened, the release of accumulated renin in circulation can lead to development of acute hypertension.
- *Procedure:* Sprague-Dawley rats, weighing sround 300 g are anesthetized with intraperitoneal hexobarbital sodium (100 mg/kg). After tracheal cannulation and the blood pressure is measured by the introduction of a pressure transducer into the carotid artery. The jugular vein is cannulated for administration of potential test compound. A polyvinyl chloride (PVC) coated clip is placed in the kidney's left hilum and fixed to the back muscles. The renal artery is occluded for around 4 h, after which the renal arterial clip is removed. Release of clip causes elevated plasma renin levels, which in turn increases the blood pressure. Test drug can be administered through intravenous route and the percent reduction in blood pressure measured and compared with pre-treatment values.

22.5.1.2 One Kidney One Clip Model (Chronic Renal Hypertension in Rats)

- *Principle:* The renal artery is constricted on one side, and the contralateral kidney is removed. The clip is left for 4–5 weeks, and the increase in blood pressure due to renal ischemia is measured and the changes produced on administration of test compound are compared with pre-treatment values.
- *Procedure:* Anesthetize the Sprague-Dawley rats with intraperitoneal pentobarbitone sodium (50 mg/kg). The left renal artery is dissected through a incision made in the flank, in the left lumbar area. The artery is constricted with a silve clip applied close to the aorta. Following this, remove the contralateral right kidney after tying off the renal pedicle. The clip is left intact for 4–5 weeks duration, and the blood pressure is measured. Potential test compounds are given for the next 3 days, and the change in blood pressure is measured and compared with the controls and also with the pre-treatment values.

22.5.1.3 Two-Kidney Two Clip Method [Chronic Renal Hypertension in Rats]

- *Principle:* In this method, both renal arteries are constricted with a clip causing severe renal ischemia, which activates the renin-angiotensin system and sympathetic nervous system. All these changes coupled with a profound increase in serum vasopressin can cause an increase in blood pressure. Percent reduction in blood pressure values after intravenous administration of test compound can be compared to pre-treatment values.
- *Procedure*: Anesthetize the Sprague-Dawley rats, weighing around 300 g, with intraperitoneal pentobarbitone sodium. Cannulate the trachea, jugular vein, and carotid artery. Both the renal arteries are clipped close to their origin from the abdominal aorta, with a U-shaped silver clip. Potential test compound is administered via jugular vein and the blood pressure changes are monitored using the pressure transducer placed in carotid artery.

22.5.2 Dietary Hypertension Models in Rats

Dietery manipulation of animals with special diet containing high salt and high sugar/fat can lead to development of hypertension on a long run after long term exposure. The increased sugar/fat in the diet can enhance the production of reactive oxygen species which in turn can induce oxidative stress and nitic oxide inactivation. This coupled with high salt diet can significantly elevate the blood pressure in these animal models of hypertension that in many ways reflects as a replica of human hypertension.

22.5.2.1 Fructose-Induced Hypertension in Rats

- *Principle*: High intake of sugar (glucose or sucrose, or fructose) can lead to development of spontaneous hypertension in rats. Fructose, in addition is also known to induce increased insulin levels, insulin resistance and increased triglyceride levels in normal rats.

- *Procedure:* Wistar rats (200–250 g) are placed in water (includes 10% fructose solution) and chow diet *ad libitum*. Various parameters, including water intake, food intake, body weight, blood pressure and pulse rate are monitored carefully every week. The pulse rate and blood pressure are measured non-invasively by the tail-cuff method. Blood samples are collected every 2 weeks after treatment with the test drug, and the plasma levels of glucose, insulin, and triglycerides are measured and compared with pre-treatment values.

22.5.2.2 High Salt-Induced Hypertension in Rats

- *Principle*: Long tern dietary manipulation with a high-salt diet can induce hypertension in rats and these animal models can morphologically resemble human hypertension.
- *Procedure:* Wistar rats (200–250 g) are given a chow diet *ad libitum*. High salt intake is ensured by using 1–2% sodium chloride solution for drinking water. The animals are monitored and parameters including pulse rate, blood pressure are measured by the tail-cuff method. Blood samples are collected before and every second week after treatment with the test drug.

22.5.3 Neurogenic Hypertension Model in Rats

Primarily neural damage can induce a permanent increase in blood pressure, establishing the role of central nervous system in the induction of hypertension. Denervation of sino-aortic baroreceptors is the most commonly used neurogenic model of hypertension.

Blood pressure in pithed rats:

- *Principle:* The pithed rat model is devoid of neurogenic reflex, which may alter the effect of primary drug and help with evaluating the action of potential test compounds on the cardiovascular system.
- *Procedure:* Male Wistar rats, weighing around 300 g are initially anesthetized using halothane. The trachea, jugular vein, and carotid artery are cannulated for artificial ventilation, injection of test compounds, and monitoring blood pressure using a pressure transducer, respectively. The rats are pithed by introducing a steel rod through the orbit into the foramen magnum down the spinal canal through its entire length. The air inlet of the ventilation pump is attached to a T-piece which can be utilized to increase the oxygen enrichment of the inspired air. Around 0.3 ml of blood is withdrawn from the carotid cannula 30 min after pithing and is analyzed for pH, pCO_2, pO_2 and bicarbonate concentration using a blood gas analyzer. After the potential test compound is given intravenously, the agonist dose-response curves are recorded and the potency ratios are calculated. The effect of α_1 and α_2 antagonism can be tested using this model, after administration of selective α_1 and α_2 agonists.

22.5.4 Endocrine Hypertension

Mineralocorticoids can cause salt and water retention, which can induce hypertension by increasing the plasma volume. Selye et al. demonstrated that deoxycorticosterone acetate (DOCA) administration in rats coupled with a high salt diet could cause hypertension attributed to increased sympathetic activity and increased vasopressin levels.

22.5.4.1 DOCA Salt-Induced Hypertension in Rats

- *Principle:* Deoxycorticosterone acetate (DOCA) salt is a mineralocorticoid that causes salt and water retention, increase in plasma volume and subsequently elevated blood pressure. Simultaneously administration of DOCA salt with high salt diet and unilateral nephrectomy can result in a low-renin hypertension which can be compared with an experimentally induced high-renin form of hypertension.
- *Procedure:* Male Sprague Dawley rats, weighing around 300 g are anesthetized with ether. A flank incision is made to remove the left kidney. DOCA salt (20 mg/kg) is dissolved in olive oil and injected subcutaneously into rats twice weekly for 4 weeks. DOCA pellets or implants can also be used to avoid repeated injections. Drinking water is replaced with 1% NaCl solution, and the changes in blood pressure are recorded before and after the potential test drug is given, to evaluate the anti-hypertensive efficacy.

22.5.5 Psychogenic Hypertension Models in Rats

- Repeated exposure to stress can increase the risk of persistent hypertension. Stress can be induced in different forms including certain form of emotional stimuli, immobilization stress, psychosocial stress or an electrical stimuli. This form of hypertension is not renin-dependent, and hence the plasma renin levels can be normal or suppressed. The degree and stability of hypertension are not comparable to other standard forms of hypertension. This model is more often used to study the pathophysiology of hypertension.
- *Air-jet stimulation-induced hypertension:*
- *Principle: Air*-jet stimulation, used as a stressor agent, can increase the systolic blood pressure in borderline hypertensive rats exposed for a long duration. Exposure to stress can alter the basic auto-regulatory stress response elements in the hypothalamic-pituitary-adrenal (HPA) axis, sympathetic adrenal medullary system and the RAAS axis inducing an increase in blood pressure.
- *Procedure:* Borderline hypertensive rats are exposed to short (20 min) or long (120 min) daily sessions of air-jet stimulation. The blood pressure is monitored non-invasively by the tail-cuff method. Animals exposed to a long period of stress have a significantly higher systolic blood pressure than those exposed for a short duration.

22.5.6 Genetic Models of Hypertension in Rats

- The role of genetics in the development and pathophysiology of hypertension can be studied in certain strains of salt-sensitive or spontaneously hypertensive rats.
- *Salt-sensitive Dahl rats:*
  - *Principle:* Salt-sensitive Dahl rats, when fed on the high salt diet, develop severe and fatal hypertension compared to salt-resistant strains (on a high salt diet).
  - *Procedure:* Sprague-Dawley rats (250–300 g) are maintained in a high Dahl salt diet (made by mixing salt with regular diet) and 8% NaCl as drinking water. Their blood pressure is monitored and the changes observed after giving the potential test compound is also recorded. After the experimental duration, the animals are sacrificed for the measurement and comparison of total cardiac mass, size and mass of left and right ventricles among the test drug and control group to study the drug's ability to reverse these changes.
- *Spontaneously hypertensive rats:*
  - *Principle:* Spontaneously hypertensive rats (SHR) can be used to study the pathophysiological features of hypertension including the end-organ damage that manifests in different organs such as ventricular hypertrophy, cardiac failure, and renal dysfunction. But gross vascular changes, including macrovascular atherosclerosis or vascular thrombosis, do not develop in these strains, and hence the occurrence of stroke is not a standard feature with these strains.
  - *Procedure:* Okamoto and Aoki described a strain of spontaneously hypertensive rats (SHR) obtained by breeding a strain of spontaneously hypertensive Wistar rats with female rats having slightly elevated blood pressure. A subset of this is the SHR stroke-prone strain (SHR-SP) which is prone to develop very high blood pressure levels and has a strong tendency to die from stroke.

22.6 Dog Models of Hypertension

22.6.1 Chronic Renal Hypertension

- *Principle*: After wrapping one kidney with cellophane, the contralateral kidney on the other side is removed by ligating the renal pedicle including the renal artery, vein, and ureter. Ischemia leads to the development of chronic renal hypertension, and the various test drugs are assessed.
- *Procedure:* Dogs (8–12 kgs) are anesthetized with intravenous thiopental 15 mg/kg. One kidney is wrapped with cellophane and placed back in the abdomen, and the other kidney is removed after ligating the pedicel. After 6 weeks, the blood pressure is measured by the indirect tail-cuff method or by direct measurement

through the carotid artery. Test drugs are administered for 5 days, and blood pressure measurements are recorded.

22.6.2 Neurogenic Hypertension

Principle: The carotid sinus and aortic arch baroreceptors play an very important and essential role in regulating the blood pressure. Sectioning of baroreceptors can induce a sustained increase in blood pressure, a form of neurogenic hypertension.

Procedure: Adult dogs (10–15 kg) are anesthetized using 15 mg/kg thiopental, 200 mg/kg sodium barbital and 60 mg/kg sodium pentobarbital. The carotid sinus nerves are carefully dissected, ligated and sectioned. This is followed by bilateral vagotomy to induce neurogenic hypertension. Test drugs are given through femoral vein cannula, and change in left ventricular pressure is recorded using Miller microtip pressure transducer inserted through the common carotid artery. Changes in hemodynamic parameters are expressed as a percentage of the values before and after administering the drugs.

22.7 Monkey Models of Hypertension

22.7.1 Renin Inhibition in Monkeys

Principle: Renin-angiotensin pathway is one of the most critical pathways involved in the regulation of blood pressure. Renin is an aspartyl protease enzyme that converts angiotensinogen into angiotensin I, which is then converted into Angiotensin II by the angiotensin-converting enzyme. Since the renin inhibitors developed so far were found to have high specificity for primates, the commonly used small laboratory animals such as mice, rats or dogs are not suitable for in-vivo screening of compounds with potential renin inhibiting activity.

Procedure: Marmosets (Callithrix jacchus) of 300–400 g, fed on pellet diet and fruits are used. 2 days before the experiment, they are anesthetized, and a catheter is implanted in the femoral artery for blood pressure recording. Cannualte the lateral tail vein for injection of the potential test compounds. Frusemide (5 mg/kg) is given intravenously to stimulate renin release. They are sedated with diazepam (0.3 mg/kg, intraperitoneally) and are restrained. The test compound and standards are given, and the changes in blood pressure are recorded.

22.8 Transgenic Models

22.8.1 Transgenic Rats Overexpressing Mouse Ren-2 Gene [TGR (mRen 2) 27]

Principle: Overexpression of mouse *Ren-2* gene in rats leads to the development of severe hypertension that can be potentially lethal to homozygous rats. Though it is a genetic model of hypertension, the exact mechanism underlying hypertension remains elusive, probably due to increased renin activity. The genetic background of rats [TGR (mRen2) 27] used for breeding contributes significantly to the severity of the hypertension.

End organ damage: Around 70% of heterozygous rats survive till 5 months of age. By that time, they develop marked hypertrophy of cardiac tissue and impairment of endothelium-dependent relaxations. This depends on increased local angiotensin II formation, and hence this form of hypertension is exquisitely sensitive to renin-angiotensin inhibition.

22.8.2 Knockout Models

- Genes for atrial natriuretic peptide and nitric oxide-synthase can be knocked out in rats to induce hypertension. ANF knockout rats result in salt-sensitive hypertension, whereas knockout of type A receptor for ANF can result in salt-independent hypertension and can be used to study various anti-hypertensive agents.

Bibliography

Dornas WC, Silva ME. Animal models for the study of arterial hypertension. J Biosci. 2011;36(4): 731–7. Review

Gupta SK. Drug screening methods. 3rd ed. Jaypee Brothers Medical Publisher (P) Ltd: Jaypee Brothers Medical Publishers; 2016.

Lerman LO, Chade AR, Sica V, Napoli C. Animal models of hypertension: an overview. J Lab Clin Med. 2005;146(3):160–73. Review

Vogel H, editor. Drug discovery and evaluation: pharmacological assays. 3rd completely rev., updated and enlarged ed. 2008th ed. Berlin; New York: Springer; 2007.

Screening Methods for the Evaluation of Antiarrhythmic Drugs

23

Saranya Vilvanathan

Abstract

Abnormalities associated with impulse generation or conduction can manifest as conduction block or re-entry of cardiac impulse and arrhythmias. Screening of anti-arrhythmic agents can be done with *in-vitro or in-vivo* models. The *in-vitro* Guinea pig models include Langendorff technique where in the isolated heart is perfused in retrograde direction from aorta at a constant pressure or flow. The other more simple and accurate method would be using an isolated papillary muscle from the right ventricle, mounted in organ bath, to assess the developed tension, excitability and effective refractory period. Acetylcholine or potassium induced arrhythmias in dissected atria of New Zealand white rabbits, can evaluate the potency and efficacy of anti-arrhythmic agents. In-vivo arrhythmic models can be chemically induced with agents such as aconitine, digoxin, strophanthin, adrenaline or calcium or can be electrical, mechanical or exercise induced models. Inherited ventricular arrhythmias were observed in a colony of German Shepherd dogs that predisposed to sudden death during sleep or at rest after exercise and can be used as genetic models to screen anti-arrhythmic effects of compounds.

Keywords

Impulse generation · Conduction abnormalities · Guinea pig models · Acetylcholine · Digoxin induced arrhythmias · Aconitine

S. Vilvanathan (✉)
Department of Pharmacology & Clinical Skills, Medical University of the Americas, Nevis, West Indies
e-mail: s.vilvanathan@mua.edu

M. Lakshmanan et al. (eds.), *Introduction to Basics of Pharmacology and Toxicology*, https://doi.org/10.1007/978-981-19-5343-9_23

23.1 Introduction

Anti-arrhythmic agents are drugs used to prevent or treat cardiac rhythm abnormalities that can be due to irregularities or impairment in cardiac conduction or automaticity. Abnormalities associated with impulse generation are considered impairment in automaticity, and anomalies related to impulse conduction can manifest as conduction blocks or re-entry of cardiac impulses. Anti-arrhythmic agents screening can be done in two different types of models: *In-vitro* models and *In-vivo* models and are summarized in Table 23.1.

23.2 In Vitro Models

23.2.1 Isolated Guinea Pig Papillary Muscle

- *Principle:* Guinea pig papillary muscle from the right ventricle of the heart is used to measure the developed tension (DT), excitability (EX) and effective refractory period (ERP). It is a non-microelectrode in-vitro model that is simple, precise and accurate to identify and classify the potential anti-arrhythmic agents belonging to different classes.
- *Procedure:* Guinea pig of either sex is sacrificed. The heart is removed and placed in a warm, oxygenated physiological salt solution. The right ventricle is opened and the tendinous end of the papillary muscle is ligated with a silk thread after which the chordae tendinae are freed from the ventricle. Mount the preparation in an organ bath while the required experimental conditions are maintained. To

Table 23.1 Experimental models of anti-arrhythmic agents

In-vitro models	*In-vivo* models
Isolated guinea pig papillary muscle	1. Drug/Chemical induced arrhythmia: Aconitine antagonism inducing arrythmias in rats Digoxin induced arrhythmias in guinea pigs Strophanthin/Ouabain induced arrhythmia Adrenaline induced arrhythmia Adrenaline- Halothane induced arrhythmia Calcium-induced arrhythmia
Langendorff technique	2. Electrical induced arrhythmia Ventricular fibrillation electrical threshold Programmed electrical stimulation-induced arrhythmia
Acetylcholine and potassium induced arrhythmia	3. Exercise-induced ventricular fibrillation Sudden coronary death models in dogs
	4. Mechanically induced arrhythmia A] Reperfusion arrhythmia in rats B] Reperfusion arrhythmia is Dogs C] Two-stage coronary ligation in Dogs
	5. Genetically induced arrhythmia

connect the muscle to force transducer, a silk thread is used and the muscles are field stimulated to induce isometric contraction at a duration of 1 ms. Pulses are delivered using a constant voltage stimulator. The developed tension (DT) is measured and recorded using a polygraph recorder. The force-frequency curve is obtained by measuring the developed-tension over a range of stimulus frequencies (0.3, 0.5, 0.8, 1.0, 1.2 Hz). The percentage change in the developed tension at 1 Hz is compared, post-treatment versus pretreatment, and is used to quantitate the ionotropic effect of the given test drug.

- *Evaluation:* Change in the effective refractory period, degree of shift in the strength-duration curve, percentage change noted in post-treatment developed tension, duration of the action potential, contraction force is calculated and is used to classify the agents into different classes of anti-arrhythmic agents. Drugs affecting the sodium channels will affect the excitability of the papillary muscle; those affecting the calcium channels will cause a change in the contraction force. Those affecting the potassium channels will affect the effective refractory period.

23.2.2 Langendorff Technique

- *Principle:* This is an In-vitro model, where the heart perfusion is maintained in a retrograde direction using oxygenated saline solution infused from aorta either at constant pressure or constant flow that closes the aortic valve during diastole, just in situ. The perfusate is then displaced into the open right atrium and coronary sinus.
- *Procedure:* Guinea pigs of either sex (300–500 g) are sacrificed. The heart is removed as soon as possible and placed in a Petri dish containing Ringer Lactate solution at 37 °C.

  Locate the aorta and make a small to cut below the division to insert a cannula and tie it in place. The heart is then perfused with Ringer lactate solution and transferred to double wall plexiglass perfusion apparatus maintained at 37 °C. To occlude the perfusion and induce arrhythmias, a ligature is placed around the left anterior descending coronary artery. The ligature is tightened to occlude the artery for around 10 min and maintained by reperfusion. Inject the test compound in the perfusion medium either before or after the occlusion. An epicardial ECG electrode is used for pulsatile stimulation and induction of arrhythmia. Rectangular pulses 0.75 ms duration, 10 V at the frequency of 4001800 shocks/min. Attach a small steel hook with string to the heart's apex, and the heart rate is measured with a chronometer. The incidence and duration of ventricular fibrillation and ventricular tachycardia in control and test group is recorded.

23.2.3 Acetylcholine or Potassium Induced Arrhythmias

- *Principle:* Acetylcholine and potassium can induce arrhythmias in isolated atrial muscle, and the anti-arrythmic potency of the test compounds to control the arrhythmias can be measured.
- *Procedure:* New Zealand white rabbit, weighing around 1–3 kg, are sacrificed, and their hearts are removed. The atria are dissected from other tissue and placed in Ringer Lactate solution and are attached to an electrode. Fibrillation is produced when the atria are exposed to acetylcholine or potassium chloride solution. After 5 min of exposure to acetylcholine or potassium chloride, the atria are stimulated with around 10 V pulses for a duration of 0.75 ms, and fibrillations induced are recorded on Kymograph. Control arrhythmia is induced for 6–10 min, followed by 30 min rest period. The test compound is added to the bath to check for its anti-arrhythmic potency.
- *Evaluation:* The test drug is found to be effective if fibrillation disappears immediately or within 5 min. If the fibrillation doesn't disappear in 8–10 min, it is inferred that the test compound doesn't work, and the preparation is washed.

23.3 In-Vivo Models

23.3.1 Chemically Induced Arrhythmias

- *Aconitine antagonism in rats:*
  - *Principle:* Aconite persistently activates sodium channels. Administering test compound 5 min before aconite infusion [aconite (5 mg/kg) dissolved in 0.1 N HNO_3] continuously in the saphenous vein and monitoring lead II ECG every 30 s can give an estimate of the anti-arrhythmic potency of the test compound.
  - *Procedure:* Male Ivanova rats, weighing around 0.3–0.4kg, are anesthetized intraperitoneal injection of urethane at a dose of 1.25 g/kg. Infuse aconitine dissolved in 0.1 N HNO3 continuously into the saphenous vein at the rate of 0.1 ml/min. The test compound is given 5 min before aconitine infusion and the lead II ECG is recorded every 30 s.
  - *Evaluation:* The dosage requirement of aconitine to induce arrhythmia in the test group (in which the test compound is administered) and control group is compared. The index of anti-arrhythmic activity is given by comparing the requirement of increased dosages of aconitine in test group compared to control group. The amount of aconitine required per 100 g of animal, based on infusion duration, is also calculated and can also be used to assess the anti-arrythmic potential of the test compound.
- *Digoxin induced arrhythmia:*
  - *Principle*: Cardiac glycosides, such as digoxin, when infused at higher doses, can induce arrhythmias, including ventricular extrasystoles, ventricular fibrillation and cardiac arrest. Test compounds with potential anti-arrhythmic

activity can decrease the dosage of digoxin required to induce arrhythmias and cardiac arrest.

- *Procedure:* Male Guinea pigs are anesthetized. Cannulate their trachea and catheterize their jugular vein and carotid artery. Digoxin, at dosage of 85 mg/kg, is infused into the jugular vein at a rate of 0.266 ml/min. The test compound is administered only in the test group, either orally for 1 h or intravenously for 1 min, before digoxin infusion. And the control group is infused only with digoxin. ECG recordings are traced in the lead III and compared between both the groups.
- *Evaluation:* The time taken for the onset of ventricular extrasystoles, ventricular fibrillation, and cardiac arrest is recorded in both test and control groups. The total amount of digoxin required to induce arrhythmias is measured and noted. Using the Student *t*-test, the doses of digoxin needed to cause arrhythmias in the test and control groups are statistically compared to analyse the anti-arrhythmic potential of the test compound.

• *Strophanthin or ouabain induced arrhythmia in dogs:*

- *Principle:* Strophanthin K, a cardiac glycoside, has a potential to induce ventricular tachycardia or multifocal ventricular arrhythmia in dogs. This potential of the compound can be used to assess the anti-arrhythmic potential of various test compounds.
- *Procedure:* Dogs of either sex are initially anesthetized and kept on artificial respiration. Locate two peripheral veins and cannulate the veins for administration of test compounds and chemicals (strophanthin K) to induce arrhythmia. Also the duodenum is cannulated for intraduodenal administration of test compounds. In one of the peripheral veins cannulated earlier, Strophanthin K is administered by continuous intravenous infusion at a rate of 3 mg/kg/min and the heart is monitored. When the induced arrhythmias are stable for 10 min, the test substance is administered intravenously through the other cannulated peripheral vein (5 mg/kg) or intraduodenally (10 and 30 mg/kg). ECG recordings are traced at 0.5, 1,2, 5- and 10-minutes following administration of test drugs, and further duration is increased if required.
- *Evaluation:*

  Intravenous route: The disappearance of strophanthin-induced arrhythmias after the administration of test compound at increasing dosages is evaluated. Once the test compounds reverse the induced arrhythmias, the next higher dosage of the test compound is given after the reappearance of stable arrhythmias. To reestablish the normal sinus rhythm, anti-arrhythmic medications such as ajmaline, quinidine or lidocaine at dosage of 1, 3 and 10 mg/kg are used.

  Intraduodenal route: The test compound is said to have an anti-arrhythmic effect if strophanthin K-induced extra systoles disappear within 15 min after administration of the chemical through intraduodenal route. If the test compound doesn't reverse strophanthin-K induced arrhythmias within the next 60 min, it is said to have no anti-arrhythmic potential.

• *Adrenaline induced arrhythmia:*

- Principle: Adrenaline at a high dose may precipitate arrhythmia.
- *Procedure*: Dogs, weighing around 10kg, are initially anesthetized with intraperitoneal pentobarbitone, given at dosage of 30–40 mg/kg. Locate and cannulate the femoral vein to infuse adrenaline at a rate of 2–2.5 mg/kg via the femoral cannula. The test drug is administered 3 min after adrenaline infusion, and Lead II ECG is recorded.
- *Evaluation:* The anti-arrhythmic potential of the test compound is evaluated by its potential to control adrenaline-induced arrhythmias. The immediate disappearance of the extrasystoles after the administration of the test compound will clearly indicate the anti-arrhythmic potential of the compound.

• *Halothane-adrenaline induced arrhythmia:*
  - *Principle:* The sympathetic and vagal influences on the pacemaker activity of the heart can be utilized to induce arrhythmias. Stimulation of the sympathetic innervation of the heart using sympathomimetic drugs and certain anesthetic agents including halothane which is known to sensitize the myoacardium to the action of catecholamines, can decrease the threshold for arrhythmias.
  - *Procedure:* Dogs or Guinea pigs can be used for this experiment. The animal is anesthetized using 1% halothane vaporized by 100% oxygen. Halothane is also infused at the rate of 2–3 μg/kg/min, along with adrenaline. The infusion speed is adjusted to produce stable ventricular tachycardia lasting for 20 min, and the drugs are given by i.v bolus.

• *Calcium-induced arrhythmia:*
  - *Principle:* injection of calcium chloride can induce arrhythmias, ventricular fibrillation, and flutter in rats
  - *Procedure:* Wistar Albino rats (60–130 g) are anesthetized with Nembutal (60 mg/kg) administered by intraperitoneal injection. Ventricular fibrillation and flutter are induced by administering 2 ml/kg of 10% aqueous calcium chloride through the femoral vein. Two percutaneous precordial clamps attached to cardioscope can be used to monitor the cardiac rhythm and behavior. The test drug is administered 2 min before calcium infusion.
  - *Evaluation:* Isolated ventricular premature systoles, short and long-run ventricular fibrillation and flutter are recorded and compared in the test and control groups.

23.3.2 Electrically Induced Arrhythmias

• *Ventricular fibrillation electrical threshold:*
  - *Principle:* Electrical stimulation of heart can induce ventricular fibrillation and the increase in ventricular fibrillation threshold (VFT) caused by the anti-arrhthmic potential of the test compounds can be assessed when the heart is electrically stimulated through different types of impulses including single pulse stimulation, train-of-pulse stimulation or continuous 50 Hz stimulation.

VFT is the minimum intensity of the electrical pulse that is required to induce sustained ventricular fibrillation in the anesthetized animal.

- *Procedure:* Adult dogs of around 10 kg are anesthetized using sodium pentobarbital at the dosage of 35 mg/kg and are maintained on artificial respiration with Harvard respiratory pump. Blood pressure is monitored, and the body temperature is maintained using a thermal blanket. The chest cavity is opened to suspend the heart in a pericardial cradle. Once after the sinus node is crushed and electrical impulses are delivered to the left ventricle with an Ag-AgCl stimulating electrode embedded in a Teflon disc sutured to the anterior surface of the ventricle. The test drug or standard is given through the femoral vein, and the heart is electrically stimulated for 400 ms through the driving electrode while recording the ECG in lead II.
- *Evaluation:* Ventricular Fibrillation Threshold (VFT) is determined first, using a 0.2–1.8 s train of pulses which are of around 50 Hz, delivered for a duration of 100 ms after every 18th basic driving stimulus. The intensity of the electrical stimuli is increased from the diastolic threshold. When ventricular fibrillation is recorded in the electrocardiogram, the heart immediately defibrillated and allowed to recover for the next 20 min. The increase in threshold to induce ventricular fibrillation using the electrical impulse before and after administration of test compounds is assessed and recorded at various time intervals.

• *Programmed electrical stimulation-induced re-entry arrhythmia:*
  - *Procedure:* Dogs, weighing around 10kg, are anesthetized with pentobarbital sodium given intravenously at dosage of 30 mg/kg and maintained on artificial respiration. After the left external jugular vein is cannulated, a thoracotomy incision is performed to expose the heart. The left anterior descending artery is identified and isolated, and a 20-gauge hypodermic needle is placed inside. A ligature is then tied around the artery and the needle inside for a while and then removed. When the needle is removed, it leads to a reactive critical stenosis of the blood vessel. The artery is then perfused for 5 min and then occluded for 2 h with a silicone rubber snare, which causes ischemic injury to the myocardium. After suturing an epicardial bipolar to the interventricular septum, the chest wall is closed and subcutaneous silver disc electrodes are placed on the chest wall for monitoring the ECG. After a week, the chest is re-opened, and programmed electrical stimulation is performed through the electrode implanted in the non-infarcted zone. The pacing stimuli are set to 200 ms, and after every 15 pacing stimuli, an extra stimulus is delivered. The test drug is administered 30 min after the stimulus.
  - *Evaluation:* The minimum electrical pulse intensity required to induce and sustain ventricular fibrillation is recorded before and after administering test drugs. The experiment is repeated for a minimum of around 10 times and the mean values are compared using the student *t*-test.

23.3.3 Exercise-Induced Ventricular Fibrillation

- *Sudden coronary death model in dogs:*
  - *Principle:* Physical exercise (stress) induced coronary vasoconstriction can be experimentally recreated in animals, which can very closely resemble the situation in coronary patients.
  - *Procedure:* Mongrel dogs (weighing around 15–20 kg) are anesthetized using intravenous sodium pentobarbitone at a dosage of 10 mg/kg. The chest cavity is opened and the heart is suspended in a pericardial cradle. Using a doppler flow inducer, measure the blood pressure. Dissect around the left coronary artery, and using a hydraulic occluder, the left coronary artery is occluded near the circumflex artery. Following this a pair of insulated silver-coated insulated wires are sutured to the left and right ventricles and are used to monitor the heart rate (Gold biotachometer). Insert a solid-state pre-calibrated pressure transducer in the left ventricle. Following this, a two-stage occlusion procedure, that is partial occlusion for around 20 min and tie off to completely occlude, is performed in the left anterior descending artery. The animal is to be maintained on analgesics and antibiotics throughout the procedure. 4 weeks after the procedure, the animals are made to walk on a motor-driven treadmill and are trained to quietly lie down on a table (without restraint) during the recovery phase. Susceptibility to exercise induced arrhythmias, is recorded by increasing the intensity of exercise given using the motor-driven treadmill.
    The treadmill protocol routinely starts with a 3 min warm-up period when the animal is made to run at a speed of 6.4 km/h. Following this warm-up, the speed intensity of the treadmill is increased every 3 min from 0 and 4 and 8 and 12% and then finally to 16%. During the last minute of exercise, complete occlusion of the left coronary artery is done and maintained for another 1 min after the treadmill is stopped, making the total occlusion time of 2 min. An electrical defribrillator is used when the animal becomes unconscious and the occlusion is released when ventricular fibrillation occurs.

    After the control group recordings are measured, the test compound is administered to the test group and the exercise plus ischemia experiment is repeated. The change in coronary blood flow and the effective refractive period are monitored, measured and recorded after pre-treatment with the test compound.
  - *Evaluation:* All hemodynamic data including the rate of change in left ventricular pressure, the effective refractory period, reactive hyperemia response to each occlusion is recorded and the mean value is obtained from a set of recording and the data is analyzed using analysis of variance (ANOVA). Chi-square test with Yate's correction is used to assess and determine the effect of drug intervention on induction of arrhythmia.

23.3.4 Mechanically Induced Arrhythmias

- *Coronary artery occlusion and reperfusion induced arrhythmia in rats:*
  - *Principle:* Occluding the coronary artery by ligation can lead to infarction of the myocardial tissue and subsequent reperfusion induced arrhythmias. An electrocardiogram recording is performed during the ligation and subsequent reperfusion procedure. Staining the myocardial lesions with p-nitro-blue tetrazolium chloride can help measure the amount of infarcted tissue in the myocardial lesions.
  - *Procedure:* Sprague Dawley rats (350–400 g) are anesthetized with intraperitoneal injection of pentobarbitone sodium, given at dosage of 60 mg/kg. Maintain the anesthetized rats on artificial respiration and cannulate the jugular vein for administration of test compounds. A pressure transducer inserted in the carotid artery and connected to a polygraph, is utilized to record the blood pressure of the animal. Open the chest cavity and expose the heart. Carefully dissect the left coronary artery, insert a thin silk thread around the artery and keep it ready for ligation during appropriate time. After an equilibrium period of around 45 min, once the animal is stable, administer the test substance intravenously through the jugular vein. Five minutes later, tighten the ligature around the left coronary artery for around 15–90 min followed by reperfusion for the next 30 min. Recordings of the blood pressure and ECG tracings are taken. After reperfusion, the animal is sacrificed and the heart is dissected and cut out in transverse sections of 1 mm thickness, from apex till the base. The slices obtained are stained with p-nitro-blue tetrazolium chloride to visualize the infarcted tissue within the myocardial lesions.
  - *Evaluation:* The ECG tracings are carefully analyzed to record the intensity and number of ventricular premature contractions, ventricular tachycardia and fibrillation episodes during the occlusion and reperfusion periods. Hemodynamic parameters are measured. After the staining procedure, the infarct size is determined.
- *Reperfusion arrhythmia in Dogs:*
  - *Principle:* Ligating the coronary artery in dogs, under controlled settings, can increase increase the heart rate, blood pressure, left ventricular contractility, left-ventricular end-diastolic pressure as well the risk of ventricular arrhythmias during the reperfusion period.
  - *Procedure:* Dogs, weighing around 23 kg, are anesthetized with intraperitoneal injection of thio-butobarbital sodium (30 mg/kg). Anesthesia is maintained with intravenous administration of chloralose (20 mg/kg) and urethane (250 mg/kg) followed by subcutaneous 2 mg/kg morphine. The animal is then kept on artificial respiration. Locate and cannulate the peripheral saphenous vein to inject the test compounds. Locate and cannulate the femoral artery and insert a pressure transducer to record the blood pressure. Attach ECG leads and the tracings are monitored in lead II. The left ventricular pressure curves are analyzed to determine the heart rate and left ventricular

end-diastolic pressure. Twenty minutes after administering the test compound, the coronary artery is ligated for 90 min.

- *Evaluation:* Changes in parameters (mortality, hemodynamic, and arrhythmia) in drug-treated animals are compared to vehicle controls.

- *Two-stage Coronary Ligation in Dogs:*
  - *Principle:* Two-stage coronary artery ligation can induce and sustain arrhythmia and is as effective as one-stage ligation procedure in dogs with a decrease in mortality risk.
  - *Procedure:* Dogs (8–11 kg) are anesthetized by intravenous injection of methohexitone sodium (10 mg/kg). After anesthesia, the animal is kept and maintained on artificial respiration. The chest is opened to expose the heart. Carefully dissect the left coronary artery and ligate the artery in two stages; after placing a 21-gauge needle inside the artery, a ligature is tied around the artery, which is then removed. After 30 min, the second ligature is tied around the artery tightly. The chest is closed in layers 30 min after the second ligature is placed, and the animal is recover from anesthesia. After 24–48 h of ligation, arrhythmia develops and abates within 3–5 days.
  - *Evaluation:* Lead II ECG, atrial electrogram, and mean blood pressure is measured. Changes in parameters (mortality, hemodynamics, and arrhythmia) in test animals (drug treated) are compared to vehicle controls.

23.3.5 Genetically Induced Arrhythmia

- *Genetic arrhythmia:*
  - *Principle*: Inherited ventricular arrhythmias observed in a colony of German Shepherd dogs predisposed them for sudden death that occurs very often during sleep or at rest after exercise. These dogs can be used to screen for potential anti-arrhythmic effects of compounds.
  - *Procedure:* The dogs are screened for ECG abnormalities. The dogs with the inherited disorder are more likely to have polymorphic ventricular tachycardia following long R-R intervals. These genetically induced arrhythmia are likely triggered by early depolarizations in the Purkinje system. The density of the transient outward current (I_0) channels and the time of constant inactivation is found to be reduced in the epicardial myocytes of these animals which are genetically predisposed to arrhythmias. There can be deficiencies in the cardiac sympathethetic innervations which can explain the predisposition to arrhythmias as well. The type of arrhythmias commonly observed in these animals include polymorphous ventricular tachycardia (Torsade de Pointes) which makes them counterpart for patients who have been reported to have similar type of arrhythmias with normal QT interval.

Bibliography

Gupta SK. Drug screening methods. 3rd ed. Jaypee Brothers Medical Publisher (P) Ltd: Jaypee Brothers Medical Publishers; 2016.

Vogel H, editor. Drug discovery and evaluation: pharmacological assays. 3rd completely rev., updated and enlarged ed. 2008th ed. Berlin; New York: Springer; 2007.

Screening Methods for the Evaluation of Cardiotonic Drugs

24

Abialbon Paul

Abstract

Various experimental methods are available to assay the cardiac activity of drug molecules in preclinical search for newer cardiac drugs. In vitro techniques include the ouabain binding assays, isolated hamster heart preparations and isolated cat papillary muscle preparations. In vivo techniques include mouse models, rat models and larger animal models like guinea pigs, hamsters, rabbits and dogs. Coronary artery ligation and banding in rats are useful models to test drugs that provide benefit in myocardial infarction and injury. Salt sensitive rats and spontaneously hypertensive rats are used to assay antihypertensive drugs. Larger animal models are more reliable as they are more similar to the human heart in structure and physiology. Dog models include the chronic pacing model, volume and pressure overload models and coronary artery ligation models. Rabbit models have an advantage over dog models in that they are less expensive, easier to maintain and the beta myosin heavy chain is the predominant myosin chain similar to human hearts. Genetic models include the Myosin Lim protein knockout mice and the TNF-alpha overexpressing mice. This chapter includes a brief overview of these models, their advantages and disadvantages.

Keywords

Drug assay · Coronary artery ligation · Cardiac models

A. Paul (✉)
Assistant Professor Pharmacology & Clinical Skills, Medical University of the Americas, Nevis, Saint Kitts and Nevis
e-mail: a.paul@mua.edu

M. Lakshmanan et al. (eds.), *Introduction to Basics of Pharmacology and Toxicology*, https://doi.org/10.1007/978-981-19-5343-9_24

24.1 In vitro & Ex vivo Methods

24.1.1 Ouabain Binding Assays

Principle: Cardiac glycosides can be characterized by its binding kinetics (association, dissociation, and equilibrium processes) on the ouabain receptor when incubated with ouabain.

***Description*:**

- Membrane preparation of cardiomyocytes required for this procedure is obtained by collagenase digestion from the hearts of dogs or rats.
- Tritium labeled ouabain with radioactivity of 20 C_i/mmol is incubated with the ligand to be tested along with $MgCl_2$, inorganic phosphate, and Tris-HCl buffer with a pH of 7.4. This is prepared into 10 ml of binding medium.
- To measure the association process, 200 μg of membrane preparation is added to a solution containing the binding medium with ouabain concentrations at 10 or 100 nM. At various time points, 4.5 ml of the medium is removed and rapidly filtered under vacuum on HAWP Millipore filters. The radioactivity that is bound to the filters is measured.
- To measure the equilibrium binding, varying concentrations of [3H]-Ouabain are added to 40 μg of membrane preparations. After 30 min, 4.5 ml is removed from the medium and filtered.
- To measure the dissociation kinetics, Mg^{2+} plus Pi-Tris-HCl with unlabeled ouabain is added to the previous solution which is already at equilibrium to initiate the dissociation of radiolabelled ouabain. At various time intervals, 0.9 ml is removed and rapidly filtered.
- Scatchard plots are used to analyze the various binding kinetics parameters.

24.1.2 Isolated Hamster Hearts

Principle: Drugs with cardiotonic activity increase the contractile force and alters the coronary blood flow in isolated hamster heart preparation.

***Description*:**

- Syrian Hamsters of either sex and age of at least 50 weeks are chosen for this study with similar matched controls.
- After pretreating with intraperitoneal heparin (5 mg/kg), the heart is prepared for the method proposed by Lagendorff and is perfused with Ringer's solution.
- The heart is kept at the isolated state for an hour at (30–32 °C) with a preload of 1.5 g.
- The drugs to be tested are added along with the perfusing solution and the various parameters (contractile force, coronary blood flow, and heart rate) are measured and compared with the controls.

24.1.3 Isolated Cat Papillary Muscle

Principle: Repeated stimulation of the papillary muscle results in diminishing contractions. The addition of a cardiac glycoside or drugs with action similar to cardiac glycosides restore the contractile force.

Description:

- This method uses the papillary muscle from the right ventricle as described by Catell and Gold.
- Cats of either sex are anesthetized with ether and the heart is exposed through a left thoracotomy. The papillary muscle from the right ventricle is isolated and fixed in an organ bath with a force transducer.
- Electrical stimuli of 4–6 V are applied at the rate of 30 per minute for about an hour until the contractile force decreases.
- The drug is added to the bath and the restoration of the contractile force is compared with that of ouabain.

24.2 In Vivo Methods

24.2.1 Mouse Models

The left coronary ligation model (myocardial infarction model) and the transverse aortic constriction model (pressure overload model) are very similar to their counterparts in rats (described below). However, because performing surgery in mice is difficult and the specimen size is small, these models are less preferred than rats. Genetic models are useful mice models. (Described under genetic models).

24.2.2 Rat models

24.2.2.1 Models of Myocardial Injury

The initial models used the subcutaneous injection of isoproterenol and electrocautery to the surface of the heart to impose myocardial injury. However, after Pfeffer and his team introduced the left coronary ligation model, it became the standard model for heart failure induced by myocardial injury.

24.2.2.2 Rat Left Coronary Artery Ligation Model

Principle: Complete obstruction of the left anterior descending artery, a branch of the left coronary artery results in myocardial infarction and overt heart failure. The incomplete obstruction results in chronic heart failure due to chronic ischemia. The infarct results in dilation of the left ventricle increased diastolic filling pressures, and reduced cardiac output.

***Description*:**

- Male Sprague Dawley rats weighing 250–300 g are chosen. They are anesthetized with 200 mg/kg of hexobarbital. The trachea is cannulated and the animal is put on artificial respiration. The chest cavity is opened and the LAD is identified and ligated.
- Four weeks after the procedure, the animal is again opened up and the carotid artery and the jugular vein are cannulated. All hemodynamic measurements are made with the control and with varying doses of the drug and noted.
- After the experiment, the hearts are isolated and homogenized to study the calcium content, ATPase levels, etc.

***Advantages*:**

- Multiple parameters can be assessed
- Larger size of the heart compared to mice makes surgery easier and more tissue for further analysis

***Disadvantages*:**

- Various similarities exist in the normal physiology of rat heart with the human heart.
  - The action potential is very short and lacks a plateau phase
  - The role of Na^{+}/Ca^{2+} exchanger is limited
  - The myocardium composition of the myosin chains is different.
  - The resting heart rate is high and the force-frequency relationship is inverse in rats.

24.2.2.3 Rat Aortic Banding Model (Pressure Overload Model)

Principle: Obstruction to the aorta results in left ventricular hypertrophy and hypertension. A subset of rats will develop cardiac failure.

***Description*:**

- Sprague-Dawley rats (250–280 g) are anesthetized with 200 mg/kg hexobarbital and the aorta is exposed. A cannula is placed close to the aorta and a silk ligature is used to tie the aorta and the cannula. After the ligature is applied, the cannula is removed. Hence, the diameter of the aorta will be the diameter of the cannula.
- The advantages and disadvantages are similar to the left coronary ligation model.

24.2.2.4 Dahl Salt-Sensitive Rats

Principle: Dahl salt-sensitive rats develop hypertension on the consumption of a high-salt diet. Hypertension results in hypertrophy which transitions into heart failure.

***Description*:**

- Sprague-Dawley rats (250–300 g) are selected and administered a high-salt diet with 1% NaCl solution for drinking. The test drug and the control are administered orally for a month. After the completion of the experiment, the rat is sacrificed and various cardiac parameters are studied.

***Advantages*:**

- No surgery is required.
- The development of heart failure is gradual and hence might mimic the clinical scenario

***Disadvantages*:**

- Time-consuming and expensive to maintain the colonies.

24.2.2.5 Spontaneous Hypertensive Rats and Spontaneous Hypertensive-Heart Failure Rats (SH-HF Rats)

***Description*:**

- Spontaneously hypertensive rats and SH-HF rats are genetic models that develop hypertension and cardiac failure at about 18–24 months of age. Various intracellular parameters are studied with this model (mRNA levels of various calcium transporters, apoptotic mechanisms, etc.)
- The advantages and disadvantages are similar to the Dahl salt-sensitive rats.

24.3 Large Animal Models

While mice and rats reduce the time and cost of maintenance of animal models, large animal models are required for the following reasons.

Smaller animals have a higher heart rate than humans.

- The myocardial contractile fiber composition is different.
- The oxygen consumption of the heart and the ratio of beta receptors are different.
- The dependence on Na^+/Ca^{2+} transporter and mechanisms of calcium handling in the failing myocardium is different.

24.3.1 Dog Models

24.3.1.1 Chronic Rapid Pacing Model

Principle: Pacing the heart of dogs at rates above 200 bpm is associated with a reversible form of cardiac failure within 4 weeks.

***Description*:**

- Adult male dogs (18–25 kg) are anesthetized with pentobarbital (30 mg/kg) intraperitoneally. The animal is put on an artificial respirator and the heart is exposed by a long-thoracotomy. The ventricular apex is identified and a pacing lead is placed that paces the heart at rates of 240–260 bpm. The animals develop

hypertrophy of both ventricles and cardiac failure. This technique has also been tried in sheep and pigs.
- The test drugs and the control are administered subcutaneously or intramuscularly for 14 days.

24.3.1.2 Volume Overload Model

Principle: Increased volume in the ventricles slowly results in a dilated form of cardiomyopathy and leads to cardiac failure. The volume overload can be created by damage to the mitral valve or by creating an artificial AV fistula.

***Description*:**
- Dogs are anesthetized and put on artificial respirators similar to the chronic pacing model. The chordae or leaflets of the mitral valve are identified and abolished without opening the heart through a catheter.
- Alternatively, an AV fistula could also be created which is a side-to-side anastomosis of the femoral artery and the vein. Cardiac failure takes about 3 months to develop. The test and control drugs are given either s.c. or i.m. for 14 days.

24.3.1.3 Pressure Overload Model

Similar to the rat-pressure-overload model, various techniques have been tried to create pressure overload models in dogs. They include aortic banding, renal artery stenosis, and aortic stenosis. However, large animal models are generally not used for the pressure overload model.

24.3.1.4 Coronary Artery Ligation and Microembolization Model

Similar to the coronary artery ligation model in rats, myocardial ischemia can be induced by coronary artery ligation or by injection of Polystearyl microspheres in the left atrium.

However, this method generally has more disadvantages and hence less commonly preferred. Due to variations in collateral circulation, the infarct and the results might vary drastically. Further, this model is time-consuming, expending, and requires more skill.

24.3.2 Rabbit Models

Advantages of Rabbit Models

- Less expensive compared to dog models.
- The beta myosin heavy chain is the predominant myosin chain similar to humans.
- The Na^{+}/Ca^{2+} channels contribute to about 30% of the calcium exchanger activity.
- The force-frequency relationship is positive similar to humans.

24.3.2.1 Volume and Pressure Overload Model

***Description*:**

- Rabbits are anesthetized with phenobarbitone (35 mg/kg) and the trachea is cannulated and connected to an artificial respirator. The thorax is opened and the heart is exposed. The aortic valve is destroyed by a catheter introduced through the carotid artery to produce a volume overload lesion. The aorta can be clamped below the diaphragm with PVC clamps to produce a pressure overload lesion. Both can be performed in the same animal and this increases the chance and speed of the animal going into failure.

***Advantages*:**

- The Beta-receptors in the heart and various other mechanisms closely resemble the end stage of the failing human myocardium and hence is very useful to study.

24.3.2.2 Tachycardia Pacing Model

A cardiac catheter that paces the heart of the rabbit at rates 350–400 bpm is placed surgically. After 4–6 weeks, the animals develop both hemodynamic and neurohumoral evidence of cardiac failure. The test and the control should be administered intraperitoneally for 14 days.

24.3.2.3 Doxorubicin Cardiomyopathy Model

Principle: Doxorubicin can produce both acute and chronic forms of myocardial damage. Injection of doxorubicin can be used to cause cardiac failure in rabbits and other species. The cardiotoxicity is mediated through free radical generation, lipid peroxidation, and inhibition of mRNA synthesis and protein synthesis.

***Description*:**

- Rabbits of either sex weighing 5–6 kg are chosen and 1 mg/kg of doxorubicin is administered intravenously twice a week for 6–9 weeks. After that, the animals are surgically opened and various parameters are measured.

***Advantages*:**

- Chronic doxorubicin toxicity causes a change in the expression of the RYR2 receptors (ryanodine receptors). Hence, this model is well suited to study the consequences of alterations in the ryanodine receptors in heart failure.

24.3.3 Guinea Pig Models

24.3.3.1 Cardiac Insufficiency Model

***Description*:**

- Male guinea pigs weighing 250–400 g are anesthetized with ether. The thorax is opened and the heart is brought out and held with a clamp without shutting off the circulation. A ligature is run along the apical third of both ventricles and constricted. The constriction should be partial and must not result in necrosis.

- These guinea pigs will develop congestive heart failure within 1 day. The physiology of heart failure is similar to humans in many ways including myosin composition, calcium cycling, etc.

24.3.3.2 Heart Failure Associated with Sudden Cardiac Death

Sudden cardiac death is one of the important complications of heart failure. This model is developed by surgically constricting the aorta and daily isoproterenol injections intraperitoneal. These animals will have an early onset of decompensated heart failure and a very high incidence of sudden cardiac death.

24.3.3.3 DOCA-Salt Associated Heart Failure

Guinea pigs are around 5 weeks weighing 300 g are chosen for this procedure. Left nephrectomy is performed surgically and these guinea pigs are administered DOCA-salts by intramuscular injection as well as added to the drinking water. They develop high blood pressure and left ventricular hypertrophy which progresses into heart failure by 10 weeks.

24.3.4 Syrian Hamster Models

24.3.4.1 Cardiomyopathic Syrian Hamsters

***Description*:**

- Cardiomyopathic strains of Syrian hamster is chosen and reared for about 10 months. Most of these will develop signs of cardiomyopathy within 7–10 months.
- The cardiomyopathy progresses in various phases. A pre-necrotic phase that cannot be detected, a fibrosis phase, a phase of calcium deposition, reactive hypertrophy, and a decline in cardiac performance. Advances in instrumentation have enabled identifying various lesions in this model.
- The drugs to be tested are administered for about 14 days. The inheritance pattern is autosomal recessive.

24.4 Genetic Mice Models

It is common to have genetic mice models of heart failure to study the progression of the disease and therapies that might change the course of the disease. Genetic models are preferred in mice as surgical techniques are difficult to perform on their small bodies.

24.4.1 Myosin Lim Protein Knockout Mice

Myosin Lim protein is an actin-based protein that regulates myocyte differentiation. They develop left ventricular dilated hypertrophy and heart failure.

24.4.2 TNF-Alpha Overexpressing Mice

This is also a model of dilated cardiomyopathy. This model is used to study TNF-alpha lowering therapies like the TNF-alpha sequestering soluble receptors.

Myogenic factor 5 knockout mice, Overexpressing beta-receptor kinase, and overexpressing tropomodulin have also been established.

24.5 Newer Target Identification in CHF and Other Modalities of Treatment

- Various new targets are being identified due to advances in gene therapy and stem cell research.
- The gene therapy is targeted towards correcting the alterations in calcium handling in the failing myocardium. Various models have been used to study gene therapy including smaller animal models like mice and larger animals like sheep, pigs, and dogs.
- Stem cell research aims at reducing post-MI remodeling of the left ventricle by injecting stem cells either in the coronary artery or systemically.
- Mice models are useful but significant phenotypic differences exist between the stem cells of mice and humans. Large animal models including dogs, sheep, and pigs are also used with external devices and mechanical assist.
- Large animals are often chosen to study the effectiveness of various pacemakers, LV assist devices, and other modalities of treatment.

Bibliography

Liu T, Demazumder D, O'Rourke B. A guinea pig model for heart failure-associated sudden cardiac death. Biophys J. 2015;108(2):273a.

Tiritilli A. DOCA-salts induce heart failure in the guinea pig. Eur J Heart Fail. 2001;3(5):545–51.

Vogel H, editor. Drug discovery and evaluation: pharmacological assays. 3rd completely rev., updated and enlarged ed. 2008th ed. Berlin; New York: Springer; 2007. p. 2068.

Gupta SK. Drug screening methods. 3rd ed. Jaypee Brothers Medical Publisher (P) Ltd: Jaypee Brothers Medical Publishers; 2016. p. 744.

Dixon JA, Spinale FG. Large animal models of heart failure: a critical link in the translation of basic science to clinical practice. Circ Heart Fail. 2009;2(3):262–71.

Patten RD, Hall-Porter MR. Small animal models of heart failure: development of novel therapies, past and present. Circ Heart Fail. 2009;2(2):138–44.

Screening Methods for the Evaluation of Antiplatelet Drugs

25

Ilanthamizhan Jayakumar

Abstract

Thrombosis is the key event in major cardiovascular and cerebrovascular events resulting in significant morbidity and mortality. Antiplatelet drugs like aspirin, clopidogrel inhibits platelet aggregation and inhibits thrombus formation. It is used widely in the primary and secondary prophylaxis of major cardiovascular and cerebrovascular events. Use of appropriate animal and model is utmost importance in preclinical evaluation because of the species difference variation in the event of a thrombosis. Various animals like rats, dogs, rabbits and cynomolgus monkey are used in the preclinical evaluation of antiplatelet drugs. Recently Zebrafish have been evolved as a useful model because of easiness in handle & care and resemblance of its genetics to humans. This chapter will discuss in details about various invitro and invivo model in the evaluation of the antiplatelet drugs.

Keywords

Thrombosis · Antiplatelet · Animal model · Aspirin · $P2Y_{12}$ antagonist

25.1 Introduction

- Platelets are anucleate cells that are primarily involved in the pathogenesis of homeostasis. Platelets are dynamic but remain inactive unless there is damage to the subendothelial surface of the blood vessels.
- Platelets have a major role in the pathogenesis of thrombosis in cardiovascular and cerebrovascular disorders. Platelet adhesion and aggregation are essential

I. Jayakumar (✉)
Department of Clinical Pharmacology, JIPMER, Puducherry, India

M. Lakshmanan et al. (eds.), *Introduction to Basics of Pharmacology and Toxicology*, https://doi.org/10.1007/978-981-19-5343-9_25

events in the formation of thrombus. Apart from that, platelet also plays a vital role in various conditions like inflammation, cancer, and organ transplantation rejection.

- Platelet was discovered by G. Bizzozero in 1882, but the idea of targetting platelets as drug targets was not established for a long time until they found out the role of platelets in thrombosis in the 1960s, which leads to the approval of the first antiplatelet drug, Aspirin in 1988 by US FDA.

25.2 Targets for the Antiplatelet Drug

- Platelets have two important granules, namely alpha and dense granules. Alpha granules contain proteins like GP IIb/IIIa, Von Willebrand factor (vWf), fibrinogen, and adhesion molecule P-Selectin on the membrane, which is involved in the coagulation cascade. Dense granules also contain a variety of molecules like catecholamine, calcium, serotonin, adenosine 5′ diphosphate (ADP), and adenosine triphosphate (ATP), which acts as an agonist for platelet aggregation.
- Damage to the vascular endothelial endothelium initiates the coagulation cascade. The initial response after damage to endothelium is vasoconstriction of the blood vessel and activation of platelet plug formation by various internal and external stimuli.
- An initial critical step in platelet aggregation is the interaction between the GP Ib-IX–V complex in the platelet membrane and the A1 domain of vWF in the exposed sub-endothelium.
- The secretion of the contents from the stored granules occurs along with the change in the shape of the platelet from smooth disc to spiky edges with a tremendous increase in the surface area of platelets. Release of granule contents like ADP and other numerous factors triggers platelet activation, which is accompanied by an alteration in the GP IIb- IIIa complex causing translocation of phosphatidylserine to the membrane surface which favors binding the calcium that activates the assembly of the coagulation cascade. Alternatively, platelet activation is also done by special G-protein coupled receptors called protease-activated receptors (PAR). Activated platelet release molecules like thromboxane A2, which again favors platelet aggregation.
- Later changes in GP IIb/IIIa complex favor platelet binding with fibrinogen that connects the adjacent platelets forming initial reversible aggregation. Parallel activation of thrombin binds to these complexes and causes platelet contraction.

25.3 Antiplatelet Drugs

- Aspirin was the first antiplatelet drug approved, which acts by irreversible inhibition of COX inhibitors. Dipyridamole acts by inhibiting the phosphodiesterase enzyme and inhibits the uptake of ADP.

- Later $P2Y_{12}$ inhibitors were introduced. These drugs inhibit the Gi $P2Y_{12}$ purinergic receptor-mediated adenylyl cyclase inhibition irreversibly (Clopidogrel, ticlopidine) and reversibly (Ticagrelor, cangrelor)
- GP IIb/IIIa receptor antagonist like abciximab, tirofiban inhibits GP IIb/IIIa complexes which are involved in platelet aggregation activated by all type of agonist.
- Vorapaxar, PAR -1 antagonist, is the new antiplatelet drug approved by the US FDA in 2014 to reduce thrombosis in patients with h/o cardiovascular events and peripheral arterial disease.

25.4 Screening for Antiplatelet Drugs

The selection of appropriate animals and models is of utmost importance in the preclinical evaluation of antiplatelet drugs. This is because of species differences and variation in response to the antiplatelet drugs. For example, evaluation of GP IIb/IIIa antagonist in the rat model is limited in use because of variation in response to GP IIb/IIIa in rats. Hence, large animals like dogs were used later because they resembled GP IIb/IIIa activity in humans. Later modifications were done, and platelets rich plasma was obtained from drug-treated rats and untreated dogs, which reduced the use of large animals.

25.4.1 Invitro Screening Method

- For the in-vitro screening method, blood from healthy volunteers is obtained, and 3.8% sodium citrate is added as an anticoagulant. Other agents like EDTA, heparin are not used as it interferes with the platelet aggregation
- Platelet aggregation assay is done by a device called aggregometer, a type of photometer that works on the principle of change in light transmission after adding an agonist for platelet aggregation. Change in the light transmission intensity reflects the platelet aggregation index.

25.4.1.1 Plate Aggregation in Platelet-Rich Plasma (PRP) or Washed Platelets (WP)

***Purpose & Principle*:**

- This method was first introduced by Born in 1962. Hence, it is also known as the "Born method."
- It is used to measure quantitatively the effect of the antiplatelet activity of the new compounds on induced platelet aggregation
- The platelet aggregation pathway is activated by compounds like PRP, washed platelet, ADP, arachidonic acid, serotonin, epinephrine, or PAF, and the evaluation of antiplatelet activity is determined by measuring the change in the optical density following platelet aggregation

Animal used: Mice/rats/guinea pig of either sex.

***Procedure*:**

- The test drug is administered either by oral or intravenous route. Two samples are taken, one just before administering the test drug and the second sample at the end of the absorption time. The blood sample is obtained through caval puncture following administration of phenobarbitone and xylazine. In the case of large animals, the sample can be obtained through cardiac puncture.
- Hirudin and Acid Citrate Dextrose (ACD) are added to the blood sample to obtain PRP.
- The PRP supernatant is centrifuged at 1500 rpm for 10 min to obtain platelet-poor plasma (PPP) and then diluted with PRP to obtain a platelet count of 3×10^8/ml before starting the platelet assay
- To obtain washed platelet, eight volumes of human blood is mixed with 1.5 volume of ACD and centrifuged to get the PRP. Then it is acidified by adding ACD and then suspended with Tyrode solution to obtain a platelet count of 3×10^8/ml.
- To 40 μl of the test solution, 320 μl of WP or PRP is added from the untreated subject and inserted into an aggregometer. Light transmission of PPP is to be set at 100%. Then the sample is incubated at 37° C followed by magnetic stirring at 1000 rpm, and 40 μl of the aggregating agent is added. The change in the optical density is noted down after adding the aggregating agent.

***Observation*:**

- Percentage inhibition of platelet aggregation in the drug-treated sample compared to the pretreatment sample
- Percentage of platelet aggregation

$$\%\text{inhibition} = \frac{10\ \text{min platelet count} \times 100}{\text{Initial platelet count}}$$

- IC_{50}, i.e., 50 percentage inhibition of platelet aggregation value (log dose concentration vs. percentage inhibition of platelet aggregation)

Disadvantages:

- The use of PRP does not reflect the exact in-vivo platelet activity as it lacks erythrocyte and leucocytes
- The complexity of sample preparation
- High level of expertise technique
- Time-consuming

25.4.1.2 Platelet Aggregation by Laser Scattering Mmethod

Purpose & Principle:

Conventionally light transmittance method is used to determine the antiplatelet activity. To improve the sensitivity for detecting the antiplatelet activity, a new aggregometer AG-10, Kudo, was introduced in Japan that uses laser light scattering to measure the antiplatelet activity.

Procedure:

- The PRP supernatant is centrifuged at 1500 rpm for 10 min to obtain platelet-poor plasma (PPP) and then diluted with PRP to obtain platelet of 3×10^8/ml before starting the platelet assay.
- To obtain washed platelet, eight volumes of human blood is mixed with 1.5 volume of ACD and centrifuged to get the PRP. Then it is acidified later by adding with by adding ACD and then suspended with Tyrode solution to obtain a platelet count of 3×10^8/ml.
- An aggregating agent like ADP is added and transferred into the cuvette. A laser beam measuring 40 μm is generated using a 20–uW diode laser, and it is passed through the PRP.
- The light that scatters from the cuvette is detected by a photocell array. The light intensity corresponds to the platelet aggregate size

Observation:

The light intensity corresponds to the aggregate size.

Disadvantage:

Only a PRP or washed platelet sample can be used in this method. Whole blood is not recommended.

25.4.1.3 Platelet Aggregation After Gel Filtration

Purpose & Principle:

- Use of Gel filtration platelet (GFP) is recommended when a weak agonist of platelet aggregation (Serotonin, epinephrine) is used where an exogenous addition of fibrinogen is required for platelet aggregation.
- GFP is used to evaluate platelet aggregation inhibition by light transmittance aggregometer by measuring the optical density
- This technique is particularly used in the drugs that are targeting GP IIb/IIIa complex

Procedure:

- Venous blood is withdrawn from the healthy volunteers, and 3.8% sodium citrate is added as the anticoagulant.
- To obtain PRP, 8.4 ml of the test sample is mixed with 1.6 ml of ACD and centrifuged at 120 g for 10 min.
- Then the PRP is adjusted to a pH of 6.5 with ACD, and the resulting pellet is suspended in Tyrode's solution, and later the platelet suspension is prepared using Sepharose gel filtration CL 2B column and platelet count adjusted to 3×10^8 /ml. The platelet suspension is kept at room temperature for 1 h before the test is started.

- The platelet suspension is incubated with 0.5 mM concentration of $CaCl_2$, and 20 μl of the test solution is added and kept at room temperature for 2 min, and 20 μl of platelet agonist is added. The percentage inhibition of platelet aggregation is measured by the aggregometer.

Observations:
- Percentage inhibition of platelet aggregation
- The IC_{50} value of the antiplatelet activity

25.4.1.4 Microchannel Array Flow Analyzer (MC-FAN)
Purpose& Principle:
MC-FAN consists of silicon chip channels of length 10–100 μm and diameter similar to the human capillaries. The whole blood sample is passed through the chamber at a negative pressure of 20 cm H_2O after adding the platelet aggregating agents, and the evaluation of antiplatelet activity is done. This method is more suitable to evaluate the antiplatelet activity in the capillaries and other microvasculature.

Procedure:
- In this method, whole blood is preferred over the platelet-rich plasma.
- Blood from the human volunteers with no h/o of bleeding disorder, hyperlipidemia, and no h/o of medication for the last 3 months is obtained and mixed with anticoagulant 3.8% sodium citrate.
- Then aggregation-inducing agents such as ADP or ristocetin are added and passed through siliconized microchannels of 7 mm width and 30 mm length at a negative pressure of 20 cm of H_2O, and the average transit time of blood through the cylinder is determined and compared with the control for 5 min.
- The chip is taken out later and fixed with 2.5% glutaraldehyde in 0.1 M phosphate buffer and examined under the scanning electron microscope (SEM).

Observation:
- Mean blood flow rate/average transit time through the cylinder
- IC_{50} value through the dose-response curve
- Microscopic examination under SEM of the thrombi formed

Disadvantages:
- Results are highly variable
- Lacks sensitivity

25.4.1.5 Platelet Function Aanalyzer-100 (PFA-100)
Purpose and Principle:
- It is commonly used as a screening tool in point of care testing before initiating antiplatelet agents and also for the preclinical evaluation of antiplatelet drugs.
- It is developed as an alternative technique for measuring bleeding time and light transmission platelet aggregometry platelet function test.

- It works on the principle that platelets undergo aggregation under shear stress in the presence of an agonist for aggregation.

Procedure:

- Whole blood is obtained from the healthy volunteer, and 3.8% sodium citrate is added as the anticoagulant
- PFA – 100 contains two types of disposable cartridges, namely collagen plus ADP (CADP) and collagen plus epinephrine (CEPI) coated with collagen and an ADP agonist.
- The whole blood sample is transferred and incubated at 37 °C in the cartridge. The cartridge has a small aperture, and the time taken for the clot formed with shear stress and ADP agonist to occlude the aperture is noted down. This is called closure time (CT). Time for the closure of the aperture in the cartridge is observed for a maximum of 5 min.

Observation

- Closure time by the clot formed.

25.4.2 In-Vivo Models

25.4.2.1 Chemically Induced Thrombosis Model

FeCl3 Induced Thrombosis

Purpose & Principle: It is the most commonly used chemically induced thrombosis method in the evaluation of antithrombotic agents. Topical application of 25% FeCl3 was found to produce occlusive thrombus. It was initially described by Reimunn – Hunziger in 1944 in Inferior vena cava in rabbits. Later it was adapted in the Murine model by Kurz et al. in 1990.

Animal used: Carotid artery in Wistar rats weighing 250–300 g.

Procedure: Rats are anesthetized with Inactin (100 mg/kg), and tracheostomy is done to facilitate breathing. Catheters are placed in the femoral artery and jugular vein to record arterial blood pressure and administration of test agents, respectively. Then the carotid artery is exposed, and an Ultrasonic flow probe is placed inside the carotid artery. The test is administered prior through intravenous route administration. Thrombus formation is induced by applying the 25% Fecl3 solution topically through Whatman filter paper (2 × 5 mm) and leftover for 10 min.

***Observation*:**

- Blood flow through the carotid artery before and after 60 min of induction of thrombus
- Weight of the thrombus on filter paper
- Time taken for occlusion of the vessel.

Advantage: This method can be used to evaluate both antithrombotic and anticoagulant drugs.

25.4.2.2 Photochemical Model for Thrombosis

Purpose & Principle: Intravenously administered fluorescein dye can cause endothelial damage by producing singlet oxygen and reactive oxygen species leading to platelet adhesion and platelet aggregation.

Animal used: Wistar rats weighing 250–300g.

Procedure: Animal are weighed and anesthetized, the right femoral vein is cannulated to inject the dye. Dye like sodium fluorescein or fluorescein isothiocyanate dextran 70 is injected intravenously, followed by excitation of the mesenteric artery with ultraviolet light at a wavelength of 490–510 nm to induce thrombosis.

Observations: Time for initiation and complete thrombus formation and blood flow rate through the mesenteric artery are monitored.

25.4.2.3 Rose Bengal Induced Thrombosis

Purpose& Principle: Rose bengal, calcein acetoxymethyl ester is a photosensitizer dye which on excitation by a xenon filtered light lamp at 540 nm, produce singlet oxygen and reactive oxygen species followed by thrombosis. Hence, this model can be used to evaluate the drug, which acts as an antithrombotic drug by inhibiting the reactive oxygen species.

Animal used: Wistar rat weighing 300 g.

Procedure: After anesthetizing the rats with 1% isoflurane, the femoral vein is exposed to inject the 1% Rose Bengal and the left common carotid artery is dissected &exposed, and a Light-emitting diode (LED) is implanted. Then the rose bengal is injected intravenously at 10 mg/kg followed by irradiation with LED at 540 nm for 5 min. Blood flow is monitored by a pulse doppler flow meter.

***Observation*:**

- Detection and imaging of thrombotic plaque via Ultrasonography and Magnetic resonance angiography.
- Monitoring the flow and stable occlusion by thrombosis.

25.4.2.4 Laser-Induced Thrombosis Model

Purpose and principle: Argon laser-induced endothelial injury may cause stimulation of ADP mediated platelet aggregation leading to thrombus formation. Hence, this model can be used to evaluate the antithrombotic agents which act at the later phase of platelet aggregation.

Animal used: New Zealand Rabbit mesenteric artery or Sprague Dawley rats.

Procedure: Before inducing thrombosis, animals are pretreated with atropine and intraperitoneally anesthetized with ketamine 100 mg/kg and xylazine 4 mg/kg. Animals are pretreated with antithrombotic agents. Argon laser of 514 nm is used, which is coupled with the compound microscope to direct the laser into the target organ. The blood flow is monitored continuously by a video analyzer software package.

***Observation*:**

- The blood flow rate is measured.
- The number of laser excitation required to produce a thrombus and the total number of thrombus formed.
- Histological and immunohistochemical analysis of mesentery artery after sacrificing the animal.

25.4.2.5 Arteriovenous Shunt Model for Thrombosis

Purpose & Principle: The formation of an arteriovenous shunt in larger animals like rabbits, dogs, and cats have been used as a model for studying the antithrombotic potential of the drugs.

Animal used: Japanese White rabbits weighing 2.2–3 kg.

Procedure: Rabbits are weighed and anesthetized. Then the left carotid artery and the right jugular veins are isolated and cannulated with a shunt catheter made of polypropylene tubes. Later, thrombosis is induced by placing a silk thread with a thrombogenic agent soaked with recombinant human tissue factor. The test drug is injected through a femoral vein 30 min prior, and occlusion by the silk thread is released to allow blood flow through the shunt. The blood sample can be collected for further studies.

***Observation*:**

- Weight of the thrombus attached to the silk thread.
- Invitro platelet assay

25.4.2.6 Electrical Induced Thrombosis Model

Purpose and principle: This model was initially described in rabbits by Wong et al. later, this model is used in animals like dogs, rats, and cynomolgus monkeys for evaluation of the antithrombotic effect. Electrical stimulus application can be made in two ways either by external application on the vessel wall through hook-like electrodes or application of electrical stimulus inside the vessel wall in the intimal layer.

Animal used: Wistar rats weighing 250–300 g.

Procedure: Rats are anesthetized with 1% Isoflurane. The right femoral vein is exposed to inject the test drug, and the right carotid artery is dissected and separated from the surrounding tissue. A pair of the electrode is placed inside the lumen of the vessel wall. Test drug injected is injected 30 min before the application of the electrical stimulus. An electrical stimulus of 2 mA is applied for 5 min for the formation of the thrombus. A cannula is inserted in the left carotid artery to record the blood pressure. The ultrasonic Doppler flow meter is connected to measure the blood flow. At the end of the experiment, rats are sacrificed, and the carotid artery is examined for thrombus formation under the scanning electron microscope.

***Observation*:**

- Time for occlusion of the vessel wall
- Blood flow through the vessel wall before and after 60 min of the experiment

- Patency of the vessel wall
- Histological examination of the vessel wall

Disadvantages: Damage to all layers of the vessel wall and damage to the surrounding tissue by the electrical stimulus.

25.4.2.7 Mechanically Induced Thrombosis Model/Stenosis Induced Thrombosis Model

Purpose and principle: Thrombosis from the stenosed blood vessel is one of the common reasons for the development of myocardial infarction, angina, and sudden death. The acute stenosis model was described by Folts et al., who described this model in Dog's Left descending coronary artery. This model was used to study thrombosis with modifications in animals like pigs, baboons, and monkeys.

Animal used: Canine dogs/ Baboons/ monkeys.

Procedure: Cynomolgus monkey is weighed and sedated with ketamine at 10 mg/kg. Then it is anesthetized by intravenous sodium phenobarbital 20 mg/kg. The right carotid artery is exposed from the surrounding tissue, and the femoral vein is dissected to inject the test drug and withdraw the blood sample during the procedure. Then a constrictor is applied to the carotid artery until it produces a 50% reduction in the blood flow in the coronary artery. Cyclic blood flow reductions occur with the stenosed vessel due to embolization of the thrombus and reformation of the thrombus in the blood vessel. An ultrasonic flow probe connected to the blood vessel is used to measure the blood flow inside the carotid artery. Blood samples are obtained via the femoral vein for in-vitro studies during the procedure.

***Observations*:**

- Reduction in the number of cyclic flow reduction (CFR).
- Platelet aggregation study by whole blood aggregometer.
- Blood flow rate inside the carotid artery before and after 60 min.
- Time for occlusion of the carotid artery.

25.4.3 Zebrafish as a Thrombosis Model

Recently Zebrafish model gained popularity in the thrombosis model because of the similarity in the gene coding for coagulation and platelet activation pathway in humans and zebrafish. Although it has nucleated platelets, its function is similar to human platelets. The similarity in the vascular endothelium and easiness in handle and care encourages the use of Zebrafish as alternatives. Currently, FeC13 and laser-induced thrombosis models were developed and use in the preclinical evaluation of the antithrombotic agents.

25.4.4 Genetic Models

Various genetic knock-out models have been tried in the preclinical evaluation of antiplatelet drugs to identify and validate the novel targets. Some of the genes knock out tried were P-selectin, PAR-1, PAR −3, Thromboxane A2, and P2Y1.

Bibliography

Andre P, Bal ditSollier C, Bonneau M, Pignaud G, Hainaud P, Azzam K, Drouet L. Which experimental model to choose to study arterial thrombosis and evaluate potentially useful therapeutics? Haemostasis. 1996;26(4):55–69.

Andrews RK, Berndt MC. Platelet physiology and thrombosis. Thromb Res. 2004;114:447–53.

Bodary PF, Eitzman DT. Animal models of thrombosis. Curr Opin Hematol. 2009;16:342–6.

Chintala M, Strony J, Yang B, Kurowski S, Li Q. SCH 602539, a protease-activated receptor-1 antagonist, inhibits thrombosis alone and in combination with cangrelor in a folts model of arterial thrombosis in cynomolgus monkeys. ArteriosclerThrombVasc Biol. 2010;30:2143–9.

Dyszkiewicz-Korpanty A, Olteanu H, Frenkel EP, Sarode R. Clopidogrel anti-platelet effect: an evaluation by optical aggregometry, impedance aggregometry, and the platelet function analyzer (PFA-100). Platelets. 2007;18:491–6.

Gremmel T, Frelinger AL 3rd, Michelson AD. Platelet physiology. Semin Thromb Hemost. 2016;42:191–204.

Hajjarian Z, Tshikudi DM, Nadkarni SK. Evaluating platelet aggregation dynamics from laser speckle fluctuations. Biomed Opt Express. 2017;8:3502–15.

Jagadeeswaran P, Cooley BC, Gross PL, Mackman N. Animal models of thrombosis from zebrafish to nonhuman primates: use in the elucidation of new pathologic pathways and the development of antithrombotic drugs. Circ Res. 2016;118:1363–79.

Kamada H, Okamoto T, Hayashi T, Suzuki K. An in vitro method for screening anti-platelet agents using a microchannel array flow analyzer. Biorheology. 2010;47:153–61.

Koltai K, Kesmarky G, Feher G, Tibold A, Toth K. Platelet aggregometry testing: molecular mechanisms, techniques and clinical implications. Int J Mol Sci. 2017;18:1803. https://doi.org/10.3390/ijms18081803.

Kweon OJ, Lim YK, Kim B, Lee MK, Kim HR. Effectiveness of platelet function analyzer-100 for laboratory detection of anti-platelet drug-induced platelet dysfunction. Ann Lab Med. 2019;39: 23–30.

Majithia A, Bhatt DL. Novel antiplatelet therapies for atherothrombotic diseases. ArteriosclerThrombVasc Biol. 2019;39:546–57.

Matsuno H, Tokuda H, Ishisaki A, Zhou Y, Kitajima Y, Kozawa O. P2Y12 receptors play a significant role in the development of platelet microaggregation in patients with diabetes. J Clin Endocrinol Metab. 2005;90:920–7.

McFadyen JD, Schaff M, Peter K. Current, and future antiplatelet therapies: emphasis on preserving haemostasis. Nat Rev Cardiol. 2018;15:181–91.

Mizuno T, Tsukiya T, Takewa Y, Tatsumi E. Differences in clotting parameters between species for preclinical large animal studies of cardiovascular devices. J Artif Organs. 2018;21:138–41.

Ogawa T, Hashimoto M, Niitsu Y, Jakubowski JA, Tani Y, Otsuguro K, Asai F, Sugidachi A. Effects of prasugrel, a novel P2Y(12) inhibitor, in rat models of cerebral and peripheral artery occlusive diseases. Eur J Pharmacol. 2009;612:29–34.

Olson JD, Fass DN, Bowie EJW, et al. Ristocetin-induced aggregation of gel filtered platelets: a study of von Willebrand's disease and the effect of aspirin. Throm Res. 1973;3:501–51.

Paniccia R, Priora R, Liotta AA, Abbate R. Platelet function tests: a comparative review. Vasc Health Risk Manag. 2015;11:133–48.

Pérez P, Alarcón M, Fuentes E, Palomo I. Thrombus formation induced by laser in a mouse model. Exp Ther Med. 2014;8:64–8.
Salzman EW, Rosenberg RD, Smith MH, Lindon JN, Favreau L. Effect of heparin and heparin fractions on platelet aggregation. J Clin Invest. 1980;65:64–73.
Shinozawa E, Kawamura M. Anti-thrombotic effect of a factor Xa inhibitor TAK-442 in a rabbit model of arteriovenous shunt thrombosis stimulated with tissue factor. BMC Res Notes. 2018;11:776. https://doi.org/10.1186/s13104-018-3886-4.
Westrick RJ, Winn ME, Eitzman DT. Murine model of thrombosis (Eitzman series). AteriosclerThombVasc Biol. 2007;27:2079–93.
Wong PC, Watson C, Crain EJ. The P2Y1 receptor antagonist MRS2500 prevents carotid artery thrombosis in cynomolgus monkeys. J Thromb Thrombolysis. 2016;41:514–21.
Yeh JC, Huang KL, Hsiao YC, Hsu YH, Lin YH, Lou SL, Lee TH. A rat model of thrombosis in common carotid artery induced by implantable wireless light-emitting diode device. Biomed Res Int. 2014;2014:724134. https://doi.org/10.1155/2014/724134.
Yen TL, Wu MP, Chung CL, Yang WB, Jayakumar T, Geraldine P, Chou CM, Chang CY, Lu WJ, Sheu JR. Novel synthetic benzimidazole-derived oligosaccharide, M3BIM, prevents ex vivo platelet aggregation and in vivo thromboembolism. J Biomed Sci. 2016;23:26. https://doi.org/10.1186/s12929-016-0245-4.

Screening Methods for the Evaluation of Diuretics

26

Nishanthi Anandabaskar

Abstract

Diuretics are drugs that promote increase production of urine by augmenting the excretion of sodium and water. Although they are used in a wide range of diseases like essential hypertension, congestive cardiac failure, renal failure, ascites etc.; they are associated with adverse drug reactions especially due to fluid and electrolyte disturbances. Thus, there is a need for development of newer drugs with diuretic activity and less adverse effects. Various *in vitro* and *in vivo* screening methods have been used for evaluating the diuretic activity of drugs. The various *in vitro* techniques include carbonic anhydrase inhibition *in vitro*, patch clamp technique in kidney cells, perfusion of isolated kidney tubules and isolated perfused kidney technique. The various *in vivo* techniques include Lipschitz test, saluretic activity in rats, micropuncture techniques in the rat, diuretic and saluretic activity in dogs, clearance methods and stop-flow technique. Each of these techniques has been discussed in detail in this chapter.

Keywords

Diuretic · Lipschitz test · Natriuretic · Saluretic

26.1 Introduction

Diuretics are drugs that promote increased production of urine by augmenting the excretion of sodium and water. They are used commonly for the management of various diseases like essential hypertension, congestive cardiac failure, renal failure, ascites, edema of varied origin, and glaucoma. However, they are associated with

N. Anandabaskar (✉)
Department of Pharmacology, Sri Manakula Vinayagar Medical College and Hospital, Puducherry, India

M. Lakshmanan et al. (eds.), *Introduction to Basics of Pharmacology and Toxicology*, https://doi.org/10.1007/978-981-19-5343-9_26

Table 26.1 Screening methods involved in the evaluation of diuretics

In vitro methods	*In vivo* methods
1. Carbonic anhydrase inhibition *in vitro*	1. Diuretic activity in rats (LIPSCHITZ test)
2. Patch-clamp technique in kidney cells	2. Saluretic activity in rats
3. Perfusion of isolated kidney tubules	3. Micropuncture techniques in the rat
4. Isolated perfused kidney	4. Diuretic and saluretic activity in dogs
	5. Clearance methods
	6. Stop-flow technique

$$CO_2 + H_2O \underset{\text{Carbonic anhydrase}}{\rightleftharpoons} H_2CO_3$$

Fig. 26.1 Activity of carbonic anhydrase enzyme

adverse effects especially due to fluid and electrolyte disturbances like hypokalaemia, dilutional hyponatremia, hypocalcemia, and magnesium depletion. This poses the need for the development of newer drugs with diuretic potential and lesser adverse effects. Various *in vitro* and *in vivo* screening methods have been employed in the evaluation of drugs with diuretic activity. They are enumerated in Table 26.1

26.2 *In Vitro* Methods

26.2.1 Carbonic Anhydrase Inhibition *In Vitro*

26.2.1.1 Principle

- Test compounds with possible diuretic activity are **assessed for carbonic anhydrase inhibition**
- This method is used to find the activity of sulfonamide diuretics
- Carbonic anhydrases (CAs) belong to a family of zinc metalloenzymes, which catalyze the conversion of CO_2 and H_2O to form H_2CO_3 (refer Fig. 26.1). This is a reversible reaction and the process is called hydration/hydroxylation
- H_2CO_3 undergoes non-enzymatic dissociation to form HCO_3^- and H^+
- The carbonic anhydrase inhibiting the activity of the test compounds is tested by the method described by Maren in 1960 (Micro method), which is highly reliable

26.2.1.2 Procedure

- Materials required
  - Phenol red indicator solution
  - 1 M sodium carbonate/bicarbonate buffer
  - Enzyme: Carbonic anhydrase (Source: dog blood; it is sampled and filled into a tube containing heparin; and diluted 1:100 with deionized water)
  - Equipment: Reaction vessel, Monostatbench-mounted flowmeter, and 30% CO_2

- Assay
  - CO_2 flow rate is adjusted to 30 (45) ml/min
  - Phenol red indicator solution, enzyme, H_2O, and bicarbonate buffer are added to the reaction vessel

26.2.1.3 Evaluation

- Determination of the following parameters are performed in duplicate samples:
  - Tu = (uncatalyzed time) = time taken for change in color, in the absence of enzyme
  - Te = (catalyzed time) = time taken for change in color, in the presence of enzyme
  - Tu–Te = enzyme rate
  - Ti = enzyme rate in the presence of various concentrations of inhibitor
- Percent inhibition of carbonic anhydrase is estimated based on the following formula:

$$\%\text{Inhibition} = \frac{1 - (T_u - T_e) - (T_i - T_e)}{(T_u - T_e)} \times 100.$$

- Standard data for acetazolamide and chlorothiazide are given in Table 26.2

26.2.2 Patch-Clamp Technique in Kidney Cells

26.2.2.1 Principle

- This test aids in the **investigation of ion channels**, which perform vital roles in the kidney cell functioning
- The different modes of patch-clamp technique include cell-attached, cell-excised, and whole-cell mode
- Also, this method helps to explore other electrogenic transport mechanisms (e.g., sodium-coupled alanine transport)

26.2.2.2 Procedure

- The patch-clamp technique can be used in the following cell types
  - Cultured kidney cells
  - Freshly isolated kidney cells
  - Cells of isolated perfused kidney tubules
- The procedure with the use of cells of isolated perfused kidney tubules are described in detail below

Table 26.2 Standard data for carbonic anhydrase inhibition *in vitro*

Compound	IC50 [M]
Acetazolamide	9×10^{-9}
Chlorothiazide	9×10^{-7}

- Segments of late superficial proximal tubules of rabbit kidney are dissected
- With the aid of a perfusion system, it is perfused from one end
- With the help of optical control, the patch pipette is moved through the open end into the tubule lumen and is brought in contact with the brush border membrane
- On application of slight suction of the patch electrode, gigaseals are produced immediately and single potassium or sodium channels can be recorded in the cell-attached or inside-out cell-excised mode
- Cotransport systems—can only be investigated by the whole-cell method of the patch-clamp technique. Freshly isolated cells of rabbit proximal tubules are used to study the sodium-alanine cotransport system.

26.2.2.3 Evaluation

- In isolated perfused kidney tubules, patch-clamp technique can be used to obtain the concentration-response curves of drugs inhibiting ion channels
- In isolated cells of proximal tubule, patch-clamp technique (whole-cell mode of it) helps in assessing the sodium-alanine cotransport system

26.2.3 Perfusion of Isolated Kidney Tubules

26.2.3.1 Principle:

- This is the best method **to explore the site and mechanism of action of a drug that has demonstrated action on kidney function** in clearance and micropuncture studies.
- The various tubule segments having different functional properties and thus, needs to be studied are as follows:
  - Proximal tubule (PT)
  - Descending thin limb of the loop of Henle (DTL)
  - Ascending thin limb of the loop of Henle (ATL)
  - Thick ascending limb of the loop of Henle (TAL)
  - Distal convoluted tubule (DCT)
  - Connecting tubule (CNT)
  - Cortical collecting duct (CCD)
  - Medullary collecting duct (MCD)
  - Papillary collecting duct (PCD)

26.2.3.2 Procedure

- Kidney tubule segments of different species are used—man; rabbit; rat; mouse; hamster; snake; birds, etc.
- All the tubule segments (PT, DTL, ATL, TAL, DCT, CNT, CCD, MCD, PCD) are dissected from thin kidney slices (<1 mm thickness) and transferred into the perfusion chamber having an inverted microscope (20–400X)
- The perfusion chamber and the bath perfusate are kept at 37 °C. The selection of bath perfusate depends upon the tubule segment that is to be studied.

- The following parameters are measured
  - Flux measurements
  - Transepithelial electrical measurements
  - Intracellular electrical measurements
  - Patch-clamp studies
  - Fluorescent dyes in isolated perfused tubule (fluorescent dyes can be used for the monitoring of Na^{+}, K^{+},Cl^{-}, Ca^{2+}, and pH)

26.2.3.3 Evaluation

- One or several of the parameters mentioned above can be estimated in the presence or absence of test drug and their concentration-response curves can be obtained
- Intracellular measurements help in delineating the mechanism of action of test drug
- Number of preparations required for screening–3
- Number of preparations required for obtaining concentration-response curves–10

26.2.4 Isolated Perfused Kidney

26.2.4.1 Principle

- Isolated kidney—useful in evaluating function of proximal tubule but of limited use in assessing distal tubule function
- The kidney can be perfused in situ and isolated *in vitro*
- An isolated kidney can be perfused with blood or plasma-like solutions by a pump
- The disadvantage of blood-perfused dog kidney *in vitro* model—instability (especially when compared to *in situ* models)
- Isolated perfused rat kidney also can be used in this method (its procedure is detailed below)

26.2.4.2 Procedure

- Source of kidneys—anesthetized male rats with a bodyweight of 300–400 g
- Donor animals—kept for overnight fasting before surgery; however, no water restriction is enforced
- The abdominal cavity of the animal is opened
- A polyethylene tubing is used to cannulate the right ureter
- Vena cava is cannulated and heparin (500 u/kg body weight) is injected
- The renal artery is cannulated
- A perfusion solution is used to continuously perfuse the kidney
- The kidney is then dissected out from the animal and placed in a plexiglass chamber
- Speed of the perfusion pump is adjusted to maintain a perfusion pressure of 80–90 mm Hg in the renal artery

26.2.4.3 Evaluation

- Clearance periods of 20 min are used after the equilibration period
- Assessment of global renal function—a collection of urine samples and perfusate got at the midpoint of the clearance period
- Estimation of glomerular filtration rate (GFR) and fluid transport -^{3}H-labelled polyethylene glycol is added to a modified Krebs-Henseleit bicarbonate buffer
- Measurement of electrolytes in urine—by standard flame photometry
- Calculation of fractional excretions of water, electrolytes, and test compounds done

26.3 *In Vivo* Methods

26.3.1 Diuretic Activity in Rats (LIPSCHITZ Test) (Refer Fig. 26.2)

Lipschitz et al. described the technique for evaluation of diuretic activity in rats, in the year 1943. It is a standard method for screening test compounds with potential diuretic activity.

26.3.1.1 Principle

- Excretion of sodium and water are compared between rats given test drug and rats given high dose of urea (control group), and Lipschitz- value is calculated
- "The Lipschitz- value is the quotient between excretion by test animals and excretion by the urea control"

$$\text{Lipschitz-value} = T/U$$

Where, T is the response produced with test compound, and U that with urea treatment.

26.3.1.2 Procedure

- Animals used—Male Wistar rats (weight range of 100–200 g)
- The animals are placed in metabolic cages which have a wire mesh bottom and a funnel for collecting the urine
- The funnel contains stainless-steel sieves which aid in retaining the feces and allowing the urine to pass-through
- Diet and water—the rats are fed with standard diet and water ad libitum
- Food and water are withdrawn 15 h before the experiment
- One metabolic cage houses 3 animals
- Test group—two groups of 3 animals each (thus, a total of 6 rats) are used for a single dose of the test drug (especially for screening procedures). The test drug is given orally at a dose of 50 mg/kg in 5 ml water/kg body weight
- Control group—two groups of 3 animals each(thus, a total of 6 rats) given urea orally at a dose of 1 g/kg body weight

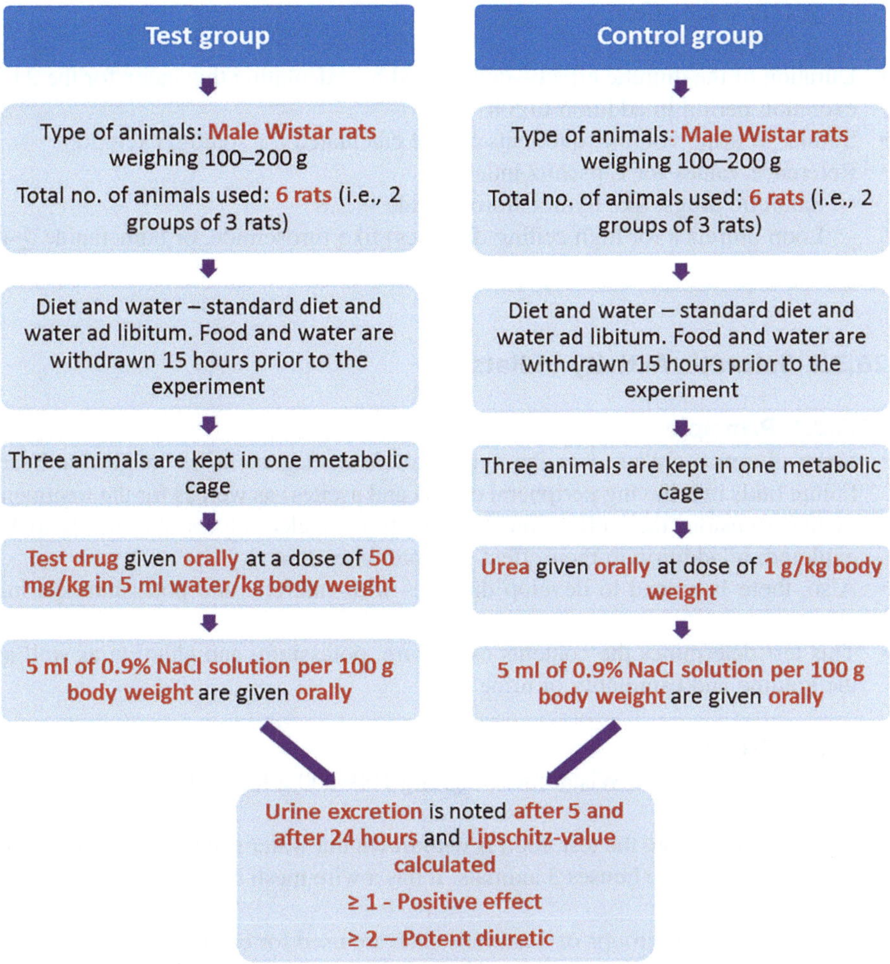

Fig. 26.2 Methodology of evaluation of diuretic activity in rats by Lipschitz test

- In addition, 5 ml of 0.9% NaCl solution per 100 g body weight are given orally by gavage
- Urine excretion is noted after 5 and after 24 h
- Flame photometry is used to assess the sodium content of the urine

26.3.1.3 Evaluation

- For each group, we find the volume of urine excreted per 100 g body weight
- Lipschitz-value, i.e., the ratio *T*/*U* (where *T* and *U* are the response produced by the test compound and urea control respectively) is calculated
- Interpretation of Lipschitz-value

  - $\geq$ 1–Positive effect
  - $\geq$ 2–Potent diuretic
- Duration of the diuretic effect—determined by calculating this index for the 24 h excretion period in addition to 5 h
- Similar to urine volume, quotients can be calculated for sodium excretion
- Reference values for Lipschitz index
  - Saluretic drugs, like hydrochlorothiazide $\approx$ 1.8
  - Loop diuretics (or high ceiling diuretics) like furosemide, or bumetanide $\geq$4

26.3.2 Saluretic Activity in Rats

26.3.2.1 Principle

- Since diuretics are used in the symptomatic management of congestive heart failure (aids in relieving peripheral edema and ascites) as well as for the treatment of hypertension, their effect on the excretion of electrolytes also needs to be explored, in addition to their effect on water excretion
- Also, there is a need to develop diuretics with saluretic and potassium-sparing effects
- This test determines the contents of sodium, potassium, and chloride as well as the volume and osmolality of urine

26.3.2.2 Procedure

- Animals used—Male Wistar rats weighing 100–200 g fed with standard diet and water ad libitum
- Fifteen hours before the test, food is withdrawn but water is made freely available
- One metabolic cage houses 3 animals. It has a wire mesh bottom and a funnel for collecting the urine
- Test group—Two groups of 3 animals each are used for each dose of a test drug. The dose of the test compound is 50 mg/kg orally in 0.5 ml/100 g body weight starch suspension
- Standard group—Two groups of 3 animals each are used for standard drugs. Standard drugs include furosemide (25 mg/kg p.o.), hydrochlorothiazide (25 mg/kg p.o.), triamterene (50 mg/kg p.o.), or amiloride (50 mg/kg p.o.). (Note: "p.o" stands for "per oral" administration of the drugs)
- Urine excretion is recorded hourly up to a duration of 5 h
- The 5-h urine is analyzed for the following contents
  - Sodium
  - Potassium
  - Chloride
- However, 24 h of urine is collected and analyzed for evaluation of compounds with prolonged effects

Table 26.3 Saluretic activity in rats—calculations performed for the estimation of saluretic, natriuretic, and carbonic anhydrase inhibiting activity

Type of activity	Calculation
Saluretic activity	Sum of Na^+ and Cl^- excretion
Natriuretic activity	Na^+/K^+ ratio Note: – Values >2 indicate a favorable natriuretic effect – Values >10 indicatea potassium-sparing effect
Carbonic anhydrase inhibiting the activity	$\frac{Cl^-}{Na^+ + K^+}$ ratio Note: • Ratios between 1 and 0.8—No carbonic anhydrase inhibition • With decreasing ratios—slight to strong carbonic anhydrase inhibition is present

26.3.2.3 Evaluation

The calculations performed for the estimation of saluretic, natriuretic, and carbonic anhydrase inhibiting activity are described in Table 26.3. Comparison of the two in vivo models - evaluation of diuretic activity in rats (Lipschitz test) and saluretic activity in rats is described in table 26.4.

26.3.3 Micropuncture Technique in the Rat (Refer Fig. 26.3)

26.3.3.1 Principle:

- This test is used to study the effect of diuretics on functioning of a single nephron
- The mechanism of action of the diuretic can be assessed by studying the following parameters:
  - Tubular fluid reabsorptive rates
  - Electrolyte concentrations
- In rats, the proximal and distal tubules as well as collecting ducts are easily accessible for micropuncture and hence used as the animal model of choice in this experiment

26.3.3.2 Procedure:

- Animals used—rats (weight range of about 250 g)
- Anesthetic used—thiopentone 50 mg/kg (intraperitoneal route)
- Rats are made to undergo 16 h of fasting before starting the experiment but they are given free access to water
- After anesthesia, the rats are kept on a thermostatically heated table
- Then, rats are tracheotomized and the carotid artery and jugular vein are cannulated and used for the following purposes:
  - Carotid artery—used for blood pressure recording
  - Jugular vein—used for blood sampling, and infusion of compounds

Table 26.4 Comparison of the two *in vivo* models—evaluation of diuretic activity in rats (Lipschitz test) and saluretic activity in rats

Parameter	Evaluation of diuretic activity in rats (Lipschitz test)	Evaluation of saluretic activity in rats
Animals used	Male Wistar rats (weight range of 100–200 g)	
Food and water	Rats are fed with standard diet and water ad libitum	
Restriction of food before the experiment	Withdrawn 15 h before the experiment	
Restriction of water before the experiment	Water is withheld 15 h before the experiment	No
No. of animals placed per metabolic cage	Three animals are kept in one metabolic cage	
No. of animals used in test and control/standard group	Test group—Two groups of 3 animals each (thus, a total of 6 rats) Control/standard group—Two groups of 3 animals each (thus, a total of 6 rats)	
Test drug dosing	The test drug is given orally at a dose of 50 mg/kg in 5 ml water/kg body weight. Additionally, 5 ml of 0.9% NaCl solution per 100 g body weight are given orally	The test drug is givenorally at a dose of 50 mg/kg in 0.5 ml/100 g body weight starch suspension
Control/standard drug dosing	Oral urea at a dose of 1 g/kg body weight. Additionally, 5 ml of 0.9% NaCl solution per 100 g body weight are given orally	Furosemide (25 mg/kg p.o.), hydrochlorothiazide (25 mg/kg p. o.), triamterene (50 mg/kg p.o.), or amiloride (50 mg/kg p.o.) can be used as standards
Time of measurement of urine excretion	After 5 and after 24 h	Hourly up to 5 h and after 24 h
Parameters analyzed in urine	Urine volume and sodium content	Content of sodium, potassium, and chloride as well as volume and osmolality of urine
Evaluation and interpretation	• Results are given as the "Lipschitz-value", i.e., the ratio T/U (where T and U are the response produced by the test compound and urea control respectively) • Interpretation of Lipschitz-value ≥1–Positive effect ≥2–Potent diuretic	• Saluretic activity = Sum of Na^+ and Cl^- excretion • Natriuretic activity = Na^+/K^+ ratio (values >2 indicate a favorable natriuretic effect and values >10 indicate a potassium-sparing effect) • CA inhibiting activity = $\frac{Cl^-}{Na^+ + K^+}$ ratio (values between 1 and 0.8 exclude CA inhibition and with decreasing ratios slight to strong CA inhibition can be assumed)

Abbreviations: p.o, peroral administration; CA, carbonic anhydrase

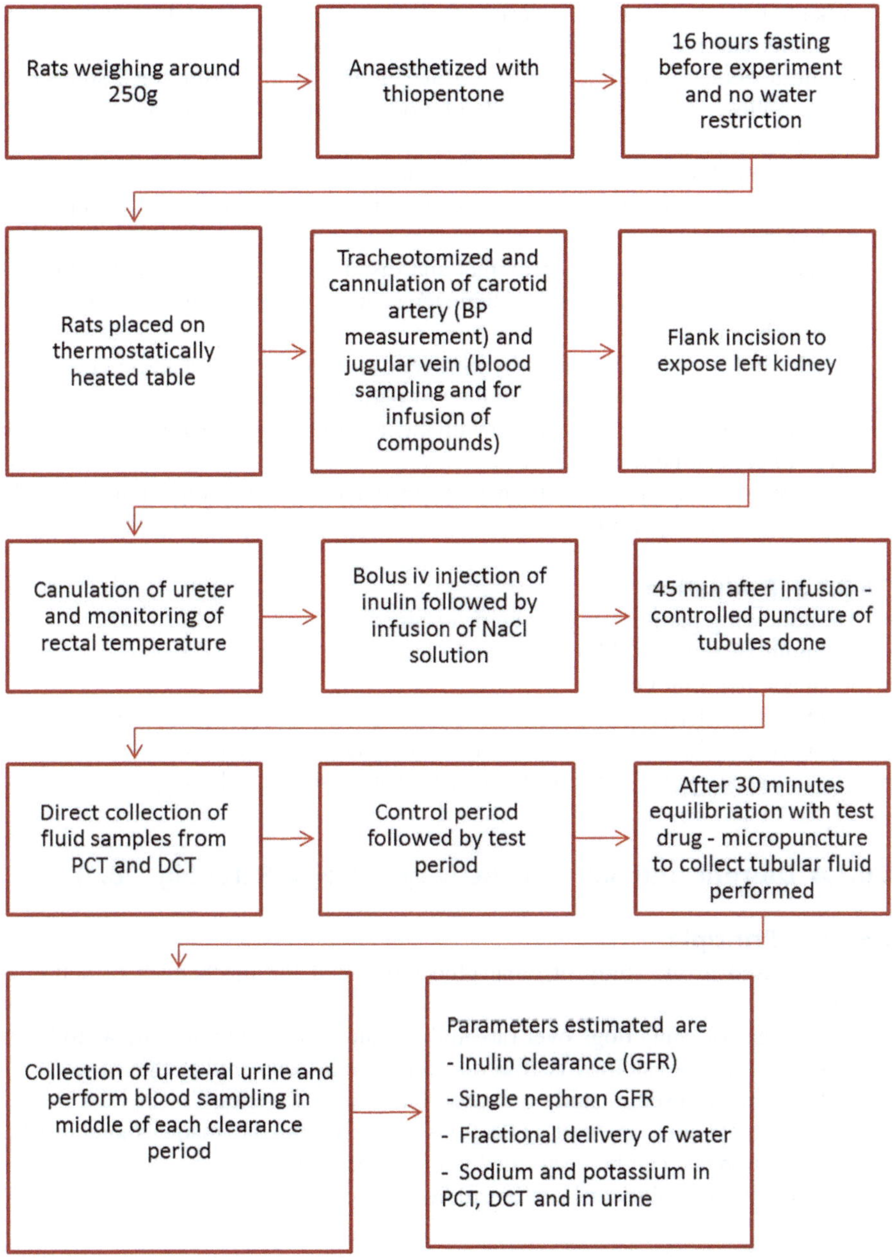

Fig. 26.3 Methodology of micropuncture technique in the rat. Abbreviations: BP, blood pressure; PCT, proximal convoluted tubule; DCT, distal convoluted tubule; GFR, glomerular filtration rate

- A flank incision is made to expose to the left kidney and a cannula is inserted into the ureter and rectal temperature monitored continuously
- An intravenous bolus injection of 75 μCi inulin ^{3}H in 0.7 ml NaCl solution is given, followed by a 0.85% NaCl solution at a rate of 2.5 ml/min/100 g body weight
- The sustained infusion delivers 75 μCi inulin ^{3}H per hour
- After 45 min of starting the intravenous infusion, control puncture of tubules is performed
- Tubular fluid samples from proximal and distal tubules are collected with glass capillaries of 8–10 μm external diameter using a micromanipulator and microscopic observation
- Distal tubules can be identified by intravenous injection of lissamine green
- The control period is followed by the test period
- After a 30 min equilibration period with the test drug, micropuncture is performed again to collect the tubular fluid
- Then, removal of urine in ureter is performed and blood is sampled in the middle of each clearance period

26.3.3.3 Evaluation

- The following parameters may be determined:
  - Inulin clearance (GFR)
  - Single nephron GFR
  - Fractional delivery of water
  - Sodium and potassium in proximal and distal tubules and in urine
- The test group values are compared with that of the control group

26.3.4 Diuretic and Saluretic Activity in Dogs (Refer Fig. 26.4)

26.3.4.1 Principle

- Dogs—used in the study of renal physiology and the mechanism of action of diuretics
- Advantages of using dogs over rats in the evaluation of diuretics are as follows:
  - Dog's renal physiology is closer to that of man compared to that of rats
  - Dogs can be used to study oral absorbability of diuretics
  - Using catheters, interval collections of urine can be made (more reliable measurements can be performed in dogs, than rats)
  - Also, pharmacokinetics can be studied by withdrawing blood samples

26.3.4.2 Procedure:

- Animals used—Beagle dogs of either gender undergone intensive training for accepting gavage feeding and hourly catheterization without any resistance
- The dogs are kept in metabolic cages
- A minimum of 4 dogs are used as

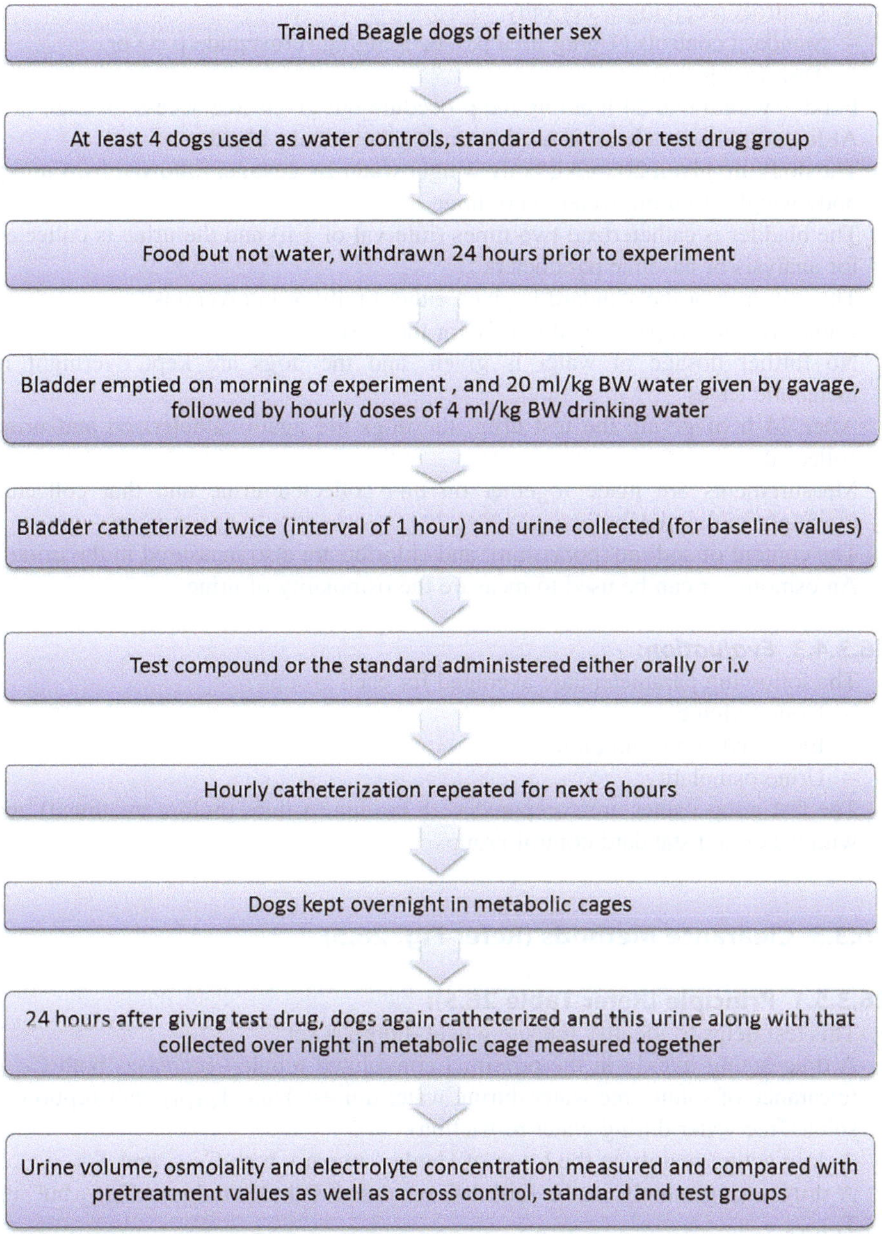

Fig. 26.4 Methodology of diuretic and saluretic activity in dogs. p.o.: peroral; BW: body weight; i. v: intravenous

  - Controls receiving water only,
  - Standard controls (1 g/kg urea p.o. or 5 mg/kg furosemide p.o.) or
  - Test drug group
- Food is withdrawn 24 h before the procedure but given free access to water
- At first, a plastic catheter is used to empty the urinary bladder
- The dogs are given 20 ml/kg body weight water by gavage, followed by 4 ml/kg body weight drinking water every hour
- The bladder is catheterized two times (interval of 1 h) and the urine is collected for analysis of baseline parameters
- The test drug or the standard is given either orally or intravenously
- Catheterization is performed hourly for the next 6 h
- No further dosage of water is given, and the dogs are kept overnight in metabolic cages
- After 24 h of giving the test drug, the dogs are again catheterized and urine collected
- Measurements are made together on this collected urine and that collected overnight in the metabolic cage
- The content of sodium, potassium, and chloride are also measured in the urine
- An osmometer can be used to measure the osmolality of urine

26.3.4.3 *Evaluation:*

- The following parameters are averaged for each group:
  - Urine volume
  - Electrolyte concentrations
  - Urine osmolality
- The test goup values are compared with baseline values (before treatment) and with water and standard control groups

26.3.5 Clearance Methods (Refer Fig. 26.5)

26.3.5.1 Principle (Refer Table 26.5):

- This test helps to identify the site where diuretics act
- A drug acting mainly in the proximal convoluted tubule—increases both C_{H2O} (clearance of solute-free water during water diuresis) and T_{CH2O}(reabsorption of solute-free water during water restriction)
- A drug acting mainly in the Loop of Henle—impairs both C_{H2O} and T_{CH2O}
- A drug acting mainly in the Distal Convoluted Tubule -reduces C_{H2O} but not T_{CH2O}

26.3.5.2 Procedure

- Animals utilized—conscious or anesthetized beagle dogs (conditions of experiment -water diuresis and hydropenia)

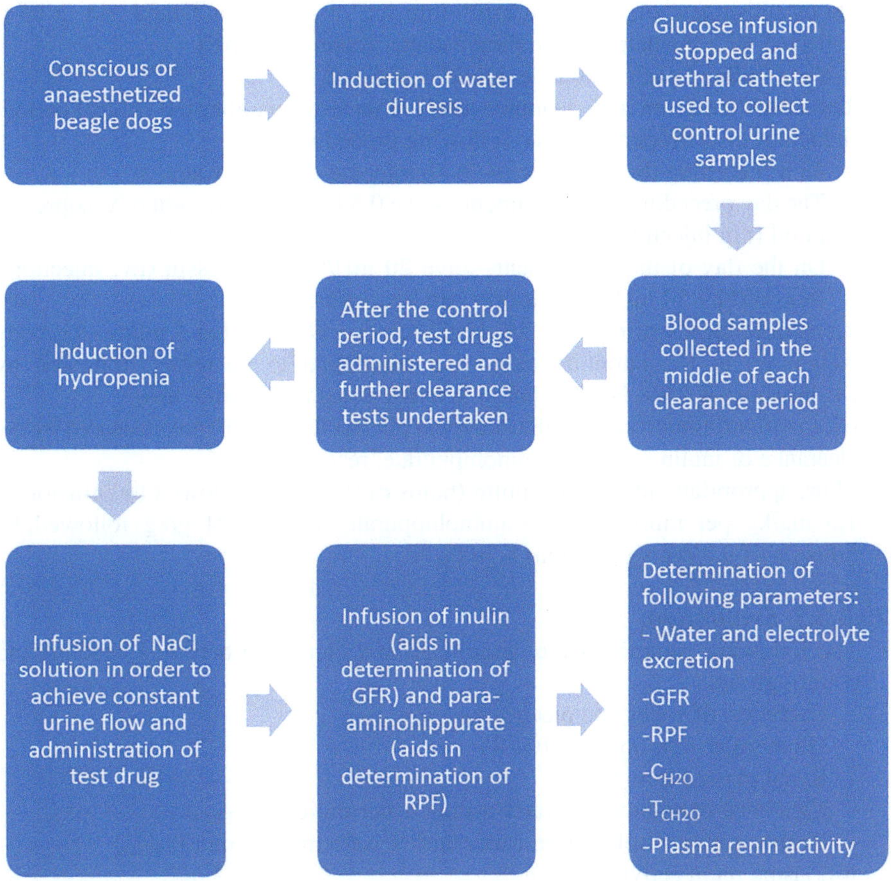

Fig. 26.5 Methodology of clearance test. Abbreviations: GFR, Glomerular filtration rate; RPF, renal plasma flow; C_{H2O}, clearance of solute-free water during water diuresis; T_{CH2O}, reabsorption of solute-free water during water restriction

Table 26.5 Identification of the site of action of diuretics based on its influence on C_{H2O} and T_{CH2O} in the clearance test

Site of action of the drug	Influence on C_{H2O} and T_{CH2O}
Proximal convoluted tubule	Increase both C_{H2O} and T_{CH2O}
Loop of Henle	Impair both C_{H2O} and T_{CH2O}
Distal Convoluted Tubule	Reduce C_{H2O} but not T_{CH2O}

C_{H2O}: clearance of solute-free water during water diuresis; T_{CH2O}: reabsorption of solute-free water during water restriction

- Induction of water diuresis—Oral water administration (50 ml/kg body weight); followed by continuous infusion of 2.5% glucose solution and 0.58% NaCl solution into the jugular vein (infusion rate: 0.5 ml/min/ kg body weight)

- On establishment of water diuresis—infusion of glucose is halted and urine is collected with the help of a urethral catheter (control sample)
- In the middle of each clearance, blood is sampled
- Test drugs are given and clearance tests are undertaken, after the control period
- Induction of hydropenia–by the following procedures:
  - 48 h preceding the experiment—withdraw the drinking water
  - The day preceding the experiment—give 0.5 U/kg body weight of vasopressin in oil (i.m injection)
  - On the day of the experiment—give 20 mU/kg vasopressin (i.v. injection), followed by 50 mU/kg/h vasopressin (i.v. infusion)
- To achieve steady urine flow, 5% NaCl solution is infused at 1 ml/min/kg body weight up to i.v. administration of a compound to be tested, followed by i.v. infusion of 0.9% NaCl solution at a rate equal to the urine flow
- Glomerular filtration rate (GFR) and renal plasma flow (RPF) are measured by the clearance of inulin and para-aminohippurate, respectively
- Thus, appropriate infusion of inulin (bolus of 0.08 g/kg followed by infusion of 1.5 mg/kg per min) and para-aminohippurate (bolus 0.04 g/kg followed by infusion of 0.3 mg/kg per min) is initiated

26.3.5.3 *Evaluation:*

- The following parameters are determined and compared between standard and control group:
  - Excretion of water and electrolyte
  - Glomerular filtration rate (GFR)
  - Renal plasma flow (RPF)
  - Clearance of solute-free water during water diuresis (C_{H2O})
  - Reabsorption of solute-free water during water restriction (T_{CH2O})
  - Plasma renin activity

26.3.6 Stop-Flow Technique (Refer Fig. 26.6)

26.3.6.1 Principle

- This test is useful for discovering the location of transport processes along the length of the nephron
- When ureter is clipped, GFR is drastically diminished
- Also, there is an increase in the time period of contact for the tubular fluid in the respective renal tubular segments
- On release of the clamp, the tubular fluid passes quickly and produces only a minor alteration in the fluid's composition
- The initial samples are from the distal nephron segment, and the final ones correspond to glomerular fluid

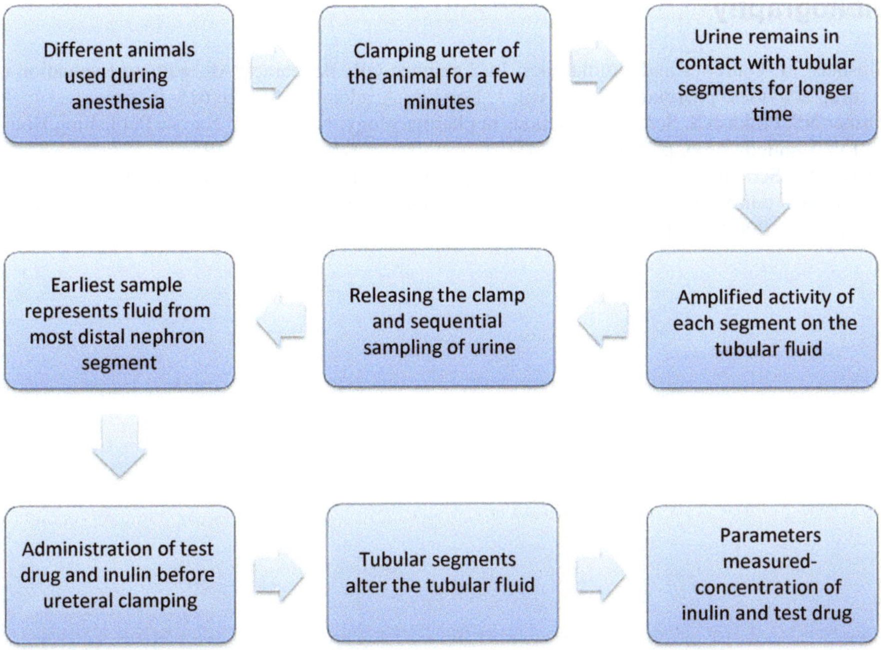

Fig. 26.6 Methodology of the stop-flow technique

26.3.6.2 Procedure

- Animals used—different animals can be used during anesthesia
- For the animal going through intense osmotic diuresis, its ureter is occluded with a clamp, for a few minutes
- Thus, a moderately static section of urine is in contact with the various tubular segments for prolonged periods
- This leads to amplification of the activity of each segment on the tubular fluid
- Then, the ureter is unclamped and the urine is collected in a sequential manner
- The fluid in contact with the most distal nephron segment comes out first
- Test drugs are given along with inulin before applying the ureteral clamp
- When the fluid moves from the proximal renal tubular segments downstream, the distal segments may alter the tubular fluid

26.3.6.3 Evaluation

- In each sample, the following parameters are estimated
  - Inulin concentration (a glomerular marker)
  - Test drug concentration
- Fractional value of the excretion of the test drug and the glomerular marker are plotted versus the aggregate urinary volume

Bibliography

Ginsburg M. Diuretics and antidiuretics. In: Laurence DR, Bacharach AL, editors. Evaluation of drug activities: pharmacometrics, vol. 1. London: Academic; 1964. p. 615–9.

Parmer NS, Prakash S. Screening methods in pharmacology. New Delhi: Narosa Publishing House Pvt. Limited; 2006. p. 241–5.

Turner RA. Screening methods in pharmacology. New York: Academic Press Inc; 1965. p. 251–4.

Vogel HG, editor. Drug discovery and evaluation: pharmacological assays. 2nd ed. Berlin: Springer; 2002. p. 317–28.

Screening Methods for the Evaluation of Antihyperlipidemic Drugs

27

Bertilla Benjamin

Abstract

Statins are the first line agents to combat hyperlipidemia. However, there is a need for newer drugs as there have been cases of statin resistance and a few untoward adverse effects. The following chapter gives insight into a few commonly used screening models both *in-vitro* and *in-vivo*. The *in-vitro* models like Caco −2 cell lines helps in initial screening to see if the drug has any effect on lipoprotein levels. Inhibition of isolated HMG-CoA reductase inhibitors and ACAT inhibitory models helps further in elucidating the mechanism of action. *In-vivo* models used include rats, rabbits, hamsters and genetic models. Recently, transgenic mice are also in use. Hyperlipidemia is induced either through diet like in cholesterol diet induced atherosclerosis in rabbits or induced with chemicals such as triton or propyl-thiouracil. Hypolipidemia action in Syrian hamsters closely mimics human hyperlipidemia. Chronic Model or high fat diet (HFD) induced hyperlipidemia is the one most widely used. Genetic models help in screening of drugs with newer mechanism of action. However, no model is ideal and the model chosen should be relevant to the drug screened.

Keywords

Hyperlipidemia · Hypercholesterolemia · Hypolipidemia · Anti-atherogenic

B. Benjamin (✉)
Community Health Centre, Irikkur, Kannur, Kerala, India

M. Lakshmanan et al. (eds.), *Introduction to Basics of Pharmacology and Toxicology*, https://doi.org/10.1007/978-981-19-5343-9_27

27.1 Lacunae with Current Anti-Hyperlipidemics

The most widely used anti-hyperlipidemics are statins. Statins are safe and efficacious. However, in some groups of patients significant residual cardiovascular risk remains despite statin therapy. Some patients are also unresponsive to statins. The major danger with this class of drugs is the development of myopathy and abnormal liver function tests. Moreover, there are a lot of controversies regarding the use of statins such as it hastens the development of diabetes in patients with impaired glucose tolerance. Thus, there is a need for a new class of drugs beyond statins to improve the treatment of dyslipidemia and halt atherosclerosis.

Newer classes like Pro-protein-Convertase-Subtilisin/Kexin-type 9 (PCSK9) inhibitors, Microsomal triglyceride transport protein (MTP) inhibitors and antisense oligonucleotides against ApoB have shown modest efficacy and are now in clinical trials.

27.2 How to Choose a Relevant Model for Screening?

The following criteria help to decide whether a model is relevant for screening a drug:

- The model should be sensitive in a dose-dependent manner to standard compounds possessing the desired therapeutic property.
- The relative potency of active agents in the model must be comparable to the relative potency in clinical use.
- The model must be selective. The effects of known agents for a therapeutic indication should be distinguishable from the effects of drugs for other indications.
- If it is a first of kind drug for a disease for which no drugs are available, then the model should have a similar pathological status of the disease.

27.3 Classification of Screening Methods for Anti-Hyperlipidemias

***In-Vitro* Screening Methods**

(1) Inhibition of isolated HMG-CoA reductase inhibitors
(2) ACAT inhibitory model
(3) Caco-2 cell lines

***In-Vivo* Screening Methods**

(1) Triton Wistar rat induced hyperlipidemia
(2) Cholesterol diet-induced atherosclerosis in rabbits
(3) Chronic Model or high-fat diet (HFD) induced hyperlipidemia

HMG CoA $\xrightarrow{\text{HMG CoA reductase}}$ Mevolonate

Fig. 27.1 Mechanism of action of HMG CoA reductase. HMG CoA: 3-hydroxy, 3-methyl glutacyl CoAHMG CoA reductase inhibitors reduce cholesterol synthesis and also reduce the expression of LDL receptors on the liver, this is NADPH dependent

(4) Propylthiouracil (PTU) induced hyperlipidemia
(5) Fructose induced hyperlipidemia in rats
(6) Hypolipidemia action in Syrian hamsters
(7) Hereditary hypercholesterolemia in rats
(8) Transgenic animal model (Fig. 27.1)

27.3.1 *In-Vitro* Screening Methods

27.3.1.1 Inhibition of Isolated HMG-CoA Reductase

HMG CoA reductase inhibitors reduce cholesterol synthesis and also reduce the expression of LDL receptors on the liver, this is NADPH dependent (Fig. 27.1). HMG CoA assay kits are available. These kits measure the quantity of NADPH that has been utilized. This is obtained by observing the decrease in absorbance at 340 nm. The percentage of inhibition of the enzyme is calculated. The enzyme can also be purified and quantified.

27.3.1.2 In Vitro Assay Using Caco-2 Cell Lines

Caco-2 cell lines are colon cell lines which are differentiated to human intestinal epithelium like cells. These cells secrete lipoproteins like chylomicrons, VLDL, LDL, and HDL when sodium oleate is added into the culture medium. Thus, the effect of various drugs on lipoprotein levels can be tested.

27.3.1.3 ACAT Inhibitory Model

Cholesterol is present as free cholesterol and as cholesterol esters. Cholesterol esters are synthesized by the acyl-CoA cholesterol acyltransferase (ACAT) enzyme. Thus this enzyme has been an important target for developing newer agents as it is implicated in the synthesis of lipoproteins and foam cells. Lecithin cholesterol acyltransferase (LCAT) is another enzyme involved in cholesterol ester synthesis and plays a role in reverse cholesterol transport.

Procedure:

Liver microsome preparation is made from the liver of male Sprague Dawley rats and it is homogenized to obtain liver homogenate (LH). Other reagents required include:

- Iodoacetic acid (LCAT inhibitor)
- Reagent 1 (R1): Cholesterol esterase, Cholesterol oxidase, Phenol, 4-amino antipyrine

- Lignocaine (cholesterol esterase inhibitor)
- Reagent 2 (R2): Standard (STD) Cholesterol
- Herbal drugs/test compound

The reagents are added in a stepwise manner and the absorbance is measured in each step (absorbance value gives total cholesterol). The steps are repeated in the presence of an LCAT inhibitor and then the amount of free cholesterol is calculated.

27.3.2 *In-Vivo* Methods

27.3.2.1 Triton-Induced Hyperlipidemic Rat Model

- This test studies effect of drugs on the synthesis and excretion of cholesterol. Though simple and rapid, it needs validation with other methods.
- Triton is a surfactant that causes a biphasic elevation of cholesterol and triglycerides.
- The methodology is depicted in Fig. 27.2 below.
- Inference
  - Drug interfering with synthesis produces action in phase I.
  - Drug interfering with metabolism and excretion shows effect in phase II.

27.3.2.2 Cholesterol Diet-Induced Hyperlipidemia

Methodology:

- Swiss albino mice male/female 20–30 g are taken.
- Cholesterol diet given for 7 days, it consists of cholesterol 1%, sodium cholate 0.5%, sucrose 30%, cascien 10%, butter 5%, standard chow 53.5%.
- The mice are divided into 3 groups—Control, test, and standard.

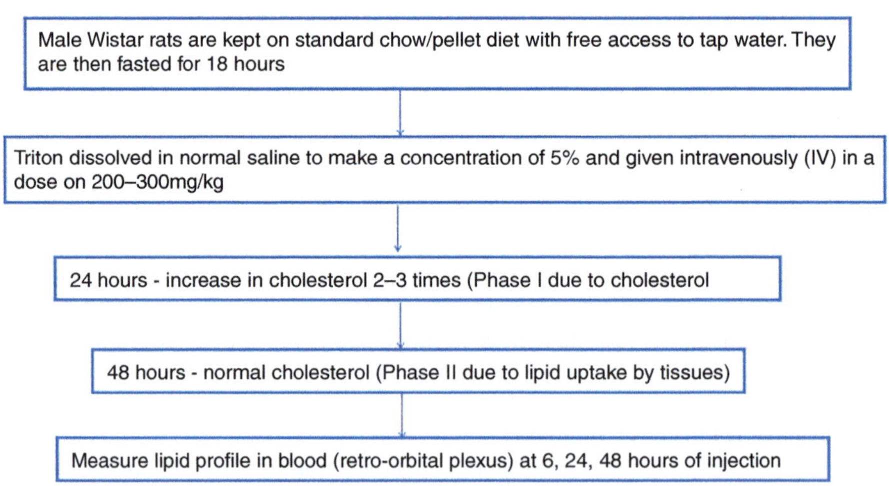

Fig. 27.2 Flowchart of methodology of triton induced hyperlipidemia

- The experiment is done in 2 stages; in the first stage, the lethal dose is calculated; and in the second, three graded doses are given and ED_{50} is calculated.

27.3.2.3 Chronic Model or High-Fat Diet (HFD) Induced Hyperlipidemia

This model is used widely as it mimics human hyperlipidemia. It also induces atherosclerosis and cardiovascular disease.

Methodology:

- Rats are divided into 4 groups namely Control (receiving normal diet), Standard-drug group, test-drug group and high fat diet group.
- An increased amount of cholesterol is mixed with vegetable oil and given to all groups except the control group.
- At 30th day, the blood sample is collected and rat is sacrificed and organs are taken for histopathological examination

27.3.2.4 Propylthiouracil (PTU) Induced Hyperlipidemia

PTU causes hypothyroid state and also increases cholesterol and triglycerides.

Methodology: In this method, around 32 rats are taken and are divided into 5 groups. The procedure is outlined in the flow chart below (Fig. 27.3).

27.3.2.5 Fructose Induced Hyperlipidemia in Rats

This method has a limited role but used for screening nowadays.

- *Principle:*
  Rats are maintained on a diet low in carbohydrates and rich in proteins are suddenly shifted to high fructose diet leads to **hypertriglyceridemia**.
- *Methodology:* The methodology has been illustrated in Fig. 27.4.

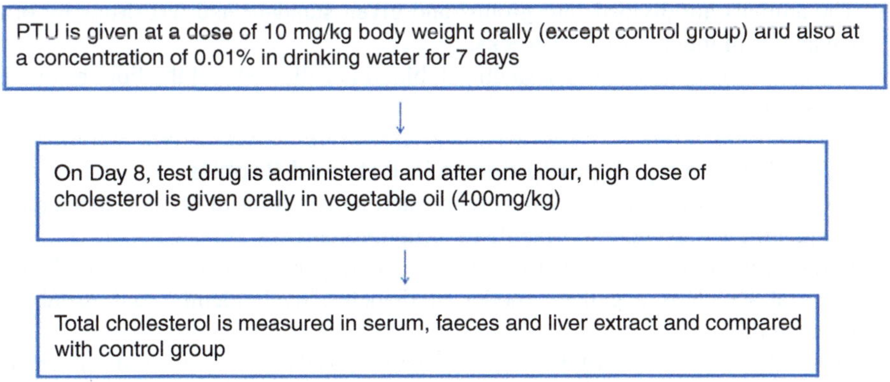

Fig. 27.3 Flow chart of methodology of PTU induced hyperlipidemia

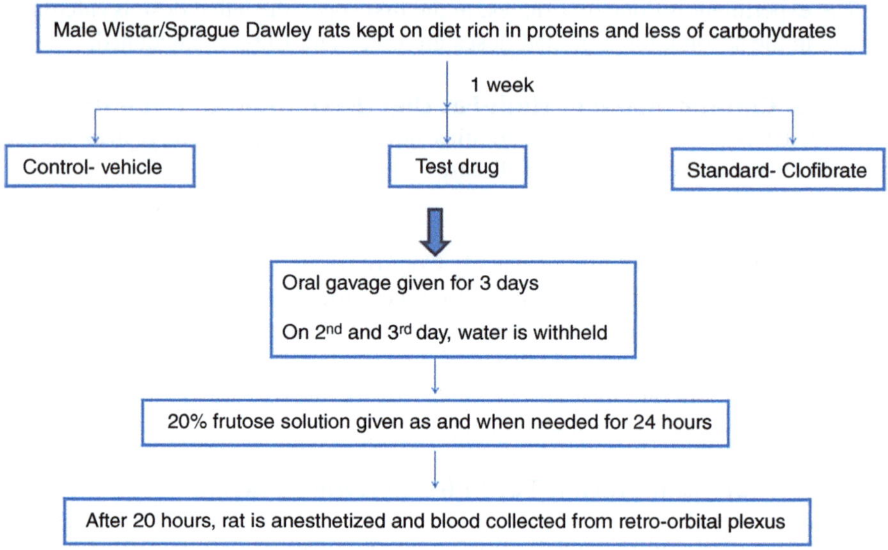

Fig. 27.4 Flowchart of methodology of fructose induced hyperlipidemia

27.3.2.6 Hypolipidemic Activity in Syrian Hamsters

- This is a widely used model. Statins, ezetimibe, and cholestyramine have shown efficacy in this model.
- Lipoprotein and bile acid metabolisms are more similar to human metabolism.
- Hamsters have cholesteryl ester transfer protein (CETP) activity similar to humans which is absent in rats, rabbits, and mice.

Procedure:

- Hyperlipidemia is induced by giving a diet containing 0.2% cholesterol and additional fat like coconut butter (10%) for 2–3 weeks.
- The animals are divided into groups and given standard and test drugs by oral gavage or in the diet for 1–3 weeks.
- After this, the animals are anesthetized, blood is collected and the lipid profile is measured.

27.3.2.7 Hereditary Hyperlipidemia in Rats and Rabbits

- RICO strain is a genetically hypercholesterolemic strain of rat due to increase production and decreased catabolism of lipoproteins.
- Beta-cyclodextrin was found to be effective in this model. Others include Zucker fatty rat which is a genetic model of diabetes, dyslipidemia, and metabolic syndrome.
- In rabbits, Watanabe-heritable hyperlipidemic (WHHL) strain is used for screening of drugs. Rabbits develop atherosclerosis by 10–14 months of age.

Lomitapide an MTP inhibitor which is a newer agent was evaluated in this model. It caused a reduction in ApoB lipoprotein

27.3.2.8 Transgenic Animal Models

- Transgenic mice are developed by integrating different human apolipoprotein genes.
- Apo E knockout mouse has been mostly used. The mouse develops spontaneous hypercholesterolemia and atherosclerosis when on normal chow diet
- Integrating Apo-I causes raised HDL whereas B-48 and 100 causes atherogenic changes.

Bibliography

Ahn CH, Choi SH. New drugs for treating dyslipidemia: beyond statins. Diabetes Metab J. 2015;39: 87–94.

Jijith US, Jayakumari S. Screening methods for antihyperlipidemic activity: a review. DIT. 2018;10:257–9.

Parmar V, Gaurang SG. Screening of antihyperlipidemic herbal drugs for their ACAT (acyl coenzyme: a cholesterol acyltransferase) inhibitory activity. IJSR. 2017;6:1406–12.

Parmer NS, Prakash S. Screening methods in pharmacology. U.K: Alpha science international; 2006.

Ramesh M, Kumar PV, Chaganti M. Pharmacological screening of antihyperlipidemic, anti-stress, and anxiolytic activity of polyherbal formulation in rodents. Int Res J Pharm App Sci. 2013;3: 238–43.

Vogel HG. Drug discovery and evaluation: pharmacological assays. 3rd ed. New York: Springer-Verlag Berlin Heidelberg; 2008.

Screening Methods for the Evaluation of Antiulcer Drugs

28

B. Maharani

Abstract

Peptic ulcer disease encompasses both gastric and duodenal ulcer. Though different group of drugs like antacids, H_2 blockers, Proton Pump Inhibitors (PPIs), etc are utilised for the treatment, each one has its own advantages and disadvantages. Hence, the search for newer and safer antiulcer drugs are in process. Various in vitro methods analyse the acid neutralising capacity of test drug. In vivo methods evaluate the anti-ulcer activity of test drugs on acute, subacute and chronic peptic ulcers that are induced by various methods like pyloric ligation, stress, drugs and chemicals. In this chapter, each method is discussed in detail.

Keywords

Gastric ulcer · Duodenal ulcer · Gastric acid · NSAIDs · Cysteamine · Histamine · Antiulcer drugs · Ulcer index

28.1 Introduction

- Peptic ulcer disease encompasses both gastric and duodenal ulcers. The morbidity and mortality due to peptic ulcers are increasing world-wide.
- *Etiology:*
  - *H. pylori* infection and NSAIDs use are the major etiological factors.
  - Psychosomatic factors, alcohol consumption, chronic smoking, and gastric irritant drugs are also additional etiological factors.
- *Pathogenesis:*

B. Maharani (✉)
Department of Pharmacology, Indira Gandhi Medical College and Research Institute, Puducherry, India

M. Lakshmanan et al. (eds.), *Introduction to Basics of Pharmacology and Toxicology*, https://doi.org/10.1007/978-981-19-5343-9_28

  - An imbalance between aggressive factors (pepsin and HCl) and defensive factors (mucus and bicarbonate).
  - Free radicals mediated damage to the gastric and duodenal mucosa.
- *Management:*
  - Irrespective of the etiology, neutralization, or reduction of acid secretion is the main approach in the management of peptic ulcer.
  - Drugs like antacids, H_2 blockers, proton pump inhibitors (PPIs), ulcer protectives, and anti-*H. pylori* drugs are employed in the management of peptic ulcer.
  - Each drug group has its advantages and disadvantages. PPIs are potent acid suppressants and the most commonly used drug in the current scenario. But long-term administration is associated with accelerated osteoporosis in the elderly, compensatory hypergastrinemia, atrophic gastritis, and increased risk of enteric and respiratory infections.
  - Hence, the search for newer and safer antiulcer drugs is in process. Various methods employed to screen antiulcer drugs are discussed below.

28.2 *In Vitro* Methods

28.2.1 Neutralization Effect on Artificial Gastric Acid

- Sodium chloride (2 g) and pepsin (3.2 mg) are dissolved in 500 ml distilled water. To this, 7 ml HCL with adequate water is added to prepare 1 l of artificial gastric juice with a pH of 1.2.
- Artificial gastric juice (100 ml, pH 1.2) is added to ninety ml of the test solution. The neutralizing effect of the test solution is determined by a change in pH value.
- The same procedure is repeated for the standard drug (the combination of aluminum hydroxide and magnesium hydroxide or sodium bicarbonate) and analyzed.
- *Application*: Used to evaluate drugs neutralizing gastric acid.

28.2.2 Vatier's Artificial Stomach Apparatus Model

- This model has three parts, artificial stomach, peristaltic pump, and pH recording system
- Artificial Stomach has three subparts:
  - *secretary flux*
  - *reservoir*
  - *gastric emptying flux*
- *Peristaltic pump:* to create peristalsis similar to human
- *pH recording system:* to determine the duration of acid neutralization

- In the reservoir portion of the stomach, artificial gastric juice (100 ml, pH 1.2) is added to ninety ml of the test solution at 37 °C. The solution is stirred continuously with a magnetic stirring apparatus maintained at 30 rpm speed.
- Concurrently 3 ml of solution is pumped out and 3 ml of artificial gastric juice is added to the reservoir portion of the stomach. The duration of the neutralization of artificial gastric juice by test drug is determined with the return of pH to 1.2.
- The same procedure is repeated for the standard drug and analyzed.
- *Application:*This method evaluates the duration of the neutralizing capacity of the test drugs.

28.2.3 Fordtran's Model

- Ninety ml of sample (standard or test drug) is placed in a 250 ml beaker at 37 °C and stirred continuously to imitate stomach movements maintained at 30 rpm speed.
- Standard and test samples are separately titrated with artificial gastric juice to achieve the pH 3.
- Quantity (V) of artificial gastric juice consumed is measured.
- Total consumed hydrogen ion (mmol) = 0.063096 (mmol/ml) × V (ml). Test and standard drugs are analysed separately.
- *Application:*Used to evaluate drugs neutralizing gastric acid.

28.3 *In Vivo* Method: Gastric Ulcer Evaluation

- *Animals used for evaluation:*
  - *Rats:*

    Ideal animal for screening anti-ulcer drugs

    Continuous acid secretion in the stomach

    The stomach is analogous to man

    Being omnivorous resembles man nutritionally
  - *Guinea pigs:* They are used to evaluate anti-ulcer drugs when histamine is used to induce gastric ulcer.

28.3.1 Acute Gastric Ulcer Evaluation

28.3.1.1 Rats: Pylorus Ligation Method (Shay Rat Model)

- *Principle:* Pylorus ligation for a certain period induces the formation of gastric ulcers.
- *Procedure:*
  - Female Wistar rats (150–170 g) starved for 48 h with water ad libitum are utilized for the experiment. To avoid coprophagy, they are housed separately in cages with the raised bottom of wide wire mesh.

– The animals are anesthetized with ether. A midline abdominal incision is made. Pylorus is identified and ligated without traction or any damage to the blood supply. The stomach is replaced and the abdominal wall is sutured after completing the procedure.
– After the procedure, test or standard drugs are administered either orally or subcutaneously
– After 17–19 h of pyloric ligation, rats are sacrificed with CO_2 anesthesia.
– Contents of the stomach are emptied into the centrifuge tube and the acidity of the contents is determined with 0.1 N NaOH by titration method.
– The stomach is opened along the greater curvature and pinned on a cork plate.
– Lesions occur mainly in the rumen and the antrum of the stomach as the mucosal protective mechanisms against gastric acid is less in these sites.
– The number of ulcers and severity is recorded with ulcer scores.
  0 = No ulcer
  1 = Superficial ulcers
  2 = Deep ulcers
  3 = Perforation• *Evaluation:* By calculating the ulcer index. Ulcer index (UI) = $U_N + U_S + U_P \times 10^{-1}$ wherein U_N = Average number of ulcers per animal, U_S = Average of severity score, and U_P = Percentage of animals with ulcers. Ulcer index and acidity of the gastric content of the treated animals are compared with the controls.

• *Application:* Used to evaluate antiulcer drugs that act by different mechanisms

28.3.1.2 Stress Ulcer Model in Rats: Restrain-Induced Ulcers (Immobilization Stress), Hanson and Brodie

• *Principle:* Psychogenic factors including stress play a major role in the development of a peptic ulcer. The method is employed only for final drug evaluation.
• *Procedure*
– Female Wistar rats (150–170 g) are used. The animals should fast for 24 h before the experiment with no access to food and water.
– Test drug or standard drug is administered either orally or subcutaneously.
– Animals are immobilized for 24 h by molding a galvanized steel window screen around the animal and held in place with wire staples.
– Limbs of animals are restricted with adhesive tape to avoid movement.
– Rats are horizontally suspended in the darkroom for 24 h at 20 °C and then sacrificed with CO_2 anesthesia.
– The stomach is removed; number, the severity of gastric lesions, acidity of gastric contents, and ulcer index (UI) are determined for the control group, standard, and test group animals.
– The number and severity of gastric lesions are graded as follows (Buyukcoskun et al, 2007)

0–No lesion
1–Mucosal edema and petechiae
2–1 to 5 small lesions (1–2 mm)
3–more than 5 small lesions or 1 intermediate lesion (3–4 mm)
4–2 or more intermediate lesion or 1 gross lesion (>4 mm)
5–Perforated ulcers

$$\%\text{Inhibition} = [1 - (\text{UI pre} - \text{treatment group}/\text{UI non} - \text{pre} - \text{treatment group})] \times 100$$

- *Application:* Used to evaluate antacids, H_2 blockers, PPIs, and also neuroleptics (since ulcers are produced due to stress). Used only for final drug evaluation.

28.3.1.3 Stress Ulcer Model in Rats: Cold Water Immersion Restraint (WRS) Stress, Takagi et al

- *Principle:* Cooling the rats during the immobilisation period, hastens the occurrence of gastric ulcers and the immobilization time is reduced from 24 to 1 h.
- *Procedure:*
  - Adult Wistar female rats (150–200 g) are used for the experiment. The animals should fast for 24 h
  - After the administration of the test/standard drug orally or SC, they are immobilized in a restraint cages and placed vertically in water up to the level of the xiphoid process at 22 °C for 1 h.
  - The animal is removed from water and Evan's blue (30 mg/kg) dye is injected into the tail vein.
  - The animals are sacrificed after 10 min with CO_2 anesthesia.
  - The completely ligated stomach is removed, filled with formal-saline, and stored overnight
  - To examine the ulcerative lesions, the stomach is opened with an incision along the greater curvature and washed with warm water.
  - Total lesion score = Summation of the length of the longest diameter of the lesions
  - The mean count for each group is calculated and compared (Range for control rat is 20–28).
- *Application:* Used to evaluate antacids, H_2 blockers, proton pump inhibitors, and also neuroleptics. Used only for final drug evaluation.

28.3.1.4 Stress Ulcer Model in Rats: Cold Restraint stress, Senay and Levine (1967); Das et al (1993)

- *Principle:* Cooling the rats during the immobilisation period hastens the occurrence of gastric ulcers and reduces the immobilization time.
- *Procedure:*
  - Adult female Wistar rats (150–200 g) are utilized for the experiment. They are fasted for 24 h.

– After the test or standard drug administration by the oral or subcutaneous route, the rats are immobilized in restraint cages and subjected to cold stress at 4 ± 1 °C without visual contact for 3.5 h.
– Animals are then sacrificed with ether anesthesia; the stomach is incised along the greater curvature, opened, and gently rinsed with water.
– The number, severity of gastric lesions, acidity of gastric contents, and ulcer index (UI) are determined for control, standard, and test group animals and suitably analyzed.

- *Application:* Used to evaluate antacids, H_2 blockers, PPIs, and also neuroleptics. Used only for final drug evaluation.

28.3.1.5 Stress Ulcer Model in Rats: Stress and NSAID-Induced Ulcers

- *Principle:* Stress plays a major role in the development of a peptic ulcer. NSAIDs also cause gastric mucosal damage by inhibiting prostaglandin production. The combination of both methods increases the chance of gastric mucosal injury.
- *Procedure:*
  – Adult female Wistar rats weighing 150–200 g are fasted for 24 h.
  – Test/standard substances are administered orally, simultaneously NSAID is administered intraperitoneally (*Indomethacin* 0.2–4 mg/kg or *diclofenac* 1.25–12 mg/kg or *aspirin* 3.25–25 mg/kg).
  – The rats are also subjected to cold water immersion stress.
  – Animals are sacrificed and ulcer index is evaluated.
  – The dose of NSAID required to increase the ulcer index to 100% against control (stress alone) is compared with the dose of NSAID to produce a 100% increase in gastric erosion under the protective effect of the test drug.
- *Application:* This model is suitable for studying the dose-dependent effect of antiulcer drugs.

28.3.1.6 Stress Ulcer Model in Rats: Haemorrhagic Shock-Induced Gastric Ulcers

- *Principle:* Nonlethal hemorrhagic shock induces severe gastric epithelial necrosis and gastric erosion.
- *Procedure:*
  – Male Wistar rats weighing 250 g are fasted for 12 h. Drinking water is replaced with a 5% sucrose solution.
  – The sucrose solution is withdrawn 2 h before the experiment.
  – Standard or Test drug is administered 2 h before the procedure intraperitoneally.
  – Under urethane anesthesia, the volume of blood equivalent to 2% of the animal weight is removed from the inferior vena cava in 2 min.
  – After 15 min, the animals are sacrificed, the stomach is removed and the intensity of the ulcer is graded.
- *Application:* This model is suitable for evaluating drugs that protect gastric mucosa during major trauma, surgery, or thermal burns.

28.3.1.7 Histamine-Induced Ulcer in Guinea Pigs

- *Principle:* Acute gastric or duodenal ulcer develops with either single intraperitoneal or multiple intramuscular administration of histamine respectively.
- *Procedure:*
  - Male guinea pigs are fasted for 14–16 h with a plastic collar around the neck to prevent coprophagy.
  - Depending upon the experiment, gastric erosions are produced by single-dose intraperitoneal administration of histamine phosphate (5 mg/kg) or duodenal erosions are produced by injecting histamine phosphate 5 mg/kg intramuscularly every 30 min for 3–4 h.
  - Standard or Test drug is administered 1 h before and 2 h after starting the experiment.
  - Animals are sacrificed by cervical dislocation 3 h after a single intraperitoneal injection or 30 min after the last intramuscular injection.
  - The stomach and proximal small bowel are removed and cut open and examined for mucosal erosions and ulcerations.
  - Stomach ulceration is graded similarly to Shay et al method.
  - Duodenal ulcer is graded based on the severity

    1–erosion <5 mm

    2–erosion >5 mm, no hemorrhage evident

    3–erosion >5 mm, hemorrhage evident

    4–a perforated ulcer
  - The acidity of gastric contents is analysed by the titration method. The acidity of gastric contents and the severity of duodenal ulcers are compared between standard and test drug-treated animals.
- *Application:* This method is used to evaluate H_2 blockers.

28.3.1.8 NSAIDs-Induced Gastric Ulcer

- *Principle*: NSAIDs like aspirin, indomethacin, and ibuprofen induce gastric ulcers by inhibiting the constitutive COX enzyme which in turn inhibits gastric prostaglandin synthesis and leads to the formation of gastric ulcer.
- *Procedure:*
  - Male Wistar rats weighing 150–200 g are used.
  - Test drug or standard drug is administered orally in 0.1% Tween 80 solution 10 min before oral indomethacin (20 mg/kg) dissolved in 0.1% Tween 80 solution.
  - After 6 h, the animal is sacrificed and the stomach is removed. The completely ligated stomach is filled with 2% formal-saline and left overnight.
  - The next day, formal-saline is emptied, the stomach is washed with warm water and examined for ulcer.
  - Total lesion score = Summation of the length of the longest diameter of the lesions.
  - Mean count for each group is calculated and compared (Range for control rat is 20–28)

- *Modification of the Method:* As an alternative to indomethacin, aspirin (150–200 mg/kg, animals are sacrificed after 4 h of aspirin administration), phenylbutazone (100 mg/kg, 2 doses at 15 h interval, animals are sacrificed 6 h after the second dose) or ibuprofen (400 mg/kg, animals are sacrificed after 5 h of ibuprofen administration) can be utilized for inducing gastric ulcers
- *Application:* This model is used to evaluate the antiulcer, cytoprotective, and anti-secretory activity of test drugs.

28.3.1.9 Ethanol-Induced Mucosal Damage in Rats

- *Principle:* Gastric lesions are produced by intragastric administration of irritants like absolute ethanol.
- *Procedure:*
  - Male Wistar rats (250–300 g) are utilized for the experiment. They are fasted for 18 h with access to water alone.
  - Appropriate standard or test drug is administered intragastrically. After 30 min, 1 ml of absolute ethanol is administered intragastrically.
  - Animals are sacrificed 1 h after ethanol administration. The stomach is incised along the greater curvature and ulcer scores are analyzed by comparing standard and test drug animals.
  - A circular full-thickness piece of tissue of 13 mm in diameter from the glandular and non-glandular portion of the fundus of the stomach is examined under Aristo Model T-16 cold cathode transilluminator (38 × 38 cm) containing W-45 blue white lamp for optical density.
  - A lower optical density value indicates damage while a higher optical density value indicates little or no damage.
- *Modification of the Method:* As an alternative to ethanol, irritant agents like alcohol, HCL, NaOH or hypertonic NaCl can be utilized for inducing gastric ulcers.
- *Application*: This model is used to evaluate cytoprotective compounds used in the management of gastric ulcers.

28.3.2 Sub-Acute Gastric Ulcer Evaluation

- *Principle:* Sub-acute gastric ulcer is induced by using thin steel wire with a needle tip of 1.2 mm diameter.
- *Procedure:*
  - Female Wistar rats (120–150 g) are utilized for the experiment. They are fasted for 24 h with free access to water only.
  - After anesthetizing the rat with ether, a catheter made of polyethylene with a thin steel wire having a needle tip of 1.2 mm diameter is introduced into the stomach.
  - Once the cannula reaches the stomach wall, the upper part of the steel wire is pressed to puncture the gastric wall.

- Standard drug or test drug is administered orally 30 min and after 24 h of the procedure.
- After 2 h of the procedure up to the end of the experiment, the animals are allowed for free access to food and water.
- The animals are then sacrificed with ether overdose at the definitive interval after the procedure.
- The stomach is removed and incised along the lesser curvature and rinsed with tap water. It is fixed around a polyethylene tube of 10 mm diameter with the punched ulcer in the center and suspended in a beaker containing physiological saline.
- The pressure in the tube is gradually increased with a valved rubber wall which is tonometrically calibrated up to 1 bar. The tensile strength (mm of Hg) at which bubbles appear at the gastric ulcer is recorded.
- Healing rate of gastric ulcer = (A-B)/C (mm Hg/h)

    A = tensile strength at C time point after the puncture

    B = tensile strength 30 min after puncture (average value is 143 mm Hg)

    C = time course (h) of the experiment
- The healing rate of standard and test drug are compared and analyzed.

• *Application:* This method is used to evaluate drugs like H_2 blockers, PPIs, and cytoprotective drugs.

28.3.3 Chronic Gastric Ulcer: Acetic Acid-Induced Gastric Ulcer

• *Principle:* Chronic gastric ulcer is induced in rats by subserosal injection of *acetic acid.*

• *Procedure:*
  - Male Wistar rats weighing 200–225 g are fasted for 4 h.
  - Under ether anesthesia, laparotomy is performed with midline epigastric incision. The stomach is exposed and then 0.05 ml of 1 or 10 or 30% acetic acid is injected into the subserosal layer in the glandular portion of the anterior wall without disturbing the blood vessels.
  - A thumb is placed at the injection site to prevent leakage of injected acid. The abdomen is closed and animals are fed normally with the test or standard drug administered orally for 2 weeks.
  - After 2 weeks, animals are sacrificed and the stomach is removed, filled with 10 ml of 1% formalin solution, and immersed in a 1% formalin solution for 5 min to fix the outer layer.
  - The stomach is cut opened and examined for ulcers and ulcer index (UI) is calculated.
  - Curative ratio = (Control UI–Test drug UI) × 100 / Control UI

• *Application:* This method is used for evaluating the antisecretory and cytoprotective effects of drugs used in the treatment of chronic gastric ulcers.

28.4 *In Vivo* Method: Duodenal Ulcer Evaluation

28.4.1 Cysteamine-Induced Duodenal Ulcers

- *Principle:* Administration of cysteamine either by oral or subcutaneous route induces experimental duodenal ulcers in rats. It inhibits mucus secretion in the proximal duodenum, stimulates gastric acid secretion rate, and increases serum gastrin concentration.
- *Procedure:*
  - Male Wistar fed rats weighing 200–225 g are used.
  - *Cysteamine HCL* (10% solution in normal saline) is administered either orally or subcutaneously at an interval of 3–4 h to induce acute duodenal ulcers. Standard or test drug is administered based upon their respective dosage
    Oral dose–28 mg/100 g × 3, sacrifice the animal 28 h after the first dose,
    Oral dose–40 mg/100 g × 2, sacrifice the animal 48 h after the first dose,
    Oral dose–75 mg/100 g × 1, sacrifice the animal 48 h after the dose (or)
    Subcutaneous dose–20 mg /100 g × 2, sacrifice the animal 18 h after the dose
  - Duodenal ulcers develop 2–4 mm away from the pylorus on the anterior wall sometimes penetrating the liver and a small ulcer develops on the posterior wall of the duodenum in the control animal.
  - The intensity of ulcers produced in control, standard, and test groups are compared and analyzed
    0 = no ulcer
    1 = superficial mucosal erosion
    2 = deep ulcer with transmural involvement
    3 = perforated or penetrated ulcer
  - Chronic duodenal ulcer is induced by an acute ulcerogenic regimen on the first day followed by the addition of 0.2 or 0.05 or 0.01% *cysteamine HCL* to the drinking water for 21–60 days. Standard or test drug is administered based on the respective dosage.
  - In addition to ulcers, granulation tissue and dense fibrous connective tissue are observed around the ulcer.
  - The intensity of ulcer produced in control, standard, and test groups are compared and analyzed.
- *Application:* This model is used to evaluate anticholinergic agents, antacids, H_2 receptor antagonists, and prostaglandin analogs used in the treatment of duodenal ulcers.

28.4.2 Chemical-Induced Duodenal Ulcers

- *Principle:* Dulcerozine, L-Methyl-4-Phenyl-tetrahydropyridine (MPTP), indomethacin + histamine, pentagastrin, and carbachol are also utilized to induce duodenal ulcers in rats. Dimaprit is used to induce duodenal ulcers in guinea pigs.

- *Procedure:*
  - *Dulcerozine*–300 mg/kg is administered subcutaneously to 24 h fasted rats. Animals are euthanized 18 h after the dose.
  - *MPTP*–20 mg/kg, 3 doses/day for 4 days is administered subcutaneously to fed rats. Animals are euthanized on the 5th day of MPTP administration.
  - *Indomethacin*–5 mg/kg is administered subcutaneously to 24 h fasted rats. After 30 min, histamine HCL - 40 mg/kg, 3 doses at an interval of 2.5 h is administered. Animals are sacrificed 8 h after the last dose of histamine.
  - *Pentagastrin* 1 μg/kg/min and *carbachol* 0.25 μg/kg/min are administered by subcutaneous infusion to overnight fasted rats at a rate of 13 ml per 24 h for 48 h. Animals are sacrificed after the infusion.
  - *Dimaprit*–0.09–0.18 mg/kg, 8 doses are administered intramuscularly at an hourly interval or 1.8–3.6 mg/kg, a single dose is administered intraperitoneally to 24 h fasted guinea pigs. Animals are sacrificed 1 h after the last injection. The intensity of ulcers produced in control, standard, and test groups are compared and analyzed.
- *Application:* The above models are used to evaluate drugs effective in the treatment of duodenal ulcers.

Bibliography

Azhari H, King JA, Underwood F, Coward S, Shah S, Ho G, et al. A258 the global incidence of peptic ulcer disease at the turn of the 21st century: a study of the organization for economic co-operation and development (OECD). J Can Assoc Gastroenterol. 2019;2:504–7.

Cho CH, Pfeiffer CJ. Gastrointestinal ulceration in the guinea pig in response to dimaprit, histamine, and H_1- and H_2-blocking agents. Dig Dis Sci. 1981;26:306–11.

Dekanski D, Janićijević-Hudomal S, Ristić S, Radonjić NV, Petronijević ND, Piperski V, et al. Attenuation of cold restraint stress-induced gastric lesions by an olive leaf extract. Gen Physiol Biophys. 2009;28:135–42.

Fordtran JS, Morawski SG, Richardson CT. In-vivo and In-vitro evaluation of liquid antacid. N Engl J Med. 1973;288:923–8.

Hohnke LA. Gastric acid secretion and ulcerogenesis in histamine-treated guinea pigs. Gastroenterology. 1974;66:1161–7.

Keshavarzian A, Wibowo A, Gordon JH, Fields JZ. MPTP-induced duodenal ulcers in rats. Prevention by reuptake blockers for serotonin and norepinephrine, but not dopamine. Gastroenterology. 1990;98:554–60.

Kurebayashi Y, Asano M, Hashizume T, Akashi A. Dulcerozine-induced duodenal ulcers in rats: a simple, highly-reliable model for evaluating anti-ulcer agents. Arch Int Pharmacodyn Ther. 1984;271:155–68.

Landeira-Fernandez J. Analysis of the cold-water restraint procedure in gastric ulceration and body temperature. Physiol Behav. 2004;82:827–33.

Robert A, Stout TJ, Dale JE. Production by secretagogues of duodenal ulcers in the rat. Gastroenterology. 1970;59:95–102.

Szabo S. Duodenal ulcer disease. Animal model: cysteamine-induced acute and chronic duodenal ulcer in the rat. Am J Pathol. 1978;93:273–6.

Takagi K, Okabe S, Saziki R. A new method for the production of chronic gastric ulcer in rats and the effect of several drugs on its healing. Jpn J Pharmacol. 1969;19:418–26.

Takeuchi K, Furukawa O, Tanaka H, Okabe S. A new model of duodenal ulcers induced in rats by indomethacin plus histamine. Gastroenterology. 1986;90:636–45.

Thabrew MI, Arawwawala LDAM. An overview of in vivo and in vitro models that can be used for evaluating anti-gastric ulcer potential of medicinal plants. Austin Biol. 2016;1:1007–16.

Urakawa T, Nagahata Y, Azumi Y, Itoh A, Sano I, Takeda K, et al. The mechanism of acute gastric ulcer after induced hemorrhagic shock. - PubMed - NCBI. Scand J Gastroenterol. 1989;24:193–201.

Vatier J, Malikova SE, Vitre MT, Mignon M. An artificial stomach-duodenum model for the in-Vivo evaluation of antacids. Aliment Pharmacol Ther. 1992;6:447–58.

Vogel HG. Drug discovery and evaluation: pharmacological assays. 2nd ed. New York: Springer; 2002. Chapter J, Activity on the gastrointestinal tract; p. 833–905

Screening Methods for the Evaluation of Antidiarrheal Drugs

29

B. Maharani

Abstract

Diarrhoea can be due to infective etiology, secretory diarrhoea, stress-induced diarrhoea, or adverse effect of some medications. The routinely used antidiarrheal drugs act by either decreasing the secretory function of intestinal mucosa or by reducing the intestinal peristalsis or by both of these mechanisms. Antimotility drugs may not be beneficial in infective diarrhoea. Another limitation of antidiarrheal drugs is short duration of action. Hence, the search for effective and safe drug is in process. The various in vitro methods employed for screening anti-diarrheal drugs are inhibition of acetylcholine, histamine or serotonin-induced contraction on isolated segments of small intestine of rat or rabbit by test drugs. The various in vivo methods evaluate the ability of test drugs to inhibit diarrhoea or enteropooling induced by chemicals, drugs or stress. Other in vivo methods include test drug effect on cecectomised rats and gastrointestinal transit time. In this Chapter, in vivo methods to evaluate antidiarrheal activity of test drugs are discussed in detail.

Keywords

Antidiarrheal drugs · Magnesium sulphate · Castor oil · Charcoal meal · Enteropooling

B. Maharani (✉)
Department of Pharmacology, Indira Gandhi Medical College and Research Institute, Puducherry, India

M. Lakshmanan et al. (eds.), *Introduction to Basics of Pharmacology and Toxicology*, https://doi.org/10.1007/978-981-19-5343-9_29

29.1 Introduction

- Diarrhoea is defined by WHO as passage of 3 or more, loose or watery stools in 24 h period. It is one of the major causes of morbidity and mortality. Diarrhoea may be due to infective aetiology or secretory diarrhoea or stress-induced diarrhoea or adverse effect of some medications.
- Proper rehydration and treating the underlying cause are the mainstay in the management of diarrhoea.
- The routinely used antidiarrheal drugs act by either decreasing the secretory function of intestinal mucosa or by reducing the intestinal peristalsis or by both the mechanisms. Majority of antimotility drugs decreases the peristalsis and reduces intestinal secretions. It may not be beneficial in infective diarrhoea. Racecadotril—an enkephalinase inhibitor decreases intestinal hypersecretion without affecting intestinal motility. Nevertheless, it has a limitation of a short duration of action. Hence the search for an effective and safe drug is in process. Various methods employed to screen antidiarrheal drugs are discussed below.

29.2 *In Vitro* Methods

The various *in-vitro* methods employed for screening anti-diarrheal drugs are:

- Effect of test drugs on isolated segments of the small intestine of rat or rabbit with acetylcholine or histamine or serotonin-induced contraction.
- Cell lines with Intestinal chloride channel derived from Fischer rat thyroid epithelium or human colonic biopsies are used to assess the percentage of inhibition of intestinal chloride channel by the test drug.

29.3 *In-Vivo* Methods

Rats and mice are the commonly used laboratory animals to screen antidiarrheal agents.

29.3.1 Chemical-Induced Diarrhoea

29.3.1.1 Castor Oil-Induced Diarrhoea

- *Principle:* Castor oil hydrolysis yields ricinoleic acid. It sensitises the intramural neurons of the gut, causes local inflammation and alters water and electrolyte transport in the gut leading to a hypersecretory response.
- *Procedure:*
  - Female Wistar rats (210–230 g) are utilized and they are fasted overnight.
  - The rats are housed individually without water.

- Standard (loperamide 2–5 mg/kg, p.o.) or test drug (in graded dose) is administered orally.
- One hour after the drug administration, 1 ml of castor oil is orally administered.
- After castor oil administration, stools are collected up to 24 h in non-wetting paper sheets.
- Early and late diarrhoeal excretions are recorded.
  Net stool weight is recorded every 15 min during the first 8 h by draining off the urine by gravity—*Early diarrhoeal excretion*
  Net stool weight occurring between 8 and 24 h after castor oil administration is recorded—*Late diarrhoeal excretion*
- The diarrhoea free period and acute diarrhoeal phase are also recorded.
  Diarrhoea free period—Time in minutes between castor oil administration and occurrence of the first diarrhoeal output
  Acute diarrhoeal phase—Time between the first and last diarrhoeal output of the 8 h observation period
- Anti-diarrhoeal effect of test drug is evaluated with the dose-response curve for the decrease of hypersecretion (stool weight) and increase in diarrhoea free period.

• *Application:* Drugs inhibiting electrolytes secretion in the gut can prevent castor oil-induced diarrhoea.

29.3.1.2 Magnesium Sulphate Induced Diarrhoea

- *Principle:* Magnesium sulphate is an osmotically active agent, increases electrolyte secretion and prevents electrolyte and water reabsorption from the gut.
- *Procedure:* Similar to Castrol-oil induced diarrhoea method.
- *Dose:* 1 h after the administration of standard or test drug, 2 g/kg of magnesium sulphate is administered orally through orogastric tubes.
- *Application:* Agents that prevent increase electrolyte secretion in the gut can be screened by this method.

29.3.1.3 Serotonin Induced Diarrhoea

- *Principle:* Serotonin regulates gastrointestinal motility (direct activation of duodenal smooth muscles and indirectly through neural activation of ileal smooth muscles), pancreatic fluid secretion and intestinal chloride secretion.
- *Procedure:* Similar to Castrol-oil induced diarrhoea method.
- *Dose:* Serotonin (600 μg/kg) is administered intraperitoneally 30 min after administration of standard or test drug.
- *Application:* Drugs that are effective in reducing intestinal motility, chloride secretion in the gut and the drugs effective in modulating toxin-induced diarrhoea can be screened by this method.

29.3.2 Chemical-Induced Enteropooling

29.3.2.1 Castor Oil-Induced Enteropooling

- *Principle:* Castor oil alters fluid and electrolyte secretion in the gut. The ability of the drug to alter the accumulation of fluid in the intestine (enteropooling) is assessed.
- *Procedure:*
  - Female Wistar rats weighing 210–230 g are fasted overnight.
  - The rats are housed in individual cages with no access to water.
  - Standard (loperamide 2–5 mg/kg, p.o.) or test drug (in graded dose) is administered orally.
  - One hour after the drug administration, 1 ml of castor oil is administered orally.
  - After 1 h of castor oil administration, animals are sacrificed by cervical dislocation, abdomen cut open with a midline incision, the intestine is exposed by careful dissection, pyloric end and ileocecal end of the small intestine are tied with the thread to avoid spilling out of the contents and dissected out.
  - The contents of the intestine are emptied in a graduated cylinder and the volume is measured
  - Weight of intestine before and after emptying the contents are measured to calculate the percentage inhibition of intestinal secretions

    $\%\ \text{inhibition} = (A - B)/A \times 100$

    A—Average volume or weight of intestine in the control group

    B—Average volume or weight of intestine in the test group
- *Application:* Drugs inhibiting electrolyte secretion in the gut can be screened by this method.

Other substances used in enteropooling test: PGE_2, magnesium sulphate, bile, taurocholate, tauro-cheno-desoxy-cholate.

29.3.2.2 PGE_2 Induced Enteropooling

- *Principle:* Prostaglandin (PGE_2) acts on EP1 receptors in the gut and causes contraction of intestinal smooth muscles and secretion of water and electrolytes. *Procedure:* It is similar to castor oil-induced enteropooling, except PGE_2 is administered orally in the dose of 100 µg/kg immediately after test or standard drug administration in a suitable vehicle. After 1 h. of PGE_2 administration, animals are sacrificed and intestinal content is measured similar to castor oil-induced enteropooling.
- *Application:* Drugs that act by inhibiting the activity of prostaglandins in the intestine can prevent its enteropooling effect are screened by this method.

29.3.3 Gastrointestinal Transit Time Using Charcoal Meal

- *Principle:* The charcoal meal is not absorbed and it is used to measure the gastrointestinal transit time based on the intestinal transport of charcoal meal. Lesser the peristalsis, lesser will be the distance travelled by charcoal meal.
- *Procedure:*
  - Mice or rats are used in this experiment. Before the experiment, animals are fasted for 18 h with only free access to water.
  - Test drug or standard drug (loperamide 2–5 mg/kg p.o. or atropine sulfate 3–5 mg/kg IP) is administered to the animals either by subcutaneous route (15 min before charcoal meal administration) or oral route (60 min before charcoal meal administration). 0.2 ml charcoal meal (4% suspension of charcoal in 2% carboxymethylcellulose solution) is given orally in this experiment.
  - Animals are sacrificed at intervals of 20, 40, 60 and 120 min after charcoal meal administration.
  - The entire extent of intestine from the pylorus to caecum is removed instantly and immersed in 5% formalin to pause peristalsis followed by washing in running water.
  - Transit of charcoal meal from the pylorus to caecum is measured with a help of a ruler.
  - Movement of charcoal in the intestine is expressed as a peristaltic index (PI).

$$\mathrm{PI} = \mathrm{A/B} \times 100[\mathrm{A} - \text{Distance travelled by the charcoal meal, B} - \text{full length of intestine}]$$

$$\text{Percentage of inhibition} = [(\mathrm{APIC} - \mathrm{APIT})/\mathrm{APIC}] \times 100\ [\mathrm{APIC} - \text{average PI of control, } \mathrm{APIT} - \text{average PI of test group}]$$

- *Application:* Drugs that have a spasmolytic activity and capable of reducing peristalsis can be screened by this method.

29.3.4 Antidiarrheal Effect in Cecectomized Rats

- *Principle:* Rat has a distinct caecum, it temporarily stores intestinal contents and holds excess intestinal fluid. It has a high concentration of various microbial flora that helps in digestion of carbohydrates, cellulose and peptides by microbial fermentation. Cecectomised rats are employed to study the antidiarrheal effect of test compounds to eliminate the confounding factor. This model mimics secretory diarrhoea in man.
- *Procedure*
  - Rats (200–250 g) are carefully chosen and cecectomy is performed.
  - The animals are caged immediately and allowed for 48 h recovery period with free access to food and water.

- Animals are then placed in wire bottomed cages without access to food and water. Sheets of paper are placed below the cages to assess the faecal output.
- Rats are acclimatized for 2 h in the cages and those rats passing out stools other than pellets are not included in the study.
- After acclimatization, 16, 16 dimethyl prostaglandin E2 (0.3 mg/kg) in 3.5% ethanol or carbachol (15 mg/kg) in water or cholera toxin (0.5 mg/kg) in aqueous vehicle of 2% NaHCO3 plus 2% casamino acids is administered orally to induce diarrhoea.
- Substances with antidiarrheal activity are administered orally after the commencement of diarrhoeal episodes.
- Diarrhoea is evaluated at 15 min interval for diarrhoea induced by carbachol, 30 min interval for diarrhoea induced by PGE2, and hourly interval for diarrhoea induced by cholera toxin. Consistency of stools and faecal output index are calculated as follows. Consistency of stools is scored as follows:

  1 = normal pelleted stools

  2 = soft—formed stools

  3 = watery stool and/ or diarrhoea

  Faecal output index is the summation of the number of defecation periods and their ranked consistency score which is expressed as mean ± S. D• *Application:* Drugs effective in the management of secretory diarrhoea are evaluated by this method.

29.3.5 Antidiarrheal Effect in Cold-Restrained Rats

- *Principle:* Stressful situation can produce diarrhoea in humans. In this model, rats are subjected to cold restraint stress and the antidiarrheal effect of test drugs are evaluated by determining the fluid content of faecal pellet.
- *Procedure:*
  - Rats weighing 260–310 g pre-treated with the test or standard drugs are subjected to cold restrain (wire mesh restraining cylinders placed at 4 °C) with free access to food and water.
  - Faecal output (no. of pellets expelled) is measured at 1 and 3 h.
  - Faecal pellets are weighed before and after drying them in an oven at 37 °C and the fluid content of faecal pellet is determined.
  - The dose of the drug that reduces cold restrain induced increased faecal output by 50% or the maximum percentage decrease in fluid content is calculated.
- *Application*: Drugs that are effective in stress-induced diarrhoea can be evaluated by this method.

Bibliography

Ghosh MN. Fundamentals of experimental pharmacology. 6th ed. Kolkatta: Hilton & Company; 2015. Chapter 17, Some standard isolated muscle preparations; p.97-106

Khan MA, Khan NA, Qasmi IA, Ahmad G, Zafar S. Protective effect of Arque-Ajeeb on acute experimental diarrhoea in rats. BMC Complement Altern Med. 2004;4:8.

Kinuthia DG, Muriithi AW, Mwangi PW. Freeze dried extracts of bidens biternata (Lour.) Merr. and Sheriff. show significant antidiarrheal activity in-vivo models of Diarrhea. BioRxiv. 2016: 046185.

Kopic S, Geibel JP. Toxin mediated diarrhea in the 21st century: the pathophysiology of intestinal ion transport in the course of ETEC, V. cholerae and rotavirus infection. Toxins (Basel). 2010;2 (8):2132–57.

Maniyar Y, Bhixavatimath P, Agashikar NV. Antidiarrheal activity of flowers of Ixora Coccinea Linn. in rats. J Ayurveda Integr Med. 2010;1(4):287–91.

Pandhare RB, Balakrishnan S, Bangar GD, Dighe PD, Deshmukh VK. Antidiarrheal activity of Clitoria ternatea Linn. (Fabaceae) ethanol leaf extract in rats. Ayu. 2018;39(1):40–5.

Saikia B, Barua CC, Haloi P, Patowary P. Anticholinergic, antihistaminic, and antiserotonergic activity of n-hexane extract of Zanthoxylum alatum seeds on isolated tissue preparations: An ex vivo study. Indian J Pharmacol. 2017;49(1):42–8.

Schulzke JD, Andres S, Amasheh M, Fromm A, Günzel D. Anti-siarrheal mechanism of the traditional remedy Uzara via reduction of active chloride secretion. PLoS One. 2011;6(3): e18107.

Sisay M, Engidawork E, Shibeshi W. Evaluation of the antidiarrheal activity of the leaf extracts of Myrtus communis Linn (Myrtaceae) in mice model. BMC Complement Altern Med. 2017;17: 103.

Thiagarajah JR, Ko E-A, Tradtrantip L, Donowitz M, Verkman AS. Discovery and development of antisecretory drugs for treating diarrheal diseases. Clin Gastroenterol Hepatol. 2014;12(2): 204–9.

Vogel HG. Drug discovery and evaluation: pharmacological assays. 2nd ed. New York: Springer; 2002. Chapter J, Activity on the gastrointestinal tract; p. 833–905

Screening Methods for the Evaluation of Antiemetics

30

B. Maharani

Abstract

Nausea and emesis are the components of a protective reflex that helps to get rid the gastrointestinal tract from toxic substances and prevents further ingestion. Various potent antiemetic drugs like 5-HT_3 antagonist, D_2 blockers, and H_1 blockers are utilised in the treatment of chemotherapy and radiotherapy- induced nausea and vomiting, post-operative nausea and vomiting, and others. Though potent antiemetics are available, each has its own advantages and disadvantages. Hence, the search for effective antiemetic agent with good therapeutic effect and minimal side effect is in process. Experimental drugs with possible antiemetic activity with different mechanism of action are screened by using various isolated muscle preparations like guinea pig ileum, taenia caecum, colon; rat duodenum, colon, uterus; and rabbit ileum and jejunum. In vivo methods evaluate the antiemetic activity of test drug in emesis induced by chemicals, drugs, motion and by radiation. In this chapter, in vivo methods employed in the screening of antiemetic drugs are discussed in detail.

Keywords

Emesis · Radiation-induced · Motion-induced · Delayed emesis

30.1 Introduction

Nausea and emesis are the components of a protective reflex that helps to rid the gastrointestinal tract from toxic substances and prevents further ingestion. Various receptors and pathways are involved in the pathogenesis of vomiting. The most

B. Maharani (✉)
Department of Pharmacology, Indira Gandhi Medical College and Research Institute, Puducherry, India

M. Lakshmanan et al. (eds.), *Introduction to Basics of Pharmacology and Toxicology*, https://doi.org/10.1007/978-981-19-5343-9_30

common receptors involved in the process of nausea and vomiting are 5-HT_3, D_2, M_1, H_1, opioid μ, CB_1 and NK_1 and are located in Chemoreceptor Trigger Zone (CTZ). Irritant impulses arising from the gut and blood vessels mediate vomiting through 5-HT_3 receptors. Impulses from the vestibular apparatus utilize M_1 and H_1 receptors and send information to the emetic centre. Other receptors located in CTZ and NTS act as a relay centre for emetic signals arising from cortex and peripheral sites and they are the potential targets for antiemetic drug action.

Various potent antiemetic drugs like the 5-HT_3 antagonist, D_2 blockers and H_1 blockers are utilised in the treatment of chemotherapy and radiotherapy-induced nausea and vomiting, postoperative nausea and vomiting, etc. Though potent antiemetics are available, each has its advantages and disadvantages. Hence the search for an effective antiemetic agent with good therapeutic effect and the minimal side effect is in process. The various animal models used in the screening of antiemetic agents are discussed in this section.

30.2 *In-Vitro* Methods

- New drugs with potential antiemetic activity can be screened by using *in-vitro* methods. Experimental drugs with possible antiemetic activity with a different mechanism of action can be screened by using isolated muscle preparation.
- Some standard muscle preparations that can be used are:
  - Guinea pig ileum for H_1 blockers,
  - Guinea pig taenia caecum for muscarinic antagonist,
  - Guinea pig colon for 5-HT_3 antagonist,
  - Rabbit ileum or jejunum is sensitive to muscarine and can be used for screening muscarinic antagonists,
  - Rat duodenum and colon are sensitive to substance P and are used for screening its antagonist and rat uterus for screening 5-HT_3 antagonist.

30.3 *In-Vivo* Methods

- Commonly used laboratory animals like mouse, rat, rabbit, guinea pig and hamster are not utilised to evaluate antiemetic agents.
- Reduced muscularity of diaphragm, the geometry of stomach which is not well structured to move the contents towards the oesophagus, and absent brain stem neurological component for vomiting has led to the absence of vomiting reflex in rodents. They on exposure to emetogenic agents or motion have a typical character of increased intake of non-nutritive substances like kaolin (pica).
- Pica is an illness-response behaviour similar to vomiting in other species. Rabbits have a thin-walled stomach, which may rupture when emesis is induced and it is not used to evaluate antiemetic agents. Other mammalian species like dogs,

monkeys, cats, ferrets, pigs and vertebrates like birds, reptiles, fishes and amphibians exhibit vomiting reflex similar to humans.

- Dogs, cats, monkeys, ferrets, house musk shrew (*Suncus murinus*), least shrew (*Cryptotis parva*), gerbils and pigeons are the animal species used to evaluate antiemetic activity. Ferret is the commonly accepted animal model for emesis. They are used to test 5-HT_3 antagonists and NK_1 receptor antagonists.
- Dogs, cats and monkeys are not utilized for repeated testing of antiemetic activity since they have good learning abilities. Chemoreceptor trigger zone in cats has α_2 receptors. Hence xylazine or dexmedetomidine produces a better emetic response in cats.

30.3.1 Chemicals/Drugs Induced Emesis Model

- Most commonly used chemicals/drugs to induce emesis are apomorphine, morphine, nicotine, copper sulphate, and anticancer drugs like cisplatin, methotrexate, cyclophosphamide, doxorubicin, etc.

30.3.1.1 Apomorphine Induced Emesis

- *Principle:* Apomorphine acts directly on dopamine receptors in CTZ and induces vomiting. It is the emetic of choice for dogs because of its rapid action and ability to reverse its effect. Apomorphine is used to induce emesis in dogs, ferret and rat. It causes excitation in the cat; hence it is not used to induce emesis in cats.
- *Procedure:*
  - *Dog model:*
    - Beagle dogs (7.3–9.5 kg) of either sex are utilised for the study.
    - Dogs are fed with dry pellet food with water ad libitum and maintained in a temperature and humidity-controlled rooms with a 12:12 h light, dark cycle for 1 week.
    - The animals are shifted to observation cages and administered with apomorphine (0.1–0.3 mg/kg) subcutaneously to induce emesis.
    - The number of vomiting and retches, each episode separated by 1 min are observed over a period of 1 h.
    - If the animal fails to vomit or retch, the latency period is equal to the period of observation (60 min).
    - The same procedure is repeated after two weeks with prior administration of the test or standard drug and the latency of onset of vomiting or retching and number of episodes of vomiting or retching is recorded and analysed.
  - *Ferret model:*
    - Adult male ferrets weighing 1–1.5 kg are utilised for the study.
    - Ferrets are fed with dry pelleted cat chow with water ad libitum and maintained in a temperature and humidity-controlled environment with 12:12 h light, dark cycle for 1 week.
    - The animals are also allowed to acclimatise in the observation cages 2 h every day.

  - On the day of the experiment animals are kept for 30 min in the observation cage and observed for spontaneous movement, episodes of retching and/or vomiting, lip licking, burrowing, rearing, curling-up, backward walking, and defecation.
  - Ferrets are then administered with test drug/standard antiemetic drug in the appropriate route and after 30 min, Inj. apomorphine hydrochloride (0.25 mg/kg) is administered subcutaneously.
  - Animals are then observed for 60 min for any behavioural changes or emetic responses and suitably analysed.
- *Rat model:*
  - Male Wistar rats weighing 150–200 g are utilized for the study
  - They are housed in individual cages with freely available food, water and kaolin pellets and maintained in 12:12 light, dark cycle for 3 days. Kaolin pellets are placed in the cages to allow the animals to adapt to its presence. Everyday kaolin pellets are weighed in a specified time to obtain the correct value of kaolin consumption by the rats.
  - Preparation of kaolin pellets–Kaolin (China clay, hydrated aluminium silicate) is mixed with 1% gum acacia in distilled water to form a thick paste and was shaped similar to the food pellets and dried at room temperature. (Mitchell et al).
  - On the day of the experiment, rats are administered with test drug/standard antiemetic drug in the appropriate route and after 30 min, Inj. apomorphine hydrochloride (1–10 mg/kg) is administered intraperitoneally. Control animals are administered with only apomorphine.
  - After 24 h of drug administration, kaolin intake is measured and compared between groups and analysed.

30.3.1.2 Copper Sulphate Induced Emesis

- *Principle:* Copper sulphate on oral administration induces nausea and vomiting by causing gastrointestinal irritation. It is used to induce emesis in dogs, cats, ferrets, *Suncus murinus* and chick.
- *Procedure:*
  - *Dog model:*

    Selection and acclimatization of the animals to the experimental set up is similar to apomorphine-induced emesis.

    Animals are fasted overnight, test or standard antiemetic drug in the appropriate route is administered and after 30 min, copper sulphate solution (Copper sulphate pentahydrate in distilled water) at a dose of 100 mg/kg is administered through the orogastric tube.

    The number of vomiting and retches experienced by the different group animals, each episode separated by 1 min are observed over a period of 1 h and suitably analysed.
  - *Ferret model:*

After prior experimental selection and adaptation of ferrets, test or standard antiemetic agent is administered in a suitable route followed by oral administration of copper sulphate solution at a dose of 40 mg/kg.

Animals are then observed for 60 min for any behavioural changes or emetic responses and suitably analysed.

- *Cat model:*

  Male and female cats (2–4 kg) are used for the study.

  Cats are housed individually in observation cages and are fed with standard cat food.

  Copper sulphate is administered orally in a dose of 20 mg to 40 mg/cat once a week to find out the threshold dose for each cat by observing the cat for emesis for 45 min.

  Those cats which do not exhibit vomiting with the dose of more than 40 mg or cats with the latency period to the vomit of less than 5 min or more than 45 min are excluded.

  After a week, selected cats are pretreated with standard or test antiemetic drug and the threshold dose of copper sulphate which induced vomiting is administered and the animals are observed for vomiting for 45 min and suitably analysed.

- *Suncus murinus model:*

  Male and female Suncus murinus (>10 weeks old, 30–70 g BW) are utilised for the study.

  They are housed individually with 12:12 light, dark cycle and with free access to pellet chow.

  Animals are then acclimatized in the observation cages, 30 min after the administration of standard/test drug, copper sulphate solution (40 mg/kg) is administered orally.

  The animals are observed for 2 h for emesis and suitably analysed. Emesis is characterized by the rhythmic contraction of the abdomen with or without expulsion of materials from the gastrointestinal tract.

30.3.1.3 Cisplatin-Induced Emesis

- *Principle:* Cisplatin, an anticancer drug is cytotoxic to enterochromaffin cells and releases serotonin which in turn acts on 5-HT3 receptors in the periphery and in the vomiting centre which induces vomiting (acute phase of cisplatin-induced vomiting) in first 24 h of drug administration. Cisplatin-induced emesis model is used to screen drugs with 5-HT3 antagonist property. Cisplatin solution is prepared by dissolving cisplatin powder in normal saline at 70 °C followed by slow cooling to 40 °C.
- *Procedure:*
  - The various animals that are used to screen antiemetic drugs by utilising cisplatin-induced emesis model are dogs, cats, ferrets, rats, suncus murinus, and pigeon.

- Animals pretreated with standard or test drug are administered with cisplatin in the dose of 3.2 mg/kg/ml for dogs, 3–7.5 mg/kg for cats, 10 mg/kg for ferrets, 20 mg/kg for *Suncus murinus* and 3–10 mg/kg for rats. In all the animal models except rat, cisplatin is administered intravenously. In rats, it is administered intraperitoneally. The animals are observed for a period of 4 h for the emetic response or behavioural changes or pica (observed for 24 h) and suitably analysed.
- *Pigeon model*
  - Male adult pigeons (350–550 g) are used for the study.
  - They are maintained at a constant temperature, light-dark cycle and humidity in individual stainless-steel cages.
  - The experiments are conducted during illuminate phase of the light-dark cycle.
  - Pigeons pre-treated with test/standard drug are administered with 4–10 mg/kg cisplatin intravenously into the wing vein.
  - Animals are observed for emesis for a period of 4 h and suitably analysed.

30.3.1.4 Cisplatin-Induced Delayed Emesis

- The delayed phase of emesis is characterised by the release of substance-P from the damaged cells that activates NK-1 receptors in CTZ and induces vomiting.
- Cisplatin is administered at a dose of 5 mg/kg intraperitoneally to ferrets and the animals are observed for a period of 72 h for the delayed phase of emesis.
- The animals are pre-treated with test/standard antiemetic drug with periodic administration of the same depending upon the half-life of the drug. It is always better to observe the animals with a video camera enabled with night photographic system to observe the animals continuously.

30.3.1.5 Methotrexate Induced Delayed Emesis

- Methotrexate is administered at a dose of 2.5 mg/kg I. V to dogs and the animals are observed for a period of 72 h in the observation cages which are enabled with night photographic video recording system.
- The efficacy of antiemetic drugs to prevent delayed phase of emesis which develops after 24 h of methotrexate administration can be evaluated by pre-treatment and periodic administration of test/standard drug depending upon the half-life of the drugs.

30.3.1.6 Other models of drug-induced emesis

- Inj. Xylazine (0.4–0.5 mg/kg) I.M is used to induce emesis in cats.
- Inj. Dexmedetomidine (0.6–10 μg/kg) I.M is used to induce emesis in cats.
- Inj. Morphine (0.1–2.5 mg/kg) S.C is used to induce emesis in ferrets.
- Syrup Ipecac (1–3 ml/kg) is used to induce emesis in pigeons.
- Inj. Emetine Hcl (1–20 mg/kg) or m-(chlorphenyl)-biguanide (mCBG) (0.32–5 mg/kg) or ditolylganidine (DTG)(5.6 mg/kg) injected intramuscularly in to the pectoralis muscle is to induce emesis in pigeons.

30.3.2 Emesis Induced by Motion

- *Principle:* Laboratory animals, exposed to linear horizontal movement in a controlled environment displays salivation or other signs like nausea/discomfort with or without emesis which is similar to motion sickness in humans involving histaminergic receptors. Most commonly used lab animals to study motion-induced emesis are dogs, cats, rat and *Suncus murinus*.
- *Procedure:*
  - *Cat model:*

    Cats weighing 1.5–3 kg are utilised for the experiment. They are acclimatised to the laboratory environment. Each cat should be used only 3 times at an interval of 2 weeks apart to avoid adaptation to motion stimulus.

    Horizontal oscillation model—Crampton and Lucot (1985) developed a motorized device similar to amusement park Ferris wheel. Two ventilated plexiglass boxes (for horizontal balancing) are suspended from the ends of a beam that rotated around a horizontal axle at 17 rpm. The boxes are large enough to accommodate one cat and allowed its free movement. Pretreated or control animals are kept in the boxes and exposed to 30 min of rotation followed by observation for 5 min. The degree of protection by the test drug against motion-induced emesis is assessed by the number of repetitive licking, salivation and or vomiting episodes between control and pre-treated cats.

    Vertical oscillation model—Cats are placed in a plexiglass box with holes at the base to reduce resistance during oscillation. The boxes are suspended with springs from the ceiling. Motion is induced by a gentle push to initiate and maintain vertical oscillation at 0.3 Hz with a distance of 75 cm vertically. Pre-treated or control animals are kept in boxes and are exposed to motion for 30 min followed by 5 min observation period followed by suitable analysis.
  - *Suncus murinus model:*

    Male (BW range 55–90 g) and female (BW range 35–50 g) *Suncus murinus* are used for the study. Each animal can only be used twice at an interval of 1 week to avoid adaptation to motion stimulus.

    Pre-treated or control animals are positioned in Perspex chamber (11 cm length and breadth, 22 cm long) attached to a platform of the shaker and accustomed for 3 min.

    The animals are then exposed to 4 cm linear horizontal movement at 1 Hz frequency for 5 min along the axis of the chamber.

    The number of emetic episodes and timing (a brief period of rapid retch followed by vomiting) are recorded and then suitably analysed.
  - *Rat model*

    Male Wistar rats (150–200 g) are utilised for the experiment.

    Motorized Ferris wheel developed by Crampton and Lucot (1985) is utilized for the study purpose.

The animals are positioned in plexiglass chambers with the long axis of the body perpendicular to the horizontal rotation rod.

The device is rotated in a clockwise and counterclockwise direction at an angular velocity of 120/s with 1 s pause. Clockwise, pause and counter-clockwise motion should last for 21 s.

Pretreated and control animals are subjected to rotation stimulation in complete darkness for 2 h. The number of faecal pellets in the plexiglass chamber during stimulation is recorded.

Immediately, the animals are subjected to spontaneous locomotion activity testing measured by animal behaviour test system and suitably analysed.

30.3.3 Emesis Induced by Radiation

- *Principle:* Laboratory animals exposed to radiation in a controlled environment exhibits salivation or other signs like nausea/discomfort with or without emesis. Radiation delivered to the particular site of body induces emesis by damage to the cells and release of neurotransmitters that activates receptors in the emetic centre and induces nausea and vomiting.
- Similar to chemotherapy-induced nausea and vomiting, acute phase of radiation-induced emesis is caused by damage to enterochromaffin cells and release of serotonin which activates serotonergic receptors in the periphery and the vomiting centre.
- The delayed phase is caused by the release of substance-P from damaged cells which activates the NK-1 receptor. The most commonly used laboratory animals are dogs, ferrets and rats. Cats are not employed as the radiation needed to induce emesis is very high.
- *Procedure:*
  - *Dog model:*

    Adult male dogs weighing 12–20 kg are utilised for the study.

    Selection and acclimatisation of the animals are done according to standard procedure.

    Animals are fed with 0.4 kg canned dog food, one hour before the experiment.

    Test group animals are pretreated with the drugs 30 min before the procedure.

    Dogs are positioned in special acrylic boxes with rods to minimise animal movement.

    The animals are irradiated with 60Co gamma radiation, 258 rad for 3–11 min. Entire body except the head is exposed to the radiation.

    The dogs are then placed in observation cages and observed continuously for eight hours.

Latency and onset of general behavioural changes, salivation, and emetic episode are recorded and suitably analysed.

- *Rat model:*
  Male Wistar rats of 180–200 g are utilized for the experiment.
  The animals are housed in individual cages in a controlled environment with free access to water and food with kaolin pellets for 3 days.
  Test group animals are pretreated with drugs administered intraperitoneally 30 min before irradiation.
  Control and test group animals are placed individually in acrylic cages designed in a way that limits free movement of animals.
  The animals are irradiated with 4 Gy (Gray) of 4 MV X-rays. The entire body is irradiated using a medical linear accelerator.
  The animals are then placed in cages with kaolin pellets (prepared using standard procedure) and observed for radiation-induced pica.
  Kaolin intake for 24 h after irradiation is measured and compared between groups and analysed.
- *Ferret model:*
  Adult male ferrets (0.9–1.4 kg) are utilised for the experiment.
  The animals are maintained under standard laboratory conditions in stainless steel cages with free access to dry cat chow.
  Test group ferrets are pretreated with drugs 30 min before the procedure.
  Control and test group ferrets are then placed in plastic cages and irradiated with gamma rays emitted by 60Co source delivered at a dose rate of 50–100 cGy/min.
  Latency and time for the onset of retching and number of retches and or emetic responses are observed, recorded and suitably analysed.

Bibliography

Agarwal R. Antiemetic agents. In: Gupta SK, editor. Drug screening methods. 3rd ed. New Delhi: Jaypee Brothers Medical Publishers; 2016. p. 567–77.

Andrews PLR, Horn CC. Signals for nausea and emesis: Implications for models of upper gastrointestinal diseases. Auton Neurosci. 2006;125(1–2):100–15.

Beleslin DB, Jovanović-Mićić D, Japundžić N, Nikolić SB, Samardžić R. Differential inhibition by ethanol of norepinephrine- and clonidine-induced emesis in cats. Alcohol. 1992;9(5):389–93.

Cubeddu LX. Serotonin mechanisms in chemotherapy-induced emesis in cancer patients. OCL. 1996;53(Suppl. 1):18–25.

Davies RR, Rees Davies JAE. Rabbit gastrointestinal physiology. Veterinary Clinics of North America: Exotic Animal Practice. 2003;6(1):139–53.

Fukui H, Yamamoto M. Methotrexate produces delayed emesis in dogs: a potential model of delayed emesis induced by chemotherapy. Eur J Pharmacol. 1999;372(3):261–7.

Fukui H, Yamamoto M, Sasaki S, Sato S. Possible involvement of peripheral 5-HT4 receptors in copper sulfate-induced vomiting in dogs. Eur J Pharmacol. 1994;257(1–2):47–52.

Ghosh MN. Fundamentals of experimental pharmacology. 6th ed. Kolkatta: Hilton & Company; 2015. Chapter 17, Some standard isolated muscle preparations; p.97–106

Hickman MA, Cox SR, Mahabir S, Miskell C, Lin J, Bunger A, et al. Safety, pharmacokinetics and use of the novel NK-1 receptor antagonist maropitant (Cerenia TM) for the prevention of emesis and motion sickness in cats. J Vet Pharmacol Ther. 2008;31(3):220–9.

Horn CC, Kimball BA, Wang H, Kaus J, Dienel S, Nagy A, et al. Why can't rodents vomit? A comparative behavioral, anatomical, and physiological study. Covasa M., editor. PLoS One. 2013;8(4):e60537.

King GL. Characterization of radiation-induced emesis in the ferret. Radiat Res. 1988;114(3):599.

Lau AHY, Ngan M-P, Rudd JA, Yew DTW. Differential action of domperidone to modify emesis and behaviour induced by apomorphine in the ferret. Eur J Pharmacol. 2005;516(3):247–52.

Mattsson JL, Cordts RE, Yochmowitz MG, Hardy KA. Prevention of radiation emesis in dogs by combinations of drugs. International Journal of Radiation Oncology*Biology*Physics. 1984;10 (7):1067–72.

Nakayama H, Yamakuni H, Higaki M, Ishikawa H, Imazumi K, Matsuo M, et al. Antiemetic activity of FK1052, a 5-HT3- and 5-HT4-receptor antagonist, in suncus murinus and ferrets. J Pharmacol Sci. 2005;98(4):396–403.

Rudd JA, Jordan CC, Naylor RJ. The action of the NK1 tachykinin receptor antagonist, CP 99,994, in antagonizing the acute and delayed emesis induced by cisplatin in the ferret. Br J Pharmacol. 1996;119(5):931–6.

Takeda N, Hasegawa S, Morita M, Matsunaga T. Pica in rats is analogous to emesis: an animal model in emesis research. Pharmacol Biochem Behav. 1993;45(4):817–21.

Thawley VJ, Drobatz KJ. Assessment of dexmedetomidine and other agents for emesis induction in cats: 43 cases (2009–2014). J Am Vet Med Assoc. 2015;247(12):1415–8.

Vogel HG. Drug discovery and evaluation: pharmacological assays. 2nd ed. New York: Springer; 2002. Chapter J, Activity on the gastrointestinal tract; p. 833–905

Wynn RL, Essien E, Thut PD. The effects of different antiemetic agents on morphine-induced emesis in ferrets. Eur J Pharmacol. 1993;241(1):47–54.

Yamamoto K, Takeda N, Yamatodani A. Establishment of an animal model for radiation-induced vomiting in rats using pica. J Radiat Res. 2002;43(2):135.

Yoshino Furukawa T, Nakayama H, Kikuchi A, Imazumi K, Yamakuni H, Sogabe H, et al. Antiemetic effects of a potent and selective neurokinin-1 receptor antagonist, FK886, on cisplatin- and apomorphine-induced emesis in dogs. Biol Pharm Bull. 2013;36(6):974–9.

Zhou W, Wang J, Pan L, Qi R, Liu P, Liu J, et al. Sex and age differences in motion sickness in rats: the correlation with blood hormone responses and neuronal activation in the vestibular and autonomic nuclei. Front Aging Neurosci [Internet] 2017. Feb 15 [cited 2020 Mar 28];9. Available from: https://www.ncbi.nlm.nih.gov/pmc/articles/PMC5309225/

Screening Methods for the Evaluation of Hepatoprotective Agents

31

Jayapriya Balamurugan and Lourdu Jafrin

Abstract

Liver being the major site of metabolism of xenobiotics, it is predisposed to the harmful effects of the toxic metabolites generated during biotransformation. Drugs designed to prevent this damage are referred to as hepatoprotective agents. To determine the safety and efficacy of the potential hepatoprotective agents, a plethora of screening methods are available both in vitro and in vivo. Some of the commonly used hepatoprotective agents are N-acetyl cysteine, d-penicillamine, s-adenosyl methionine (SAM), melatonin, L-carnitine etc. Plant derivatives like Silybum marianum, Curcuma longa, Camellia sinensis and Glycyrrhiza glabra are also used frequently to name a few.The in vitro screening methods utilize fresh hepatocytes, primary hepatocyte culture and immortalized cell lines for screening hepatoprotective agents. Cell viability test is a in vitro method while precision cut liver slices (PCLS) and isolated perfused liver are ex vivo screening methods.The in vivo screening methods frequently use the Sprague dawley rat for pretreatment with hepatotoxins. The hepatotoxins used are acryl amide, alcohol, adriamycin, cadmium chloride, carbon tetrachloride, erythromycin stearate, galactosamine, paracetamol, tamoxifen, t-BHP and olanzapine.Other models used are chloroform model, hypoxia model, diclofenac model, isoniazid and rifampicin model, concavalin A & lipopolysaccharide model. The hepatoprotective effect is assessed by the morphology, biochemistry and histopathological parameters.

J. Balamurugan
Department of Pharmacology, Government Medical College, Virudhunagar, Tamil Nadu, India

L. Jafrin (✉)
Department of Pharmacology, Indira Gandhi Medical College and Research Institute, Puducherry, India

M. Lakshmanan et al. (eds.), *Introduction to Basics of Pharmacology and Toxicology*, https://doi.org/10.1007/978-981-19-5343-9_31

Keywords

Hepatoprotective agents · Screening method · In vivo · In vitro · Rats and hepatotoxins

31.1 Introduction

Liver is the main site of metabolism of the xenobiotics consumed by humans. This predisposes it to the detrimental effects of the toxic metabolites generated during biotransformation. So, there has always existed a need for drugs which can protect the liver from the toxic metabolites.

Any substance which leads to damage of liver cells is called a *hepatotoxin*; any substance which antagonizes the effects of hepatotoxin is an *antihepatotoxic agent* and any substance that helps to prevent liver damage is called *hepatoprotective agent*. Many allopathic drugs and medicinal plant preparations have been tried and also are in the pipeline for protecting the liver from damage. In order to ascertain the therapeutic efficacy and safety, these novel formulations have to be screened by in vitro and in vivo screening techniques before proceeding to clinical trials in humans.

31.2 Classification of Liver Diseases

- *Based on duration of illness:*
  - Acute <6 week
  - Sub-acute 6 weeks to 6 months
  - Chronic >6 months
- *Based on etiology:*
  - Alcoholic liver disease—steatosis, acute and chronic hepatitis, cirrhosis
  - Non-alcoholic steatohepatitis (NASH)
  - Viral liver disease—acute and chronic viral hepatitis, cirrhosis
  - Drug induced liver disease (DILI)—hepatitis, cholestasis, steatosis
  - Herbal induced liver disease (HILI)—acute hepatocellular injury, acute and chronic hepatitis, acute liver failure, cirrhosis
  - Herbal and dietary supplements (HDS) induced liver injury
  - Metabolic liver disease like Wilson's disease, Hereditary hemochromatosis, alpha 1 antitrypsin deficiency
  - Autoimmune disease—hepatitis, cirrhosis
  - Cholestatic disease—primary biliary cirrhosis, primary cholestatic hepatitis, secondary cholestatic hepatitis
  - Liver damage due to other diseases (Refer 31.3)
  - Liver tumors
- *Based on histopathology:*

- Hepatitis
- Steatosis
- Cirrhosis
- Cancer

31.3 Causes of Liver Disease

(a) Hepatotoxins
(b) Viral—parenteral exposure through IV drug use, blood transfusion, travel to areas endemic for hepatitis virus
(c) Secondary to other diseases
    - Hemochromatosis—diabetes mellitus, skin pigmentation
    - Right sided heart failure—congestive hepatopathy, arthritis, hypogonadism and dilated cardiomyopathy
    - Obesity—Non-alcoholic fatty liver disease (NAFLD),
    - Pregnancy—gallstones,
    - Inflammatory bowel disease—primary sclerosing cholangitis,
    - Celiac disease—Celiac hepatitis ranging from mild asymptomatic elevation of transaminase levels to severe liver failure
    - Thyroid disease

    Generally, adults and females are more susceptible to DILI. Children are more susceptible to liver injury with propylthiouracil. Drug-drug interaction, alcohol use disorder and malnutrition predispose to DILI with acetaminophen.

31.3.1 Hepatotoxins

- Hepatotoxins can be classified as
  Intrinsic/Predictable agents—Constant latent period between exposure and liver damage and is dose dependent. E.g. Carbon tetrachloride, thioacetamide, acetaminophen, ethanol etc.

  Idiosyncratic agents—Unpredictable non—dose dependent liver damage with variable latent period and is non reproducible in animals. E.g. Halothane, sulfonamides, isoniazid etc.
- Exposure can be occupational, environmental, domestic—accidental, homicidal or suicidal
- Accounts for 10% of all acute hepatitis, 50% of acute jaundice, 50% of acute liver failure (out of which 16% are idiosyncratic) is due to DILI. 32% of drug withdrawal form market is due to DILI. 15–20% of DILI could be due to herbal and dietary supplements.
- Almost all the NSAIDs cause elevation of liver enzymes. Acetaminophen is the most common reason for DILI associated acute liver failure, followed by

antibiotics (Amoxicillin/clavulanate), Amiodarone, Allopurinol, Anti-seizure medications, Isoniazid, Azathioprine, Alcohol, iron overdose, etc.

- For further information, this link below can be used to search the database of hepatotoxic drugs, herbal and dietary supplements. (https://www.ncbi.nlm.nih.gov/books/NBK547852/)
- Other hepatotoxins include excess alcohol, aflatoxin, carbon tetrachloride, chlorinated hydrocarbons, industrial chemicals like vinyl chloride, amatoxin from mushrooms—Amanita phalloides and Amanita verna.
- Inorganic compounds—arsenic, phosphorus, copper and iron
- Organic compounds—plant toxins like pyrrolizidine alkaloids, mycotoxins, bacterial toxins and thioacetamide

31.3.2 General Mechanism of Action of Hepatotoxins

It could be direct hepatotoxicity or immune mediated injury.

- *Direct hepatotoxicity:* The general mechanism of action of hepatotoxins is formation of reactive metabolites which interact with cellular lipids, proteins and nucleic acids. This leads to lipid peroxidation, DNA damage and oxidative stress. Also, they may disrupt ionic gradients and intracellular calcium stores leading to mitochondrial dysfunction.
- *Immune mediated:* Hepatic injury also induces both innate and adaptive immune reaction. Stimulation of innate immune cells like Kupffer cells, NK cells etc. These cells produce both proinflammatory mediators, chemokines and cytokines like TNF–α, IFN–Υ, IL–1 β which promote liver injury and also IL-10, IL-6 and prostaglandins which protect liver cells. The balance between proinflammatory and protective mediators determine the susceptibility or adaptation to liver injury (Table 31.1 and Fig. 31.1).

  Kupffer cells, hepatic stellate cells and endothelial cells are hypothetically more exposed or sensitive to oxidative stress-related molecules. Kupffer cells release ROS, cytokines and chemokines which induce neutrophil extravasation and activation, inflammation and apoptosis. Hepatic stellate cells proliferate and synthesize collagen in response to oxidative stress triggered lipid peroxidation.

31.4 Liver Function Tests

Many of the liver enzymes are present in the cytosol of the hepatocyte and they reach the circulation because of damage to the cell membrane. These are detected as increased levels in the serum and used as markers for hepatotoxicity. Liver function tests can be used to differentiate hepatocellular injury, cholestasis and loss of synthetic function of liver.

Markers of liver cellular injury—Alanine aminotransferase ALT(SGPT), Aspartate aminotransferase AST(SGOT), alkaline phosphatase and bilirubin

Table 31.1 Hepatotoxins used in rat models

Hepatotoxin	Dose	Mechanism of action
Acryl amide	6 mg/kg, IP daily for 15 days	• Carcinogenic in rodents • Oxidized to epoxide glycinamide by CYP2E1
Alcohol	7.9 g/kg body weight/day IP for 6 weeks	• Production of reactive oxygen species leading to oxidative stress • Membrane lipid peroxidation leading to loss of integrity of cells and increase in the membrane bound enzyme GGT. • Inhibition of glutathione peroxidase • Reduced activity of catalase and dismutase superoxide
Adriamycin	10 mg/kg body weight IP single dose	• Undergo redox cycling between semiquinone and quinone radicals
Alpha–naphthyl-iso-thiocyanate (ANIT)	75 mg/kg IP. Assessed at 24 h	• Injures bile duct epithelium and liver parenchymal cells
Cadmium chloride	3 mg/kg /d orally for 3 weeks	• Enhances peroxidation of membrane lipids leading to interaction of cellular organelles with metal ions • Depletes glutathione and sulfhydryl groups leading to production of reactive oxygen species like superoxide ions, hydroxyl ions and hydrogen peroxides.
Carbon tetrachloride	1 ml/kg body weight, i.p., 1: 1 v/v mixture of CCl4 and olive oil (0.2–2 ml/kg single dose for acute liver damage; 1.5–5 ml / kg in divided doses over a period of 1 week for chronic (reversible), 12–20 ml /kg for 5–12 weeks for irreversible damage.	• CYP2E1 mediated metabolism of CCl4 to trichloromethyl radicle CCl_3O^- • CCl4 inhibits antioxidants, reacts with protein, cause membrane lipid peroxidation • Reduction in CYP 450 enzymes leads to functional failure and hence decrease in proteins and fatty liver (accumulation of triglycerides)
Erythromycin stearate	100 mg/kg body weight for 14 days IP	Generates free radicals

(continued)

Table 31.1 (continued)

Hepatotoxin	Dose	Mechanism of action
Erythromycin esolate	800 mg/kg/d for 15 days IP	
Galactosamine – • Highly selective hepatotoxin compared to acetaminophen and CCl4, as hepatocytes have high levels of galactokinase and galactose-1-uridyltranseferase	800 mg/kg IP single dose	• Diffuse hepatitis like human viral hepatitis • Disrupts synthesis of uridine nucleotides and in turn synthesis of RNA and proteins, leading to cell death • Reduces the number of viable hepatocytes and oxygen consumption • Damage to bile ducts and ductules leading to cholestasis
Lead acetate Lead nitrate	550 ppm for 21 days in drinking water 5 mg/kg body weight daily for 30 days	• Lipid peroxidation • Production of reactive oxygen species leading to disturbance of prooxidant—antioxidant balance
Microcystin- LR • cyclic heptapeptide from the blue green algae *Microcystis aeruginosa*	20 µg/kg for 28 weeks develop neoplastic liver nodules	Severe diffuse centrilobular hepatocellular necrosis and haemorrhage
Paracetamol	2 g/kg P.O.	• At toxic doses, CYP2E1 mediated production of NAPQ1 (also known as NAPBQI or N-acetyl-p-benzoquinone imine) metabolite occurs which binds covalently with sulphydryl groups of proteins leading to lipid peroxidative degradation of glutathione, mitochondrial dysfunction and oxidative stress thereby resulting in centrilobular necrosis
Phalloidin, toxin from Amanita phalloides	50 mcg/100 g body weight	• Irreversible polymerization of actin filaments leading to cholestasis
Tamoxifen citrate	45 mg/kg/d in 0.1 ml dimethyl sulfoxide and normal saline for 6 days IP	Oxygen radical formation and lipid peroxidation leading to liver cancer
Thioacetamide	200 mg/kg, IP thrice weekly for 8 weeks	• Bioactivation by cytochrome P 450 enzymes or flavin-containing monooxygenases (FMOs) into sulfine and sulfone type compounds which affects the synthesis of proteins, DNA, RNA and GGT

(continued)

Table 31.1 (continued)

Hepatotoxin	Dose	Mechanism of action
		• Decreases the volume and contents of bile
Tert—Butyl hydroperoxide (t-BHP)	0, 0.2, 0.5, 1 or 3 mmol/kg IP. Serum markers increase from 2 h and max at 8 h, with doses 0.5 mmol/kg and above	Metabolized to free radical intermediates leading to lipid peroxidation, glutathione depletion, oxidative stress and cell damage
Olanzapine	4 mg/kg dissolved in 0.9% saline daily IP for 6 weeks in wistar albino rats	

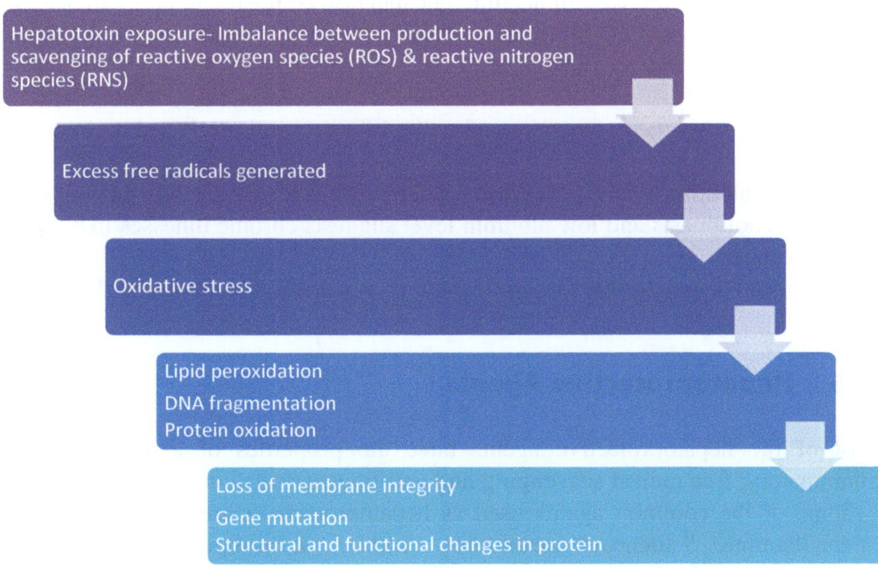

Fig. 31.1 Oxidative stress pathway

Table 31.1 ROS generators and scavengers

ROS generators	ROS scavengers
• Cytochrome P450 enzymes • Mitochondrial electron transport chain • NADPH oxidase • Lipoxygenase	• Reduced Glutathione (GSH) • Catalase (CAT) • Superoxide Dismutase (SOD) • Glutathione Peroxidase (Gpx) • Glutathione Reductase (GR) • Glutathione S Transferase (GST) • Alpha Tocopherol • Ascorbic Acid

- ALT—relatively more specific marker of liver cell injury
- AST—present in liver, cardiac muscle, skeletal muscle, kidney and brain
- LDH—used to differentiate ischemic hepatitis from viral hepatitis
- Disproportionate increase in aminotransferase with or without elevated bilirubin and abnormal liver synthetic function markers indicates hepatocellular injury.

Markers of biliary obstruction—Alkaline phosphatase, Gamma-Glutamyl Transferase (GGT)

- Alkaline phosphatase—from both liver and bones. Elevated values due to liver (cholestasis) is arrived at, by the simultaneous increase in gamma-glutamyl transpeptidase
- Disproportionate increase in alkaline phosphatase compared to aminotransferase with or without elevated bilirubin, and abnormal liver synthetic function markers indicate cholestasis.

Markers of liver synthetic function—albumin, bilirubin and prothrombin time (PT)

- Normal albumin level suggests acute illness like viral hepatitis or choledocholithiasis and low albumin level suggests chronic illnesses like cirrhosis or cancer.
- Prolonged PT may be due to significant liver dysfunction or vitamin K deficiency.

31.5 Hepatoprotective Agents

Most of the hepatoprotective agents have the properties of antioxidant, anti-inflammatory, free radical scavenging and anti-lipoperoxidation.

Some of the common agents used as hepatoprotectives are *N*-acetyl cysteine, d-penicillamine, *S*-adenosyl methionine (SAM), cardiotropin-I, antioxidants—vitamins, melatonin, beta carotene, glutathione, L-carnitine, etc.

Herbal medications—nearly 90 plants are used in 33 mono/poly herbal hepatoprotective formulations in India. Pharmacopoeia Foundation claims 160 phytoconstituents from 101 plants to possess hepatoprotective effect. Some of the commonly used herbs are:

- Silybum marianum (milk thistle),
- Picrorhiza kurroa (kutkin),
- Curcuma longa (turmeric),
- Camellia sinensis (green tea),
- *Phyllanthus* species (amarus, niruri, emblica)
- Glycyrrhiza glabra (licorice).

The active components are mostly Polyphenolic compounds,e.g. Flavonoids, Terpenoids,e.g. Carotenoids, Triterpenes,e,g. Ursolic acid.

31.6 Screening Models

31.6.1 In Vitro Models

- *Cells used:*
  - Fresh hepatocytes
  - Primary hepatocyte culture—Metabolic properties resemble normal liver cells but long-term maintenance not possible
  - Immortalized cell lines—Stable for long time, cryopreservation can be done but biochemical and metabolic properties differ from normal cells.
- *Uses:* Screening of hepatoprotective compounds
- Understand molecular and cellular level mechanism of action of compounds
- *Parameters monitored:* Hepatocyte multiplication, morphology, macromolecular synthesis and oxygen consumptions, and release of transaminases in the medium
- Hepatoprotective agents cause increase in percentage of cells, increase in oxygen consumption rate, reversal of the enzymatic values such as SGPT, SGOT.
- *Advantages:* Large-scale primary screening can be done followed by in vivo methods.
- Quick testing in 2–3 days
- Small amounts of test substance (milligrams) needed
- Experimental conditions strictly controllable
- Different samples analyzed in same test
- Cheap
- Little variability and hence reproducible
- *Disadvantages:* No cell-to-cell interaction, lack of natural exposure to absorption or distribution process of the organ system. So interpretation should be done cautiously.
- *Methodology:*
- *Cell viability test.*
- Sprague Dawley or Wistar rats of either gender is fasted and anaesthetized with ketamine and 1% sodium citrate is injected intraperitoneally to prevent blood clotting. The liver lobes are isolated after cardiac and liver perfusion with calcium, magnesium free hanks buffer salt solution for 15 min. The hepatocytes are isolated and placed in chilled HEPES (N–2–hydroxy ethyl piperazine–N–2–ethanesulphonic acid).
- These hepatocytes are incubated along with the hepatotoxins like CCL4, paracetamol, thioacetamide et cetera at 37 °C for 3 h.
- Hepatoprotective activity was measured by checking viability of cells using tryptan blue and analysis of the released enzymes like SGPT, SGOT and LDH in the medium.

31.6.2 Ex Vivo Models

Precision cut liver slices (PCLS): Because of retained cellular interactions, functioning metabolizing enzymes and biliary canaliculus, liver metabolism and cellular damage can be studied using this model. LDH released is used as a marker of hepatotoxicity. The liver slices are incubated with different treatments and the percentage of LDH release is calculated.

Isolated perfused liver: Porcine, rats, mice and rabbit livers can be used. As the structure and function of liver is preserved, can be used to study hemodynamic parameters and collect bile in real time.

Advantages: Mimics in vivo atmosphere

- Cheap
- Reproducible
- Less animal sacrifice in PCLS model

Disadvantages: Difference between human and murine liver in size, function and geometry.

31.6.3 In Vivo Models

Advantages: All biochemical and metabolic markers and histopathological parameters can be assessed. Highest correlation with humans.

Disadvantages:	Large number of animals are needed
	Long duration
	Inter individual variation is more
	There is a difference between human and animal models in pathogenesis
	Only one plant or drug can be screened at a time

Commonly used animal model is Sprague dawley rat. The dose of the hepatoprotective agent has to be decided by a pilot study.

Rats should be divided into groups.

Group I—Placebo control—Receive only distilled water

Group II—Pretreatment with distilled water and hepatotoxin (toxic dose or repeated dose)

Group III—Pretreatment with hepatoprotective agent and hepatotoxin

Daily monitoring of food and water intake, body weight, signs of any toxic effects and mortality need to be monitored. At the end of the study animals have to be sacrificed under ether anesthesia and blood collected from dorsal aorta—2 ml of blood with EDTA for hematology and 4 ml with heparin for biochemistry.

Hepatotoxins commonly used in rats are given in the Table 31.1:

Apart from these, the other available models are

- Chloroform model
- Hypoxia model
- Diclofenac model
- Isoniazid and rifampicin model:
  The incidence of hepatitis and hepatic necrosis is increased with concomitant use of INH and Rifampicin. INH metabolized to mono acetyl hydrazine (AcHz) first and then further to a toxic product by CYP enzymes.

Rifampicin—increases the production of toxic metabolite due to enzyme induction

- decreases half-life of AcHz and aids quick conversion of AcHz to toxic product,
- induces hydrolysis path of INH metabolism into hepatotoxic pathway of conversion of INH to isonicotinic acid and hydrazine, both of which are hepatotoxic

- Iron overload inducing model
- Cisplatin model
- Concanavalin A (ConA) and lipopolysaccharide (LPS) model—This model can be used in the study of cellular mechanism of autoimmune liver disease
- Nicotine is not used as experimental model for liver injury but for its prooxidant mechanism and evaluation of antioxidant and protective properties of natural compounds

31.7 Evaluation of Hepatoprotective Effect

- Morphological—liver weight and volume
- Biochemical markers of liver injury or biliary obstruction—Liver enzymes Alanine aminotransferase (SGPT), Aspartate aminotransferase (SGOT), alkaline phosphatase, Gamma glutamyl transpeptidase (GGT), 5′ nucleotidase, Lactate dehydrogenase (LDH)
- Liver synthetic function: Serum proteins, Bilirubin, Prothrombin time
  - Bilirubin also reflects the detoxifying ability of liver
- Cholerectic activity
- Functional—Phenobarbitone and hexabarbitone sleeping time
- Histopathological—Necrosis, fatty degeneration and cirrhosis
- Survival rate

31.8 Conclusion

There is a necessity for hepatoprotectives (including complementary & alternative medicines) as the incidence of liver diseases are increasing because of wrong dietary habits, food adulteration, water contamination and drug abuse. Plants are used for the treatment of liver ailments, as synthetic drugs are inadequate or sometimes can cause serious adverse effects. Herbal preparations with appropriate screening seem to have lower side effect profile and are relatively less costly. They have been shown to have antioxidant, anti-inflammatory, anticarcinogenic, antifibrotic and antiviral properties when assessed with a few chemically induced liver damage models in rats. The limited number of RCTs and the toxicological evaluation of herbal drugs need to be improved.

Many herbal preparations claim to be effective without adverse effects but only a few of them are evaluated properly. Since there are many types of liver diseases, the most effective drug for each kind of liver disease must be selected by discrete screening assessment.

Bibliography

Afeefa CM, Kumar MS, Rasheed A, Rajesh V, Babu G, Anjukandan A. A comprehensive review on in vitro and in vivo models used for hepatoprotective activity. J Pharm Sci. 2016;8(6):1.

Ahmad F, Tabassum N. Experimental models used for the study of antihepatotoxic agents. J Acute Dis. 2012;1(2):85–9.

Approach to the patient with abnormal liver biochemical and function tests: UpToDate [Internet]. [cited 2021 Nov 28]. Available from: https://www.uptodate.com/contents/approach-to-the-patient-with-abnormal-liver-biochemical-and-function-tests?

Areefa Shaik A, Elumalai A, Eswaraiah MC, Swathi S. An updated review on hepatoprotective medicinal plants. J Drug Delivery Ther. 2012;2(2) Available from: http://jddtonline.info/index.php/jddt/article/view/97

Asadi-Samani M, Kafash-Farkhad N, Azimi N, Fasihi A, Alinia-Ahandani E, Rafieian-Kopaei M. Medicinal plants with hepatoprotective activity in Iranian folk medicine. Asian Pac J Trop Biomed. 2015 Feb;5(2):146–57.

Delgado-Montemayor C, Cordero-Pérez P, Salazar-Aranda R, Waksman-Minsky N. Models of hepatoprotective activity assessment. Med Univ. 2015 Oct;17(69):222–8.

Drug-induced liver injury: UpToDate [Internet]. [cited 2021 Nov 29]. Available from: https://www.uptodate.com/contents/drug-induced-liver-injury?

Farghali H, Kgalalelo Kemelo M, Wojnarová L, Kutinová CN. In vitro and in vivo experimental hepatotoxic models in liver research: applications to the assessment of potential hepatoprotective drugs. Physiol Res. 2016 Dec;28:S417–25.

Hepatotoxicity due to herbal medications and dietary supplements: UpToDate [Internet]. [cited 2021 Nov 29]. Available from: https://www.uptodate.com/contents/hepatotoxicity-due-to-herbal-medications-and-dietary-supplements.

Ilyas U, Katare D, Aeri V, Naseef P. A review on hepatoprotective and immunomodulatory herbal plants. Phcog Rev. 2016;10(19):66.

Sengupta P, Bagchi C, Sharma A, et al. Olanzapine-induced hepatopathy in albino rats: a newer model for screening putative hepatoprotective agents, namely silymarin. Indian J Pharmacol. 2010;42(6):376–9. https://doi.org/10.4103/0253-7613.71924.

Simeonova R, Kondeva-Burdina M, Vitcheva V, Mitcheva M. Some *in vitro/in vivo* chemically-induced experimental models of liver oxidative stress in rats. Biomed Res Int. 2014;2014:1–6.

Screening Methods for the Evaluation of Antiobesity Drugs

32

Lourdu Jafrin and Mageshwaran Lakshmanan

Abstract

Obesity has become a global pandemic affecting all age groups and is related to significant morbidity and mortality. About 600 million adults are obese, i.e., a body mass index (BMI) greater than 30 kg/m^2. As of date, there are only six US-FDA-approved drugs available for long-term management of obesity in adults. The ideal anti-obesity drug is expected to produce sustained weight loss with minimal adverse drug effects. It is important that animal models are selected with high face construct and predictive validity which resemble the human disease pathophysiology to ensure successful drug development. The screening models for the evaluation of drugs involve the development of good animal models via diet-induced, surgical, chemical, or drug-induced methods. The role of various genes and intermediate pathways are now being elucidated using novel genetic animal models of obesity like Lep ob/ob mouse, Japanese KK mouse, cp/cp rat, Zucker fatty rat, OLEFT rat, etc., Invertebrate models like drosophila has also been now developed successively for rapid and large volume screening of molecules with potential anti-obesity activity. The principles, methods, pros, and cons of each model are discussed in this chapter.

Keywords

Obesity · Animal models · Fatty rat · Fatty mouse · Anti-obesity drug · Drug screening

L. Jafrin
Department of Pharmacology, Indira Gandhi Medical College and Research Institute, Puducherry, India

M. Lakshmanan (✉)
Department of Pharmacology, Thanjavur Medical College, Thanjavur, Tamil Nadu, India

M. Lakshmanan et al. (eds.), *Introduction to Basics of Pharmacology and Toxicology*, https://doi.org/10.1007/978-981-19-5343-9_32

32.1 Introduction

- Obesity has become a global pandemic affecting all age groups and is related to significant morbidity and mortality. It is a chronic metabolic disorder resulting from multiple factors embracing physiological influence, epigenetics, and the environment. About 600 million adults are obese, i.e., with a body mass index (BMI) greater than 30 kg/m^2.
- Obesity is an imbalance between energy expenditure and energy intake, with an increased quantity and duration of intake and probably decreased energy expenditure leading to excessive storage. An understanding of the pathways controlling appetite, satiety, and food intake are needed for the development of new therapeutic drugs for the management of this condition.
- As of date, there are nearly six US-FDA-approved drugs available for long-term management of obesity in adults namely, phentermine (sympathomimetic amine), orlistat (intestinal lipase), phentermine/topiramate ER, bupropion/naltrexone, liraglutide (GLP-1), and semaglutide (GLP-1). Likewise, there are classic examples of antiobesity drugs that were withdrawn due to safety concerns such as rimonabant (depression and suicidal thoughts), fenfluramine-phentermine (primary pulmonary hypertension and heart-valve defects), fenfluramine and dexfenfluramine (primary pulmonary hypertension and heart-valve defects), sibutramine (CV risks), and lorcaserin (cancer).
- The ideal anti-obesity drug is expected to produce sustained weight loss with minimal adverse drug effects. Initially, drugs that target pathways in metabolic tissues, such as adipocytes, liver, and skeletal muscle, showed potential in preclinical studies but were not very effective. The current understanding of peptidic signaling of hunger and satiety from the gastrointestinal tract mediated by ghrelin, cholecystokinin, peptide YY and glucagon-like peptide and mechanisms related to leptin and its pathways in the hypothalamus have opened new avenues for drug development.
- It is important that animal models are selected with high face construct and predictive validity which resemble the human disease pathophysiology to ensure successful drug development. The screening models for the evaluation of drugs with anti-obesity effects are discussed below.

32.2 Animal Models for Inducing Obesity

32.2.1 High Fat Diet-Induced Obesity

- This involves consumption of a high-fat diet
- The following results are seen with consumption or rather overconsumption of high-fat diets
  - low satiation
  - alteration of enzymes and hormones involved in energy balance

  - (hyperleptinemia and hyperinsulinemia, reduction in suppression of ghrelin release)
- C5 7 BL/6 mice are genetically more susceptible to high-fat diet-induced obesity, impaired glucose tolerance, and type 2 diabetes mellitus.
- The method involves the administration of a high-fat diet to normal rats or mice over a period of 3–4 months.
- The parameters measured are
  - altered blood pressure
  - insulin resistance
  - plasma leptin level elevation
  - total cholesterol LDL and triglycerides level
- The advantages of a high-fat diet are that it is cost-effective and is appropriate for the study of lifestyle-induced obesity
- The disadvantage is that a longer period is required to induce obesity leading to a longer study duration

32.2.2 Cafeteria Diet-Induced Obesity

- The cafeteria diet mimics the Western diet of humans (high-fat and high salt).
- The animals demonstrate an increase in adiposity and inflammation of the white and brown fat as well as the liver. In this method, insulin resistance and hypertriglyceridemia can be induced rapidly
- The rodents are administered with a mix of sweets, savory, fat, and high sugar solid foods. These four items are provided for 15 days for four different rat strains.
- *Junk food diet:* A mild modification of cafeteria diet method in which processed foods and snacks are provided along with lard chow mix for both sexes before mating. The diet will be continued for the offspring also.
- Perinatal exposure of the offspring to high-fat and high sugar diets increases their fat intake from the time of weaning. This manipulation results in a change in the expression of genes in the mesolimbic reward system also.
- Reward mechanisms also contribute to the development of obesity via the cafeteria diet-induced obesity method. When male Wistar rats are fed with a cafeteria diet for 18–23 h/day for 40 days, downregulation of striatal dopaminergic transmission resulting in compulsive feeding behavior can occur.

32.2.3 Fat or Sugar Choice Diet-Induced Obesity

- This is a simplified form of a cafeteria diet. The 'free choice of a high-fat/high sugar diet' gives a choice for the rats to choose between saturated fat, a 30% sucrose solution, and standard chow.
- The rodents become hyperphagic and obese when compared to the other group which was fed 'no choice remix off fat, sugar, and chow'. This model proves that

when given a choice there is overconsumption and subsequently leading to obesity.

32.2.4 Meal Feeding-Induced Obesity

- This model involves feeding the animal for specific periods determined by the experimenter which can be denoted as "set meals".
- This model mimics the patterns of human food intake. This imposed meal feeding has the advantage of standardization of the physiological state of the animal before, during and after the meal.
- The amount of energy consumed and the timing of meals is defined. This method has been used to study hunger and satiety-induced brain response to food ingestion more than obesity.
- Binge-type feeding: Binge eating is defined as the consumption of more food than is normal in a short period, with a lack of control and eating when not hungry. In this model, the animals receive continuous access to chow which is supplemented for 2 hours with vegetable fat.

32.2.5 Surgically-Induced Hypothalamic Obesity

- Hypothalamus interacts with a lateral feeding center and a medial satiety center. The ventromedial hypothalamus (VMH) is responsible for appetite control, satiety, and energy homeostasis. Thus, ventromedial hypothalamic lesions cause increased food intake leading to increased body weight and obesity.
- Female Sprague Dawley rats, weighing about 190 g are fed with a high-fat diet for 5–9 days. Later after fasting and under anesthesia bilateral knife cuts are made stereotactically using standard coordinates to reach the hypothalamus of the rat. Sham-operated rats serve as controls.
- *Evaluation*: Parameters like food consumption and increase in body weight, brown adipose tissue enzymes, and guanosine diphosphate binding to brown adipose tissue mitochondria and noradrenaline turnover in various organs are measured and compared between the groups.

32.2.6 Ovariectomy-Induced Obesity in Rat

- Surgically removing both ovaries causes an abrupt fall of estrogen levels in rats. This ovariectomy leads to the development of leptin resistance and obesity in rats.
- Initially, the leptin level falls, and later on, it increases due to leptin resistance.
- However, this method is less preferred as the animal may develop obesity with a wide range of duration.

32.2.7 Castration-Induced Obesity in Male Rat

- In male rats, testosterone is responsible for fat homeostasis. Castration causes an abrupt fall in testosterone levels leading to fat accumulation and obesity.
- Excess visceral fat deposition, increased fasting glucose level in the blood, and increased abdominal adipose tissue accumulation are noted after castration in the male rat.
- Fat accumulation is mainly subcutaneous than visceral in this model. Similar to the ovariectomy model in female rats, this model is also less preferred as the first choice due to wide variation in the duration of development of obesity in animals.

32.2.8 Gold-Thioglucose-Induced Hypothalamic Obesity

- Six-weeks old Swiss albino mice of either sex are given a single intraperitoneal injection of gold-thioglucose (30–40 mg/kg).
- The injected thioglucose reaches the brain and the glucose moiety present in the gold-thioglucose reduced blood circulation to the hypothalamus by altering the permeability in the ventromedial portion of the hypothalamus. Necrosis of tissue occurs leading to the development of lesions in the ventromedial portion of the hypothalamus. The animals exhibit obesity in 15 days due to hyperphagia.
- *Evaluation* – body weight is recorded for 3 months and compared with control.

32.2.9 Monosodium Glutamate-Induced Obesity

- Male Charles River mice are injected daily with monosodium glutamate subcutaneously in the doses of 2–4 g/kg for 5 consecutive days in the early stages of life. Normal saline is injected for the control group.
- The injected mice pups exhibit vagal hyperactivity and or sympathoadrenal hypoactivity leading to hyperinsulinemia and increased weight fat.
- *Evaluation*: Food consumption and weight gain are recorded and a comparison is made.

32.2.10 Drug-Induced Obesity

- Reduction in serotonin, dopaminergic activity, and adrenergic transmission in the central nervous system leads to weight gain. Thus, drug groups like anti-psychotic and anti-depressants are commonly associated with weight gain.
- Thus, animals can be administered with these groups of drugs to induce obesity. However, this method has the disadvantage of possible interactions with the test drug.

32.3 Genetically Obese Animals

- Obesity, diabetes, and genetic influences are intertwined together and reflected in the genetically obese animals used for screening.
- Genetic models utilized for the evaluation of anti-obesity drugs are summarized in Tables 32.1 and 32.2.

32.4 Uncommon Models Used in Obesity

32.4.1 Obesity Induction in Macaques

- Rhesus macaques monkeys developed late-onset obesity around 10–15 years mainly due to overeating.
- This large laboratory animal model resembles the metabolic changes seen in humans, such as elevated basal insulin levels, impaired glucose tolerance, elevated serum triglycerides and cholesterol, and an increase in intra-abdominal fat.

32.4.2 Drosophila Model for Obesity

- The *'brummer'* gene in drosophila is homologous to the *ATGL* gene (Adipocyte triglyceride lipase) in humans. *ATGL* gene is responsible for the control of lipid storage in adipocytes in humans. Similarly in drosophila, the *brummer* gene is responsible for encoding the 'Lipid droplet-associated TG lipase' enzyme.
- Mutation in the *'brummer'* gene leads to the development of obesity in drosophila.
- Thus, the drosophila model linked with 'green fluorescent protein' serves as a potential model for rapid and simultaneous screening of anti-obesity drugs.

32.5 In-vitro Assay Methods for Evaluation of Anti-Obesity Activity

32.5.1 Measurement of GDP Binding in Rat's Brown Adipose Tissue

- *Principle:* Non-shivering thermogenesis in the brown adipose tissue promotes weight loss. The thermogenesis activity is indicated by the binding of GDP to the 'thermogenin' protein in the brown adipose tissue. Thus, the test drug with potential anti-obesity effect and its ability to induce weight loss by non-shivering thermogenesis in brown adipose tissue can be evaluated by measuring the GDP binding in thermogenin.
- *Procedure:*
  - Adult male Fatty Zucker rats weighing around 450 g are fed with varying concentrations of test drug and normal saline (control) for 3 weeks. The body

Table 32.1 Genetic mouse models of obesity

Mouse model	Features	Comment
Lep ob mice (commonly referred to as ob or ob/ob)	• Homozygous for the obese spontaneous mutation • Suffer from hyperglycemia, atrophy of pancreatic islets, and premature death similar to diabetes mellitus • Obesity is seen in addition to severe hyperphagia, transient hyperglycemia, glucose intolerance, and elevated plasma insulin • Leptin-deficient	Used to assess drugs to counter hyperphagia related obese phenotype
db/db mice	• Have high plasma insulin levels at 10–14 days, are deficient in leptin, and develop obesity at 4–5 weeks • They demonstrate Polyphagia, proteinuria, glycosuria, polyuria, polydipsia, and hyperinsulinemia	---
The New Zealand obese mouse (NZO)	• It is an inbred polygenic strain that exhibits obesity at 4 to 5 weeks • It also exhibits Type 2 diabetes mellitus due to insulin resistance and improper leptin signaling • Reaches about 50–70 g by 6–8 months • Affected by renal disease and autoimmune disorders by 6 months of age	Moderate hyperphagia, reduced energy expenditure, and reduced voluntary activity lead to adiposity in this model
Fat mouse (fat/fat or CPE Fat/CPE fat)	• Autosomal recessive mutation on chromosome 8 which codes for carboxypeptidase E, which is involved in the processing of insulin in the late-stage • Obesity is exhibited between 6 and 8 weeks of age and reaches the bodyweight of 60–70 g by 6 months. • The mouse has hyperinsulinemia, obesity, and infertility.	Used as a model for the late-onset type of obesity
Tubby mouse	• The tubby colony was bred from the C5 7 BL/6 J male mouse as autosomal recessive mutation • obesity is noticed only in the third month of age and reaches an average weight of about 50 g at 6 months. • Associated with sensorineural hearing loss and retinal detachment	Slow onset obesity
Japanese K K mouse	• Most suitable polygenic mouse model • Obesity display around 2–3 months of age and reaches 40–50 g by 6–9 months.	Late-onset of obesity

Table 32.2 Genetic rat models of obesity

Rat models	Features	Comment
Zucker fatty (fa/fa) rat	• The Sherman and Merck stock M rats were mated and inherited as autosomal recessive mutation. • obesity, hyperinsulinemia, and insulin resistance are noted • obesity is noticed by 3–5 weeks of age	• most widely used rat model
Obese spontaneous hypertensive rat (OSHR)	• Spontaneous hypertensive female rats (Kyoto–Wistar strain) were mated with normotensive Sprague–Dawley male rats over many generations. • Obesity, hypertension, and hyperlipidemia noted	----
Otsuka–long–Evans-Tokushima–fatty (OLETF) rats	• Selective breeding of Long Evans rats maintained at the Tokushima Research Institute, Japan • develop mild obesity, hyperglycemia, polyuria, and polydipsia	---
Spontaneously diabetic Torri fatty rat	• After weaning, both sexes show spontaneous hyperphagia and obesity • Initial hyperinsulinemia is seen and later on insulin level falls	Used as an obesity model for an initial duration and as a diabetic model for later ages
Cp/cp rat	• When the cp/cp gene is homozygous, the rat develops hyperphagia, hyperglycemia, obesity, and insulin resistance. • The rats are normotensive initially even though they are derived from the spontaneously hypertensive strains. • Renal lesions are noted at a later age and develop hypertension besides obesity due to renal lesions.	NASH like lesions develop after 23 weeks of age
WBN/Kob fatty rat	• They are a novel congenic strain for the 'fa' homozygous allele of the leptin receptor gene. • Hyperinsulinemia is seen till 8 weeks and later on level falls. • NASH and microangiopathy are not seen with this model	A good model for obesity with type-2 diabetes

weight and food intake are measured daily. After 3 weeks, rats are sacrificed and brown adipose tissues are collected.

– The binding of GDP to thermogenin is measured by the radioimmunoassay method. Radiolabelled sucrose ([^{14}C]-Sucrose) and GDP ([^{3}H]-GDP) are used in this method.

- Using a homogenizer, the tissue pellet is prepared from the brown adipose tissue and suspended in sucrose buffer (albumin free). Now the mitochondria in the tissue pellets are incubated with unlabelled sucrose, unlabelled GDP, and radiolabelled sucrose and radiolabelled GDP along with other standard reagents for 10 min. The reaction is terminated by filtration.
- The same step is performed in absence of unlabelled sucrose and GDP and the radioactivity is measured using a scintillation counter for both steps.
- The entire procedure is repeated for all groups of rats administered with varying concentrations of test drug and control.

• *Calculation:*
  - Total binding is estimated by measuring the amount of ^{125}[C]-Sucrose and ^{3}[H]-GDP bound in absence of unlabeled sucrose and GDP. Non-specific binding is estimated by measuring the amount of ^{125}[C]-Sucrose and ^{3}[H]-GDP bound in presence of unlabeled sucrose and GDP. Specific binding is estimated by the difference between total binding and non-specific binding.
  - Using radioactivity values of ^{125}[C]-Sucrose, the correction for the trapped medium is done and the final ^{3}[H]-GDP binding is estimated.
  - Increased ^{3}[H]-GDP binding indicates increased non-shivering thermogenesis. The values are compared between the groups and control using two-way ANOVA.

32.5.2 Measurement of Uncoupling Protein-1 in Rat's Brown Adipose Tissue

• Along with thermogenin, the uncoupling protein-1 (UCP-1) is also expressed in brown adipose tissue and is involved in non-shivering thermogenesis. Thus, measurement of UCP-1 levels by using northern blot or western blot techniques is used for assaying the anti-obesity action of the test drug.

32.5.3 β3-Adrenoceptor Binding in the Chinese Hamster Ovary Cells

• *Principle:* Stimulation of β3-adrenoceptor in the adipose tissue produces increased thermogenesis (brown adipose tissue), reduction of serum leptin level and its gene expression, and promotion of lipolysis (white adipose tissue) leading to weight loss without affecting food intake. The Chinese hamster ovary (CHO) cells transfected with the human β3-adrenoceptor gene can be used to study the agonistic activity of test drugs on these receptors.

• *Procedure:*
  - The CHO cells that express β3-adrenoceptor stably are selected. Using the electroporation method, these cells are transfected with cAMP-response-element-Luciferase (CREL) plasmids. After successful transfection, the CHO

cells are kept in 96-microwell plates at a density of approximately 40,000 cells/well and grown for 20 h.
  - After 20 h, varying concentrations of test drugs (between 10^{-11} M and 10^{-4} M) are added into the wells and incubated for another 4 h. Isoproterenol, at a concentration of 10^{-6} M, is used as a positive control.
  - Now, the cells are lysed thoroughly, and using a luminometer, the Luciferase activity is measured. The amount of Luciferase activity is directly proportional to the agonistic activity of the test drug on the β3-adrenoceptor.
- *Calculation:*
  - The analysis is done with duplicate or triplicate samples for each concentration. The Luciferase activity of each concentration of test drug is compared with the standard.

32.5.4 Determination of mRNA Level Of Leptin in Rat's Adipose Tissue

- *Principle:* Leptin, secreted mainly by white adipose tissue, causes weight loss by acting on the hypothalamus via promotion of energy expenditure and reduction of food intake. Leptin causes reduced lipogenesis, thermogenesis in brown adipose tissues, and increased lipolysis resulting in weight reduction. Leptin levels are high in obese individuals due to leptin resistance. Thus, measuring the mRNA level expression of leptin in rats administered with test drug will provide insight into the efficacy of test drug for obesity.
- *Procedure:*
  - Male Wistar rats are treated with test drugs for at least 14 days and the tissues like brown adipose tissue (intracapsular), white adipose tissue (from epididymis), and liver are harvested. The tissues are immediately frozen using liquid nitrogen.
  - Using the 'guanidinium-thiocyanate- chloroform/phenol' method, the mRNA of the leptin is extracted and fractionated by 'horizontal-gel-electrophoresis'.
  - Using standard 'northern-blot-analysis', the mRNA levels are estimated. Alternatively, 'competitive-RTPCR' or 'capillary-electrophoresis-with-laser-induced-florescence-detection' techniques can be used to estimate the mRNA levels of leptin.
- *Calculation:* The quantity of mRNA of leptin of various groups is summarized as mean with SD and using 'two-way ANOVA' the variance between the groups is compared.

32.5.5 Plasma Leptin Level Determination in Rats

- *Principle:* Refer to sect. 32.5.4.
- *Procedure:*
  - The male Wistar rats are treated with test drug for at least 14 days and the blood samples are collected via retro-orbital venous plexus or from rat tail vein in fasting state and after well-fed state.
  - The plasma levels of leptin are measured using standard 'radio-immunoassay' using standard rat-leptin-specific-antibodies.
- *Calculation:* The quantity of plasma leptin levels of various groups is summarized as mean with SD and using 'two-way ANOVA' the variance between the groups is compared.

32.5.6 Binding Assay of Neuropeptide Y (NPY) Receptor in Pigs

- *Principle:* NPY is an orexigenic endogenous peptide distributed widely in the body. It promotes weight gain by a reduction in thermogenesis and increased food intake. After fasting, the levels of NPY and its mRNA are increased in the hypothalamus. NPY exerts its action via six different receptors namely, Y1, Y2, Y3, Y4, Y4, Y5, and Y6. Thus, the antagonist of the NPY receptor is a potential anti-obesity target.
- *Procedure:*
  - The receptor binding assay of the test drug for NPY receptors is done by following steps. First, NPY are radiolabelled with ^{125}[I] and filtered using reverse phase-HPLC and stored at −20 °C.
  - Second, synaptosomal membranes are prepared by standard method from the hippocampal tissues of the Yorkshire pigs. Subsequently, the membranes are diluted with suitable buffer (final concentration of 2 g/l) and stored at −80 °C.
  - Finally, the receptor binding assay is done and the radioactivity is measured using a scintillation counter.
- *Calculation:*
- Total binding is estimated by measuring the amount of ^{125}[I]-NYP peptide bound in absence of unlabeled peptide.
- Non-specific binding is estimated by measuring the amount of ^{125}[I]-NYP peptide bound in presence of unlabeled peptide.
- Specific binding is estimated by the difference between total binding and non-specific binding.
- Percent inhibition: 100–specific binding as a percentage of the control value.
- Using log-probit analysis, the IC_{50} calculations are performed. At least 3–4 different concentrations of the test compounds in triplicate should be used to determine the IC_{50}.

32.5.7 Binding Assay of Orexin-A and Orexin B Receptors In Hamster Ovary Cells

- *Principle:* Orexins A and B, found mainly in the hypothalamus and locus coeruleus, stimulate food intake by multiple mechanisms. Besides food intake, it also regulates sleep cycles, modulates stress, influences cardiovascular, sexual, and behavioral activities, and stimulates insulin release. Weight gain is noted in animals that are injected with orexins A and B and weight loss occurs with orexin-antagonists. All these effects are mediated via two Orexin receptors—OX_1 and OX_2. Thus, the antagonist of the Orexin receptor is a potential anti-obesity target.
- *Procedure*
  - The receptor binding assay of the test drug for orexin receptors is done by following steps. First, orexin-A are radiolabelled with ^{125}[I] at the tyrosine amino acid in the 17th position and purified using reverse phase-HPLC
  - Second, Chinese hamster ovary cells (CHO cells) are transfected with OX_1 and OX_2 genes and grown in a standard medium. Stable expression of orexin receptors is confirmed and incubated with radiolabelled orexin-A with unlabeled competitors at 20 °C for 90 min. The reaction is terminated by washing and lysing of cells.
  - Finally, the radioactivity is measured using a scintillation counter.
- *Calculation:*
  - Total binding is estimated by measuring the amount of ^{125}[I]-orexin A bound in absence of unlabeled peptide.
  - Non-specific binding is estimated by measuring the amount of ^{125}[I]-orexin A bound in presence of unlabeled peptide.
  - Specific binding is estimated by the difference between total binding and non-specific binding.
  - Percent inhibition: 100–specific binding as a percentage of the control value
  - Using log-probit analysis, the IC_{50} calculations are performed. At least 3–4 different concentrations of the test compounds in triplicate should be used to determine the IC_{50}.

32.5.8 Binding Assay for Galanin Receptors in Rats

- *Principle:* Galanin, isolated first from the pig intestine, stimulates food intake and increases body weight. Galanin mediates the effects via three GPCR receptors-GAL_1, GAL_2, and GAL_3. Glucose-induced insulin release is blocked by galanin. The release of various hormones like dopamine, prolactin, and growth hormones is also influenced by galanin. Antagonists of galanin have been shown to reduce food intake.

- *Procedure:*
  - First, the radiolabelled galanin is prepared by iodination using radioactive iodine ^{125}[I]. Following the reaction, the fractions of ^{125}[I]-galanin are pooled and stored at −18 °C.
  - Hypothalamic tissues from male rats (Sprague-Dawley) are obtained and tissue is homogenized for making standard membrane preparations. Samples are incubated with radiolabelled ^{125}[I]-galanin for 30 min at 37 °C.
  - The reaction is terminated by rapid filtration and the radioactivity is measured using a scintillation counter.
- *Calculation:*
  - Total binding is estimated by measuring the amount of ^{125}[I]-galanin bound in absence of unlabeled galanin peptide.
  - Non-specific binding is estimated by measuring the amount of ^{125}[I]-galanin bound in presence of unlabeled galanin peptide.
  - Specific binding is estimated by the difference between total binding and non-specific binding.
  - Percent inhibition: [100]–[specific binding as a percentage of the control value].
  - Using log-probit analysis, the IC_{50} calculations are performed. At least 3–4 different concentrations of the test compounds in triplicate should be used to determine the IC_{50}.

32.5.9 Other In-vitro Assays

- Similar to the principle described above, the assay for the following is also being done for the evaluation of the anti-obesity activity of test compounds
  - mRNA level measurement of adipsin in mouse
  - Serum level measurement of adipsin in mouse
  - Measurement of agouti-related peptide expression in the hypothalamus of rats
  - Measurement of Melanin-concentration-hormone (MCH) and Cocaine-Amphetamine-regulated transcript (CART) level in mouse

32.6 In-Vivo Methods for Evaluation of Anti-obesity Activity

32.6.1 Measuring Resting Metabolic Rate in the Mouse

- *Principle:* Resting metabolic rate is one of the main factors for the determinant of the development of obesity. An obese individual has a lower resting metabolic rate than the normal one.

- *Procedure:*
  - 'Female C57B1' or 'yellow KK mouse' are used in this experiment. They are fed with commercial powdered chow and water as *ad-libitum.* The test drugs of varying concentration are administered (preferably parenteral route) for at least 14 days one group is kept as control. The daily food intake and body weight are measured.
  - Using a closed-circuit-metabolic system, O_2 consumption and CO_2 production are measured. Initially, the animals are allowed to stabilize for 30 min and the parameters are measured for 1 hour at the temperature of 22 °C.
  - The resting metabolic rate between the test group and control is then compared using Two-Way ANOVA.

32.6.2 Measurement of Food Consumption in Rats

- *Principle:* Anorectic activity of potential anti-obesity drugs can be measured by the amount of food consumption in rats. This procedure can be done using normal or genetically obese rats as acute experiments. Changes in body weight from baseline can be measured when the experiment is designed as semi-chronic.
- *Procedure:*
  - Adult female rats (Zucker) weighing around 250 g are selected and divided into respective drug groups. After adaptation to the new environment, the baseline food intake and body weight are measured and noted.
  - The drugs are administered to the respective groups orally without spillage or intraperitoneally for 7 days. The food intake is measured everyday morning between 8 and 9 a.m. regularly by measuring the amount of food left and spillage.
  - If the food chow spillage is soaked with urine or water, it should be air-dried and weighed.
  - Bodyweight is also measured if the drug is administered for 14 days and the experiment is designed as 'semi-chronic'
- *Evaluation:*
- The mean reduction of food intake and body weight from the baseline for different groups is calculated and compared using ANOVA.

Bibliography

Aswar U, Patil R, Bodhankar S. Review on the induction of obesity in laboratory animals. Diabesity. 2019;5(4)

Cho H, Lee JY, Choi SY, Lim C, Park MK, An H, et al. Identification of a new chemotype of anti-obesity compounds by ensemble screening. ACS Omega. 2020;5:4338–46.

Crowley VE. Overview of human obesity and central mechanisms regulating energy homeostasis. Ann Clin Biochem. 2008;45:245–55.

Jordi J, Guggiana-Nilo D, Bolton AD, Prabha S, Ballotti K, Herrera K, et al. High-throughput screening for selective appetite modulators: a multibehavioral and translational drug discovery strategy. Sci Adv. 2018;4:eaav1966.

Miller GD. Appetite regulation: hormones, peptides, and neurotransmitters and their role in obesity. Am J Lifestyle Med. 2017;13:586–601.

Mujwar S, Kumar V. Computational drug repurposing approach to identify potential fatty acid-binding protein-4 inhibitors to develop novel antiobesity therapy. Assay Drug Dev Technol. 2020;18:318–27.

Rodgers RJ, Tschöp MH, Wilding JP. Anti-obesity drugs: past, present, and future. Dis Model Mech. 2012;5:621–6.

Suissa K, Schneeweiss S, Kim DW, Patorno E. Prescribing trends and clinical characteristics of patients starting antiobesity drugs in the United States. Diabetes Obes Metab. 2021;23:1542–51.

Absorption Studies

33

Muthu Meera Ramasubramanian

Abstract

Extent of absorption of drugs into the systemic circulation plays a major role in determining their pharmacological action. Starting from the phase of drug development, studying the extent of drug absorption is of much importance. Studies to evaluate the extent of drug absorption can be broadly classified into in vitro, in vivo and in silico studies.

In vitro studies involve the use of isolated intestinal tissue, cultured colon carcinoma cells, etc. In vivo studies involve the use of anaesthetized live animals to which the drug is administered followed by measurement of drug level or the pharmacological response. In silico study is like a bridge between in vitro and in vivo studies in which the blood supply and innervation of the tissue are preserved. Among the three, in vivo studies provide a closer picture regarding the absorption profile of the drug.

Absorption studies are very useful while screening large libraries of compounds about their absorption profile, as part of the high throughput screening process. The appropriate experimental technique must be opted for; also, the findings of the study must be interpreted carefully.

Keywords

Absorption · Bioavailability · Pharmacokinetics

M. M. Ramasubramanian (✉)
Department of Pharmacology, Velammal Medical College Hospital and Research Institute, Madurai, Tamil Nadu, India

M. Lakshmanan et al. (eds.), *Introduction to Basics of Pharmacology and Toxicology*, https://doi.org/10.1007/978-981-19-5343-9_33

33.1 Introduction

Absorption refers to the movement of drug molecules from the site of administration to the systemic circulation. Several factors influence the rate and extent of absorption of a drug into the systemic circulation. They include drug formulation, particle size, disintegration rate & dissolution rate of the drug, pH of the stomach, gastrointestinal motility, concomitantly taken drugs & food, and concurrent diseases. Physiological changes like pregnancy and aging can also influence the extent of drug absorption.

33.2 Importance of Drug Absorption

The rate and extent of absorption of a drug into the systemic circulation is defined as 'bioavailability.' Bioavailability plays an essential role in determining the therapeutic usefulness. The route of administration plays a vital role in determining bioavailability. Bioavailability is 100% for the intravenous route. It is less than 100% for all the other routes. Increasing the extent of absorption of a drug increases bioavailability, thereby enhancing its therapeutic usefulness.

33.3 Classification of Drug Absorption Studies

Starting from the phase of drug development, studying the extent of drug absorption is of vital importance. Studies to evaluate the extent of drug absorption can be broadly classified into:

- Invitro studies
- Invivo models
- In situ study

33.4 In vitro Methods to Evaluate Drug Absorption

33.4.1 Partition Coefficient

33.4.1.1 Log-P

The partition coefficient between an oil and water phase is an essential property of a drug molecule that can be determined. It is denoted as 'log-P.' It is the ratio of concentrations of a compound in a mixture of two immiscible solvents. One of the two solvents is water, whereas the other is a hydrophobic solvent like octanol. Octanol has properties similar to that of biological membranes. It is a well-known fact that the higher the lipophilicity, the higher the extent of drug absorption. Here, Log-P is a measure of the lipophilicity of drugs. It predicts the amount of drug that can cross the biological membranes in the human body. Here the term 'biological membrane' refers to the barriers that separate the various compartments of the body.

33.4.1.2 Log-D

Another commonly used measure is log-D. Log-D is the distribution coefficient. It takes into account the ionization of the drug at a particular pH. It is the ratio of the concentrations of both ionized and unionized fractions of a drug distributed between the lipid and aqueous phases. The lipid-aqueous partition coefficient is an important factor influencing drug permeation across biological membranes. It is important to note that log-P does not consider the degree of ionization, and hence log-D is used. Log-D measured at intestinal pH will give a much better idea about the extent of drug permeability across intestinal membrane than log-P.

33.4.2 Parallel Artificial Membrane Permeability Assay (PAMPA)

- In this model (Fig. 33.1), the artificial membrane simulates the phospholipid bilayers in the human body through which drug permeation occurs.
- This model consists of a lipid-infused artificial membrane, a donor compartment, and an acceptor compartment.
- The drug is added to the donor compartment, and the amount of drug permeated through the artificial membrane into the acceptor compartment is determined by a suitable assay.
- This technique is mainly used to study the extent of passive absorption of the drug through the intestinal membrane.

33.4.3 Brush Border Membrane Vesicles (BBMV)

- Brush border refers to the epithelium found in microvilli of the small intestine. It increases the surface area for absorption.
- Both human and animal brush border tissue can be used in this model for studying drug absorption.

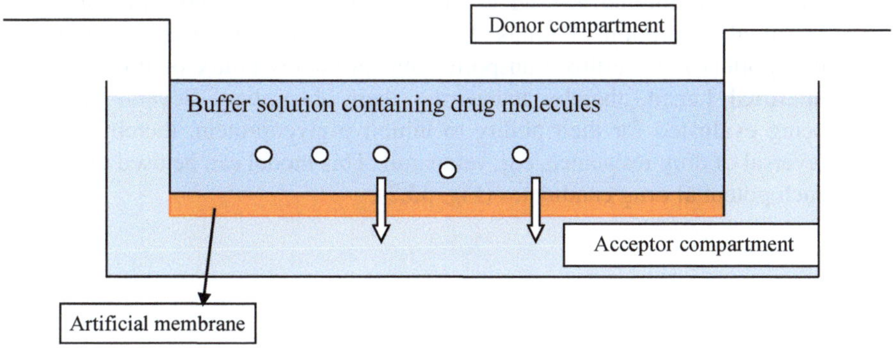

Fig. 33.1 Parallel artificial membrane permeability assay

- The brush border epithelium is isolated and purified by treatment with calcium chloride, enabling separation of brush border fragments from microsomal fragments. This is followed by centrifugation and the formation of vesicles.
- These BBMVs are then mixed with the drug in a buffer solution, and the extent of uptake is assayed.
- Again, this technique is helpful to study the intestinal absorption of drugs.

33.4.4 Isolated Intestinal Cells

- Intestinal cells are obtained from genetically related animals like rats/pigs.
- The isolated intestinal cells are added to the buffer solution to which the drug has been previously added.
- After a specific time, the cells are then separated, and the extent of drug absorption is assayed.
- This model has been employed to study phase 1 of drug metabolism also.

33.4.5 Everted Small Intestinal Sac Technique

- In this technique, the intestinal segment is isolated from rats and everted so that the serosa faces inside and mucosa faces outside. Sutures are tied at the proximal and distal ends.
- The segment is then immersed in an Erlenmeyer flask with a buffer solution containing the drug that has been incubated at a specific temperature and adequately oxygenated.
- At specified time intervals, the serosal solution is removed, and the amount of drug in it is determined, which is indicative of the extent of drug absorption.
- Applications:
  - This technique is used to find out the segment of the small intestine that permits optimal absorption of drugs, particularly in the case of actively absorbed drugs.
  - Also, this technique is used to evaluate the role of p-glycoprotein in the intestinal absorption of drugs. P-glycoprotein (multidrug resistance type-1 transporter) is an efflux transporter that promotes efflux of drugs into the intestinal lumen, thereby hampering their absorption. Several drugs are being evaluated for their ability to inhibit p-glycoprotein, thereby enabling reversal of drug resistance, e.g. verapamil. This model can be used to evaluate such potential drug candidates (Fig. 33.2).

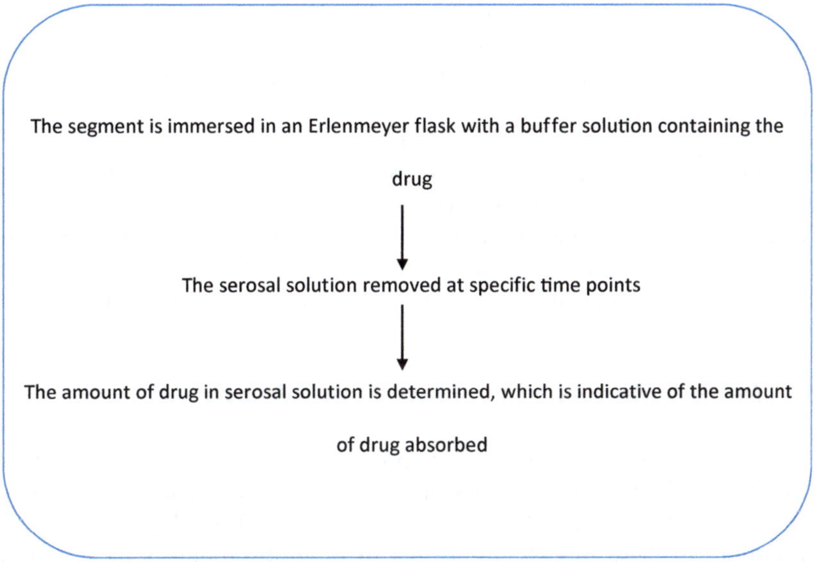

Fig. 33.2 Everted small intestinal technique for evaluating drug absorption

33.4.6 Everted Sac Modification Technique

- In this method, the test animal is kept under fasting for 24 h. The animal is sacrificed, and the entire small intestine is everted and cut into small segments. One segment is selected.
- The distal end of the segment is tied with a suture, and the proximal end is attached to a cannula. The segment is suspended in a drug solution. The serosal surface is exposed to a drug-free buffer solution via the attached cannula.
- The serosal solution is removed from the sac at specific time intervals with the help of a syringe and replaced with a fresh buffer solution. The amount of drug that permeates the intestinal mucosa is measured and plotted against time. Thus, the absorption profile of the drug can be described.
- This modified technique will be more efficient than the conventional technique since the serosal surface is kept exposed to the drug-free buffer solution, and thus the measurement of the extent of drug absorption across the intestinal barrier tends to be more accurate.

33.4.7 Cell Culture Techniques

- The commonly used cell lines for studying drug absorption are colon carcinoma cells. The human CaCo-2 (colon carcinoma) cell system serves as an ideal in vitro model of the intestinal epithelial barrier.

- In this technique, the drug solution is placed in contact with a monolayer of cultured CaCo-2 cells. The portion of the drug that permeates through the cellular barrier and reaches a 'latter compartment' (representing the systemic circulation) is determined.
- Nevertheless, it is necessary to be careful while interpreting the conclusions from these models due to their undesirable phenotype and functionality compared to the native intestinal tissue. Compared with the monolayer culture of CaCo-2 cells, a three-dimensional model has more physiologically relevant characteristics.
- This technique is beneficial in screening large libraries of new chemical entities concerning their permeability properties following oral administration.

33.5 *In-Vivo* Models to Evaluate Drug Absorption

- The *in-vivo* models to evaluate drug absorption more closely resemble the drug absorption under actual physiological conditions. They help in understanding the influence of physiological factors like gastrointestinal motility and pH on drug absorption. In these models, the extent of drug absorption is measured either directly or indirectly.
- In the direct method, the blank blood or urine of the animal is collected before administration of the test drug. This is followed by the test drug administration. Then, samples are collected at specific time intervals, and the amount of drug in them is measured by sensitive analytical procedures. The commonly used experimental animals for this purpose include rats, rabbits, pigs, and dogs.
- In situations when it is not possible to directly measure the drug level in blood/urine, an indirect method is employed. In this method, the pharmacological response of the drug is taken as an indirect measurement of the amount of drug in the body, which in turn depends upon the extent of drug absorption.

Some of the commonly employed in vivo models to evaluate drug absorption are as follows:

33.5.1 Doluisio Method

- This method is used to study absorption from the small intestine.
- Rat is the preferred experimental animal. Rats are fasted for about 24 h.
- The animal is anesthetized, and a midline abdominal incision is made. The small intestine is identified. The proximal and distal ends of the small intestine are connected to two separate syringes via cannulas that are secured with sutures. The intestine is replaced in the abdominal cavity.
- The drug solution is passed through the intestinal lumen via the first syringe. Then, the same solution is collected via the second syringe attached to the distal segment, and the amount of drug in it is measured. This indirectly evaluates the amount of drug that has been absorbed already in the intestinal segment.

33.5.2 Intestinal Loop Technique

- Adult male rats are kept under fasting before the experiment. The animal is anesthetized, a midline abdominal incision is made, and the small intestine is exposed. Ligatures are made in the proximal and distal segments of the small intestine separated by a distance of about 8 cm.
- The intestinal loop thus created is replaced in the abdomen. The drug solution is introduced into the intestinal segment via a syringe. A proximal ligature secures the syringe. Unlike the Doluisio method in which a second syringe is used to collect the drug solution, here, the intestinal loop is excised, and the amount of unabsorbed drug is determined.
- The above two techniques are used to study the extent of intestinal drug absorption. The extent of absorption of the drug through the stomach is limited because several drugs get ionized/get inactivated by gastric acidity. However, the amount of drug absorbed in the stomach can be studied by a similar technique- by ligating the cardiac end & pylorus of the stomach, followed by introducing the drug solution into this created pouch and measuring the amount of drug in this pouch that has been left over after absorption.

33.6 *In-Situ* Method to Evaluate Drug Absorption

- *The in-situ* method is the bridge between in vitro and in vivo models.
- In the *in-situ* rodent model, the animal is anesthetized, and a specific segment of the small intestine is cannulated and perfused with a drug solution.
- The mesenteric vein draining the intestinal segment is cannulated, and blood samples are collected from it at specific time intervals to determine the amount of drug absorbed into the blood. The values can be plotted against time to describe the absorption profile of the drug.
- *The in-situ* method is peculiar in that the blood supply and innervation are preserved. Thus, it creates an experimental condition that very closely resembles in vivo environment Major differences between the three types of drug absorption studies have been tabulated in Table 33.1.

33.7 Conclusion

Drug absorption studies provide vital information regarding the pharmacokinetic profile of the drug. The appropriate method has to be selected depending upon the study drug and available experimental conditions. Also, interpretation of findings has to be done carefully. These studies are very useful while screening large libraries of compounds concerning their absorption profile as part of the high throughput screening process.

Table 33.1 Comparison of in-vitro, in-vivo, and in-situ drug absorption studies

In-vitro absorption studies	In-vivo absorption studies	In-situ absorption studies
Involves use of isolated intestinal tissue, cultured colon carcinoma cells, etc	Involves the use of anesthetized live animals (rats) to which the drug is administered, followed by measurement of drug level/ pharmacological response in the live animal	It is a bridge between *in-vitro* and *in-vivo* studies. In this technique, the blood supply and innervation are preserved.
It does not involve the collection of blood samples	Involves collection of blood samples	Blood samples are collected from the mesenteric vein draining the intestinal segment.
Lack of resemblance to *in-vivo* conditions	*In-vivo* studies give a better picture regarding the absorption profile of the drug in live animal	More closely resembles *in-vivo* conditions

Bibliography

Alam MA, Al-Jenoobi FI, Al-Mohizea AM. Everted gut sac model as a tool in pharmaceutical research: limitations and applications. J Pharm Pharmacol. 2012;64:326–36.

Barthe L, Bessouet M, Woodley J, Houin G. The improved everted gut sac: a simple method to study intestinal P-glycoprotein. Int J Pharm. 1998;173:255–8.

Bonnefille P, Sezgin-Bayindir Z, Belkhelfa H, Arellano C, Gandia P, Woodley J, et al. The use of isolated enterocytes to study phase I intestinal drug metabolism: validation with rat and pig intestine. Fundam Clin Pharmacol. 2011;25:104–14.

Stappaerts J, Brouwers J, Annaert P, Augustijns P. In situ perfusion in rodents to explore intestinal drug absorption: challenges and opportunities. Int J Pharm. 2015;478:665–81.

Wilson G. Cell culture techniques for the study of drug transport. Eur J Drug Metab Pharmacokinet. 1990;15:159–63.

Screening Methods for the Evaluation of Antiasthmatic Agents

34

Neel Jayesh Shah

Abstract

Anti-asthma agents range from bronchial smooth muscle relaxants, anti-inflammatory agents, leukotriene inhibitors, and monoclonal antibodies. Hence, a diverse set of models are needed to test anti-asthma efficacy. While in-vivo models can test the efficacy in the presence of all physiological factors at play, in-vitro models can measure the contraction of airway smooth muscles directly. Testing monoclonal antibodies present an additional challenge as cytokines from animals and humans have a genetically different make-up.

Keywords

Models for asthma · In-vivo anti-asthma models · In-vitro anti-asthma models · OVA sensitization · OVA challenge

34.1 Introduction

It is now well understood that the treatment of asthma consists not only of bronchodilation but also addressing underlying inflammation. This can be seen in the latest update to the guidelines, which now recommend the use of a low-dose inhaled corticosteroid with formoterol for as-needed treatment instead of salbutamol. Asthma in humans is mediated by T_H2 (predominantly), T_H1, and T_H17 pathways of inflammation. IL-4, IL-5, and IL-13 are cytokines that belong to the T_H2 pathway. The end result is the spasm of the bronchial smooth muscles, inflammation, and in the long term, airway remodeling. Models which mimic every step of this pathogenesis are used as screening tools for anti-asthma agents.

N. J. Shah (✉)
Quality & Compliance Services Inc., Mississauga, ON, Canada

M. Lakshmanan et al. (eds.), *Introduction to Basics of Pharmacology and Toxicology*, https://doi.org/10.1007/978-981-19-5343-9_34

34.2 *In-Vitro* Models

34.2.1 Checking for Smooth Muscle Relaxant Action

Guinea pig trachea can be isolated after sacrificing the animal. The trachea can be dissected into isolated rings. Each ring can be mounted into an organ bath, and bioassay apparatus can be set up by suspending the tissue under tension and amplifying movement to measure the responses. Contraction of the guinea pig tracheal smooth muscle can be elicited using carbachol or histamine. This contracted tissue can now be exposed to the test drug. The spasmolytic/ relaxant action can be observed by plotting a concentration-response curve.

34.2.2 Testing Anti-inflammatory Action

Fibroblasts can be isolated from the healthy tissue portion from a patient who is undergoing lobectomy. Confluent fibroblast layers in 96-well plates should be pre-incubated with the test drug (steroids have been proven to demonstrate efficacy in this model). Modulators of chronic inflammation like interleukin-1β (IL-1 β) or TNF-α must be then added. In the control wells, IL-1β or TNF-α will upregulate ICAM-1 and VCAM-1 expression. This expression can be quantified by measuring the binding of anti-ICAM and anti-VCAM antibodies by ELISA. Also, IL-1β causes the production of GM-CSF (Granulocyte-macrophage colony-stimulating factor) and IL-8 by human fibroblasts. Such an experiment showed that budesonide prevented the upregulation of ICAM/ VCAM and production of GM-CSF and IL-8. Fibroblasts also have β-2 receptors. Hence, the above model has also been proven to check the effect of long-acting beta-agonists like salmeterol and formoterol.

34.3 Animal Models

Since the 1990s, the most common animals used as asthma models are mice. Apart from being easy to maintain and a variety of reagents available to measure cellular response, some strains of mice even are more T_H2 predisposed such as the BALB/c strain.

34.3.1 The Murine Allergic Airway Inflammation Model

Sensitization phase: Ovalbumin (OVA) is the main protein found in egg white. Mice can be sensitized by intraperitoneal (i.p.)/ subcutaneous (s.c.)/ intra-tracheal injection of 10–50 mcg of OVA on days 1 and 14. Mixing with an adjuvant-like alum provides a greater IgE response.

AHR challenge phase: In the next phase, airway hyper-responsiveness can now be induced by nebulizing mice placed in a chamber with OVA for 20 min. This is repeated for 4–8 consecutive days. This creates an increase of eosinophils, bronchospasm, mucus secretion, and release of mediators by mast cells such as histamine (early response), rise in IL-4, IL-5, and IL-13 (late response, as measured in the bronchoalveolar lavage (BAL) fluid and anti-OVA IgE in serum). The model is now ready, and AHR can be also be induced by nebulized methacholine.

Disadvantages of the above acute model using OVA:

- A disproportionately high number of eosinophils in BAL compared to human asthma
- Inflammation of bronchioles greater compared to humans
- If a large amount of OVA is used to challenge, it can deposit in the airways and cause granulomatous inflammation
- The inflammation and hyper-responsiveness resolves after few days
- To resemble human asthma better, OVA can be replaced by antigens such as dust mite (more immunogenic), ragweed (typically causes pollen allergy and hay fever), cotton dust, or cockroach extracts.

Chronic model: When low-level OVA is administered by nebulization repeatedly over 8–12 weeks, it can cause persistent inflammation, airway remodelings such as smooth muscle hypertrophy, goblet cell hyperplasia, and subepithelial fibrosis. The chronic model resembles asthma in humans more closely.

34.3.2 Other Animal Models

- Guinea pigs were amongst the first animal models due to the following advantages: easy to sensitize and challenge, strong response to human antigens, and similar autonomic nervous system innervation of lungs. Disadvantages include longer gestation time, no transgenic models, and lack of reagents to measure cellular responses. This lack of reagents is also a problem with rat, cat, dog, and horse models.
- Regarding the cytokines involved in inflammation, felines demonstrate IL-4, sheep demonstrate IL-4 and IL-13, while equine inflammation is mediated by T_H17 cytokines, like CXCL13 and IL-17. Dogs, sheep, and horses have an advantage as they are naturally susceptible to allergens, develop atopy (dogs) and immediately respond to inhaled allergens (sheep), but are too costly to be used in experiments.

34.4 Measuring Outcomes

(1) Immunological: measurement of IgE, IgG, and cytokines in blood and BAL
(2) Histopathological: pattern of inflammatory infiltrate in the airway

(3) Functional: lung function measured by plethysmography. A better way to quantify lung function is by using a forced oscillation technique (FOT). Under the effect of an anesthetic and muscle relaxant, a cannula is inserted into the trachea, and a computer-controlled piston ventilates the lungs in an oscillatory manner. Pressure volume changes are converted to digital waveforms. The resistance offered to the piston by airways can be measured. Examples of output parameters that can be obtained by FOT include pulmonary function resistance, dynamic compliance, forced expiratory volume, peak expiratory flow, and forced vital capacity.

34.5 Example of Published Trial Testing Anti-muscarinic in an Acute Model of Asthma

Tiotropium was tested in guinea pigs that were prior sensitized to OVA by i.p. injections on days 1, 3, and 6. After giving tiotropium by tracheal insufflation, guinea pigs were challenged with OVA by the same route (Fig. 34.1). This causes bronchoconstriction, which was worsened by electrical stimulation of the vagus nerve in the control animals. However, in animals given tiotropium, there was no bronchoconstriction, neither by OVA challenge nor by vagal stimulation. Hence, it means that tiotropium prevented antigen-induced bronchoconstriction (mimicked by OVA challenge) and airway hyperresponsiveness (mimicked by vagal stimulation) as well.

34.6 Example of Published Trial Testing a Chronic Model for Late-Phase Response

Four monoclonal antibodies are currently approved to treat severe asthma: omalizumab, reslizumab, mepolizumab, and benralizumab. The latter three drugs inhibit IL-5 induced differentiation and recruitment of eosinophils. Cynomolgus

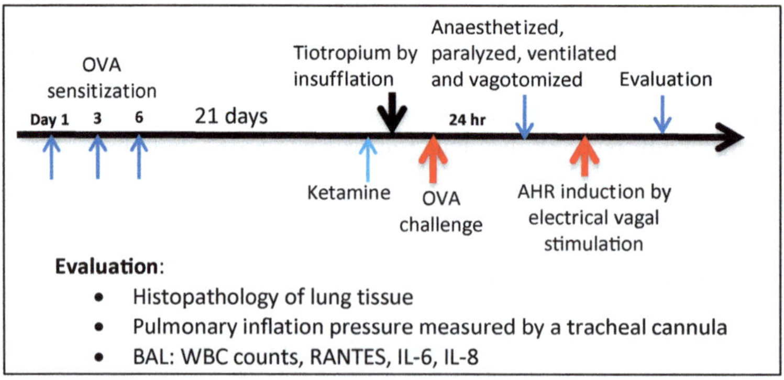

Fig. 34.1 Testing tiotropium in guinea pigs. *AHR*, Airway hyper-responsiveness

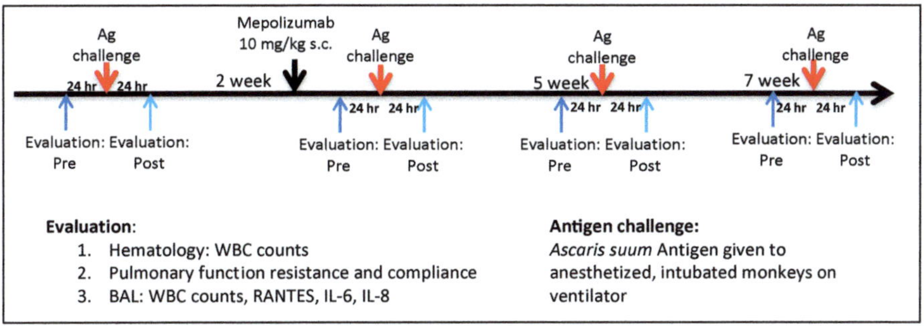

Fig. 34.2 Testing of mepolizumab in Cynomolgus monkeys

monkeys were determined to be the only pharmacologically relevant species, as observed by the USFDA review of the drug mepolizumab. Monkey IL-5 differs from human IL-5 by two amino acids, and mepolizumab can inhibit both of them.

In Fig. 34.2, the trial of mepolizumab in cynomolgus monkeys is described. The monkeys were selected based on positive bronchoconstrictor response to the Ascaris antigen; no prior sensitization had to be done. In the mepolizumab group, the evaluation done 24 h after giving antigen challenge had a significantly reduced amount of eosinophils, cytokines, and RANTES (a chemokine) compared to the control group. However, pulmonary function resistance and dynamic compliance had no improvement, suggesting that this drug works in the late phase, not the acute phase.

Bibliography

Aun MV, Bonamichi-Santos R, Arantes-Costa FM, Kalil J, Giavina-Bianchi P. Animal models of asthma: utility and limitations. J Asthma Allergy. 2017;10:293–301.

Buels KS, Jacoby DB, Fryer AD. Non-bronchodilating mechanisms of tiotropium prevent airway hyperreactivity in a guinea-pig model of allergic asthma. Br J Pharmacol. 2012;165(5):1501–14.

Chu X, Xu Z, Wu D, Zhao A, Zhou M, Qiu M, et al. In vitro and in vivo evaluation of the anti-asthmatic activities of fractions from Pheretima. J Ethnopharmacol. 2007;111(3):490–5.

Hart TK, Cook RM, Zia-Amirhosseini P, Minthorn E, Sellers TS, Maleeff BE, et al. Preclinical efficacy and safety of mepolizumab (SB-240563), a humanized monoclonal antibody to IL-5, in cynomolgus monkeys. J Allergy Clin Immunol. 2001;108(2):250–7.

Kumar R, Herbert C, Foster P. The “classical” ovalbumin challenge model of asthma in mice. Curr Drug Targets. 2008;9:485–94.

McGovern TK, Robichaud A, Fereydoonzad L, Schuessler TF, Martin JG. Evaluation of respiratory system mechanics in mice using the forced oscillation technique. J Vis Exp JoVE. 2013 [cited 2021 Apr 29];(75). Available from: https://www.ncbi.nlm.nih.gov/pmc/articles/PMC3684007/.

Pelaia G, Vatrella A, Maselli R. The potential of biologics for the treatment of asthma. Nat Rev Drug Discov. 2012;11(12):958–72.

Spoelstra FM. Additive anti-inflammatory effect of formoterol and budesonide on human lung fibroblasts. Thorax. 2002;57(3):237–41.

Screening Methods for the Evaluation of Antitussives and Expectorants

35

Lakshmi Balasundaram

Abstract

Cough is one of the reflexes that is protective in nature, however it can be distressing if untreated. The number of animal models to study antitussives are limited and are classified based on the techniques used to induce cough such as by chemical stimuli, mechanical stimuli and by nerve stimulation.

While models for studying antitussives are based on evaluating the test drug's ability to suppress cough, animal models of expectorants are based on studying mucus secretion, which can be performed by acute or chronic cannulation studies. This chapter aims to provide the readers an insight on the existing models, modifications and their criticisms.

Keywords

Cough · Antitussives · Expectorants · Mucus secretion · Cannulation studies

35.1 Introduction

Cough: It is a protective reflex that occurs due to a variety of causes. It is imperative to address the etiology of the cough rather than treating it as a symptom. For example, antibiotics help relieve cough in bacterial pneumonia; likewise, inhaled corticosteroids are beneficial in asthmatic cough. The cough reflex arc consists of the afferent pathway (carried by sensory nerve fibers of vagus nerve branches), cough center (upper brain stem and pons), and the efferent pathway (carried by vagus, phrenic and spinal motor nerves to abdominal muscles and diaphragm). This reflex occurs when the receptors, predominantly in the bifurcation in the trachea, get

L. Balasundaram (✉)
Department of Pharmacology, Jawaharlal Institute of Postgraduate Medical Education & Research (JIPMER), Puducherry, India

M. Lakshmanan et al. (eds.), *Introduction to Basics of Pharmacology and Toxicology*, https://doi.org/10.1007/978-981-19-5343-9_35

stimulated. Stimuli can be mechanical or chemical (inhalation of irritants). Animal models that mimic acute or chronic cough are limited, and the models can be categorized as those using:

- Chemical stimuli
- Mechanical stimuli
- Nerve stimulation

35.2 Cough Induced by Chemical Stimuli

35.2.1 Citric Acid-Induced Coughing in Guinea Pigs

This is a simple yet the most effective model to test antitussive agents. The instrument consists of a cylindrical glass chamber containing an inlet and an outlet tube. The inlet tube allows the entry of citric acid aerosol, and the outlet tube serves as a vent for the efflux of the aerosol. The outlet tube has a side tube connected to a tambour that allows measurement of changes in pressure. A pinch-clamp connected beyond the sidearm permits measurement of changes in the pressure caused, such as during a coughing episode; this improves the instrument's sensitivity. Changes in pressure during normal respiration are not recorded. Steps involved in the procedure are given in the flowchart below:

Guinea pigs of either sex weighing 300-400g are placed in a cylindrical chamber containing an inlet tube and outlet tube

↓

Aerosolised 7.5% citric acid in water is introduced through the inlet tube for 10 minutes

↓

Control response is recorded as the number of tussive response in each animal

↓ After 60 minutes

Test drug is administered by subcutaneous/oral route

↓ After 30 minutes

Number of coughs are recorded in 10 minutes

Evaluation: The number of coughs in the treatment period is expressed as a percentage of that in the control period. ED_{50} can also be calculated by using different doses of the test drug.

Other chemical stimuli: Capsaicin inhalation, nicotine, histamine, ammonia, nebulized sulfuric acid, sulfur dioxide have been tried in place of citric acid in cats and dogs. Aerosolised antigens have also been used in sensitized guinea pigs.

35.3 Cough Induced by Mechanical Stimuli

This procedure involves the use of anaesthesia in guinea pigs. Care is taken to lightly anaesthetise the animals so that there is no respiratory depression. Steps involved in this procedure are given in Fig. 35.1:

Evaluation: Students paired t-test is used to check statistical significance for paired data.

Modifications of this method:

- Electrical stimulation of trachea by bronchoscope
- Electrical stimulation of implanted copper electrodes
- Nylon- bristled stimulator thrust in the trachea of dogs
- The vibration of an iron slung by the magnet in the trachea of dogs

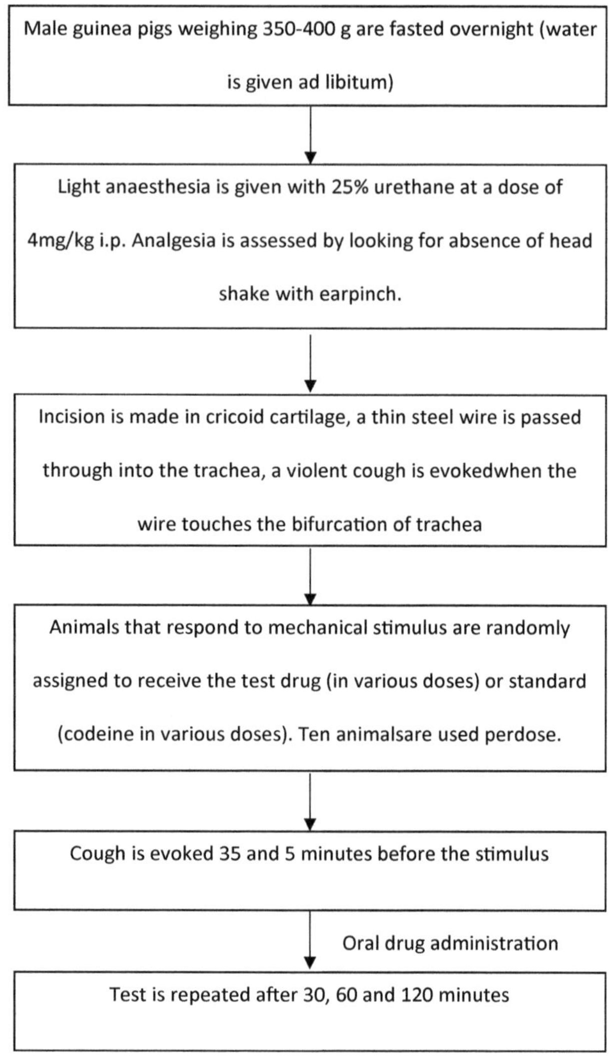

Fig. 35.1 Cough-induced by mechanical stimuli

35.4 Cough Induced by Stimulation of Superior Laryngeal Nerve

Efferent pathway in the cough reflex is constituted by the superior laryngeal nerves (branch of the vagus nerve) and is involved in the closure of glottis resulting in cough. Centrally acting antitussives are generally effective in this model.

Cats are utilized in this procedure, and it involves the administration of light anesthesia so that there is no respiratory depression. The procedure is described in Fig. 35.2:

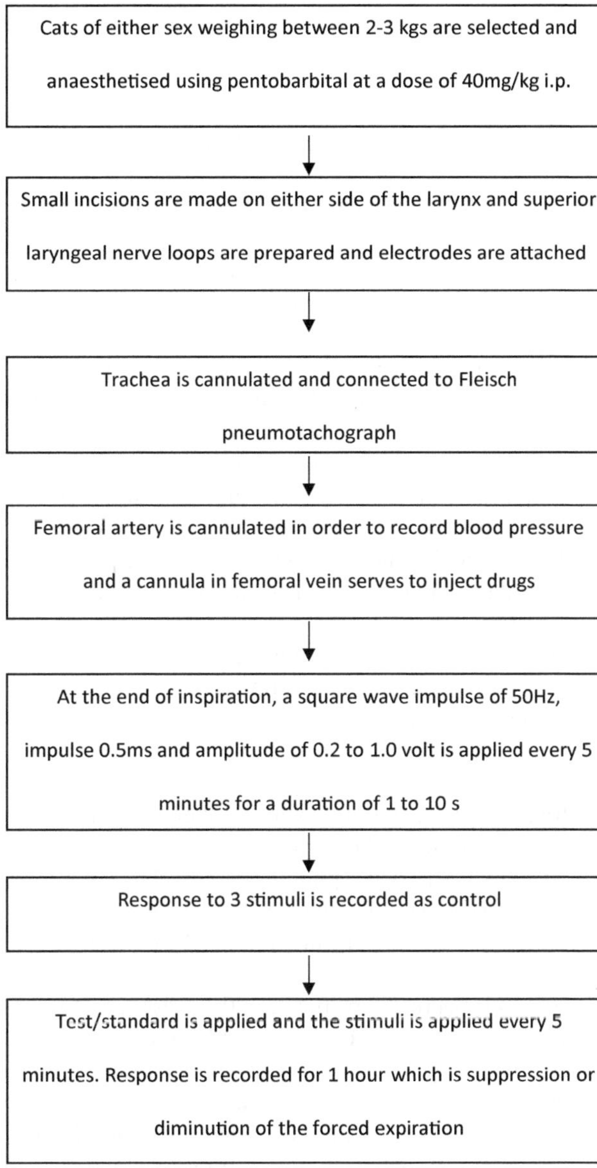

Fig. 35.2 Cough induced by stimulation of superior laryngeal nerve

Evaluation: Suppression of forced expiration, either partial or complete, can be expressed as a percentage of the control.

Modifications of this method:

- Stimulation of the dorsolateral region of the medulla was used to produce cough response in decerebrated cats.

- Cough was produced by stimulating the vagus nerve in the loop in conscious dogs.

35.5 Screening of Expectorants

35.5.1 In Vitro Studies of Mucus Secretion

Analysis of mucus secretion is done to screen the activity of expectorants in isolated tracheal preparations in dogs and ferrets.

- Ferrets weighing 0.6–1.5 kg of either sex are anesthetized with sodium barbital.
- The trachea is cannulated, following which a midline incision is made in the chest, and the trachea with the carina is dissected out.
- The tissue with its laryngeal end downwards and submucosal side facing outside is placed in Kreb's Henseleit solution with 0.1% glucose solution. Temperature is maintained at 37^0 C and aerated with 95% O_2 and 5% CO_2.
- The tracheal lumen is filled with air. A plastic catheter is passed through the cannula and made to form an airtight seal with the mucosa; this helps to collect the secretions.
- Drugs are added to the submucosal surface.
- The procedure helps to estimate the volume of secretions and changes in the volume of tissues.
- Mucus secretion is stimulated by passing electrical field stimulation at 50–100 V, 20 Hz for 1–2 m.
- Radioactive tantalum powder adheres to airway mucus on the surface of the trachea and traps the secretions, and forms hillocks shaped as hemispheres.
- At regular intervals, photographs of the surface are taken through a dissecting microscope. The diameter of hillocks can be measured using a microscope, and secretion volume per unit area can be estimated.

Evaluation: Secretory response in the presence or absence of drug can be recorded at various time intervals 45, 90, and 135 min.
Modifications of this method: Cat tracheal tissue has been used similarly, wherein the serosal side of the isolated trachea was bathed in ringer solution while the epithelial surface was covered with water-saturated paraffin oil. After stimulation of mucus, tiny droplets of secretions were observed under a dissecting microscope on the epithelial surface. Timed collection of the droplets along with volume estimation enables the calculation of the rate of secretion.

35.5.2 In Vivo Studies of Mucus Secretion

35.5.2.1 Acute Studies

The steps involved in assessing mucus secretions in a rabbit are given in Fig. 35.3:

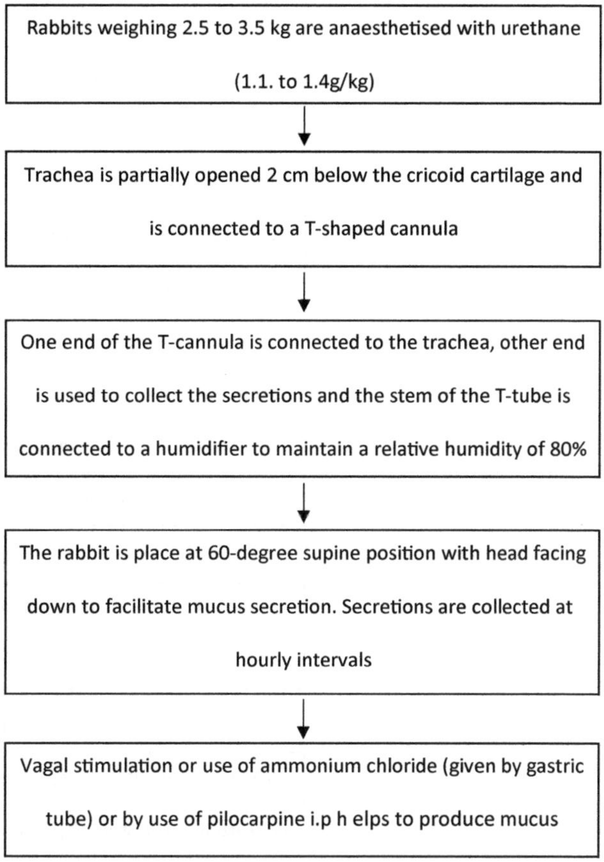

Fig. 35.3 Acute studies for mucus secretion

Evaluation: Animals are divided into two groups: treated and untreated. Time response curves are obtained after stimulating mucus secretion and are compared between both groups.

35.5.2.2 Chronic Cannulation Studies

This model utilizes beagle dogs wherein a small portion of the trachea is brought out like a pouch, and secretions can be collected. The procedure is described in Fig. 35.4:

Evaluation: Pressure changes in the balloon can be compared between baseline and after the test drug use.

Criticisms of this method: This method does not help to understand the pharmacological effects of drugs. Instead, the model is more suitable for physiological studies.

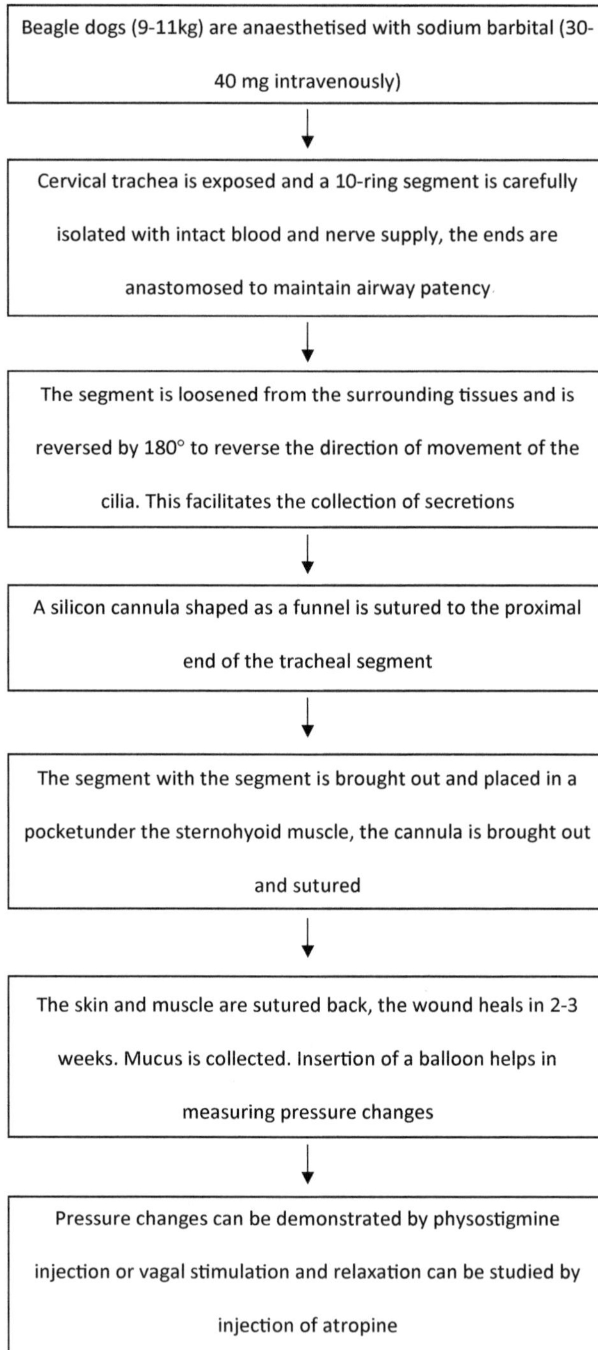

Fig. 35.4 Chronic cannulation studies

Bibliography

Hock FJ. Drug discovery and evaluation: pharmacological assays. 4th ed. Cham, Springer; 2016.

Polverino M, Polverino F, Fasolino M, Andò F, Alfieri A, De Blasio F. Anatomy and neuropathophysiology of the cough reflex arc. Multidiscip Respir Med. 2012 Jun 18;7(1):5.

Xu X, Chen Q, Qiu Z, Shi C, Ding H, Wang L, et al. Association of cough hypersensitivity with tracheal TRPV1 activation and neurogenic inflammation in a novel guinea pig model of citric acid-induced chronic cough. J Int Med Res. 2018 Jul;46(7):2913–24.

Screening Methods for the Evaluation of Antidiabetic Drugs

36

Avinash Arivazhahan

Abstract

Diabetes mellitus is a very common disorder that has seen the emergence of various drugs, day in and day out. Animal models are specific for each type of diabetes mellitus – type 2 or type 1. In addition, there are also models available for assessing the diabetic complications, both micro- and macro-vascular. These models, as with the other systems, can be in vitro or in vivo in nature. Among the in vivo models, we have various inducing techniques – chemical, hormonal, genetic, microbiological, etc. We will focus on the commonly used animal models for screening drug with antidiabetic potential.

Keywords

Streptozotocin · Alloxan · Quinolones · Tubby · NODD

36.1 Introduction

As we are aware, diabetes mellitus is of two broad types that are commonly seen in clinical settings—type 2 diabetes mellitus (T2DM) and type 1 diabetes mellitus (T1DM). There are other types of diabetes mellitus that are less common and are usually interspersed in the spectrum of clinical features. Animal models have always been a talking point when it comes to diabetes mellitus and its discovery. The earliest recorded mention is that of Joseph von Mering in the early 1890s, who documented that removal of the pancreas can lead to symptoms of polyuria and polydipsia in dogs (which would later become part of the classical triad of insulinopenic symptoms, along with polyphagia). However, the dog that has hogged the limelight

A. Arivazhahan (✉)
Sanofi India Ltd., Chennai, Tamil Nadu, India

M. Lakshmanan et al. (eds.), *Introduction to Basics of Pharmacology and Toxicology*, https://doi.org/10.1007/978-981-19-5343-9_36

in the history of diabetes mellitus was Marjorie, one of the dogs that Banting and Best used to test the effects of insulin, thus making Marjorie a common name in the field of diabetes mellitus and insulins. Ever since the animal models for diabetes mellitus have come a very long way in terms of their advancements and ease of use.

The screening methods and models are divided into 3 types:

- Models for T1DM
- Models for T2DM
- Models for diabetic complications

These can further be divided into *in-vivo* and *in-vitro* models.

36.2 *In-Vivo* Models for T1DM

The animal models that induce a state of total insulin deficiency (as in classical T1DM) are listed here. They can be broadly divided into the following.

- Chemical models
- Hormonal models
- Antibody-mediated models
- Microbiological models
- Surgical models
- Genetic models

36.2.1 Chemical Models for T1DM

Various chemicals have been found to induce a state of T1DM in animals. A few worthy mentions are streptozotocin, alloxan, dithizone, monosodium glutamate, gold thioglucose, etc.

36.2.1.1 Streptozotocin-Based model

- Streptozotocin is a broad-spectrum antibiotic molecule that comes from *Streptomyces achromogens.*
- Its chemical name is [2-deoxy-2-(3-methyl-3-nitrosourea)-1-D-glucopyranose].
- It induces a state of T1DM by causing damage to the pancreatic beta-cells through 3 means: by methylation, by generating free radicals, and by producing nitric oxide.
- While streptozotocin has been studied in various species like mice, rats, and dogs, it is most commonly carried out in rats and mice. Guinea pigs and rabbits are resistant to its action.
- The dose of streptozotocin required to induce T1DM depends on the species used: 175–200 mg/kg i.p. or I .v. in mice; 50–60 mg/kg i.p. or I .v. in rats; 10–15 mg/kg for 3–5 days in dogs.

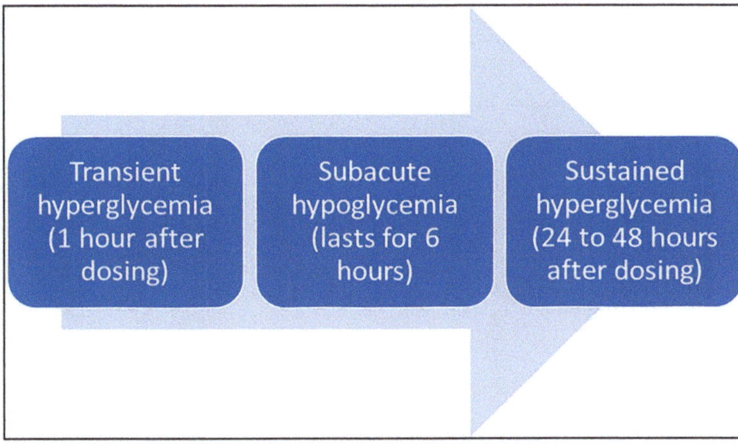

Fig. 36.1 Triphasic response seen with streptozotocin model for T1DM

- A classical triphasic response is usually resultant, expressed as in Fig. 36.1.
- In addition to dosing streptozotocin in isolation, it may also be combined with cyclosporin-A, Freund's adjuvant, etc. for additive effects.
- Streptozotocin is currently the most preferred chemical agent (and has superseded the alloxan-based model) owing to a few clear advantages: low mortality rates, ability to induce an irreversible state of diabetes, and high selectivity for pancreatic beta cells.

36.2.1.2 Alloxan-Based Model

- Alloxan is very rarely used in modern experimental pharmacology, although it was among the earliest chemicals used to induce an irreversible (or reversible) diabetes mellitus.
- Alloxan is chemically a cyclic urea analog.
- It induces a state of diabetes mellitus due to the following mechanisms: production of free radicals and induction of cell necrosis.
- As with streptozotocin, the dosage of alloxan also depends on the species being tested: 100–175 mg/kg s.c. for rats, 150 mg/kg i.v. infusion for 10 min for rabbits, 60 mg/kg i.v. for dogs, etc.
- Following the administration of alloxan, the animals will receive glucose and regular insulin for the next 7–10 days.
- After 7–10 days, a single daily dose of insulin (28 IU) is to be given s.c.
- A triphasic response, as seen with streptozotocin, is seen with alloxan as well (depicted in Fig. 36.2).
- However, when compared to the streptozotocin-based model, this model has a lower success rate of around 70% (the remaining 30% animals either become reversibly hyperglycemic or die).
- Also, alloxan may induce ketosis in a few animals.
- Guinea pigs are resistant to both alloxan and streptozotocin.

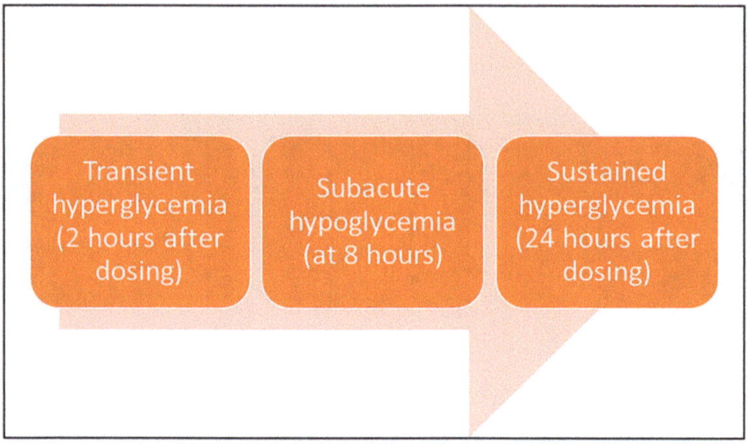

Fig. 36.2 Triphasic response seen with streptozotocin model for T1DM

36.2.1.3 Other Chemical Models

- As mentioned earlier, dithizone (40–100 mg/kg i.v.) can induce a state of permanent hyperglycemia after 24–72 h of dosing. While a triphasic response is seen with dithizone dosing, a state of hypoglycemia as seen with streptozotocin and alloxan is replaced by a state of normal blood glucose level.
- Other agents used include gold thioglucose, monosodium glutamate, quinolones, etc.

36.2.2 Hormonal Models for T1DM

- Corticosteroids are known to induce hyperglycemia in normal or diabetic individuals in clinical scenarios.
- Corticosteroids stimulate the adrenal cortex, resulting in diabetes (reversible or irreversible) due to the ensuing hormonal imbalance.
- Dexamethasone is commonly used as an inducing agent when administered intraperitoneally at 2–5 mg/kg BID dose for a few days in series.
- Rats, guinea pigs, and rabbits are preferred for this model (guinea pigs are resistant to the chemical models for diabetes).

36.2.3 Antibody-Mediated Models for T1DM

- As T1DM is pathophysiologically an auto-immune disorder, immune models in animals have also shown success.
- Guinea pigs are initially used for this model, where bovine insulin is administered with complete Freund's adjuvant to induce anti-insulin antibodies.

- Rats are then subjected to 0.25–1 ml of guinea pig anti-insulin serum (slow i.v. infusion or i.p.)
- This induces hyperglycemia, which is claimed to be dose-dependent.
- While lower doses (or shorter courses) lead to reversible hyperglycemia, higher doses (or prolonged courses) can lead to irreversible damage and persistent hyperglycemia.

36.2.4 Microbiological Models for T1DM

- Among the multifaceted pathophysiology of T1DM, viruses also find a significant place.
- The mechanisms by which viruses can induce hyperglycemia include the destruction of pancreatic beta cells, eliciting auto-immune reactivity, and the causation of non-beta cell-mediated systemic effects.
- Viruses used in animal models include, but are not limited to, Coxsackievirus B4, RNA picornaviruses, encephalomyocarditis (D or M variants), reovirus, lymphocytic choriomeningitis virus (Armstrong variant), etc.
- These viral isolated are inoculated into suckling mice, depending on their susceptibility, as shown in Table 36.1.

36.2.5 Surgical Models for T1DM

- Surgical excision of the pancreas induces diabetes mellitus in animal models.
- This excision can be total or partial. However, in partial pancreatectomy, at least 90% of the organ needs to be removed to have a lasting effect.
- The duration and intensity of induced diabetes depend on the degree of pancreatectomy performed.
- The major demerit of this model is that there is also a loss of alpha cells due to pancreatectomy, which results in the abolition of counter-regulatory responses. In addition to this, pancreatic enzymes are also not secreted by pancreatectomized animals.

Table 36.1 Mouse strains used for each viral isolate

Virus & strain	Mouse strain used
Coxsackie B4	SJL/J SWR/J
Encephalomylocarditis (D or M variants)	SJL/J SWR/J DBA/1 J DBA/2 J BALB/cCUM
Reovirus	SJLj
Kilham rat virus	BB-DR

- A technical challenge is that complete or total pancreatectomy is difficult to perform in most animals.

36.2.6 Genetic Models for T1DM

- A few popular genetic rodent models are as follows:
  - Non-obese diabetic (NOD) mice
  - Bio-breeding (BB) rats
  - Wistar Bonn/Kobori (WBN/KOB) rats
  - Cohen diabetic rats
- The salient features are enlisted in Table 36.2.

36.3 *In-Vivo* Models for T2DM

- As with the models for T1DM induction, there are various broad categories of animal models for T2DM, as below.
  - Chemical models
  - Genetic models
  - Miscellaneous models
- Of these, the genetic and chemical models are more commonly preferred. Also, a few of the models used for T1DM can be extrapolated to induce T2DM if used in different doses or when dosed in the neonatal stage.

Table 36.2 Genetic models for T1DM

Genetic model	Salient features
NOD mice	• Among the most preferred genetic models for diabetes • Developed in Japan • Derived from in-breeding of Jcl:ICR Swiss mice • Gross hyperglycemia and glucosuria are seen in female mice of this strain
BB rats	• Developed in Canada (named after the Bio-breeding lab) • Not in-bred as with most other models (but closed out-bred) • Frank hyperglycemia, glucosuria, ketosis, and weight loss are classical features • Highly prone to diabetic ketoacidosis and death • Both genders develop symptoms
WBN/KOB rats	• Less commonly used • Pancreatic beta-cell degeneration is observed at around 3 months of age • Diabetic symptoms have onset at around 5 months of age
Cohen diabetic rats	• Used as models for both T1DM and T2DM since these rats also show significant insulin resistance (preferred for T2DM these days) • More prone to chronic diabetic complications than other models

36.3.1 Chemical Models for T2DM

- *Streptozotocin-induced T2DM model*
  - Neonatal rats are used for this model.
  - Streptozotocin at a dose of 80–100 mg/kg i.p. is administered at birth or within the first 5 days of delivery.
  - A major distinction, when compared to the T1DM model that uses streptozotocin, is that the rats here do not have complete and severe beta-cell destruction.
  - There is a transient period of hyperglycemia during which studies can be conducted.
  - Following this period of hyperglycemia, there is a regeneration of beta cells, which leads to normoglycemia.
  - The duration of hyperglycemia rests on several factors like species used, age of the animals, the dose of streptozotocin, etc.
- *Other chemical models*
  - Other agents that are used for inducing a state of T2DM are adrenaline, quinolones, EDTA, diazoxide, etc.
  - Adrenaline is dosed at 0.1 mg/kg s.c. and is preferred for the screening of oral antidiabetic drugs.
  - Rats, mice, and rabbits are also susceptible to the hyperglycemic effects of agents including thiazides, diazoxide, furosemide, and others.

36.3.2 Genetic Models for T2DM

Genetic models can be monogenic or polygenic in nature.

36.3.2.1 Monogenic Models

- The monogenic models used for screening drugs for T2DM usually have a characteristic phenotypic profile of sustained hyperglycemia, obesity, hyperinsulinemia, and dyslipidemia, the features of metabolic syndrome.
- A few of the commonly employed monogenic models include the Agouti mice, Tubby mice, Zucker diabetic fatty rats, etc. A few of their characteristics are listed in Table 36.3.

36.3.2.2 Polygenic Models

- Since the development of T2DM and obesity is multifactorial, there are several polygenic models that involve multiple genetic changes.
- Hence, in a way, it is widely agreed that polygenic models for T2DM and obesity are closer to human pathogenesis than the monogenic models. Strains of rodents employed as polygenic models are detailed in Table 36.4.

Table 36.3 Monogenic models for T2DM

Monogenic model	Key characteristics
Agouti mice	• Also known as yellow mice • Developed in China • Named after *Dasyprocta agouti*, a South American rodent • They exhibit hyperglycemia, insulin resistance, and hyperinsulinemia from around 4 weeks of age
Tubby mice	• Developed as a result of spontaneous mutation (tub mutation, hence the name) • Hyperglycemia is not usually very prominent • Other features like hyperinsulinemia and obesity are well-defined • These mice may go into occasional hypoglycemic episodes
ob/db mice	• Two mutations, ob (obesity) and db (diabetes) are seen in these mice (autosomal recessive) • They have obesity, insulin resistance (usually severe), mild hyperglycemia
Fat mice	• This mutation is also autosomal recessive • There is obesity associated with a transient period of hyperglycemia (particularly in males) at 8 weeks, following which there is a return to normoglycemia
Zucker diabetic fatty rats	• Male rats are prone to obesity, insulin resistance, and persistent hyperglycemia • Female rats are prone to obesity and insulin resistance, but usually do not become hyperglycemic
Wistar fatty rats	• Transfer of gene from Zucker rat to Wistar Kyoto rats has developed this monogenic model • They are characterized by obesity, insulin resistance, impaired glucose tolerance, and dyslipidemia

36.3.2.3 Transgenic and Knockout Models

- Apart from the monogenic and polygenic animals, there is also active manipulation of genes controlling the different aspects of insulin secretion and action, which can result in transgenic and knockout animals. A few of these genes that can be manipulated are mentioned in Table 36.5.

36.3.2.4 Miscellaneous Models

- In addition to the above-mentioned genetic models, there are a few models developed from unclear or unestablished genetic or situational components.
  - Spiny mice (*Acomys russates* and *Acomys cahirinus*)
  - Sand rats (*Psommomys obesus*)
  - Tuco-tuco rodents (*Ctenomys talarum*)

Table 36.4 Polygenic models for T2DM

Polygenic model	Key characteristics
db/db mice	• Autosomal recessive mutation of the leptin receptor gene, which is spontaneous in nature • Since leptin is affected, the mice are obese with hyperglycemia and hyperinsulinemia • They are prone to developing nephropathy at a relatively young age
New Zealand obese mice	• The NZO mice are typically obese with hyperglycemia and hyperinsulinemia • They also have moderate to severe insulin resistance
Japanese KK mice	• KK (Kasukabe-Kasukabe) mice exhibit hyperglycemia and hyperinsulinemia, along with insulin resistance
Nagoya-Shibata-Yasuda mice	• Males are preferred, which have obesity and impaired glucose tolerance • Not commonly used because the onset of symptoms is quite late (beyond 32–36 weeks of age)
Otsuka-Long-Evans-Tokushima fatty rats	• OLEFT rats are developed from Long Evans rats that are inbred to produce off-springs that develop permanent hyperglycemia (late-onset but sustained) • Besides, they are also obese and dyslipidemic
Goto-Kakisaki rats	• These are inbred hyperglycemic Wistar rats that develop symptoms and feature very closely similar to the human presentation • In addition to the features of hyperglycemia and insulin resistance, they also develop complications of diabetes mellitus (particularly micro-vascular) • Obesity is not very prominent
Others	• Chinese hamsters/*Cricetulus griseus* • South African hamsters/*Mystromys albicandatus* • Siberian hamsters/*Phodopus sungorus*

Table 36.5 Genes targeted to produce transgenic and knockout animals

Genes controlling insulin resistance	Genes controlling insulin secretion
Insulin receptor Glucose transporters (GLUTs) Hexokinase-2 TNF-alpha	GLUT-2 Glucokinase Islet amyloid polypeptide

36.4 Models for Diabetic Complications

- A major reason for stressing on timely diagnosis and management of diabetes mellitus is the impact that diabetes exerts on various end organs in the long run. The complications of diabetes mellitus can be of two types: micro- and macro-vascular.
  - Micro-vascular complications: Nephropathy, neuropathy, and retinopathy
  - Macro-vascular complications: Atherosclerosis, myocardial infarction, stroke

- Some experimental models have been developed to study and assess these complications. A few of these models are enlisted here.

36.4.1 Models for Diabetic Nephropathy

- Akita mice (with spontaneous mutation of the Ins-2 gene, leading to protein misfolding and beta-cell toxicity)
- eNOS-deficient mice (endothelial nitric oxide-deficient mice that develop renal vascular disorder)
- BTBR ob/ob mice (insulin-resistant diabetic mice with early-onset diabetic nephropathic features)

36.4.2 Models for Diabetic Retinopathy

- Streptozotocin-induced diabetic rats
- Galactose-induced diabetic rats
- Oxygen-induced retinopathy (non-diabetic)
- Retinal occlusion model (non-diabetic)
- Angiogenesis model/VEGF-induced model (non-diabetic)
- Zebrafish models (hypoxia-induced/glucose-induced/angiogenesis models)

36.4.3 Models for Diabetic Neuropathy

- Streptozotocin-induced diabetic rats/mice
- The L-fucose-induced diabetic neuropathic rat model
- Partial sciatic nerve ligation model
- Surgical model

36.4.4 Models for Diabetic Cardiac Disease

- Akita diabetic mice
- OVE26 diabetic mice

36.5 *In-Vitro* Models

The *In-vitro* methods are also available for screening of drugs with antidiabetic efficacy. The important in vitro methods are highlighted in Table 36.6.

Table 36.6 In vitro models for antidiabetic drug screening

Isolated organs/tissues	Cell lines to study insulin secretion	Assays for glucose transport
• Isolated rat pancreas • Isolated rat islets • Isolated rat liver • Isolated rat hepatocytes • Isolated perfused rat hind limb	• Pancreatic cell lines (HIT, INS-1, INS-2 cell lines) • Muscle cell lines (BC3H1, L6, C2C12 cell lines)	• Rat diaphragm assays • Rat adipose tissue assays

Bibliography

Caccetta R, Salami HA. Screening for antidiabetic activities. Methods Mol Biol. 2013;1055:207–18.

Eddouks M, Chattopadhyay D, Zeggwagh NA. Animal models as tools to investigate anti-diabetic and anti-inflammatory plants. Evid Based Complement Alternat Med. 2012;2012:142087.

Fang J, Lin C, Huang T, Chuang S. In vivo rodent models of type 2 diabetes and their usefulness for evaluating flavonoid bioactivity. Nutrients. 2019;11(3):530.

Gupta SK. Antidiabetic agents. In: Drug screening methods. 3rd ed. Jaypee Brothers Medical Publishers; 2016. p. 613–44.

Islam MS. Animal models of diabetic neuropathy. J Diabetes Res. 2013;2013:149452.

King AJF. The use of animal models in diabetes research. Br J Pharmacol. 2012;166(3):877–94.

Screening Methods for the Evaluation of Antithyroid Drugs

37

Suja Xaviar

Abstract

Hyperthyroidism is a systemic disease resulting from excessive thyroid hormone production or release from the thyroid gland. Thyroid hormone (TH) signaling is crucial for growth, development and function of multiple organs in humans and animals. The negative feedback system by TH is modulated centrally through the hypothalamic-pituitary-thyroid axis. Antithyroid drugs are the mainstay for the treatment of hyperthyroidism. They either inhibit the TH biosynthesis and/or decrease the conversion of T4 to T3. The identification of potential therapeutic targets for hyperthyroidism requires models that can be either in vitro or in vivo. The various assays and experimental models for thyroid hormones with emphasis on screening methods in evaluating antithyroid drugs is discussed.

Keywords

Antithyroid · Hyperthyroidism · Thyroid hormone · Animal models

37.1 Introduction

Triiodothyronine (T_3) and thyroxine (T_4) are the cell metabolism regulators that are associated with different biologic processes in vertebrates. Thyroid dysfunctions are one of the common endocrinopathies in both humans and animals. Hence, experimental studies associated with thyroid dysfunctions are based either on suppression of hormone production or administration of thyroxine in high doses resulting in iatrogenic hyperthyroidism. This chapter mainly focuses on the screening methods involved in the evaluation of antithyroid drugs.

S. Xaviar (✉)
Department of Pharmacology, Jawaharlal Institute of Postgraduate Medical Education & Research (JIPMER), Puducherry, India

M. Lakshmanan et al. (eds.), *Introduction to Basics of Pharmacology and Toxicology*, https://doi.org/10.1007/978-981-19-5343-9_37

37.2 Thyroid Hormones

The thyroid gland is the major endocrine gland that controls the biological functions of our body. The important hormones synthesized by it include T_3 and T_4 which are essential for normal growth and development, and the calcitropic hormone calcitonin, involved in calcium phosphate regulation.

The main effects of T_3 and T_4 include:

- biological- they are involved in the differentiation of all the cells in our body
- calorigenic- they increase the basal metabolic rate
- cardiovascular- they increase the sensitivity of the heart to catecholamines
- metabolic- they increase lipid, carbohydrate, and collagen metabolism
- Thyroid hormone (TH) analogs and metabolites can be tested based on the above effects.
- Thyroid hormones help in the uptake and utilization of iodine by the thyroid. This activity is blocked by antithyroid compounds.
- An increase in T_3 and T_4 levels causes feedback inhibition of TSH from the pituitary. Various mechanisms acting at different sites of the hypothalamic-pituitary-thyroid axis can cause changes in thyroid hormone homeostasis, as depicted in Fig. 37.1.

An overactive thyroid gland results in a condition called hyperthyroidism, where there is a sustained increase in synthesis and/or secretion of thyroid hormones. The commonly used antithyroid drugs are thioamides, iodides, radioactive iodine, and

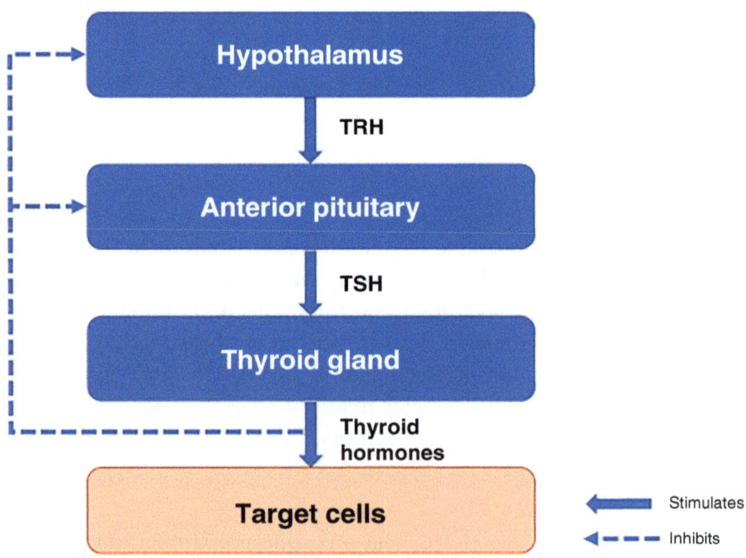

Fig. 37.1 Hypothalamic- pituitary- thyroid axis

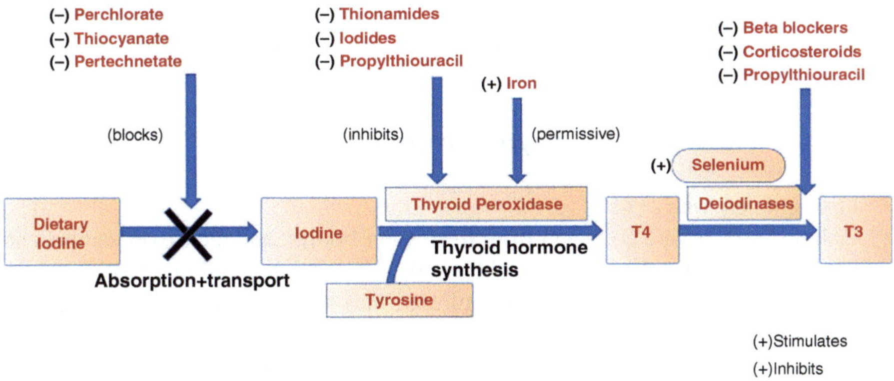

Fig. 37.2 Overview of thyroid hormone synthesis

adrenergic blockers. An overview of the thyroid hormone synthesis with the site of action of antithyroid agents is illustrated in Fig. 37.2.

37.2.1 Historical Assays

- The role of thyroid hormones on amphibian morphogenesis was studied. It is an ideal model to understand the action of the hormone. Morphogenesis is a developmental process that fails to occur without thyroid hormones but can be induced precociously by exogenous TH. The frog thyroid glands are located between the eyes. Administration of exogenous thyroid hormones to tadpoles induces premature metamorphosis, where the tadpole is transformed into a small frog with the simultaneous growth of limbs, resorption of the tail, and gills with development of lungs.
- Treatment of tadpoles with antithyroid compounds causes retardation or complete blockade of the metamorphosis process.
- Effects of exposure of pre and pro-mctamorphic larvae to methimazole, propylthiouracil(PTU), and T4 were tested. Methimazole and PTU (thyroid hormone synthesis inhibitors) caused a delay in larval development and morphological changes in the thyroid gland. However, a concentration-dependent increase in larval development was seen with T_4.
- An example of a study object where metamorphosis was induced by thyroid hormones is the axolotl (*Ambystoma mexicanum* or *tigrinum*). This animal loses its gills and develops lungs, and simultaneously changes the shape of its tail.
- Amphibian thyroid glands change continuously during metamorphosis, making it difficult to evaluate. Histopathological examination of the thyroid gland has demonstrated increased sensitivity.
- T_3 and T_4 also cause metabolic activation and increased energy expenditure. Thyroid preparations were standardized using weight loss of guinea pigs after

1 week of treatment. A guinea pig unit is a dose that causes at least 10% reduction within 7 days in the weight of guinea pigs with an initial weight of 250–300 g.

37.2.2 In Vitro Tests for Thyroid Hormones

- *Lipogenic enzyme assay*
- Fishes (18–22 g) are collected from freshwater bodies, and their liver samples are pooled and minced with ice-cold Hank's balanced solution (HBSS). After centrifuging the samples, they are cultured into separate flasks containing C-14 acetate in 2 ml of HBSS and various test and standard concentrations (10^{-7}, 10^{-8} or 10^{-9} M). The control group contains a medium without hormones. Incubation of the samples is done at 30° C in a shaking water bath for 8 h. Following incubation, any unbound C-14 acetate is removed by washing the samples in HBSS. Samples are stored at -20 °C, and incorporation of C-14 acetate into lipids is analyzed. Lipogenic enzymes are assayed by conducting similar experiments without C-14 acetate. Major lipogenic enzymes are measured in the liver samples. The values are expressed as IU/mg protein. Absorbance is recorded using a UV-visible spectrophotometer. Lipids are extracted from the liver tissue and separated by thin-layer chromatography (TLC). Lipids of interest are collected from the TLC plate and placed in separate scintillation vials containing 5 ml of scintillate fluid and 0.2 g of 1,4 bis (2–15 – phenyloxazolyl) benzene/liter. Activity is counted in a scintillation counter and expressed as CPM/mg tissue.
- *Tadpole tail culture method*
- The structure of T_3 and T_4 in tadpoles is similar to the human thyroid hormone. Metamorphosis in tadpoles depends on the availability of thyroid hormones. Tadpole tails are cultured. Staged tadpoles are treated with Steinberg's solution (10 mm Hepes, 60 mM NaCl, 0.67 mM KCl, 0.34 mM $Ca(NO_3)_2$, 0.83 mM $MgSO_4$, pH 7.4) containing gentamicin (70 μg/ml) and streptomycin (200 μg/ml). After 24 h, tadpoles are anesthetized, and 6 mm of the tail is cut. Drug treatment is started after 24 h, and the culture medium is changed every 48 h. The length of each tail is measured every 24 h and compared between test and control groups.

37.2.3 In Vivo Tests for Thyroid Hormones

Animal models are used to evaluate the screening of different diseases due to the biological similarity between animals and humans, shorter life cycles, and controlled environments.

- *Thyroidectomy*
- Thyroidectomized rats were used for screening thyroid hormones and their analogs.

- In this model, the endogenous thyroid hormones are entirely removed from the circulation. It carries the disadvantage that it is an invasive procedure with repercussions on calcium metabolism as the parathyroid glands are also removed.
- Thyroidectomy is performed under pentobarbital anesthesia. The fur of the neck is clipped, and the median incision is made. Salivary glands and maxillary lymph nodes on either side are pushed aside to expose the hyoid muscle covering the trachea. This muscle is split to expose the isthmus of the thyroid that connects the two lobes. Using blunt forceps, the lobes and isthmus are separated from the trachea and the blood vessels ligated. Parathyroid glands are also usually removed.
- To prevent hypocalcemia postoperatively, calcium gluconate 1% in drinking water is advised.
- *Oxygen consumption*
- Thyroid hormones produce a wide array of physiological effects in our body, like an increase in basal metabolic rate, oxygen consumption, and carbon dioxide production.
- One of the hallmarks of thyroid hormone action has been to alter oxygen consumption in the body. This has been put to diagnostic use in humans and screening thyroid hormones and their derivatives in animals.
- Individual mice were placed in tightly closed jars, and their survival time was measured. This method is now obsolete.
- A reliable and straightforward automatic closed-circuit system is used to measure the oxygen consumption of small laboratory animals. The apparatus consists of an animal chamber with a removable cover plate at one end, which provides access to the inside of the chamber and has connections for oxygen and pressure lines. A rubber gasket placed between the cover and the chamber provides an airtight compartment. The animal is placed on a wire grid on a layer of carbon dioxide adsorbent such as soda lime. An automatic syringe dispenser delivers predetermined volumes of oxygen into the chamber. The dispenser is inactivated when the chamber pressure exceeds the ambient by 3 mm H_2O. It is reactivated when the pressure drops below the threshold level.
- With each dispenser activation, a fixed volume of oxygen is delivered. The rate of oxygen consumption can be observed by obtaining the pump rate.
- *Inhibition of iodine release*
- Iodine is a vital element in thyroid hormone synthesis and secretion. The thyroid gland has a strong affinity for iodine.
- Radioisotopic iodine (I-131 or I-125) is used to evaluate thyroidal uptake of iodine. Release of I-131 from the thyroid gland in rats is inhibited by thyroxine. The inhibition is dose-dependent. This phenomenon is used to compare the potency of thyroid hormone derivatives with standard thyroxine.
- Male Sprague Dawley (SD) rats (180–240 g) are fed with commercial laboratory chow with or without supplementation of 0.03% propylthiouracil (PTU). PTU inhibits the thyroid peroxidases, which catalyzes the iodination of tyrosine residues in thyroglobulin and the oxidative coupling of iodinated tyrosine molecules. Intraperitoneal injection of I-131 or I-125 is given after withholding

food for 8 h before and 24 h after the injection. After 40 h, radioactivity is measured over the thyroid region under light ether anesthesia. This reading is taken as time zero, and after this reading, the diet is changed to a feed containing 0.03% PTU. Following this, several doses of standard and test preparations are injected subcutaneously at 24-hour intervals for a total of four doses. After the last four doses, the percentage of time zero counts of I-131 remaining in the thyroid is plotted against the logarithm of dose, and from these dose-response curves, potency ratios are calculated.

- Quantitative estimation of thyroid hormone has been replaced by analytical methods.

Short-term uptake of I-125 or I-131 can be used as a parameter to assess thyroid peroxidase inhibition by antithyroid drugs.

- *Anti-goitrogenic activity*
- Administration of exogenous goitrogenic compounds blocks thyroid hormone synthesis resulting in reduced levels of thyroid hormone in circulation. The negative feedback inhibition of T_3 and T_4 is reduced, resulting in TSH release from feedback inhibition. TSH controls the weight and size of the thyroid in rats. With an increase in TSH, thyroid enlargement, and increase in weight due to hyperplasia of thyroid follicles is noticed with features of goiter. By administration of thyroxine, triiodothyronine, or thyroid hormone analogs, thyroid hyperplasia can be prevented.
- Male SD rats (150–180 g) in groups of eight to ten are used. During treatment, 0.1% PTU is added to the drinking water or food, and the baseline thyroid weight is obtained. Rats are administered the test compound or the standard drug (thyroxine) subcutaneously in a dose of 10–40 μg/kg over a period of 2 weeks. Animals in the control group receive PTU diet and saline injection only. Rats are sacrificed on day 14. Thyroid glands are dissected and weighed. Weight of the thyroid gland is increased by 2 to 3 times by PTU diet and is reversed by dose-dependently to normal values by thyroid active substances. Dose-response curves are plotted, and the potency ratio is calculated.
- *Reduction in tensile strength*
- High doses of thyroid hormones can reduce the tensile strength of connective tissue in a dose-dependent manner. This phenomenon can be used to evaluate thyroid hormone derivatives.
- The strength of connective tissue increases with short-term treatment with corticosteroids. This activity is antagonized by thyroid hormones. The biphasic effect is seen with thyroid hormones. The dose-dependent tensile strength of epiphyseal cartilage, tail tendon, and skin strips decreases with short-term treatment with thyroid hormones, whereas treatment exceeding 10 days increases the tensile strength.
- Male SD rats (100–120 g) are administered thyroid hormones subcutaneously. The animals are sacrificed after 24 h, and the tensile strength of distal femoral epiphyseal plates, tail tendons, or skin strips is measured.

- Dose-response curves of test and standard are plotted, and potency ratios are calculated.

37.3 Antithyroid Drugs

- They lower the basal metabolic rate by interfering with the synthesis, release, and/or peripheral action of the thyroid hormone.
- They are cornerstones in the treatment of hyperthyroidism. Reduction in T_4/T_3 levels releases TSH from inhibition by the pituitary gland, thereby increasing TSH secretion leading to induction of goitrogenic response.
- This response was used to detect the presence of antithyroid drugs and has been widely used for screening procedures.

37.3.1 In Vivo Model for Antithyroid Activity

- *Inhibition of iodine uptake in rats*
- PTU and other antithyroid drugs may inhibit thyroid hormone synthesis by inhibiting thyroid peroxidase. This results in reduced iodine uptake by the thyroid gland by inhibiting the organic binding of iodine. This phenomenon is seen with graded doses of PTU. However, higher doses of PTU are required to observe an increase in thyroid weight in rats.
- Measurement of the iodine content of the thyroid gland was used as a parameter previously. Currently, uptake and release of I-131 are noted.
- Male Wistar rats, age 26–28 days of weight 40–45 g are used. They are fed a regular diet. Potassium iodide is added to their drinking water. While doing toxicology studies, the test compounds are added to the drinking water for 10 days, following which the amount ingested by each rat is calculated from the total food consumed. This is expressed in milligram daily per kilogram body weight.
- The rats are sacrificed after 10 days, and the thyroid weight and iodine content are determined. It is observed that PTU decreases the iodine content in a dose-dependent manner. However, higher doses are necessary to note an increase in thyroid weight.
- Evaluation is done using plotting of dose-response curves of test compounds and reference standard followed by calculation of potency ratios.
- Modification of the above method involved using subcutaneously implanted pellets of PTU to alter the thyroid status in rats. Here, the uptake of labeled iodine is measured. Protirelin (TRH) injection may be administered to stimulate the release of labeled iodine.
- The inverse relation between PTU and thyroid function is observed.

37.3.2 Antithyroidal Effects in Animal Assays

- This is a historical bioassay that is modified for demonstrating antithyroid activity.
- The oxygen consumption in iodine-treated mice is used as a parameter to measure antithyroid activity.
- It is observed that oxygen consumption is increased in potassium iodide-treated mice, resulting in decreased asphyxiation time.
- Antithyroid compounds, when given in a dose-dependent manner, are seen to antagonize this effect. They reduce the metabolic rate, and hence the time to convulsions is prolonged due to the reduced metabolic rate.
- Initially, antithyroid activity was measured based on thyroid weight. Rabbits were exclusively fed on goitrogenic compounds such as cabbage. There was a drastic increase in thyroid weight. Iodine treatment reversed these effects.

37.4 Conclusion

The studies conducted in experimental models play an important role in understanding the disease pathology and identifying the best therapeutic options for its treatment. The selection of an in vivo experimental model where the pathologic processes could be likened to the same phenomena occurring in humans is essential.

Bibliography

Argumedo GS, Sanz CR, Olguín HJ. Experimental models of developmental hypothyroidism. Horm Metab Res. 2012 Feb;44(02):79–85.

Broadhead GD, Pearson IB, Wilson GM. The effect of prolonged feeding of goitrogens on thyroid function in rats. J Endocrinol. 1965 Jul 1;32(3):341–51.

Miyata K, Ose K. Thyroid hormone-disrupting effects and the amphibian metamorphosis assay. J Toxicol Pathol. 2012;25(1):1–9.

Tata JR. Amphibian metamorphosis as a model for the developmental actions of thyroid hormone. Mol Cell Endocrinol. 2006 Feb 26;246(1–2):10–20.

Varghese S, Shameena B, Sudhakaran PR, Oommen OV. Short-term effects of thyroid hormones on lipogenic enzymes and 14C-acetate incorporation into various lipid classes: in vivo and in vitro studies. Indian J Exp Biol. 2001;39:–431, 5.

Walker JS, Levy G. Induction of experimental thyroid dysfunction in rats with implantable pellets of thyroxine or propylthiouracil. J Pharmacol Methods. 1989 May 1;21(3):223–9.

Screening Methods for the Evaluation of Antifertility Drugs

38

Selvalaxmi Gnanasegaran and Mangaiarkkarasi Adhimoolam

Abstract

The discovery and development of a new antifertility drugs relies heavily on the preclinical use of animal models to establish efficacy and safety prior to trials in humans. Animal models have played a fundamental role in advancing our understanding of basic mechanisms and preclinical development of novel contraceptive drugs. Both in-vivo and in-vitro animal models are currently used for the screening of anti-fertility drugs. Presently, the in-vivo models are intended to screen the estrogenic, anti-estrogenic, progestational, anti-progestational, anti-ovulatory, anti-implantation activity, androgenic, anti-androgenic and anabolic activity in animals. These animals models which includes rats and mice used in in-vivo studies are either ovariectomized or immature. They should be maintained under controlled environmental conditions. Test preparations should be fresh, either in solution or in suspension in a suitable vehicle. In-vitro models are used to screen the assessment of sperm motility and count, its viability and morphology, estrogen and progesterone receptor binding activity, aromatase and 5-alpha reductase inhibition activity. Identification of drugs having anti-fertility activity is the need of current time and hence series of drug discoveries and development process is carried out. This chapter encompasses the comprehensive review of various animal models used for screening of various anti-fertility activities.

S. Gnanasegaran (✉)
Department of Pharmacology, Vinayaka Mission's Medical College and Hospital, Karaikal, Puducherry, India

M. Adhimoolam
Department of Pharmacology, Sri Venkateshwaraa Medical College Hospital and Research Centre, Puducherry, India

M. Lakshmanan et al. (eds.), *Introduction to Basics of Pharmacology and Toxicology*, https://doi.org/10.1007/978-981-19-5343-9_38

Keywords

Antifertility screening · Preclinical models · Contraceptives · Estrous cycle · Fertilization

38.1 Introduction

Contraceptive substances range from natural agents available as herbal derivatives to synthetic chemical moiety. The wide range of natural and synthetic substances have been shown to intercept pregnancy during different stages of pregnancy. They are found to be active in females when it hinders fertilization, ovulation, implantation and destroys the zygote or causes abortion. In males, it inhibits testosterone production by preventing spermatogenesis. The search for new therapies with better efficacy and tolerability remains an essential goal in discovering antifertility drugs.

38.2 Physiology of Female Sexual Organs

The female reproductive system consists of ovaries, fallopian tubes or uterine tubes, uterus, and vagina. Oogenesis is the process of the development of ova in the ovaries. In the middle of each menstrual cycle, a single ovum is expelled from an ovarian follicle into the abdominal cavity near the open fimbriated ends of the two fallopian tubes. This ovum then passes through one of the fallopian tubes into the uterus; if sperm has fertilized it, it gets implanted in the uterus, developing into a fetus. The reproductive system of women shows regular cyclic changes for preparation of fertilization or pregnancy every month. The cycle length is variable in women, but an average period is 28 days from the start of one menstrual period to the next. This is called the female monthly sexual cycle or **menstrual cycle.** The most conspicuous feature of the menstrual cycle is the shedding of the blood along uterine mucosa through the vagina, with an average duration of the menstrual flow ofabout 3–5 days. Only 25% of menstrual blood is of venous origin; the remaining predominant flow is arterial. The fibrinolysin present in the menstrual blood lyses clots, so that menstrual blood does not usually contain clots unless the flow is excessive. Cyclical changes co-occur in the ovary (ovarian cycle) and the uterus (uterine cycle) during the menstrual cycle. The anterior pituitary gland's gonadotrophic hormones FSH and LH secreted in a pulsatile manner govern the ovaries from puberty until menopause; the ovaries remain inactive after menopause. The time of the first menstrual cycle is called menarche.

38.2.1 Follicular Phase or Proliferative phase

The first phase of the menstrual cycle occurs from day zero to day 14, based on the average duration of 28 days. In this phase, 17-beta-estradiol supports increasing growth of the endometrial layer of the uterus and increased stromal and glandular secretion. It also creates an environment that is friendly and helpful for the possible entry of sperm. Simultaneously, follicular growth occurs with moderate enlargement of the ovum with the growth of additional layers of granulosa cells in some of the follicles known as primary follicles. The concentration of both FSH and LH increase in the initial days of the cycle, which causes accelerated growth of 6–12 primary follicles every month.

Along with the proliferation of the granulosa cells, spindle cells derived from the ovary interstitium give rise to thesecond mass of cells called the theca. This is divided into two layers, theca interna and theca externa. The distended follicle ruptures at about the 14th day of the cycle, and the ovum is extruded into the abdominal cavity. This is the process of ovulation**.** Only one follicle fully matures each month, and the remainder undergoes atresia. Even with the abundance of FSH, ovulation does not occur without LH, which is essential for final follicular rupture and ovulation.

38.2.2 Luteal or Secretory Phase

This phase lasts from day 14 to day 28 of the cycle. Once the ovum is released out of the follicle, the remaining granulosa and theca cells undergo the luteinization process forming corpus luteum. The corpus luteum is highly secretory, producing a copious amount of both progesterone and estrogen. These hormones have negative feedback effects on the anterior pituitary gland to reduce the secretion of both FSH and LH, which causes complete degeneration of the corpus luteum, a process called *involution* of the corpus luteum. Final involution occurs around the 26th day of the normal female sexual cycle. At this time, the feedback inhibition of the anterior pituitary gland vanishes; starts secreting FSH and LH, which initiate the growth of new follicles, beginning a new ovarian cycle. The thickness of the endometrium may reach upto 5–6 mm at the peak of the secretory phase. Until the ovum gets implanted (7–9 days after ovulation), the uterine secretions provide nutrition for the early dividing fertilized ovum. Once the ovum implants in the endometrium, the trophoblastic cells on the surface of the implanting ovum (in the blastocyst stage) begin to digest the endometrium and absorb the endometrial stored substances, thus making plentitude of nutrients available to the early implanting embryo. If the ovum is not fertilized, the corpus luteum in the ovary involutes, followed by menstruation.

38.3 Estrous Cycle in Rodents

The reproductive cycle in rodents is referred to as the estrous cycle. It consists of four phases – proestrus, estrus, metestrus, and diestrusthat repeat every 4–5 days unless interrupted by pregnancy, pseudopregnancy, or anestrus. The short length of the estrus cycle makes the rodents an ideal model for screening the newer compounds, which makes changes during the reproductive cycle. Both ovaries exhibit the same

Table 38.1 Length of various phases of the estrous cycle in Rodents

EstrousPhase	Cycle length (h)	
	Rat	Mice
Proestrus	14	<24
Estrus	24–48	12–48
Metestrus	6–8	8–24
Diestrus	48–72	48–72

morphological changes throughout the cycle and ovulate bilaterally at the same time. Since the number of ova released is relatively large in rodents (approximately 12–14 overall), many follicles and corpora lutea can be observed in the ovaries. The follicles develop rapidly until ovulation in proestrus phase. At this stage, two to four large follicles can be observed. When estrus starts at ovulation, follicles are transformed into corpora lutea, the luteal cells become more basophilic and smaller than in later phases of the cycle. These corpora lutea begin to regress at metestrus and continue regressing in diestrus. This pattern of development and regression repeats in the next cycle of estrous.

The proestrus phase corresponds to the human follicular stage where there is rise in circulating estradiol concentrations and little surge in prolactin which causes increased release of LH and Follicle Stimulating Hormone (FSH). The peak in FSH concentration with an associated rapid decline in estradiol levels correlates to ovulation and estrus phase. Metestrus and diestrus are homologous to human early and late secretory stages of the reproductive cycle, respectively, with high progesterone levels. Whereas in rabbits, estrus cycles are not regular, and they may have periods of anestrus, and in every 4–17 days, they have 1–2 day periods of non-receptivity (Table 38.1).

38.4 Experimental Animal Models

38.4.1 Invivo Methods

38.4.1.1 Methods for Females

Test for Anti-Ovulatory Activity

- HCG Induced Ovulation in Rats
  - *Principle:* Young female albino rats do not exhibit spontaneous ovulation. Ovulation is induced in 2 days by injecting human chorionic gonadotropin (HCG). Test compounds with anti-ovulatory activity inhibit the HCG-induced ovulation.
  - *Assay Procedure:* Ovulation is induced in 24–26 days old immature female albino rats by priming with human chorionic gonadotropin (HCG). Two days before HCG administration, these animals are treated with various doses of test drugs, and they are sacrificed after twodays of HCG administration, and their ovaries are dissected for histological examination.

- *Assessment:* Compounds with anti-ovulatory activity inhibit HCG-induced ovulation.

- Cupric Acetate- Induced Ovulation in Rabbits
  - *Principle:* As rabbits are the reflex ovulators, they ovulate within few hours after mating, intravenous administration of certain chemicals, and/ or mechanical stimulation of the vagina. In this method, cupric acetate is used to induce ovulation in rabbits. The administration of test compounds inhibit the same, proving its anti-ovulatory activity.
  - *Assay Procedure:* Non-pregnant female rabbits weighing 3-4 kg are treated with the test drug. Intravenous injection of cupric acetate (0.3 mg/kg, using 1% copper acetate in 0.9% saline) is given to the rabbits after 24 h of administering the test drug. The animals are then sacrificed, and ovaries are examined 18–24 h later.
  - *Assessment:* The total number of ovulation points on both ovaries isrecorded for each animal. Fewer ovulation points are seen in test groups than in the control group, proving the test drug's anti-ovulatory action.

38.4.2 Test for Estrogenic Activity

- Vaginal cornification
  - *Principle:* This assay for estrogenic activity is based on vaginal epithelial proliferation with associated changes in the secretion of hormones. This test is otherwise known as the Allen-Doisy test, named after those (Allen and Doisy) who observed vaginal cornification in rodents (1923) which is again based on the observations of Stockard and Papanicolaou (1917), who observed the same in guinea pigs. The estrus cycle of rodents is classified into proestrus, estrus, metestrus, and diestrus. The compounds that change the animal to the estrous stage skipping the other stages, are considered estrogenic potential.
  - *Assay Procedure:* Ovariectomized female Sprague-Dawley or Albino rats are kept on a standard laboratory feed and water ad libitum for 1 week. The standard compound – estradiol at various doses 0.02, 0.1 and 0.5 μg and the test compound at several doses in 0.5% solution of carboxymethyl-cellulose are given orally or subcutaneously to the standard group and the test group animals, respectively. The drugs are administered twice daily at the same time for three consecutive days. On the fourth day, vaginal smears are taken with cotton swabs dipped in saline. In normal animals, the smears contain cornified cells, leukocytes, and epithelial cells. Whereas in test compound treated animals, irregular and large cornified cells suggest maximum growth of the vaginal mucosa indicating the positive estrous smear.
  - *Assessment:* The number of positive estrous animals in each group is found, indicating the compound administered's estrogenic property. ED_{50} values are calculated and compared with the standard estradiol group.

- Assay For Water Uptake
  - *Principle:* The principle of this assay is responsiveness of the uterus to estrogens by increased uptake and retention of water with its peak uptake at 6 h after administration.
  - *Assay Procedure:* Ovariectomized immature 18-day-old mice or 22-day-old rats are obtained and randomly grouped as control group and estrogen control group. The control group is injected subcutaneously with 0.1 ml of cottonseed oil. The estrogen control group is given a range of doses (0.01–0.1 *mcg)* to establish a dose-response curve. The test compound is given to groups in the initial test at a high and a low dose. All doses are given in 0.1 ml of cottonseed oil. After 5 h of treatment, the animals are killed by cervical dislocation followed which quick removal of uteri. The uterus is picked up with forceps and severed from the vagina. The uterine horns are separated from the connecting tissues and then cut at the constriction seen near the ovary (the ovary, of course, the ovary will not be seen in ovariectomized animals). The uteri are kept damp by placing them on damp (not wet) filter paper and covering them with the same. They are then rapidly weighed to the nearest 0.1 mg. The uteri are dried in an oven at 100 °C for 24 h and are reweighed.
  - *Assessment*: The percent increase in water uptake over control can be calculated and compared with other groups' values. No less than tenanimals per test group should be included, and the Student "t" test should be used to determine the significance of the difference of the test group from the control.
- Vaginal Opening Assay
  - *Principle:* This qualitative test for estrogenic activity is based on the principle that vaginal opening occurs in immature female animals by treating them with estrogenic compounds.
  - *Assay Procedure:* The animals used are immature female albino mice – 18-day old mice or 21-day old rats. The test and standard drugs are administered to the animals intramuscularly in cottonseed oil.
  - *Assessment:* The time of complete vaginal opening can be observed as a sign of estrogenic activity.
- Uterine weight Assay
  - *Principle:* Repeated administration of estrogenic compounds triggers protein synthesis causing myometrial/endometrial proliferation, which leads to a dose-dependent increase in uterine weight of ovariectomized female rats.
  - *Assay Procedure:* The test drug or standard (estradiol) are administered intramuscularly to the ovariectomized albino rats for thee days. The animals are then sacrificed by cervical fracture on the fourth day and the uteri are dissected out rapidly from the abdomen. The uterine contents are cleared and uterine weight is measured immediately in the wet state. The uterus is placed in the oven at 100 °C, dehydrated and again weighed to calculate the increase in dry weight.
  - *Assessment:* An increase in uterine weight is an indication of the estrogenic property of the test drug.
- EChick-oviduct method

- *Principle:* This is one of the early methods for screening estrogenic compounds. A dose-dependent increase in weight of the oviduct of young chicken proves estrogenic potential of the compound.
- *Assay Procedure:* Pullet chicks are injected subcutaneously, twice daily with various doses of test compound and 17 β-estradiol as standard for 6 days. Animals are sacrificed 24 h after the last injection, and the weight of the body and oviduct are measured.
- *Assessment:* Increase in weight of the oviduct suggests estrogenic property of the test drug.

38.4.3 Test for Anti-Estrogenic Activity

- *Antagonism of Physiological Effects of Estrogen*
  Principle: Anti-estrogenic compounds inhibit the physiological effect of estrogen, such as
  (a) water uptake of the uterus
  (b) uterotrophy and
  (c) vaginal cornification
  This principle is used for the screening of anti-estrogenic activity. In this method, antiestrogenic compounds inhibit the estradiol-stimulated increase in uterine weight in castrated female rats. The assays used for anti-estrogens are modifications of the above estrogenic assays.

  Assay procedure: Groups of 5–10 female immature Sprague-Dawley rats weighing about 55 g are ovariectomized. They are injected with estradiol 0.03–0.06 g per animal s.c. for seven consecutive days. The test compound is administered in 0.5% carboxymethylcellulose solution or cottonseed oil either orally or injected subcutaneously. On the eighth day, the animals are sacrificed, and the uterine weight is determined.

  Assessment: Percentage inhibition of uterine weight is calculated using the following equation:

$$\text{Percentage inhibition in ovarian weight} = [1 - \text{XE} - \text{C})]/\text{E} - \text{C} \times 100.$$

where C indicates mean uterine weight from rats treated with.
vehicle, E for estradiol and XE indicates the mean uterine.
weight of rats treated with extract and estradiol.

38.4.4 Test for Progestational Activity

Determination of the progestational activity of compounds depends on:

- changes in the uterine endometrium of rabbits or rodents
- effects on ovulation in rats or rabbits
- the intrusion of corpus lutea formation in the ovary
- Effects on implantation and disturbance in the maintenance of pregnancy in rabbits, rodents, or hamsters.

38.4.4.1 Clauberg-McPhail Method

Principle: Development of endometrial glands in estrogenpretreated rabbits after progestin administration was first described by Clauberg in 1930 and examined by McPhail in 1934.

Assay procedure: Immature female rabbits (Yellow Silver or New Zealand strain-800–1000 g) are given a daily subcutaneous injection of 5.0 μg of 17β-estradiol dissolved in sesame oil for six days. For the next fivedays, different doses of test drug dissolved in sesame oil – 0.02, 0.08, and 2 mg progesterone per day or 0.01, 0.02, and 0.04 mg Medroxyprogesterone orally are given for each group of animals. Each group comprises 3–4 animals. The control group receives either the vehicle or 17β-estradiol only. Twenty-four hours after the last injection, animals are killed, uteri are dissected out, and frozen sections of the segment of the middle portion of one horn are prepared and examined for histological interpretation.

Assessment: Rabbits treated with 17β-estradiol have an increased uterine weight. For interpretation of progestational proliferation of the endometrium, the following scores are established:

0–Ramification of the uterus mucosa, but no proliferation (estrogen treatment only)
1–Slight proliferation of the uterus mucosa
2–Medium proliferation of the uterus mucosa, slight additional ramification
3–Pronounced proliferation of the uterus mucosa
4–Very pronounced proliferation of the uterus mucosa, pronounced ramification.

The increase in weight of the uterine horns is an unreliable index of progestational activity.

38.4.4.2 Endometrial Carbonic Anhydrase Activity

Principle: Carbonic anhydrase activity in the uterine endometrium depends upon the functioning of a corpus luteum, and hence, there exists a linear dose-response relationship between the dose of progestogens and carbonic anhydrase activity in rabbit endometrium. This principle is used for the screening of progestational compounds.

Assay Procedure: Estrogen primed immature female albino rabbits are injected with progesterone daily for fivedays. These animals are killed by cervical dislocation, and uteri are opened longitudinally. The endometrium is dissected, weighed, and homogenized with ten-fold volume in a glass homogenizer.

Assessment: After centrifugation for 15 min in 4 °C, carbonic anhydrase activity is estimated using the calorimetric method of Philpot and Philpot (1936), using bromothymol as an indicator. Thus carbonic anhydrase activity of the test and the standard can be established.

38.4.4.3 Deciduoma Reaction in Rats

Principle: Progestational compounds induce cell division and deciduoma formation in the traumatized uterus of ovariectomized rats. It is considered a classical assay of progestogens.

Assay Procedure: Ovariectomized adult female Sprague Dawley or albino rats weighing between 150 and 200 g areestrogen-primed with 0.5–1 μg of estradiol once daily s.c for fourdays. Then progesterone or the test compound is given to the rats for ninedays. On day 5, one uterine horn is exposed, and 1 mg of histamine dihydrochloride is injected into the lumen. Twenty-four hours after the last dose of the drug, animals are killed, uterine horns are cut off, weighed, and histologically examined.

Assessment: The degree of deciduoma formation is evaluated by the percent increase in the weight of the histamine-injected uterine horn as compared with the control horn. Dose-response curves and potency ratios are calculated.

38.4.4.4 Assays causing changes in ovulation

Principle: Progesterone has a biphasic effect – facilitatory and inhibitory effect on ovulation by affecting the pituitary gonadotropins.

Assay procedure: Facilitatory effect – Immature female rats (22 days old) are injected with 15 IU of pregnant mare serum gonadotrophin(PMS) s.c in 0.5 ml of saline followed by progesterone injection twodays later. The next day, rats are sacrificed, oviducts are dissected, and the number of ova estimation is done using a binocular microscope. Thus the number of rats ovulating is more in progesterone treated group compared to the group treated with PMS alone.

Inhibitory effect – In sexually mature female rabbits, ovulation occurs after mating, mechanical stimulation of the vagina, or after injecting copper acetate. In progesterone-treated mature female rabbits, the above-mentioned stimulus isgiven 24 h later. The animals are then sacrificed, and the number of ovulation points on both sides isestimated, which is found to be reduced.

Assessment: The effect of facilitation or inhibition of progesterone ovulation can be correlated with the effects of progesterone on the release of luteinizing hormone from the pituitary gland.

38.4.4.5 Pregnancy Maintenance Test

Principle: Maintenance of pregnancy requires the presence of progestogen. Ovariectomy done in rats during the first trimester causes pregnancy termination, whereas the same done in the second trimester does not cause abortion because of the placental source of progestin and estrogen.

Assay procedure: Mature Sprague Dawley female rats are made pregnant after mating with male rats overnight. Rabbit, Mouse, and Hamster can also be used for this test. After examining the uterus for pregnancy, the rats are ovariectomized on the eighth day of pregnancy. Immediately after ovariectomy, the test compound, progesterone 5–10 mg/day, is injected subcutaneously once daily with estradiol (0.1 μg/day) administration for potentiating progesterone effect.

Assessment: Prior to term, the animals are sacrificed, the uteri are examined for live young, and the number of implantation sites using which efficacy of the drug can be determined. The net success index is estimated for each group of animals.

$$\text{Net success index} = [\text{No.of live young/no.of mothers x 11}] \times 100.$$

38.4.5 Test for Anti-Progestational Activity

- McGinty Test
- *Principle:* Anti-progestational compounds inhibit the physiological effects of progesterone-like deciduoma formation in rats and increased proliferation of estrogen-primed endometrium of rabbits.
- *Assay procedure:* A local progestational test involving the direct injection of progesterone into a uterine horn of sexually immature female rabbits (750–900 g) primed with subcutaneous estradiol for sixdays. The positive control group and the test group receive estradiol 1 μg/kg/animal subcutaneously, whereas the negative control group is given only the vehicle (0.5% carboxymethylcellulose-CMC). On the seventh day, the uterus is exposed by laparotomy. The upper-middle segment of each uterine horn is ligated without disturbing its blood supply. The right uterine horn of all the groups receives only the progesterone 0.1 mg/ml. The left horn of the positive and negative control group receives only the vehicle, while the test group's left uterine horn receives the vehicle along with progesterone. The animals are allowed to recover, and threedays after the procedure, the animals were sacrificed. The uteri were dissected out and examined histologically for McPhail score.
- *Assessment:* Inhibition of progesterone-induced proliferation of estrogen-primed endometrium indicates anti-progestational activity.

38.4.6 Test for Anti-Implantation Activity /Abortifacient Activity

Assay procedure: Female rats of proestrus or estrous phase are kept with male rats of proven fertility in the ratio of 3:1. The female rats are examined onthe following morning for evidence of copulation. The female rats showing thick clumps of spermatozoa in vaginal smear will be separated from the male rats, and that day designated as day 1 of pregnancy. These rats are administered with the test drug orally from 1 to 7 days of pregnancy. Control rats received the vehicle (distilled water). On day 10 of pregnancy, the animals are laparotomised, and the uterine horns are examined for the number of implants and the number of corpus lutea(CL). After the complete gestation period of 21–23 days, the number of litters delivered iscounted.

Assessment: The anti-implantation activity is expressed as the percentage decrease in the number of uteri implants on day 10 of pregnancy. The number of resorbed implants from the existing number of implants will be recorded on day 18 for evaluating the early abortifacient activity.

Percentage of Pre-post- and anti-implantation activity are calculated using:

% Pre-implantation loss = [(No. of CL – No. of implants) / No. of CL] × 100% **Post-implantation loss** = [(No. of implants – No. of litters)/ No. of implants] × 100% **Abortifacient activity** = (Number of resorptions/number of corpus luteum) × 100.

38.4.7 Methods for Males.

38.4.8 Test for Androgenic and Anabolic Activity:

- Chicken comb method
- *Principle:* Growth of the chick comb has been identified as androgenic activity and extremely useful for isolation and structural elucidation of natural androgens.
- *Assay procedure:* The comb is measured using a ruler before the assay. Test substances and standard substances are given orally, intramuscularly, or subcutaneously for fiveconsecutive days. Ten chicks per dose level are an adequate number for this assay.
- *Assessment:* The chicks are killed using ether 24 h after the last dose of the test substance, and the combs are remeasured.
- *Modification of the method:* With the same array of procedures, capons can also be used for this assay with the increase in its weight and the length of its comb as evaluation. Nevertheless, the advantages of using chicks are
- Castration is not necessary
- Housing the bird for a prolonged period is not necessary
- The accuracy in weighing is greater than that of measuring with calipers or a ruler.
- Accessory sex organs of the Rodents
- *Principle:* Androgenic substances increase the seminal vesicles' weight, ventral prostate, and levator ani muscle in orchidectomized rodents.
- *Assay procedure:* The mouse test was more sensitive than the rat test. Nevertheless, the dissecting and weighing of target tissues is easier in rats than in mice. Hence 21–23 days old immature male Sprague-Dawley rats are used for this assay. Ten animals are used for each group- test, standard, and control group. The test group receives 0.5 ml of 0.5% carboxymethylcellulose or 0.2 ml sesame oil suspension subcutaneously daily for tendays. The standard group receives testosterone in 0.02, 0.1, and 0.5 mg per animal subcutaneously. At the same time, the control group is given vehicle only. The rats are killed with ether 24 h after the last dose. The ventral prostate, seminal vesicles, and levator ani muscle are dissected out gently. Their fluid contents are squeezed out.
- *Assessment:* The above-mentioned target organs are weighed, and they are compared with other groups of animals. Comparison is made between before and after assay. An increase in weight of the ventral prostate is a more sensitive index of androgenic activity than the increase in weight of the seminal vesicles.

The increase in weight of levator ani muscle is indicative of anabolic activity than androgenic activity.

- Nitrogen –balance
- *Principle:* This method is based on the principle that anabolic agents induce positive nitrogen balance in living organisms.
- *Assay procedure:* 25 day old castrated rats are placed on a restrictive diet to maintain the constant weight and given casein and brewer' yeast as nitrogen sources.
- *Assessment:* Total urinary nitrogen is measured before and after drug administration. The amount of nitrogen in the urine drastically decreases after the test drug administration, indicating the anabolic or nitrogen-retaining activity.

38.5 Test for Anti-Androgenic Activity

The same assay techniques used for the evaluation of androgenic compounds are used to demonstrate anti-androgenic activity. The antiandrogen is given simultaneously with the androgen and androgenic action like increasing the growth of capon's and chick's comb and increasing the growth of ventral prostate, seminal vesicle, and levator ani muscle in rodents are inhibited.

38.6 Invitro Methods

38.6.1 Methods for Females

38.6.1.1 Test for Estrogenic Activity

- Estrogen receptor binding assay
  - ***Principle:*** The estrogenic activity of the test compound is determined by their binding ability with the estrogen receptors. Estradiol-17β is used as reference compound. Mouse uteri or human endometrium serve as sources of estrogen receptors.
  - ***Assay Procedure:*** 18-day old female Swiss mice uteri' are removed and homogenized with 0.25 M sucrose buffer at 0 °C in 1:50 (w/v) of 10 mM Tris-HCl (pH 7.4), in a conical homogenizer. Menopausal women's endometrium is frozen within 2 h of hysterectomy and frozen with liquid nitrogen until use. This frozen endometrium is pulverized and homogenized in 1:5 (w/v) Tris-sucrose buffer and centrifuged for 1 h.

38.6.1.2 Techniques used to determine various parameters

- *Dextran-coated charcoal (DCC) adsorption technique:* A 100 μl aliquot of incubated cytosol is stirred for 10 min at 0 °C in a micro-titer plate with 100 μl of Dextran-coated charcoal(DCC) suspension (0.625% Dextran 80,000, 1.25% charcoal Norit A) and then centrifuged for 10 min at 800 *g*. The concentration of bound steroid is determined by measuring the radioactivity in a 100 μl aliquot of supernatant.

- *Determination of specific binding in mouse uterus cytosol as a function of steroid concentration, incubation time, and temperature.*
  - Triplicate aliquots of 125 μl of cytosol are incubated with 5 or 25 nM labelled steroid (estradiol-17β) either for 2 or 24 h at 0 °C or for 2 or 5 h at 25 °C in the absence (total binding) or presence (non-specific binding) of a 100-foldexcess of radio-inert steroid. The bound steroid is measuredby DCC adsorption.
- *Measurement of association rate at 0 °C.* 1 or 5 nM concentration of labeled steroid is added to cytosol maintained at 0 °C. Every 5 min for 1 h after adding the labeled steroid, a 100 μl aliquot is transferred into a 5000 nM radio-inert steroid solution in a microtiter plate to stop the reaction. Bound radioactivity is determined by DCC adsorption.
- *Measurement of dissociation rate at 25 °C.* The dissociation rate is measured by the isotopic dilution technique. Radio-inert steroid (2500 nM) is added to crude cytosol previously incubated with 5 nm labeled steroid for 15 h at 0 °C. After different incubation times at 25 °C, 100 μl samples are treated with DCC at 0 °C in order to determine bound radioactivity. Specific binding is evaluated by subtracting non-specific binding from total binding.

  Assessment:

  Relative binding affinity is calculated by plotting the percentage of radioligand bound in the presence of a competitor compared to that bound in its absence against the concentration of unlabeled competing steroid. The relative affinity of a test compound is established as the ratio of unlabeled radioligand concentration to competitor concentration, at 50% competition. This ratio is multiplied by 100.

  Association rate ($k + 1$) is calculated by the slope of the line.

$$k + 1\, t = (2.3/E0 - R0) \log (E\, R0/R\, E0).$$

where $E0$ and E represent free steroids and $R0$ and R.

free receptor at time $t = 0$ and time t, respectively.

Dissociation rate ($k - 1$) is calculated from the slope of the line.

$k - 1 = -2.3 \log B/B0$.

where $B0$ and B represent bound steroids at time $t = 0$.

and time t, respectively.

38.6.2 Test for Anti-Estrogenic Activity

- Aromatase inhibition
  - *Principle:* Inhibiting the enzyme aromatase would be the wise approach in suppressing the estrogen biosynthesissince that is the enzyme thatconverts androgen to estrogen. Test compounds which not only potently but the one which selectively inhibits aromatase enzyme are the ones considered as an optimal anti-estrogenic agent. Selective inhibition of estrogen biosynthesis

occurs when aromatase is inhibited without inhibiting the activity of other steroidogenic enzymes like the side-chain cleaving enzyme converts cholesterol to pregnenolone (the precursor for progesterone) would lead to a suppression of gonadal and adrenal steroidogenesis.

- *Assay procedure:* Adult female golden hamsters of 80–120 g kept in groups of 4–8 animals under controlled laboratory conditions were used for this experiment. Prior to the experiment, three consecutive fourdays estrus cycles are monitored bythe presence of a conspicuous vaginal discharge onday 1 of the cycle (= day of ovulation); day corresponded to proestrus. After decapitation, the ovaries were removed and quartered. The quarters are transferred into plastic incubation flasks with 2 ml of Kreb's Ringer bicarbonate salt (KBR) solution pH 7.6, containing 8.4 mM glucose. The flasks are gassed with O2/CO2 (95%/5%) tightly closed and placed in a shaker/water bath (37 °C) to incubate the fragments. The incubation media are replaced with fresh KBR after pre-incubation for 1 h. Estrogen production was stimulated by incubating adult female hamster ovarian tissuewith ovine LH. The production rates of estrogens (E), testosterone (T), and progesterone (P) weredetermined using radioimmunoassays to measure the amount of these steroids released into the incubation medium over a 4-hour incubation period.
- *Assessment:* The selectivity of aromatase inhibition was assessed by determining the IC_{50} s with which each inhibitor inhibited the production of E (end product), T (immediate precursor of E), and P (an early precursor of E). The data of control and test groupsare compared with suitable statistical analysis.

38.6.3 Test for Progestational Activity

- Progesterone receptor binding assay
  - *Principle:* Test compounds with progesterone activity can be assessed by theirability to bind with progesterone receptors. Progesterone receptors can be obtained from uteri tissue of estrogen primed rabbits and human uteri obtained after hysterectomy or from the cultured cells. The standard used is progesterone.
  - *Assay procedure:* The frozen homogenized human uteri obtained after hysterectomy are stored at −80 °C until use. Thehomogenates are then centrifuged at 105,000 *g* at 4 °Cfor 30 min in ice-cold buffer composed of 10 mM KH2PO4, 10 mM K2HPO4, 1.5 mM EDTA, 3 mM NaN3, 10% glycerol, pH 7.5 (PENG buffer). The supernatant is taken as cytosol. Thecytosol preparations are incubated with 3H-R 5020 asradioligand at a concentration of 8 nmol/l and increasingconcentrations (1 × 10–10 to 1 × 10–5 mol/l) of thecompetitor steroids overnight at 4 °C. Then, unboundsteroids are adsorbed by incubating with 0.5 ml of DCC(0.5% Norit A, 0.05% dextran T400 in PENG buffer) for

10 min at 4 °C. After centrifugation (10 min at1500 *g* at 4 °C), 0.5 ml of the supernatant is withdrawn and counted for radioactivity.

Modifications of the method: For screening procedures, homogenates of rabbit uteri may be used.

- *Assessment:* Relative binding affinity is calculated by plotting the percentage of radioligand bound in the presence of a competitor compared to that bound in its absenceagainst the concentration of unlabeled steroid. The ratio of unlabeled radioligand andcompetitor for 50% competition multiplied by 100 iscalculated for relative binding affinity.

Association rate:

(k + 1) is calculated by the slope of the line.

$$k + 1\, t = (2.3/E0 - R0) \log (E\, R0/R\, E0).$$

where $E0$ and E represent free steroids and.

$R0$ and R, free receptor at time $t = 0$ and time t, respectively.

Dissociation rate:

(k–1) is calculated from the slope of the line.

$$k - 1 = -2.3 \log B/B0.$$

where $B0$ and B represent bound steroids at time $t = 0$.

and time t, respectively.

38.6.4 Methods for Males

38.6.4.1 Emergent Spermatozoa Made Non-Functional

- Spermicidal activity

  Using eosin/nigrosin stain, a viability study was done, accounting for the percentage of live spermatozoa. Spermicidal drugs are diluted with normal saline, and serial dilutions are made in 0.2 ml of human seminal fluid with 1 ml of spermicidal solution and incubated at 37 °C for 30 min. Thin smears were then prepared and observed under a light micro- scope at ×400 magnification to assess the viability of the sperm. Viable sperm remained colorless while non-viable sperm stained red. Effective spermicidal agents kill the sperms.
- Immobilisation assay

  Semen ejaculates from normal subjects after 72–96 h of sexual abstinence are liquefied at 37 °C and subjected to routine semen analysis. Sperm count above 100 million/ml with normal morphology and progressive motility isused for the test. Adult male sheep' (Ram') cauda portion of the epididymis is isolated and filtered through a piece of cheese paper to get sperm suspension after mincing in 0.9% saline solution. In a 1:1 ratio, either ram' epididymal sperm suspension or human ejaculates are mixed thoroughly with different concentrations of drugs and incubated at 37 °C. The percentage of progressive sperm motility was evaluated visually at 400 × magnification. Motility estimates were performed from five different fields in each sample.

- Sperm revival test
  This test evaluates the spermicidal and immobilization capacity of the drugs by checking the sperm revival. The spermatozoa are washed in physiological solution and incubated at37 °C for 30 min after completion of the immobilization assay. Then the reversal of sperm motility is observed.

38.6.5 Test for Anti-Androgenic Activity

- Inhibition of 5-alpha reductase
  - *Principle:* The enzyme 5α- reductase converts testosterone to 5-dihydrotestosterone (DHT) in androgen-target tissues such as seminal vesicles, epididymis, and skin. In androgen-sensitive tissues, such as the skeletal muscles and central structures, testosterone mediates the androgenic stimulus.
  - *Assay procedure:* Prostate from various species such as humans, dogs, and rats with benign prostatic hyperplasia serves as the source of 5α- reductase enzyme source. Frozen human prostates obtained from benign prostatic hyperplasia patients are thawed, minced, and homogenized in 3 tissue volumes of medium A (20 mM potassium phosphate, pH 6.5, containing 0.32 M sucrose, 1 mM dithiothreitol, and 50 μM NADPH) with a Brinkmann Polytron and a glass homogenizer. The homogenate is centrifuged at 140,000 *g* for 60 min, and the pellets are washed with approximately threetissue volumes of medium A. In the same way, Mongrel dog prostatic particulates are also prepared. Male Sprague-Dawley rats' prostatic tissue is prepared as human tissue except for that medium A without NADPH is used throughout the procedures since the rat enzyme is stable without the co-enzyme.
  - Reaction solutions are prepared in duplicate tubes. Test compounds or standard inhibitors are added in 5 μl ethanol at concentrations between 10–9 and 10–5 M. The control tubes receive the same volume of ethanol. The reactions for the rat and dog enzymes are started by the addition of the prostatic particulates. The human prostatic particulate is premixed with NADPH before starting the reaction. The reactions are linear for at least 1 h at 37 °C and carried out for 10–30 min. After centrifugation at 1000 *g* for 5 min, the ethyl acetate phase is transferred to a tube, and The solutions are applied to Whatman LK5DF silica plates which areair-dried, and the chromatography is repeated. Nonradioactive steroid standards are located by UV and by spraying with 1% CeSO4/10% H2SO4 solution followed by heating. The radioactivity profiles are determined by scanning the plates and counting in a scintillation counter. 5α -Dihydrotestosterone is the only radioactive product for the rat and human enzymes. With the dog enzyme, 5α-dihydrotestosterone, 3–17β-androstanediol, androstane-3-17-dione, and androstenedione are formed. The radioactivities of the first threeproducts are combined for the calculation of the 5α -reductase activity. Androgenic activity of the test compounds isdetermined by theirability to bind with androgen receptors. Test compounds inhibiting the activity of 5α- reductase have anti-androgenic action.

– *Modifications of the method* Using human genital skin fibroblasts and simian COScells, specific inhibition of 5α-reductase type 1 has been observed.

Assessment: IC-50 (Inhibitory concentration) values are calculated based on at least fivedilutions of test preparations or standards.

Appendix

Phases of Menstrual Cycle

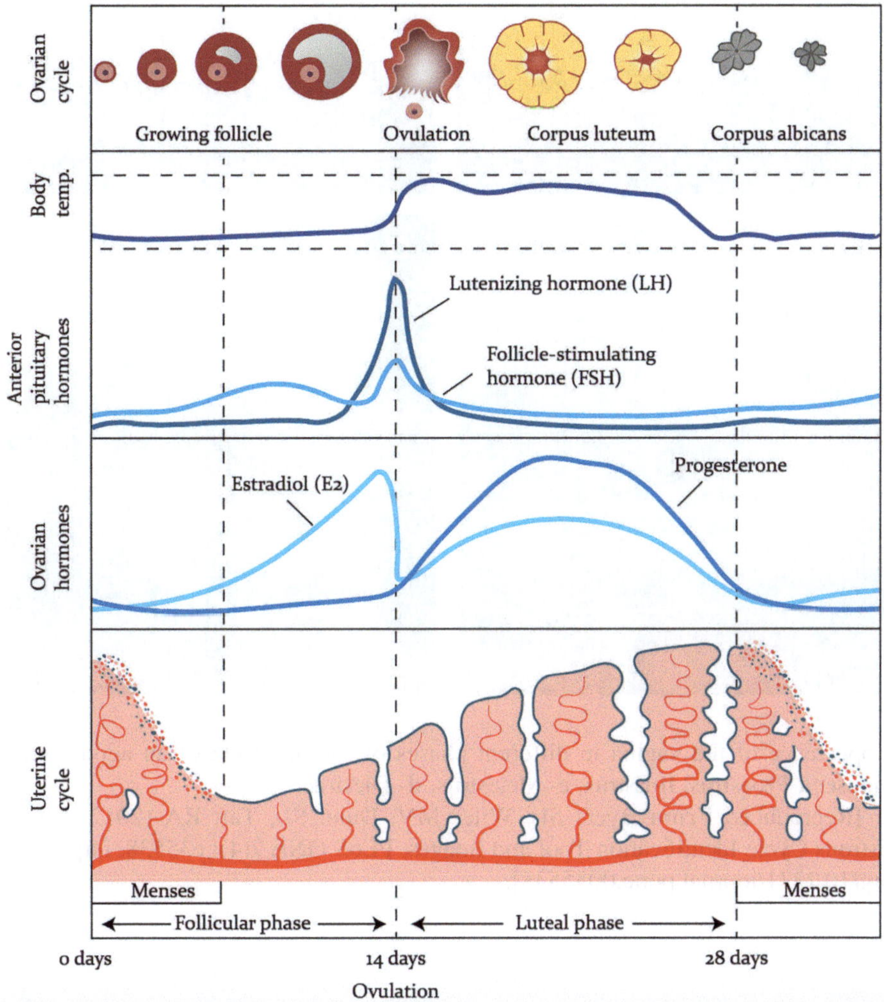

Image Source: Vios Fertility Institute.

Vaginal Opening Assay.

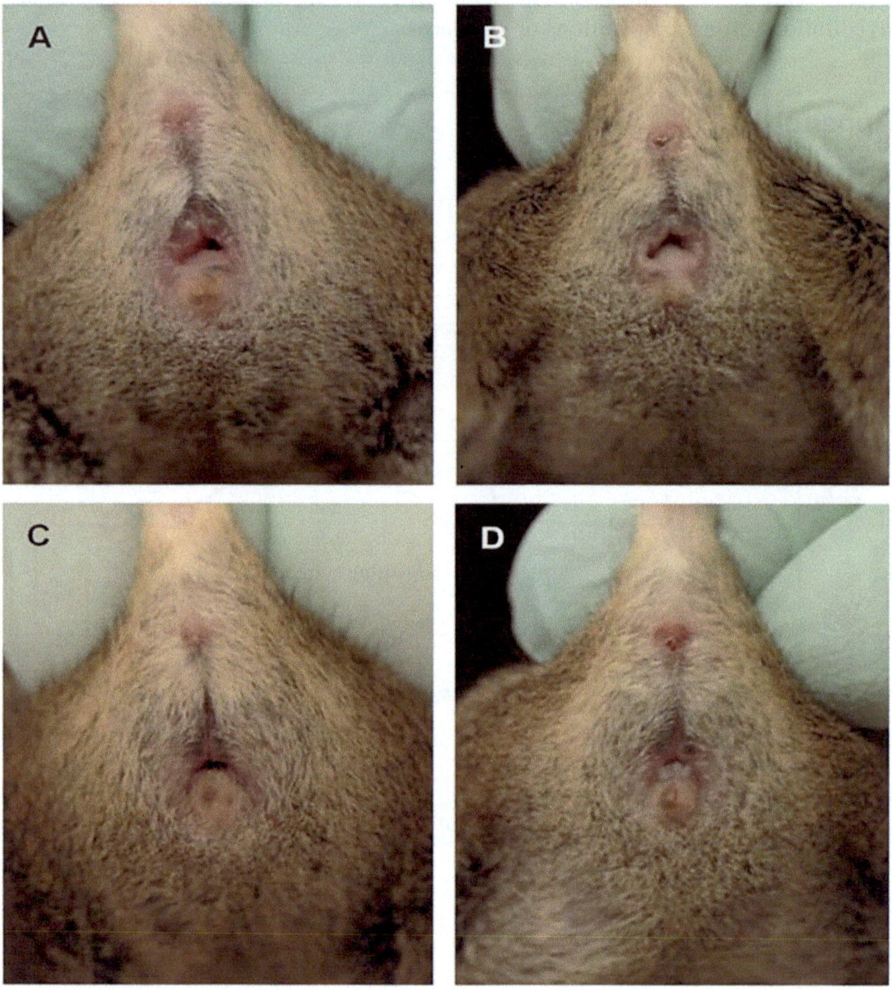

Appearance of the vagina in different phases of estrous cycle of an agouti strain mouse. **a**-Proestrus, **b**-Estrus, **c**- Metestrus, **d**- Diestrus.

[Reproduced **From** Byers SL, Wiles MV, Dunn SL, Taft RA (2012) Mouse Estrous Cycle Identification Tool and Images. PLoS ONE 7(4): e35538. https://doi.org/10.1371/journal.pone.0035538].

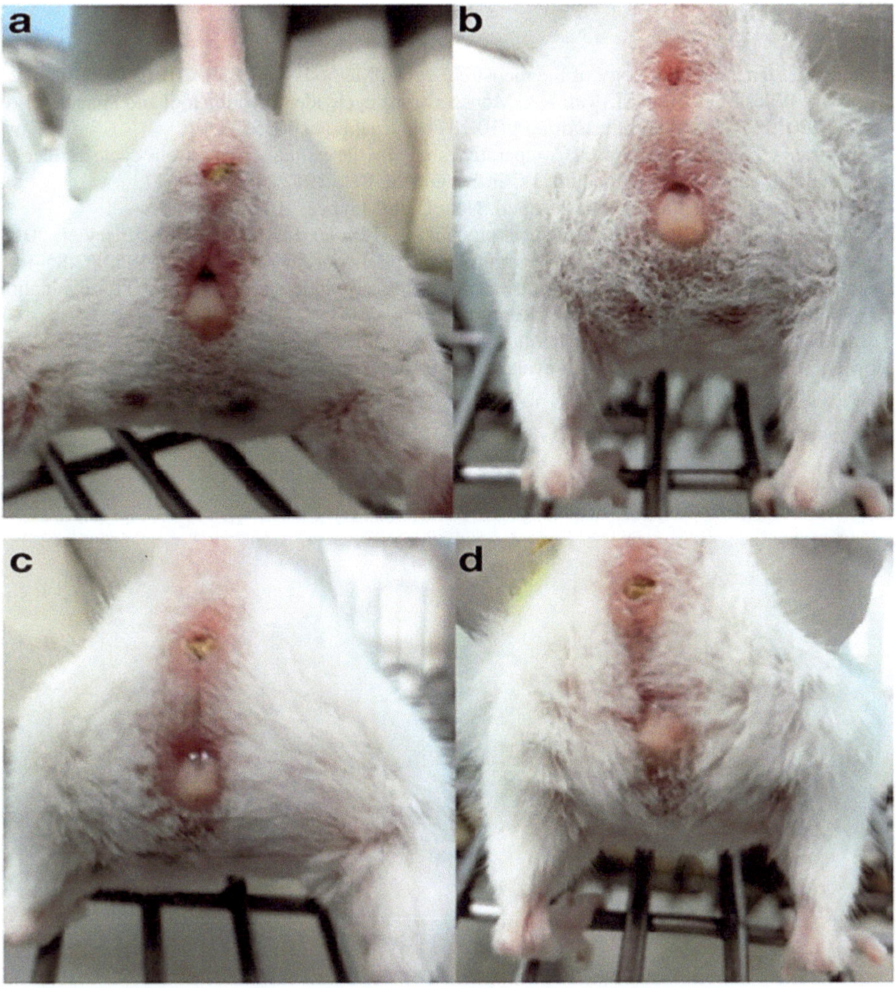

Appearance of the vagina in different phases of estrous cycle of a Swiss albino strain mouse. a-**Proestrus,** b-**Estrus,** c- **Metestrus,** d- **Diestrus**

[Reproduced From Ekambaram G, Sampath Kumar SK, Joseph LD. Comparative Study on the Estimation of Estrous Cycle in Mice by Visual and Vaginal Lavage Method. J Clin Diagn Res. 2017 Jan;11(1):AC05-AC07. doi: 10.7860/JCDR/2017/23977.9148. Epub 2017 Jan 1. PMID: 28273958; PMCID: PMC5324403.]

Bibliography

Barbieri RL. The endocrinology of the menstrual cycle. Methods Mol Biol. 2014;1154:145–69.

Byers SL, Wiles MV, Dunn SL, Taft RA. Mouse estrous cycle identification tool and images. PLoS One. 2012;7(4):e35538. https://doi.org/10.1371/journal.pone.0035538.

Ekambaram G, SKS K, Joseph LD. Comparative study on the estimation of estrous cycle in mice by visual and vaginal lavage method. J Clin Diagnostic Res. 2017;11(1):AC05–7.

Gupta SK. Drug screening methods. 3rd ed. New Delhi: Jaypee; 2016.

Guyton AC, Hall JE. Textbook of medical physiology. 12th ed. Philadelphia, PA: Elsevier; 2011.

Long JA, Evans HM. Determination of estrous cycle phases of rats. Braz J Biol. 1952;62:85–9.

Shah M, Singh R, Shah R, Kakar S. An overview of the current methodologies used for evaluation of anti-fertility agents. Asian Pac J Reprod. 2016;5:175–8.

Turner RA. Screening methods in pharmacology. New York: Academic; 1971. p. 75–119.

Vogel HG. Drug discovery and evaluation: pharmacological assays. 4th ed. New York: Springer; 2016.

Screening Methods for the Evaluation of the Drugs Acting on Posterior Pituitary, Adrenal Steroid, Testicular, Parathyroid, Ovarian, and Thyroid Hormones

39

Sushil Kiran Kunder

Abstract

Hormonal or endocrine screening is an important attribute of animal model screening. This chapter looks at the various endocrine screening techniques, both in vivo and in vitro – for example, tadpole tail culture and thyroidectomy models for drugs acting on thyroid system, sebum secretion and chicken comb methods for drugs acting on androgenic system, vaginal cornification and uterine weight methods for drugs having an effect on estrogenic system, McPhail or Clauberg test for drugs acting on the progesterone system, milk ejection method for the drugs acting on posterior pituitary system, adrenalectomy and liver glycogen testing models for the adrenal system, and electrolyte excretion method for the parathyroid system.

Keywords

Tadpole tail · Sebum secretion · Chicken comb · McPhail · Clauberg

39.1 Screening of Drugs Acting on the Thyroid System

A few of these techniques are detailed below (Table 39.1).

39.1.1 Tadpole Tail Culture Method

Structures of T3 and T4 of tadpole are similar to human T3 and T4, making this a good model for in vitro screening. The process of metamorphosis is dependent on

S. K. Kunder (✉)
Novo Nordisk India Ltd., Bengaluru, Karnataka, India

M. Lakshmanan et al. (eds.), *Introduction to Basics of Pharmacology and Toxicology*, https://doi.org/10.1007/978-981-19-5343-9_39

Table 39.1 Screening models for drugs acting on the thyroid system

In vitro methods	In vivo methods
Lipogenic enzyme assay	Thyroidectomy model
Tadpole tail culture technique	Iodine release inhibition
	Anti-goitrogenic activity
	Reduction in tensile strength

thyroid hormones. Tadpoles are anesthetized, and 0.6 cm of their tail is cut. The length of the tadpole tail is checked and compared between the test and control groups.

39.1.2 Thyroidectomy Model

Thyroidectomized animal is the preferred model for evaluating thyroid hormones and analogs. Rats, mice, and guinea pigs are preferred. The fur on the neck of the animal is shaved, and the neck is disinfected. A median incision is done. The maxillary lymph nodes and salivary glands are pushed to the side. Muscles covering the trachea are split. The isthmus is separated, and blood vessels are ligated. The thyroid gland is removed.

39.1.3 Iodine Release Inhibition

Thyroxine inhibits the release of I-131 from the thyroid gland. Male SD rats are administered I-131. Radioactivity is determined after 40 hs under the influence of ether anesthesia. This forms the baseline. Diet is changed to diet having propylthiouracil (0.03%). Test and standard are administered daily for 4 days. I-131 is plotted versus logarithmic dose followed by calculation of potency ratios.

39.1.4 Anti-Goitrogenic Activity

Secretion of thyroid hormones can be blocked using exogenous compounds possessing goitrogenic activity. Reduced thyroid hormones stimulate TSH secretion, causing thyroid gland enlargement. Six-week-old Wistar rats are treated for 7 weeks with propylthiouracil and low iodine chow to induce goiter. The weight and size of the thyroid gland are measured and compared. Thyroidal substances prevent goitrogenesis.

39.1.5 Reduction in Tensile Strength

Thyroid hormones in the short term reduce the tensile strength of skin, tail tendons, and epiphyseal cartilage. Reduction in tensile strength is dose-dependent. On the other hand, long-term treatment increases these measures. Sprague Dawley rats are injected with thyroid hormones subcutaneously. After 1 day, animals are sacrificed, and tensile strengths are measured.

39.2 Screening of Drugs Acting on the Androgenic System (Table 39.2)

39.2.1 Sebum Secretion in Rats

Sebum secretion from the sebaceous gland is directly proportional to the dose of androgens. Anesthetized castrated male rats are suspended into the solvent. Ears should be above the level of solvent. After a minute, they are moved up and down and then shifted to the second solvent. Rats are cleaned and put in their cages. Solutions are weighed and dried. The amount of lipid extracted is calculated.

39.2.2 Castration Technique

Young male rats are anesthetized, following which a small transverse incision is made over the symphysis. Testes are pushed into the abdominal cavity, and the vessels and ductus deferens are crushed (using artery forceps). Testes are then cut with the epididymal fat tissue.

39.2.3 Chicken Comb Model

It is an important model to demonstrate the androgenic activity of a substance as androgens promote the growth of chicken comb. White Leghorn chicks (single comb) are divided into groups and injected with test and standard preparations parenterally once in 3 days. Body weights are noted at weekly intervals. In the end, the white leghorn chicks are sacrificed, and their combs are measured for size and weight.

Table 39.2 Screening models for drugs acting on the androgenic system

In vitro methods	In vivo methods
Sebum secretion technique	Castration model
Receptor binding assays	Chicken comb method
	Nitrogen retention test
	Growth of secondary sex organs
	Change in beta-glucuronidase activity

39.2.4 Nitrogen Retention Assay

Adult male Albino rats are castrated and fed with a laboratory diet till they weigh around 350 g. They are then put on a liquid diet with 350 g of nitrogen as one of the ingredients. Once the rats are accustomed to this diet, urine samples are collected and assayed for nitrogen. There is usually an increase in the bodyweight of the rats. In addition, there would be nitrogen retention, as shown by decreased nitrogen content in the urine.

39.2.5 Growth of Secondary Sex Organs

The action of androgen on secondary sex organs is dose-dependent. Male immature castrated Sprague Dawley rats are divided into the test, control, and standard (testosterone) groups. They receive the preparation daily for 10 days, following which the rats are sacrificed, and the levator ani, ventral prostate, and seminal vesicles are dissected out. Organ weight to body weight ratio is calculated and compared.

39.2.6 Change in Beta-glucuronidase Activity

The alteration in testosterone levels is associated with changes in renal beta-glucuronidase activity in experimental animals. Therefore, renal beta-glucuronidase activity can be used for androgen bioassay.

39.3 Screening of Drugs Acting on the Estrogenic System (Table 39.3)

39.3.1 Castration of Female Rats

Sprague Dawley rats are used. Rats are anesthetized using ketamine, acepromazine, and xylazine through the intraperitoneal route. A dorsal midline incision is given, and bilateral ovariectomy is done. This model is commonly used to analyze the tissue-specific effects of selective estrogen receptor modulators (SERMs).

Table 39.3 Screening models for drugs acting on the estrogenic system

In vitro methods	In vivo methods
Alkaline phosphatase assay	Castration model
Receptor binding assays	Vaginal cornification
	Uterine weight assessment
	Chick oviduct method

39.3.2 Vaginal Cornification

Immature female Wistar rats are ovariectomized. Standard and test drugs are administered for 7 days orally. During these 7 days, the vaginal opening is observed daily. After a week, the vaginal smear is taken, and after 24 h, rats are anesthetized; the uterus is removed, cleaned, and weighed. An increase in the size of the vaginal opening indicates that the test substance has estrogenic activity. In addition, if only cornified epithelial cells are present, it is indicative of estrogenic activity.

39.3.3 Increased Uterine Weight

Female Sprague Dawley rats are used. Their ovaries are removed. Estradiol-treated rodents show uterine hypertrophy and an increase in uterine weight.

39.3.4 Chick Oviduct Method

One-week-old pullet chicks are used and administered test/standard drugs for 6 days. Chicks are sacrificed, and the oviducts are weighed. Estrogens increase the weight of the chick oviduct.

39.4 Screening of Drugs Acting on Progesterone System (Table 39.4)

39.4.1 McPhail/Clauberg Test

This model demonstrates the effect of progestational effects of the test compound on the uterus. The endometrial cells undergo proliferation so as to aid the implantation of the embryo. Female immature rabbits are primed with estrogen for 6 days, followed by administration of standard/test compounds for 5 days. Animals are sacrificed 4 days later; uteri are removed and prepared for histological examination. Scores are mentioned below:

- 0–uterine mucosal ramification, no proliferation is seen.
- 1–minimal mucosal proliferation.
- 2–moderate mucosal proliferation.

Table 39.4 Screening models for drugs acting on the progesterone system

In vitro methods	In vivo methods
Receptor binding assays	McPhail / Clauberg test
	Rat decidualization test
	Rat uterine C3 model
	Rat ovulation inhibition model

- 3–pronounced mucosal proliferation.
- 4–pronounced proliferation and ramification.

Each group is scored. DRC is plotted. The potency ratio is calculated.

39.4.2 Rat Uterine C3 Model

Female ovariectomized Sprague Dawley 60-day old rats are used. Rats are administered standard or test compound or vehicle for 2 days. On the second day, additionally, they are treated with ethinylestradiol orally. The animals are sacrificed a day later. The uterus is removed. Mesentery and fat are stripped. The uterus is weighed and frozen. RNA is isolated, and Northern blot analysis is done. C3 is quantified.

39.4.3 Rat Decidualization Test

Female ovariectomized Sprague Dawley 60-day-old rats are used. Test and standard compounds are given for 3 days. A day later, one horn of the uterus is decidualized (D). This is done by scratching of anti-mesometrial epithelium with the help of a blunt needle. The other horn (unscratched) serves as control (C). Rats are sacrificed, and weights of control and decidualized horns are checked. D/C ratio is measured, which gives the decidual response.

39.4.4 Rat Ovulation Inhibitionspiepr146 model

Progestins inhibit luteinizing hormone (LH) surge during ovulation, and this principle is utilized in this model. Female mature rats are used. They are synchronized to the estrous stage using luteinizing hormone releasing hormone (LHRH). 8 days later, test compounds are given for 4 days. Oviducts are removed. They are placed on glass slides and visualized under the dissecting microscope. Ova are counted.

39.5 Screening of Drugs Acting on Posterior Pituitary/Oxytocin System

39.5.1 Milk Ejection Method

Compounds with oxytocic properties lead to the milk ejection. This method is based on this principle. Rabbits/rats can be used in this method. Rabbits (female) are anesthetized, and one of the ducts of their nipples is cannulated. This cannula is attached to the polygraph. Standard/test drugs are administered through the jugular vein, and potency ratios are calculated for test substances.

39.5.2 Chicken Blood Pressure Method

Oxytocin is known to cause a transient reduction in the blood pressure of birds. White Leghorn chicken is anesthetized with pentobarbital sodium. The ischiadic vessels and crural veins are exposed. The ischiadic artery is cannulated, which is connected to a transducer. The test and standard drugs are administered through the crural veins. Potency ratios are calculated.

39.5.3 Isolated Uterus Method

Guinea pig or rat uterus is used in this method. Animals were administered estradiol benzoate prior to this experiment. Uterine horn is dissected and suspended in a tissue bath with DeJalon's solution at a temperature of around 32 °C. DRC of test / standard compounds are recorded, and potency ratios calculated.

39.6 Screening of Drugs Acting on the Adrenal System

The adrenal cortex releases steroids into circulation. These steroids may be classified as glucocorticoids and mineralocorticoids.

39.6.1 Screening of Glucocorticoids (Table 39.5)

39.6.1.1 Adrenalectomy in Rats

Male Wistar Rats weighing around 200–240 g are used. Ether anesthesia is administered. The retroperitoneal cavity is opened (transverse incision at costovertebral line 5 mm in length). Adrenal glands are exposed and removed entirely.

39.6.1.2 Atrophy of Adrenal, Thymus Gland, and Spleen and Reduced Lymphocyte Count in Rats

High-dose cortisol in newborn rats produces significant atrophy of the adrenal and thymus gland along with the marked reduction in lymphocytes. Different corticoids produce different effects. While betamethasone and dexamethasone produce atrophy of the thymus gland and spleen, deoxycorticosterone and cortisol reduce the corticosterone content of the adrenals.

Table 39.5 Screening models for glucocorticoids

In vitro methods	In vivo methods
Receptor binding assays	Adrenalectomy in rats
	Organ atrophy and reduction in lymphocyte count in rats
	Liver glycogen test

39.6.2 Screening of Mineralocorticoids (Table 39.6)

39.6.2.1 Electrolyte Excretion

Mineralocorticoids increase potassium excretion and sodium retention. Sprague Dawley male rats are adrenalectomized. After the fourth day of adrenalectomy, water is administered via Ryle's tube, followed by oral NaCl (0.9%). The test compound is then injected, and urine is collected and measured. They are then diluted and analyzed using a flame photometer for sodium. The potency ratio is calculated by comparing the standard and test dose-response curves.

39.7 Screening of Drugs Acting on the Parathyroid System (Table 39.7)

39.7.1 Serum Calcium Increase Model

This model can be performed in dogs, rabbits, and rats. Male Wistar rats are anesthetized using pentobarbital, and parathyroidectomy is done by cauterization. A week later, a basal reading is taken. Test/control and standard are administered. About 21 hours later, blood samples are taken. Calcium is measured using flame photometry. An increase in calcium levels is measured for each rat.

39.7.2 Serum Phosphate Decrease Model

Parathyroid hormone administration leads to a decrease in serum phosphate level, which forms the principle of this model. Male Wistar rats are used. Only water is given to the rats during the experiment phase. Baseline and post-treatment (test/ control/ standard) phosphate levels are checked. DRC is plotted, and potency ratios are calculated.

Table 39.6 Screening models for mineralocorticoids

In vitro methods	In vivo methods
Receptor binding assays	Electrolyte excretion

Table 39.7 Screening models for drugs acting on parathyroid system

In vitro methods	In vivo methods
Receptor binding assays	Serum calcium increase model
	Serum phosphate decrease model
	cAMP release model

39.7.3 cAMP Release Model

Parathyroid hormone releases cAMP from bones, and this forms the basis of this model. Wistar rats are anesthetized, and their femur is removed. Below the neck of the femur, a hole is made. A needle is inserted through this hole. Bone is placed in the liver perfusion apparatus and perfused using Krebs' Ringer Bicarbonate. cAMP levels are measured by radioimmunoassay before and after administration of test/ standard. DRC is plotted, and potency ratios are calculated.

Bibliography

Gallo D, Zannoni GF, Fabrizi M, De Stefano I, Mantuano E, Scambia G. Comparative effects of 17β-estradiol and phytoestrogens in the regulation of endometrial functions in the rodent uterus. J Endocrinol Investig. 2008;31:48.

Gawienowski AM, Stadnicki SS, Stacewicz-Sapuntzakis M. Androgenic properties of gibberellic acid in the chick comb bioassay. Experientia. 1977;33(11):1544–5.

Gupta SK. Drug screening methods. 3rd ed. Jaypee Brothers Medical Publishers; 2016. p. 585–8.

Jawaid T, Awasthi A, Kamal M. Estrogenic activity of a hydro-alcoholic extract of *Bambusa arundinaceae* leaves on female Wistar rats. J Adv Pharm Technol Res. 2015;6(1):19–24.

Laezza C, Mazziotti G, Fiorentino L, Gazzerro P, Portella G, Gerbasio D, et al. HMG-CoA reductase inhibitors inhibit rat propylthiouracil-induced goiter by modulating the ras-MAPK pathway. J Mol Med. 2006;84:967.

Lennon HD, Saunders FJ. Anabolic activity of 2-oxa-17α-methyldihydrotestosterone (oxandrolone) in castrated rats. Steroids. 1964;4(5):689–98.

Lundeen SG, Zhang Z, Zhu Y, Carver JM, Wnneker RC. Rat uterine complement C3 expression as a model for progesterone receptor modulators: characterization of the new progestin trimegestone. J Steroid Biochem Molecul Biol. 2001;78:137–43.

Malarvizhi D, Mathur PP. Renal beta glucuronidase activity is a bioassay of serum testosterone levels in rats. Indain J Exp Biol 1996;34(6):582–583.

Ronsini ML, Collares EF. Effects of adrenalectomy and adrenal enucleation on liquid gastric emptying in rats. Braz J Med Biol Res. 2000;33(9):1047–51.

Vogel HG, editor. Drug discovery and evaluation: pharmacological assays. 3rd ed. Berlin: Springer; 2008. p. 1772.

Screening Methods for the Evaluation of Analgesics, Anti-Inflammatory Drugs, and Antipyretics

40

Avinash Arivazhahan

Abstract

Several drugs and agents in clinical practice have been found to possess mild to moderate analgesic, anti-inflammatory and antipyretic activity. This makes it essential for pharmacologists to understand the various pre-clinical models and methods available for such screening. Numerous in vitro methods have been highlighted in this chapter. In addition, focus is maintained on the in vivo models – hot plate method, tail flick technique, tail immersion method for analgesic activity assessment to mention a few, paw oedema method, granuloma pouch method, cotton wool granuloma technique for anti-inflammatory screening, and Brewer's yeast suspension method for antipyretic activity screening.

Keywords

Hot plate · Tail flick · Tail immersion · Paw oedema · Carrageenan

40.1 Screening for Analgesic Activity

Pain is arguably the most common symptom that instigates the patient to look for treatment options. Pain can be of various types and grades clinically. Since pain is a common phenomenon in most animals, testing the same in animals may be considered homologous to the human reaction in terms of time and intensity. As with other disease models, pain models can also be of in vitro or in vivo types. They can further be divided into models for central and peripheral analgesic activity.

A. Arivazhahan (✉)
Sanofi India Ltd., Chennai, Tamil Nadu, India

M. Lakshmanan et al. (eds.), *Introduction to Basics of Pharmacology and Toxicology*, https://doi.org/10.1007/978-981-19-5343-9_40

40.1.1 In Vitro Models for Analgesic Screening

In vitro models for analgesic activity mostly pertain to the mediators of central pain pathways like opioids, cannabinoids, and vanilloids.

40.1.1.1 ^{3}H-Naloxone Binding Assay

The whole-brain (except cerebellum) of Wistar rats is removed after decapitation and centrifuged to get a homogenized tissue isolate. Different concentrations of the test compound are then made to react with the tissue to determine the binding potency. This analysis gives us the binding levels to the opioid receptors. To check for the type of binding, sodium is added to the centrifuged isolate. Sodium has the property to enhance the binding of receptor antagonists and reduce the binding of receptor agonists. Hence, the amount of shift happening when sodium is added gives the estimate of the type of binding. The higher the shift, the higher is the agonistic property of the agent.

40.1.1.2 ^{3}H-Dihydromorphine Binding Assay

This technique is similar to the ^{3}H-naloxone binding assay. However, this test can be considered highly specific for the μ receptors (distributed mainly in the supraspinal areas). Compounds that displace or prevent the binding of ^{3}H-dihydromorphine to the receptors are those with significant μ-receptor binding action and thus possess analgesic activity. The tissue used is the same (rat brain except for the cerebellum).

40.1.1.3 ^{3}H-Bremazocine Binding Assay

This test is a specific binding assay for the kappa receptors (distributed chiefly in the spinal cord). Guinea pig cerebellum is preferred for this technique since the sensitivity and specificity for kappa receptors are very high.

40.1.1.4 Enkephalinase Inhibition Assay

Rat kidney or rat striatal tissues are preferred for this method. The homogenized isolates are subjected to different concentrations of the agent to be tested. A comparison of data with the standard is made to evaluate the binding potential.

40.1.1.5 Nociceptin Receptor Binding Assay

Nociceptin is an endogenous agonist of the ORL_1 (Opioid-Receptor-Like) receptor. The downward cascade pathways of the activation of this receptor are similar to those of opioid receptor activation. Hence, the nociceptin receptor is sometimes included under the opioid receptor subgroup under the name of OP_4 receptor. Guinea pig brain (without cerebellum) is preferred for this binding assay. Chinese hamster ovary cells may also be employed for this assay. Comparison versus the standard gives the estimate of the actual binding potency.

40.1.1.6 Cannabinoid Receptor Binding Assay

Cannabinoid receptor agonists function as analgesic agents similar to opioid agents but can be safer since they do not affect respiratory function. Binding assays can be non-specific or specific to CB_1 and CB_2 receptors. The rat brain cortex is used for these binding assays. Guinea pig brain is an alternative. To test individual receptor subtypes, mice lacking one of the two receptors can be in-bred and used for the assays.

40.1.1.7 Vanilloid Receptor Binding Assay

Capsaicin has recently gained a lot of importance for its analgesic properties. This analgesic property has been identified to be mediated via vanilloid receptors. Chinese hamster ovary cells transfected with rat vanilloid receptors are used for this binding assay. Rat spinal cord cells have also been employed for this method.

40.1.1.8 VIP Receptor Binding Assay

Vasoactive Intestinal Polypeptide (VIP) is a neuropeptide molecule that plays a role in both central and peripheral algesia mediation. Its function is supplemented by the analogous pituitary adenylate cyclase-activating peptide (PACAP). Chinese hamster ovary cell lines that express the different VIP and PACAP receptors are used for binding assay.

40.1.1.9 Bioassay Techniques for Peripheral Mediators of Central Analgesia

Bioassay of nociceptin in the peripheral tissues can be evaluated by bioassay techniques using tissues like vas deferens of rats, mice, or rabbits, ileum, or renal pelvis of guinea pigs.

40.1.2 In Vivo Models for Analgesic Screening

40.1.2.1 In Vivo Models for Central Analgesic Activity

(i) *Hot plate method*

The hot plate method is one of the most commonly used methods for evaluation or screening of centrally acting analgesic agents (peripherally acting analgesics are not sensitive to this method). Although this method was first proposed in 1944 by Woolfe and MacDonald, it was popularized by Eddy and Leimbach in 1953, thus giving it the name "Eddy's hot plate method." This method is based on the principle that paws of rodents like rats and mice are sensitive to high temperatures, characterized by jumping, withdrawal, or licking of the paws. Time taken for these reactions before and after test drug administration is noted to evaluate the analgesic property. The hot plate (or Eddy's analgesiometer) is maintained at around 55 °C. The animals are placed on the surface with pre-determined cutoff durations (usually set at 15 s to avoid damage to the animal's paws). Animals that do not react within 6–8 s at baseline are excluded from the test (Fig. 40.1).

False-positive results may be obtained with drugs that possess significant muscle relaxant or sedative properties. Also, drugs with mixed opiate

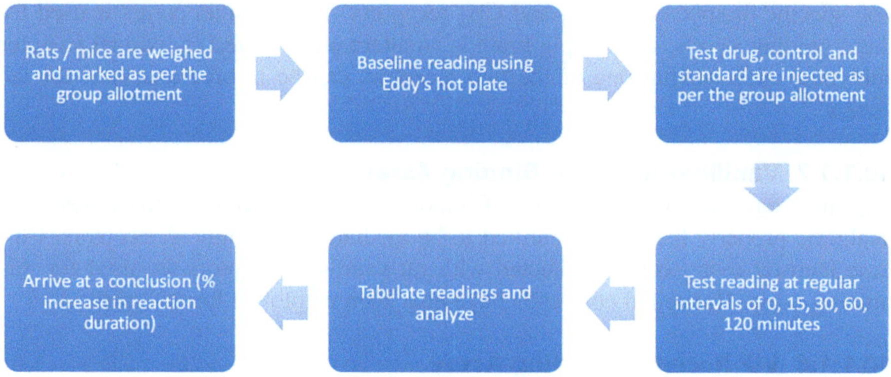

Fig. 40.1 Methodology of a hot plate technique

agonistic-antagonistic activity may show unreliable and inconsistent results (they can be ideally tested at a lower temperature of around 49–50 °C).

A popular modification of this hot plate method is that the temperature is continuously increased in phases instead of maintaining it at a constant temperature of 55 °C.

(ii) *Radiant heat method*

The tails of rats and mice function as temperature regulators, in general. The radiant heat method employs a heat source focused on the animals' tail, thus inciting flicking of the tail or an escape reaction. Restrainers or cages are used to contain the rat or mouse's body, leaving the tail exposed. A light source beam is focused on the proximal one-thirds of the animal's tail, thus creating a localized heat source. The rat or mouse usually exhibits a biphasic response: an early tail-flick (which is considered a spinal reflex) and a late escape reaction (which can be genuinely called a central reflex). In the most real sense, drugs that abolish this late escape reflex can be termed as those with significant central analgesic activity (Fig. 40.2).

The same technique is also employed rarely on rat ears, rat paws, and rabbit ears.

(iii) *Tail flick method*

Tail flick analgesiometers (or tail-flick meters) have been designed to assess tail flicking reactions to mechanical or electrical stimulation. Again, a cutoff duration of 10–20 s is maintained to avoid damage to the tail (Fig. 40.3).

As described earlier, the tail-flick cannot be considered as a genuine central reaction, as it can also be elicited by peripherally acting analgesic agents when given at high doses. However, this test can be considered a preliminary screening technique as it is simple, rapid, and easy to perform.

(iv) *Tail clip method*

The tail clip method (described by Haffner) is one of the earliest methods elucidated for analgesic activity screening. It is a straightforward method wherein a Haffner's artery clip is placed at the root of the animal's tail

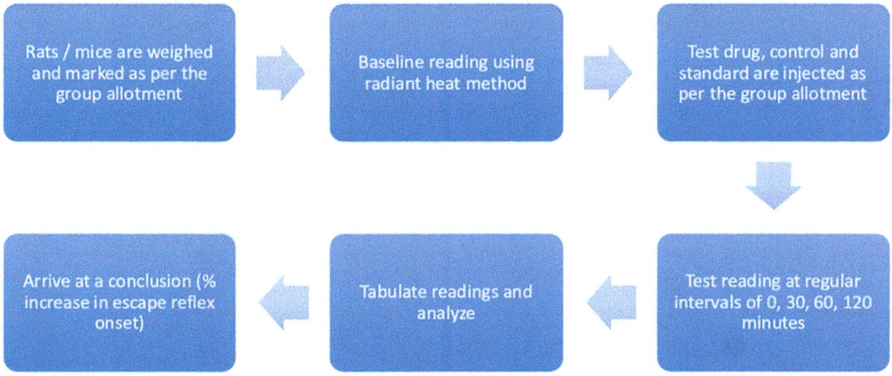

Fig. 40.2 Methodology of the radiant heat method

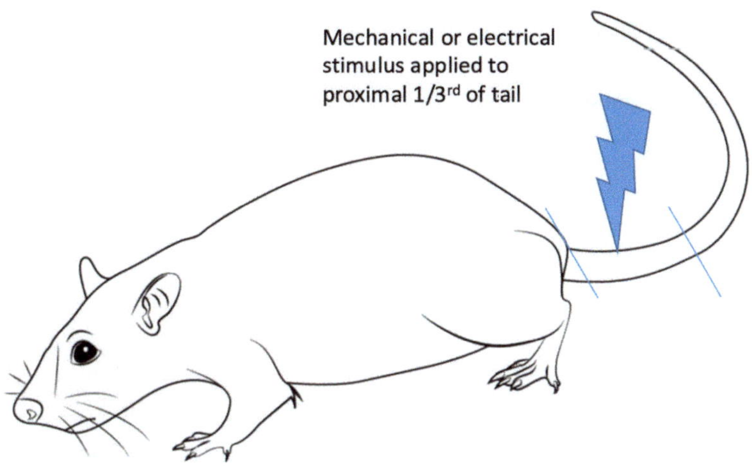

Fig. 40.3 Tail flick method

(mice or rats). The rat or mouse responds to this painful stimulus by biting the clip or the tail close to the clip. This positive reaction is prolonged once an analgesic agent is administered. Peripherally acting analgesics do not have any effect in this technique (Fig. 40.4).

(v) *Tail immersion method*

Here, the rats or mice are kept in a restrainer with their tails hanging outside. The tails (distal one-third) are dipped in a hot water bath (heated to °C). The reaction from the animal is the quick withdrawal of the tail (similar to a flick). However, central analgesics tend to prolong or inhibit this withdrawal. A cutoff time of 15 s is maintained to prevent injury to the tail. Tabulated data with time durations recorded before and after administration of the test agent are compared, and results are drawn.

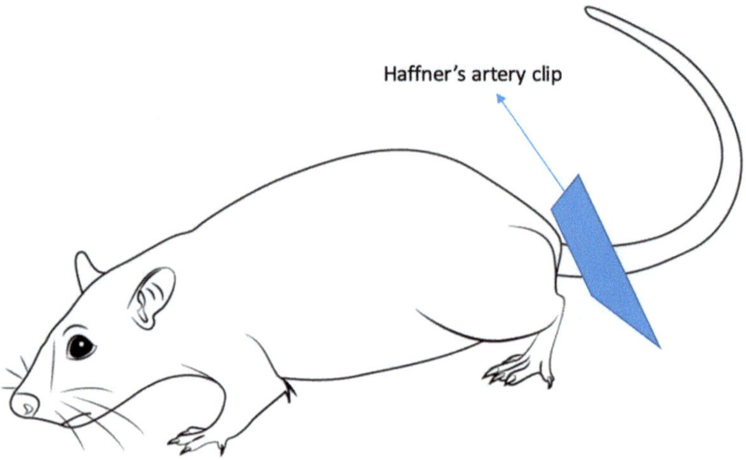

Fig. 40.4 Haffner's tail clip method

Fig. 40.5 Electric grid shock method

Another modification of this method is that cold stimulus (ethylene glycol) is used instead of heat. However, the sensitivity was found to be lower for cold stimulus as compared to heat.

(vi) *Grid shock method*

A closed box (with a transparent glass or plastic top) is divided into grids created by stainless steel wires. The grids are designed to deliver increasing intensities of electric shock. The animals (rats or mice) placed in the grid box react to the electric stimulus by either vocalizing the pain or increasing mobility, or jumping. The threshold at which this reaction is elicited is displayed on the attached oscilloscope (Fig. 40.5).

Centrally acting analgesics are known to increase the threshold time observed at baseline. An advanced version of this method is the "learned grid shock method," wherein the boxes are provided with a lever or a switch to regulate the electric current's intensity. Once the rats learn that pressing the lever brings more comfort to them, they start pressing it once the electric current crosses the threshold. This helps to understand the tolerance of the animals to the effect of central analgesics.

(vii) *Tooth pulp stimulation method*
Rabbits are used for this technique, wherein the pulp cavities are exposed, and electrodes are placed. Stimulation of the pulp tissue results in licking, chewing, and the flick of the head. The threshold at which these reactions are exhibited is increased in response to the administration of centrally acting analgesics. The threshold time duration before and after drug administration is measured, and results are drawn.

(viii) *Formalin paw method*
The formalin paw model is considered as a chronic pain model for centrally acting analgesic agents. Here, a small volume (roughly 0.05 ml) of formalin solution (10%) is injected into one of the rat's paws. Mice may also be used instead (5% formalin at a dose of 0.02 ml is preferred). Positive reactions are licking or biting the injected paw or sparing that paw when walking around. Successive observations are done after the administration of the test and control agents. Agents that can prevent these positive reactions or alleviate the same are those with significant analgesic activity. Some experts report that the pain reaction obtained in rats or mice can be termed biphasic. While central analgesics inhibit both phases, peripheral analgesics may possess the ability to inhibit the second phase of pain. Instead of injecting in the paw, upper lip injections have also been demonstrated in rats.

(ix) *Monkey shock titration model*
This model cannot be called a screening test in essence since it is considered a confirmatory model with higher specificity. The tail of the monkey is shaved in two spots, where electrodes are attached. A pedal or switch is provided for the monkeys to self-titrate the electric current's intensity. The test compound's ability to increase the threshold confirms the analgesic activity present in the compound compared with control or standard.

(x) *Migraine pain models*
Rats and mice are not considered to be ideal models for migraine-like pain. Instead, cats are preferred. Anesthetized and ventilated cats are subjected to craniotomy, following which electrodes are placed in the superior sagittal sinus and cervical spinal cord (at the level of C2). Drugs with potential utility in managing migraines are known to inhibit 'spontaneous' and 'evoked' firing from the cells.

Other models for migraine pain include constriction of carotid arteriovenous anastomosis (in anesthetized pigs), trigeminal neurogenic plasma extravasation (in anesthetized rats), inhibition of CGRP release (in anesthetized rats), and genetic models (like R192Q-FHM knock-in mice).

(xi) *Neuropathy models*
Chronic nerve constriction techniques are the most commonly used model for neuropathy. Rats are anesthetized, and both sciatic nerves are exposed. One sciatic nerve is ligated, and the other is just mobilized. The skin is closed on

both sides, and the rats are monitored over the next few days to check for the onset of paw or foot drop. Once this is seen, the animals are then tested for pain sensitivity using radiant heat produced using a high-intensity halogen lamp (thermal hyperalgesia is a typical neuropathy feature in animals). Withdrawal of the paw in response to radiant heat is a positive reaction. Agents with analgesic activity will prolong or inhibit this reaction time. This prolongation is known as paw withdrawal latency time.

While sciatic nerve ligation is a standard method employed, vincristine-induced neuropathy is another model that simulates drug-induced neuropathic pain. Vincristine is administered to adult rats for 10–14 days at a daily dose of 100 mcg/kg. Although hyperalgesia starts manifesting right from day 2 of dosing, testing (using mechanical or thermal stimuli) can be done during week 2 of dosing. Drugs that have analgesic properties typically delay or inhibit the response to mechanical or thermal stimuli.

Diabetic neuropathy is a prevalent complication seen in human diabetics. An animal model that is used to simulate diabetic neuropathy in the rat model of streptozotocin-induced diabetes. Streptozotocin is administered chronically to rats, following which there is blunting of responses to mechanical, thermal, or chemical stimuli at around 2–4 weeks. The response time duration before and after drug administration can be compared to assess analgesic activity.

(xii) *Cancer pain models*

The in vivo models for cancer pain pertain to the bone metastases that occur with different cancers since metastasis to the bones is considered to be very painful. Rats and mice are generally used here, wherein MRMT-1 tumor cell line (mammary), Walker-256 tumor cells (mammary), NCTC-2472 tumor cells (sarcomatous), or HCa-1/2 tumor cells (hepatocellular) are injected into the long bones. 1–2 weeks are allowed to pass following injection for the bone invasion to occur. Once the bone invasion is confirmed, analgesic activity is tested for mechanical, thermal, or chemical means.

(xiii) *Post-operative incision pain model*

A 1 cm-long incision is made longitudinally on the plantar surface of the hind-paw of anesthetized rats such that the incision penetrates the muscle layer. Responses to von Frey filaments are done for 1 week following this incision and compared with pre-incision readings. A control and a test group of rats can be used to assess the test drugs' analgesic activity. As an alternative to the von Frey filaments, mechanical pressure-based techniques may also be used. However, there have been data to suggest the increased sensitivity when von Frey filaments are utilized.

(xiv) *Thoracotomy pain model*

Conventionally, pain following thoracotomy is said to persist for 45–60 days on average. Hence, it is ideal for providing good analgesic cover for such post-thoracotomy patients. Screening for such agents using this model requires expertise on the part of the investigator. Anesthetized rats are surgically operated on to incise the pleura under their fourth and fifth ribs, following which a retractor is used to retract the tissue by around 0.8 cm. At least half of these rats develop allodynia if retraction is maintained for an hour. The test drugs that reduce the incidence of such allodynia can be claimed to possess significant analgesic activity. A control group of rats can be used, wherein pleura is opened, but no retraction is applied.

40.1.2.2 In Vivo Models for Peripheral Analgesic Activity

In the real classical sense, there are no isolated tests to check exclusively for peripheral analgesic activity. There are likely to be variable degrees of central action involved. However, a few of the widely accepted models of screening for peripheral analgesic activity are described below.

(i) *Writhing test*
The basic principle of this method is to induce irritant-related pain in the test animals. Mice are preferred for the test, although rats are also used frequently. The irritant is injected into the peritoneal cavity, which induces a typical reaction in rodents called "writhing." Writhing is the phenomenon wherein the mice or rats exhibit a characteristic stretching of their trunk. Although any irritant can produce this reaction in rodents, validated irritants include acetic acid, phenylquinone, ethacrynic acid, etc. Test drugs are administered prior to injecting the irritant. 5–10 min following the irritant's injection, the rodents are observed for the development of writhes. Several scoring systems exist to rank the writhes, but positive writhe can be one that has the characteristic stretching of the abdomen along with the extension of at least one hind limb. Test drugs that inhibit the writhing reflex can be said to have analgesic activity. Since this test is non-specific for peripheral and central analgesia, this is usually employed as an early screening test. Further tests to confirm the site of action can be performed for greater sensitivity and specificity.

(ii) *Randall-Selitto test*
The Randall-Selitto method is based on the principle that the usual threshold for pain is lowered in times of inflammation, which can be elevated again by drugs with significant peripheral or central analgesic property. In this method, local inflammation is induced, following which the threshold of pain is checked using mechanical or other pain induction methods. Inflammation is usually caused using Brewer's yeast or carrageenan. Other alternatives include prostaglandin E2, Freund's adjuvant, etc. The inflammation-inducing agent is injected into the plantar surface of one of the hind paws, following which the response to mechanical or other modes of painful stimuli (analgesiometers) is now available to simplify the process and encourage objectivity of findings) is evaluated. Test drugs that can raise the pain threshold are said to have significant analgesic activity.

(iii) *Modified Randall-Selitto test*
There exist several modifications to the original Randall-Selitto method. The most commonly used modification involves the deletion of mechanical pressure on the paws of the affected rodents. Instead, the animals are made to walk on the gridded metal pad, and the examiner observes their gait. A scoring

system is applied to the animals based on the gait they develop, as shown below:

- Score 1 for normal gait.
- Score 0.5 for limping gait.
- Score 0 for hopping gait (the paw is not allowed to come in contact with the grid).

  The cumulative scores are calculated as a percentage of the total scores assessed. The higher the score, the higher is the analgesic activity of the test substance.

(iv) *Mechanical visceral pain test*

The duodenal distension model is the most popular visceral pain analysis method that is used experimentally. Male rats are anesthetized with general anesthesia, and an abdominal incision is made below the left costal margin, through which the stomach is accessed. A longitudinal incision is made 1–2 cm above the pylorus on the rat stomach's greater curvature. Gastrostomy is performed following a purse-string closure of the incision. A catheter is designed with a distensible balloon that can hold about 1.5 ml volume attached to it. This catheter is inserted through the gastrotomy opening, thus created and made sure to progress into the first part of the duodenum. Following this, the purse string s closed around the inserted catheter, and sutures are tightened at the dermal level around the catheter to stabilize the same. A recovery period of 3–6 days is given depending on the extent and time taken for the procedure.

After this initial recovery period, the duodenal distension volume is measured by determining the threshold at which writhing is produced (ranges between 0.5 and 0.7 ml). Test, standard, and control groups are compared to draw conclusions. A scoring system exists to ensure the objectivity of the findings:

- Score 0–Normal behavior (resting, exploration, escape, etc.)
- Score 1–Mild modification in the behavior (lack of exploration, excessive grooming attempts, hyperventilation, the chattering of teeth, shaking like a wet dog, etc.)
- Score 2–Moderate modification in the behavior (retching, hunching, grooming of the abdomen, hindlimb immobility, etc.)
- Score 3–Severe modification in the behavior (hindlimb stretches, dorsiflexion of hind paws, etc. in addition to features of score 1 and 2).
- Score 4–Intensive modification in the behavior (intense writhing movements in addition to features of score 1, 2, and 3).

  The lower the final mean score, the greater is the analgesic potential of the test drug.

  Several modifications of this distension model are in use: colorectal distension, renal pelvis distension, etc. However, these modifications are complex to perform and also carry a higher mortality risk. Hence, the mechanical duodenal distension model still remains the most common method of this group.

(v) *Chemical visceral pain tests*

These tests are very similar to the mechanical visceral pain models, the only difference being the agents used to induce pain. Chemical agents are injected into the hollow viscera to induce pain here. The two widely accepted models are the sigmoid colon distension pain model and the inflammatory uterine pain model.

In the sigmoid colon distension model, 5% formalin is used as the inciting agent, injected into the test rodents' sigmoid colon. This is said to produce a classical biphasic pain response. The first phase involves excessive stretching of the flanks or the entire body. The second phase involves grooming of the abdomen. Comparisons of such behaviors with standard and control groups can help estimate the test agents' analgesic potential.

The inflammatory uterine pain model is a clear reflection of the human uterine inflammation and is commonly used to estimate visceral pain. Here, irritants like mustard oil or castor oil are injected into one of the rodents' uterine horns to induce inflammation. Sham operations are conducted in control animals. The rodents are then observed over the next week for the presence or absence of abnormal movements and behaviors. Usually, a biphasic response is seen in most animals. The first half of the week shows the rats exhibiting abnormal behavior in the form of frequent hunching or arching of the back, abdominal grooming, stretching of the body, grazing of the abdomen against the floor, etc. While the same may be decreased over the second half of the week, most of these flank muscle hyperalgesia features are evident till the end of the week. As with other models, a comparison of the scores versus the test and control (or standard) is made to conclude the test agents' analgesic potential.

(vi) *Bradykinin-based tests*

Bradykinin is one of the key mediators in analgesic pathways. However, administration of bradykinin via the intravenous or oral route is ineffective since enzymes rapidly degrade it. Hence, intra-arterial administration is preferred.

In the most common and accepted model, male rats are anesthetized, following which a polyethylene catheter is inserted into the right carotid artery in the centripetal vector. After allowing for a brief recovery period, bradykinin is injected via the catheter. This injection would provoke dextro-rotation of the head along with forelimb flexion. Squeaking may also be occasionally seen, but not in all cases. Since the bradykinin dose required to elicit such responses may vary among the rodents, this preliminary analysis is done to determine the minimal dose of bradykinin necessary for each animal. Once this initial testing is done, the test agents are injected at least 15 min before bradykinin injection. Sequential injections of bradykinin are administered in intervals of 5–10 min until the bodily effects reappear.

This method's modifications include injecting bradykinin in the femoral artery or the subcutaneous space of guinea pigs. Besides, bradykinin antagonists can also be employed as a negative control to enhance the test's sensitivity.

(vii) *Spinal neuronal test*

While the peripheral analgesics are believed to majorly produce their action by inhibiting the well-studied arachidonic acid pathways, there are other minor mechanisms for their analgesic action, including the action at the spinal level. This procedure is a relatively less common one that involves the usage of cats.

Cats are anesthetized using ketamine and chloralose and then ventilated mechanically (pancuronium may be used for the same). Then, the sartorius tendon is cut to expose the medial articular nerve. The cat's thigh is fixed to a board so that flexion and extension of the lower leg can be performed. The saphenous nerve is cut at around the inguinal fossa to enable recording. Bipolar electrodes are placed at the medial articular nerve. Recordings are drawn from the saphenous nerve from single medial articular nerve units. To facilitate spinal cord neuronal recording, T12 to L7 segments are exposed by performing surgical laminectomy. Spinal cord transection is performed at the lower thoracic region to prevent mechanical axonal activation in the long spinal tracts. Ascending tract neurons are identified and used for extracellular recording.

Roughly 3–6 h before electrical stimulation and recording, acute arthritis is induced in the cat knee using irritants like kaolin (4%) and carrageenan (2%). On average, a duration of 1–3 h is allowed to pass for the acute arthritic phenomenon to set in.

Recordings are started, and action potentials are read. An initial lag time of 40–60 min is set to stabilize the recordings. Following this lag time, the test agent (and the control/standard agents, if needed) are administered intravenously to the cats. Analgesic activity is then estimated using mechanical (or sometimes, chemical) stimuli. Comparison of the activity before and after test drug administration (or between test and control) is made to arrive at the test agent's analgesic activity.

The Han-Neugebauer modification of this method employs rats, wherein a computer-based algorithmic analysis of the rats' vocalizations (both audible and ultrasonic) of the rats are used. Other modifications of this method include neonatal rat spinal cord, lumbar puncture, injection in mice, compression of the spinal cord, etc.

(viii) *Neuronal growth factor tests*

The Nerve Growth Factor (NGF) role in perception and modulation of pain has been established across a few studies, making this one of the second-line screening procedures for analgesic activity. While the in-vitro tests are more commonly performed, there is scope for in vivo NGF tests.

Spinal hemisection is performed in rats at the level of T13, followed by laminectomy at two segments of T11 and T12. After the hemisection is confirmed, a 10-day course of intraperitoneal anti-NGF is administered to the test rats. Estimation of analgesia can be performed using any of the previously described modalities: mechanical, chemical, electrical, etc. Control groups can be maintained on saline administration instead of the anti-NGF agent.

40.2 Screening for Anti-Inflammatory Activity

Anti-inflammatory agents' screening methods are practically critical since these agents' utility is multifold – inflammatory conditions, autoimmune disorders, carcinomas, trauma management, etc. Although anti-inflammatory agents are predominantly supportive in managing most of these conditions, it is crucial to study and analyze potential drugs for their anti-inflammatory properties. The methods used for such screening may be in vitro or in vivo in nature.

40.2.1 In Vitro Methods for Anti-Inflammatory Screening

In vitro methods relate to the inflammatory mediators produced in response to any kind of inflammatory damage. These mediators include histamine, serotonin, substance P, bradykinin, prostaglandins, leukotrienes, thromboxanes, cytokines, lymphokines, and platelet-activating factor, and other minor ones. The effects of the potential anti-inflammatory agents on the levels of these mediators comprise the in vitro methods of drug screening. A few of the important methods are listed here.

(a) *Screening involving the bradykinin family*
Bradykinin's family of receptors includes three subtypes: B1 and B_2 (further divided into B_{2a}, B_{2b}, B_{2c}), and B_3. While B_3 is predominantly localized to the pulmonary system, the other two receptors are located in most tissues susceptible to inflammatory damage. Guinea pig ileum is the most common tissue used for bradykinin assay. Different concentrations of ^{3}H-bradykinin are used sequentially to check for binding after the addition of the agent to be tested. Levels of bound and free levels are then measured. The values are compared with the control chart to assess for anti-inflammatory activity. Bovine aorta, canine trachea, rat duodenum, and rabbit aorta are a few other tissues that exhibit good levels of sensitivity to bradykinin.

(b) *Screening involving the tachykinin family*
Substance P is an important mediator, which belongs to the broad family of tachykinins. The highest levels of substance P seem to be present in the brain and related neural tissue. ^{3}H-Substance-P receptor binding is a standard method, which employs pig brain. Radioactivity bound levels are measured using a liquid scintillation meter. Other tissues used include rabbit pulmonary vessels, rabbit vena cava, guinea pig ileum, rat urinary bladder, etc.

Neurokinins are the other major entities in the tachykinin family. There are three well-established neurokinin receptors – NK_1, NK_2, and NK_3. While substance P itself binds to NK_1 and NK_2, neurokinins A and B bind preferentially to NK_2 and NK_3. Chinese hamster ovary cells (CHO cells) are used for NK_1 and NK_2 assays, and guinea pig-derived cortical membranes are used for NK_3 assays. Other common tissues used are rat esophagus, rat portal vessels, rabbit iris, etc.

(c) *Screening involving arachidonic acid pathway*
Arachidonic acid and its metabolites are chief constituents of the inflammatory cascade. The two major pathways of arachidonic metabolism are the cyclo-oxygenase (COX) and 5-lipo-oxygenase (LOX) pathways. Levels of leukotriene B_4, eicosanoids, COX-1, and COX-2, are the most commonly used techniques for screening anti-inflammatory agents. This method's major demerit is that the arachidonic acid pathway is highly non-specific at a tissue level. The activity thus ascertained can only be assessed at a generalized level.

(d) *Screening involving cytokines*
Cytokines, majorly interleukin-1, 6, 8, interferons, and TNF-alpha, are measured to demonstrate the anti-inflammatory activity of drug compounds. While pro-inflammatory cytokine levels are reduced, anti-inflammatory cytokine levels are increased by drugs with significant anti-inflammatory properties. ELISA, flow cytometry, absorbancemetry are a few techniques used for evaluation.

(e) *Screening involving chemokines*
Human chemokines are more than 20 in number, and these are associated with roughly 20 different chemokine receptors. Inhibitors of these chemokines have significant anti-inflammatory actions. CCR-3 receptor binding is considered to be one of the standard methods of evaluation. Other routinely used methods include eosinophilic chemotaxis assay (migration of eosinophils for chemotactic functions), calcium flux assay (CCR-3-mediated fluctuations in intracellular calcium flux), etc.

(f) *Screening involving PPARs*
Although the peroxisome proliferator-activated receptors (PPARs) are chiefly involved in the pathogenesis of dyslipidemia and diabetes mellitus, it has been discovered that they also play a significant role in the inflammatory cascades. PPAR inhibitors have been found to inhibit the presence and actions of cytokines produced in response to inflammation. Several methods are documented to assess the PPAR-mediated inflammatory processes, like luciferase assay, RNA blot method, transient transfection assay, etc. Most of these methods are expensive and time-consuming.

(g) *Screening involving histaminergic system*
Histamine receptors (H_1 to H_4) have a role to play in the cascade of inflammation, with the H_4 subtype (expressed most commonly in the intestines and peripheral leucocytes) being the key receptor of interest for the assessment of anti-inflammatory activity in humans. 4-methyl-histamine, a highly selective H4 agonist, is used as control, and drugs with H_4 blocking properties can be deemed to possess anti-inflammatory potential.

40.2.2 In Vivo Methods for Anti-Inflammatory Screening

Inflammation cannot be entirely defined as a single entity since the actual inflammatory process is spread across three distinct and demarcated stages, as mentioned here.

- Acute stage of inflammation, which is a transient phase, defined by localized vasodilatation and enhanced capillary permeability;
- Subacute stage of inflammation, which is a more prolonged phase, defined by leucocytic infiltration at the site of inflammation;

The chronic stage of inflammation, which is a proliferative phase, is characterized by degeneration and fibrosis of the inflamed tissue.
The models developed for assessing anti-inflammatory activity also depend on the inflammation stage and are very specific to the same. While there are specific models for testing acute (and subacute) inflammation, there is another set of models to check for chronic inflammation. Besides, several immune-mediated models have also been developed for testing the immunological factors involved in inflammation.

40.2.2.1 In Vivo Methods for Acute and Subacute Inflammation

(i) *UV erythema method*

Also known as the Wilhelmi technique, guinea pigs are traditionally used for this technique since they are one of the most sensitive animal groups. Albino guinea pigs are shaved (on the dorsum and flanks) roughly 1 day prior to the test. Chemical depilation (barium sulfide suspension applied for 15–20 min) is preferred over mechanical shaving. The test drug is administered in two sittings on the test day: one half at 30 min prior to UV exposure and the other half at 2 min following UV exposure. Once the first half is administered to the animals, the animals are covered such that only a small area of the shaved portion (around 1.5 × 2.5 cm) is exposed to the UV light. UV light (using an emitter that is pre-warmed and can emit light at wavelengths of 180–200 nm) is administered to this portion at a distance of around 15–20 cm above the animal. As mentioned earlier, following 2 min of exposure, the second half of the test agent is administered. The erythema degree is recorded at multiple intervals, with mandatory recordings at 2- and 4-hours post-exposure.

The simple scoring system that is followed is as follows.

- Score 0 – no erythema.
- Score 1 – weak or mild erythema.
- Score 2 – strong erythema.
- Score 4 – very strong or intense erythema.

As is evident from the scoring system, the assessment is very subjective and is prone to errors. Hence, the observation by multiple investigators is recommended, from which the mean scores can be obtained. Test agents that can offer protection from UV erythema (with scores 0 and 1) have significant anti-inflammatory activity. This method is a straightforward

Table 40.1 Alternative irritants (phlogistic agents) for edema induction

Alternative substance	Volume to be injected (ml)
1% formalin	0.1
1% dextran	0.1
1% egg albumin	0.1
Undiluted fresh egg white	0.05
2.5% Brewer's yeast	0.1
5% kaolin	0.1
0.25% papaya latex	0.1

screening method that is liable to several biases. Hence, different approaches are needed to confirm the results of this technique.

Modifications of this crude method include the utility of rats, mice, and even humans.

(ii) *Paw edema method*

Although the earlier-described UV erythema model is the simplest of the lot, the paw edema model is arguably the most commonly used method. This model involves the injection of irritants (phlogistics) like carrageenan, Brewer's yeast, kaolin, formaldehyde, dextran, or egg albumin into the paw of rodents to induce acute and sub-acute phases of inflammation.

The test agents are administered to rats intraperitoneally or orally. 30–60 min later, subcutaneous injection of 1% (w/v) carrageenan (0.05 ml) is administered into one of the hind paws of the test rats (administration into both hind paws will result in severe and excruciating pain and immobility, and hence should be avoided). An indelible mark is made at the lateral malleolar level of the injected paw. Paw volume is measured consecutively using a plethysmograph that is filled with mercury. The paw volumes are compared between the test and control groups to conclude the test agents' anti-inflammatory activity. The percentage in the reduction of paw edema is reflective of the activity of the test drug.

A few of the other agents that can be used instead of carrageenan are listed in Table 40.1.

(iii) *Ear oedema method*

The ear edema model is quite similar to the previous model in that it also reflects the typical acute and subacute phases of inflammation. However, this model is usually considered to be more sensitive for topical agents. Croton oil is used as the inciting substance in this model.

Rodents are preferred for this technique, wherein one ear is coated with croton oil to induce inflammation. The test drug is usually applied to the inner surface of the ear 30–60 min before croton oil application. About 6 hs after croton oil application, the rodents are subjected to euthanasia (preferably by cervical dislocation). Plugs of 6 mm diameter are removed from both ears (also the control ear) to check for differences in weight, which indicate the extent of oedematous response.

An alternative method that is less used is oxazolone-induced ear edema, wherein 2% oxazolone is applied to the ear's inner surface.

(iv) *Granuloma pouch method*

In this screening technique, rats are anesthetized and subjected to a thin needle injection to create a small pneumoderma in the dorsal skin (by injecting a pocket of air, roughly 20 ml). 1% croton oil or 20% carrageenan is injected into this air pocket thus created. The test agent is administered to the rats daily. After 2 days, the air is released from the pouch. Another day later, adhesions in the pouch are broken down. The next day (day 4), the rats are sacrificed. The exudate from the pouch is collected and compared with the control animals. Test agents that can reduce the exudate volume can be said to have significant anti-inflammatory activity.

(v) *Pleurisy test*

Pleurisy is a documented type of exudative inflammation in humans. A similar phenomenon may be induced in rodents using irritants like carrageenan, bradykinin, histamine, prostaglandins, etc.

Test rats are anesthetized, following which a minor incision is made between the seventh and eighth ribs to expose the intercostal muscle. Another incision is created in the muscle layer through which carrageenan solution (2%) is injected into the pleural cavity. One hour prior to this injection, the test drug is administered to the rats. Also, the test drug is repeated at 24 h and 48 h of carrageenan injection. Euthanasia is performed after 72 h, and the pleural cavity is accessed. Heparinized Hank's solution (a balanced salt solution) is injected into the cavity. The cavity fluid is then drawn out using a pipette and transferred into a plastic tube with gradings. The total volume of fluid collected is compared with the control group. Other parameters that can also be used to compare include WBC count, fibronectin analysis, PGE_2 analysis, etc.

40.2.3 In Vivo Methods for Chronic Inflammation

(i) *Cotton wool granuloma model*

This model is based on the principle that an implanted foreign body draws inflammatory mediators and serves as a chronic inflammation source, depicting the proliferative phase. In the cotton wool model, sterilized cotton wool pellets (weighed before the procedure) are implanted in subcutaneous pockets created on both sides of the scapular area in anesthetized rats. Following this, the test drug is administered for the next 5–7 days. On the fifth or the seventh day, the rats are euthanized, and the pellets are carefully removed from the subcutaneous pockets. These pellets are dried at around 60 °C for 18–24 h and then weighed. The final dry weight is calculated by subtracting the initial weight of the pellets. The lower the weight, the greater is the anti-inflammatory activity of the test drug. This model has also been modified into several versions, of which the groin implantation model is an accepted technique.

(ii) *Glass rod granuloma model*

Also known as the Vogel granuloma model, this method is based on a similar principle as the cotton wool model. Sterilized small glass rods (3–4 cm long) are implanted in the subcutaneous plane of the test rats' dorsum for the next 20–40 days. Following the implantation, the test agent is administered to the rats daily for the entire duration (to assess for prophylactic effects) or during the last 10 days (to evaluate therapeutic effects). At the end of the test period, the rats are sacrificed, and the rod weights are estimated. Comparisons are drawn. Further, if needed, biochemical analyses of the rods can also be performed.

(iii) *PVC sponge granuloma model*

Introduced by Saxena, the sponge granuloma method uses polyvinyl foam discs (soaked in irritants like carrageenan or dextran). The dorsal incision is made in test rats to implant the sponge discs. The test drug is then administered over the next 3–4 weeks, following which the discs are removed to check for the weight change if any. Also, the sponge exudates are subjected to biochemical and genetic analyses.

40.2.4 Miscellaneous Models of Anti-Inflammatory Activity

(i) *Models of arthritis*

Induction of arthritis in rodents can be done using adjuvants like Freund's complete antigen or irritants like papaya latex or chicken collagen (type 2).

(ii) *Models of cystitis and prostatitis*

Cystitis can be simulated in rodents by chronic administration of cyclophosphamide (acute or single dose can result in acute cystitis).

Prostatitis can be induced by injecting irritants like carrageenan (3%), formalin (5%), or capsaicin into male rats' prostate.

40.3 Screening for Antipyretic Activity

Antipyretics are arguably the most commonly prescribed and most widely consumed drugs in today's clinical practice. Although paracetamol is the drug used commonly and not many new drugs are developed for this purpose, it is essential to know the methods used for screening antipyretic activity since many drugs might have antipyretic properties as added or secondary pharmacological actions (e.g., most anti-inflammatory agents). The in vivo methods used for this purpose are listed here. Regarding the in vitro methods, they are only extensions of the methods described earlier for analgesic and anti-inflammatory activity screening.

40.3.1 Brewer's Yeast Suspension Method

This is the most classical method that is available for antipyretic screening. Brewer's yeast in a concentration of 15% suspended in normal saline has been documented to cause pyresis. Drugs with significant antipyretic activity will naturally lower the rise in temperature, thus induced by Brewer's yeast. Rats are commonly used for this method. Brewer's yeast is administered subcutaneously, following which body temperature rises by around 16–20 h. Once a rise in temperature is documented using a rectal thermometer (insertion to be done at least 2 cm into the rectum), the drugs to be screened are injected, and successive temperature measurements are done sequentially. The difference between baseline and post-drug levels is measured to get an estimate of the antipyretic activity.

40.3.2 Bacterial Lipopolysaccharide Method

Lipopolysaccharide fractions isolated from *Escherichia coli* can induce fever in animals when injected via the intravenous route. Rabbits are preferred for this technique, although rats can also be used (intraperitoneal administration can be performed in rats). Rabbits classically exhibit a bimodal pattern of pyresis, with an early peak at 1 h (rise of temperature by 1 °C or more) and a late peak at 3 h. Once the lipopolysaccharide fraction is injected into the rabbits, the drugs to be tested are administered after an hour once the fever sets in. As mentioned earlier, since there is a late peak at 3 h, observations are carried out until completion of 3 h' duration. Drugs that tend to reduce the rectal temperature by at least 0.5 °C can be considered to have antipyretic activity. This technique is associated with higher sensitivity levels compared to Brewer's yeast method done in rats.

40.3.3 Other Less Commonly Used Methods

- Circumventricular micro-injection of PGE_1 in rats has been documented to cause fever within 2 h. However, the complexity of the procedure and the associated high mortality levels are drawbacks.
- Similarly, intracerebral injection of pyrogenic substances is also used for screening antipyretic agents.
- Turpentine-induced pyresis is an inexpensive (but inconsistent) method of inducing fever in rabbits and rats.
- Microbial endotoxins can also produce fever in rats, rabbits, goats, and monkeys.
- Another screening method is an indirect one that assesses agents' ability to potentiate hypothermia induced by chlorpromazine and related drugs in rats and rabbits.

Bibliography

Huang BP, Lin CH, Chen YC, Kao SH. Anti-inflammatory effects of Perilla frutescens leaf extract on lipopolysaccharide stimulated RAW264.7 cells. Mol Med Rep. 2014;10:1077.

Koster R, Anderson M, de Beer EJ. Acetic acid for analgesic screening. Fed Proc. 1959;18:412.

Randal LO, Selitto JJ. A method for measurement of analgesic activity on inflamed tissue. Arch Int Pharmacodyn. 1957;111:409–19.

Sancillo LF, Nolan JC, Wagner LE, et al. The analgesic and anti-inflammatory activity and pharmacologic properties of bromfenac. Arzneim Forsch/Drug Res. 1987;37:513–9.

Shukla P, Shukla P, Mishra SB, Gopalakrishna B. Screening of anti-inflammatory and antipyretic activity of Vitex leucoxylon Linn. Indian J Pharmacol. 2010;42(6):409–11.

Vogel HG. Analgesic, anti-inflammatory, and antipyretic activity. In: Vogel HG, editor. Drug discovery and evaluation: pharmacological assays. Germany: Springer. p. 984–1116.

Wang G, Hu Z, Song X, Cui Q, Fu Q, Jia R, et al. Analgesic and anti-inflammatory activities of resveratrol through classic models in mice and rats. Evid-based Complementary Altern Med. 2017;5197567

Winter CA, Risley EA, Silber RH. Carrageenan-induced edema in hind paw of the rat as an assay for anti-inflammatory drugs. Proc Soc Exp Biol Med. 1962;111:544.

Screening Methods for the Evaluation of Antiglaucoma and Anticataract Drugs

41

Rupadevi Muthaiah

Abstract

Glaucoma and cataract are the most common ophthalmic conditions that can lead to visual impairment and eventually blindness, if untreated. Visual impairment or dysfunction significantly affects the quality of life, and hence there is always a need for new and potential drugs to restore vision. In drug discovery, experimental animal models are one of the mandatory tools to decipher the molecular mechanisms, pathophysiology, and applicability of therapy in diseases. Though several models are available, it is important to identify the ideal or the most appropriate pre-clinical models that mimic the features of ophthalmic conditions such as glaucoma and cataract to discover the targets for promising therapeutic agents. For a deeper comprehension, the advantages and limitations of the available pre-clinical models ranging from larger mammals to small animals for glaucoma and cataract are discussed in this section. This chapter summarizes the perspectives and various aspects to be considered for designing a study to evaluate the efficacy of drugs against glaucoma and cataract.

Keywords

Glaucoma · Ocular hypertension · Cataract · Congenital cataract · Ophthalmic drug screening · Animal models

41.1 Introduction

The evaluation of ophthalmic drugs has a major challenge in drug penetration and absorption due to:

R. Muthaiah (✉)
Department of Pharmacology & Toxicology, Jacobs School of Medicine and Biomedical Sciences, The New York State University at Buffalo, Buffalo, NY, USA

M. Lakshmanan et al. (eds.), *Introduction to Basics of Pharmacology and Toxicology*, https://doi.org/10.1007/978-981-19-5343-9_41

- presence of tissue barriers to drug penetration, namely the lipophilic corneal epithelium, the hydrophilic corneal and scleral stroma, the conjunctival lymphatics, choroidal vasculature, and the blood-ocular barriers.
- highly sensitivity of ocular tissues (e.g., the lens, uveal tract, and retina)

Thus, stringent screening methods may improve the successful transition of outcomes from benchtop to bedside.
This chapter review about the screening methods involved in the evaluation of anti-glaucoma and anti-cataract drugs.

41.2 Glaucoma

Glaucoma is a neurodegenerative disease and a leading cause of blindness worldwide. The main characteristic feature of glaucoma is a progressive retinal ganglion cell (RGC) loss and degeneration of axons in the optic nerve (ON), which eventually leads to an irreversible vision loss.

41.2.1 Types of Glaucoma

In humans, glaucoma is primarily classified into three major types based on the iridocorneal angle and developmental abnormality:

- *Primary open-angle glaucoma (POAG)*—iridocorneal angle appears open and normal but has diminished aqueous outflow.
- *Primary angle closure glaucoma (PACG)*—iridocorneal angle appears closed due to apposition of the peripheral iris to the trabecular meshwork (TM) with obstructed aqueous outflow.
- *Primary congenital glaucoma (PCG) with POAG* — an autosomal recessive disease caused by an abnormality in the anterior chamber angle during the development stage.

Glaucoma is also classified as normal tension glaucoma, autoimmune glaucoma, exfoliation syndrome, trauma-related glaucoma and pigmentary glaucoma.

41.2.2 Aqueous Humor Outflow

It is important to maintain the balance between the rate of aqueous humor production and the aqueous outflow to maintain the intraocular pressure (IOP) under normal range. Aqueous humor outflow is regulated by two mechanisms:

- Schlemm's canal collects aqueous humor from the anterior chamber and drains into episcleral and limbal vessels through the trabecular meshwork (TM)
- the uveoscleral pathway

Ocular hypertension is mainly associated with the consequential effect of diminished or impaired aqueous humor outflow and lead to the elevation of IOP.

- The association of aqueous humor production and outflow is described by the following equation:

Flow in = Flow out
Rate of formation = Pressure gradient across entire outflow pathway × Ease with which fluid can exit + uveoscleral outflow.

$$F = (P_i - P_e) \times C + U.$$

where,

F is rate of aqueous formation in µl/min.
Pi is IOP in mmHg.
Pe is episcleral venous pressure in mmHg.
C is facility of aqueous outflow in µl/min/mmHg.
U is uveoscleral outflow.

41.2.3 Intraocular Pressure Measurement

Manometry is the only available direct and invasive method to measure IOP. Briefly, anterior chamber is cannulated with a microneedle that is connected with calibrated pressure transducer through fluid filled tubing. But this method has some limitations such as it may cause inflammation and introduce other artifacts due to cannulation.

Tonometer is an indirect non-invasive method to measure IOP by estimating the turgidity of the cornea or the eye-wall. Below mentioned are the types of tonometry:

- *Applanation tonometry* — basic principle is the amount of force required to flatten portion of surface of the sphere that is directly proportional to the pressure inside the sphere. Examples are Tono-Pen, Pneumatonometer, Perkins and Goldmann tonometers
- *Rebound tonometry* — estimates IOP by the rate of speed rebound from the corneal surface

41.2.4 Designing the Study to Test the Drug Efficacy

While designing the study to evaluate the efficacy of test drug in lowering the IOP, the following factors have to be considered:

- Appropriate species selection
- Rate of non-responders in the study population
- Use of normotensive or hypertensive animals
- Secondary endpoints for the assessment of lowering IOP
- Selection of control (For example, one eye acts as control)

On the other hand the following endpoints should not be included:

- Blood draws
- Fluorescein staining
- Dilated pupil examinations
- Pupillometry
- Corneal sensitivity
- Corneal pachymetry
- Noncontact specular microscopy
- Anterior segment optical coherence tomography
- Electroretinography
- Fundus imaging
- *Species Selection*

It is essential to consider the cost, availability, ease of handling and housing requirements to select the species for the drug efficacy study.
The following approaches could be adopted for species selection:

- Target receptor distribution and concentration that resembles with humans.
- Perform pilot studies in small number of animals in different species.
- Generally nonhuman primates resemble humans in many factors.
- Similar physiological pathways to elicit a necessary response.
- The most important factor in the species selection is the size of the various parts of the eye such as cornea, anterior chamber, lens and vitreous volume which may significantly influence the pharmacokinetics and absorption, distribution, metabolism and excretion (ADME) of the drug.
- *Nonhuman Primates:*
- The most commonly used and well-studied nonhuman primates are Cynomolgus (*Macaca fascicularis*) and Rhesus macaques (*Macaca mulatta*).
- They are the most appropriate test species due their close phylogeny and high homology with humans. The responses to the test drug extrapolated from these species are well correlated with the humans.
- Whereas, differences in the eye parts such as smaller corneal surface area and overall body mass are the limitations of these species that may affect pharmacokinetics, ADME, and systemic toxicity of topically applied antiglaucoma drugs.
- Due to the aggressive nature of the animal, the use of anesthesia for the tonometry significantly influences the IOP values.

- Overall, the limitations of these species are limited availability, required special housing facilities, maintenance cost and chance of transmission of diseases, such as tuberculosis and Herpes B to humans.
- *Cats (Felis catus):*
- To investigate the mechanism of action of autonomically active drugs, cats are the ideal species due to their autonomic innervation and adnexa.
- Cats elicit response to majority of the antiglaucoma drug classes, but the magnitude of IOP reduction is not significant compared to monkeys and dogs.
- The difficulty in handling and tonometry use due to their claws and occasional aggressive nature makes them as less feasible species.
- Moreover, the variation in the eye parts such as large cornea, vertical slit pupil and small body mass may affect the pharmacokinetics, ADME, and toxicity profiles of topically applied drugs.
- Dogs (*Canis lupus familiaris*):
- They are readily available and easy to handle compared to monkeys and cats. The use of tonometry without general anesthesia produces reliable IOP values and they respond to most of the antiglaucoma drug classes.
- The disadvantages of this species are the anatomic features like larger cornea and anterior chamber compared to humans, larger housing requirements, a need for continual training to use tonometry and IOP variations in day and night.
- *Rabbit (Oryctolagus cuniculus):*
- Rabbits are the widely studied species for ophthalmic drugs and respond to most of the antiglaucoma drug classes
- The IOP can be measured using tonometry without general anesthesia
- The following ocular anatomical and physiological features need to be considered for the selection of this animal
- Low tear production
- Highly sensitive to ocular irritation
- Thin cornea and sclera which reduces ocular rigidity (causes tonometer to underestimate the real IOP)
- Fragile blood: aqueous barrier
- Large lens
- Different blood supply to the retina and optic disc
- Some strains have albinism, which may also influence the pharmacokinetics and drug efficacy due to the interference of pigment binding
- *Rodents:*
- Transgenic and knockout models are widely used to study the pathophysiology of increased IOP.
- They are convenient to use due to their low cost, minimal housing requirements and feasibility of large sample size to achieve statistical power and significance.
- The differences in the anatomical and physiological features are substantial between humans and rodents. Moreover, the responses to drugs vary from one strain to another.
- The presence of albinism in both rats and mice and background retinal degeneration in many mouse strains also can be a significant variable.

- Though it is easy to handle, the difficulty in restraining the animal for measuring IOP may introduce technical error and affect the accuracy. Hence, pre-clinical studies in rodents for antiglaucoma drugs are not sufficient to support FDA investigational new drug applications.
- *Responders:*
  In drug efficacy studies, another important factor to be considered is the percentage of responders within the study population—as the type and distribution of the target receptors varies within the species. Moreover, the rate of responders is critical for identifying the sample size to extrapolate the test results.
- *Normotensive or Hypertensive Models:*
  In normotensive rats, IOP lowering is found to be minimal and hence, ocular hypertensive models are usually being used to investigate the mechanism of action of the test compounds and to identify potentially promising new drugs.
- *Spontaneous vs Experimentally-Induced Models:*
  In various reports, spontaneous or transgenic glaucoma animal models are found to be useful in the delineation of the series of pathophysiological events involved in glaucoma. In general, experimentally-induced models of glaucoma in nonhuman primates or rabbits are used to evaluate the efficacy of antiglaucoma drugs. Rats and mice with experimentally-induced glaucoma are useful to study the molecular mechanism involved in the pathogenesis of glaucoma. They are not an ideal model for therapeutic efficacy studies, as the drug efficacy study data cannot be translated directly into humans.

41.2.5 Animal Models

- *Primate Models of Ocular Hypertension:*
  - Laser photocoagulation of the TM is the most widely used nonhuman primate model for glaucoma.
  - For this model, diode or argon laser is used and several laser treatments are required with the varying period of interval to induce an IOP elevation. One eye is injured and the other eye acts as a control in this model. The laser treatment elevates the IOP chronically and it may vary from day-to-day. Animals are trained to accept tonometry measurement while they are conscious.
  - Other glaucoma monkey models are induced by obstructing the TM using intracameral injection of latex microspheres or autologous fixed red blood cells which elevates IOP chronically.
- *Rabbit Models of Ocular Hypertension:*
  - Rabbit models of ocular hypertension can either be *acute* (transient elevation of IOP for few hours by inducing osmotic shifts in the eye) or *chronic* (persistent increase in IOP).
  - *Injection of Hypertonic Saline:* 5% sodium chloride (0.1 ml) is injected intravitreously that causes a transient IOP spike. A test compound is administered before the saline injection and the IOP is measured at 6 different

time points, i.e., pre-saline injection, 30 min, 1, 1.5, 3 and 5 h of post-saline injection. The efficacy of the test compound is evaluated based on the inhibition of increase in IOP.

- *Water Loading:* For this model, the animal is fasted for 24 h and water (60–70 ml/kg) is administered orally using a 12 French catheter. Warm water is used to prevent confounding factors due to substantial drop in body temperature. The administration of water bolus generates transient elevation of IOP from the baseline 15 mmHg to low 30s and reaches the peak at 20–30 min which may last for 2–3 h. Hence, the test compound is administered prior to water loading and IOP is measured at various intervals.
- *Glucose Infusion:* In this model, 5% glucose (15 ml/kg) is injected intravenously for over 20 s and it elevates IOP transiently. The IOP spike is generated due to the reduction of blood osmolality by this hypotonic solution that transfers the water into the eye. The IOP spike reaches the peak upto double from the baseline value by 30 min and returns to the baseline value within 50–135 min post-infusion. The test drug is usually given prior to infusion and the efficacy of the drug is evaluated by the ability to inhibit the IOP spike.
- *Corticosteroid-Induced Hypertension:* Corticosteroids increase the resistance to aqueous humor outflow and IOP as consequence of biochemical alterations in the TM. Elevation of IOP in this model is chronic compared to the osmotic models. Not all the rabbits are inclined to develop IOP elevation due to corticosteroid and the percentage of IOP elevation depends on the strain and cohort within the strain. Researchers adopted various methodologies to induce an IOP elevation in this species. Of which, common methodologies for the development of IOP elevation by corticosteroids are listed below:

  intravitreous injection of 0.1 ml at a concentration of 4 mg/ml triamcinolone acetonide.

  weekly subconjunctival ocular injection of repository betamethasone (4 mg) for 3 weeks.

  0.7 ml of betamethasone suspension (betamethasone sodium phosphate, 3 mg/ml) and betamethasone acetate (3 mg/ml) are administered subconjunctivally once a week for 4–7 weeks.

  0.1% dexamethasone phosphate is applied topically 3 times a day for 5 weeks.

  The efficacy of the test drug is evaluated by the ability to lower IOP in the corticosteroid-treated eye.

- *Intracameral α-Chymotrypsin:* A specific mechanism of action for the IOP elevation due to α-chymotrypsin injection is unclear. It is hypothesized that it may be due to lysed lens zonular proteins, inflammatory debris, or peripheral anterior synechia involved in the obstruction of the TM or blood aqueous barrier breakdown that lead to increase in the rate of aqueous humor inflow. The administration of α-chymotrypsin into the posterior chamber induces a chronic elevation of IOP to 30 mmHg and sometimes it may reach greater than 50 mmHg.

Commonly used dose varies between 45 and 150 units dissolved in different volumes of sterile saline from 100 μl to 0.5 ml.
Anesthetize the animal with intramuscular ketamine and xylazine, and constrict the pupil with pilocarpine 15 min prior to the α-chymotrypsin injection. Inject α-chymotrypsin into posterior chamber through the pupil using 27–30 g needle or a blunt cannula of a particular size. Retain the needle in the posterior chamber for at least 1 min to avoid the enzyme leakage to prevent corneal injury due to enzyme. This model requires careful examination for upto a week after the injection to avoid complications such as inflammation and intraocular damage.

- *Rodent Models*
- *Dexamethasone Treatment:* Topical application of dexamethasone in rats mimics human PAOG. This model develops IOP elevation after 2 weeks of topical application. Specifically this model is useful in studying the association of myocilin and the pathogenesis involved in the glaucoma.
- *Laser-Induced Ocular Hypertensive Models:* The most common approach to elevate IOP in animal models is to obstruct the aqueous humor outflow by laser photocoagulating the episcleral and limbal blood vessels. In Wistar rats, India ink is injected into the anterior chamber 1 week prior to laser treatment to the TM. The carbon particles accumulated in the anterior chamber absorb anargon laser energy and produce a localized scarring effect. Minimum 3 consecutive laser treatments with the interval of 7 days are required to induce an IOP elevation to at least 25 mmHg and lead to ON head cupping, thinning of the nerve fiber layer and ON degeneration. Laser-induced ocular hypertension method can be used in both mice and rats.

  The duration and extent of IOP elevation is associated with the modulation of various factors such as laser intensity, duration, and spot number.
- *Episcleral Vein Injection of Hypertonic Salin***e:** In this model, injection of hypertonic saline through the veins located in the episclera causes obstruction of aqueous outflow that lead to the sustained IOP elevation after 4 weeks of injection in Brown Norway rats. The IOP elevation varies between 7 and 28 mmHg from baseline when compared to the control eye. Of the total injected animals, 45% of rats develop IOP elevation after 2–4 weeks of the injection whereas 35% requires repeated injections. Following the IOP elevation, this model exhibit progressive RGC loss and ON degeneration.
- *Episcleral Vein Cauterization:* Cauterization of two or more episcleral veins in Wistar rats impairs the aqueous outflow from the collecting ducts of Schlemm's canal resulting in IOP elevation upto 60 mmHg. Due to the ocular hypertension, an apoptosis of RGC is observed at a rate of approximately 4% per week.
- *Injection of Substances that Induce Ocular Hypertension:* The intracameral injection (once a week for upto 9 weeks) of the TM extracellular matrix with molecules such as hyaluronic acid in rats induced a sustained ocular hypertension. Injection (6 weekly repeated injections) of latex microspheres with the size of 10 μm with or without the hydroxypropylmethylcellulose (HPM) also diminished the aqueous outflow through the blockage of the TM and increased IOP.

Though these models are considered as technically simple and cost effective, they may have side effects such as corneal abnormalities and/or inflammation. To mimic uveitic glaucoma in rodents, S-antigen in complete Freund's adjuvant is administered systemically. IOP increases up to 35 mmHg after 2 weeks of injection and this model is characterized by angle closure resulting in a reduction of aqueous outflow. This model exhibits a significant inflammation in both the anterior and posterior chambers as part of the pathogenesis.

- *Genetic Rodent Models of Glaucoma*
  - The DBA/2J mouse strain is widely used and studied model for secondary glaucoma. Gene mutations of *Tyrp1* encoding tyrosinase-related proteins and *Gpnmb* encoding glycosylated transmembrane proteins are mainly associated with this model. It is characterized by the obstruction of aqueous outflow through pigment dispersion, iris transillumination, iris atrophy and anterior synechia. This model develops ocular hypertension by the age of 9 months.
  - DBA/2NNia is another substrain of DBA/2J that also develops elevated IOP and shows RGC loss and ON degeneration approximately at 15 months of age.
  - A transgenic mouse model that expresses the Tyr413His myocilin point mutation corresponding to the human MYOC Tyr437His mutation in the drainage structures of the eye. Of the causative genes associated with POAG in humans, myocilin has been extensively studied. This model is characterized by approximately 20% of RGC loss, ON axonal degeneration, endothelial cells detachment of the TM and sustained IOP elevation at the age of 18 months.
  - Transgenic mouse model with mutation in the $\alpha 1$ subunit of collagen type I also used to study POAG. Features of this model are open angles, ON axonal loss and IOP elevation.
  - Vav2/Vav3-deficient mice also demonstrated an elevated IOP, RGC loss and ON head excavation. This model suggesting that Vav2 and Vav3 expression is associated with the iridocorneal angle closure.

41.3 Cataract

Cataract, one of the major eye diseases, causes lens opacity that ultimately leads to blindness. The incident rate of cataract is about 50% of all the other visual impairments. The causes of cataract are both acquired and hereditary; and the risk factors are aging, diabetes, gender, sunlight exposure and lifestyle change. But aging and diabetes are identified as the most common risk factors.

At present, the only therapeutic option for cataract is surgery; however, a demand for an alternative and economical option is growing in aging population. Further, the incidence rate could be reduced with appropriate preventive measures that delay onset of the disease. In order to obtain an apt prevention strategy, it is important to understand the pathophysiology of cataract and require a valid system to test toxicity and efficacy of the drug candidates, which can be achieved with various animal models.

Cataract is classified into four major forms based on the regions in the lens where cataract initiates, such as:

- Sub-capsular
- Cortical
- Nuclear
- Mixed

Cataract formation is considered as matured when red fundus reflex is no longer visible through lens and the lens appear dull white to the naked eye. Intensity of the progression of cataract formation is scored as:

- Stage 1—Clear normal lens
- Stage 2—Peripheral vesicles
- Stage 3—Peripheral vesicles with cortical opacities
- Stage 4—Diffuse central opacities
- Stage 5—Opacity involving the entire lens

41.3.1 Animal Models of Congenital Cataracts

Total 39 loci are mapped in the human genome for congenital cataracts, in which 26 loci are linked to mutations in specific genes such as crystalline, connexin and cytoskeletal proteins and membrane proteins. Majority of the mutations related to congenital cataracts are identified in the crystalline family of proteins.

- *Crystallins:*
  - The crystallins are the family of proteins in lens and maintain the structure to facilitate refraction of light. They are classified into three types based on their molecular weight–alpha, beta and gamma.
  - The α-class of proteins are encoded by *CRYAA* (αA) and *CRYAB* (αB) genes and involved in the prevention of protein accumulation in response to temperature changes. The αA-crystallin knockout mice and the double (αA and αB-crystallin) knockout mice develop microphthalmia and nuclear cataracts but not the αB-crystallin knockout mice.
  - Of the four independent point mutations in the *CRYAA* gene, a mutation at codon 54 with a transition from arginine to cysteine is observed in both human and mouse. But the mouse R54C mutation is a dominant one, and human R54C mutations possess recessive traits that indicate αA-crystallins has different functional activities in the different species.
  - The β- and γ-crystallins play a role in the maintenance of the lens structure. Two forms of human β-crystallin are identified and they are encoded by six

genes *CRYBA1,2,4* and *CRYBB1,2,3*. Human γ-crystallin is encoded by six genes, namely, *CRYGA* to *CRYGF*, encoding γA–γF-crystallin proteins.

- *Connexins*

  Connexins are gap junction proteins which play an important role in the maintenance of the osmotic and metabolic homeostasis in avascular tissues such as lens. The expression of the following connexin genes is identified in the mammalian lens.

  - Cx43 encoded by *Gja1 gene*
  - Cx46 encoded by *Gja3 gene*
  - Cx50 encoded by *Gja8 gene.*

  Mutations in the genes *Gjp3* (encoding Cx46 or α3) and *Gjp8* (encoding Cx50 or α8) are associated with cataracts in both dominant and recessive inheritance. Homozygous Cx46-null mice display mild lens opacity at 2–3 weeks of age and form complete nuclear cataract by 2 months of age, indicates the importance of Cx46 in the lens function.

  The cataract changes include crystallin degradation, lens protein aggregation, and protein insolubility and disulfide cross-linking between the proteins, which are reminiscent of human senile cataract.

  More than 15 mutations of the human Cx50 gene at various codons are found to be associated with inherited cataracts. Further, Cx50 homozygous knockout mice show microphthalmia and mild opacity at 2 weeks of age and mature by 6 months.

- *Cytoskeletal and Membrane Proteins*

  Mutations in the structural and membrane proteins are also linked to the development of cataracts. An abundant membrane protein in lens fiber cells is a major intrinsic protein (MIP) or aquaporin 0, a voltage-dependent water channel. The phenotype of the MIP mutant mouse is characterized by impairment in the fiber cell differentiation and protein solubility in the crystalline lens. Moreover, the KFRS rat model has a spontaneous frameshift mutation in the *kfrs4* gene and develops recessive cataracts by inactivating MIP.

  Cataract formation is also associated with mutation of a membrane protein, vimentin, which is highly expressed in lens fiber cells and mesenchymal tissue. It is a type-III filament and interacts with α-crystallin in lens epithelial cell and fiber cell. Vimentin transgenic mice are used to determine the association of vimentin and lens opacity in congenital cataracts. Vimentin overexpressing mice show disruption in fiber cell differentiation. Another structural protein that is associated with cataract formation is called beaded filament structural protein 2 (bfsp2). Knocking out bfsp2 in mice exhibit impaired lens fiber cell differentiation by disrupting the interaction with α-crystallins and other intermediate filaments.

41.3.2 Animal Models: Induced Cataract

- *L-Buthionine-(S, R)-Sulfoximine (BSO): Induced Cataract*

Normal human lenses are resistant to oxidative damage due to high levels of glutathione. Glutathione is a principle antioxidant and an intracellular thiol in the lens and it plays an important role in combating oxidative stress and maintains lens transparency. BSO selectively inhibits γ-glutamylcysteine synthetase that actively controls the synthesis of GSH (reduced glutathione, a major tissue antioxidant). Inhibition of γ-glutamylcysteine synthetase disrupts the lens transparency indirectly by depleting the GSH levels. Correlation of decreased GSH levels and diminished scavenging activity of lens to oxidative insult in the aged population is well established and it induces cataract formation.

Intraperitoneal injection of BSO (4 mmol/kg body weight) once daily for 3 days to the weaner mice or rats (less than 2 weeks old). Cataract formation is graded once the eyes of the animals are open (approximately 15 days) using a slit microscope and animals are sacrificed for further lens analysis.

Weaner lenses did not show effective cataract changes when it is cultured with BSO in vitro. But, the lens culture of pretreated mice with BSO demonstrated a significant cataract development that is similar to the *in vivo* model.

It is a good model to study the relationship between GSH depletion and loss of lens transparency in both in vivo and in vitro. This model demonstrates a mixed (corticonuclear) type and age-related cataract.

- *Selenite-Induced Cataract*

  Selenite cataract model is a widely used model that mimics aspects of nuclear and age-related cataracts. A high concentration of selenium can effectively induce cataracts in both in vivo and in vitro. Selenite treatment disrupts the antioxidant defense mechanism of the lens by depleting GSH levels and impairs calcium homeostasis. Consequently, calcium elevation and calpain activation lead to nuclear cataract formation.

  Sodium selenite is injected subcutaneously either as a single dose (19–30 μM/kg body weight) or repeated smaller dosage (40–50 nmol/g body weight) to suckling rats (10–14 days of age). The extent of lens opacity is examined and graded using slit-lamp microscopy after 2 weeks of injection.

- *OXYS Rats*

  Galactose over-fed Wistar rats generate OXYS strain with inherited galactosemia. This model does not develop diabetes but displays increased glucose transport, which results in glucose accumulation in the lens. Because of glucose accumulation, ROS is generated in excess in the lenses that leads to the development of nuclear cataract at 1–2 months of age. Later, cortical is involved with nuclear opacification at 6–12 months of age.

  OXYS rats are identified as an ideal model to investigate glucose transport, oxidative stress, and age-related nuclear (ARN) cataracts.

- *Naphthalene-Induced Cataract*

  Naphthalene is metabolized to naphthalene 1,2-dihydrodiol in the liver. This naphthalene metabolite reaches aqueous humor where it is converted to 1,2-dihydroxy naphthalene. In lens, this breaks into highly reactive compound naphthoquinone that generates free radicals.

To induce cataract, naphthalene is administered orally at a dose of 0.5 g/kg initially for 3 days followed by 1 g/kg for 25 days, and the opacity develops in the cortex at 4–6 weeks. Change in the lens morphology is examined using an ophthalmoscope. Further, the lens that cultured in vitro in the presence of naphthalene dihydrodiol develops opacity, but only in the deeper cortex region.

Though this model produces similar results in both in vitro and in vivo, it is not an ideal model for ARN cataract.

- *UV-Induced Cataract*
  One of the major risk factors for cataract formation is UV radiation. Intraocular penetration of UV light induces the formation of thymine dimer and ROS generation that cause DNA damage to the lens. Exposing the Sprague-Dawley rats or mice to 8 kJ/m^2 UV-B radiation for 15 minutes induces cataract formation. This model can be used to study ARN cataracts but there are discrepancies in the association of UV light and cataract induction.
- *Steroid-Induced Cataract*
  In recent years, the incidence rate of cataracts is increased with steroid therapy. Steroids such as glucocorticoids are used in the therapeutic management of inflammatory conditions. The development of posterior subcapsular (PSC) and nuclear cataracts is observed with the chronic use of glucocorticoids (GC).

  The steroid-induced cataract models are as follows:
  - To study the response to GC treatment, chick embryos are treated with dexamethasone (0.02 μmol/egg) and the lenses of chick embryo develop cataract with 48 h of treatment.
  - The Brown-Norway rats develop morphological changes similar to the features of human steroid-induced cataract when they are treated with 1% prednisolone acetate instillation daily of a total volume of 1 mg/kg or intramuscular injection of 0.8–1 mg/kg prednisolone acetate for 10 months.
  - The treatment of rat lens explants with dexamethasone and fibroblast growth factor-2 results in the features of PSC formation such as increased cell proliferation and coverage on the capsule.
- Hyperbaric oxygen (HBO)-induced cataract
  HBO treatment in older animals is an ideal model to study the mechanisms of ARN cataract formation. For the *in vivo* model, guinea pigs with the age of 17–18 months are treated with 100% O_2 at 2.5 atm three times a week, for up to 100 times. For the *in vitro* model, cultured rabbit lenses show the development of nuclear cataractswhen they are treated with HBO at pressures ranging between 1 and 100 atm for 3 h.

  The biochemical features of this model resemble the human ARN cataract. However, this model develops only haziness in the lens nucleus but not a completely mature cataract. Hence, it may not be a useful tool to study the drug efficacy on mature cataract.

41.3.3 Hereditary Animal Models

Some animal models develop cataract spontaneously at their early stages of life by inherited mechanisms. These animals are an ideal model to investigate the efficacy of drugs for the treatment of congenital cataracts. However, some animals can be used to study ARN cataracts.

- *Emory Mouse*
  Even though this model develops a hereditary cataract, it is a well-suited model for the ARN cataract. This model has an effect on the expression of the α-crystallin and ARK receptor tyrosine kinase genes in the lens. There are two types of sub-strains in this model; one develops cataract at the age of 5–6 months and the other at the age of 6–8 months. Opacification initially appears in the superficial cortex and progresses to the deep anterior cortex, and the cataract formation intensifies with age and exhibits biochemical changes similar to human ARN cataract.
  Emory cataract shows following biochemical changes:
  - Decreased GSH and protein sulfhydryl levels
  - Protein insolubility
  - Increased lipid peroxidation
  - Increased PSSG (protein-*S*-*S*-glutathione) and PSSC (protein-*S*-*S*-cysteine)
  - Increased MP24 (intrinsic membrane protein 24)
  - Decreased antioxidant enzyme activity

  This model is also helpful to study pre-cataractous changes in lenses.
- *Senescence Accelerated Mouse (SAM)*
  SAM is another genetic mouse model for the ARN cataract. There are 3 sub-strains in this model; SAMP-P/1 and SAM-P/9 are senescence prone strains and SAM-R3 is senescence resistant strain.
  SAM-P strains show the earliest changes such as ripple mark body in the lens cortex of cataract at around 3 months of age. With age, it develops into ripple rings and causes liquefaction of the lens cortex. SAM-R3 strain develops opacity at the posterior pole at 10 weeks of age and later progresses to the posterior perinuclear and cortical regions. The morphological changes of SAM cataracts are posterior pole protrusion, lens cortex liquefaction, and nuclear dislocation, and biochemical changes are decreased GSH content and increased GSSG (oxidized GSH) content. Despite the pathogenesis of accelerated senescence and neurodegenerative disorders, SAM strains are not ideal models to study the ARN cataract.
- *Ihara Cataract Rat (ICR)*
  The Ihara cataract rat develops cataract spontaneously at the age of 3–4 months and matures by 4–6 months. Initially a mild opacification is observed in the posterior subscapular region and subsequently, swelling and degeneration of fiber cells occur in the posterior cortex and nucleus at about 4 months of age. Moreover, this model shows a decreased SH groups concentration and increased disulfide bonds and serum lipid peroxide level as a very early sign before

opacification. After the initiation of opacification, Ca^{2+} content is also increased in the lens. Ihara rat is a useful model to study cortical cataracts.

41.3.4 Diabetic-Induced cataracts

The STZ rat and galactose-fed animals develop cataract acutely and possess similar features of the human diabetic cataract. Though these models are helpful to understand the association of the polyol pathway and osmotic stress, we need better diabetic models with slow-developing cataracts for efficient drug development.

The following models can be considered to study cataract development in spontaneous type 1 diabetes rats.

- Diabetes-prone BB rat
- Long Evan Tokushima Lean (LETL) rat
- LEW-IDDM rat
- Komeda diabetes prone (KDP) rat
- *Obese Models of Type 2 Diabetes*
  Type 2 diabetes with or without obesity could be either due to insulin resistance and/or β cell failure. Among the type 2 diabetes animal models with obesity, Zucker Diabetic Fatty (ZDF) rat and Otsuka Long Evans Tokushima Fatty (OLETF) rat are the ideal models that resemble human diabetic cataract due to slow-development of cataract and exhibit similar morphological and biochemical changes to the human diabetic cataract.
- *Zucker Diabetic Fatty (ZDF) Rat*
  ZDF rat strain is an outbred colony of Zucker fatty rat but less obese with more severe insulin resistance. Male ZDF rats show cataract signs such as peripheral vesicles in the lens and cortical opacities at about 15 weeks and complete formation occurs by 21 weeks of age. Morphological changes are tissue liquefaction, fiber cell swelling, and membrane ruptures in the cortex. Biochemical changes include decreased GSH level, AGEs accumulation in the epithelium, increased apoptotic factor nuclear factor kappaB (NFkB) expression in the lens.
- *Otsuka Long Evans Tokushima Fatty (OLETF) Rat*
  The OLETF rat is an obesity-induced diabetes model and reared from Long Evans rats. Early cataract changes such as swelling in the anterior and PSC lens fibers are observed at 40 weeks of age and vacuoles and cell swelling in the equatorial cortex and supranuclear fibers at 60 weeks. The morphological changes of this model are identical to galactosemic and STZ-induced diabetic rats. Lenses exhibit increased sorbitol levels, and aldose reductase (AR) and sorbitol dehydrogenase (SDH) activity. Morphological and a few biochemical changes of this model are reminiscent of human diabetic cataracts. It is a very useful model to study the chronic pathways involved in type 2 diabetes-induced cataract formations.
- *Non-obese Models of Type 2 Diabetes*
  Of the non-obese diabetic models, Spontaneous Diabetic Torri (SDT) rat and Wistar Bonn/Kobori (WBN/Kob) rat are recognized for the slow development

cataract as they exhibit similar physiological, morphological, and biochemical aspects of human diabetic cataract.

- *Spontaneous Diabetic Torri (SDT) Rat*
  In both male and female rats, lens opacity appears initially at the posterior pole and matures as cortical cataract at the age of 40 weeks or older. Morphological changes include swelling, vacuolation, lens fibers disintegration, Morgagnian globules formation, nuclear sclerosis, and opacification in the lens cortex. This model is ideal to study the efficacy of drugs on female diabetic cataracts.
- *Wistar Bonn/Kobori (WBN/Kob) Rat*
  The WBN/Kob rat is also an animal model for retinal degeneration. The male rats display cell swelling in the lens periphery and lens fibers disorder at about 12 months of age. This model is useful to study the role of calpain activity in the development of human diabetic cataracts.
- *Transgenic or Knockout Mice*
  Diabetic mice are not readily used to study diabetic cataracts due to resistance in the development of cataracts. Whereas, transgenic mouse models are useful tools to study a new pathogenic mechanism involved in the polyol pathway by genetic manipulation.
- *Aldose Reductase Transgenic (AR-Tg) Mice*
  The AR-Tg mice develop cataract signs such as vacuoles at the lens periphery at about 1–2 days after exposing to 50% galactose diet, and complete lens opacity at about 2–3 weeks. The accumulation of galactitol in the lenses is identified as the cause of galactose-induced cataract. AR-Tg mice show the appearance of cataracts after 7–14 days of STZ injection, which matured to complete lens opacification by 2–3 weeks of post-injection.
- *Sorbitol Dehydrogenase–Deficient Mice*

The SDH-deficient mouse shows a high level of sorbitol in the lens but does not affect the glucose and oxidative stress level. This indicates the high activity of the polyol pathway even in the absence of diabetes. About 70% of mice display small vacuoles followed by opacity in the lens nucleus at 16–17 months of age.
Diabetes induced AR-Tg mice and SDH deficient mice are ideal models to study the contribution of AR and SDH pathways in the development of diabetic cataract.

41.4 Conclusion

It is important to follow the three R's (i.e., Replacement, Reduction, and Refinement) while designing the drug efficacy and toxicity studies using animal models. Overall, the selection of an appropriate species with similar retinal anatomy and retinal vascular supply to humans and a reliable method to mimic the ocular disease is essential to increase study validity, decrease variability and reduce the number of animals used in research.

Bibliography

Bouhenni RA, Dunmire J, Sewell A, Edward DP. Animal models of glaucoma. J Biomed Biotechnol. 2012;2012:692609

Johnson TV, Tomarev SI. Rodent models of glaucoma. Brain Res Bull. 2010;81(2–3):349–58.

Lim JC, Umapathy A, Donaldson PJ. Tools to fight the cataract epidemic: a review of experimental animal models that mimic age-related nuclear cataracts. Exp Eye Res. 2016;145:432–43.

Miller PE. Study design and methodologies for evaluation of anti-glaucoma drugs. In: Gilger B, editor. Ocular pharmacology and toxicology. Methods in pharmacology and toxicology. Totowa, NJ: Humana Press; 2013.

West-Mays J, Bowman S. Animal models of cataracts. Animal models of ophthalmic diseases. Essent Ophthalmol. 2015;11–29

Screening Methods for the Evaluation of Drugs Affecting Peripheral Nerve Functions

42

Avinash Arivazhahan

Abstract

Drugs affecting peripheral nerve function are commonly skeletal muscle relaxants and local anaesthetic agents. Apart from the dedicated drugs that possess such activity, there are drugs used for other purposes that also exhibit activity on the peripheral nerves. Screening methods for drugs affecting peripheral nerve function include in vitro and in vivo techniques. The most routinely conducted experiments for skeletal muscle relaxant screening include the inclined plane method, rotarod method, palmar grip strength method, etc. Similarly, for screening of potential local anaesthetics involves methods like corneal reflex suppression, wheal reaction, tooth pulp method, etc. This chapter aims to discuss these various screening methods and models in brief.

Keywords

Skeletal muscle relaxants · Local anaesthetics · Rotarod · Inclined plane

42.1 Screening of Skeletal Muscle Relaxants (Neuromuscular Blocking Agents)

Neuromuscular blocking agents are those agents that act at the myoneural junctions and cause skeletal muscular paralysis, either by acting pre-synaptically or post-synaptically. At the presynaptic level, these agents inhibit the synthesis or release of acetylcholine, while at the post-synaptic level, they inhibit the action of acetylcholine at the receptors. As discussed in detail in Volume 2 of this review series,

A. Arivazhahan (✉)
Sanofi India Ltd., Chennai, Tamil Nadu, India

M. Lakshmanan et al. (eds.), *Introduction to Basics of Pharmacology and Toxicology*, https://doi.org/10.1007/978-981-19-5343-9_42

neuromuscular blockers can be generally of two types: depolarising (non-competitive) and non-depolarizing (competitive).

42.1.1 Scope and Limitations of Neuromuscular Screening Tests

Neuromuscular screening tests are extensive and can be used to assess several distinct properties of neuromuscular blocking agents. A few of these properties are as follows.

- Potency and efficacy of the agent
- Type of blockade
- Onset of action
- Duration of action
- Nature of reversibility of blockade
- Adverse effects associated with the blockade

However, these screening tests come with a major limitation that gross extrapolation cannot be done with the results from one screening test population to another. There have been differences observed between species and also within the same species. Hence, it is always ideal to screen the drugs in humans under ideal conditions before clinical approval of the same.

42.1.2 In-Vitro Screening Tests

The in-vitro methods are said to be more sensitive to changes in the environment, and hence it is possible to identify and nullify various external factors that could hamper the results in direct screening tests. Further, internal factors like hemodynamics and nutrient dynamics can also be masked.

42.1.2.1 Isolated Nerve-Muscle Preparations

These are the most widely applied in-vitro screening tests. The classical nerve-muscle preparations consisted of amphibian tissue, with mammalian tissues also being tested later on. These are typically conducted in a liquid bath medium, which comprises modified Kreb's solution (with glucose). The isolated tissue is kept immersed in the medium, with the proximal end of the nerve exposed outside the medium. A stimulus is applied directly to the muscle or to the exposed nerve via electrodes and stimulator device, to generate impulses that are recorded. The neuromuscular blocking agent to be studied is added to the medium, and the recording is done.

As stated earlier, frog nerve-muscle preparation is the most commonly used tissue. Gastrocnemius (the earliest tissue used), sartorius, or rectus abdominis are tissues that are routinely employed.

Mammalian tissues like the phrenic nerve—diaphragm of young mammals (rats or Guinea pigs) are also used for this purpose. Young animals are selected to ensure that the tissue is permeable enough for oxygen in the medium. It is to be noted that rat preparation is resistant to the action of the decamethonium class of agents. Hence, to study this class of agents, the Guinea pig tissue may be preferred. Lumbrical-nerve preparation of rabbits is also uncommonly used for screening (chiefly for non-competitive blockers). Limitations of this tissue preparation are that the lumbricals may be functionally inert in rabbits, thus becoming unpredictable in screening neuromuscular blocking agents. Also, this preparation is more expensive and needs higher levels of expertise, when compared with the other lower mammalian tissues described earlier.

42.1.2.2 Single Muscle Fiber Study

In this method, the muscle to be evaluated is mounted in a bath that is partly illuminated to facilitate viewing via a microscope. Electrodes are used to stimulate the muscle. This method is easy and quick and also enables comparison among different fibers of a muscle. Amphibian sartorius is commonly used.

42.1.2.3 Ionophoretic Testing Method

Also known as Thelseff's method, this technique uses cat tenuissimus nerve-muscle preparation. The change in potential that occurs due to depolarisation is studied using microelectrodes inserted into the muscle close to the site of drug application. This is a very rapid method that is helpful in also assessing the extent of depolarization.

42.1.2.4 Human In Vitro Preparations

Fetal phrenic nerve-diaphragm preparations and intercostal nerve-muscle preparations have been used for screening neuromuscular blockers.

42.1.3 Screening Tests in Anesthetized Animals

In these methods, the animal is first anesthetized and then incubated, following which the drugs are screened in different ways. These tests can be used to assess the presence, degree, and duration of neuromuscular blockade. A few of them are listed here.

- Lower limb study involving gastrocnemius, tibialis anterior or soleus muscles is performed in rats, cats, and dogs
- Phrenic nerve-diaphragm testing
- Facial nerve stimulation test
- Reversal of succinylcholine-induced apnea in mice or rats

42.1.4 Screening Tests in Intact Non-Anesthetized Animals

Here, the animals are unanesthetized so as to study the effects of neuromuscular blockade in an intact physiological milieu. These tests are arguably the most common screening tests that we use for neuromuscular blockers.

42.1.4.1 Inclined Plane Method

In this method, rats or mice are placed on a board that is inclined at an angle relative to the horizontal plane. The time duration for which the animals can hold on to the board without falling is noted before and after drug administration. A cutoff time of 20–30 min is maintained. Animals that fall off the board before this cutoff period are termed as positive reactors. Pre-screening of animals is necessary to avoid false positive or false negative responses. A less frequently used modification of this method is the inverted grid method, which has similar utility (the cutoff time is kept as 15–20 min here).

42.1.4.2 Rotarod Method

The rotarod apparatus comprises a set of rods elevated from the base that rotate on their own axes at set speeds. Rats or mice are made to stand on these rods to check for the time duration that they can balance themselves. As with the previous method, the animals are tested before and after drug administration. A major advantage of this method is that multiple animals can be tested simultaneously.

42.1.4.3 Testing of Righting Reflexes

Rats, mice, and rabbits can be used for this method. The righting reflex is a normal phenomenon where the animal tries to restore its physiological posture when it is inverted on its back. Once neuromuscular blockade sets in, this righting reflex is lost. The time for onset and potency of the agents can be studied by this crude method.

42.1.4.4 Head Drop Technique

Rabbits are preferred for this method although rats, mice, dogs, and monkeys can also be tested. When an agent causing neuromuscular blockade is administered, there is the phenomenon of a head drop, wherein the animal's head drops down to the surface to the extent that it cannot be reverted back even when the animal is stimulated by a tap on the back.

42.1.4.5 Others

While the methods listed above are the ones that are routinely used, there are a few other methods that are documented to demonstrate neuromuscular blockade in intact lower animals.

- Avian muscle contracture method
- Bernard's amphibian lymph sac method

42.1.5 Screening Tests in Anesthetized Human Beings

A point to be noted before moving on to the actual methodology is that there is always a risk of hypersensitivity and respiratory paralysis associated with testing these agents in humans, anesthetized or non-anesthetized. Hence, all these tests should be performed in a center that has the personnel and the equipment required for adequate and timely resuscitation and management. Cumulative effects and tachyphylaxis of neuromuscular blocking agents can be studied in detail in anesthetized subjects.

The two methods that have been used in anesthetized human subjects are the evaluation of muscle twitches in response to stimulation of nerves, and assessment of pulmonary functions (tidal volume, inspiration and expiration pressures, etc.) in response to administration of the agents.

42.1.6 Screening Tests in Intact Non-Anesthetized Human Beings

Following the screening of the agents in animal models, it is essential to also screen them in healthy human volunteers since there is always a high chance of inter-species variations with most neuromuscular blocking agents. A few of the methods in non-anesthetized human subjects are listed below.

(a) *Palmar grip strength method:* In this method, the human volunteers are asked to squeeze (by applying the maximal possible exertion) an elastic bulb-like device (ergograph), which is linked to another measuring device (dynamometer) that measures the force exerted on the bulb. Sequential testing and estimation of forces applied to the bulb are recorded after the administration of the agents to be screened. Analysis of the results will tell us about the potency, onset, duration, and efficacy of the agents.
(b) *Hand fatiguability method:* This method can be considered as an extension of the grip strength method. The duration and extent to which the ergograph can be squeezed without developing fatigue are estimated after drug administration.
(c) *Muscular twitch method:* Here, indirect stimulation of a nerve induces twitches to the muscles innervated by the nerve. This method is more objective and eliminates participant bias.
(d) *Voluntary power estimation:* This is a very simple and crude method that takes into account only the voluntary muscle power as exhibited by the participant after drug administration. High chances of error are associated with this method.
(e) *Electromyographic evaluation:* The contractions produced at the level of individual muscles or muscle groups are estimated in an objective setting. This can also be done using local and restricted intra-arterial drug administration, which is usually not preferred for safety reasons.
(f) *Pulmonary function analysis:* Here, as in the anesthetized subjects, the pulmonary parameters are evaluated following drug administration.

42.2 Screening of Local Anesthetics

Every potential local anesthetic agent has its own potency and limitations in different forms of local anesthesia like surface, conduction, infiltration, etc. Each of these techniques of anesthesia has a unique set of screening techniques, which are described in brief in this section.

42.2.1 Screening for Surface Anesthesia

(a) *Blockade of corneal reflex:* Rabbits are routinely used for evaluation of the surface anesthetic activity. In this method, the agents are applied to the rabbit eye and the corneal reflex is tested sequentially. One eye can be used as a control, while the other is used for testing. The test can also be done in guinea pigs, although it is not considered to be a standard method.
(b) *Suppression of sneezing reflex:* Rabbits have a well-established sneezing reflex, which makes them ideal for this method. The loss of this reflex after the local application of the agent is taken as positive for local anesthetic activity.

42.2.2 Screening for Infiltration Anesthesia

(a) *Wheal reaction test:* This test is the most common screening technique, and can be claimed to be a standard method for evaluation of the local anesthetic activity. The wheal reaction test is almost exclusively performed in guinea pigs. Pinpricks are used to check for local anesthetic action in areas where the agents are infiltrated (a wheal is marked around the injection area). The pricks done outside the wheal area are taken as control. If there are no squeals when pricked within the wheal, then the agents have significant local anesthetic activity.
(b) A modification of this method is the squeak test done in rats, wherein the rats are subjected to electrical stimulation in the marked area. When the rats fail to produce a squeak in response to electrical stimulus, it is taken as a positive reaction.

42.2.3 Screening for Conduction Anesthesia

(a) *Sciatic nerve technique:* Also known as Ther's method (or Bulbring-Wajda method), this method is done in decapitated frogs. Cotton balls or pads soaked in the local anesthetic agent are placed around the exposed sciatic nerves. After about 1–2 min, the pads are removed. The legs are then pinched using forceps to check for responses. The time of onset and duration of anesthesia can be estimated by this technique. A similar model in rats can be employed using the sciatic or trigeminal nerves.

(b) *Analgesic model extensions:* Screening methods used for analgesic activity can also be used to screen local anesthetic compounds. Rat or mouse tail is subjected to radiant heat or tail clip constriction before and after injection of the local anesthetic. Tail flick and hot plate techniques can also be used for this purpose.
(c) *Tooth pulp assay:* This is one of the most common and specific methods for evaluation of the local anesthetic activity. New Zealand rabbits are the animals of choice for this technique. Pulp chambers of the central incisors are exposed and subjected to electric current after injection of local anesthetic agents. A major advantage of this method is that the same animals can be used on a chronic repetitive basis, with adequate washout times in between. The drilled pulp openings usually remain patent for at least 2–3 weeks.
(d) *Retrobulbar and infra-orbital block methods:* Rats are preferred for the infra-orbital block method (analogous to humans although rats do not have a closed orbit), while dogs are used for the retrobulbar blockade. In the infra-orbital technique, a lack of response to pinching of the upper lip is taken as the onset of the effect of local anesthetic agents. The retrobulbar block is estimated by checking for pupillary, corneal, and lacrimation reflexes.
(e) *Isolated tissue methods:* Isolated tissues including frog sciatic nerve, rabbit vagus, peripheral nerves of frogs, rats, guinea pigs, and rabbits can be evaluated using electrode-based stimulation to screen for local anesthetic activity. These methods are less sensitive and highly non-specific.

42.2.4 Screening for Spinal and Epidural Anesthesia

The agents to be evaluated are injected into the intrathecal or epidural spaces to study the onset of action, duration, and extent of anesthesia. Rats, guinea pigs, rabbits, dogs, etc. have been used for these tests. Usually, the larger the animal, the easier it is to inject the drugs into the right spaces. More recently, ultrasound-guided administration is also being carried out for better accuracy and reliability. Following the injection into the spaces, the anesthetic activity is evaluated most commonly using pinpricks. However, other methods like tail-flick and hot plates can also be employed here.

Bibliography

Chakraborty A, Devi BR, Sanjebam R, Khumbong S, Thokchom IS. Preliminary studies on local anesthetic and antipyretic activities of *Spilanthes acmella* Murr. In experimental animal models. Indian J Pharmacol. 2010;42(5):277–9.

Prabhakar A, Kaye AD, Wyche MQ, Salinas OJ, Mancuso K, Urman RD. Novel drug development for neuromuscular blockade. J Anaesthesiol Clin Pharmacol. 2016;32:376–8.

Sivaji K, Benerji GV, Farid Babu M, Rekhakumari D. Effect of local anesthetics on animal models. Ind J Basic Appl Med Res. 2015;4(4):60–5.

Mutagenic Toxicity Testing

43

Kiruthika Sivagourounadin

Abstract

Genetic toxicology is a subspecialty of toxicology, which deals with identification, testing and analysis of substances which can cause damage to genetic material of living organisms. With the increasing burden of human genetic disorders and further for the safety evaluation of various routinely used substances like pharmaceutical products, cosmetics, agrochemicals, industrial compounds, food additives, it is mandatory to detect genotoxic agents among them. Genetic toxicology testing includes several in vitro and in vivo tests, which are used to evaluate two major endpoints, namely mutagenicity and genotoxicity. These assays are targeted to identify potential mutagens, clastogens and aneugens among the test substances. New chemical entities can cause genetic damage through multiple mechanisms and thereby it is recommended to use a battery of mutagenicity assays for assessing genotoxicity. Several global regulatory guidelines are available for genotoxicity testing, which provide a detailed description about the principles and methods to be followed for performing these assays and interpreting the data obtained. This chapter outlines the several in vitro and in vivo assays that are commonly used to evaluate genotoxicity.

Keywords

Ames test · Genetic toxicology · Genotoxicity · Mutagenicity · Mutagenicity assays

K. Sivagourounadin (✉)
Department of Pharmacology, Jawaharlal Institute of Postgraduate Medical Education & Research (JIPMER), Karaikal, Puducherry, India
e-mail: kiruthikas.kkl@jipmer.edu.in

M. Lakshmanan et al. (eds.), *Introduction to Basics of Pharmacology and Toxicology*, https://doi.org/10.1007/978-981-19-5343-9_43

43.1 Introduction

In this growing era of urbanization, a great threat faced by us is our exposure to potential chemical, biological or physical genotoxic agents, either at environment or workplace. Even though multiple repair mechanisms are induced in a cell, as a response to genetic insult, in some circumstances these defense strategies tend to fail leading to development of various medical conditions like cancer, cardiovascular, neurodegenerative and immune related disorders.

Genetic toxicology deals with the identification, testing and assessment of substances which can interact and cause damage to genetic material of living organisms. A genotoxic agent has direct physical interaction with genetic material and causes genetic damage.

The characteristic feature of a genotoxic agent is to bring out an alteration in genetic material by any of the following means (Fig. 43.1).

Genetic toxicology testing includes assays for the assessment of two major endpoints, namely mutagenicity and genotoxicity.

Mutagenicity is a subset of genotoxicity. Mutagenicity leads to events causing irreversible changes in DNA sequence, number and structure of chromosome. Unlike genotoxicity, mutagenicity can be transmitted to subsequent generations.

Genotoxicity includes mutagenicity and also the events leading to DNA damage and DNA repair, the cell's response to genetic damage. Genotoxicity encompasses a broad spectrum of events of genetic alteration, which may or may not cause permanent alterations in a cell. Thus, tests for genotoxicity include assays to assess mutagenicity as well as those assays for DNA damage and DNA repair, to measure the early, potentially reversible effects caused by test agent on genetic material.

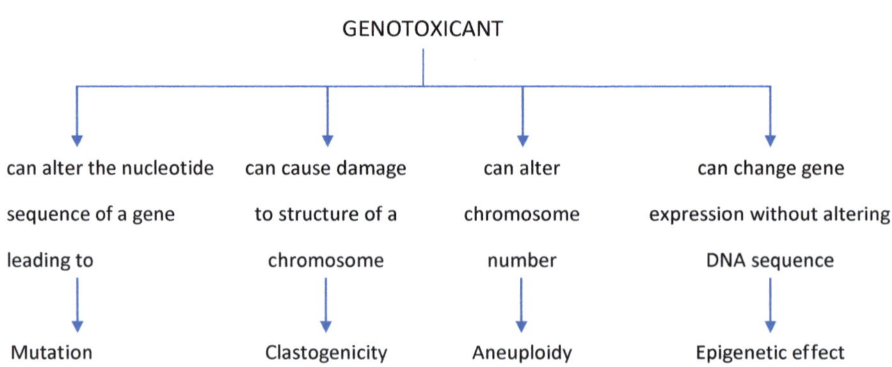

Fig. 43.1 Types of genetic damage caused by a genotoxicant

43.2 History of Genotoxicity Testing

Genetic toxicology traces its beginning to early 4320s when H.J. Muller identified the mutagenic effect of X-rays in fruit fly, *Drosophila*. In the early 4370s, a breakthrough occurred with the development of a bacterial assay by Bruce Ames, using *Salmonella typhimurium* to detect mutations induced by chemical mutagens. At present more than 200 mutagenicity assays are available to screen potential agents which might induce genetic alteration. In the recent years the field of genetic toxicology, has undergone a paradigm shift from identification and risk assessment of genotoxic agents to applying these data for developing strategies in the risk management of these substances.

43.3 Introduction to Genetic Toxicology Assays

The main objectives of various genotoxicity assays used for the assessment of a test agent is:

- To identify mutagens and to thereby attenuate the health hazard encountered by humans on exposure to these substances.
- To corroborate and describe the mechanisms through which these substances produce mutagenesis.
- To analyse and establish dose–response relationship between the test agent and its mutagenic effect.
- To utilize all the information obtained above for decisions on safety evaluation of these substances by regulatory agencies.

At present more than 200 assays are available in a wide range of species like bacteria, fungi, cultured mammalian cells, plants, insects, and mammals. Even though such a vast array of in vitro and in vivo assay is available, but relatively few tests are recommended and routinely used for the purpose of screening and assessment of genotoxic agents.
Table 43.1 depicts distinguishing features between the in vitro and in vivo genotoxicity assays.

- The use of selection techniques in gene mutation assays will simplify and expedite the process of detecting mutation. This can be done by creating an experimental environment with conditions and prerequisites which will allow only the growth of mutant cells. For example, in mouse lymphoma assay the medium contains trifluorothymidine and only cells with mutant *thymidine kinase TK*gene can grow in the presence of this inhibitory chemical.
- Based on type of mutation detected, mutagenicity assays can identify either of the following gene mutation:

Table 43.1 Comparison between the in vitro and in vivo genotoxicity assays

In vivo genotoxicity assays	In vitro genotoxicity assays
Here intact animals are treated with the test agent and are sacrificed, the appropriate tissues and organs are isolated and analyzed.	These assays are performed in cultured cells and microorganisms.
They follow complex design making them expensive and time consuming.	They are comparatively easy, inexpensive.
They provide several additional information like the effect of route of exposure, in vivo metabolism of agents can be obtained.	They can screen a larger number of agents and can also detect mutant cells which occur at low frequency.
Promutagens and procarcinogens are activated by endogenous metabolic pathways.	Exogenous metabolic activation system is needed to detect the promutagens and procarcinogens.
Relevance for human genetic risk assessment is more than in vitro assays	Relevance for human genetic risk assessment is comparatively less than in vivo assays.

Gene mutation

Forward mutation	Reversion or back mutation
A genetic change occurs in a wild-type gene leading to loss of its function and causes an alteration in phenotype	A genetic change leads to gain of function of a gene and reverts the gene to its wild-type.
Eg: In vitro mouse lymphoma assay detects forward mutation of thymidine kinase*TK*gene	Eg: Ames test detects histidine revertants in *Salmonella*

43.3.1 Metabolic Activation System

- Promutagens and procarcinogens are substances which need metabolic activation to be converted into mutagens and carcinogens, prior to this activation processthey are not mutagenic and carcinogenic.In intact animals, usually metabolism and the process of activation occurs in liver, but microorganisms and mammalian cell cultures lack this function. Hence microbial and cell culture assays must be provided with an exogenous metabolic activation system to detect promutagens and procarcinogens.Eg: Dimethylnitrosamine and benzopyrene are promutagens which induce mutations in bacteria only after the addition of mammalian liver homogenate(exogenous metabolic activation system) to the bacterial assays.

- The metabolic activation system which is commonly used is a rat liver homogenate, from which a postmitochondrial supernatant is obtained and then combined with suitable buffers and cofactors. Rat liver S9 mixture is considered as the standard liver metabolic activation system. Ames et al. first prepared this rat liver homogenate by centrifuging thesupernatant at 9000 g (hence it is termed as S9 mixture). Microsomes and cytosols containing the enzymes which are needed for phase I and II metabolic reactions are present in the subcellular fraction of this mixture. The enzymes like cytochrome P450 which are used for metabolic activation are induced by pre-treating the rats with chemicals like polychlorinated biphenyl mixture or phenobarbital-plus-β-naphthoflavone. Apart from rat liver S9, homogenates from other species like mice, hamster, guinea pig and monkey have also been used as exogenous metabolic activation system in some assays. These preparations tend to exhibit a varying efficiency in detecting mutagens when compared with rat liver S9. However, the disadvantage faced with the use of exogenous metabolic activation system is that it cannot be a precise substitute to in vivo metabolism of intact animals. This is due to the following differences between exogenous metabolic activation system and the in vivo metabolism occurring in intact animals:
- normal intestinal microbial flora can also cause metabolism of some compounds in intact animals
- even within a same species, variation tend to occur among several tissues involved in metabolism of xenobiotics
- sometimes the chemicals used to induce the metabolic enzymes might lead to an alteration in the metabolism of xenobiotics leading to a different mechanism in vitro

Recently, using recombinant DNA technology, genes coding for enzymes involved in xenobioticmetabolism inhumans can be incorporated into microorganisms and mammalian cell cultures. These microbes and cell lines will express cytochrome P450 enzymes and other enzymes involved in phase I and II reactions of metabolic activation. Hence, these genetically engineered microbial and mammalian cell culture assays can detect promutagens and procarcinogens without the use of an exogenous metabolic activation system.

43.3.2 Genotoxicity Testing Endpoints

The strategies used for genotoxicity testing must include a broad spectrum of endpoints, in order to measure events which might cause early reversible as well as irreversible damage to DNA in a somatic cell or germ cell. Hence for thorough evaluation of the mutagenic potential of a chemical the following endpoints must be assessed:

- gene mutation
- chromosome aberration (clastogenicity)

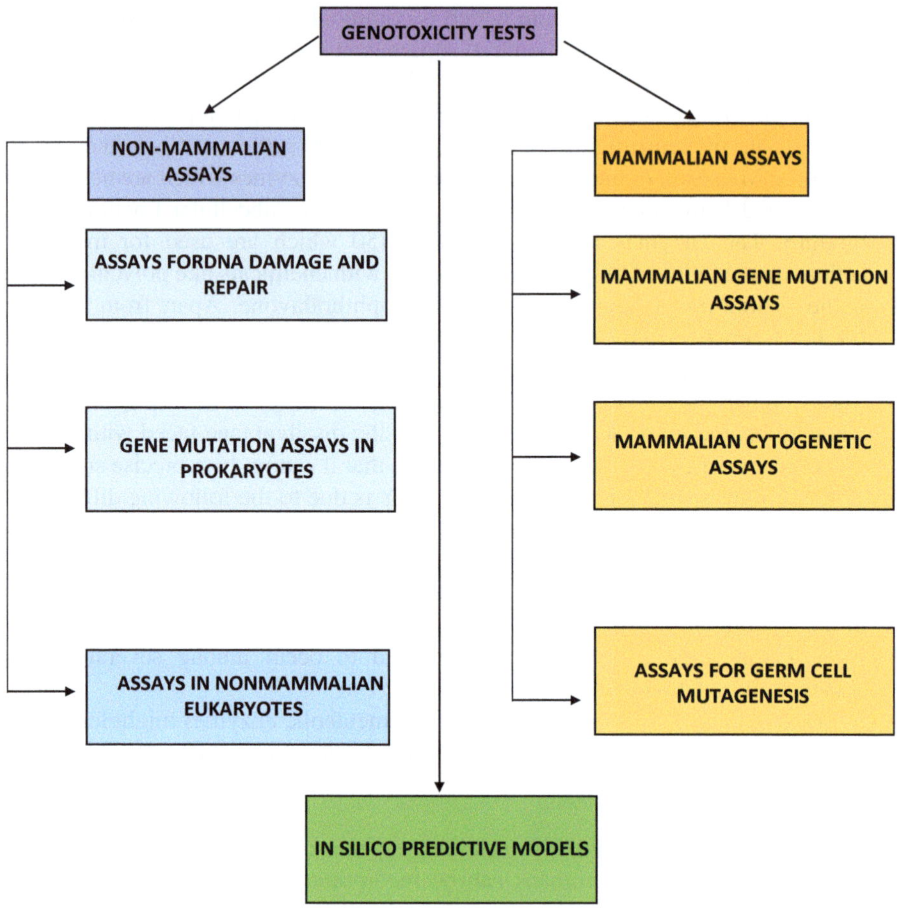

Fig. 43.2 Flow chart depicting the assay principles of various genotoxicity tests

- chromosome loss or gain (aneuploidy)
- DNA damage and DNA repair, genotoxic stress

Figure 43.2 depicts the principles of various genotoxicity tests, which are briefly described in the subsequent sections of this chapter.

43.4 Predictors of Genotoxicity: Structural Alerts and In Silico Assays

An analysis of chemical structure of a substance helps us to predict about to its mutagenic potential. The electrophilic sites in a chemical substance tend to readily react with the nucleophilic sites of DNA. Thus, the presence of electrophilic sites in a compound serves as a structural alert and indicates its chances of being a potential mutagen.

In silico assays are computer-based systems which can predict the genotoxic potential of a molecule depending on its chemical structure.

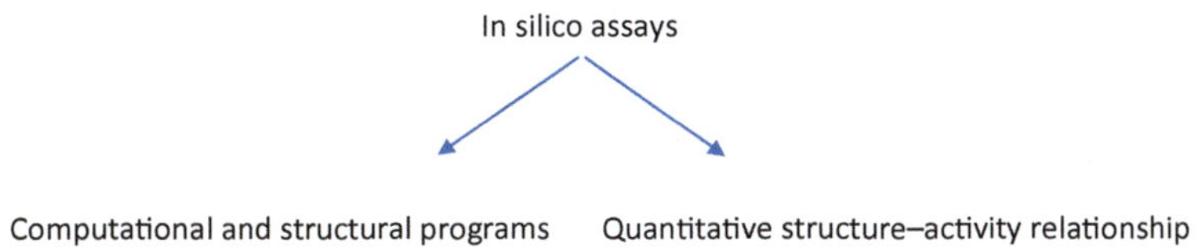

They can be used as predictors of genotoxicity and as add on to the existing mutagenic assays. It is optimal to use these assays along with other standard battery of in vitro and in vivo tests to improve its reliability.

In the near future, in silico assays might evolve as a cost effective, resource and time saving tool to screen large number of compoundsfor genotoxicity testing.

43.5 Assays for DNA Damage and Repair

DNA damage and DNA repair assays are grouped as indicator tests, since they mainly assess DNA damage and may or may not identify the effects of mutation. DNA damage can be assessed directly by measuring DNA adducts or strand breakage. DNA damage can be assessed indirectly also by measuring DNA repair processes which happens as a response to genetic insult. (Fig. 43.3).

43.5.1 Comet Assay for DNA Strand Breakage

It is also known as single cell gel electrophoresis assay (SCGE). It is a sensitive indicator of DNA damage and it is widely used for genotoxicity testing.

In comet's assay, after exposure to test agent, the cells are suspended on to agarose plated slides, cells are lysed and DNA is released. This is followed by electrophoresis at alkaline pH >13 and then for observing and analysing the image, DNA is stained with fluorescent dye. This entire procedure facilitates the detection of strand breaks in DNA. During the migration process, the large pieces of DNA settle near the head whereas the DNA fragments which are formed due to DNA damage move rapidly towards to tail. These pieces of DNA fragments which move towards to the tail resemble a comet. The extent of DNA damage can be assessed based on the intensity of the comet tail relative to the total intensity (intensity of head and tail).

Rat or mice liver, stomach and bone marrow are the preferred tissues for conducting in vivo comet assay. For in vitro comet assay, cell lines usually used

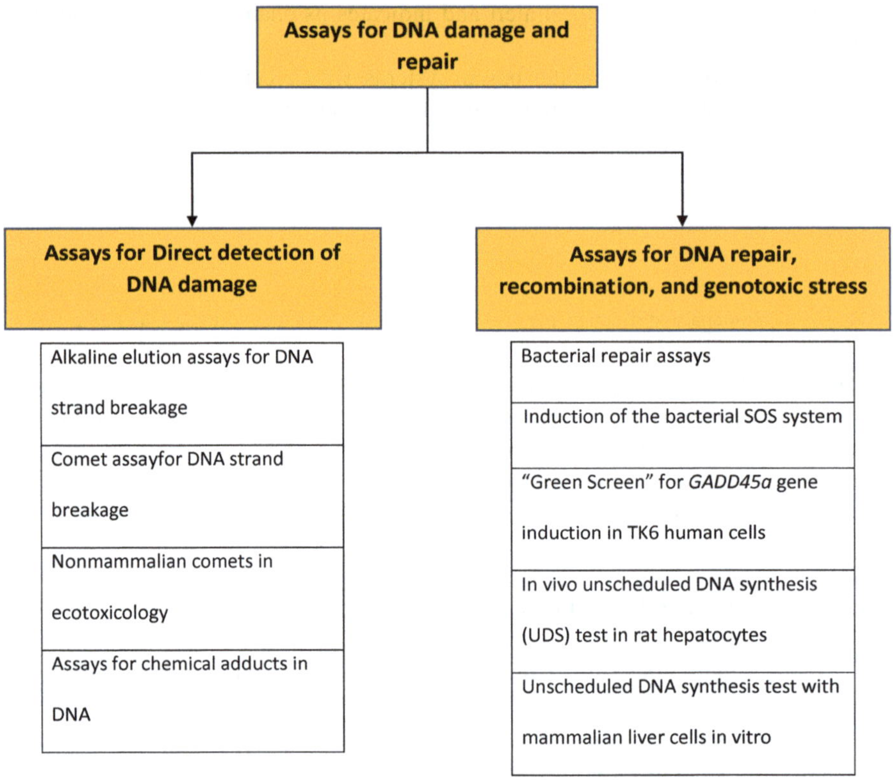

Fig. 43.3 Genotoxicity assays detecting DNA damage, DNA repair and genotoxic stress

are mouse lymphoma cells, TK6 human lymphoblast cells and human peripheral lymphocytes.

Comet assay can also be performed in different species like humans, drosophila, plants, fish, worms and amphibians. Different comet assay procedures are available, hence to facilitate the regulatory decision process, it is necessary to validate and standardise the existing comet assay techniques.

43.5.2 DNA Repair Assays

DNA damage can be measured indirectly by assessing DNA repair processes and other genotoxic stress response mechanisms which gets activated in a cell due to DNA damage.

43.5.2.1 In Bacteria

A genotoxic compound can cause increased toxicity in bacterial strains deficient of DNA repair mechanisms than the wild type strains that are capable of DNA repair.

Bacterial repair assays have been developed based on the principle of this differential killing of repair-deficient and wild type bacteria. For example, bacterial repair assay developed in *rec*$^+$ and *rec*$^-$ strains of *Bacillus subtilis*. However, at present these bacterial repair assays are not commonly used for mutagenicity testing.

43.5.2.2 Unscheduled DNA Synthesis in Cultured Mammalian Cells and Intact Animals

The DNA repair processes occurring in a cell due to DNA damage can be assessed by measuring unscheduled DNA synthesis. UDS can be evaluated by assays in cultured mammalian hepatocytes, germ cells, hepatocytes, other tissues and organs isolated from rodents and other mammals. Even though UDS is an indicator of DNA damage, in cases of irreparable DNA damage, unscheduled DNA synthesis might not occur. Hence, with the absence of UDS in a cell, it cannot be concluded that DNA damage has not occurred.

43.5.3 Assays Based on DNA Stress Response Mechanisms

When exposed to a genotoxicant, apart from DNA repair processes, the stress response mechanisms also get induced in a cell and these genotoxic stress responses can serve as an indirect indicator of DNA damage. Assays have been developed in microorganisms and cultured mammalian cells for the assessment of genotoxic stress response.

In bacteria, DNA repair cascades called "SOS response" gets activated following DNA damage and it can be detected by phage induction, colorimetry, or green fluorescent protein (GFP) reporter induction.

43.5.3.1 GreenScreen Assay or GADD45a-GFP Assay

In mammalian cells, stress responses activated due to DNA damage, cause increased expression of growth arrest and DNA damage gene (*GADD45a*), which is known to play a key role in DNA repair cascades. Green screen assay is an in vitro green fluorescent protein-based assay developed in human lymphoblastoid TK6 cell line to assess the levels of *GADD45a* gene when exposed to a genotoxicant.

43.6 Gene Mutation Assays in Prokaryotes

Among the in vitro genotoxicity tests available for detecting gene mutation, Ames test is widely used and preferred.

Auxotrophsare microbial strains which have a unique nutritional requirement that is distinct from wild type strains of a species. Detection of back mutations or reversion of the auxotrophs to wild type, forms the basis of microbial assays like Ames test. Dr.Bruce and his colleagues developed the Ames test, which quantifies histidine and tryptophan revertants in *Salmonella typhimurium* and *Escherichia coli*respectively.

Table 43.2 Various types of bacterial tester strain used and mutation detected by Ames test

Bacterial species	Type of strain	Mutations detected
Salmonella typhimurium	TA1535	Base pair substitution mutation
	TA1537	Frameshift mutation due to addition
	TA1538	Frameshift mutation due to deletion
	TA 100	Base pair substitution mutation not detected by TA1537
	TA 98	Frameshift mutation
Escherichia coli	WP2 uvrA	Base pair substitution mutation

The specific strains of *Salmonella*and *E. coli*, used for Ames test are mutated at histidine(*his*)and tryptophan (*trp*) biosynthesis gene locus and are not capable of producing histidine and tryptophan respectively. Thus, they are dependent on exogenous source of these amino acids in growth medium for cell division.

In Ames test, after exposure to different concentrations of test compound, the bacterial strains are plated on agar medium with trace amount of histidine, which is necessary for the initial cycles of cell division. When histidine available in the medium gets exhausted, bacteria will revert to wild type and will produce histidine and thus they will be able to form colonies. On the other hand, non-mutated bacteria will die. If the test compound is a mutagen, it will cause dose related increase in colonies in revertants than the spontaneous revertants in control plates.

Since bacterial strains do not metabolize promutagens in the same way as mammalian tissues, the assay is usually performed in the presence and absence of a rat liver S9 metabolic activation system. The use of different tester strains in the assay, helps to assess the specific mutagenic mechanism of the compound. Each tester strain can predict different types of mutation. Table 43.2 lists the routine tester strains used and the type of mutation detected by Ames test.

The other assays developed in bacteria for detecting gene mutation are *E. Coli LacZ* specific reversion assay and *E. Coli LacI* forward mutation assay.

43.7 Assays in Nonmammalian Eukaryotes

Fungi, plants and insects were used for experimental purposes during the early phase of mutagenicity testing. But now their use as an efficient tool for genotoxicity testing has been largely replaced by assays in bacteria and mammalian cells and most of these assays are of historical importance with limited use at present.(Fig. 43.4).

43.7.1 Sex-Linked Recessive Lethal Test

The sex-linked recessive lethal test developed in fruit fly *Drosophila melanogaster*is an in vivo forward mutation assay. It can provide information about germ cell

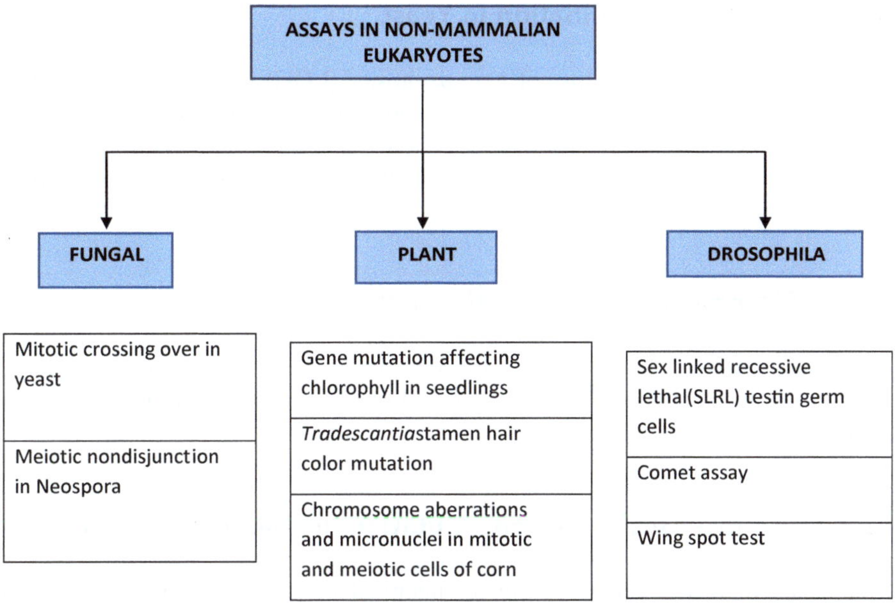

Fig. 43.4 Genotoxicity assays developed in non-mammalian eukaryotes

mutagens by identifying recessive lethal mutations at about 800 different loci on the X chromosome. The wild type male fruit fly is treated with the test agent and is mated with one or more females. The males of subsequent F_2 generation are screened for the presence or absence of wild type phenotype. If sex-linked recessive lethal mutation has occurred, wild type males will be absent and there will be substantial increase over the frequency of spontaneous SLRLs in the lineages derived from treated males. The disadvantage encountered is the need to screen large numbers of fruit fly offspring and also the genetic risk relevance to humans of this test in insects, is uncertain due to differences in mechanisms of metabolism and gametogenesis between the two species.

43.7.2 Plant Assays

Chromosomal aberration assays and micronucleus test in *Tradescantia* and specific locus tests in corn are some of the genotoxicity assays developed in plants. Even though plant assays are having very limited use at present for genotoxicity testing, they can be used for monitoring and evaluating the environmental pollutants and also for analysing the mechanisms involved in metabolism of promutagens by agricultural plants.

43.7.3 Mitotic Recombination in Yeast

Studies have reported that loss of heterozygosity (LOH) of one allele of tumour suppressor gene p53 can play a vital role in carcinogenesis. LOH tends to occur due to recombination of DNA sequences between chromosomes. Assays to detect mitotic crossing over and mitotic gene conversion have been developed in yeast for assessing mitotic recombination.

43.8 Mammaliangene Mutation Assays

Gene mutation assays have been developed using various mammalian cells and they are divided into three broad categories based on the test system which is used for genotoxicity testing. (Fig. 43.5).

43.8.1 In Vitro Assays for Gene Mutation in Mammalian Cells

Mutagenicity assays donein cultured mammalian cells for identifying gene mutations, detect forward mutations of reporter gene, leading to the development

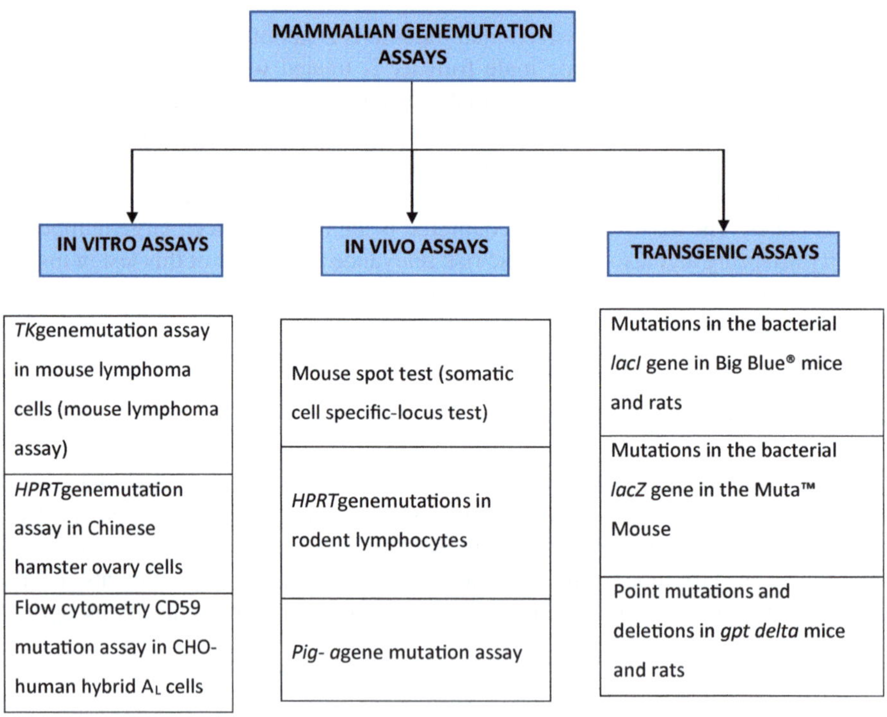

Fig. 43.5 Genotoxicity assays developed in mammalian for detecting gene mutations

of resistance to a toxic chemical. The two widely used in vitro mammalian cell gene mutation assays are

- Hypoxanthine guanine phosphoribosyl transferase (HPRT)assay
- Thymidine kinase (TK) assay

Forward mutations occurring at autosomal heterozygous loci of *TK* gene and at X chromosome linked lociof *HPRT* gene, inactivates the wild-type gene and results in loss of function of the gene products. Hence, forward mutations of *HPRT* and *TK* gene lead to development of resistance to the cytotoxic effects of the purine analogue, 6-thioguanine and pyrimidine analogue, trifluorothymidine respectively. Thereby the mutant cells can proliferate in the presence of toxic chemical.

In HPRT assay, $HPRT^{+}$ proficient cells are sensitive to 6-thioguanine and further cell division cannot occur. On the other hand, $HPRT^{-}$deficient mutant cells can proliferate in presence of purine analogues. Similarly, in TK assay, TK^{-}deficient mutant cells can proliferate in presence of pyrimidine analogues. The mutagenic effect of the test agent can be assessed based on the presence of mutant colonies in the culture plates.

Chinese hamster ovary (CHO) cells and mouse lymphoma L5178Y cells are used for HPRT and TK assay respectively. Studies have reported that human cells can also be used for the both these assays.

The need of using selective media in the above two assays, can be overcome with the use of immunological techniques like flow cytometry, as described in the flow cytometry CD59 mutation assay. It is also an in vitro mammaliancell gene mutation assay, performed using CHO-human hybrid A_L cell line.*CD59* gene on human chromosome 11 codes for glycosylphosphatidylinositol GPI-linked surface protein, CD59. CHO-human hybrid A_L cell line has been designed by incorporating CHO cells with a single copy of human chromosome 11.The hybrid cells will express CD59 and the normal CHO cells will not express this protein, since they lack *CD59* gene. If a mutation occurs in hybrid cells at *CD59* gene, the surface protein CD59 will not be expressed. In flow cytometry CD59 mutation assay, CHO-human hybrid $A_L$cells are stained with fluorescent monoclonal antibodies against CD59 and flow cytometry is used to estimate the number of $CD59^{-}$mutant cells.

43.8.2 In Vivo Assays for Gene Mutation in Mammalian Cells

43.8.2.1 Mouse Spot Test

It is a mutagenicity test of historical importance and it is used for detecting gene mutations in somatic cells. If inutero mutation of coat colour genes occurs in embryos offemale mice, treated with test agent, it leads to visible spots of altered phenotype (coloured coat spots) in heterozygous mice progeny. In spite of its simple design, large number of animals are required for this test and it has limited its routine use for genotoxicity testing. The mouse specific-locus test is the counterpart of mouse spot test used for detecting gene mutations in germ cells.

43.8.2.2 In Vivo HPRT Assay

Similar to in vitro CHO cells HPRT assay, in vivo HPRT assay can be conducted in lymphocytes derived from humans, rodents and monkeys.

43.8.2.3 *Pig-a* Mutation Assay

The phosphatidylinositol glycan class A gene (*Pig-a)* is a single copy gene located on the X chromosome and it encodes for an enzyme which is involved in the initial steps of GPI anchor synthesis. Even though around 30 other genes are essential for its synthesis, only *Pig-a*is a single copy gene and thereby a single inactivating mutation of*Pig-a*gene will result in alteration of cell phenotype.

Pig-a gene mutation will disrupt GPI anchor synthesis, leading to loss of GPI anchors and its associated cell surface protein markers like CD55, CD59, CD24, which are present on surface of mammalian hematopoietic cells.

In *Pig-a* gene mutation assay, peripheral red blood cells derived either from rats, mice, monkeys or humans, are stained with fluorescent labelled monoclonal antibodies against specific GPI-anchored protein (usually CD59 in humans, rats and CD24in mice). Flow cytometry is used to estimate the frequency of *Pig-a* mutant cells, which will bed eficient in GPI-anchored proteins.

Several studies conducted using *Pig-a* gene mutation assay, have reported that it can be used as an in vivo follow up test for agents with positive mutagenic results in in vitro assays. Currently the assay procedures are being standardised and validated, for its inclusion in the standard battery of tests by regulatory authorities.

43.8.3 Transgenic Assays

Transgenic animals are produced by incorporating foreign DNA sequence to the host genome, using recombinant DNA technology.

"MutaMouse" assay and "Big Blue" assay are the widely used mammalian Transgenic Rodent (TGR) mutation assays. In "Big Blue" assay target gene for mutagenesis is *lacI*, while *lacZ*is the target gene in "MutaMouse" assay.

The lac operon of *E. coli*contains genes which are involved in lactose metabolism and they are one of the most commonly used reporter genes in the field of molecular biology. The *lacZ* gene encodes for the enzyme β -galactosidase, which cleaves lactose into glucose and galactose and gets activated only in the presence of lactose. *LacI* gene encodes for a lactose repressor protein, which normally inhibits the transcription of β -galactosidase. The presence of lactose releases this block and allows the production of β -galactosidase.

The activity of the *lacZ*gene can be detected by using X-gal, a chromogenic substrate which gets cleaved by β -galactosidase and forms a blue coloured precipitate.

In "Big Blue" assay, mutation of *lacI* gene inactivates the lactose repressor protein and allows the expression of β -galactosidase. X Gal gets cleaved by β-galactosidase enzyme and the mutant colonies will be blue in colour. On the other hand, the normal plaques will be colourless due to presence of lactose repressor

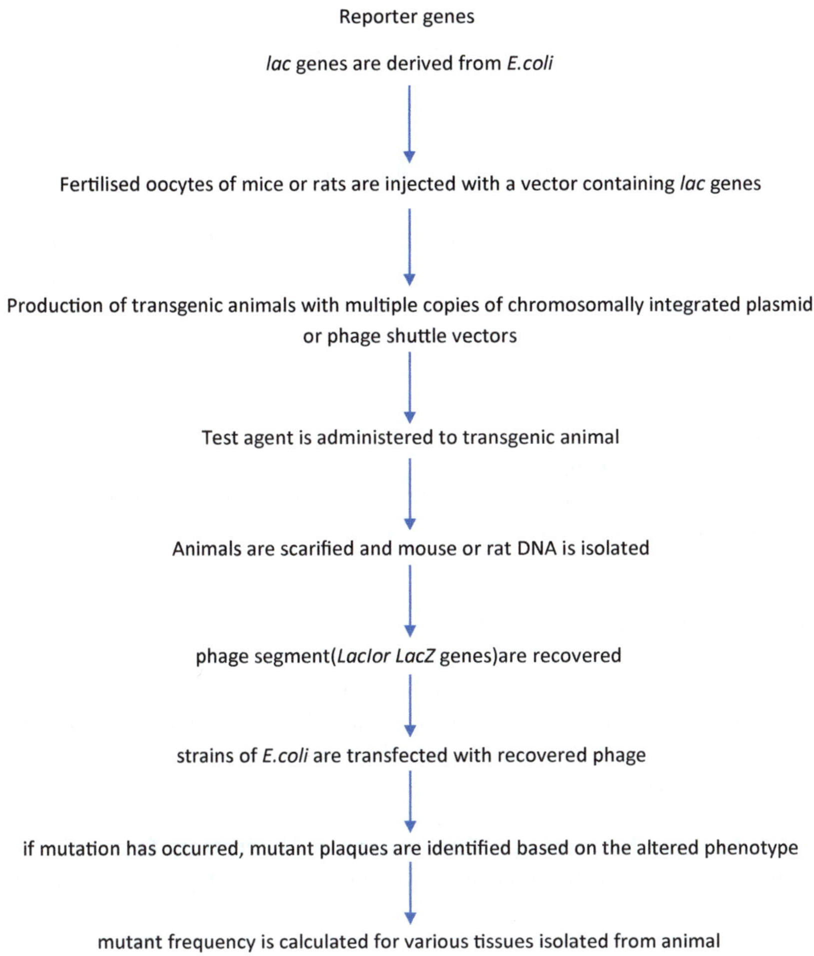

Fig. 43.6 Flowchart depictingthe sequence of events followed in transgenic rodent mutation assays

protein. In the "MutaMouse" assay, if *lacZ* gene is mutated, β-galactosidase enzyme will not be produced and hence X Gal will not be cleaved. This results in the formation of colourless mutant plaques. Whereas the normal colonies will appear blue due to presence of β-galactosidase enzyme. For both systems, the mutant frequency will be estimated based on the principle of colorimetric detection of mutants. (Fig. 43.6).

The transgenic animals used for detecting resistance to 6-thioguanine are *gpt delta* mice and rats, to identify point mutations in *gpt* gene. The use of transgenic animals for mutagenicity testing provides the advantage of combining in vivo metabolic activation and pharmacodynamics with in vitro microbial detection systems. The advantage with use of TGR mutation assays is that any rodent tissue yielding high quality DNA can be used for genotoxicity testing. However, some studies have observed TGR mutation assays to be less sensitive in detecting clastogenic events.

43.9 Mammalian Cytogenetic Assays

Cytogenetic assays use direct methods like microscopy for observing genetic change that has occurred in a cell, whereas the gene mutation assays depend on the alterations in phenotype for analysing genetic damage.

The various endpoints assessed by cytogenetic assays are the following:

- chromosome aberrations
- micronuclei
- sister chromatid exchanges
- aneuploidy—changes in chromosome number, which can be hyperploidy (gain of chromosomes) or hypoploidy (loss of chromosomes).

43.9.1 Chromosome Aberrations

Chromosomal aberration test can be performed in vitro using cultured mammalian cells like Chinese hamster or human cells and in vivo using mammalian bone marrow cells. For detecting chromosomal aberrations, metaphase analysis is widely followed in cytogenetics.

43.9.1.1 In Vitro Mammalian Chromosome Aberration Test

The important factors to be considered while designing cytogenetic assays are the following:

- To select and obtain cells suitable for treatment and analysis.
  Chinese hamster cells and human peripheral lymphocytes are preferred because they have stable, well-defined karyotype and low chromosome number with large chromosomes.
- To provide adequate time interval between exposure to test agent and the sampling of target cells for analysis.

The cell cultures are treated with test agent and colchicine (metaphase arresting compound) is used to arrest mitosis at metaphase stage. These cells are collected, stained and analysed microscopically for identifying chromosome aberrations. The extreme culture conditions might lead to development of artifacts, which can interfere with analysis of chromosomal aberrations.

43.9.2 In Vivo Mammalian Chromosome Aberration Test

In vivo assays for detecting chromosomal aberrations have been developed using cells derived from rodent or hamster bone marrow. After exposure to test agent, animals are treated with colchicine 3–5 h before their sacrifice. The cells are collected, stained and analysed microscopically for identifying chromosome aberrations. Automated techniques like Fluorescent in situ hybridization (FISH) can be used to aid in identifying, scoring and analysis of chromosomal aberrations.

43.9.3 Micronucleus Assay

The micronucleus assay is a simpler alternative to metaphase analysis, latter being more time consuming and labour-intensive. Micronucleus are membrane-bound structures which contain chromosomal fragments or whole chromosomes, that were not transferred to daughter nuclei during the mitotic process. Micronucleus assay can serve as a simple indicator of chromosomal damage and can be performed either as an in vitro or in vivo assay. Initially, mammalian lymphocytes were used for in vitro micronucleus assay, but later other cell lines like Chinese hamster, human TK6, and mouse lymphoma cells are being used. For in vivo assay, tissues like bone marrow, testes, bladder, buccal mucosal cells, stomach, liver, lung and spleen are utilised. The erythrocytes derived from rodent bone marrow are most commonly used for in vivo micronucleus assay.

Cytokinesis-blocked micronucleus (CBMN) assay is the most commonly used technique for detecting micronuclei. After exposure to a mutagenic substance, cells that have completed one nuclear divisionare treated with cytochalasin- B, which inhibits cytokinesis without blocking mitosis. This results in the formation of binucleated and multinucleated cells. Binucleated cells alone are selected for the micronuclei scoring. Even though microscopy is the most conventional technique for visualizing micronuclei, nowadays FISH technique is used for a precise analysis of micronuclei formed.

The micronucleus assay is a rapid, highly reliable tool for detecting potential clastogens and aneugens, especially those which might cause genetic toxicity due to occupational exposure. The test protocol and procedures of micronucleus assay have been accepted and validated by several international regulatory agencies. It is also a part of the standard battery of tests for genotoxicity testing.

43.9.4 Sister Chromatid Exchange

In sister chromatid exchanges (SCE), a reciprocal exchange of segments occurs between two chromatids of a chromosome. SCE can be induced in cultured cells and in intact animals by genotoxic agents and the exchange between chromatids can be visualised cytologically by differential staining of chromatids. In spite of the simple design of SCE assays, doubt prevails about its biological and toxicological significance and therefore, SCE assays are considered as nonspecific indicators of mutagen exposure.

43.9.5 Aneuploidy

Assays designed for detecting aneuploidy are targeted towards agents which disrupt the mitotic and meiotic process. The endpoints usually included for aneuploidy assays are chromosome counting, presence of micronuclei with kinetochores, occurrence of abnormal spindle in cells. Even though, assays for aneuploidy are not well established, studies have reported that effects of aneuploidy can significantly affect human health. FISH based assays for detecting aneuploidy have been developed in somatic cells and germ cells (sperms).

43.10 Germ Cell Mutagenesis Assays

Assays developed in mammals for assessment of germ cell mutagenesis serve to be more relevant indicator of genetic risk to humans. Thus, in spite of their limitations like being complex, time consuming and expensive, germ cell mutagenesis assays are of unique importance in genotoxicity evaluation.

Studies have reported that some chemical mutagens specifically affect certain germ cell stages and also it has been observed that late stages of spermatogenesis are more sensitive to mutagens.

43.10.1 Mouse Specific-Locus Test

Mouse specific-locus test has been designed to identify recessive mutations in gene, which can cause alteration in phenotype that is visibly identified like changes in coat colour and ear size. In this assay, wild type mice of either sex are treated with test compound and are mated with tester strains of mice, which express homozygous recessive genes at specific loci. If the progeny is of wild type, then it suggests the absence of mutation. If a recessive phenotype is present in the offspring, it indicates the occurrence of mutation in mice exposed to test substance.

43.10.2 Cytogenetic Assays

Assays have been developed in rodent spermatogonia, spermatocytes and oocytes for detecting chromosomal aberrations, micronuclei. These include the following assays namely —Micronucleus assay in rodent spermatids

- Chromosomal aberrations test in mouse spermatogonia and spermatocytes
- Cytogenetic analysis of mice oocytes
- FISH based techniques to detect aneuploidy in sperms of mice and rats.

43.10.3 Mouse Heritable Translocation Assay

Mouse heritable translocation assay has been developed to detect translocation, which is complete exchange of genetic material between two chromosomes. In this assay, after exposure to test agent, male mice are mated with normal females. Male offspring are mated with group of female mice. If translocation has occurred, then it would result in complete or partial infertility in first generation male offspring, which can be easily identified from the litter size of the subsequent generation.

43.10.4 Dominant Lethal Assay

The dominant lethal assay is an in vivo assay developed in mice, rats and it serves as a reliable tool to screen clastogens and aneugens affecting germ cells. In this assay, male mice or rats are treated with test agent, mated with normal females and group of females are replaced with new groups at selected intervals. If a dominant lethal mutation has occurred in the parent germ cells, it will cause early foetal death. Before parturition, pregnant female rats or mice are sacrificed and early embryonic deaths are estimated.

43.10.5 Assays for DNA Damage

Unscheduled DNA synthesis in rodent germ cells and comet assay for DNA strand breaks are some of the assays developed in rodents for detecting DNA damage in germ cells.

43.11 Regulatory Guidelines for Genotoxicity Testing

- Initially, a binary (yes/no) approach was followed widely for evaluating the outcome of mutagenicity assays. In that way, the information gained, indicated whether a chemical has mutagenic potential or not. Gradually over several years, some of the various genotoxic mechanisms, through which chemicals and other substances affect human health were established. This led to a paradigm shift in the field of genetic toxicology, from mutagenic risk identification to a further complete mechanistic approach for genetic risk and hazard assessment.
- The data obtained from several mutagenicity assays can be evaluated using a two-step approach. In this approach, first a weight-of-evidence (WOE) analysis is

established by integrating all the available information relevant to the genotoxic potential of a compound. Then the second step is to determine the mutagenic mechanism leading to adverse health outcome through quantitative dose–response analysis. Thus, detailed information about the mutagenic potential of a chemical substance can be obtained, facilitating the process of regulatory decision-making.

- A series of accurately conducted, standardized, validated mutagenicity assays with high sensitivity are the basic prerequisites for regulatory submissions of a product. To ensure the safety of human participants, genotoxicity testing of a new drug candidate is a mandatory requirement by most of the drug regulatory agencies.
- A mutagen can cause genotoxicity through multiple mechanisms and thereby it is recommended to use a battery of mutagenicity assays. Further, by conducting this battery of tests, the chances of missing out a potential genotoxicant can be minimized. It is optimal to follow the methods and procedures described in internationally accepted guidelines for performing the genetic toxicology assays. Since, these guidance documents have been prepared by international regulatory bodies, after considering the various strengths and limitations of the routine mutagenicity tests.
- Depending on the type of substance to be tested and the country to be registered, guidelines and specific requirements for mutagenicity testing may differ significantly. Various global and regional regulatory guidelines are available for mutagenicity testing of a substance based on its use as pesticide, food additive, pharmaceuticalsand medical use products.
- Some of the global regulatory guidelines for genotoxicity testing are:
  - Organisation for Economic Co-operation and Development (OECD) Guidelines for the Testing of Chemicals
  - International Conference on Harmonization ICH S2(R1)—Guidance on genotoxicity testing and data interpretation for pharmaceuticals intended for human use
  - International Workshops on genotoxicityTesting (IWGT)
  - The International Programme on Chemical Safety (IPCS) under auspices of the World Health Organization (WHO)
  - United States Environmental Protection Agency (EPA) Guidelines for Mutagenicity Risk Assessment
  - In India, "New Drugs and Clinical Trials Rules, 2043" deals with the prerequisites to carry out the clinical trials of a new drug before its marketing.
- These guidance documents provide a detailed description about the standard battery of assays used for screening and follow-up testing. They also describe the steps to interpret the results of various in vitro and in vivo genotoxicity tests.
- The following factors regarding the test agent should be considered, as they might influence the choice of mutagenicity test:
  - Chemical structure and physicochemical properties of the test agent, for probing on the possible structure–activity relationships

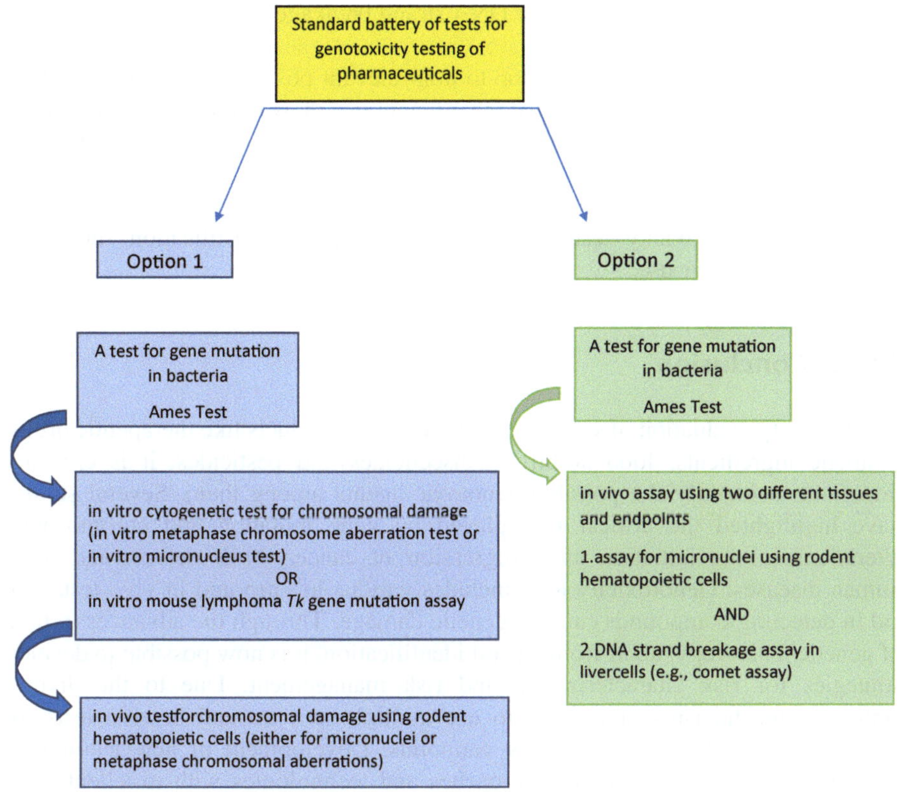

Fig. 43.7 Schematic representation of ICH S2(R1) guidelines standard battery of tests for genotoxicity testing of pharmaceuticals

  - Chemical and biological reactivity of the test agent, and its associations with established genotoxic compounds
  - Routes of exposure, bioavailability, expected routes of metabolism and target organs.
- ICH S2(R1) guidance on genotoxicity testing for pharmaceuticals, revised in 2008, replaces and combines the ICH S2A and S2B guidelines. This revised guidance recommends the use of the following two options as a standard test battery for genetic toxicology testing of drugs. (Fig. 43.7).
- The first option includes Ames test, in vitro cytogenetic assay using mammalian cells such as in vitro metaphase chromosome aberration test or in vitro micronucleus test or in vitro mouse lymphoma *Tk* gene mutation assay can be used as an alternative. Finally, an in vivo test for chromosomal damage using rodent hematopoietic cells.
- The second option includes a bacterial mutation assay (Ames test) and an in vivo test with two different endpoints, e.g., assay for micronuclei using rodent

hematopoietic cells and assay for DNA strand breakage in liver cells (e.g., comet assay)

- To minimize the delay caused due to non-relevant positive results, obtained for in vitro genotoxicity tests and for the efficient use of available animal resources, the second option of standard test battery was included in the revised ICH S2 (R1) guidelines.
- If inconclusive results are obtained from the standard battery of tests, other validated genotoxicity tests can be used and proper justification should be provided for their use.

43.12 Conclusion

For the safety evaluation of various commonly used products like therapeutic drugs, cosmetic ingredients, food additives, insecticides and pesticides, it is crucially important to identify the possible genotoxic agents among them. Several studies have highlighted the critical role played by gene mutation and chromosomal alterations in the initiation and progression of cancer, birth defects and other human diseases. Genotoxicity tests includes various in vitro and in vivo tests that aid in detecting compounds causing genetic damage. Through the advances in field of genetic toxicology, apart from hazard identification, it is now possible to develop strategies for risk characterization and risk management. Due to the diverse genotoxic mechanisms, it is ideal to use multiple in vitro and in vivo tests for analysing the different genotoxicity endpoints. Development of new automated, cost effective, and time saving approaches and technologies with excellent data-mining tools to screen large number of compounds is the need of the hour. Further by integrating and streamlining the different old and new strategies, a tremendous change is set to happen in the near future in the field of genetic toxicology testing.

Bibliography

Ames BN, McCann J, Yamasaki E. Methods for detecting carcinogens and mutagens with the salmonella/mammalian-microsome mutagenicity test. Mutat Res Mutagen Relat Subj. 1975;31 (6):347–63.

Bajpayee M, Pandey AK, Parmar D, Dhawan A. Current status of short-term tests for evaluation of genotoxicity, mutagenicity, and carcinogenicity of environmental chemicals and NCEs. Toxicol Mech Methods. 2005;15(3):155–80.

Bode G. International guidelines. In: Vogel HG, Maas J, Hock FJ, Mayer D, editors. Drug discovery and evaluation: safety and pharmacokinetic assays. Berlin: Springer Berlin Heidelberg; 2013 [cited 2020 Jun 1]. p. 1235–1263. Available from: http://link.springer.com/10.1007/978-3-642-25240-2_53.

Brusick DJ, Fields WR. Genetic toxicology. In: Hayes AW, Kruger CL, editors. Hayes' principles and methods of toxicology. 6th ed. New York: CRC Press; 2014. p. 1173–200.

Dhawan A, Bajpayee M, editors. Genotoxicity assessment: methods and protocols. (Methods in molecular biology; Vol. 2031). New York, NY: Springer New York; 2019 [cited 2020 Jun 1]. Available from: http://link.springer.com/10.1007/978-1-4939-9646-9.

Fenech M. The in vitro micronucleus technique. Mutat Res Mol Mech Mutagen. 2000;455(1–2): 81–95.

International Conference on Harmonisation of Technical Requirements for Registration of Pharmaceuticals for Human Use (ICH). 2020aGuidance on genotoxicity testing and Data interpretation for Pharmaceuticals intended for human use S2(R1) [Internet]. [Cited 2020 Jun 1]. Available from: https://database.ich.org/sites/default/files/S2%28R1%29%20Guideline.pdf.

International Conference on Harmonisation of Technical Requirements for Registration of Pharmaceuticals for Human Use (ICH).2020b. Assessment and control of DNA reactive (mutagenic) impurities in pharmaceuticals to limit potential carcinogenic risk. M7 (R1) [Internet]. [Cited 2020 Jun 1]. Available from: https://database.ich.org/sites/default/files/M7_R1_Guideline.pdf.

Klapacz J, Gollapudi BB (2043) Genetic toxicology. In: Klassen CD, editor. Casarett & Doull's toxicology: the basic science of poisons. 9th ed McGraw-Hill Education, New York. Available from: https://accesspharmacy.mhmedical.com/content.aspx?bookid=2462&sectionid=202672669.

Kumaravel TS, Vilhar B, Faux SP, Jha AN. Comet assay measurements: a perspective. Cell Biol Toxicol. 2009;25(1):53–64.

Lambert IB, Singer TM, Boucher SE, Douglas GR. Detailed review of transgenic rodent mutation assays. Mutat Res Mutat Res. 2005;590(1–3):1–280.

Miura D, Dobrovolsky VN, Kasahara Y, Katsuura Y, Heflich RH. Development of an in vivo gene mutation assay using the endogenous Pig-A gene: I. Flow cytometric detection of CD59-negative peripheral red blood cells and CD48-negative spleen T-cells from the rat. Environ Mol Mutagen. 2008;49(8):614–21.

OECD genetic toxicology test guidelines and updates. [Internet].2020. [Cited 2020 Jun 1]. Available from:: https://www.oecd.org/publications/overview-on-genetic-toxicology-tgs-9789264274761-en.htm.

Parry JM, Parry EM, editors. Genetic toxicology: principles and methods. (Methods in molecular biology; vol. 817). New York, NY: Springer New York; 2012 [cited 2020 Jun 1]. Available from: http://link.springer.com/10.1007/978-1-61779-421-6.

Preston RJ, Hofmann GR. Genetic toxicology. In: Klaassen CD, Watkins JB, editors. Casarett & Doull's essentials of toxicology. 3rd ed. New York: McGraw-Hill Education; 2015. p. 135–47.

Ross CD, Lim C-U, Fox MH. Assay to measure CD59 mutations in CHO AL cells using flow cytometry. Cytometry A. 2005;66A(2):85–90.

Sukumaran SK, Ranganatha R, Chakravarthy S. High-throughput approaches for genotoxicity testing in drug development: recent advances. Int J High Throughput Screen. 2016;6:1–12.

Teaf CM, Middendorf PJ. Mutagenesis and genetic toxicology. In: Williams PL, James RC, Roberts SM, editors. The principles of toxicology: environmental and industrial applications. 2nd ed. New York: Wiley-Interscience; 2000. p. 239–64.

Turkez H, Arslan ME, Ozdemir O. Genotoxicity testing: progress and prospects for the next decade. Expert Opin Drug MetabToxicol. 2017 Oct 3;13(10):1089–98.

Wang T, Jacobson-Kram D, Pilaro AM, Lapadula D, Jacobs A, Brown P, et al. ICH guidelines: inception, revision, and implications for drug development. Toxicol Sci. 2010;118(2):356–67.

Screening Methods for the Evaluation of Anticancer Drugs

44

A. Vimala Ananthy

Abstract

Drug screening method for anticancer agents is rather a recent entity in the field of medicine. Most of the drugs which are already approved for various types of cancers possess the limitation of being more cytotoxic and acting through a specific type of mechanism. Key statistics of cancer worldwide in the past decade shows that there will an 80% increase in cancer incidence especially for low and middle income countries. It is observed that the increased rates are due to greater exposure to risk factors and longer life span. The need for novel drugs for the treatment of different cancers is still under great pressure. For a drug to be successful in treating the target population it requires competent preclinical studies using animal models and in-vitro studies which can exactly simulate the disease in patients. There are various types of models developed for studying the anticancer property which include both in-vitro and in-vivo models. Meticulous use of these methods specific to the cancer subtype can help to decrease the rate of failure of drug action in clinical phases. This chapter highlights the various in-vitro and in-vivo models for anticancer drug discovery with due mention of their strengths and limitations.

Keywords

Anticancer drugs · Antineoplastic · In-vitro assays · Cancer

A. V. Ananthy (✉)
Department of Pharmacology, MGMCRI, Puducherry, India

M. Lakshmanan et al. (eds.), *Introduction to Basics of Pharmacology and Toxicology*, https://doi.org/10.1007/978-981-19-5343-9_44

44.1 Introduction

Cancer is a disease with unregulated cell proliferation and a tendency to spread throughout the body. It has become a significant cause of death in recent days due to the inherent nature of the disease of having uncontrolled growth. Even though numerous preclinical researches are going on for various cancers, very few drugs cross all the phases of clinical trials to get into the market. The pathogenesis of cancer is multifactorial and involve genetic malfunction with proto-oncogene and tumor suppressor gene. Current screening methods include various *in-vitro* and *in-vivo* techniques with their respective pros and cons. This chapter provides insight into the different preclinical screening methods used to study the effect of anticancer agents and their advantages and disadvantages.

44.2 *In-Vitro* Methods

In-vitro tumor models not only help in searching new compounds showing antitumor activity but also for assessing their effectiveness. These models facilitate more comprehensive primary screening antitumor drugs, thus stopping those with deficient antitumor activity from entering preclinical animal testing. However, the use of animal models has certain demerits, such as high cost, feasibility issues, and varied responses between species (Table 44.1) and (Fig. 44.1).

44.2.1 Functional Assays (Cell Viability Assays)

- *List of cell viability assays:*
  - Tetrazolium reduction assays
    MTT [3-(4,5-Dimethylthiozol-2-yl)-2, 5-diphenyltetrazolium bromide] assay
    MTS [3-(4,5-Dimethylthiozol-2-yl)-5-(3-carboxymethoxyphenyl)-2-(4-ulphophenyl-2Htetrazolium]

Table 44.1 Merits and demerits of in-vitro anticancer models

Merits	Demerits
Less time consuming	More false positive or false negative results
Cost-effective	The role of pharmacokinetics in drug response cannot be studied
A small quantity of drugs is needed	The solid tumor geometry will be different in monolayer cultures
A large number of drugs can be tested	
Easier to manage	
Minimal experimental errors	

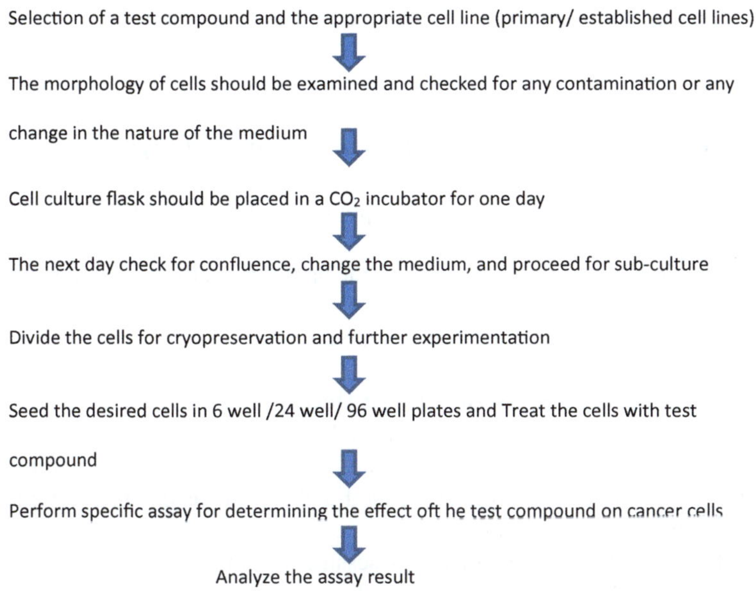

Fig. 44.1 General protocol for *in vitro* anticancer drug screening

XTT (2,3-Bis-(2-Methoxy-4-Nitro-5-sulfophenyl)-2H-Tetrazolium 5-Carboxanide

WST-1 (4[3-(4-iodophenyl)-2-(4-Nitrophenyl)- 2H-Tetrazolio]-1-3-benzene disulfonate)

- Resazurin reduction assay
- Protease viability marker assay
- ATP assay

- *Microculture tetrazolium test (MTT) assay:*
  - *Principle:* The conversion of MTT into purple formazan crystals by metabolically active cells with the help of mitochondrial dehydrogenase enzyme is the principle used by this assay.
  - In the presence of cytotoxic agents, dead cells are unable to cleave MTT; thus, the intensity of formazan production corresponds to the viable cell number and is inversely proportional to the degree of cytotoxicity.
  - The percentage of inhibition is calculated by using the following formula.

$$\text{Percentage of inhibition} = [(\text{mean OD of control} - \text{mean OD of test})/\text{Mean OD of control}] \times 100$$

  - Factors influencing the sensitivity of MTT assay

    Concentration of MTT

    Incubation period of culture plate

    Percentage of viable cells and their metabolic activity

Altered pH or depletion of essential nutrients in culture
Confluence of cells

44.2.2 Non-Functional Assays (Morphological Assays)

- The specific mechanism of the test compound can be studied with non-functional assays.
- The intracellular signaling or the site of the apoptotic pathway getting affected by the cytotoxic drug can be identified using this assay.
- The following are the various non-functional assays

44.2.2.1 Assessment of Mode of Cell Death

- *Acridine orange-ethidium bromide (AO-EB) staining:*
  - Apoptosis due to anticancer drugs can be detected by AO-EB staining.
  - The cell membrane changes associated with apoptosis due to cytotoxic agents are responsible for the staining.
  - Both quantitative and qualitative assessments can be made using this assay.
  - Principle: AO penetrates normal and early apoptotic cells with intact membrane, binds to DNA and emits fluorescence green.
  - EB can only enter cells that have lost membrane integrity, such as late apoptotic and dead cells, and bind to concentrated DNA fragments or apoptotic bodies, and emit orange-red fluorescence.
- *Hoechst staining:*
  - Both live and dead cells can be identified in the rapid assay using Hoechst staining, and hence it is being used. In apoptotic cells, chromatin condensation and nuclear fragmentation can be better examined through the double staining with Hoechst / propidium iodide stain using fluorescence microscope analysis.
  - *Principle:* Hoechst stains nuclei of both live and dead cells, which can be examined by fluorescence microscopy.
  - Hoechst is a blue fluorescing dye that stains chromatin DNA.
  - The red fluorescing dye propidium iodide is only permeable to dead cells and cannot enter the intact plasma membrane of living cells.
  - Cells are scored apoptotic if their nuclei show chromatin condensation and marginalization of nuclear beading or other apoptotic morphology.
  - Often apoptotic nuclei fragment into smaller structures.

44.2.2.2 Assessment of Cell Death Markers

- Annexin V assay:
  - Apoptosis at an early stage can be identified using Annexin V than assays based on nuclear changes such as DNA fragmentation.
  - It precedes the loss of membrane integrity which accompanies the latest stages of cell death resulting from either apoptosis or necrosis caused by anticancer agents

  - This assay does not distinguish between apoptotic or necrotic cell death.
  - *Principle:* Annexins are a group of calcium-dependent phospholipid-binding proteins that bind to phosphatidylserine (PS) with high affinity.
  - Under functional conditions, PS is mainly found in the inner leaflet of the plasma membrane.
  - Upon initiation of apoptosis, PS is translocated to the extracellular membrane leaflet.
  - Annexin V may be conjugated to fluorochrome for flow cytometric analysis.
- JC-1 (5,5′,6,6′-Tetrachloro-1,1′,3,3′-tetraethyl-imidacarbocyanine iodide) staining
  - The changes in the transmembrane potential of mitochondria can be quantitatively measured using JC-1 staining.
  - Principle: apoptosis begins with changes in the mitochondrial permeability during which there is loss of electrochemical gradient across the membrane.
  - In apoptotic cells, the cationic dye JC-1 stays as a green fluorescent monomeric form in the cytoplasm.
  - In healthy cells, due to intact mitochondrial membrane potential, the dye collects into the mitochondrial matrix and stains bright red.

44.2.2.3 Assessment of Genotoxicity of Compounds

- Comet assay: comet assay (single-cell gel electrophoresis) is a rapid and sensitive technique to measure DNA strand breaks in individual cells.
- In addition to measuring DNA single-strand breaks, it can also measure the cross-links, base damage, and apoptotic nuclei.
- Staining is done with fluorescent dye.
- It is named a "comet" since the stained image looks like a comet with the head formed by the intact DNA and the tail by the broken pieces of DNA
- The steps involved in comet assay are as follows:
  - Living cells are obtained from tissue, culture media, or blood
  - These cells are subjected to DNA damage through chemical methods, UV or GAMMA radiation
  - Damaged cells are embedded on agarose gel and lysed
  - These cells undergo electrophoresis and fluorescent staining generating the comet-shaped image
  - Interpretation: The analysis of DNA damage can be done by visual scoring or image analysis using software (e.g., CASP software).
  - There are various software packages available that compute fluorescence parameters for comments.
  - The most commonly used parameters for measuring and scoring DNA damage are

    The DNA tail length

    Relative fluorescence intensity of head and tail

    Tail movement

44.2.3 Molecular Techniques in *In-Vitro* Screening

- The molecular techniques are useful alternative tools for understanding the mechanism of anticancer agents on various biological functions.
- These techniques help to assess the effect of the drug candidate on a specific pathway or protein involved in tumor cells.
- The following are the commonly performed molecular methods in *in-vitro* anticancer drug screening protocols.
- Immunohistochemistry
- Polymerize Chain Reaction
- Western Blotting

44.3 *In-Vivo* Screening Methods for Anticancer Drugs

44.3.1 Chemical Carcinogen Models

- The most common type of carcinogen known in humans is a chemical carcinogen. All type of carcinogens requires activation for inducing carcinogenesis involving multiple steps. The following are the steps involved in experimental carcinogenesis:
  - Initiation
  - Promotion
  - Malignant conversion
- Anticancer drug screening in animals involves several chemical carcinogens and their respective animal models.

(a) DMBA induced mouse skin papilloma:
  - This is a two-stage experimental carcinogenesis model where DMBA (dimethylbenz[a]anthracene) acts as an initiator and TPA (12-O-tetradecanoyl-phorbol-13- acetate) act as a promotor.
  - A Single-dose of 2.5 μg of DMBA and 5–10 μg TPA in 0.2 ml of acetone is given twice weekly to mice topically.
  - After 8–10 weeks, papilloma begins to appear.
  - Tumor incidence & multiplicity can be compared between the control and treatment groups.
  - The drug is given either through an oral or topical route.
  - The percentage reduction in tumor incidence observed in treatment groups will provide the anticancer property of the drug.

(b) DMBA induced mouse mammary gland carcinogenesis:
  - Female Sprague rats are chosen for inducing mammary gland carcinogenesis.
  - Through intragastric injection, 12/kg DMBA is given at 50 days of age.
  - Within 120 days of injection, the rats develop mammary tumors such as adenocarcinoma, fibroadenoma, or adenoma.

- Drug efficacy is studied by the reduction in adenoma incidence, multiplicity, and latency time for tumor development in the treatment and control group.

(c) MNU-induced tracheal squamous cell carcinoma in Hamster:
- Male Syrian golden hamsters are chosen for this model.
- Specific area of the trachea is exposed to 5% of MNU using a special catheter once a week for 15 weeks.
- Tumor incidence occurs within 6 months of exposure.
- Percentage reduction in the tumor incidence is used for studying the drug efficacy.

(d) N, N-Diethylnitrosamine (DEN) induced lung adenocarcinoma in Hamster:
- Male Syrian hamsters at the age of 7–8 weeks are used for inducing carcinogenesis with DEN. Through the subcutaneous route, 17.8 mg /kg DEN is given twice weekly for 20 weeks.
- The origin of the tumor is from pulmonary Clara cells or endocrine cells present in the lungs. Percentage reduction in the tumor incidence is used for studying the drug efficacy.

(e) 1,2-dimethylhydrazine (DMH) induced colorectal tumors:
The procarcinogen DMH is most commonly used for inducing colorectal tumors in rats and mice (Box 44.1).

(f) Azoxymethane induced aberrant crypt foci in the rat:
Single-dose of azoxymethane at 30 mg/kg body weight administered to rats.
↓
Induce potential precancerous lesions exhibiting dysplasia.
↓
The animal is sacrificed, and the lesion is studied under microscopic examination

(g) N-Butyl-N-(4-hydroxybutyl)-nitrosamine(OH-BBN)-induced bladder carcinoma in the mouse:
50 days old male BDF mice strain are selected.
↓
7.5 mg OH-BNN is given through intragastric installation at eight weekly doses.
↓
After 180 days of the initial dose, this model induces invasive transitional cell carcinoma in the urinary bladder

(h) DMBA – induced oral cancer in Hamster:
Male Syrian hamsters are selected.
↓
0.5% solution of DMBA in liquid paraffin is taken.
↓
Painted in the right buccal mucosa three times a week for 16 weeks.
↓
Tumor size and number can be compared between the anticancer drug and the control.

(i) Fibrosarcoma tumor model in Mouse:
3-methylcholanthrene (MCA) is used to induce fibrosarcoma in the mouse.
↓
Single-dose of 200 μg of MCA is injected subcutaneously into the mouse thigh.
↓
Tumor incidence occurs within 7 weeks. The tumor incidence is observed every week till 15 weeks of post-injection.

(j) Skin tumors in the mouse:
MCA is used to induce skin tumors in the shaved back of ddY mice.
↓
0.05 ml of 0.3% MCA is used as the twice-weekly application.
↓
Tumor incidence is observed for a period of 15 weeks in control and drug groups.

(k) Benzopyrene induced Forestomach tumors in the mouse:
- Benzopyrene is administered through gavage to mice twice weekly for 4 weeks at the dose of 1 mg in 0.1 ml of peanut oil.
- It is found to produce a 100% incidence of forestomach tumors in controls which can be compared with drug groups.

Box 44.1. Procarcinogen DMH Is Most Commonly Used for Inducing Colorectal Tumors in Rats and Mice

1,2-dimethylhydrazine (DMH) (carcinogen) ⟹ (Activated) azomethane ⟹ (Activated) methylazoxymethanol

Rat model

7 wks old F-344 male rats
↓
Single subcutaneous dose of 30mg/kg
↓
Within 40 wks colon adenomas and adenocarcinomas appear

Mice model

9-11 wks old Female CF1 mice
↓
Intraperitonial injection of MAM
↓
4 times in 11days (low dose)
8 times in 22 days (high dose)
↓
Within 38 wks colon adenomas and adenocarcinomas appear

44.3.2 Viral Infection Models

- Mouse Mammary Tumor Virus (MMTV) was the first virus to be studied to cause tumorigenesis in mice.
- The mechanism involved was the insertion of the proviral genome which resulted in protooncogene activation.
- At present, engineered viruses are used in laboratories to induce carcinogenesis in animals to study anticancer drug efficacy.

44.3.3 Methods Involving Cell Line/Tumor Pieces Implantation

- Chemical carcinogen model lack specificity to a certain extent which can be bridged by using cell lines for inducing carcinogenesis.
- By inoculating a particular cell line into sensitive animal strain, the tumor can be developed in a more rapid way.
- Commonly used cell lines are L-1210 and P-388 cell lines derived from mouse lymphocytic leukemia.
- The animal implanted with the cell line will have a 100% growth fraction and dies when the tumor burden reaches ten cells.
- The effectiveness of the drug can be studied by its potential to decrease the tumor burden and increase the life span of the animal.
- The other cell lines used are Lewis lung carcinoma, sarcoma-180, and B-16 melanoma cells.
- The parameters studied are mean survival time and tumor inhibitory activity.
- The implantation of cells into the experimental animals can take place by one of the following methods:
- *Hollow fiber technique:*
  - This technique uses hollow fibers made up of plastic, polyvinylidene fluoride at dimensions of 1 mm diameter and 2 cm length.
  - Tumor Cells are carried by these hollow fibers and inserted into the mouse skin or body cavity.
  - Anticancer agents are tested at two dosages and against at least 12 tumor cells.
  - The duration of the test is 4 days.
  - If the anticancer agents show promising results, they will be carried to the next level of testing.
- *Use of xenografts:*
  - This method employs direct injection of the tumor into the mouse skin.
  - The anticancer agents that are found effective in the hollow fiber technique will be used to test the ability to kill or retard the growth of tumors but cause only minimal toxicity to the animal.
  - The total duration of testing with xenograft is 30 days.
- *Nude mouse model:*
  - The nude mouse model has a genetic mutation resulting in the absence of the thymus.

  - Thus it lacks helper T cells and suppressor T cells and also any antibody response to transplanted material or antigen.
  - It is widely used to test the tumorigenicity of cells and screening of anticancer agents.
  - Melanoma and colon carcinomas grow very well, whereas leukemias and prostate carcinoma grow poorly in this model
  - The disadvantages of this model are, it requires a very large number of cells to be implanted and expensive maintenance.
- *Newborn rat:*
  - A cost-effective alternative for nude mice is the newborn rat model, especially for neural tumors.
  - C6 glioma cells are transplanted into the left side of the newborn rat model of Sprague-Dawley strain, and a palpable tumor will appear in 15–20 days.

44.3.4 Genetically Engineered Mouse Models

- The major pathogenesis in cancer is the mutation of either the tumor suppression gene or protooncogene.
- Such mutation can be induced by pronuclear injection artificially using genetic engineering, and mouse models can be created to study both the disease as well as the gene therapy of the disease.
- Metamouse is a genetically engineered mouse in which tumor sections of patients are implanted into the organ of primary growth.
- *Hepatocellular carcinoma knockout models:*
  - Inflammation-associated hepatocellular carcinoma model can be induced by MDR-KO mice.
  - This knockout model lacks P-glycoprotein, which is responsible for the transport of phosphatidylcholine across the canalicular membrane.
  - Lack of phospholipids results in bile regurgitation and portal inflammation which leads to dysplasia followed by carcinoma of hepatocytes.
  - Anticancer drugs can be studied in such models by observing the anti-mitotic and anti-oxidant properties.
  - These models have the advantage of being more suitable for screening anti-cancer drugs and having more consistent results.
- *Pancreatic cancer models:*
  - The exocrine pancreatic cancer model of a genetically engineered mutant mouse is one of the recently developed mouse models for screening anticancer drugs.
  - The models are characterized histologically and immunohistochemically.
  - Pancreas epithelium-specific endogenous Kras activation leads to murine pancreatic intraepithelial neoplasia (mPanIN).
  - Further inactivation of p16, p53, or transforming growth factor-β signaling coverts mPanIN to invasive pancreatic ductal adenocarcinoma (PDAC) with abundant stromal expansion and marked fibrosis (desmoplasia).

Bibliography

Kumar S, Bajaj S, Bodla RB. Preclinical screening methods in cancer. Indian J Pharmacol. 2016;48: 481–6.

Dhanya KC, Menon A, Rai LS. In-vitro models in anticancer screening. In: Kumar S, Egbuna C, editors. Phytochemistry: an in-silico and in-vitro update. Singapore: Springer; 2019. p. 251–65.

Zips D, Thames HD, Baumann M. New anticancer agents: in vitro and in vivo evaluation. In Vivo. 2005;19:1–7.

Meerloo JV, Kaspers GJL, Cloos J. Cell sensitivity assays: the MTT assay. Methods Mol Biol. 2011;731:237–45.

Prakash J, Singh JN. Chapter ten: anticancer agents. In: Gupta SK, editor. Drug screening methods (Preclinical evaluation of new drugs). 2nd ed. New Delhi: Jaypee Brothers Medical Publishers; 2009. p. 166–84.

Williams O. Flow cytometry-based methods for apoptosis detection in lymphoid cells. Methods Mol Biol. 2004;282:31–42.

Ijichi H. Genetically-engineered mouse models for pancreatic cancer: advances and current limitations. World J Clin Oncol. 2011;2:195–202.

Screening Methods for the Evaluation of Drugs for Benign Prostatic Hyperplasia

45

S. C. Santosh Kumar

Abstract

Benign prostatic hyperplasia (BPH) affects majority of the older male population. Lower urinary tract symptoms are commonly seen in patients suffering from BPH. The pathophysiology of the BPH is indefinite and ageing, hormonal imbalance, epithelial -mesenchymal interactions play role in the evolution of BPH. Pharmacologically BPH is treated with alpha 1a adrenergic antagonists acts on the dynamic component and 5-alpha reductase inhibitors act on the static component of the BPH. Different *invitro, invivo, exvivo*, transgenic, xenograft and spontaneous BPH screening models are developed to study the pathophysiological insights of the BPH and to develop novel drug molecules for the treatment of BPH. WPMY-1, NRP152 and NRP-154 human prostatic cell lines are used for *invitro* screening. Testosterone and sulpiride induces BPH in rats, hemorrhagic cystitis by cyclophosphamide, partial ligation of urethra of the bladder in rats and intraprostatic injection in mongrel dogs are used as invivo models. Isolated prostate gland, detrusor muscle and urethral contractility studies are used for exvivo screening. Transgenic BPH models are developed by transfection using probacin prolactin, murine mammary tumor virus (MMTV)-Int2 and MMTV-keratinocyte growth factor genes. Human BPH specimen are implanted in immunocompromised rats and mice are used as xenograft models. Naturally occurring BPH in old dogs is referred as spontaneous BPH model.

Keywords

Benign prostatic hyperplasia · Dihydrotestosterone · Lower urinary tract symptoms · Castration · Hemorrhagic cystitis · Transgenic mice · Overactive bladder

S. C. Santosh Kumar (✉)
Department of Pharmacology, Jawaharlal Institute of Postgraduate Medical Education and Research (JIPMER), Puducherry, India

M. Lakshmanan et al. (eds.), *Introduction to Basics of Pharmacology and Toxicology*, https://doi.org/10.1007/978-981-19-5343-9_45

45.1 Introduction

- The elderly male population is at higher risk of developing benign prostatic hyperplasia (BPH) associated with lower urinary tract symptoms (LUTS). Health-related quality of life is adversely affected in BPH patients due to bothersome LUTS. Pathogenesis of BPH is elusive, and it is thought to be multifactorial. Ageing, hormonal imbalance and epithelial-mesenchymal interactions play a crucial role in developing and advancing BPH. The α-adrenergic antagonists decrease the smooth muscle tone of the prostatic urethra and the bladder neck. The 5α-reductase inhibitors block the conversion of testosterone to the more active form dihydrotestosterone (DHT). For BPH patients who do not respond to pharmacological treatment, transurethral resection of the prostate (TURP) is preferred as a surgical intervention.
- Experimental models are developed to thoroughly understand the pathophysiology and also to identify and screen new drug entities. *In vitro* models are usually used to study cellular functions' mechanisms and delineate the downstream pathways involved. However, *in vitro* study results should be substantiated by the *in vivo* experimental studies for further validation.
- Transgenic mice models are used to study the role of oncogenes and other regulatory proteins involved in the pathogenesis and progression of the disease. Xenograft models can be used to study the etiology and evaluate the short-term efficacy of some lead compounds. Spontaneous models are the best and ideal models because they reciprocate the pathophysiology of the disease more naturally but are limited to very few species, namely dogs and chimpanzees. Ethical and financial issues also limit the use of these spontaneous models.

45.2 Pathophysiology and Pharmacology of BPH in Brief

- BPH is the most common chronic condition seen in aged male and autopsy studies shows the histological prevalence of 50% in 50–60 years of men and 90% in men aged 80 years or above. Clinical studies suggest that BPH develops in the transitional and central zone of the prostate, whereas prostate cancer is from the peripheral zone.
- Increased proliferation of connective tissue, stromal smooth muscle cells, and glandular epithelial cells associated with the initiation and progression of BPH. An increase in prostatic volume (static component) and stromal smooth muscle tone (dynamic component) physically compress the urethra resulting in bladder outlet obstruction, which in turn present clinically as LUTS, urinary tract infections, acute urinary retention (AUR), renal failure, haematuria, and bladder calculi.
- Patients of BPH experience obstructive symptoms such as weak stream, hesitancy, intermittency, incomplete bladder emptying, and storage symptoms such as frequency, urgency, and nocturia. The involvement of androgen and its

receptors in BPH development is well known as prostate growth is androgen dependent.

- Estrogens play an essential role in BPH-LUTS disease because estrogen receptors are expressed throughout the male lower urinary tract and other reproductive organs, namely the prostate, epididymis, testes, urethra, and bladder.
- Exogenous administration of estrogens has increased the proliferation of prostatic epithelial, urothelial, and bladder fibroblast via estrogen receptor alpha.
- Estrogen receptor alpha is over-expressed in the prostatic epithelium of symptomatic BPH patients compared to men without LUTS.
- Mild chronic inflammation is commonly seen in clinical BPH patients with the presence of inflammatory infiltrates such as T and B lymphocytes and macrophages.
- Prostatic hyperplasia occurs in regions associated with chronic inflammation.
- Obesity and associated abnormalities in glucose homeostasis influence prostate growth and might play a role in benign prostatic hyperplasia.
- The α_{1a} adrenergic blockers, namely tamsulosin, doxazosin, alfuzosin, silodosin, are commonly used because of their efficacy and fast action as they relieve symptoms (dynamic component) of LUTS/BPH by relaxing the smooth muscle contraction around the bladder neck and intra-prostatic urethra.
- 5-α reductase inhibitors, namely finasteride, and dutasteride are used in BPH, which reduces the prostate size (static component) and thus inhibits the progression of the disease. The 5-α reductase inhibitors are slow slow-acting, and the prolonged use of this class of drugs for a period of 6 months shows a clinically beneficial effect.
- The FDA approved the single capsule formulation of tamsulosin and dutasteride in June 2010 to treat symptomatic BPH patients with an enlarged prostate.
- Muscarinic antagonists, namely tolterodine, solifenacin, and darifenacin, are commonly employed to treat OAB with storage symptoms and as add-on therapy for symptoms of BPH. Muscarinic antagonists relieve voiding symptoms by inhibiting detrusor overactivity, decreases smooth muscle tone and size of the prostate.
- Transurethral resection of the prostate (TURP) is the surgical intervention used in BPH patients with severe urinary retention and who are refractory to pharmacological treatment. TURP is effective in relieving obstructive symptoms when compared to irritative symptoms.

45.3 *In Vitro* Models

- Human prostate epithelial cell lines are used to study the mechanisms of prostate cell function and investigate the hormonal regulation of the prostate and as a control for the studies involving BPH and prostate cancer.
- WPMY-1 cell line derived from stromal cells of the peripheral zone of the histologically normal adult prostate. WPMY-1 cell line is used for studies on paracrine and stromal to epithelial interactions.

- NRP152 is a rat prostatic epithelial cell line used for studies involving the normal physiological function of the prostate as well regulation by steroidal hormones and growth factors.
- NRP-154 is a tumorigenic epithelial cell line derived from the preneoplastic dorsal-lateral prostate of rats used for studying the molecular mechanisms of carcinogenesis and related apoptosis pathways.
- BPH cell lines are usually derived from patients who underwent open prostatectomy or transurethral resection of the prostate. BPH1 is a human hyperplasic epithelial cell line that shows androgen-dependent growth.

45.4 *In Vivo* Models

45.4.1 Testosterone Induced Benign Prostatic Hyperplasia in Rats

- Male Wistar or Sprague-Dawley rats weighing 200–250 g are initially acclimatized to laboratory conditions for 1 week.
- It is always necessary to consider the rats for experiments that have almost equal weight.
- The rats are individually caged and surgically castrated to avoid the influence of endogenous testosterone. The rats are allowed to recover for 7–10 days post castration.
- The rats are randomly divided into different groups ($n = 6$), and testosterone propionate in corn oil is administered subcutaneously at 3–5 mg/kg body weight (BW) and the standard drug or test drug daily for 28 days (Fig. 45.1)

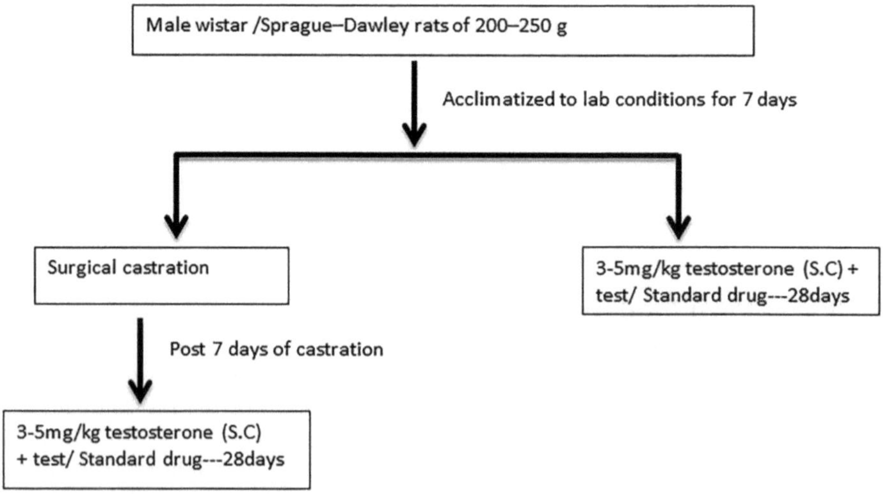

Fig. 45.1 Testosterone induced benign prostatic hyperplasia in rats

- The testosterone administration to the rats can be carried out without surgical castration also. It is always wise to carry out the surgical castration initially to avoid the influence of endogenously produced testosterone.
- On the 29th day, 4 ml of blood is collected by intracardiac puncture, and the rats are sacrificed 72 h. after the last dose of testosterone dose by cervical dislocation, and immediately prostate gland will be dissected out for assessing different parameters.
- The following parameters are commonly assessed:
  (a) Bodyweight measurement of all rats at weekly.
  (b) Prostate weight to body weight ratio measurement after sacrificing rats
  (c) Determination of serum testosterone/dihydrotestosterone levels by ELISA
  (d) Histopathology of Prostate gland
  (e) Percentage of recovery
  Based on mean prostatic weight (P) and P/BW ratios, percentage recovery in the P/BW ratio by test group compared with the model group will be calculated. The increase induced by testosterone alone will be considered to be 100%, and all other test groups are compared with this reading.
  The formula is used as follows:

$$\text{Percent recovery by the test sample} = [A - \text{B}]$$

A: percentage of increase in prostatic weight induced by testosterone.
B: percentage of increase in prostatic weight induced by the test sample.
  (f) Percentage of Inhibition (PI)of increase in prostate weight is calculated as follows:

$$\begin{aligned} PI = 100 - & PW \text{ of } treated\ group - PW \text{ of } negative\ control \\ & \times PW \text{ of } model\ group - PW \text{ of } negative\ control \\ & \times 100 PW \text{ of } model\ group - PW \text{ of } negative\ control * 100 \end{aligned}$$

- To study disease pathogenesis at the molecular level and to study the mechanisms of new drug entities-
  (a) Biomarkers role by Immunohistochemistry
  (b) Protein expression studies by western blotting
  (c) Gene expression studies

45.4.2 Testosterone + Estrogen Induced Benign Prostatic Hyperplasia in Rats

- Estrogens have a role in the maintenance and function of the prostate, and they also play an essential role in the development of BPH since it is involved in the growth and differentiation of the prostate.
- Sprague-Dawley rats of 120–130 gm body weight are usually used for the study. Initially, rats are acclimatized to laboratory conditions for a week.

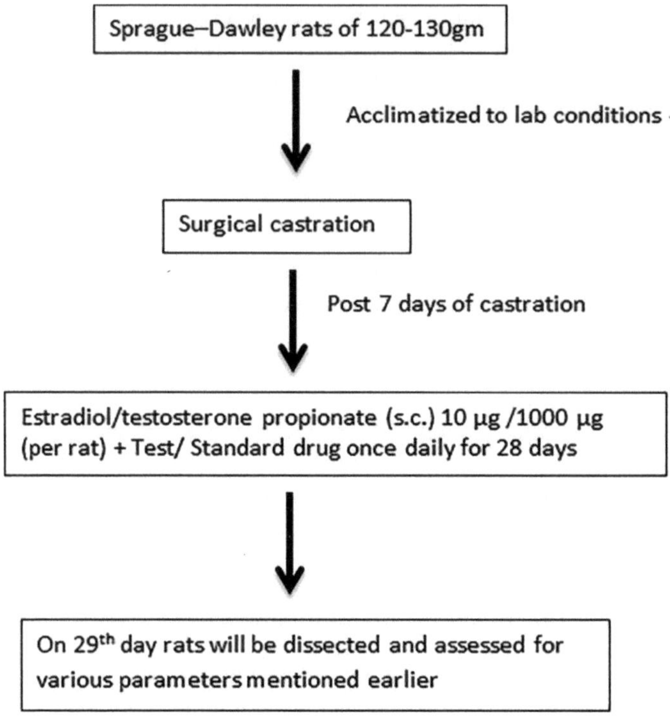

Fig. 45.2 Testosterone + Estrogen induced benign prostatic hyperplasia in rats

- The rats are then surgically castrated to avoid the influence of endogenous testosterone. After 7 days of recovery from castration, the rats are co-administered subcutaneously with estradiol 10 μg and 1000 μg for 120–130 gm bodyweight rats along with test or standard drugs for 28 days (Fig. 45.2)
- The rats are euthanized, and the prostate gland is dissected out and weighed 48 h after the last injection and assessed for different parameters, as mentioned earlier.

45.4.3 Sulpiride-Induced BPH in Rats

- Sulpiride-induced BPH is one of the most commonly used models for studies involving BPH.
- Sulpiride induces prostatic hyperplasia in both the dorsal and lateral lobes. Sulpiride acts as a dopamine antagonist at the level of the pituitary gland and enhances the secretion of prolactin.
- Sulpiride induced BPH model closely resembles that of human BPH, since the prolactin levels of serum in human increases with age
- Male Wistar rats weighing 240–280 gm are used for the study. Once the rats are acclimatized to laboratory conditions for 7 days, they are divided into different groups (n = 6).

- Sulpiride at 30 mg/kg body weight is administered intraperitoneally for 28 days. The test/standard drug is administered to rats three times a week for 30 days from the fifth day of sulpiride administration.

45.4.4 Bladder Outlet Obstruction of Hemorrhagic Cystitis by Cyclophosphamide in Rats

- Patients with benign prostatic hyperplasia experience overactive bladder (OAB) symptoms due to restructure of the bladder wall, neuronal enlargement, hypoxia, hypersensitivity to the locally released acetylcholine.
- These symptoms are also observed after administration of cyclophosphamide due to hemorrhagic cystitis because of its metabolite acrolein which induces severe oxidative stress.
- 6-weeks-old male Wistar rats are acclimatized to laboratory conditions for 1 week. Four doses of intraperitoneal injections of cyclophosphamide (75 mg/ kg b.w) are administered on alternative days (days 1, 3, 5, and 7).
  After 2 h of administration of cyclophosphamide (75 mg/kg b.w) 4 doses of standard/test drug is administered intraperitoneally four times (Fig. 45.3).
- Parameters assessed
  (a) Measurement of urine volume
  (b) Assessment of edema of the urinary bladder tissue

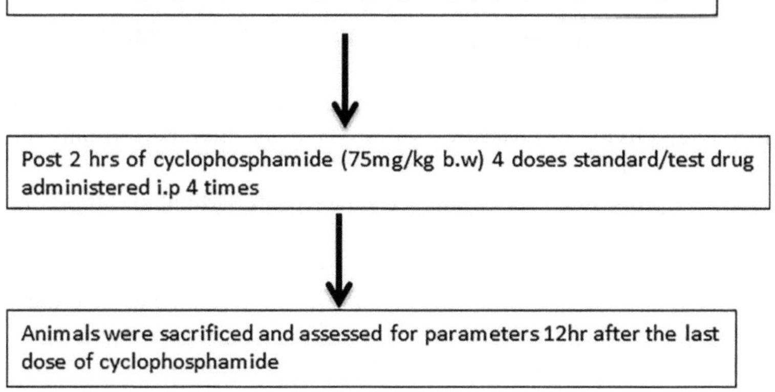

Fig. 45.3 Bladder outlet obstruction of hemorrhagic cystitis by cyclophosphamide in rats

After dissection, edema of the urinary bladder is determined by analyzing the weight of the urinary bladder and then divided by the bodyweight of rats.

(c) Histopathology studies

Bladders are removed, fixed in formalin, embedded in paraffin, and cut into 5 μm sections, which are stained with hematoxylin and eosin for morphological analysis.

Dissected urinary bladder tissue is fixed with formalin and then embedded in paraffin. Approximately 5 μm sections are cut, and morphological analyses are carried out using hematoxylin and eosin stain.

A scoring system may be used to assess the severity of the urinary bladder cystitis by analyzing the mucosal erosion and ulceration, edema, infiltration of leukocytes, and hemorrhage. Any changes of the above said parameters can be expressed as normal, mild, moderate, and severe.

Gene expression, protein expression by western blotting, and immunohistochemistry studies may be performed.

45.4.5 Single Dose of Cyclophosphamide Induced Hemorrhagic Cystitis in Rats of Bladder Outlet Obstruction Model

- 100/200 mg of a single dose of cyclophosphamide per kg body weight is administered to male Wistar rats (220–270 g,) and the test drug/vehicle is administered 1 h prior to cyclophosphamide administration.
- Mesna (standard) (40 mg/kg, 1 ml/kg) is administered to rats intraperitoneally, and after 5 min, cyclophosphamide (100/200 mg) is given, followed by administration of mesna at 4 and 8 h, respectively.
- Rats are sacrificed after 24 h of cyclophosphamide injection and assessed for different parameters, as mentioned previously in a bladder outlet obstruction of hemorrhagic cystitis by cyclophosphamide in rats.

45.4.6 Partially Ligated Urethra of Bladder Outlet Model

- Previous studies suggest that a condition clinically corresponding to BPH with bladder overactivity is observed within 15 days after performing the partial ligation of the urethra.
- The surgical procedures are conducted under pentobarbital anesthesia at 40 mg/kg b.w.
- Over the projection of the urinary bladder, a medial incision is made, Urethra is catheterized, and the proximal part of the urethra is ligated around the catheter, and finally, the catheter is removed.
- After that, the abdominal integument and skin are sutured.
- To minimize the risk of post-surgical infection, antibiotics (neomycin) will be sprayed over the skin. Sham surgery is also performed.

- To avoid infection after surgery, antibiotics (neomycin) are sprayed over the sutured skin.
- Sham surgery to be carried out.
- Test drug will be administered 2 days post-surgery for 4/8 weeks.
- Animals are sacrificed at 4/8 weeks and assessed for various parameters, as mentioned earlier.

45.4.7 Intraprostatic Injection Model

- Mongrel dogs of weight 13–17 kg are used in this model.
- To administer perioperative antibiotics and sedatives such as pentobarbital, the dogs are put in the lithotomy position.
- Test drug/standard will be administered to the prostate by transperineal injection under transrectal ultrasonography (TRUS) guidance.
- Normal saline is injected into the prostates of the dogs, which belongs to the control group.
- Parameters assessed:
  - Standard/test drug or normal saline is administered for one to 3 months, and then the prostate gland is dissected out for histological evaluation.
  - The prostatic tissues are used to study apoptosis with TUNEL stain.

45.5 Advantages of *In Vivo* Models

- *In vivo* models are helpful for studying the mechanisms of hormonal regulation on the prostate.
- The dorsolateral lobe of the rodent prostate is ontogenetically comparable to humans.
- It is convenient to use small laboratory animals like rats and mice in a well-controlled environment.
- It is quite easy to develop BPH in small laboratory animals.
- *In vivo* models, especially bladder outlet obstruction, represents the clinical condition seen in BPH patients.
- Intra prostatic injection model is useful for studying the role of new drug molecules and their role in apoptosis of the prostate gland.

45.6 Limitations of *In Vivo* Models

- It is not sure whether the rodent models can simulate the human BPH because the rodent prostate has three lobes ventral, dorsolateral, and anterior lobes. In contrast, four distinct regions are seen in man: anterior fibromuscular stroma and the central, peripheral, and transitional zones.

- As the age progresses, the testosterone levels decrease, and it is also observed in clinical BPH patients, but to develop BPH in experimental animal's testosterone is used.
- Increased prostate size in rodent models may not necessarily cause bladder obstruction, which is the commonest symptom observed in BPH patients.
- Testosterone-induced BPH models in rats show epithelial hyperplasia, whereas human BPH shows more stromal hyperplasia of the prostate gland.
- Spontaneous BPH is not seen in rodent models. It is confined to dogs and chimpanzees, which is a more appropriate and suitable model.

45.7 *Ex Vivo* Models

45.7.1 Isolated Prostate Gland Contractility Model (Flow Chart Given Below)

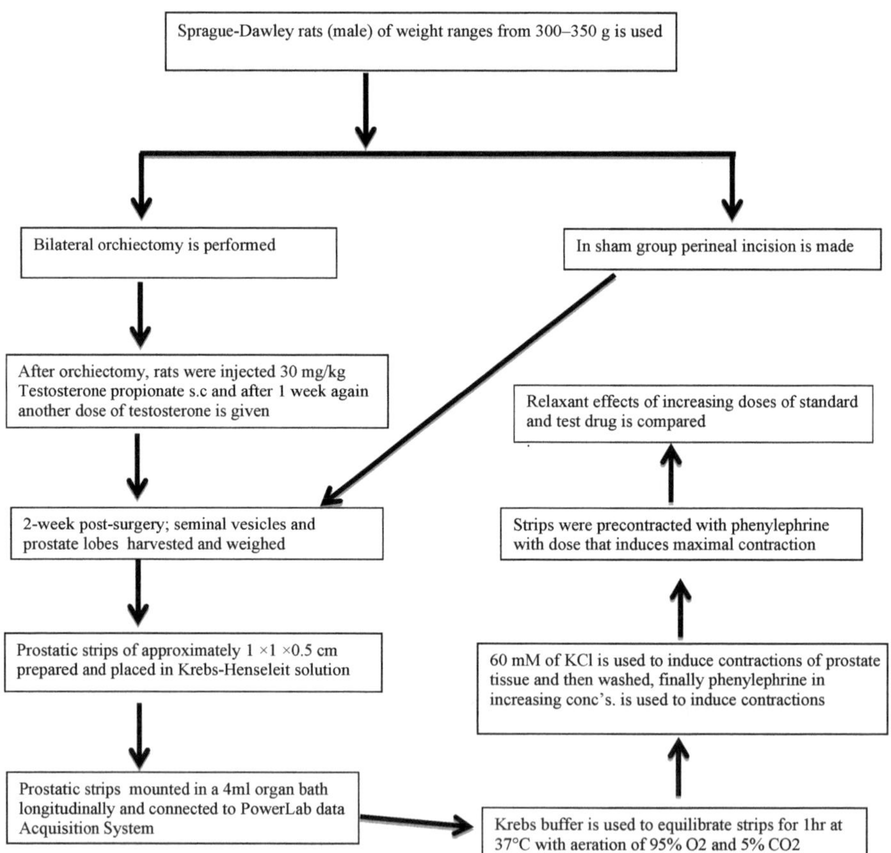

- In the previous methods, castration is not performed, and the rats are sacrificed directly by cervical dislocation, and the prostate gland is dissected out and placed in Krebs-Henseleit solution and mounted in a 30 ml organ.
- The dose-response curve for both acetylcholine and norepinephrine is established. Tissue is washed with bath solution and then allowed to recover for 15 min.
- Tissue is exposed to test compound for 3 min every time, and second concentration-response curve is recorded in the presence of test compound.
- Isolated prostate gland contractility studies can be modified based on the study objectives and performed to serve the purposes.
- These experimental procedures are done to understand the normal physiology and the role of different cell types in smooth muscle contraction and to study the role of new therapeutic agents in smooth muscle contraction.
- These kinds of studies can be executed in rats, guinea pigs, rabbits, and even in higher animals like canines.

45.7.2 Isolated Detrusor Muscle and Urethra Contractility Model

- *Ex vivo* studies involving detrusor smooth contractility can be performed to assess the function of bladder smooth muscle tone.
- To explore the mechanisms for its contractility as well as for evaluating the therapeutic potential of promising new drug molecules.
- For studying the function of bladder smooth muscle, usually, rats, rabbits, and guinea pigs are preferred, and some studies have used caprines also.
- Procedure:
  - The rat bladders are longitudinally cut into approximately 2 × 2 × 6-mm strips, and then the strips are suspended in Krebs solution.
  - The tissue will be mounted in a 10 ml organ bath and maintained at 37 °C with 95% O_2 and 5% CO_2.
  - The maximum response induced by KCl is considered 100%, and the contractile effect induced by different concentrations of carbachol is referred to this value.
  - The relaxant effect induced by the test compound is compared with carbachol.
  - The urethra of rats is also used for isolated tissue preparations.

45.8 Advantages of *Ex Vivo* Models

- With the use of ex vivo models, the role of different receptors in smooth muscle contraction and relaxation can be studied.
- Mechanisms of potential therapeutic drugs can be delineated, and the particular receptor responsible for the drug's action can be traced out.

- The response of the tissue to the drug is very well appreciated in ex vivo studies
- The dose range data obtained can be used for further studies.

45.9 Limitations With the Use of *Ex Vivo* Studies

- Ex vivo models do not actually represent the physiological condition.
- *Ex vivo* models are not suitable for studying the hormonal regulation of BPH, which plays an important role in pathogenesis.
- With the use of *ex vivo models,* the different genes and proteins involved in the disease development and progression cannot be determined.

45.10 Transgenic Models

45.10.1 Probasin Prolactin Transgenic Mouse Model of Benign Prostate Hyperplasia

- The probasin-prolactin mouse model showed increased expression of rat prolactin, which is controlled by the prostate-specific probasin (Pb) minimal promoter.
- The Probasin-prolactin model represents many human BPH features, including prostate enlargement (all lobes), marked stromal hyperplasia, and ductal dilatation.
- This model is also suitable for studying the role of inflammatory mediators and the drug targets affecting BPH.
- The transgenic probasin-prolactin mouse model is generated by transfection using Pb minimal promoter and the rat prolactin genes in the mice.
- At 4–5 weeks of age, this model represents the gene expression in the prostate luminal epithelial cells in a developmentally and hormonally regulated fashion.
- Rat prolactin transgene expression could not be detected in other tissues, and its expression is limited to dorsal, lateral, ventral, and anterior prostate.
- This transgenic mouse model shows marked prostate enlargement from 10 weeks of age and increases further with age.

45.10.2 MMTV (Murine Mammary Tumor Virus) -Int2 (Proto-Oncogene) Transgenic Mice Model

- Muller et al. developed this model. MMTV acts as a promoter/enhancer for Int2, a fibroblast growth factor 3 (fgf3).
- When MMTV-int2 genes are introduced into the mice and expressed shows a significant enlargement of prostate resembling glandular hyperplasia of human and dog BPH.
- This model is suitable for studying the growth factor-induced epithelial hyperplasia, and it is hormonally sensitive.

45.10.3 MMTV-Keratinocyte Growth Factor (fgf7)-Induced Hyperplasia of the Transgenic Model

- Transgenic mice are developed using a keratinocyte growth factor as a transgene gene under the control of the MMTV promoter/enhancer.
- This model has similar characteristic features to the MMTV-Int2 transgenic model.
- Male mice have developed hyperplasia in the prostate, seminal vesicle, and vas deferens.

45.11 Advantages of Transgenic Models

- The creation of transgenic models has helped researchers to understand pathological insights of the disease more precisely.
- Using the transgenic model, the role of a particular gene/protein and the consequences of its downregulation and expression in disease pathogenesis can be explained.

45.12 Limitations of Transgenic Models

- Concerns about the ethical issues involved.
- Transgene integration is apparently random; hence neighboring sequences might interfere, both qualitatively and quantitatively.
- Transgenic models are the best suited and most common model for studying the pathogenesis of different cancers and may not be suitable for BPH.

45.13 Xenograft Models

- In xenograft models, cells derived from primary surgical specimens and human tissue culture are transplanted into the immunocompromised rats or mice for the studies involved in understanding pathogenesis and studying the efficacy of new drug molecules for the disease.
- D. Love et al. implanted the human prostate xenografts in severe combined immunodeficiency (SCID) mice and studied the genes involved in androgen regulation.
- This model represents the biologically relevant interactions of epithelium and stroma that occur *in vivo,* and it allows for the manipulation of androgens which can be done either by castration or by hormonal supplementation of the host.
- The xenograft model is useful to understand the epithelial and stromal interactions *in vivo,* and it allows for androgen manipulation by administering testosterone to castrated and non-castrated animals.

- De-identified human prostate tissue from the right and left transition zone and peripheral zone obtained from radical prostatectomies, and histological analyses are performed on frozen sections.
- 2–3 mm thick sections of the transition zone of the prostate cut, and 4–6 sections implanted under the renal capsules of the male SCID mice.
- Standard group mice are then implanted with a 2.5 cm length of Silastic tube containing 25 mg of testosterone subcutaneously to comp the effects of with and without testosterone on xenografts.
- The xenografts are preserved for 1 month. Harvested xenografts are dissected immediately under magnification, snap-frozen is carried out using liquid nitrogen, and tissues are stored at -80 °C.
- Human BPH tissues are implanted subcutaneously in male beige nude mice and maintained for up to 16 weeks.
- The implanted tissues are maintained without any evidence of degeneration up to 6 weeks, as shown by histological investigation, and thereafter gradually, tissue became atrophic. Immunohistochemistry data showed continuous expression of prostate-specific antigen and prostatic acid phosphatase except in severe atrophy.

45.14 Advantages With the Use of Xenograft Models

- Human BPH specimens transplanted into the mice maintain biology similar to that of original patient tissue. The histology of the gland is well preserved.
- The xenograft model preserves the stromal epithelial interactions as seen in BPH patients.
- Xenografts have been shown to preserve prostate-specific antigen expression and prostatic acid phosphatase in the epithelial cells but gradually lose androgen expression.
- The xenograft model helps study the expression of human genes within a primarily *in vivo* human context, which strongly influences gene expression and cell growth.
- Response to experimental therapies can be evaluated.
- Proteomic and pharmacogenomic data can be obtained.

45.15 Limitations with the Use of Xenograft Models

- Access and availability of human tissue specimens require a coordinated approach between clinicians and researchers.
- Contamination of transplanted tissue by the host vasculature and inflammatory cells.

45.16 Spontaneous BPH Model

- An Animal model that represents the naturally occurring disease condition as observed clinically in patients is known as the spontaneous model.
- BPH is an age-related disease occurs in elderly male, and the spontaneous models for BPH have many advantages compared to other rodent models but seen only in dogs and chimpanzees.
- Most aged dogs develop BPH and hyperplastic changes seen in all dogs histologically at 6 years of age.
- When compared to the human prostate, the canine prostate is not fixed anatomically and may expand in all directions leading not only to urinary obstruction because rectal compression may lead to constipation as well.
- Still, many anatomical and pathological differences exist between human and dog BPH, but the natural reason for hyperplasia in both species remains the same.
- Ethical issues and the lack of commercially available aged dogs make the use of this spontaneous model a distant reality.

45.17 Conclusion

BPH is a multifactorial age-related disease observed commonly in the elderly male. The main objective in treating men of BPH is to relieve the symptoms of LUTS. Drugs used in the treatment of BPH act either on the static or the dynamic component of the disease. Several different experimental models of BPH have been developed to understand the pathogenesis and screen novel drug molecules. Depending on the objectives of the study-specific the experimental model can be selected. *In-vitro*, *In-vivo*, and *Ex-vivo* models are more commonly used because of their ease of access and financial and ethical benefits. In vitro models are used to study epithelial-mesenchymal interactions, cellular transformations. The *in-vitro* data obtained is further explored and validated in the in vivo models. The *In-vivo* models can be exploited to study both the pathogenesis and progression of BPH and to analyze test compounds.

The selection of the specific in vivo model should be done carefully, which fulfills the needs of the study. A testosterone-induced rat model can be selected to test the compounds that act on the disease's static component. In contrast, partial bladder outlet models or *ex-vivo* models such as the isolated prostate gland and detrusor contractility models can be chosen for testing drugs acting on the dynamic component of BPH. Transgenic models are used to understand the involvement of a particular gene in the disease pathogenesis and to explore its downstream pathway. Xenograft models mimic the human BPH pathophysiology of the disease condition more precisely out the human body; hence it helps understand the pathology of the disease. Spontaneous models are valuable tools, but the ethical and financial issues and the limited availability of the animals have made use of these models less practical.

Bibliography

AUA Practice Guidelines Committee. AUA guideline on management of benign prostatic hyperplasia (2003). Chapter 1: diagnosis and treatment recommendations. J Urol. 2003;170:530–47.

Bernichtein S, Pigat N, Camparo P, Latil A, Viltard M, Friedlander G, et al. Anti-inflammatory properties of Lipidosterolic extract of Serenoa repens (Permixon®) in a mouse model of prostate hyperplasia. Prostate. 2015;75:706–22.

Chuang YC, Tu CH, Huang CC, Lin HJ, Chiang PH, Yoshimura N, et al. Intraprostatic injection of botulinum toxin type-A relieves bladder outlet obstruction in human and induces prostate apoptosis in dogs. BMC Urol. 2006;6:12.

Danielpour D. Transdifferentiation of NRP-152 rat prostatic basal epithelial cells toward a luminal phenotype: regulation by glucocorticoid, insulin-like growth factor-I and transforming growth factor-beta. J Cell Sci. 1999;112:169–79.

Dawson C, Whitfield H. ABC of urology. Bladder outflow obstruction. BMJ. 1996;312:767–70.

Debiec-Rychter M, Asgari K, Jones RF, Grignon DJ, Haas GP, Wang CY. Xenografts of human benign prostatic hyperplasia tissues in the nude mouse. In Vivo. 1994;8:449–50.

Dmochowski RR. Bladder outlet obstruction: etiology and evaluation. Rev Urol. 2005;7:S3–13.

Dobrek Ł, Baranowska A, Skowron B, Thor PJ. The influence of piroxicam, a non-selective cyclooxygenase inhibitor, on autonomic nervous system activity in experimental cyclophosphamide-induced hemorrhagic cystitis and bladder outlet obstruction in rats. Acta Pol Pharm. 2014;71:497–507.

Elkahwaji J. The role of inflammatory mediators in the development of prostatic hyperplasia and prostate cancer. Res Rep Urol. 2012;5:1–10.

Hennenberg M, Stief CG, Gratzke C. Pharmacology of the lower urinary tract. Indian J Urol. 2014;30:181–8.

Izumi K, Mizokami A, Lin WJ, Lai KP, Chang C. Androgen receptor roles in the development of benign prostate hyperplasia. Am J Pathol. 2013;182:1942–9.

Jiang X, Chen Y, Zhu H, Wang B, Qu P, Chen R, et al. Sodium tanshinone IIA sulfonate ameliorates bladder fibrosis in a rat model of partial bladder outlet obstruction by inhibiting the TGF-β/smad pathway activation. PLoS One. 2015;10:1–17.

Jk P, Patel N. Epidemiology and etiology of benign prostatic hyperplasia and bladder outlet obstruction. Indian J Urol. 2014;30:170.

Kitsberg DI, Leder P. Keratinocyte growth factor induces mammary and prostatic hyperplasia and mammary adenocarcinoma in transgenic mice. Oncogene. 1996;13:2507–15.

Lesovaya EA, Kirsanov KI, Antoshina EE, Trukhanova LS, Gorkova TG, Shipaeva EV, et al. Rapatar, a nanoformulation of rapamycin, decreases chemically-induced benign prostate hyperplasia in rats. Oncotarget. 2015;6:9718–27.

Lin PH, Pan Z, Zheng L, Li N, Danielpour D, Ma JJ. Overexpression of Bax sensitizes prostate cancer cells to TGF-beta induced apoptosis. Cell Res. 2005;15:160–6.

Lopez-Barcons LA. Human benign prostatic hyperplasia heterotransplants as an experimental model. Asian J Androl. 2010;12:157–63.

Love HD, Booton SE, Boone BE, Breyer JP, Koyama T, Revelo MP, et al. Androgen regulated genes in human prostate xenografts in mice: relation to BPH and prostate cancer. PLoS One. 2009;4:1–15.

Mahapokai W, Van Sluijs FJ, Schalken JA. Models for studying benign prostatic hyperplasia. Prostate Cancer Prostatic Dis. 2000;3:28–33.

McLaren ID, Jerde TJ, Bushman W. Role of interleukins, IGF and stem cells in BPH. Differentiation. 2011;82:237–43.

Moraes JP, Pereira DS, Matos AS, Santana DG, Santos CA, Estevam CS, et al. The ethanol extract of the inner bark of *Caesalpinia pyramidalis* (Tul.) reduces urinary bladder damage during cyclophosphamide-induced cystitis in rats. Sci World J. 2013; https://doi.org/10.1155/2013/694010.

Muller WJ, Lee FS, Dickson C, Peters G, Pattengale P, Leder P. The int-2 gene product acts as an epithelial growth factor in transgenic mice. EMBO J. 1990;9:907–13.

Nicholson TM, Moses MA, Uchtmann KS, Keil KP, Bjorling DE, Vezina CM, et al. Estrogen receptor-α is a key mediator and therapeutic target for bladder complications of benign prostatic hyperplasia. J Urol. 2015;193:722–9.

Parsons JK, Carter HB, Partin AW, Windham BG, Metter EJ, Ferrucci L, et al. Metabolic factors associated with benign prostatic hyperplasia. J Clin Endocrinol Metab. 2006;91:2562–8.

Raja A, Hori S, Armitage JN. Hormonal manipulation of lower urinary tract symptoms secondary to benign prostatic obstruction. Indian J Urol. 2014;30:189–93.

Rick FG, Abi-Chaker A, Szalontay L, Perez R, Jaszberenyi M, Jayakumar AR, et al. Shrinkage of experimental benign prostatic hyperplasia and reduction of prostatic cell volume by a gastrin-releasing peptide antagonist. Proc Natl Acad Sci. 2013;110:2617–22.

Roehrborn CG. Male lower urinary tract symptoms (LUTS) and benign prostatic hyperplasia (BPH). Med Clin North Am. 2011;95:87–100.

Shin IS, Lee MY, Jung DY, Seo CS, Ha HK, Shin HK. Ursolic acid reduces prostate size and dihydrotestosterone level in a rat model of benign prostatic hyperplasia. Food Chem Toxicol. 2012;50:884–8.

Ventura S, Oliver V, White C, Xie J, Haynes J, Exintaris B. Novel drug targets for the pharmacotherapy of benign prostatic hyperplasia (BPH): drug targets for BPH. Br J Pharmacol. 2011;163: 891–907.

Vignozzi L, Filippi S, Comeglio P, Cellai I, Morelli A, Maneschi E, et al. Tadalafil effect on metabolic syndrome-associated bladder alterations: an experimental study in a rabbit model. J Sex Med. 2014;11:1159–72.

Vyas B, Desai N, Patel P, Joshi S, Shah D. Effect of Boerhaavia diffusa in experimental prostatic hyperplasia in rats. Indian J Pharmacol. 2013;45:264.

Wang C, Du X, Yang R, Liu J, Xu D, Shi J, et al. The prevention and treatment effects of tanshinone IIA on oestrogen/androgen-induced benign prostatic hyperplasia in rats. J Steroid Biochem Mol Biol. 2015;145:28–37.

Wang X, Liao J, Yin D, Zhan F, Dai S, Xie G, et al. Establishment of a novel model for studying the effects of extracts of Chinese herb medicine on human type II 5alpha-reductase in vitro. Yakugaku Zasshi. 2010;130:1207–14.

Webber MM, Trakul N, Thraves PS, Bello-DeOcampo D, Chu WW, Storto PD, et al. A human prostatic stromal myofibroblast cell line WPMY-1: a model for stromal-epithelial interactions in prostatic neoplasia. Carcinogenesis. 1999;20:1185–92.

Zhang X, Zang N, Wei Y, Yin J, Teng R, Seftel A, et al. Testosterone regulates smooth muscle contractile pathways in the rat prostate: emphasis on PDE5 signaling. AJP Endocrinol Metab. 2012;302:243–53.

46 Screening Methods for the Evaluation of Drugs Acting on Autonomic Nervous System

Dhivya Elango

Abstract

Autonomic nervous system (ANS) which is maintaining the internal homeostasis, broadly classified into sympathetic and parasympathetic system. Screening of drugs acting on ANS includes initial and specific screening. Initial screening of unknown compounds for neuro pharmacological activity was given by Irwin. Irwin screening provides the direction of activity whether sympathetic or parasympathetic predominant. In Irwin method, unknown compounds are subjected to various tests and scored on each activity for example, action on the smooth muscle, heart rate. According to the score, compounds are taken to in *vitro* and in *vivo* experiments to analyse the specific activity. In *vitro* analysis will focus on the receptor activity or the mechanism of action. Alpha 1 activity of the compounds is tested against rat heart ventricles with radio ligand ^{3}H prazosin. Then the isolated organ model provides the quantification of tissue activity without the interference of neuro humoral reflexes. For example Cholinergic/ anticholinergic compounds can be evaluated by Guinea pig ileum model. In *vivo* experiments provide the biological actions of the compound. For example invasive blood pressure measurement in rats evaluates the sympathomimetic and sympatholytic activity with intact reflexes similar to human body. These screening methods give insights of newer compounds developed to act on ANS.

Keywords

Receptor binding assay · heart and uterus model · Ellman assay · Guinea pig ileum model

D. Elango (✉)
Department of Pharmacology, Jawaharlal Institute of Postgraduate Medical Education and Research (JIPMER), Karaikal, Puducherry, India

M. Lakshmanan et al. (eds.), *Introduction to Basics of Pharmacology and Toxicology*, https://doi.org/10.1007/978-981-19-5343-9_46

46.1 Introduction

Screening of drugs for any systemic or local activity includes evaluation and scanning. Scanning of a substance refers to a group of tests to detect the physiological activity. That is to understand how newer unknown substance exerts its action on our body. Evaluation is the confirmatory test to justify the physiological property of the unknown substance. Screening of unknown substance generally includes a neurological, behavioural and autonomic profile. In this chapter will focus on autonomic profile and specific screening methods for the autonomic nervous system.

46.2 Screening of Unknown Substance for ANS Activity (Irwin Method)

Screening is organised in the following manner:

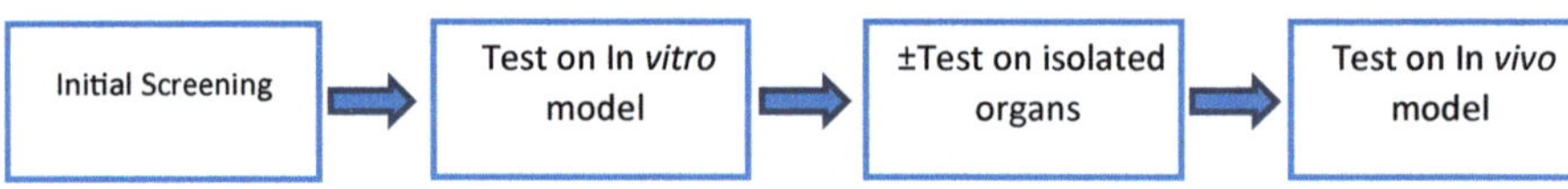

Initial screening will help to understand the following questions,

1. Whether the unknown compound is active?
2. What is the principal activity of the compound?
3. What are the systemic effects of the unknown compound?
   - Animal: Mice of the same strain, 4–5 groups, each group has 4–6 animals, with equal gender distribution.
   - Analysis: Mice were given the unknown drug in varying concentration (1, 3, 10, 30, 100, 300, 1000 mg/kg in Irwin method). General observation (weight, sex, age etc) and autonomic profile were observed and scored. The score ranges from 0 to 8, <4 is subnormal and > 4 is supernormal, 4 is considered as the base value. Example of Irwin's profile and scoring is given below (Table 46.1)
   - Each parameter is tested and scored. Each indicates specific autonomic action. E.g increased score in exophthalmos indicates sympathomimetic activity. Non-specific indicators like heart rate, are analysed after comparing all the three profile.
   - If the unknown drug has a high autonomic profile, further analysis is done on isolated organs and in *vivo* experiments

The methods available for the evaluation of substances for ANS activity is summarized in Fig. 46.1.

Table 46.1 Model of autonomic profile activity given by Irwin

Species				Weight			Sex			
				Temperature			pH			
Test drug				Route			Solvent			
	Autonomic profile Score 0–8									
Dose mg/kg	Pupil Size	Palpebral opening	Exophthalmos	Urination	Salivation	Pilo erection	Hypo thermia	Skin colour	Heart rate	Respiratory rate
1										
3										
10										
30										
100										
300										
1000										

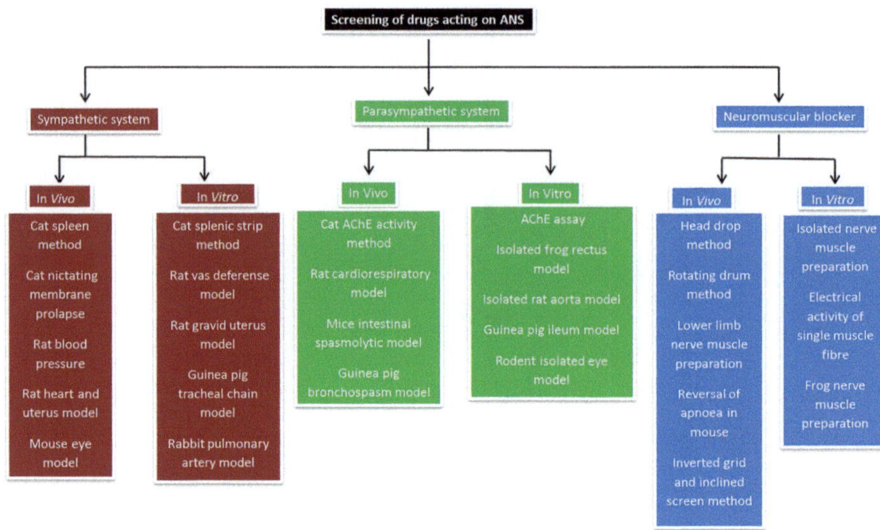

Fig. 46.1 Methods available to screen and evaluate substances for ANS activity

46.3 Sympathetic System

- Norepinephrine is the predominant neurotransmitter released at postganglionic sympathetic nerve terminals. It activates the adrenergic receptors namely alpha and beta receptors. The sympathetic system gets activated during the stressful condition, flight response. Sympathomimetics and sympatholytics are agonist and antagonist at adrenergic receptors respectively.
- Sympathomimetic drugs are evaluated by observing any of the following effects:
  - Mydriasis
  - Increased heart rate or stroke volume
  - Bronchial smooth muscle relaxation
  - Pulmonary artery constriction
  - Inhibition of gastric acid secretion
  - Spleen and uterine contraction

46.3.1 *In-Vitro* Assays

- *Alpha 1 receptor binding assay:*
  - Alpha 1 receptors are activated by both norepinephrine and epinephrine. Activation results in cardiac stimulation, activation of glucose metabolism, smooth muscle contraction and increased neuronal electrical activity.The plasma membrane of ventricles obtained from rat heart consists only alpha 1 receptors. So the membrane from Sprague Dawley rat is incubated with a

Table 46.2 Characteristics of binding assay done for adrenergic receptors

Receptor type	Principle	Animal	Tissue preparation	Standard drug
Alpha 1	Unknown drug competes with the radioligand known blocker /agonist for the binding site. Radioactivity measured by scintillation counter.	Rat	Plasma membrane of the heart ventricles	^{3}H prazosin
Alpha 2		Rat	Brain cortical tissue	^{3}H clonidine
Beta 1 & 2		Bovine	Heart ventricles	^{3}H-dihydroalprenolol
Beta 1		Rat	Brain cortical tissue	^{3}H-dihydroalprenolol
Beta 2		Rat	Lungs	^{3}H-dihydroalprenolol and propranolol

known concentration of radioligand bind ^{3}H prazosin and varying concentrations of an unlabelled unknown drug. The unknown drug competes with the prazosin for the binding site.

- Affinity to the alpha1 receptors as well as the potency of the unknown drug is determined in this test.
- The final calculation based on,

  Alpha 1 specific binding = Total binding minus Nonspecific binding

- **Alpha 2 receptor binding assay**
  - Similar to the above mentioned binding assay, here radioligand clonidine is used on a homogenised cortical tissue obtained from Sprague Dawley rat. This assay will assess the clonidine like mechanism and alpha 2 binding affinity. Alpha agonist causes displacement of ^{3}H clonidine.
  - This assay can be used to determine alpha 2 binding capacity of other classes of drugs e.g. antidepressants
  - Principle: cortical homogenates are incubated at 25 °C with a known concentration of ^{3}H clonidine and varying concentration of unknown drug for 20 min.
  - The filtered and buffer washed solution was measured for radioactivity through the scintillation counter.
  - Specific binding = Total binding minus binding in the presence of clonidine
  - Modification: Radioligand yohimbine can be used instead of ^{3}H clonidine
- **Beta receptor binding assay**
  - Bovine heart ventricles consist of 75% of beta 1 and 25% of beta 2 receptors. A bovine heart can be easily obtained from the slaughterhouse and simultaneous study on beta 1 and 2 receptors are also possible.
- The characteristics of binding assay done for adrenergic receptors are summarized in Table 46.2.

46.3.2 Isolated Organ Models

- Compared to in *vivo* methods, the isolated organ model provides accurate measurement of tissue responses
- Reflexes, neurohumoral activation and drug distribution are minimised
- *Alpha antagonist (Rat vas deferens model)*
  - *Principle:* Activation of alpha receptors elicits strong rapid contraction of vas deferens and rapid relaxation on washing the drug. Vas deferens is innervated by the autonomic system exclusively. Vas deferens of rat and guinea pig both are used commonly; a rat is preferred due to the abundance of alpha receptors
  - *Procedure:* Rat is sacrificed and vas deferens is dissected out. The tissue is kept in a suitable bioassay system with a Tyrode solution at 35 °C. A simple isotonic lever is used to record the contractile response. Cumulative DRC is obtained with various doses of norepinephrine. Phentolamine is used as a standard. The test drug is used along with norepinephrine to assess the response. Type of antagonism is determined by Schild's plot.
  - *Modification:* Guinea pig hypogastric nerve isolated organ can be used to assess the alpha antagonist activity
- *Beta agonist/antagonist (Rat heart and uterus model)*
  - Simultaneous study of beta 1 and beta 2 receptors are possible in this model. Beta 1 and 2 receptors are present in the heart and uterus respectively.
  - Animal: Sprague Dawley rats
  - Anaesthesia: IP pentobarbital 60 mg/Kg
  - *Preparation:* The anesthetized animal is connected to an artificial respirator. The left carotid artery is cannulated to monitor blood pressure. Heart rate is measured through rate meter. A midline incision is made in the abdomen to expose uterine horns. After releasing from the surrounding tissue, the uterus is mounted in the organ bath. The tissue is aerated with 5% CO_2 and 95% O_2 and the temperature is maintained at 37 °C.
  - IV isoprenaline, norepinephrine and salbutamol are given to the animal and the tissue bath. The response is noted. Later, a test drug is administered to evaluate the activity. If the test drug is an antagonist, it inhibits the action of isoprenaline on the heart and uterine tissue. The dose-response curve is obtained and compared.

46.3.3 *In Vivo* Models

- *Invasive blood pressure measurement in rat*
  - *Principle:* Sympathomimetic and sympatholytic actions are assessed simultaneously in this model by observing the effect on BP and heart rateAnimal: Sprague Dawley adult male rats
  - Anaesthesia: Urethane 1G/Kg S.C

  - *Procedure:* Rats are anaesthetised and left femoral vein and right carotid artery are exposed. The trachea is cannulated for maintaining an airway. Arterial cannula placed in carotid to measure BP and HR. The venous cannula is placed in the femoral vein to inject drugs. An arterial cannula is connected to the pressure transducer and then the recorder. Catheters are flushed with heparin. Standard and test are injected one by one and the effects are monitored. Adrenaline is used as a standard drug. To elicit a specific receptor action, animals are pre-treated with antagonist before giving agonist.
- *Cat Spleen model*
  - *Principle:* Stimulating splenic nerve by sympathomimetic drug results in the release of norepinephrine, which is measured by collecting venous effluent or observed by splenic contraction.
  - Animal: Cat
  - Anaesthesia: Chloralose
  - *Procedure:* Abdomen is cut open after giving the anaesthesia. From the mid duodenum to the colon, the intestine is removed. Spleen is released from the omental attachments. Splenic nerve is dissected out and tied. Splanchnic nerves and the adrenal glands were removed to avoid false-positive results. The portal vein was ligated proximal to the splenic vein. A cannula is placed in the superior mesenteric vein to collect the venous effluent which is diverted from the portal vein.
  - Test drugs are injected intravenously. The effluent was collected during drug administration and 20 s post-administration. Splenic contraction is also observed.
  - This model helps to understand the mechanism of action of drugs that acts by neither increasing the release of norepinephrine. The same experiment can be done in *vitro* by using cat splenic strips.

46.4 Parasympathetic System

46.4.1 *In Vitro* Model

- *Acetylcholine esterase activity by spectrophotometric Ellman assay*
  - Acetylcholinesterase (AChE) is found in nerve endings, smooth muscle, skeletal muscle, and red blood cells. AChE activity is determined by observing the change in the absorbance at 412 nm.
  - Homogenized corpora striata obtained from the brain of the male Wistar rats is used
  - The homogenized mixture is added to various concentration of test solution along with phosphate buffer, DTNB, and the substrate
  - Ellman's reagent or DTNB is a chemical compound used to quantify thiol group concentration in a sample

- Acetylthiocholine is used as substrate.
- Absorbance is measured by spectrophotometer at 412 nm after adjusting for the blank
- The percentage inhibition is obtained from the standard curve

46.4.2 Isolated Organ Model

- *Acetylcholine assay using Guinea pig ileum*
  - Animal: Guinea pig, Sacrificing method: Stunning
  - *Procedure:* The abdomen is cut open through a midline incision. The intestine is soaked with Tyrode solution. The ileum was identified by tracing the ileocaecal junction. Mysentry is removed. Thread is tied near the junction to identify the distal end. Tissue around two centimetres is taken. Tissue cleaned with Tyrode solution. The tissue is mounted in an organ bath containing Tyrode solution and aerated with room air, temperature is maintained at 37 °Celsius. The proximal portion of the tissue was attached to a clamp and then the frontal writing lever. Contractions induced by ACh are recorded with the lever on a rotating drum.
  - A standard dose-response curve was obtained by injecting various doses of ACh into the tissue bath. Then the test compound was injected to the tissue bath. The test compound can either produces contraction or relaxation.
  - The test compound can be considered as a parasympathomimetic if it produces contraction similar to ACh or parasympatholytic if it causes smooth muscle relaxation.
- *Acetylcholine assay using Isolated frog rectus muscle*
  - Animal: Frog, Sacrificing method: Pithing
  - *Procedure:* After pithing, the skin over the anterior abdominal wall was cut and extended laterally till the limb. Rectus abdominis muscle was identified as a whitish muscle originating from the pubic bone and inserted into the sternum. Muscle is dissected from its attachments and placed in a dish containing frog ringer solution. Muscle is cut into a small piece and mounted in an organ bath. The tissue is kept at 37 °Celsius and aerated with oxygen. The proximal portion of the tissue is connected to the isotonic lever and to the rotating drum to record the contractions.
  - A standard dose-response curve was obtained by injecting various doses of ACh into the tissue bath similar to Guinea pig ileum. Then the test compound was injected to the tissue bath.
  - The test compound can either produces contraction or relaxation so that the activity can be determined.

46.4.3 *In Vivo* Model

- *Cardiorespiratory model*
  - Principle: Cholinergic and anticholinergic activity of the unknown drug is evaluated by observing the effect on cardiorespiratory function in rats. Cholinomimetic drugs decrease heart rate and blood pressure and increase bronchial secretions
  - Animal: Sprague Dawley rats
  - Anesthesia: 1% chloralose and 25% urethane in an equal mixture, 5 ml/kg
  - Procedure: The trachea of the anesthetized rat is cannulated and connected to a pneumotachograph (measures airflow). The femoral artery is cannulated to record BP and the femoral vein is cannulated to inject drugs. Heart rate monitoring is done by recording a pulsatile pressure signal. Acetylcholine and atropine are used as standard drugs. Cholinergic activity is assessed by injecting test drug at varying doses intravenously after administering atropine, to observe reverting action on HR, BP and respiration.
- *Inhibition of Lacrimation*
  - This experiment is to demonstrate the anticholinergic activity by inhibiting lacrimation
  - Animal: Male Wistar rats
  - Procedure: Under the optimal laboratory setting, animals are maintained at 36 °C for 10–15 min to produce vasodilation. Test compounds are injected through the tail vein. Test solution volume was 1 mL/Kg. After the test compound, intraperitoneal carbamylcholine is injected. Carbamylcholine injection is repeated at pre-specified time interval (i.e. 15, 45, 75, 105 min). A piece of cotton is applied over the eye.
  - Presence of coloured spot within 3 min of carbamylcholine injection is a positive response. Absence of coloured spot indicates test compound has parasympatholytic property.

Bibliography

Ellman GL, Courtney KD, Andres V, Featherstone RM. A new and rapid colourimetric determination of acetylcholinesterase activity. Biochem Pharmacol. 7:88–95.

Greengrass P, Bremner R. Binding characteristics of 3Hprazosin to rat brain α-adrenergic receptors. Eur J Pharmacol. 1979;55:323–6.

Krasilnikova A, Mittal R, Gupta R. Animal models for cardiotonic agents. In: Gupta SK, editor. Drug screening methods. 3rd ed. New Delhi: The Health Sciences Publisher; 2016. p. 346–67.

Mann DE. In: Turner RA, editor. Screening methods in pharmacology. New York: Academic Press Inc.; 1965. p. 135–51.

Minneman KP, Hegstrand LR, Molinoff PB. The pharmacological specificity of beta-1 and beta-2 adrenergic receptors in rat heart and lung in vitro. Mol Pharmacol. 1979;16:21–33.

Vogel GH. In: Vogel GH, editor. Drug discovery and evaluation. 2nd ed. Berlin: Springer; 2008. p. 26–40.

47 Screening Methods for the Evaluation of Anti-Infective Agents

Sushil Kiran Kunder and Avinash Arivazhahan

Abstract

Anti-infective agents are among the most commonly prescribed and consumed drugs worldwide, particularly in developing countries like India due to the overwhelming nature and burden of infections. The various methods and models used for antifungal, anthelminthic, antibacterial and antiviral drug screening are explained in this chapter. Antifungal screening methods commonly used in experimental pharmacology include broth dilution and agar diffusion techniques, among others. Anthelminthic drugs can be divided into phenotypic screening and mechanistic screening methods. Antibacterial drug screening methods are of four broad types: diffusion methods, chromatographic methods, dilution methods and others. Antiviral drugs can be screened or studied using HIV models and non-HIV models.

Keywords

Antimicrobial · Anti-infective · Agar diffusion · Broth dilution

47.1 Screening of Antifungal Drugs

As a general principle, anti-fungal activity of drugs can be assessed in filamentous fungi based on the hyphal radial growth rate and in yeast, it is assessed using yeast growth rate.

S. K. Kunder
Novo Nordisk India Ltd., Bengaluru, Karnataka, India

A. Arivazhahan (✉)
Sanofi India Ltd., Chennai, Tamil Nadu, India

M. Lakshmanan et al. (eds.), *Introduction to Basics of Pharmacology and Toxicology*, https://doi.org/10.1007/978-981-19-5343-9_47

Fungal strains of *Trichoderma* species, *Penicillium notatum* and *Aspergillus niger*, cultured for about a week at 30 °C produces spores. A suspension of spores is placed at the centre of a petri-dish containing the test drug. It is maintained at a humid environment at 30 °C to prevent the drying-out of the growth medium. Mycelial radial growth is measured 5 days after inoculation. Growth inhibition is then calculated as a percentage of total growth.

In order to study antifungal activity against yeasts, broth dilution techniques and agar diffusion techniques are used.

In broth dilution techniques, minimal inhibitory concentration (MIC) is determined. 96-well microplate is taken, and each well is filled with broth malt extract and the drugs to be tested (in serial dilutions). These wells are then inoculated with yeast extracts. The microplate is then incubated for 24 h at 30 °C. Growth of yeast was quantified. The lowest concentration of test substance to which no growth of yeast was seen is defined as MIC.

In agar well diffusion assays, inoculum containing yeast is added to bacto-agar and malt broth medium. This mixture is poured to a petri-dish that is maintained at room temperature. Using a cork borer, wells are cut into the agar plate. The test substance is loaded into the small wells. The petri-dish is then inverted for 2 h at 4 °C to facilitate uniform diffusion. After pre-incubation, plates are re-inverted and aerobically incubated for 2 days at 30 °C. The inhibition zone diameter in recorded in 4 directions.

In paper disk diffusion assay, 6 sterile disks of paper impregnated with known concentrations of the test drug are placed on the agar surface of the petri-dish. A negative control disk is included. After 2 days, the inhibition zone diameter is measured in 4 different directions.

There exist screening tests specific to a particular mechanism of action of potential antifungal agents. These are listed below.

(a) *Test for leakage of substance:*
    Cryptococcus neoformans is incubated and cultured. The inoculum is transferred to microtubes containing the test drug and incubated at 37 °C. The microtubes are then centrifuged, and absorbance is measured at 260 nm. The absorbance due to leakage is considered.
(b) *Sorbitol Protection Assay:*
    MIC of test substance is determined in the absence or presence of sorbitol after 24 and 72 h of incubation at 35 °C. Thus, the protective effect of sorbitol can be estimated.
(c) *Ergosterol Effect Assay:*
    Drugs having affinity for ergosterol form complexes with ergosterol (exogenous), thereby preventing their interactions with fungal membrane ergosterol, resulting in an increase in the MIC value. MIC of the test drug is determined in the presence or absence of ergosterol. Amphotericin B is usually taken as the control drug. MIC is determined after 24 and 72 h of incubation at 35 °C.

These 3 tests are called "mechanism-of-action assays" as they can evaluate the effect of the test substance on the fungal cell membranes, based on established antifungal mechanisms.

47.2 Screening of Anthelminthic Drugs

Helminths are widely prevalent parasites that are said to be one of the most common groups of organisms causing diseases, particularly in developing parts of the world. In general, three strategies are regularly employed in anti-parasitic drug discovery. The most successful strategy has been introduction of the parasites in experimental animals and treating them with the test and control agents. This has been the basis of discovery of most of the anthelminthic drugs available today. The next strategy of discovery is based on creation of model parasites that are subjected in culture, to the test and control agents. The decrease in viability of these model parasites is taken as a sign of anthelminthic activity. The third and final strategy has depended purely on molecular targets for drug discovery. This is more of a mechanism-based screening that can assist in discovery of me-too drugs of a particular class, once the mechanism of action of the prototype agent is well established. In line with these strategies, anthelminthic screening can be of two broad types: phenotypic screening and mechanism-based screening.

(a) *Phenotypic screening tests:*

These tests are based on the principle that anthelminthic agents reduce the viability of the worm in vitro. The most common technique that is used is the MTT assay, which is a colorimetric assay. The MTT compound is converted to a purple-colour-yielding formazan compound by cellular oxidoreductases. While the intensity of the purple colour reflects the inherent viability of the worm (and indirectly gives an estimate of the anthelminthic activity of the test drug), it is only a rough estimate, and has to be employed in conjunction with microscopic examination for worm motility, intestinal motility and tanning of cuticle.

Caenorhabditis elegans is commonly considered as the model organism for these in vitro procedures. *Haemonchus contortus* is another worm that may be used as an alternative. Earthworms (*Pheretima posthuma*) have also been used as a model organism. As the technique is considered to be time-consuming and expertise-requiring, newer video imaging and computer software have been programmed.

As mentioned earlier, this technique is only a mere screening method, and cannot be taken as a confirmatory test. As a testament to this statement, none of the currently available anthelminthic agents have been discovered by phenotypic screening techniques. This has further emphasized the need for the existence and development of mechanism-based screening techniques.

(b) *Mechanism-based screening tests*

When compared with the previously described phenotypic screening tests, the mechanism-based tests are less time-consuming, more economical and carry a greater amount of success in discovery of the right agents. These tests are based on the principle that identification of a particular anthelminthic mechanism and isolation of agents that target this mechanism. In theory, this sounds fruitful, but not so in real world practice.

47.3 Screening of Antibacterial Drugs

With the growing threat of antimicrobial resistance, there seems to be a never-ending need for newer and more effective antibiotics in order to treat or prevent diseases in humans. As with other screening techniques, antibacterial drug screening can also be done in vitro or in vivo. While in vitro tests are the more preferred entity, in vivo techniques are also used as a confirmatory test, although ethical regulations have curbed the use of such techniques. In vivo tests are self-explanatory in the sense that the test animals (usually mice or rats) are inoculated with the pathogenic organisms (at the median lethal dose) and then the test drugs are administered. Following this, blood, tissue, urine or fecal samples are collected to check for the presence or absence of bacteria. Test agents that can significantly reduce the number of bacterial colonies are said to have antibacterial activity, when compared with the control and standard groups of animals.

The various in vitro methods used are listed in Table 47.1.

- *Diffusion methods:*

  Diffusion methods are the most commonly used in vitro techniques to check for antibacterial activity of a given substance.

  Among the various diffusion techniques, the agar disc diffusion method is the most popular. This method is also known as the "antibiogram" in clinical practice for routine antibacterial susceptibility testing. A typical agar disc diffusion plate (after growth) is shown in Fig. 47.1. In this method, the disc plates are inoculated with the bacterium, following which the test, control and standards are placed on top of the plate at specific points using filter papers. After this, the agar plates are incubated to encourage bacterial growth. 24–48 h following incubation, the plates are checked for bacterial growth. The control substance usually has no inhibitory property, while the standard substance demonstrates significant inhibition. Comparison of the test substance is done to check for antibacterial activity.

  Antibacterial grading method is another diffusion technique (also uses dilution technique described later in this chapter) that helps to determine the minimal

Table 47.1 In vitro screening methods for antibacterial agents

Diffusion methods	Thin layer chromatography-based methods	Dilution methods	Other methods
Agar disc diffusion, Antibacterial grading method, Agar well diffusion, Agar plug diffusion, Cross streak method	Agar diffusion, Direct bioautography, Agar overlay	Agar dilution, Broth dilution	Flow cytofluorumetry, ATP bioluminescence, Time-kill method, etc.

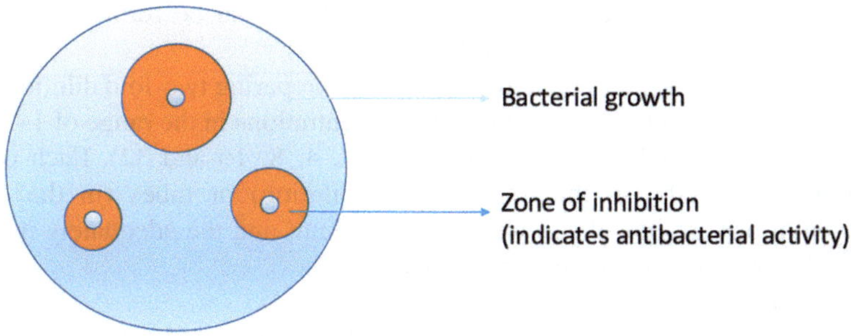

Fig. 47.1 A typical Agar disc diffusion plate

inhibitory concentration (MIC) of an antibacterial agent. Further, the effects of co-administering two antibacterial agents can also be determined using this method.

Agar well diffusion method is similar to the agar disc diffusion method, and uses microwells instead of plates in the agar medium. The principle and methodology remain the same. Agar plug diffusion method is also highly similar, the only difference being the use of plugs instead of wells or discs.

Cross streak diffusion method is a relatively faster technique that can be employed as an initial screening tool to check for antibacterial efficacy. An agar plate is taken, and a streak is made with the bacterium to be studied. Further streaks can be made following incubation. The antimicrobial efficacy can be made out by the quantum of the zone of inhibition produced by the antibacterial agents.

- *Thin layer chromatography-based methods*

  Direct bioautography is the most popular and the most commonly used tests among this class. In this method, a thin layer chromatography (TLC) plate that contains the test substance is sprayed with the microorganisms and then incubated adequately. Tetrazolium salts are then sprayed on the plate to enhance visualization of the bacterial growth (formation of formazan enables this visualization).

  TLC-based agar diffusion method is a less commonly used method. It is similar to the previously described direct bioautography technique. However, the antibacterial test substance is transferred from the TLC plate to an agar plate that contains the inoculated bacterium.

  Agar overlay diffusion technique is a combination of these two methods. The TLC plate is immersed in an agar medium to enable bacterial growth and interaction with the antibacterial agents. This method is also known as "TLC agar immersion method".

- *Dilution methods*

  Dilution methods are quantitative techniques that are most commonly used for estimation of the MIC values of antibacterial agents. Broth dilution and agar dilution are the two methods that are used. They can be further divided into microdilution (<2 ml) and macrodilution (>2 ml).

These methods are based on the principle of preparing two-fold dilutions of the test agents. This would give us multiple concentrations in the range of 1–32 mcg/ml (in the sequential dilution values of 1, 2, 4, 8, 16 and 32). Each of these dilution microwells (in the case of microdilution) or tubes (in the case of macrodilution) are inoculated with the bacterium, and the adequately incubated to check for growth or inhibition of growth.

- *Other methods*

  Flow cytofluorometry can be used to screen for antibacterial activity, although it is deemed to be an expensive method. Propidium iodide is commonly used as a fluorescent dye that also serves as an intercalating agent.

  ATP bioluminescence assay is another expensive method for antibacterial screening. Luciferase, an enzyme that helps convert D-luciferin to oxyluciferin (which emits luminescence), is used to measure the amount of light generated (using a luminometer). Greater the luminescence, greater is the viability of the microbial cell, and thus, lesser is the antibacterial efficacy of the test agent.

  The time-kill test helps to derive the relationship between the bacterium and the antibacterial agent. It can help to differentiate between a concentration-dependent versus a time-dependent mechanism of action.

47.4 Screening of Antiviral Drugs

With more and more viruses being discovered, the need for antiviral agents is rapidly increasing. However, it is a daunting task as there are significant challenges to overcome. Some of these challenges are insufficient unique targets for the drugs to act, development of drug resistance and limited availability of screening models. Screening of antiviral agents can be broadly divided into HIV-related and non-HIV-related screening methods.

47.4.1 HIV-Related Screening Methods

They may be of in-vivo or in-vitro nature, as detailed in Table 47.2.

Table 47.2 HIV-related screening methods

In-vivo models	In-vitro models
Non-Human Primates	Mono Mac 1 cell line
Severe Combined Immunodeficiency Mice	Maedi Visna Virus model
Humanized Mice	Blood Brain Barrier models

In-Vivo Methods

(i) *Non-human primate models:* Non-human primates are close to humans phylogenetically, having similar immunology, physiology and pharmacology. Chimpanzees have a more than 98% genetic identity to humans. Chimpanzees are the source of Simian Immunodeficiency Virus (SIV), which resembles acquired immunodeficiency virus (AIDS). It is considered to be the best model for AIDS research. Wild chimpanzees infected by SIV have a very high mortality rate, but very few animals infected experimentally with SIV/HIV-1 develop the disease. Another drawback is the low availability, ethical concerns and high prices.

Pig tailed macaques can be used, as HIV-1 infection can be induced in them. They are used to study the various modes of transmission like anal, oral, intravenous and vertical. Compared to chimpanzees, these species are not endangered, and their maintenance is inexpensive. However, replication of virus is poorly sustained due to which disease is not observed.

HIV-2 infections can be induced in baboons and several macaque species. The disease progression in baboons mimic the human chronic HIV picture, due to which this model is used to study AIDS pathogenesis like viral host interactions, viral latency, disease progression mechanisms and the acquired immune responses. Baboons are suitable for vaccine studies as they have immune responses like humans. Baboons species are not endangered, and sufficient numbers are available.

Many African non-human primates like sooty Mangabeys and African green monkeys are infected by SIV. SIV resemble HIV-2 more closely as compared to HIV-1. Despite high levels of replication, the hosts rarely develop the disease. This is due to evolutionary adaptation. This adaptation includes preserved mucosal immunity, limited activation of immune system and phenotypic modifications of CD4T cells.

SIV infection in Asian Macaques on the other hand results in disease that mimics AIDS in humans. SIV macaque models have limitations. There are minor differences between SIV and HIV. Furthermore, these models are expensive.

Simian human immunodeficiency viruses (SHIV) development has been advantageous as it provides a practical research aid to study the properties of HIV-1 infection. They include the characteristics of HIV 1 envelope. This aids the evaluation of vaccines and anti-HIV agents that target the envelope and reverse transcriptase. Despite their limitations, these models are still considered as primary models due to similarities in host, virus and disease pathogenesis. Non-human primates have become indispensable to study SIV/HIV pathogenesis and the development of therapeutic and prophylactic interventions to avoid transmission of the disease.

(ii) *Severe Combined Immuno-deficiency mice (SCID):* SCID mice lack both B and T lymphocytes. Human thymus tissue, foetal liver and peripheral blood leukocytes are reconstituted in them. Non-obese diabetic SCID mice are better than SCID mice as they additionally have defects in macrophage and natural killer function. But these models have limitations. The SCID thymus/liver mice have low human cells in peripheral organs and blood. The SCID-hu PBL mice lack human lymphoid organs and develop severe graft-versus-host disease, mediated by xeno-reactive donor T cells. In contrast, the SCID-hu Thy/Liv mice have very low levels of human cells in the blood and peripheral organs. Collectively, the lack of human cells in the peripheral lymphoid organs and the inability to mount functional immune responses limit the applicability of these early humanized models.

(iii) *Humanized mice models:* Recent years have marked the advent of chimeric mice, which have humanized immune systems. Human immune cells of different lineage develop in immuno-deficient mice, post-transplantation of human stem cells. These mice exhibit cellular and humoral immune responses upon microbial infection and immunization. HIV-1 can infect humanized mice causing depletion of CD4+ T cell depletion and non-specific activation of immune system. This mimics the pathology of HIV-1-infected humans. These models are ideal for studying immunopathogenesis and for developing immune therapies. The various human immune cells implicated in HIV induced pathogenesis can be investigated using humanized mice.

In-Vitro Methods

(i) Mono Mac 1 cell line: For the past few years, scientists have used in-vitro models like continuous cell lines for evaluation of HIV-1 immunopathogenesis. However, monocytoid malignant cells, the most commonly used cell lines do not closely mimic the human macrophages with respect to retro viral infection. Mono Mac cell line is considered as one of the most ideal models for understanding the immunological, genetic and biochemical functions human monocytes/macrophages. Flow cytometry has shown that Mono Mac 1 cells are positive for HIV-1 primary (CD4) as well as co-receptors (CCR3, CCR5 and CXCR4). Mono Mac 1 cell line supports infection with macrophage as well as dual tropic HIV-1 isolates. Virus production in differentiated Mono Mac cells are high probably due to raised nuclear translocation of nuclear factor kappa beta. Mono Mac 1 cell line aids the study of complex interplay between HIV-1 and monocyte/macrophages.

(ii) *Maedi Visna Virus Model:* Maedi Visna virus is known to cause inflammation of CNS and lungs in sheep. Like SIV and HIV, Maedi Visna virus belong to the genus Lentivirus. Mode of replication, latency, virus host interaction and genome organization are common to both Maedi Visna virus and HIV. Both the viruses infect monocytes and macrophages but unlike HIV, Maedi Visna virus does not infect T lymphocytes. These viruses can be used as a model to study the CNS infections.

(iii) *Blood Brain Barrier (BBB) models:* Loss of integrity of BBB plays a key role in advent of CNS destruction by HIV. In order to study this, various in-vitro BBB models have been developed, which mimic the biochemical and physical properties of human BBB. Macaque BBB model contains microvascular brain endothelial cells of rhesus macaque. In another model of BBB, endothelial cells of human umbilical cord and human cerebral astrocytes are co-cultured.

47.4.2 Non-HIV-Related Screening Methods

They can be classified into in-vitro and in-vivo screening models, as listed in Table 47.3.

In-Vitro Methods

(i) *HCV replicon system:* It was virtually impossible to study replication of HCV in cell cultures until the advent of this system. RNA or DNA, which replicate are termed replicons. Sub-genomic replicons autonomously replicate within human hepatoma cell line (Huh-7). Sub-genomic replicons are avirulent. A major disadvantage of this system is that being an incomplete viral particle, it does not represent viral life cycle completely.

(ii) *Cytotoxicity Assay: (also known as MTT (tetrazolium dye) assay):* Serial dilutions of the test drug are prepared in maintenance medium. They are added to cultures that are 24-hours-old. The incubation time depends on the type of cell line. Post incubation, staining is done using MTT. Percentage of optical density of test treated cultures to untreated ones gives the cell viability.

(iii) *Cell culture systems:* Huh-7 (human hepatoma cell line) is the specific cell culture system for Hepatitis C virus (HCV). These cells support replication of HCV. However, since they have lost certain features, they are not normal liver cells, which is a drawback of this cell culture system for screening anti-viral activity of drugs. Vero cells are a continuous cell line, isolated originally from renal epithelial cells of green African monkey. It is an extensively used cell line as it is easy to maintain. Interferons are not produced, which makes it an attractive screening model for anti-viral agents. Replication of various viruses like Japanese encephalitis virus, Influenza A and B viruses, Dengue virus, Reovirus and Rabies virus is seen with Vero cell lines.

In-Vivo Methods

(i) *Chimpanzees:* Due to the close genetic link to humans, chimpanzees have contributed extensively to biomedical research. They are susceptible to human pathogens and present a similar clinical picture. They are common animal models for Hepatitis A, B and C research. Research on them has led to the advent of hepatitis A and B vaccines, which protect millions of humans

Table 47.3 Non-HIV-related screening methods

In-vitro models	In-vivo models
HCV Replicon System	Non-human primates
Cytotoxicity Assay	Others – Mice, ferrets, guinea pigs, etc.
Cell Culture System	

worldwide. Chimpanzees have also contributed to development of newer therapies for hepatitis B virus and have evolved our knowledge of T-cell response for hepatitis C virus. They are also good models for PK/PD studies of new drugs and to study pathogenesis. However, due to ethical reasons, their use has been dramatically curtailed.

(ii) *Other Non-human primates:* These animal models help us understand the pathogenesis of disease in humans. They also aid in the development of vaccines and therapies for preventing and managing viral infections. Especially, the adaptive and innate responses that these animals develop against viruses are quite like that of humans. Asian Macaques like *Macaca nemestrina* (pigtail macaque), *Macaca fascicularis* (cynomolgus macaque) and *Macaca mulatta* (rhesus macaque) are commonly used for immunological and viral studies. African non-human primates like African green monkeys (*Chlorocebus pygerythrus* and *Chlorocebus sabaeus*) and Sooty mangabeys (*Cercocebus atys*) are used quite frequently in SIV-related research as they are natural hosts. Baboons are rarely used to study viral and host interactions.

(iii) *Other animal models:* Mice models are commonly used for studying HSV and influenza virus. Since most human viruses do not replicate, transmit efficiently or produce classical clinical signs in immunocompetent normal mice, transgenic and humanized mice are commonly needed to be used. Humanized mice express human factors and do not produce murine proteins. Transgenic mice/ genetically modified mice have certain genes disrupted. If a single gene is disrupted/ removed, they are termed knock-out mice.

Guinea pigs are often used as alternatives to mice models and are susceptible to viruses like HSV, filovirus, influenza virus and to study viral hemorrhagic fevers. Ferrets are used as animal models to study Nipah virus, corona virus, influenza virus and morbillivirus. The lung physiology of ferrets is similar to that of humans. They exhibit typical signs of influenza seen with human disease. Syrian Golden Hamsters are used as animal models for flavivirus, SARS corona virus and bunyavirus. Rats are uncommonly used for influenza A virus studies.

Bibliography

Andresdottir V. Maedi-visna virus as a model for HIV. Icel Agric Sci. 2018;31(1):23–47.

Balouiri M, Sadiki M, Ibnsouda SK. Methods for in vitro evaluating antimicrobial activity: a review. J Pharm Anal. 2016;6(2):71–9.

Blight KJ, Norgard EA. HCV replicon systems. In: Tan SL, editor. Hepatitis C viruses: genomes and molecular biology. Norfolk (UK): Horizon Bioscience; 2006. Chapter 11.

Cermelli C, Cuoghi A, Scuri M, Bettua C, Neglia RG, Ardizzoni A. In vitro evaluation of antiviral and virucidal activity of a high molecular weight hyaluronic acid. Virol J. 2011;8(141):1–8.

Cruz KS, Lima ES, Amazonzs da Silva M, Simplicio de Souza E, Montoia A, Pohlit AM. Screening and antifungal activity of a β-carboline derivative against *Cryptococcus neoformans* and *C. gattii*. Int J Microbiol. 2019:1–8.

De Senna A, Lathrop A. Antifungal screening of bioprotective Isolates against *Botrytis cinerea*, *Fusarium pallidoroseum* and *Fusarium moniliforme*. Fermentation. 2017;3:53.

Estes JD, Wong JW, Brenchley JM. Nonhuman primate models of human viral infections. Nature Rev Immunol. 2018;18:390–404.

Geary TG, Sakanari JA, Caffrey CR. Anthelmintic drug discovery: into the future. J Parasitol. 2015;101(2):125–33.

Genois N, Robichaud GA, Tremblay MJ. Mono Mac 1: a new in vitro model system to study HIV-1 infection in human cells of the mononuclear phagocyte series. J Leukoc Biol. 2000;68(6): 854–64.

Holden-Dye L, Walker RJ. Anthelmintic drugs and nematicides: studies in *Caenorhabditis elegans*. In: WormBook: the online review of *C. elegans* biology. [Internet]. Pasadena (CA): WormBook; 2005–2018.

Lanford R, Walker C, Lemon S. The chimpanzee model of viral hepatitis: advances in understanding the immune response and treatment of viral hepatitis. ILAR. 2017;58(2):172–89.

Quiroga EN, Sampietro AR, Vattuone MA. Screening antifungal activities of selected medicinal plants. J Ethnopharmacol. 2001;74:89–96.

Rompay K. The use of nonhuman primate models of HIV infection for the evaluation of antiviral strategies. AIDS Res Hum Retrov. 2012;28(1):16–35.

Zhang L, Su L. HIV-1 immunopathogenesis in humanized mouse models. Sci China Life Sci. 2010;53(2):195–203.

Preclinical Toxicity Studies

48

Mageshwaran Lakshmanan

Abstract

Preclinical toxicity studies involve testing the newly screened compounds to determine the safety of that compound in at least two different animal species. Based on the time & duration of the study and its objectives, the preclinical toxicity studies can be classified into Acute toxicity studies, Sub-acute toxicity studies, Sub-chronic toxicity studies, and Chronic toxicity studies. Based on the number of doses administered, the toxicity studies can be classified into single-dose toxicity studies (acute toxicity study) and repeated dose toxicity studies (sub-acute, sub-chronic, and chronic toxicity studies). Similarly, based on the area of exposure of interventional drugs in animals, toxicity studies can be divided into systemic toxicity studies (comprises of acute, sub-acute, chronic, sub-chronic, and reproductive toxicity studies) and local toxicity studies (Dermal toxicity, Ocular toxicity studies). This chapter will give a concise review of the selection of animals, selection of dose, duration of the study, types of protocol, and various issues related to each type of toxicological study.

Keywords

Acute toxicity study · Chronic toxicity study · Local toxicity study · Subacute toxicity study · Dermal toxicity · Ocular toxicity

M. Lakshmanan (✉)
Department of Pharmacology, Thanjavur Medical College, Thanjavur, Tamil Nadu, India

M. Lakshmanan et al. (eds.), *Introduction to Basics of Pharmacology and Toxicology*, https://doi.org/10.1007/978-981-19-5343-9_48

48.1 Introduction

Claude Bernard, Father of Physiology and The Prince of Vive sectors, said, "*Have we the right to make experiments on animals and vivisect them? It would be strange indeed if we recognized man's right to make use of animals in every walk of life, for domestic service, for food, and then forbade him to make use of them in medicine*". He had a strong view on animal testing and commented, "*Experiments on animals, with deleterious substances or in harmful substance, are very useful and entirely conclusive for the toxicology and hygiene of man. The effects of the substances tested are the same on man as on animals, save for differences in degree*". Though his arguments in favor of animal experimentation were criticized ethically, to date, animals are used in various industries to test the safety of selected compounds. As the term implies, Preclinical toxicity studies involve testing the newly screened compounds aiming to determine the safety of that compound in at least two different animal species.

- Based on the time & duration of the study and its objectives, the preclinical toxicity studies can be classified into Acute toxicity studies, Sub-acute toxicity studies, Sub-chronic toxicity studies, and Chronic toxicity studies
- Based on the number of the dose administered, the toxicity studies can be classified into single-dose toxicity studies (acute toxicity study) and repeated dose toxicity studies (sub-acute, sub-chronic and chronic toxicity studies).
- Similarly, based on the area of exposure of interventional drugs in animals, toxicity studies can be divided into systemic toxicity studies (comprises of acute, sub-acute, chronic, sub-chronic, and reproductive toxicity studies) and local toxicity studies (Dermal toxicity, Ocular toxicity studies).

Hence, based on the objectives, the preclinical toxicity studies vary accordingly. This chapter will give a concise review about the selection of animals, selection of dose, duration of the study, types of protocol, and various issues related to each toxicological study.

48.2 Acute Toxicity Study

The initial step in assessing the substance's toxic nature is establishing the dose that causes mortality in 50% of animals in a group called medial lethal dose 50 (LD_{50}). Hence the data from the acute toxicity study may be useful in:

1. Establishing the initial information regarding the mode of toxicity by the drug
2. Determining the dose required in subsequent animal studies
3. Calculating LD_{50} values that provide various indices of types of drug activity.

Hence acute toxicity study is the initial assessment of toxic manifestation of the drug and initial screening procedure conducted for all compounds.

48.2.1 Aim of Acute Toxicity Study

The following are the main aims of any acute toxicity study:

1. To determine the median lethal dose 50 and maximum tolerated dose.
2. To determine the ratio between the lethal dose and pharmacologically effective dose in the same species, i.e., the therapeutic index
3. To extrapolate the LD_{50} and MTD derived from acute toxicity study to the other species using predetermined relative body surface area calculations.

48.2.2 Definition

Acute toxicity study is defined as a study in which the animals are subjected to a single drug administration or drug exposure less than 24 h and observed for 2 weeks to determine the maximum tolerated dose and LD_{50}.

48.2.3 Animal Selection

For acute toxicity studies, the OECD, FDA, CDSCO, and various guidelines recommend using two different species. Usually, one rodent species and one non-rodent species should be used as they differ in response to toxic agents. When the effect is observed in both species, it indicates that a common physiological mechanism is involved, and the chances of toxic effects in humans are high. On the other hand, when toxicity is noted in only one species, then the effect is peculiar to that particular species, and hence chances of toxicity in the third species (human) are less likely.

In many situations, the preferred rodent species will be a rat, although other rodents like rabbits, guinea pigs can be used. For non-rodents, cats or dogs are usually selected.

48.2.4 Number of Animals

The OECD guidelines recommend three animals per group per dose in acute toxicity testing, though up to 5 animals per group are recommended by Schedule Y of Drugs and Cosmetics Act in India. The animals should be tested group-wise, i.e., a group of higher doses should be studied after completing a group administered with lower doses. Studying toxicity in multiple group-parallel modes requires many animals and is discouraged by various guidelines.

48.2.5 Sex of Animals

Previously either sex was selected and tested in acute toxicity testing. Nowadays, the majority of the guidelines recommend the use of non-pregnant young female animals. This is because the literature surveys of LD_{50} tests indicated that females are slightly more sensitive to the toxins than the male species. Hence utilization of the female sex provides a slightly accurate LD_{50} value than the male species. However, when the testing drug is utilized only for males in humans, or the possibility of interactions of female sex hormones with the testing drug is considered, then the male sex can be selected.

48.2.6 Age of the Animals

Usually, mature young adults are chosen for acute toxicity studies. However, when the testing substances are intended to use in infants, then the compound can be administered in the newborn rodents of 24 h age will be chosen. Moreover, the newborn rodents' data should be compared with the mature young adults to assess any difference in sensitivity due to age.

48.2.7 General Care of the Animals

In all the steps of the studies, the investigators should strictly follow the Good Laboratory guidelines. Standard care like allotting unique identification numbers to the animals and group, providing adequate space while caging, controlling humidity and lightings should always be maintained. In addition to this, the animals should be kept so that observation regarding the development of toxic effects could be noted easily.

48.2.8 Administration of the Testing Compound

The substance intended for testing should be a pure compound. As a rule of thumb, the drug sample volume administered to the selected animal species should always be kept constant. The concentration of the drug sample can be altered. An example of this is illustrated in Table 48.1.

In the above scenario, the volume of drug administered is kept constant, i.e., 2 ml. This is because to avoid the inadvertent results obtained by pharmacokinetic alterations by administering drug samples in excess. The maximum volume of drug administration for particular species is given in Table 48.2. and should not be exceeded under any circumstances.

Usually, in rodents, the volume of drug administered should not exceed 1 ml/ 100 g of body weight. One should also take the physical nature of drug volume administered into account. Using an aqueous drug for administration is highly

Table 48.1 Illustration of constant volume—different concentrations of drug administration for various doses

S. No	Dose (mg/kg)	Weight of animal (g)	Amount required (mg)	Volume of sample (ml)	Concentration achieved (mg/ml)
1	5	200	1	2	0.5
2	50	200	10	2	5
3	100	200	20	2	10
4	200	200	40	2	20

Table 48.2 Recommended dose volumes for common laboratory animals

		Oral route (ml/kg)		IV route (ml/kg)		IP route (ml/kg)	
S. No.	Type of the species	Ideal	Maximum	Ideal (as bolus)	Maximum (as slow injections	Ideal	Maximum
1	Mouse	10	20	5	25	5–10	20
2	Rat	10	20	5	20	5–10	20
3	Hamster	10	20	5	20	5–10	20
4	Guinea Pig	10	20	1	5	10	20
5	Rabbit	10	20	1–5	10	3–5	10 (rare)
6	Dog	5–8	15	1–5	10	1	20
7	Cat, Ferret	10	15	5	10	5	20

desired, followed by suspension, an emulsion in oil, and the least desired solution in the new vehicle. For the compounds dissolved in a vehicle, a separate group of animals should be tested for that vehicle's toxic effects. The drug can be administered as a single dose or repeated doses within 24 h.

48.2.9 Route of Drug Administration

The test substance should be ideally administered to the animals in the same route intended to be administered in humans. Usually, oral and intravenous routes are the preferred routes of drug administration. In most of the situation, though the drug is intended to be administered via oral route in humans, the regulatory authorities demand the toxicity data of intravenous route of administration. This is because to compare the LD_{50} values and Maximum tolerated dose between the two routes and determine the safety of the drug. A separate group of animals should be used for testing different routes of administration.

The value of LD_{50} depends on the type of route of administration. When the different drug administration routes are compared, the intravenous route has the highest LD_{50} value, followed by the intraperitoneal, subcutaneous, with the least oral route. The intravenous route will always be preferred over the intraperitoneal route to avoid the liver's first-pass metabolism. Moreover, the intraperitoneal route limits drug administration with an irritant nature for which the intravenous route is preferred.

48.2.10 Selection of Dose Levels

The recommended steps in selecting the dose levels for acute toxicity studies are as follows:

- Select a minimum of three animals for each step
- The recommended dose levels as the starting dose by OECD guidelines are 5, 50, 300, and 2000 mg/Kg body weight.
- Start with 5 mg/Kg in 3 animals and observe the number of death.
- If the number of moribund or dead animals is 2–3, then the dose range will be fixed between 0.5 and 5 mg/kg.
- If no death or one animal is dead, then the same dose should be repeated in another group of 3 animals.
- If 2–3 animals are dead in the second time, then the dose level will be fixed between 0.5 and 5 mg/kg. If there is no death or one death even on the second time, then the dose level will be up-scaled to 50 mg/kg.
- The same steps are repeated for 50 mg/kg groups.

Including the animal groups of the next higher dose should be delayed until one is confident about the survival of previously dosed animals. Administration of dose more than 5000 mg/Kg body weight should require clearance from the ethical and regulatory authorities.

48.2.11 Observation Period

After the administration of the testing substance, the animal should be observed as:

- Continuously for the first 30 min
- Periodically for the first 24 h with special attention in the first 4 h (every hour or every 2 h)
- And daily for a total of 14 days

The time of appearance and disappearance of toxic effects, time of excessive stress and time of death are noted.

Throughout the observation period, the following parameters are noted:

1. Changes in skin and fur
2. Changes in eyes and mucous membrane
3. Alterations in circulatory, respiratory, autonomic, and CNS systems
4. Behavior modifications and changes in somatomotor activity.
5. The appearance of tremors, convulsions, salivation, and diarrhea
6. Development of lethargy, inappropriate sleep, or coma

Animals that died during the test should be subjected to the autopsy immediately, and all the organs should be assessed for toxic effects. The autopsy should be performed within 16 h of death in order to avoid autolysis. Surviving animals will be sacrificed, and the same should be repeated.

48.2.12 Calculation of Medial Lethal Dose 50 (LD_{50})

The determination of Medial lethal dose 50 (LD_{50}) is the important reason for conducting the acute toxicity studies. The LD_{50} indicates the magnitude of toxicity of the testing substance.

The following are the various methods implemented commonly to calculate LD_{50} from the acute toxicity studies.

1. Staircase or 'up-down' method
2. Karber's arithmetical methods
3. Miller and Tainter's Graphical method
4. Lorke's method of LD_{50} determination

The various methods mentioned above are discussed as follows with an appropriate example with their advantage and limitations.

48.2.12.1 Staircase or Up-Down Method of LD_{50} Calculation

- Two animals should be injected with a particular dose and observed for 2 h for any death.
- In case of no death, another two animals should be used with a dose increment of 1.5.
- In case of death, the dose should be decreased by a factor of 0.7
- By this method, with 10–12 animals, the maximum tolerated dose and a minimum lethal dose can be determined
- Once the dose range is fixed, the LD_{50} assay will be planned using four dose levels between maximum tolerated and minimum lethal dose.

48.2.12.2 Karber's Method for Calculation of LD_{50}

- The dose of the drug administered in each group and the number of the dead should be noted and tabulated
- The dose difference between the group should be calculated

Table 48.3 Hypothetical data for calculating LD_{50} based on Karber's arithmetical method

Group No	Dose (mg/kg)	No of animals dead	Dose difference (a)	Mean mortality between current and previous group (b)	a X b	Sum of the productΣ (a Xb)
1	Vehicle	0	--	--	--	170
2	10	0	0	(0 + 0)/2 = 0	0	
3	20	0	10	(0 + 0)/2 = 0	0	
4	40	1	20	(1 + 0)/2 = 0.5	10	
5	60	2	20	(2 + 1)/2 = 1.5	30	
6	80	2	20	(2 + 2)/2 = 2	40	
7	100	2	20	(2 + 2)/2 = 2	40	
8	120	3	20	(3 + 2)/2 = 2.5	50	

- The average number of dead between the two successive dose groups (interval mean) should be calculated
- The product of dose difference and interval mean should be obtained
- The sum of the product should be divided by the number of animals in the group, and the resultant value should be subtracted from the least lethal dose
- The resultant value is the Medial Lethal dose 50 (LD_{50})

The following example illustrates Karber's arithmetic method of LD_{50} calculation. Hypothetically, when an extract "XYZ' was administered to various groups in different doses, and the number of animals dead was noted and tabulated (Table 48.3).

From the above table,

The least lethal dose = 40 mg/kg.

Sum of product = 170.

The LD_{50} can be calculated by using the following formula.

$$\begin{aligned} LD_{50} &= \text{Least Lethal dose} - \left(\sum (a \times b)/N\right) \\ &= 40 - (170/7) \\ &= 40 - 24.3 \end{aligned}$$

$$LD_{50} = 15.7\,\text{mg/kg}$$

The major disadvantage of this method is that this formula could not be obtained for the data which follows non-normal distribution (data with skewed distribution due to outlier value). In reality, the data of acute toxicity studies tend to non-normal, which makes Karber's method of LD_{50} calculation inaccurate. In addition to this, this method utilizes more animals than recommended by the guidelines.

48.2.12.3 Miller and Tainter's Graphical Method of Calculation of LD_{50}

- The different groups of animals should be administered with different doses, and mortality percentage should be tabulated.

Table 48.4 Hypothetical data for calculating LD_{50} based on Miller and Tainter's graphical method

Group No	Dose (mg/kg)	Log2 dose	No of animals dead	Dead %	Corrected %	Probit
1	Vehicle	--	0/5	0	--	
2	10	3.321	0/5	0	5	3.36
3	20	4.321	0/5	0	5	3.36
4	40	5.321	1/5	20	20	4.16
5	60	5.906	2/5	40	40	4.75
6	80	6.321	2/5	40	40	4.75
7	100	6.643	2/5	40	40	4.75
8	120	6.906	3/5	60	60	5.25

- The obtained percentage should be converted to probit using online software tools or a standard table invented by Fisher and Yates.
- The obtained probit values are plotted against the log dose in the graph, and a line of straight fit should be drawn.
- The dose corresponding to 50% or 0.5 probits should be taken as LD_{50}.

Table 48.4 is provided for illustrating Miller and Tainter's graphical method of calculation of LD_{50}.
The corrected % should be applied if there is 0% mortality observed in a group by using the formula below.

$$0\%\text{dead} = 100 \times 0.25/\text{N}.$$

With the help of an online graphical tool or an excel sheet or manual method, the 'line of straight fit' is drawn, and the log dose corresponding to 0.5 probits is noted. Conversion of this log dose will provide the LD_{50} value. This is illustrated in Fig. 48.1.

From the above graph, the line intersecting log 2 dose axis at 0.5 probit value is 6.6. Hence conversion of log2 dose value of 6.6 provides the LD_{50} value of 97.5 mg/Kg.

It is evident from the LD_{50} value obtained by this method that it differs significantly from the LD_{50} obtained by Karber's method. (15.7 vs 97.5 mg/kg). This indicates that Karber's method becomes inaccurate when data are non-normal.

48.2.12.4 LD_{50} Calculation by Lorke's Method

- This method is conducted in two phases. In phase I, Three groups of three animals each should be selected. One dose should be given to each group intraperitoneally. The treated animals should be observed for 1 day, and mortality should be noted
- In phase 2, another fresh three animals per group should be administered with test substance based on the findings in phase 1.

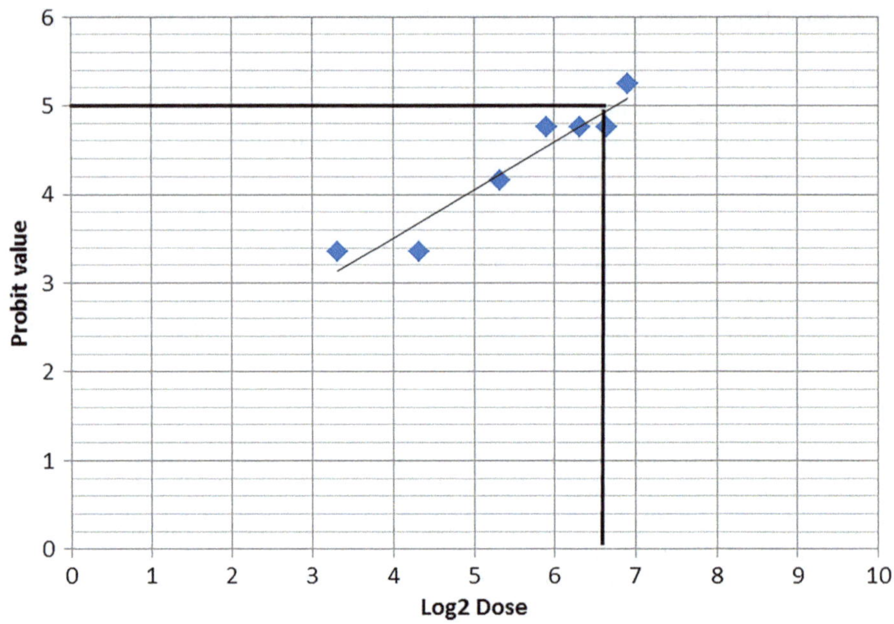

Fig. 48.1 Graph with hypothetical data to calculate LD_{50} values

- The geographic mean of the least dose that caused mortality and the highest dose that did not cause mortality should be taken as LD_{50}.
- The main advantage of this method is that it requires less time, resources, and animals. However, the accuracy, reliability of LD $_{50}$ calculated by this method is doubtful.

48.2.13 Summary of Acute Toxicity Studies

- LD_{50} and MTD are the two crucial data obtained from acute toxicity studies
- FDA, Schedule Y, and OECD guidelines nowadays recommend using one rodent and one non-rodent species
- Unless contraindicated, only female sex are used at present to determine toxic effects in acute toxicity studies
- Various methods are available to calculate LD_{50} values, and each method has its advantages and limitation.

48.3 Sub-Acute Toxicity Studies (Repeated 14–28 Days Toxicity Study)

48.3.1 Aim

- To determine the 'No observable Adverse effect level' (NOAEL) in animals which provides preliminary insight into fixing the safe-starting dose in Phase 0 clinical trial.
- To determine the dose levels that are to be used in sub-chronic and chronic toxicity studies.
- To evaluate the potential ADRs of an investigational drug following an administration period of 14–28 days.
- To provide supportive data for initial phases of clinical trials where investigational drug administration is required up to 28 days.

48.3.2 General Considerations

- One rodent of 6–10 Nos/sex/group and One non-rodent strain of 2–3Nos/sex/group is required for the sub-acute toxicity study.
- Usually, sub-acute toxicity is the continuation of acute toxicity studies where the animals are observed for a longer period. The animals alive after 14 days of acute toxicity studies (irrespective of the dose administered) are included in the sub-acute toxicity studies.
- All the parameters measured in acute toxicity studies like food intake, body weight, overt effects, histopathology, and hematological parameters are also measured in sub-acute toxicity studies.
- The highest, mid-level, and the lowest lethal dose observed in the acute toxicity studies (at the end of the 14th day) are administered to the animal groups (rodent and non-rodent) and observed for another 14 days.
- After the completion of 14 days (i.e., at the end of the 28th day), if any single mortality is recorded with that particular dose, then it is taken as sub-acute toxicity of that dose.

48.3.3 Calculation of Sub-Acute Lethal Dose

- After completion of 28 days, the highest dose that did not produce the mortality (M_0) and the lowest dose that produced mortality (M_1) should be estimated.
- Using the above values, the sub-acute lethal dose is calculated by using the formula: $LD_{\text{Sub-acute}} = (M_0 + M_1)/2$

48.4 Sub-Chronic (90 Days Repeated Dose) Toxicity Study

48.4.1 Aim

- To determine the dose range for future chronic toxicity study
- To estimate the NOAEL with better precision for various toxicology endpoints
- To provide supportive toxicological data on identified target organs for future studies.

48.4.2 General Considerations

- Usually, the sub-chronic study is conducted for 90 days. However, it can be extended up to 12 months if required.
- One rodent of 15–30 Nos/sex/group and one non-rodent of 4–6 Nos/Sex/Group are utilized in this study. Rodents between 6 and 8 weeks of age are utilized in the study. Dogs are commonly used for non-rodent groups. Dogs aged between 4 and 6 months are used.
- Sometimes only ten rodents per group are used when the study is intended to determine only dose-range or long-term studies are foreseen in the future.
- Route of drug administration should be the same as 'route intended in clinical practice', and three graded dose levels should be selected. 'Reversal group' (group in which drug is stopped after 90 days and observed for reversal signs) and its control group are also included in sub-chronic toxicity studies besides three dose groups.
- The three to five doses are usually determined from the sub-acute toxicity parameter M_0 (the highest dose that produces no-mortality after the sub-acute test). Using the formula,
  Dose $= M_0/(1 + x)$ where in $x = 0$ to infinity with increment of 1, three doses for sub-chronic toxicity study ($x = 0$, $x = 1$ and $x = 2$ for three doses levels) can be calculated.
- All the parameters that are mentioned in the above toxicity studies are estimated in sub-chronic toxicity study also for 90 days. Focus is made on the development of chronic toxicities like neurotoxicity, immunotoxicity, nephrotoxicity, and hepatotoxicity.
- For the reversal group, half of the animals are sacrificed on the 14th day, and the remaining half is sacrificed on the 28th day after drug stoppage in 90 days and examined for reversal/recovery pathological findings.

48.4.3 Calculation of Sub-Chronic Lethal Dose

- After completing 90 days of sub-chronic toxicity study, the highest dose that did not produce the mortality (M_0) and the lowest dose that produced mortality (M_1) can be estimated.

- Using the above values, the sub-chronic lethal dose is calculated by using the formula: $LD_{\text{Sub-chronic}} = (M_0 + M_1)/2$

48.5 Chronic (180 Days – 1 Year Repeated Dose) Toxicity Study

48.5.1 Aim

- To determine the NOAEL or 'benchmark dose' (BMD) with better precision than other toxicity studies
- To provide data on the effects/health hazard due to repeated exposure to the investigational drug over a significant period of the life span of animals
- To provide supportive toxicological data on identified target organs for future studies and the possibility of drug accumulation.
- To provide supportive data for prediction of the preliminary mechanism of action of investigational drugs

48.5.2 General Considerations

- Usually, chronic toxicity studies are conducted for 6 months to 1 year. One year duration of study is often required in rodents, while in non-rodents, it will be 6 months generally.
- Similar to a sub-chronic study, three to five dose levels groups, control and reversal groups are included in the chronic toxicity study. Besides this, the study design can incorporate one or more 'interim-kill-analysis' groups (e.g., at third month, sixth month, and ninth month)
- Minimum 20Nos/sex/group is used for rodents and 4 Nos/Sex/group is used for non-rodents. Rat is the preferred rodent, and dogs/cat is preferred non-rodent for chronic toxicity studies.
- Interim-kill-analysis group can accommodate only ten rodents/sex/group. The number of interim-kill-analysis groups required should be properly justified.
- If monitoring disease status is necessary during the study, a 'sentinel group' of 5Nos/Sex/group can also be incorporated in the study design.
- Route of drug administration should be the same as 'route intended in clinical practice.' For oral route, test drug should be administered daily (all 7 days/week) for 12 months. For dermal and inhalational routes, test drugs should be administered for 6 h/day for all 7 days/week for 12 months.
- All the parameters mentioned in the above toxicity studies are estimated in the chronic toxicity study also for 12 months.

48.5.3 Calculation of Chronic Lethal Dose

- After the completion of 12 months of chronic toxicity study, the highest dose that did not produce the mortality (M_0) and the lowest dose that produced mortality (M_1) can be estimated.
- Using the above values, the chronic lethal dose is calculated by using the formula: $LD_{chronic} = (M_0 + M_1)/2$

48.5.4 Calculation of NOAEL and LOAEL

- By plotting the log dose/kg on the x-axis and the % of animals died in that dose on the y-axis, one can construct a sigmoid dose-response curve
- The highest dose at which no death was observed in the group (the point at which the sigmoid curve cuts the x-axis) is considered NOAEL. Similarly, the lowest dose at which a significant adverse event (death) was observed is taken as LOAEL.

48.5.5 Calculation of Reference Concentration (RfC) or Reference Dose (Rfd) or Acceptable Daily Intake (ADI) from NOAEL or LOAEL

- RfC/RfD/ADI can be calculated from NOAEL by using the formula

$$\mathrm{ADI/RfC} = (\mathrm{NOAEL})/(\mathrm{UF1} \times \mathrm{UF2} \ldots \mathrm{UFn}) \text{ wherein UF} = \text{Uncertainty factor.}$$

- The uncertainty factors can be many, including human variability, extrapolation of animal data to humans, using LOAEL instead of NOAEL, modifying factors, and less study duration to derive NOAEL.
- For example, from 200 days chronic toxicity study in rats, the LOAEL value is estimated to be 50 mg/kg/day. To derive RfC/ADI, one should consider the number of uncertainty factors present. In this scenario, four uncertainty factors are present, namely lesser duration of the study (200 days instead of 1 year), using LOAEL instead of NOAEL, human variability, and extrapolation of rat data to humans. Any value from 0 to 10 is assigned for each uncertainty factor. Thus accounting for all the four uncertainties, the ADI/RfC for humans is calculated as

$$\text{ADI or RfC} = \frac{50\ \mathrm{mg/kg/day}}{10 \times 10 \times 10 \times 10} = 0.005\ \mathrm{mg/kg/day}$$

48.6 Reproductive Toxicity Studies

48.6.1 General Considerations

- *Selection of species:* Ideally, the species which is intended to use for these studies should be physiologically as similar as possible to humans. Hence rats, rabbits, and mice are used frequently. The rat model has the advantage of widespread use, economical, consistently high reproductive performance, and the most importantly, the convenient size of intact animal and reproductive organs. Rabbit has specific advantages like the easy collection of semen for analysis of endocrine change and artificial insemination.
- *Dose selection:* The most important consideration regarding the selection of dose is that it should ensure the adequate detection of toxicity. Hence commonly high dose levels which produce systemic toxicity but not mortality (Maximum tolerable dose- MTD) are chosen. Also, it is crucial to determine the dose-response relationship. Hence usually, one MTD dose, 50% of MTD, and 25% of MTD are used.
- *Duration of dosing:* Regarding the male reproductive toxicity assessment, a general rule of administering the drug to an adult male for at least six cycles of seminiferous epithelium prior to mating or semen collection is followed. This is because damage to the spermatogonial stem cells would not be expressed in caudal epididymal sperm for at least 6 weeks. The cycle of seminiferous epithelium is defined as series of changes that occur in the given area of seminiferous epithelium between two successive appearances of the same cellular association. Table 48.5 represents the duration of seminiferous epithelium cycling time for some common lab species and humans.
- *Number of animals:* In general, when multiple endpoints are intended to be calculated at the end of the study, it is necessary to include 20 male animals per group. However, if tests also include endpoint for evaluation of fertility, then more than 20 male animals should be included so that 20 pregnancies per treatment group is attained.
- *Length of mating period:* Cohabitation of male and female animals is done frequently in fertility testing. In such conditions, prolonged cohabitation is

Table 48.5 Duration of seminiferous epithelium cycling and dose duration for some species

S. No	Species	Seminiferous epithelium cycling duration (C) in Days	Dose Duration (D) (D=C x 6) in Days
1	Rat	12.9	78–84
2	Mouse	8.6	49–50
3	Rabbit	10.7	62–66
4	Rhesus Monkey	9.5	60–62
5	Human	16	–

allowed (8–10 days). However, this may mask the male fertility toxicity because repeated mounting and copulation may increase pregnancy rates. To prevent this, male animals are watched carefully, and after one mating, they should be removed. The general hypothesis is that restricting the number of copulation would increase the sensitivity of fertility testing. Reversing the dark-light cycle for few days for testing will make the animals to adapt, and careful watching of male animals to prevent multiple mounting can be done during work hours. During cohabitation, females should be examined daily for mating evidence (Presence of seminal plug/vaginal lavage).

48.6.2 Protocols

- *Single generation reproductive Protocol (SGRP):*
  - This protocol involves the continuous exposure of animals to a testing substance and evaluating reproductive capability for one generation.
  - The main objective is to detect the effects of the testing substance on the integrated reproductive process and study the effects on individual reproductive organs.
  - In a typical SGRP, both male and female animals are exposed to testing substances at their pre-pubertal age (5–8 weeks of age in the case of rats) and continued throughout their pubertal period. Dosing will be given for at least 8–10 weeks, and are allowed cohabiting.
  - Cohabitation should be terminated when evidence of mating is present.
  - Male animals are then sacrificed and examined for evidence of reproductive toxicity, while female animals are allowed to continue for gestation and delivery with the dosing of testing substances. Later, both F1 and female are sacrificed and examined for any evidence of reproductive toxicity.
  - The main disadvantage of SGRP is it is insufficient to identify all potential reproductive toxicants because it would exclude detection of effects caused by prenatal and postnatal exposure. It could also miss the effects on germ cells transmitted to and expressed in the next generation. Due to the short duration of exposure for parent generation, latent adverse effects might also be missed by SGRP.
- *Multigenerational reproductive Protocol (MGRP):*
  - The entire disadvantage mentioned in SGRP could be overcome just by extending the testing substance exposure and evaluation for the next few generations. Hence it is called multi-generation reproductive toxicity testing. In MGRP, usually, two or three generations are followed up.
  - One more advantage of this MGRP is that the pubertal landmarks such as vaginal opening and preputial separation are also monitored. These two monitoring is the primary markers for any reproductive system developmental defect. In the presence of any defect in these markers, then the anogenital distance is also evaluated in F2 generation litters.

   - Many studies have reported that endpoint derivation from this MGRP is better than SGRP
- *Alternative Multi-generation Protocol:*
   - This protocol was developed by National Toxicology Program (NTP). It was originally designed as single generation reproductive protocol but is currently modified into MGRP. In this protocol, the same steps are followed like SGRP with exceptions like dosing extension into F1 generation and allowing cohabitation for 14 weeks continuously. Since 14 weeks of cohabitation is allowed, up to five batches of litters are produced, which are removed soon after birth. They are examined for viability, pup weight, sex ratio, external abnormalities and then discarded. Only the last litter batch is allowed to stay with parents and is used to study the combined effects of in utero as well as perinatal and postnatal exposures.
   - This protocol provides information on spacing changes, number, and size of the litters over 14 weeks of duration.
- *Short term reproductive Protocol:*
   - The short term reproductive toxicity testing is proposed mainly to screen the chemicals for testicular toxicity. The short term indicates that testing duration will be less than the one seminiferous cycle duration in species tested. These tests are of value in the quick identification of target sites, affected cell types, and mechanism of toxicity. Nevertheless, the serious limitation is that the effects of exposures during development will not be evaluated.
   - Examples:
     Epididymal toxicity studies
     Spermatotoxicity test protocols
     35-days reproductive assessment

48.6.3 Assessment of Male Reproductive Toxicity

- Amongst the known reproductive toxic agents, many agents are more toxic to the male reproductive system than the female reproductive system. This is due to the fact that male reproductive gonads are comparatively more exposed to the external environment than female gonads, and rapid and large quantities of male germ cell (Sperm) production are being occurred. Hence minor alteration in the hormone level or temperature and many other subtle factors strongly influence this. Hence while assessing, one must consider collecting data in as possible aspects as they can.
- Table 48.6 gives an outline of potentially useful tests of Male reproductive Toxicity assessment:

Table 48.6 Potentially useful tests of male reproductive toxicity assessment in animals

S. No	Test/Organ assessment	Parameter/Analysis
1	Testis	Size, Weight, Spermatid reserve, Histology, Tubule diameter, %of tubules with lumen sperm and % of non-functional tubules
2	Epididymis	Weight, Histology, Number of sperms in distal half, % of motile sperm at distal end, Biochemical assays, Sperm morphology
3	Accessory sex gland	Histology, Gravimetric
4	Semen	Total volume, Gel free Volume, Sperm conc., Total sperm per ejaculate, % motile sperm, Gross sperm morphology
5	Endocrine	Levels of LH, FSH, Testosterone, GnRH, TSH, T_3 and T_4
6	In vitro	Hamster egg penetration test, Incubation of sperm in agent

Table 48.7 Potentially useful tests of male reproductive toxicity assessment in animals

S. No	Test/Organ assessment	Parameter/Analysis
1	Ovary	Weight, Histology, Number of oocytes, Rate of follicular atresia, Follicular maturation, Oocyte maturation, Ovulation, and Luteal function
2	Hypothalamus	Histological studies, Neurotransmitter release studies
3	Pituitary	Histology, hormone release studies
4	Oviduct	Histology, gamete transport, Fertilization studies, Transport studies
5	Endocrine	Levels of LH, FSH, Oestrogen (E1, E2 and E3), GnRH, TSH, T_3 and T_4
6	In vitro	In vitro fertilization of superovulated eggs
7	Uterus	Cytology, Histology, Decidual response, Dysfunctional bleeding
8	Cervix/Vulva/ vagina	Cytology, Histology, mucus production, Mucus quality

48.6.4 Assessment of Female Reproductive Toxicity

- Assessment of female reproductive toxicity is more complicated than male. This is because of the involvement of multiple physiological processes, namely oogenesis, ovulation, fertilization, zygote transport, and implantation with a full sequence of fetus development. Moreover, access for monitoring and sample collection regarding female reproductive tests is more complicated than male.
- Table 48.7. gives an outline of potentially useful tests of Female reproductive Toxicity assessment

48.6.5 Endpoints and Indices in Reproductive Toxicity Studies

- To assess the data obtained from the studies comprehensively, it is obligatory to set up an endpoint earlier to start the experiment. Endpoints indicating the

Table 48.8 Specific Endpoints in male reproductive toxicity

S. No	Endpoints	Parameter
1	Organ weight	Testes, Epididymis, Seminal vesicles, Prostate, Pituitary
2	Histopathology & Visual examination	Testes, Epididymis, Seminal vesicles, Prostate, Pituitary
3	Sperm evaluation	Count and detailed morphology
4	Hormone levels	LH, FSH, TSH, T3 & T4, Testosterone, Oestrogen, Prolactin
5	Sexual behavior	Mounts, Intromission & ejaculations
6	Developmental defects	Testis descent, preputial separation, anogenital distance, Structure of external genitalia, sperm production

Table 48.9 Specific Endpoints in female reproductive toxicity

S. No	Endpoints	Parameter
1	Organ weight & Histopathology	Ovary, Uterus, Fallopian tubes, Pituitary
2	Ovary evaluation	Detailed morphology, Ovulation rate,
3	Hormone levels	LH, FSH, TSH, T3 & T4, Testosterone, Oestrogen, Prolactin
4	Sexual behavior	Estrous cycling, Preparedness for reception, vaginal cytology

measures of primary functions of fertility and reproductive behavior should always be included. Multiple endpoint setting is always encouraged during the testing.

- This is because only with multiple endpoints, potential adverse effects of a particular testing agent could be detected effectively.
- The male-specific endpoints (Table 48.8), female-specific endpoints (Table 48.9), and couple mediate endpoints and indices (Table 48.10.) are summarized below.
- *Couple mediated endpoints and indices:* The following are the few couple-mediated endpoints that are commonly set in experiments.
  - Time to mating (Time of cohabitation to time of first evidence of mating)
  - Time to pregnancy (time of first mating to time of first evidence of pregnancy)
  - Pre-implantation loss [(number of corpora lutea – number of implantation sites)/number of corpora lutea)]
  - Post implantation loss [(number of implantation sites – number of the fetus)/ number of implantation sites)]
  - Implantation number
  - Litter size
  - Birth weight of litter, Postnatal weight of litter & Offspring survival
  - Internal malformations and variations

Table 48.10 Indices used in reproductive toxicity testing

S. NO	Name of the Index	Numerator	Denominator	Multiply by
1	Mating index (Rate)	Number of males or females mating	Number of male or females cohabited	100
2	Fertility index (Rate)	Number of cohabited females becoming pregnant	Number of non-pregnant couples cohabited	100
3	Pregnancy index (Rate)	Number of females delivering live young	Number of females with evidence of pregnancy	100
4	Life birth Index	Number of live offspring	Number of offspring delivered	100
5	Sex ratio	Number of male offspring	Number of female offspring	100
6	Viability index (4-day survival index)	Number of live offspring at lactation day 4	Number of live offspring delivered	100
7	Weaning index (Lactation Index)	Number of live offspring at day 21	Number of live offspring born	100

48.7 Local Toxicity Studies

48.7.1 Dermal Toxicity Studies

- Dermal toxicity studies can be acute (assessing the effect of a test chemical on the skin following uninterrupted exposure for ≤24 hr) or chronic (when duration extends beyond 24 h with repeated application). The choice of conducting acute or chronic depends upon the clinical nature and use of the test drug. Dermal toxicities should not be conducted until all the available information of the testing chemical has been assessed in 'weight-of-the-evidence' analysis.
- *Animal species*: Adult rats (minimum 8–10 weeks of age) of either gender are the preferred animal for dermal toxicity testing.
- *Number of animals:* Two animals are used at any selected dose, and the test chemical is applied to the single animal in a sequential manner. The test chemical is applied to the second animal only after 48 h to identify any severe reaction.
- *Selection of dose:*
  - 200 mg/kg body weight—when no information available on a test drug
  - 50, 1000, and 2000 mg/kg body weight—when information is available
  - Increment of dose to next level should not be done until the data on survival of the previous dose is thoroughly determined.
- *Administration of doses:* Dorsal or flank skin of the animal is commonly selected to apply test drug. By close clipping, the fur of the animal should be removed. The area of the test chemical applied should be at least 10% of the total body surface area. Test chemical should be in contact with the skin for 24 h (preferable), and if it cannot be achieved, at least 6 h. contact time should be ensured. Test chemicals

that are liquid are used undiluted, and solid test chemicals are applied with a suitable vehicle. When the vehicle is not water, then the influence of the vehicle on skin penetration by test drug should be estimated.

- *Observation duration*: First 30 min followed by first 24 h (focus on 2–6 h) and then daily for a minimum of 14 days (acute) to 1—year (chronic).
- *Observation parameters:*
  - Changes in skin, fur, mucous membrane, and eyes
  - Development of hyperemia, discoloration, edema, and necrosis.
  - Systemic changes (CVS, RS, CNS, ANS, and somatomotor)
  - Any diarrhea, salivation, tremors/convulsion, sleep, lethargy, and coma.
  - Death occurred per group for all dose levels.
- *Calculation of dermal LD_{50}:* Same as that of LD_{50} by acute oral toxicity studies. Usually, the dermal LD_{50} is greater than oral LD_{50} values, and in 99% of the testing, the dermal LD50 values will be >2000 mg/kg. Thus when oral LD50 itself is >2000 mg/kg, then dermal toxicity testing should not be done in order to avoid unnecessary testing.

48.7.2 Ocular Toxicity Studies

- Similar to dermal toxicity studies, ocular toxicity studies should not be started until all data of eye irritation/corrosivity of test substance has been assessed in weight-of-the-evidence analysis. Relevant *in vitro* studies and *in vivo* skin should be conducted first before starting an ocular toxicity study in animals for determining the corrosive/irritant nature of the test drug.
- *Animal species:* Young healthy albino rabbit is the preferred animal. Both the eyes of the selected animal should be examined 24 h before testing to rule out any pre-existing disease conditions.
- *Number of animals:* For initial testing, only one rabbit is used at any selected dose, and the test chemical is applied to one eye, and the other eye act as a control. Once a severe irritant effect is not documented in initial testing, up to three animals can be used for a particular dose level to 'confirm' the non-irritant/ irritant nature of the testing drug.
- *Administration of test substance:*
  - 0.1 ml volume—for testing liquids
  - Maximum of 100 mg ground to a fine powder and appropriately compacted, which makes up to 0.1 mm^3 volume—for testing solids
  - Simple burst directly into widely opened rabbit eye from 10 cm distance for 1 s—pressurized aerosol formulations
  - Prior collection of expelled material followed by instillation in the eye (0.1 ml) —for simple liquid spray formulations
  - Except for solid formulations, the animals' eyes should not be cleaned for at least 24 h after administering the test drug. For solid formulation, it can be washed after 1 h of observation with Normal saline or water.

- *Observation duration:* First 30 min followed by every 6 h for the first 3 days. All animals should be observed till the completion of 21 days. In case of any severe lesions within 72 h, the animal can be killed on humane grounds, and the study should be terminated.
- *Observation parameters:*
  - Presence/absence of pain, distress, excessive blinking, abnormal tearing, and repeated rubbing of eyes.
  - General examination of the cornea, conjunctiva, iris, anterior chamber, fundus, and sclera.
  - Presence/absence of corneal ulceration and depth of ulcer (Fluorescein staining and slip lamp examination) and grades of the ocular lesion.
  - Light reflex
  - Systemic adverse effects and histopathological findings.
- *Humane endpoints:*
  - Very severe lesions not possible to reverse
  - Blood in the anterior chamber
  - Grade 4 corneal opacity
  - Sloughing/necrosis of nictitating membrane
  - Ulceration of conjunctival membrane
  - Corneal perforation
  - Absence of light reflex >72 h.
- Since albino rabbit is highly sensitive to most of the chemicals than humans, the extrapolation of data from ocular toxicity study using albino rabbit is valid only to a limited degree.

48.7.3 Inhalational Toxicity Studies

- Similar to oral toxicity studies, inhalational toxicity studies can be acute, sub-acute, chronic, and chronic depending upon the duration of exposure. The same time duration of various oral toxicity studies applies to the inhalational toxicity studies.
- The main difference in inhalational toxicity study is the determination of median lethal concentration (LC_{50}) instead of LD_{50}.
- In inhalational study can be done by two methods: Traditional protocol and 'C x t' protocol: In the traditional protocol, the animals are exposed to test chemicals for a fixed time (usually 4 h). However, In the '$C \times$ t' iprotocol ('C' stands for concentration and' t' for time-duration of exposure), animals are exposed to varying concentrations with varying duration. Both the protocols adopt the initial study followed by the main study. The initial study will determine the presence/absence of severe fatal reaction and provides insight for dose range selection. The main study will determine the LC_{50} value. The traditional protocol is now obsolete and seldom followed.

- The desired particle size of test-chemical-aerosol to be used in inhalational toxicity study is <2 micrometer, and it can be achieved in the test chemical with a concentration of 2 mg/l.
- Two animals for particular '$C \times t$' interval is used. Adult rats of either sex are preferred. Mice can also be used alternatively.
- In the $C \times t$ protocol, the testing drug is administered in 4 different concentrations with five different durations. (Hence a minimum of 20 groups is required besides control, satellite, and reversal groups)
- Exposure to test chemicals can be up to a maximum of 6 h. For mice, the maximum exposure time is 4 h.
- The traditional method adopts both 'nose only' exposure and 'whole body chamber/closed rebreathing' exposure. On the other hand, the $C \times t$ method uses only 'nose only' exposure to test chemicals.
- The nose-only exposure is preferred over the closed rebreathing chambers. The advantages are
  - A high concentration of test chemicals are quickly attained
  - Minimal handling for technical procedures
  - The reaction of test chemicals in aerosols with excreta can be avoided
  - Exposure can be interrupted at any time to avoid undue suffering of animals
  - Measures for preconditioning of closed air chambers are not required
  - Exposure by other routes like oral/dermal due to licking, pruning can be avoided
- Parameters like pulmonary function tests, vital signs, bronchoalveolar lavage, signs of distress, pain, and histopathological findings are measured.
- LC_{50} calculation is similar to LD_{50} calculation in the oral toxicity studies.

Bibliography

Brusick D, Perreault SD. Assessment of female reproductive toxicity. In: Hayes AW, editor. Principles and methods of toxicology. 5th ed. Boca Raton, FL: CRC Press; 2008.

Cohen, I. Bernard, Foreword, in the Dover edition (1957) of: Bernard, Claude, An Introduction to the Study of Experimental Medicine (originally published in 1865; first English translation by Henry Copley Greene, published by Macmillan & Co., Ltd., 1927).

Demott RP, Borgert CJ. Reproductive toxicology. In: Williams PL, James RC, Roberts SM, editors. The principles of toxicology: Environmental and industrial applications. 2nd ed. New York: Wiley-Interscience; 2000.

Foster PM, Gray LE. Toxic responses of the reproductive systems. In: Klassen CD, editor. Casarett & Doull's toxicology: the basic science of poisons. 7th ed. New York: McGraw Hill; 2008.

Hamm J, Sullivan K, Clippinger AJ, Strickland J, Bell S, Bhhatarai B, et al. Alternative approaches for identifying acute systemic toxicity: moving from research to regulatory testing. Toxicol In Vitro. 2017;41:245–59.

Holzhütter H-G, Genschow E, Diener W, Schlede E. Dermal and inhalation acute toxicity class methods: test procedures and biometric evaluations for the globally harmonized classification system. Arch Toxicol. 2003;77:243–54.

ICH, International Conference on Harmonisation of Technical Requirements for Registration of Pharmaceuticals for Human Use: Note for Guidance on Toxicokinetics: The Assessment of Systematic Exposure in Toxicity Studies S3A, 1994.

Jacobs AC. Reproductive toxicology. In: Jacobson-Kram D, Keller KA, editors. Toxicological testing handbook: principles, applications, and data interpretation. 2nd ed. New York: Informa Healthcare; 2006.

Klinefelter GR, Clegg ED, Perreault SD. Assessment of male reproductive toxicity. In: Hayes AW, editor. Principles and methods of toxicology. 5th ed. Boca Raton, FL: CRC Press; 2008.

Lorke D. A new approach to practical acute toxicity testing. Arch Toxicol. 1983;3:275–89.

Nutrition C for FS and A. Search for FDA guidance documents – guidance for industry: preparation of food contact notifications for food contact substances (toxicology recommendations) [Internet]. [cited 2018 Dec 13]. Available from: https://www.fda.gov/RegulatoryInformation/Guidances/ucm081825.htm.

OECD. Test No. 420: acute oral toxicity - fixed dose procedure. OECD guidelines for the testing of chemicals, Section 4. 2002. Available from: http://www.oecd-ilibrary.org/environment/test-no-420-acute-oral-toxicity-fixed-dose-procedure_9789264070943-en.

Pauluhn J, Thiel A. A simple approach to validation of directed-flow nose-only inhalation chambers. J Appl Toxicol. 2007;27:160–7.

Thomas MJ, Thomas JA. Toxic responses of the reproductive system. In: Klaassen CD, Watkins JB, editors. Casarett & Doull's essentials of toxicology. 1st ed. New York: McGraw Hill; 2003.

Turner R. Acute toxicity: the determination of LD50. In: Screening methods in pharmacology. New York: Academic Press; 1965a. p. 300.

Turner R. Acute toxicity: the determination of LD50. In: Screening methods in pharmacology. New York: Academic; 1965b. p. 61–3.

Ethical Issues in Animal Research

49

Gerard Marshall Raj and Rekha Priyadarshini

Abstract

Contribution of animals in biomedical research—though in varied proportions—is indispensable. Both the cases for and against the use of laboratory animals are equally debatable. Apart from fundamental biological research, animals were extensively utilized in drug toxicity testings including non-pharmaceutical chemical safety assessments and also in biomedical teaching and training. However, with the growing understanding of animal experimentations and animal ethics—particularly with greater application of the 3R principles—nowadays, the experimental procedures involving animals warrant even more judicious perusal. Whenever feasible, the principle of *replacement* (absolute or relative) is given prime importance and engagement of appropriate *alternatives* to animal experiments (non-animal testing methods) is highly recommended. *Reduction* and *refinement* (and *rehabilitation*, the 4th R) are secondary principles of humane animal experimentations. This Chapter includes discussion on the principles of animal ethics, the evolution of ethical issues in animal experiments, the 3R approach including the alternatives to animal experiments, the present status of animal experimentations, and the various guidelines related to animal research.

Keywords

Animal research ethics · Humane research · 3R principles · Alternatives · Replacement principle · Reduction principle · Refinement principle

Good animal care and good science go hand in hand—Animal Research Committee, NIH

G. M. Raj (✉) · R. Priyadarshini
Department of Pharmacology, All India Institute of Medical Sciences (AIIMS) Bibinagar, Hyderabad, Telangana, India
e-mail: gerard.pharma@aiimsbibinagar.edu.in

M. Lakshmanan et al. (eds.), *Introduction to Basics of Pharmacology and Toxicology*, https://doi.org/10.1007/978-981-19-5343-9_49

Strictly speaking, it would be more appropriate to use the terms 'human animals' and 'non-human animals' (and likewise 'human primates' and 'nonhuman primates') to distinguish between humans and other animals—Nuffield Council on Bioethics

49.1 Introduction

Though closely scrutinized by staunch animal activists and other animal enthusiasts, animal experimentation in biomedical research is inevitable in one form or other. Animals, for experimental purpose, are predominantly utilized in biomedical research (40%) followed by drug testing (26%), product safety testing (20%), and teaching (7%). Educational universities (biomedical research) and pharmaceutical companies (drug safety testings) are places wherein animal research is carried out exhaustively. And in every major medical advancement, animal models have acted as the prelude to understand the pathogenesis and management of disease conditions.

Despite the blooming up of promising alternatives to animal research pursuits, complete replacement of the same is never possible. Hence, taking utmost cognizance of the prevailing norms, the need for scrupulous monitoring of animal research activities is more than necessary.

The animals used range from small rodents including mice, rats, guinea pigs, gerbils, hamsters, to the larger non-rodents including rabbits, minipigs, monkeys (macaques, baboons, and marmosets), sheep, goats, pigs, ferrets, dogs, cats, birds (including pigeons, chickens, and quails), amphibians (including frogs and salamanders), and fishes (including the zebra fish-*Danio rerio*). Among these, mice and rats are involved in more than 90% of the animal experimentations.

49.2 Terms Related to Research in Animals: The Definitions

The term 'animal experimentation' refers to the use of animals in experimentation, education, training, and research. However, other terms like animal testing, animal research, in vivo testing, and vivisection (used commonly by animal activists) are also used in the same parlance.

'Laboratory animals' (or animals) refer to any "vertebrate animal (e.g., traditional laboratory animals, agricultural animals, wildlife, and aquatic species) produced for or used in research, testing, or teaching".

'Animal use' refers to "the proper care, use, and humane treatment of laboratory animals produced for or used in research, testing, or teaching".

'Humane care' refers to "those actions taken to ensure that laboratory animals are treated according to high ethical and scientific standards".

'Experimental endpoint' of a study "occurs when the scientific aims and objectives have been reached". 'Humane endpoint' is "the point at which pain or distress in an experimental animal is prevented, terminated, or relieved". Though different, in many experimental techniques both of these endpoints are closely aligned and this needs to identified judiciously.

'Microenvironment' of an experimental animal is "the immediate physical environment surrounding the animal, i.e., the primary enclosure such as the cage, pen, or stall". 'Macroenvironment' is "the physical environment of the secondary enclosure, such as a room, a barn, or an outdoor habitat".

'Animal biosecurity' refers to "all measures taken to identify, contain, prevent, and eradicate known or unknown infections that may cause clinical disease or alter physiologic and behavioral responses or otherwise make the animals unsuitable for research".

'Sentient' denotes "having the power of perception by the senses and is usually taken to mean *being conscious*".

'Euthanasia' or good death is "the act of humanely killing animals by methods that induce rapid unconsciousness and death without pain or distress." Euthanasia includes both physical and chemical methods.

49.3 Animal Experimentations: *To Do or Not to Do?*

Apart from the ethical viewpoint of animal activists, it is sometimes felt that experiments on animals may not always translate into what is expected in the humans. This predicament is true for toxicity testings in animals, where in—false positive and false negative results occur with poor reproducibility or extrapolation in humans.

Physical stress, mental stress, discomfort, and pain are the main concerns for usage of animals in experimentations. If a procedure in animal could reasonably cause "more than slight or momentary pain or distress in a human being" that applied procedure can be considered as 'painful' in animals.

Like in research activities, laboratory animals were considered as inevitable in teaching and imparting skills in medical schools—as they were considered as a substitute for human models in the subjects of Physiology, Pharmacology, Microbiology, and Biochemistry.

Inadequacy in the competence of the animal handlers/ investigators and poorly maintained animal care (husbandry) facilities are some of the other major concerns.

49.4 The R Principles: 3Rs, 4Rs, and so on

Ever since, Russell (a zoologist) and Burch (a microbiologist) put forth their 3R principle through their treatise "The Principles of Humane Experimental Technique" (1959), the appropriate utilization of laboratory animals combined with rational scientific advancement gained significance. The opening remark by the authors in the book was, ". . .the humanest possible treatment of experimental animals, far from being an obstacle, is actually a prerequisite for successful animal experiments." They discuss extensively about *humanity (humane or wellbeing)* and *inhumanity (inhumane or distress)* and the various forms of inhumanity such as *direct inhumanity* and *contingent inhumanity*; *individual inhumanity* and *total inhumanity*.

A Special Edition was also published by the *Universities Federation for Animal Welfare* (UFAW) in 1992. A revised, updated, and abridged version was also released under the title "The Three Rs and the Humanity Criterion" in 2009 (50th Anniversary Edition) by the *Fund for the Replacement of the Animals in Medical Experiments* (FRAME).

Major Charles Hume (Founder Director, UFAW) and Sir Peter Medawar (Chairperson, Expert Scientific and Technical Committee) were the other eminent personalities behind the structuring of the 3R principles. A British Physiologist, by name Marshall Hall (1830), had laid down similar five guiding principles (like the 3Rs framework), namely, the "lack of an alternative, a clear objective, the avoidance of repetition of work, the need to minimise suffering, and full and detailed publication of the results."

Animals are to be considered as living and sentient creatures. Hence, irrespective of their usage value, all animals should be treated with dignity and respect. The 3R principles are applicable in all contexts—including research, testing, and teaching.

49.4.1 Replacement: The 1st R...

The researchers should look for alternatives to the proposed animal experiments. When appropriate alternatives are not available, the researchers shall wait till such alternatives materialize. So, it is the responsibility of the researchers to justify their stance in not choosing an alternative—when available; and not waiting for an alternative and pursuing sooner for animal experimentation rather than waiting for an alternative.

Further, adequate opportunities should be in place throughout the lifetime of the research project for replacing the animals that are utilized with alternatives—whenever relevant and applicable.

Whenever possible, animals in the lower phylogenetic scale with lesser level of sentience (cognitive awareness) are to be selected—which is 'relative or partial replacement'. Utilization of invertebrates; foetal, embryonic, and early life forms of vertebrates; and incorporation of cells and tissues extracted from animals slayed particularly for experimental purpose are considered as partial replacement strategies.

'Absolute or full replacement' refers to when animals are totally replaced with computer models (virtual humans) [and other 'in silico' methods] or tissue−/ cell-cultures (organ-on-a-chip, OOC) ['in vitro' methods]. Replacements can also include nonliving synthetic models, cadavers, abattoir left-over material, use of historic clinical (and animal data), microdosing human studies (human volunteers), immunological assays (instead of bioassays), and other mathematical simulations (including deep learning and machine learning methodologies).

Research funding bodies are also expected to support non-animal testing methods which are also relevant to human physiology.

49.4.2 Reduction Alternative: The 2nd R...

Fewer number of animals are to be utilized to obtain adequate data for testing the research hypothesis and to maximize the data ('statistical enhancement') obtained per animal used in the experimentation [obtaining more information from the same number of animals used]—both these strategies shall effectively reduce the total number of possible animals required to conduct experiments. Reduction may also include utilization of organs, tissues, or other biological materials harvested from dead animals and sometimes, even reuse of animals (animal sharing) can be considered. Using same control group of animals for multiple experiments performed simultaneously and planning experiments in such a way that the same set of animals can act as their own controls (data- and tissue-sharing) are some of the examples of animal reuse. However, these are not leeway for unnecessary repetition (duplication) of experiments and such replication of tests in the same set of animals needs valid justification.

Another ploy is to reduce the experimental variables that are to be measured which would consequently reduce the number of animals that are required. A much-controlled experimental ambience—with genetically homogenous (genetically modified) animals, healthy animals, and consistent husbandry conditions (animal housing and study areas) may help in reducing the group size. Breeding of experimental animals should be closely managed and monitored.

Though the number of animals used should be bare minimum, the scientific rigor and the robustness of the experimental design should not be compromised. As using too fewer number of animals may invalidate the results of the experiments and additionally, amounting to wastage of involved animals.

It is also the obligation of the researchers to maintain the biological diversity of the particular stock and also the entire ecosystem. Hence, experimentation conducted in endangered and vulnerable species should be largely reduced.

Harmonization of regulatory provisions across the nations would also contribute in reducing the number of experimental animals required. The interdisciplinary field of bioinformatics has also carved into the domain of experimental pharmacology—thereby aiding in converting complex in silico study results into interpretable outputs—consequently, lowering the need for conventional animal experiments. The reduction in the number of animals used can also drastically decrease the overall expenditure of the experiment.

49.4.3 Refinement Alternative: The 3rd R...

The risk of suffering inflicted on the animals should be minimized as far as possible—only the 'absolute minimum' is acceptable. The suffering can be directly due to the procedure or indirectly while setting up for the procedure (during adaptation) or even during the post-experimentation follow-up. Therefore, it is necessary to make provisions to improve the welfare of the animals used in biological testings and extend the best quality veterinary care that is possible—these include animal wellbeing, optimal housing standards, and environmental enrichment (provision of species-appropriate environment). Prompt use of anaesthetics, analgesics, tranquilizing drugs, and other palliative measures; non-invasive imaging techniques of CT and MRI; and optimal use of aseptic measures are also part of the refinement principle.

Sufferings include pain (following physical injury); extreme temperatures (cold or heat); fear (anxiety); deprivation of food or water; other illnesses and also limitations of natural behavioral response to any impulse (maladaptive behaviors).

The professional performing the experiment should be adequately competent and equally compassionate and that the particular procedure should also be compatible in protecting the animal wellbeing. The duration of the study procedure and the confinement period of the animals for the study project should be kept to a bare minimum.

Furthermore, and more pertinently, the reliability of scientific findings depends on the welfare provided to experimental animals—the accommodating cages and handling techniques should elicit more natural (physiological) behaviors. One such example is, wherein the animals (mice) are trained to voluntarily to drink from a syringe ('cooperative') rather than forcibly pushing in the contents through the gavage ('forced') for ingesting purpose. Animals are to be housed in social groups allowing for complex interactions (companionship) rather than placing them in isolated cages which would hamper their normal growth and function. Hence, the principle of refinement entails not only minimization of distress but also maximization of wellbeing.

The influence on scientific outcomes by enhancing the animal welfare activities is also studied vigorously. Compared to the replacement and reduction principles, this refinement principle translates to lesser perceptible improvement in the humaneness of the experiments—as it does not alter the number of animals utilized.

Though animal welfare (or well-being) is defined in many different ways—broadly, the indicators of animal welfare are, "longevity, health, behaviour, physiology, immunity, reproduction, expressions, among others"—as stated by the *World Organization for Animal Health* (formerly known as the *Office International des Epizooties*, OIE). And animal welfare denotes an "an animal's perspective on the net balance between positive (reward, satisfaction) and negative (acute stress) experiences of affective states.

It is generally viewed that whenever complete *replacement* of animal experimentation is not feasible, the principles of *refinement* and *reduction* needs to be maximized. Further, it is observed that the principle of *refinement* and *reduction* go hand in hand and should be balanced on a case-to-case basis.

The above-discussed 3R principles are well received worldwide and further enhancements and add-ons of these principles are getting into practice. It should be emphasized that cost and convenience shall never take upper hand above these principles.

A UK-based independent scientific organization was formed in 2004 to support in the advancement of 3Rs, namely, the NC3Rs (National Centre for the Replacement Refinement & Reduction of Animals in Research). The organization liaises with research community across the globe in all sectors—universities, pharmaceutical industries, and regulatory bodies. One such activity is hosting of a free online tool the *Experimental Design Assistant (EDA)*—a web application for refining the design of animal experiments.

Complementing the 3Rs approach in animal research, is the 3Ss principles proposed by Carol Newton, namely, *Good Science*, *Good Sense* (the Right animal is used for the Right Reason: The Three Rs of Harry Rowsell) and *Good Sensibilities*.

The 4th R of ethical conduct of animal research is the *responsibility*—as proposed by Ronald Banks (1995). Responsible conduct of animal experimentation entails development of novel research techniques, infrastructural upgradation, and other entities—so as to perform distress- and pain-free animal experiments and also educating the general public about the welfare of laboratory animals and emphasizing the importance of shared responsibility in animal research practices.

However, more commonly, the 4th R is referred to *rehabilitation* of used experimental animals—which is the 'aftercare' rendered to animals post completion of experimentation or to animals that have been bred or retained in the laboratory (animal houses) for the purpose of biomedical research, education, and testing. Occasionally, it is also known as *rehoming*.

The previously discussed (under the sect. 49.4.2) *reuse* of animals is also sometimes considered as the 4th R.

49.4.4 Miscellaneous Principles

Researchers are responsible to gauge the benefit-risk balance (*the principle of proportionality*) and proceed accordingly with the particular animal experimentation. The risk-benefit equipoise should be weighed with utmost caution considering for both short- and long-term effects. The probable benefits accrued can be for the animals, human beings, or the environment as a whole.

The natural habitat of the animals should never be disturbed for the sake of experimentation. Like in any biomedical research, data sharing and transparency in communicating the research findings (through journals) is obligatory on the part of the researchers involved in animal experiments.

49.5 Alternatives to Animal Experimentations

It has been observed multiple times that testings in animals are not always equivalent to that in humans (poor relevance to human biology) and the results are misleading—this is an inherent scientific limitation of animal research. The same holds true for toxicity testing of new chemical entities—for which animal experimentations are considered as gold standard. And it is quoted that "humane science is better science, giving a 3-fold advantage of being precise, predictive, and pain-free".

David Smyth, a physiologist and President of the UK Research Defence Society, published the book titled, *Alternatives to Animal Experiments* in 1978 which led to the usage of this taxonym, "alternatives" more commonly.

Some of the common alternatives to animal experiments are as follows,

- Ames test
  - Strains of the bacterium *Salmonella typhimurium* are utilized to determine whether chemicals cause mutations in cellular DNA
- Local lymph node assay (LLNA)
- Limulus ameobocyte lysate (LAL) assay and human whole blood pyrogen test
- In vitro eye corrosion and irritation test
- Integrated discrete Multiple Organ Co-culture (IdMOC)
- 3D in vitro living organs (human-on-chip) [https://wyss.harvard.edu/technology/human-organs-on-chips/]
- Use of invertebrates—*Drosophila melanogaster* (fruit flies), *Caenorhabditis elegans* (nematode worm), *Aplysia* sp./ (sea slugs), *Saccharomyces cerevisiae* (baker's yeast), *Daphnia,* and *Hydra*
- Koken Rat simulator
  - For training rat tail vein blood sampling and oral gavage administration

Many such formally approved alternative methods (approaches) are listed in Table 49.1.

Though in one end there is a lookout for 'simpler' applicable alternatives, on the other end it is contemplated that these simpler alternatives may not always truly replicate the 'more complex' human milieu—with 40-odd organs and more than 400 cell types. Hence, from defining, developing, validating (repetition in multiple laboratories), acceptance (in scientific community), and implementation, these alternative approaches (methods) go through a rigorous process of evaluation.

Like in other parts of the world, the Indian regulations also vouches for the 4Rs of animal experimentation.

One such alternative measure was the introduction of 'ExPharm Pro'—a computer assisted simulation software—by Prof. (Dr.) Raveendran from JIPMER Puducherry in collaboration with Elsevier India in 2012. The software had simulated experiments of effect of drugs on rabbit eye, frog heart, frog esophagus, and guinea pig ileum.

And the establishment of the *National Centre for Alternatives to Animal Experiments* (NCAAE) at the Bharathidasan University, Tiruchirappalli, India (in 2009)—referred to as the *Mahatma Gandhi–Doerenkamp Center for Alternatives to Use of Animals in Life Science Education*—by Prof. (Dr.) Akbarsha is another example.

The complete replacement of animals in experimental research is referred to as *non-animal technologies* (NATs) or *non-animal methods* (NAMs). Sometimes, the term *new approach methodologies* (NAMs) is used to refer to NATs in evaluating drug toxicity.

Table 49.1 Alternative methods to animal experimentations

Toxicity area	Alternative methods/ approaches/ guidance/ policies	3R principle(s) applied/ proposed	Weblink[a]
Skin Sensitization [Conventional: Substance was rubbed onto the shaved skin of guinea pigs who are subjectively assessed for allergy (Buehler or the guinea pig maximization test, GPMT)]	1. Adverse outcome pathway (AOP) for skin sensitization initiated by covalent binding to proteins (2012)	–	• https://www.oecd.org/officialdocuments/publicdisplaydocumentpdf/?cote=env/jm/mono%282012%2910/part1&doclanguage=en • https://www.oecd.org/officialdocuments/publicdisplaydocumentpdf/?cote=env/jm/mono%282012%2910/part2&doclanguage=en
	2. CPSC guidance to clarify the definition of "strong sensitizer"	Reduction or replacement of animal use	https://cpsc-d8-media-prod.s3.amazonaws.com/s3fs-public/pdfs/blk_pdf_strongsensitizerguidance.pdf
	3. Defined approaches on skin sensitization (OECD Guideline No. 497: Defined Approaches on Skin Sensitisation)	Replaces animal use	https://www.oecd-ilibrary.org/environment/guideline-no-497-defined-approaches-on-skin-sensitisation_b92879a4-en
	4. Guidance document on the reporting of defined approaches and individual information sources to be used within IATA for skin sensitization (OECD GD No. 256)	–	https://www.oecd.org/officialdocuments/publicdisplaydocumentpdf/?cote=env/jm/mono(2016)29&doclanguage=en
	5. Guidance on safety evaluation of immunotoxic potential of drugs and biologics (FDA)	includes provisions for reducing or replacing animal use	https://www.fda.gov/media/135312/download
	6. Guinea pig tests for skin sensitization (OECD TG No. 406)	Update refines animal use and recommends first use of non-animal methods	https://www.oecd-ilibrary.org/environment/test-no-406-skin-sensitisation_9789264070660-en

(continued)

Table 49.1 (continued)

Toxicity area	Alternative methods/ approaches/ guidance/ policies	3R principle(s) applied/ proposed	Weblink[a]
	7. In chemico skin sensitization test (OECD TG No. 442C)	Reduction or replacement of animal use	https://www.oecd-ilibrary.org/environment/test-no-442c-in-chemico-skin-sensitisation_9789264229709-en
	8. In Vitro Skin Sensitisation [ARE-Nrf2 Luciferase Test Method] (OECD TG No. 442D)	Reduction or replacement of animal use	https://www.oecd-ilibrary.org/environment/test-no-442d-in-vitro-skin-sensitisation_9789264229822-en
	9. In Vitro Skin Sensitisation assays addressing the Key Event on activation of dendritic cells on the Adverse Outcome Pathway for Skin Sensitisation (OECD TG No. 442E)	Reduction or replacement of animal use	https://www.oecd-ilibrary.org/environment/test-no-442e-in-vitro-skin-sensitisation_9789264264359-en
	10. Local Lymph Node Assay (LLNA) (OECD TG No. 429) (a) Reduced LLNA protocol (requires 40% fewer animals by using only the high dose group) (b) Updated LLNA protocol (requires 20% fewer animals)	Reduction and refinement of animal use	https://www.oecd-ilibrary.org/environment/test-no-429-skin-sensitisation_9789264071100-en
	11. LLNA: DA for skin sensitization testing (OECD TG No. 442A)	A nonradioisotopic LLNA test method	https://www.oecd-ilibrary.org/environment/test-no-442a-skin-sensitization_9789264090972-en
	12. LLNA: BrdU (5-bromo-2-deoxyuridine)-ELISA for skin sensitization testing (OECD TG No. 442B)	A nonradioisotopic LLNA test method	https://www.oecd-ilibrary.org/environment/test-no-442b-skin-sensitization_9789264090996-en

Pyrogen Testing [Rabbit Pyrogen Test (RPT)]	1. Human peripheral blood mononuclear cell/interleukin-6 in vitro pyrogen test	Replacement of animal use	https://ntp.niehs.nih.gov/whatwestudy/niceatm/test-method-evaluations/pyrogen-tests/index.html
	2. Human whole blood/interleukin-1β in vitro pyrogen test (a) application of cryopreserved human whole blood		
	3. Human whole blood/interleukin-6 in vitro pyrogen test		
	4. Monocytoid cell line Mono Mac 6/interleukin-6 in vitro pyrogen test		
	5. In vitro monocyte activation type pyrogen test (MAT) (FDA Guidance for Industry: Pyrogen and Endotoxins Testing)	Replacement of animal use	https://www.fda.gov/regulatory-information/search-fda-guidance-documents/guidance-industry-pyrogen-and-endotoxins-testing-questions-and-answers
Ocular Corrosivity and Irritation [Conventional: Substance was placed into the eyes of live rabbits which were monitored for up to 3 weeks (Draize test)]	1. Bovine corneal opacity and permeability (BCOP) in vitro test method to identify severe eye irritants/corrosives or chemicals not requiring eye hazard classification (OECD TG No. 437)	Replacement of animal use	https://www.oecd-ilibrary.org/environment/test-no-437-bovine-corneal-opacity-and-permeability-test-method-for-identifying-i-chemicals-inducing-serious-eye-damage-and-ii-chemicals-not-requiring-classification-for-eye-irritation-or-serious-eye-damage_9789264203846-en
	2. Isolated chicken eye (ICE) in vitro test method to identify severe eye irritants/corrosives or chemicals not requiring eye hazard classification (OECD TG No. 438)	Replacement of animal use	https://www.oecd-ilibrary.org/environment/test-no-438-isolated-chicken-eye-test-method-for-identifying-i-chemicals-inducing-serious-eye-damage-and-ii-chemicals-not-requiring-classification-for-eye-

(continued)

Table 49.1 (continued)

Toxicity area	Alternative methods/ approaches/ guidance/ policies	3R principle(s) applied/ proposed	Weblink[a]
			irritation-or-serious-eye-damage_9789264203860-en
	3. Cytosensor microphysiometer (CM) in vitro test method for eye safety testing	Replacement of animal use	https://ntp.niehs.nih.gov/whatwestudy/niceatm/test-method-evaluations/ocular/in-vitro/index.html
	4. Guidance document on the collection of eyes tissues for histopathological evaluation and collection of data for identification of nonsevere irritants using in vitro methods (OECD GD No. 160)	–	https://www.oecd.org/officialdocuments/publicdisplaydocumentpdf/?cote=env/jm/mono(2011)45/rev1&doclanguage=en
	5. In vitro fluorescein leakage (FL) test method for identifying ocular corrosives and severe irritants (OECD TG No. 460)	Replacement of animal use	https://www.oecd-ilibrary.org/environment/test-no-460-fluorescein-leakage-test-method-for-identifying-ocular-corrosives-and-severe-irritants_9789264185401-en
	6. In vitro macromolecular test method for identifying chemicals inducing serious eye damage and chemicals not requiring classification for eye irritation or serious eye damage (OECD TG No. 496)	–	https://www.oecd-ilibrary.org/environment/test-no-496-in-vitro-macromolecular-test-method-for-identifying-chemicals-inducing-serious-eye-damage-and-chemicals-not-requiring-classification-for-eye-irritation-or-serious-eye-damage_970e5cd9-en
	7. Non-animal testing scheme for assessing eye irritation potential of antimicrobial cleaning products [Alternate Testing Framework for	–	https://www.epa.gov/pesticide-registration/alternate-testing-framework-classification-eye-irritation-potential-epa

	Classification of Eye Irritation Potential of EPA-Regulated Pesticide Products] (EPA)		
	8. Recommendation to perform weight-of-evidence analysis and non-animal testing strategies before in vivo testing (OECD TG No. 405)	–	https://www.oecd-ilibrary.org/environment/test-no-405-acute-eye-irritation-corrosion_9789264185333-en;jsessionid=j4innnfoodqa.x-oecd-live-02
	9. Reconstructed human Cornea-like Epithelium (RhCE) test for identification of substances not requiring ocular hazard labeling (OECD TG No. 492)	Replacement of animal use	https://www.oecd-ilibrary.org/environment/test-no-492-reconstructed-human-cornea-like-epithelium-rhce-test-method-for-identifying-chemicals-not-requiring-classification-and-labelling-for-eye-irritation-or-serious-eye-damage_9789264242548-en
	10. Short time exposure test for identification of ocular corrosives and substances not requiring ocular hazard labeling (OECD TG No. 491)	Replacement of animal use	https://www.oecd-ilibrary.org/environment/test-no-491-short-time-exposure-in-vitro-test-method-for-identifying-i-chemicals-inducing-serious-eye-damage-and-ii-chemicals-not-requiring-classification-for-eye-irritation-or-serious-eye-damage_9789264242432-en
	11. Use of anesthetics, analgesics, and humane endpoints for in vivo eye safety testing (OECD TG No. 405)	Refinement of animal use	https://www.oecd-ilibrary.org/environment/test-no-405-acute-eye-irritation-corrosion_9789264185333-en;jsessionid=j4innnfoodqa.x-oecd-live-02
	12. Vitrigel-Eye Irritancy test (EIT) method for identifying chemicals not	Replacement of animal use	https://www.oecd-ilibrary.org/environment/tg-494-vitrigel-eye-

(continued)

Table 49.1 (continued)

Toxicity area	Alternative methods/ approaches/ guidance/ policies	3R principle(s) applied/ proposed	Weblink[a]
	requiring ocular hazard labeling (OECD TG No. 494)		irritancy-test-method-for-identifying-chemicals-not-requiring-classification-and-labelling-for-eye-irritation-or-serious-eye-damage_9f20068a-en
Genetic Toxicity/ Carcinogenicity [Conventional: Substance was force-fed or injected into mice or rats for 14 days; they were then killed to look at the effects on their cells. And rats or mice were fed with the substance for two years to see if they get cancer]	1. Bacterial reverse mutation test [Ames Test] (OECD TG No. 471)	–	https://www.oecd-ilibrary.org/environment/test-no-471-bacterial-reverse-mutation-test_9789264071247-en
	2. Carcinogenicity studies (OECD TG No. 451)	Reduction of animal use	https://www.oecd-ilibrary.org/environment/test-no-451-carcinogenicity-studies_9789264071186-en
	3. In vitro mammalian cell gene mutation tests using the HPRT and XPRT genes (OECD TG No. 476)	–	https://www.oecd-ilibrary.org/environment/test-no-476-in-vitro-mammalian-cell-gene-mutation-tests-using-the-hprt-and-xprt-genes_9789264264809-en
	4. In vitro mammalian cell gene mutation tests using the thymidine kinase assay (OECD TG No. 490)	–	https://www.oecd-ilibrary.org/environment/test-no-490-in-vitro-mammalian-cell-gene-mutation-tests-using-the-thymidine-kinase-gene_9789264264908-en
	5. In vitro mammalian cell micronucleus test (OECD TG No. 487)	Replacement of animal use	https://www.oecd-ilibrary.org/environment/test-no-487-in-vitro-mammalian-cell-micronucleus-test_9789264264861-en
	6. In vitro mammalian chromosomal aberration test (OECD TG No. 473)	–	https://www.oecd-ilibrary.org/environment/test-no-473-in-vitro-

			mammalian-chromosomal-aberration-test_9789264264649-en
	7. In vivo mammalian alkaline comet assay (OECD TG No. 489)	Reduction of animal use	https://www.oecd-ilibrary.org/environment/test-no-489-in-vivo-mammalian-alkaline-comet-assay_9789264264885-en
Endocrine Disruptors	1. Aromatase (human recombinant) (EPA)	–	https://www.regulations.gov/document/EPA-HQ-OPPT-2009-0576-0004
	2. EASZY assay – detection of endocrine active substances using transgenic zebrafish embryos (OECD TG No. 250)	Reduces/replaces animal use	https://www.oecd-ilibrary.org/environment/test-no-250-easzy-assay-detection-of-endocrine-active-substances-acting-through-estrogen-receptors-using-transgenic-tg-cyp19a1b-gfp-zebrafish-embryos_0a39b48b-en
	3. In vitro H295R steroidogenesis assay (OECD TG No. 456)	Replacement of animal use	https://www.oecd-ilibrary.org/environment/test-no-456-h295r-steroidogenesis-assay_9789264122642-en
	4. In vitro VM7Luc ER TA agonist assay to identify substances that induce human ER activity (OECD TG No. 455)	Replacement of animal use	https://ntp.niehs.nih.gov/whatwestudy/niceatm/test-method-evaluations/endocrine-disruptors/bg1luc/index.html
	5. In vitro VM7Luc ER TA antagonist assay to identify substances that inhibit human ER activity (OECD TG No. 455)		
	6. Integrated testing strategy to identify chemicals with the potential to	–	https://www.epa.gov/endocrine-disruption/use-high-throughput-

(continued)

Table 49.1 (continued)

Toxicity area	Alternative methods/ approaches/ guidance/ policies	3R principle(s) applied/ proposed	Weblink[a]
	interact with the estrogen receptor (EPA)		assays-and-computational-tools-endocrine-disruptor#advancements
	7. Performance-based TG (PBTG) for human recombinant estrogen receptor binding assays (OECD TG No. 493)		https://www.oecd-ilibrary.org/environment/test-no-493-performance-based-test-guideline-for-human-recombinant-estrogen-receptor-hrer-in-vitro-assays-to-detect-chemicals-with-er-binding-affinity_9789264242623-en
	8. Stably transfected human Androgen Receptor TransActivation (ARTA) assay for detection of androgenic agonist and antagonist activity of chemicals (OECD TG No. 458)	–	https://www.oecd-ilibrary.org/environment/test-no-458-stably-transfected-human-androgen-receptor-transcriptional-activation-assay-for-detection-of-androgenic-agonist-and-antagonist-activity-of-chemicals_9789264264366-en
	9. Stably transfected transactivation in vitro assay to detect estrogen receptor agonists and antagonists (OECD TG No. 455)	Replacement of animal use	https://www.oecd-ilibrary.org/environment/test-no-455-performance-based-test-guideline-for-stably-transfected-transactivation-in-vitro-assays-to-detect-estrogen-receptor-agonists-and-antagonists_9789264265295-en
Ecotoxicity	1. Determination of in vitro intrinsic clearance using cryopreserved rainbow trout hepatocytes (RT-HEP) (OECD TG No. 319A)	–	https://www.oecd-ilibrary.org/environment/test-no-319a-determination-of-in-vitro-intrinsic-clearance-using-cryopreserved-

			rainbow-trout-hepatocytes-rt-hep_9789264303218-en
	2. Determination of in vitro intrinsic clearance using rainbow trout liver S9 (RT-S9) sub-cellular fraction (OECD TG No. 319B)	–	https://www.oecd-ilibrary.org/environment/test-no-319b-determination-of-in-vitro-intrinsic-clearance-using-rainbow-trout-liver-s9-sub-cellular-fraction-rt-s9_9789264303232-en
	3. Fish cell line acute toxicity – RTgill-W1 cell line assay (OECD TG No. 249)	Reduces/replaces animal use	https://www.oecd-ilibrary.org/environment/test-no-249-fish-cell-line-acute-toxicity-the-rtgill-w1-cell-line-assay_c66d5190-en
Developmental and Reproductive Toxicity [Conventional: Pregnant female rabbits or rats are force-fed the substance and then killed along with their unborn babies]	1. Extended one-generation reproductive toxicity study (OECD TG No. 443)	reduction of animal use	https://www.oecd-ilibrary.org/environment/test-no-443-extended-one-generation-reproductive-toxicity-study_9789264185371-en
	2. Guidance on reproductive toxicity testing for oncology radiopharmaceuticals (FDA)	Includes provisions for reducing animal use	https://www.fda.gov/regulatory-information/search-fda-guidance-documents/oncology-pharmaceuticals-reproductive-toxicity-testing-and-labeling-recommendations-guidance
Dermal Phototoxicity	1. 3T3 Neutral Red Uptake (NRU) phototoxicity test for skin photo-irritation (OECD TG No. 432) (a) application to UV filter chemicals	Replacement of animal use	https://www.oecd-ilibrary.org/environment/test-no-432-in-vitro-3t3-nru-phototoxicity-test_9789264071162-en
	2. Guidance on photosafety evaluation of pharmaceuticals (FDA)	Reduction or replacement of animal use	https://www.fda.gov/regulatory-information/search-fda-guidance-documents/s10-photosafety-evaluation-pharmaceuticals

(continued)

Table 49.1 (continued)

Toxicity area	Alternative methods/ approaches/ guidance/ policies	3R principle(s) applied/ proposed	Weblink[a]
	3. in vitro reconstructed human epidermis phototoxicity test (RhE PT) method (OECD TG No. 498)	–	https://www.oecd-ilibrary.org/environment/test-no-498-in-vitro-phototoxicity-reconstructed-human-epidermis-phototoxicity-test-method_7b2f9ea0-en
	4. Reactive oxygen species (ROS) assay for photoreactivity (OECD TG No. 495)	Replacement of animal use	https://www.oecd-ilibrary.org/environment/tg-495-ros-reactive-oxygen-species-assay-for-photoreactivity_915e00ac-en
Dermal Corrosivity and Irritation [Conventional: Substance was rubbed onto the shaved backs of rabbits, and they were killed 2 weeks later]	1. Corrositex® in vitro membrane barrier skin corrosivity test (SCT) (OECD TG No. 435)	Replacement of animal use	https://www.oecd-ilibrary.org/environment/test-no-435-in-vitro-membrane-barrier-test-method-for-skin-corrosion_9789264242791-en
	2. EpiDerm™ in vitro human skin model skin corrosivity test (SCT) (OECD TG No. 431)	Replacement of animal use	https://www.oecd-ilibrary.org/environment/test-no-431-in-vitro-skin-corrosion-reconstructed-human-epidermis-rhe-test-method_9789264264618-en
	3. EpiSkin™ in vitro human skin model skin corrosivity test (SCT) (OECD TG No. 431)		
	4. Reconstructed human epidermis in vitro test method for skin corrosivity testing (SCT) (OECD TG No. 431)		
	5. SkinEthic™ in vitro human skin model skin corrosivity test (SCT) (OECD TG No. 431)		
	6. EpiDerm™ in vitro human skin model skin irritation test (SIT) (OECD TG No. 439)	Replacement of animal use	https://www.oecd-ilibrary.org/environment/test-no-439-in-vitro-skin-irritation-reconstructed-human-

	7. EpiSkin™ in vitro human skin model skin irritation test (SIT) (OECD TG No. 439)		epidermis-test-method_9789264242845-en
	8. SkinEthic™ in vitro human skin model skin irritation test (SIT) (OECD TG No. 439)		
	9. Guidance document on an integrated approach for testing and assessment for skin corrosion and irritation (OECD GD No. 203)	Replacement of animal use	https://www.oecd.org/officialdocuments/publicdisplaydocumentpdf/?cote=ENV/JM/MONO(2014)19&doclanguage=en
	10. Rat Transcutaneous Electrical Resistance Test (TER) method in vitro skin corrosivity test (OECD TG No. 430)	Replacement of animal use	https://www.oecd-ilibrary.org/environment/test-no-430-in-vitro-skin-corrosion-transcutaneous-electrical-resistance-test-method-ter_9789264242739-en
Chronic Toxicity [Conventional: Rats (occasionally rabbits, mice, or dogs) are force-fed, forced to inhale, or have the substance rubbed onto their shaved skin every day for 28 or 90 days, before being killed]	1. Chronic toxicity studies (OECD TG No. 452)	Reduction of animal use	Https://Www.Oecd-Ilibrary.Org/Environment/Test-No-452-Chronic-Toxicity-Studies_9789264071209-En
	2. Combined chronic toxicity/carcinogenicity studies (OECD TG No. 453)	Reduction of animal use	https://www.oecd-ilibrary.org/environment/test-no-453-combined-chronic-toxicity-carcinogenicity-studies_9789264071223-en
	3. Guidance for waiving subacute avian dietary tests for pesticide registration and supporting retrospective analysis (EPA)	–	https://www.epa.gov/sites/default/files/2020-02/documents/final-waiver-guidance-avian-sub-acute-dietary.pdf
	4. Guidance on nonclinical evaluation of cancer drugs (FDA)	Includes provisions for reducing animal use	https://www.fda.gov/media/100344/download

(continued)

Table 49.1 (continued)

Toxicity area	Alternative methods/ approaches/ guidance/ policies	3R principle(s) applied/ proposed	Weblink[a]
	5. Guidance on nonclinical studies for oncology therapeutic radiopharmaceuticals (FDA)	Includes provisions for reducing animal use	https://www.fda.gov/regulatory-information/search-fda-guidance-documents/oncology-therapeutic-radiopharmaceuticals-nonclinical-studies-and-labeling-recommendations-guidance
	6. Repeated dose 90-day oral toxicity study in rodents (OECD TG No. 408)	Reduction of animal use	https://www.oecd-ilibrary.org/environment/test-no-408-repeated-dose-90-day-oral-toxicity-study-in-rodents_9789264070707-en
	7. Subacute inhalation toxicity: 28-day study (OECD TG No. 412)	Reduction of animal use	https://www.oecd-ilibrary.org/environment/test-no-412-subacute-inhalation-toxicity-28-day-study_9789264070783-en
	8. Subchronic inhalation toxicity: 90-day study (OECD TG No. 413)	Reduction of animal use	https://www.oecd-ilibrary.org/environment/test-no-413-subchronic-inhalation-toxicity-90-day-study_9789264070806-en
	9. EpiAlveolar™—a 3-D Lung Model to Predict Fibrotic Potential of Multiwalled Carbon Nanotubes	Reduction or replacement of animal use	https://pubmed.ncbi.nlm.nih.gov/32167743/
Biologics testing	1. Alternative test procedure for tuberculin, PPD Bovis, intradermic (USDA)	Reduces animal use by 65%	https://www.aphis.usda.gov/animal_health/vet_biologics/publications/memo_800_114.pdf
	2. Cryopreservation protocol for leptospiral strains for vaccine testing (USDA)	Reduces animal use	https://www.aphis.usda.gov/animal_health/vet_biologics/publications/BBAPP0011.pdf

	3. ELISA test for batch potency testing of human tetanus vaccines (FDA)	Refinement: antibody quantification	https://ntp.niehs.nih.gov/iccvam/suppdocs/feddocs/fda/fda_21_610_10.pdf
	4. Guidance on preclinical assessment of investigational cellular and gene therapy products (FDA)	Minimization of animal use	https://www.fda.gov/regulatory-information/search-fda-guidance-documents/preclinical-assessment-investigational-cellular-and-gene-therapy-products
	5. Toxin Binding Inhibition (ToBI) test for batch potency testing of human tetanus vaccines (FDA)	Refinement: antibody quantification	https://ntp.niehs.nih.gov/whatwestudy/niceatm/test-method-evaluations/biologics-and-vaccines/vaccine-testing/index.html
	6. Guidelines on use of humane endpoints and methods in animal testing of biological products (USDA)	Refinement of animal use	https://www.aphis.usda.gov/animal_health/vet_biologics/publications/notice_12_12.pdf
Acute Oral Systemic Toxicity	1. Acute toxic class method for acute oral toxicity (OECD TG No. 423)	Reduction and refinement of animal use	https://www.oecd-ilibrary.org/environment/test-no-423-acute-oral-toxicity-acute-toxic-class-method_9789264071001-en
	2. Acute toxicity in vitro starting dose procedure, 3T3 cells (OECD GD No. 129)	Reduction and refinement of animal use	https://ntp.niehs.nih.gov/iccvam/docs/acutetox_docs/brd_tmer/at-tmer-complete.pdf https://ntp.niehs.nih.gov/iccvam/suppdocs/feddocs/oecd/oecd-gd129.pdf
	3. Acute toxicity in vitro starting dose procedure, NHK cells (OECD GD No. 129)		
	4. Avian acute oral toxicity test (OECD TG No. 223)	Reduction of animal use	https://www.oecd-ilibrary.org/environment/test-no-223-avian-acute-oral-toxicity-test_9789264090897-en

(continued)

Table 49.1 (continued)

Toxicity area	Alternative methods/ approaches/ guidance/ policies	3R principle(s) applied/ proposed	Weblink[a]
	5. Fixed dose procedure for acute oral toxicity (OECD TG No. 420)	Reduction and refinement of animal use	https://www.oecd-ilibrary.org/environment/test-no-420-acute-oral-toxicity-fixed-dose-procedure_9789264070943-en
	6. Guidance on nonclinical studies of microdose radiopharmaceutical diagnostic drugs (FDA)	Includes provisions for reducing animal use	https://www.fda.gov/media/107641/download
	7. Up-and-down procedure for acute oral toxicity (OECD TG No. 425)	Reduction and refinement of animal use	https://www.oecd-ilibrary.org/environment/test-no-425-acute-oral-toxicity-up-and-down-procedure_9789264071049-en
Acute Inhalation Toxicity	1. Inhalation toxicity – acute toxic class method (OECD TG No. 436)	Reduction and refinement of animal use	https://www.oecd-ilibrary.org/environment/test-no-436-acute-inhalation-toxicity-acute-toxic-class-method_9789264076037-en
	2. Inhalation toxicity testing – fixed concentration procedure (OECD TG No. 433)	Reduction and refinement of animal use	https://www.oecd-ilibrary.org/environment/test-no-433-acute-inhalation-toxicity-fixed-concentration-procedure_9789264284166-en;jsessionid=j4innnfoodqa.x-oecd-live-02
Acute Dermal Systemic Toxicity [Conventional: Substance was rubbed onto the shaved backs of rats, and they were killed the next day]	1. Acute dermal toxicity (OECD TG No. 402)	Includes provisions for waiving test and reducing or refining animal use	https://www.oecd-ilibrary.org/environment/test-no-402-acute-dermal-toxicity_9789264070585-en
	2. Guidance for waiving acute dermal toxicity tests for pesticide formulations and supporting retrospective analysis (EPA)	Reduction of animal use	https://www.epa.gov/sites/default/files/2021-01/documents/guidance-for-waiving-acute-dermal-toxicity.pdf

	3. In vitro dermal absorption methods (OECD TG No. 428)	Replacement of animal use	https://www.oecd-ilibrary.org/environment/test-no-428-skin-absorption-in-vitro-method_9789264071087-en
Multiple Toxicities	1. Guidance document on developing and assessing adverse outcome pathways (OECD TG No. 184)	–	https://www.oecd.org/officialdocuments/publicdisplaydocumentpdf/?cote=env/jm/mono%282013%296&doclanguage=en
	2. Guidance document on the reporting of defined approaches to be used within integrated approaches to testing and assessment (OECD GD No. 255)	–	https://www.oecd.org/officialdocuments/publicdisplaydocumentpdf/?cote=env/jm/mono(2016)28&doclanguage=en
	3. Guidance for waiving or bridging of mammalian acute toxicity tests for pesticides and pesticide products (EPA)	Reduction of animal use	https://www.epa.gov/pesticide-registration/bridging-or-waiving-data-requirements
	4. Guidance on nonclinical studies for pharmaceuticals for hematologic disorders (FDA)	Includes provisions for reducing animal use	https://www.fda.gov/regulatory-information/search-fda-guidance-documents/severely-debilitating-or-life-threatening-hematologic-disorders-nonclinical-development
	5. Guidance on use of microsampling techniques in toxicokinetics studies (FDA)	Includes provisions for reducing and refining animal use	https://www.fda.gov/media/100027/download
	6. Guidance regarding replacement, reduction, and refinement of animal use for testing done under the Federal Hazardous Substances Act (CPSC)	–	https://www.cpsc.gov/Business%2D%2DManufacturing/Testing-Certification/Recommended-Procedures-Regarding-the-CPSCs-Policy-on-Animal-Testing
	7. Harmonized guidance for nonclinical safety studies for pharmaceuticals (FDA)	Reduction of animal use	https://www.fda.gov/media/82666/download

(continued)

Table 49.1 (continued)

Toxicity area	Alternative methods/ approaches/ guidance/ policies	3R principle(s) applied/ proposed	Weblink[a]
	8. Harmonized guidance for preclinical safety evaluation of biotechnology-derived pharmaceuticals (FDA)	Reduction of animal use	https://www.fda.gov/regulatory-information/search-fda-guidance-documents/s6r1-preclinical-safety-evaluation-biotechnology-derived-pharmaceuticals
	9. Process for evaluating and implementing alternative approaches to traditional in vivo acute toxicity studies for FIFRA regulatory use (EPA)	Replacement and reduction of animal use	https://www.epa.gov/pesticide-science-and-assessing-pesticide-risks/process-establishing-implementing-alternative

CPSC U.S. Consumer Product Safety Commission, *EPA* U.S. Environmental Protection Agency, *FDA* U.S. Food and Drug Administration, *GD* Guidance Document, *OECD* Organisation for Economic Co-operation and Development (OECD), *USDA* U.S. Department of Agriculture, *TG* Test Guideline

a Last accessed on 10th May, 2022.

Adapted from the Table of "Alternative Methods Accepted by US Agencies" published by the National Toxicological Program (NTP) – Interagency Center for the Evaluation of Alternative Toxicological Methods (NICEATM), U.S. Department of Health and Human Services. Available from: https://ntp.niehs.nih.gov/whatwestudy/niceatm/accept-methods/index.html?utm_source=direct&utm_medium=prod&utm_campaign=ntpgolinks&utm_term=regaccept

Other References:

Taylor K. Chapter 24 Recent Developments in Alternatives to Animal Testing. In Animal Experimentation: Working Towards a Paradigm Change, Leiden, The Netherlands: Brill. 2019 Available From: Brill https://doi.org/10.1163/9789004391192_025

Barosova H, Maione AG, Septiadi D, Sharma M, Haeni L, Balog S, O'Connell O, Jackson GR, Brown D, Clippinger AJ, Hayden P, Petri-Fink A, Stone V, Rothen-Rutishauser B. Use of EpiAlveolar Lung Model to Predict Fibrotic Potential of Multiwalled Carbon Nanotubes. ACS Nano. 2020 Apr 28;14 (4): 3941–3956. doi: 10.1021/acsnano.9b06860. Epub 2020 Mar 20. Erratum in: ACS Nano. 2020 Dec 22;14 (12):17713. PMID: 32167743.

AltTox.org. Table of Validated&Accepted Alternative Methods (2016). Available from: http://alttox.org/mapp/table-of-validated-and-accepted-alternative-methods/

Table 1 OECD test methods that have been improved in respect to animal welfare under participation of ZEBET. From: Liebsch M, Grune B, Seiler A, Butzke D, Oelgeschläger M, Pirow R, Adler S, Riebeling C, Luch 654 A. Alternatives to animal testing: current status and future perspectives. Arch Toxicol. 2011 655 Aug;85(8):841–58. Available from: https://link.springer.com/article/10.1007/s00204-011-0718-x/tables/1

European Union Reference Laboratory for alternatives to animal testing – EURL ECVAM | Alternative methods for toxicity testing. Available from: https://joint-research-centre.ec.europa.eu/eu-reference-laboratory-alternatives-animal-testing/alternative-methods-toxicity-testing/validated-test-methods_en

The *European Centre for the Validation of Alternative Methods* (ECVAM) (Italy), the *Interagency Coordinating Committee on the Validation of Alternative Methods* (ICCVAM) (US), *Japanese Center for the Validation of Alternative Methods* (JaCVAM) (Tokya, Japan), the *Korean Center for the Validation of Alternative Methods* (KoCVAM) (South Korea), Center for Documentation and Evaluation of Alternative Methods to Animal Experiments (ZEBET) (Germany), and the *Brazilian Centre for the Validation of Alternative Methods* (BraCVAM) are some of the centres for alternatives to animal experiments that are involved in the concerted efforts to apply the principles of 3Rs and *alternatives*.

The ECVAM acts as the liaising organization for all other centres for alternatives and to foster development of independently validated-yet-harmonized alternative methods, the *International Cooperation on Alternative Test Methods* (ICATM) network was created. ECVAM, ICCVAM, JaCVAM, Health Canada, KoCVAM, BraCVAM, and the Chinese Food and Drug Administration (CFDA) are partners of the ICATM network. The European Union Reference Laboratory (EURL) undertakes all the activities of ECVAM. The EURL-ECVAM hosts an online portal by name *TSAR—Tracking System for Alternative methods towards Regulatory acceptance* (https://tsar.jrc.ec.europa.eu/) which depicts the progress of various alternative and non-animal methods for testing chemicals and biological agents from preliminary submission to regulatory acceptance.

Initially setup as a permanent committee as a part of the *National Institute of Environmental Health Sciences* (NIEHS), the ICCVAM is supported by the *National Toxicology Program Interagency Center for the Evaluation of Alternative Toxicological Methods* (NICEATM). In 2018, the ICCVAM put forward a strategic roadmap for achieving ideal alternatives particularly in areas of acute systemic toxicity, eye and skin irritation, and skin sensitization.

There are also other centres that promote the 3Rs agenda like the Johns Hopkins University Bloomberg School of Public Health's *Center for Alternatives to Animal Testing* (CAAT) and its European branch namely the CAAT-Europe located at the University of Kostanz, Germany, *European Society for Alternatives to Animal Testing* (EUSAAT), and the *transatlantic think tank for toxicology* (t4). Together they also a run journal titled, "ALTEX—Alternatives to Animal Experimentation". The Institute for In Vitro Sciences, Inc. and the Danish 3R-Center are some of the other centres promulgating the 3R approach.

The Norway's centre for advancement of 3R principles is the *Norecopa* (Norwegian Consensus Platform for Replacement, Reduction and Refinement of animal experiments). Among many other activities, Norecopa hosts a database [NORINA *(A Norwegian Inventory of Alternatives)*] portraying the various alternatives to the utilization of animals in education, training, and research and a website of *International Culture of Care Network* (promoting activities of *Culture of Care*—commitment to improve animal welfare, scientific quality, care of the staff and transparency for the stakeholders).

Like in research, the various proprietary human-patient simulators are playing a pivotal role in imparting the requisite clinical training among medical graduates—and indirectly reducing the burden on animals.

Continuing education (including skill training), appropriate investment, and constant attention of the various 'R' principles of alternatives to animal experimentations is indispensable for minimizing the usage of animals (number and/ distress) in research and maximizing wellbeing of animals in research.

49.6 Evolution of Ethics in Animal Research

The use of animals in research dates back to almost the same period when the field of medicine was gaining momentum—during the times of Aristotle and Hippocrates. And the Darwin's *On the Origin of Species* which highlighted the evolutionary biology and the biological resemblances between man and animal, gave more impetus for furthering animal experimentations. Though unregulated previously, from the nineteenth century onwards regulatory measures for animal experimentation came into existence in the UK and US. For example, in the UK, the *Cruelty to Animals Act* was passed in the UK Parliament in 1876 which was later repealed by the *Animals (Scientific Procedures) Act* in 1986.

Animal experiments have led to the discovery of many remarkable drugs and vaccines such as diphtheria/ tetanus antitoxin (horses; 1901), insulin (dogs; 1922), prontosil (mice; 1939), penicillin (rats; 1941), and streptomycin (chicken; 1952). The ever-increasing concern and interest in animal welfare led to the development of rules, regulations, and laws related to animal use in biomedical research across the globe.

And the argument regarding utilization of animals in experiments dates back to seventeenth century, later through Claude Bernard's wife (nineteenth century) who was opposing vivisection (N.B.: Claude Bernard is referred to as the "prince of vivisection"), and the twenty-first century is no different with organizations such as the Society for the Prevention of Cruelty to Animals (SPCA) [nonprofit animal welfare organization], The Humane Society of the United States (HSUS; https://www.humanesociety.org/), the Animal Welfare Institute (AWI; https://awionline.org/), the Royal Society for the Prevention of Cruelty to Animals (RSPCA; https://www.rspca.org.uk/), the People for the Ethical Treatment of Animals (PETA; https://www.peta.org/), the European Coalition to End Animal Experiments (ECEAE; https://www.eceae.org/), and the Physicians Committee for Responsible Medicine (PCRM; https://www.pcrm.org/)—that are strong advocates against unwarranted use of animals for research purpose (Table 49.2).

49.7 Organisation for Economic Co-operation and Development (OECD) Guidelines

The Organisation for Economic Co-operation and Development (OECD) is a 60-year-old international organization with 38 member countries across the globe to promote the economic and social well-being along with the entities of cooperation and development—with the tagline, *"Better Policies for Better Lives"*. The goal of

Table 49.2 List of guidelines for use of animals in biomedical research

Title	Organization, Country	Year of publication[a]	Weblink[b]
Compendium of Committee for the Purpose of Control and Supervision of Experiments on Animals (CPCSEA)	Ministry of Environment, Forests and Climate Change (MoEFCC), India	2018	https://www.aaalac.org/pub/?id=DCDFC6CF-BA63-03BA-34D8-2497743ECA56
Guidelines of CPCSEA for Experimentation on Fishes	Ministry of Fisheries, Animal Husbandry and Dairying, Department of Animal Husbandry and Dairying, India; CPCSEA	2021	http://cpcsea.nic.in/WriteReadData/userfiles/file/Guidelines%20of%20CPCSEA%20for%20Experimentation%20on%20Fishes-2021.pdf
Guidelines of CPCSEA for Rehabilitation/Reuse of Large Animals Post-Experimentation – 2020	Ministry of Fisheries, Animal Husbandry and Dairying, Department of Animal Husbandry and Dairying, India; CPCSEA	2020	http://cpcsea.nic.in/WriteReadData/userfiles/file/Guidelines%20of%20CPCSEA%20for%20Reuse%20Rehabilitation%20of%20Large%20Animals.pdf
Guidelines for Constitution / Reconstitution of Institutional Animal Ethics Committees (IAECs)	Ministry of Environment, Forests and Climate Change (MoEFCC), Animal Welfare Division, India	2018	http://cpcsea.nic.in/WriteReadData/userfiles/file/IAEC%20constitution%20guidelines.pdf
Guide for the Care and Use of Laboratory Animals, 8th edition	Committee for the Update of the Guide for the Care and Use of Laboratory Animals, Institute for Laboratory Animal Research, National Research Council, US	2011	https://www.nationalacademies.org/our-work/update-of-the-guide-for-the-care-and-use-of-laboratory-animals
International Guiding Principles for Biomedical Research Involving Animals	Council for International Organizations of Medical Sciences (CIOMS) and International Council for Laboratory Animal Science (ICLAS)	2012	https://olaw.nih.gov/sites/default/files/Guiding_Principles_2012.pdf
Australian code for the care and use of animals for scientific purposes, 8th edition	Australian Research Council, Commonwealth Scientific Industrial Research Organisation, National Health and	2013 (updated 2021)	https://www.nhmrc.gov.au/about-us/publications/australian-code-care-and-use-animals-scientific-purposes

(continued)

Table 49.2 (continued)

Title	Organization, Country	Year of publication[a]	Weblink[b]
	Medical Research Council, Universities Australia, Australia		
Ethical Guidelines for the Use of Animals in Research	The Norwegian National Committee for Research Ethics in Science and Technology (NENT), Norway	2018	https://www.forskningsetikk.no/en/guidelines/science-and-technology/ethical-guidelines-for-the-use-of-animals-in-research/
The ARRIVE guidelines (Animal Research: Reporting of *In Vivo* Experiments)	The NC3Rs (National Centre for the Replacement Refinement & Reduction of Animals in Research)	2020	https://journals.plos.org/plosbiology/article?id=10.1371/journal.pbio.3000410
Guide to the Care and Use of Experimental Animals Volume 1, 2nd Edition	Canadian Council on Animal Care (Conseil canadien de protection des animaux)	2020	https://ccac.ca/Documents/Standards/Guidelines/Experimental_Animals_Vol1.pdf
Animal Welfare Act and Animal Welfare Regulations (Blue Book)	U.S. Department of Agriculture (USDA)	2019	https://www.aphis.usda.gov/animal_welfare/downloads/bluebook-ac-awa.pdf
European Directive 2010/63/EU on the protection of animals used for scientific purposes	European Union (EU)	2010	https://eur-lex.europa.eu/legal-content/EN/TXT/?uri=CELEX:32010L0063
Status Report on Non-animal Methods in Science and Regulation	European Union Reference Laboratory -European Centre for the Validation of Alternative Methods (EURL-ECVAM)	2021	https://publications.jrc.ec.europa.eu/repository/handle/JRC127780
The PREPARE (Planning Research and Experimental Procedures on Animals: Recommendations for Excellence) Guidelines	Royal Society for the Prevention of Cruelty to Animals (RSPCA)	2017	https://journals.sagepub.com/doi/10.1177/0023677217724823
Consensus statement from the European Network of 3R Centres (EU3Rnet)	European Network of 3R Centres (EU3Rnet)	2020	https://www.altex.org/index.php/altex/article/view/2143/2188

(continued)

Table 49.2 (continued)

Title	Organization, Country	Year of publication[a]	Weblink[b]
The ethics of research involving animals	Nuffield Council on Bioethics	2005	https://www.nuffieldbioethics.org/assets/pdfs/The-ethics-of-research-involving-animals-full-report.pdf

[a]Only the latest version is included
[b]Last accessed on 10th May, 2022

OECD is "to shape policies that foster prosperity, equality, opportunity and well-being for all". The OECD deals with, among many key global issues at national, regional and local levels, the assessment of potential effects of chemicals on human health and environment.

The *Sect. 4* of the *OECD Guidelines (Methods) for the Testing of Chemicals* elaborates specifically on the health effects of new chemical entities under toxicology research. The *OECD Guidelines* are a set of internationally accepted working standards which are frequently expanded and revised based on the current recent developments and come under the ambit of OECD Mutual Acceptance of Data (MAD) system. The MAD system (along with the application of the *OECD principles of Good Laboratory Practice*) allows for "the participating countries (including non-members) to share the results of various nonclinical tests done on chemicals using OECD methods and principles". This multilateral agreement greatly removes duplication of testings across the globe and saves on the resources of the government, industry, and other related stakeholders of chemical testings.

The increasingly complex and heterogenous regulatory toxicity testing and assessment approaches demand for concomitant increase in the number of animals and resources required. Hence, the OECD has come forward with the *Integrated Approaches to Testing and Assessment* (IATA) to streamline regulatory decision making—which are defined as "pragmatic, science-based approaches for chemical hazard or risk characterisation that rely on an integrated analysis of existing information in a weight of evidence assessment coupled with the generation of new information using testing strategies". IATA are flexible [from non-formalized to structured rule-based (Integrated Testing Strategy, ITS)], need-based (depending on the amount of certainty required), hypothesis-driven, and follow an iterative approach to answer a defined question in a specific regulatory context.

The previously discussed translation of results from 'simpler models' to 'more complex human forms' is now emerging through mechanistic approaches—i.e., connecting one key event to another downstream. This same sequential linking approach is followed in the Adverse Outcome Pathway (AOP) concept proposed by the OECD—in which the key events (KE) are, toxicant (chemical/ drug) exposure → molecular initiating event (MIE) → cellular response → tissue/ organ response → individual (organism) response → population response [Adverse Outcome (AO)]. The AOPs facilitate and provide the mechanistic-basis for the implementation of IATA.

49.8 Compendium of CPCSEA

The Committee for the Purpose of Control and Supervision of Experiments on Animals (CPCSEA) is a statutory body constituted under the "Prevention of Cruelty to Animals (PCA) Act, 1960" and the subsequent "Breeding of and Experiments on Animals (Control and Supervision) Rules, 1998 (Amendments, 2001 and 2006)". Though constituted in the year 1964, the body took strong foothold only in the late 1990s. The prime objective of the Committee is to ensure that animals are not subjected to unnecessary pain or suffering before, during or after the performance of experiments on them. The CPCSEA under the aegis of Ministry of Environment, Forests and Climate Change (MoEFCC) [Animal Welfare Division] put forth the guidelines on the rules and regulations on the use of animals in research in 2018.

Referred technically to as the "Compendium of CPCSEA", the 2018 manual encompasses and consolidates all the norms, guidelines, rules, and regulations set forth by the CPCSEA so as to ensure humane and ethical treatment of animals in biomedical experimentation. This compendium and its associated guidelines are to be closely followed by the Indian organizations.

Box 49.1 illustrates the ethical principles adopted by CPCSEA for use of animals in scientific experiments

Box 49.1: Ethical principles adopted by CPCSEA for use of animals in scientific experiments

Ethical principle	Explanation
Principle 1	• "Experiments on animals" (including experiments involving operations on animals) may be carried out – for the purposes of *advancement by new discovery* of physiological knowledge; or – of knowledge which is expected to be useful for *saving or prolonging human life or alleviating suffering*; or – for significant gains in the *well-being for the people of the country*; or – for *combating any disease*, whether of human being, animals or plants
Principle 2	• Animals *lowest on the phylogenetic scale* (i.e., with the least degree of sentience), which may give scientifically valid results, should be used for any experimental procedure – Phylogenetic scale: Invertebrates (e.g., cockroaches) < Birds < Rodents < Canines/Felines < Bovine/ Equines < Primates (e.g., Rhesus Macaque) < More evolved Primates (e.g., chimpanzee) • Experiments should be designed with the *minimum number of animals* to give statistically valid results at 95% level of confidence • *Alternatives not involving animal testing* should be given due and full consideration and sound justification provided, if alternative, when available, are not used

(continued)

Box 49.1 (continued)

Ethical principle	Explanation
Principle 3	• Proper use of animals in experiments and *avoidance or minimization (when avoidance is not possible) of pain and suffering inflicted on experimental animals* should be an issue of priority for research personnel, and unless the contrary is scientifically established, investigators should process on the basis that procedures that cause pain or suffering in human beings will also cause similar pain or suffering in animals • All scientific procedures adopted with animals that may cause *more than momentary or slight pain and/or suffering* should be performed with appropriate sedation, analgesia or anaesthesia
Principle 4	• Persons engaged in animal experimentation have a *moral responsibility* for the welfare of the animals after their use in experiments • Investigators are responsible for the *aftercare and/or rehabilitation of animals after experimentation* • Investigators are permitted to ***euthanize animals*** only in the following situations, – When the animal is *paralyzed* and is not able to perform its natural functions; it becomes *incapable of independent locomotion*; and/or can *no longer perceive the environment in an intelligible manner* – During the course of experimental procedure, the animal has been left with a severe recurring pain and the animal exhibits obvious signs of *long-term extreme pain and suffering* – In situations where non-termination of the animal experimented upon would be *life threatening to human beings or other animals* • *Costs of aftercare and/or rehabilitation* of animals post-experimentation are to be part of research costs and should be scaled per animal in positive correlation with the level of sentience of the animals
Principle 5	• The living conditions of animals should be appropriate for their species and contribute to their *health and comfort* • The *housing, feeding, and care of all animals* used for biomedical purposes must be directed by a *veterinarian* or other scientist in a relevant discipline who is trained and experienced in the proper care, handling, and use of the species being maintained or studied • In all circumstances, veterinary care shall be provided as necessary

Reproduced and adapted from the **"Guidelines on the Regulation of Scientific Experiments on Animals, 2007"** by the Ministry of Environment & Forests (Animal Welfare Division), Government of India

The major functions of CPCSEA are as follows (as the *Compendium of CPCSEA, 2018*):

- Registration of establishments (including educational and research institutions) conducting animal experimentation or breeding of animals for this purpose.
- Selection and appointment of nominees in the Institutional Animal Ethics Committees (IAEC) of registered establishments.
- Approval of Animal House facilities on the basis of reports of inspections conducted by CPCSEA.

- Permission for conducting experiments involving use of animals.
- Recommendation for import of animals for use in experiments.
- Action against establishments in case of violation of any legal norm/stipulation.
- Composing the Standard Operating Procedures (SOP) for Institutional Animal Ethics Committee
  - The CPCSEA had previously published (in 2010) the SOP for effective functioning of the IAEC so as to maintain the quality of ethical review mechanism for animal research projects.
- To monitor the mandatory actions for all establishments registered with CPCSEA.

The IAEC shall include five members from the institute (establishment) and three nominees from CPSCEA—hence, overall, eight members with a Chairperson and Member Secretary (from the institute). The institute members include a biological scientist (1), scientists from different biological disciplines (2), a veterinarian involved in the care of animal (1), and a scientist in charge of animal facility (house) of the institute (1). A main nominee (1), a scientist from outside the institute (1), and a non-scientific socially aware nominee (1) constitute the CPSCEA nominees. In the absence of the main nominee, a designated link nominee shall fulfill the quorum.

In India, it is stressed that ethical considerations are to be applied at all levels of experimentation of laboratory animals—individual, institutional, and national and at all timepoints—prior to, during, and after the experiments.

Two drafted bills, namely, the "Animal Welfare Act, 2011" and the "Prevention of Cruelty to Animals (Amendment) Bill, 2016"—both to repeal the 1960 PCA Act—were never passed in the Parliament.

Apart from the CPCSEA, the Indian National Science academy (INSA) and Indian Council of Medical Research (ICMR) have also laid down their own guidelines for rational use of animals in scientific research. INSA had released the "Guidelines for Care and Use of Animals in Scientific Research" in 2000 and ICMR had released the "Guidelines for use of Laboratory Animals in Medical Colleges" in 2001. In 2011, INSA held a brainstorming session titled, "Man, Animal & Science" bringing together regulatory bodies, premier scientific research organizations, pharmaceutical agencies, and animal welfare bodies to deliberate on the need for animals in biomedical research, testing, and teaching.

49.9 Guide for the Care and Use of Laboratory Animals, 8th edition

The *Guide for the Care and Use of Laboratory Animals* released in 2011 is currently expecting revisions. The *Institute for Laboratory Animal Research (ILAR)* of the *National Academy of Sciences*, US is entrusted with this mandate.

The goal of this *Guide*, "is to promote the humane care and use of laboratory animals by providing information that will enhance animal wellbeing, the quality of research, and the advancement of scientific knowledge that is relevant to both humans and animals".

The intended audience of this *Guide* is wide-ranging from the biomedical scientists, the administrators, the veterinarians, the educators and trainers of animal experimentations, the producers of laboratory animals, the accreditation bodies, the regulators, to the general public.

The *Guide* endorses the premise that use of animals in research is a privilege vested with research community so that it brings about substantial new knowledge or leads to advancement in human and/or animal well-being.

The following principles are endorsed taking into consideration the 1985 *U.-S. Government Principles for the Utilization and Care of Vertebrate Animals Used in Testing, Research, and Training:*

- alternatives to reduce or replace the use of animals
- propose and execute procedures on the basis of relevance to human or animal health, improvement of knowledge, or the common good of society
- appropriateness in species selection, quality, and number of animals
- remove or minimize discomfort, distress, and pain
- use of appropriate sedation, analgesia, and anesthesia
- incorporation of humane endpoints
- provision of adequate veterinary care
- provision of appropriate animal transportation and husbandry
- supervised and executed by qualified and experienced professionals

49.10 Animal Experimentation in India and Elsewhere: *What Is the Current Status?*

India is in line with West with regards to utilization of animals for experimental purpose. Ever since the University Grants Commission (UGC) had laid down the "Guidelines of dissection and animal experimentation in Zoology/ Life Sciences in a phased manner" in 2011 and its subsequent notification in 2014 ("Dissection and Animal Experimentation in Zoology/Life Sciences and allied disciplines in under-graduate, postgraduate and research programmes"), the number of animal experimentations have been drastically reduced in academia involved in pure teaching activities. This situation is prevalent in almost all spheres of sciences that involve animals in experiments—including medical schools that teach physiology and pharmacology.

The discrepancy and discordance with regards to animal housing facilities in academia (Universities/Teaching Colleges) compared to institutions involved in pure biomedical or animal research including the drug toxicology testings in Pharma was much felt. In early part of 2000s, there was an outcry for less stringent regulatory provisions for categorizations of animal houses in pure educational organizations which carry out animal experimentations in a smaller scale.

The Indian National Medical Commission (NMC), in its latest Gazette Notification on "Minimum Requirements for Annual M.B.B.S. Admissions Regulations, 2020", clearly mentions about the utilization of computer assisted modules for teaching the undergraduate students—reflecting the earlier directives (2009) of the Medical Council of India (MCI). Further, to have animal holding area only for the conduct of research activities or postgraduate sessions in Pharmacology.

The *Animal Welfare Board of India* (AWBI) functioning under the aegis of the Ministry of Fisheries, Animal Husbandry and Dairying (Department of Animal Husbandry and Dairying) also promotes animal welfare in the country. The Ministry had also established a *National Institute of Animal Welfare* (NIAW) in Haryana with one of the objectives to improve animal welare. The *Tamil Nadu Veterinary and Animal Sciences University* (TANUVAS) in Chennai is another such premier institute that imparts training in laboratory animal welfare.

And ICMR has established the *National Animal Resource Facility for Bio-Medical Research (NARF-BR)* in Hyderabad which would be functional in the near future. This proposed state-of-the-art world-class research facility is a transformation from the previously existent *National Center for Laboratory Animal Sciences (NCLAS)* part of the National Institute of Nutrition (NIN), Hyderabad.

Like the CDSCO, the nonprofit *American Association for Laboratory Animal Science (AALAS)* coordinates with the Institutional Animal Care and Use Committee (IACUC) existent throughout member states of the US.

The selling of cosmetics (and household chemicals) tested on animals is totally banned in the Europe Union—including Belgium, the Netherlands, and the UK. However, the Registration, Evaluation, Authorization and Restriction of Chemicals (REACH) regulation of the EU can sometimes seek for in vivo animal testing of chemicals (for stricter compliance of cosmetic safety) contravening the above legislature.

In India, the 2014 Gazette notification [G.S.R 346(E)] states that, "*Prohibition of testing of cosmetics on animals-No person shall use any animal for testing of cosmetics*". However, as per the New Drugs and Clinical Trials Rules, 2019 (Second Schedule), which is legally enforceable for all new drugs, investigational new drugs for human use, clinical trial, bioequivalence study, and bioavailability study, submission of data related to 'animal pharmacology' and 'animal toxicology' is mandatory.

49.11 Conclusion

Laboratory animals form the basis for primitive preliminary biomedical research and also regulatory drug testings. There is a need for continuous communication, mutual cooperation, and transparent collaboration across regulatory bodies, scientific organizations, funding agencies, journals, animal welfare associations, and research scientists to uplift the robustness in animal research.

Owing to the differences in sociocultural aspects, religious practices, and economic factors, the implementation of ethical practices in animal research may vary from country to country. However, the overall measures to supervise animal experimentation and structuring oversight mechanisms (programs) remain the same across regions.

The handlers of animal experiments should take utmost precaution in practicing appropriate ethical practices so as to always maintain humane-touch to experimental animals. Ethics in animal research is sometimes viewed even more seriously than that for biomedical research in humans—understandably so considering the fact that animals are largely being tested for not their individual benefits!

Bibliography

Akbarsha MA, Pereira S. Mahatma gandhi-doerenkamp center for alternatives to use of animals in life science education. J Pharmacol Pharmacother. 2010;1(2):108–10. https://doi.org/10.4103/0976-500X.72353.

Akbarsha MA, Zeeshan M, Pereira S. Alternatives discussed at Indian science congress. ALTEX. 2012;29(2):216–8. https://doi.org/10.14573/altex.2012.2.216.

Badyal DK, Desai C. Animal use in pharmacology education and research: the changing scenario. Indian J Pharmacol. 2014;46(3):257–65. https://doi.org/10.4103/0253-7613.132153.

Caloni F, De Angelis I, Hartung T. Replacement of animal testing by integrated approaches to testing and assessment (IATA): a call for in vivitrosi. Arch Toxicol. 2022. Epub ahead of print. doi: https://doi.org/10.1007/s00204-022-03299-x.

Cheluvappa R, Scowen P, Eri R. Ethics of animal research in human disease remediation, its institutional teaching; and alternatives to animal experimentation. Pharmacol Res Perspect. 2017 Aug;5(4):e00332. https://doi.org/10.1002/prp2.332.

CPCSE. Compendium of CPCSEA, 2018. Available from: http://cpcsea.nic.in

CPCSEA. Standard Operating Procedures (SOP) for IAEC, 2010 [Internet]. 2010 [cited 2022 May 6]. Available from: http://cpcsea.nic.in/WriteReadData/userfiles/file/SOP_CPCSEA_inner_page.pdf.

Doke SK, Dhawale SC. Alternatives to animal testing: a review. Saudi Pharm J. 2015 Jul;23(3):223–9. https://doi.org/10.1016/j.jsps.2013.11.002.

Festing S, Wilkinson R. The ethics of animal research. Talking Point on the use of animals in scientific research. EMBO Rep. 2007 Jun;8(6):526–30. doi: 10.1038/sj.embor.7400993. PMID: 17545991; PMCID: PMC2002542.

Hubrecht RC, Carter E. The 3Rs and humane experimental technique: implementing change. Animals (Basel). 2019 Sep 30;9(10):754. https://doi.org/10.3390/ani9100754.

ICCVAM (Interagency Coordinating Committee on the Validation of Alternative Methods). A Strategic Roadmap for Establishing New Approaches to Evaluate the Safety of Chemicals and Medical Products in the United States [Internet]. 2018 [cited 2022 April 17]. Available from: https://dx.doi.org/10.22427/NTP-ICCVAM-ROADMAP2018.

INSA. Man, Animal and Science, 2011. [Internet]. 2011 [cited 2022 May 6]. Available from: https://www.insaindia.res.in/pdf/FINAL_DOCUMENT_ON_MAN_ANIMAL_SCIENCE.pdf.

Knight A. 127 million non-human vertebrates used worldwide for scientific purposes in 2005. Altern Lab Anim. 2008;36(5):494–6.

Knight J, Rovida C, Kreiling R, Zhu C, Knudsen M, Hartung T. Continuing animal tests on cosmetic ingredients for REACH in the EU. ALTEX. 2021;38(4):653–68. https://doi.org/10.14573/altex.2104221.

Kojima H, Seidle T, Spielmann H. Alternatives to animal testing. Proceedings of Asian Congress. 2016. Available from: https://link.springer.com/book/10.1007/978-981-13-2447-5.

Liebsch M, Grune B, Seiler A, Butzke D, Oelgeschläger M, Pirow R, Adler S, Riebeling C, Luch A. Alternatives to animal testing: current status and future perspectives. Arch Toxicol. 2011 Aug;85(8):841–58. https://doi.org/10.1007/s00204-011-0718-x.

National Research Council. Guide for the care and use of laboratory animals: Eighth Edition. Washington, DC: The National Academies Press [Internet]. 2011 [cited 2022 May 11]. Available from: https://doi.org/10.17226/12910.

NMC. Minimum Requirements for Annual M.B.B.S. Admissions Regulations, 2020 [Internet]. 2020 [cited 2022 March 21]. Available from: https://www.nmc.org.in/ActivitiWebClient/open/getDocument?path=/Documents/Public/Portal/NmcGazette/Medical%20College%20MSR%20regulations%202020.pdf.

Pereira S, Tettamanti M. Ahimsa and alternatives: the concept of the 4th R. The CPCSEA in India. ALTEX. 2005;22(1):3–6.

Pereira S, Veeraraghavan P, Ghosh S, Gandhi M. Animal experimentation and ethics in India: the CPCSEA makes a difference. Altern Lab Anim. 2004 Jun;32(Suppl 1B):411–5. https://doi.org/10.1177/026119290403201s67.

Sakuratani Y, Horie M, Leinala E. Integrated approaches to testing and assessment: OECD activities on the development and use of adverse outcome pathways and case studies. Basic Clin Pharmacol Toxicol. 2018 Sep;123(Suppl 5):20–8. https://doi.org/10.1111/bcpt.12955.

Smith AJ, Hawkins P. Good science, good sense and good sensibilities: the three Ss of Carol Newton. Animals (Basel). 2016;6(11):70. https://doi.org/10.3390/ani6110070.

Stephens ML. Personal reflections on Russell and Burch, FRAME, and the HSUS. Altern Lab Anim. 2009 Dec;37(Suppl 2):29–33. https://doi.org/10.1177/026119290903702S21.

Tannenbaum J, Bennett BT. Russell and Burch's 3Rs then and now: the need for clarity in definition and purpose. J Am Assoc Lab Anim Sci. 2015 Mar;54(2):120–32.

The Norwegian National Research Ethics Committees. Ethical Guidelines for the Use of Animals in Research, 2018 [Internet]. 2019 [cited 2022 Feb 24]. Available from: https://www.forskningsetikk.no/en/guidelines/science-and-technology/ethical-guidelines-for-the-use-of-animals-in-research/.

UGC. Guidelines for discontinuation of dissection and animal· experimentation in Zoology/ Life Sciences in a phased manner [Internet]. 2011 [cited 2022 March 20]. Available from: https://www.ugc.ac.in/pdfnews/6686154_guideline.pdf.

UGC. Dissection and animal experimentation in zoology /life sciences and allied disciplines in undergraduate, postgraduate and research programmes [Internet]. 2014 [cited 2022 March 20]. Available from: https://www.ugc.ac.in/pdfnews/6819407_ugcletterzoology.pdf.

Good Laboratory Practice (GLP)

50

Alphienes Stanley Xavier

Abstract

Good Laboratory Practicc principles are framed and put-forth to ensure the quality, reliability, acceptability of data generated in non-clinical health and safety studies. The principles published by OECD have been followed and referred by most of the countries. These principles help the scientists to obtain results that are reliable, repeatable, auditable, and recognized by scientists world-wide, as well as adhering to these principles prevents from rework, rejection, and deviation. In India, National GLP Compliance Monitoring Authority is functioning with the responsibilities of monitoring, inspecting the test facilities and their GLP compliance certification.

Keywords

Good Laboratory Practice (GLP) · GLP principles · GLP compliance · GLP certification

50.1 Introduction

Assuring the validity and legibility of scientific data from laboratories are vital amidst the present research and developmental proceedings in the pharmaceutical world. During the late 1970s, the upsurge in drug discovery and development, as well as the various incidences of laboratory malpractices surfaced; there was a

A. S. Xavier (✉)
Department of Pharmacology, Sri Ramachandra Medical College and Research Institute, SRIHER (Deemed University), Chennai, Tamil Nadu, India

Clinical Pharmacology, Sri Ramachandra Medical Centre, Chennai, Tamil Nadu, India
e-mail: alphclinpharm@sriramachandra.edu.in

M. Lakshmanan et al. (eds.), *Introduction to Basics of Pharmacology and Toxicology*, https://doi.org/10.1007/978-981-19-5343-9_50

burning need for strict, universal regulations to implement Good Laboratory Practice. Understanding the importance of the situation, the United States—Food and Drug Administration (US-FDA) has drafted the GLP guidelines, which was then adopted by many other nations and frames their principles.

The Organization for Economic Co-operation and Development (OECD), an intergovernmental organization has published the GLP guidelines in 1981, which is being followed by all OECD-member countries as well as many other non-member nations.

According to OECD, the GLP has been defined as "a quality system concerned with the organizational process and the conditions under which non-clinical health and environmental safety studies are planned, performed, monitored, recorded, archived and reported."

50.2 Principles of Good Laboratory Practice

GLP is a regulation covering the quality management of non-clinical safety studies. The aim of the regulations is to encourage scientists in organizing and performing their studies, which can promote and ensure the generation of data with good quality as well as validity. The principles put-forth are followed by test facilities carrying out studies to assess the health and environmental safety of chemicals of natural or biological origin, which will be submitted to regulatory authorities. Adherence to GLP enhances the acceptability of quality test data across national boundaries.

To mention briefly the GLP principles help the scientists to obtain results that are Reliable, Repeatable, Auditable and Recognized by scientists worldwide, and also GLP prevents from Rework, Rejection, and Deviation.

GLP principles are emphasized under the following major categories:

1. Resources: Organization, Personnel, Facilities, Equipment
2. Rules: Protocols, Study plans, Standard Operating Procedures (SOPs)
3. Characterization: Test items, test systems
4. Documentation: Raw data, final report, archives
5. Quality assurance: Independence from study conduct

The principles of GLP can be discussed under the following headings.

50.2.1 Test Facility: Organization and Personnel

This section explains the responsibilities of the test facility management, Study Director, Principal Investigator, and study personnel:

50.2.1.1 Test Facility Management

Individual or individuals should be identified to fulfill the responsibilities as test facility management. The management should ensure the following

- Sufficient number of qualified personnel, appropriate facilities, equipment, and materials are available for the proper conduct of the study
- Designating individuals with appropriate, training, and experience as Study Director/Principal Investigator before study initiation and to have established procedures, documentation of replacing them during the study conduct
- Implementation and performance of the Quality Assurance Programme
- Maintenance of records related to the qualification, training, job description, experience of each professional and technical individual, as well as the master schedule, historical files of all Standard Operating Procedures, documents related to the Quality Assurance Programme
- Training of study personnel and help them to understand their job responsibilities
- Clear communication among Study Director, Principal Investigator, Quality Assurance team and study personnel
- Identification of the person responsible for managing archives

50.2.1.2 Study Director

Study Director is the single point of study control, who has the responsibility for the overall study conduct and submitting its final report.

Study director's responsibilities are:

- Approval of the study plan, its implementation, any amendments as well as to ensure the availability of the study plan to all the study personnel and quality assurance team
- Timely and effective communication with quality assurance personnel during the study conduct
- Ensure the availability of SOPs and their amendments to the study personnel
- In multi-site studies ensure the approval of the study plan and final report, as well as to define the role of Principal Investigator and test sites involved in the study conduct
- Ensure complete documentation of all generated raw data
- To sign and date the final report; Accepts the responsibility for data validity and compliance with GLP principles
- Make sure that at the end of the study, the study plan, final report, raw data and supporting material are archived

50.2.1.3 Principal Investigator

Principal Investigator is responsible to ensure that all the delegated phases of the study are conducted in compliance with GLP principles.

50.2.1.4 Study Personnel

Study personnel involved in the study should

- Have adequate knowledge in the principles of GLP related to their role in the study

- Comply with the instructions given in the study plan, SOPs. Any deviation should be documented and communicated to Study Director
- Be responsible for the prompt recording and quality of data generated
- Follow all health precautions to minimize the risk and communicate to the relevant person regarding their known medical condition which requires possible exclusion of them from the study operations

50.2.2 Quality Assurance Programme (QAP)

The study facility should have a document of the Quality Assurance Programme (QAP) to assure that the studies are performed in compliance with GLP principles throughout the conduct. QAP should be carried out by an individual or team of individuals who are familiar with the test procedures and will be designated by the test facility management. They should not be a part of the study being conducted.

The responsibilities of the QAP personnel are:

- To maintain copies of all approved study plans and SOPs, and should have access to an up-to-date copy of the master schedule
- To verify the study plan for the information related to compliance with GLP principles
- Conduct inspections and documents the records related to inspections. The QAP inspections can be study-based, facility-based, or process-based
- To confirm that final reports describe the methods, procedures, and observation accurately as well as completely
- To assure that the results reported reflecting the raw data in an accurate and complete manner
- Promptly communicate the inspection findings to the management, Study Director, and/or Principal Investigator
- Prepare and sign a statement regarding the details of inspections, and observations, which should be included in the final study report

50.2.3 Facilities

- The test facility should be of suitable size, construction, and location as per the study requirements. There should be adequate separation of different activities at the facility to ensure proper study conduct.
- Storage areas should be appropriately separated from the areas of test systems to avoid infestation, contamination, and deterioration.
- There should be adequate space and storage facilities for the safe storage of hazardous substances.
- There should be appropriate provisions for the collection, storage, and disposal of wastes, as well as the transportation, decontamination procedures.

50.2.4 Apparatus, Material, and Reagents

- Apparatus used in the facility to generate, store, or retrieve data including the computerized systems should be of appropriate design and adequate quality.
- All the apparatus should undergo a periodic inspection, cleaning, maintenance, and calibration. Calibration should be according to national or international standards.
- All the chemicals, reagents, solutions at the facility should be labeled appropriately with the information to indicate the identity, expiry date, and specific storage instructions.

50.2.5 Test Systems

- Apparatus used for the generation of physical or chemical data should be of appropriate design, adequate quality, and suitably located. The integrity of the test systems should be ensured.
- Biological systems used in the study should be properly maintained to ensure data quality. Their source, date of arrival, and arrival condition should be recorded. Before their application, the systems should be acclimatized to the test environment.
- Newly received plant and animal test systems should be isolated for assessing their health status. At the initiation of the experiment, the biological test system should be free of any disease which may interfere with study conduct. During the study course, if they become diseased, they should be properly isolated and treated according to the documented procedures.

50.2.6 Test and Reference Items

- Records regarding the characteristics, receipt details, expiry, quantities received and used should be maintained properly for the test, reference items of the study
- The procedures for handling, storing, sampling of the items to be identified, and followed to prevent any possible mix-up or contamination
- The storage containers of the test, reference items should possess labels which contain information for identification, expiry date, and any specific storage instructions
- The stability of the items under storage and test conditions should be known as well as documented
- A sample from each test item should be retained for analytical purposes, except for the short term studies

50.2.7 Standard Operating Procedures (SOPs)

SOPs are "detailed, written instructions to achieve uniformity of the performance of a specific function."

The test facility should have detailed, written, updated SOPs to ensure the quality and integrity of the data generated, which should be approved by the facility's management.

Each testing facility should have immediately available, current SOPs relevant to its' relevant activities. These SOPs can be supplemented with published textbooks, analytical methods, articles, and manuals as per the requirements.

SOPs should be updated regularly. Any deviation from the SOP should be documented as well as acknowledged by the Study Director and/or principle Investigator as applicable.

SOPs should be available for the following categories of activities:

- Test and reference items
  - Receipt, identification, labeling, handling, sampling, and storage
- Apparatus, materials, and reagents
  - Use, maintenance, cleaning, calibration of apparatus
  - Validation, operation, maintenance, security, control, back-up for computerizes systems
  - Preparation and labeling of materials. Reagents, and solutions
- Recordkeeping, reporting, storage, and retrieval
  - Study codes, data collection, preparation of reports, indexing systems, and data handling
- Test system
  - Procedure for the receipt, transfer, placement, characterization, identification, and care of the systems
  - Test system preparation, observations, examinations, and environmental room conditions
  - Collection, identification, and handling of specimens from necropsy as well as histopathology
  - Handling of test system individuals found moribund or dead during the study
- Quality assurance procedures
  - Planning, scheduling, performing, documenting, and reporting QAP inspections

50.2.8 Performance of the Study

For each study undertaken, a written plan should be there before the initiation. The study plan should be approved by the Study Director and verified for GLP compliance by QAP. It should also be approved by the test facility management, sponsor, and required authorities as per the regulations of the country. Amendments or any deviations in the study plan should be appropriately documented by the Study Director and/or Principal Investigator.

The study plan should contain the following information:

- Identification of the study, test and reference items
- Information about the Sponsor, Test Facility, Study Director, Principal Investigator
- Dates
  - Date of study plan approval by Study Director, Test facility management, Sponsor
  - The proposed experimental starting and completion dates
- Test methods (OECD or other test guidelines)
- Experimental design, materials, and methods of the study, justification, and characterization of the test system
- List of records to be maintained

50.2.9 Reporting of Study Results

The final report should be prepared for each study and signed by Principal Investigators/Scientists, and the Study Director who takes responsibility for the validity of data.

The final report should have the following information:

- Identification of the study, test and reference items
- Information about the Sponsor, Test Facility, Study Director, Principal Investigator, Scientists
- Experimental starting and completion dates
- Statement from QAP about the inspection details and observations
- Description of materials and test methods
- Results: Summary, data presentation, statistical analysis, discussion, and conclusion
- Storage: Study plan, samples of test and reference items, specimens, raw data, and final report

50.2.10 Storage and Retention of Records and Materials

The following documents should be retained at the test facility for the period as mentioned by the concerned authorities. If no retention period was insisted, the final deposition of materials should be properly documented:

- Records of qualification, training, experience, and job description of the personnel

- Study plan, raw data, samples of test and reference items, specimens, and the final report of each study
- Inspection documents of the quality assurance program
- The historical file of all Standard Operating Procedures
- Documents related to maintenance and calibration of apparatus at the test facility

The record stored at the archives should be properly indexed for the orderly storage and retrieval. The individuals who are authorized by the management of the test facility only should have access to the archives. There should be proper documentation of movements of documents in and out of the archives.

50.3 GLP Advantages and Drawbacks

50.3.1 Advantages of GLP

- Studies are better controlled
- High quality and reliable data
- Mutual acceptance of data across authorities/nations
- Increases public confidence over the data generated
- Prevents rework, rejection, the deviation for/of/from the study

50.3.2 Drawbacks of GLP

- Extensive documentation
- More manpower requirement
- Time-consuming process
- Cost of the project rises by 5–20%

50.4 National Good Laboratory Practice (GLP) Compliance Monitoring Authority (NGCMA)

Website: https://dst.gov.in/ngcma.

At present, India is a provisional member of the OECD's working group on GLP. In 2002, the Government of India established the NGCMA under the Department of Science and Technology (DST). The authority is the apex body for overseeing the national GLP program functions as per the OECD principles. Secretary, DST is the chairman of the apex body, and the Secretaries from other concerned Ministries/ Departments are the members. The authority also has constituted the Technical committee with the responsibility of helping the NGCMA in evaluating the competence of test facilities based on the inspections organized through GLP inspectors.

Some of the important responsibilities of NGCMP are the following:

- To implement the National GLP program
- To lay down procedures and coordinate with test facilities associated with the program
- To ensure the functioning of the National GLP program as per current international norms
- To process the applications received for GLP certification from test facilities, and to organize and conduct inspections and audits
- To appoint and train the GLP inspectors, and communicate to the test facilities regarding the observations of audit as well as to initiate corrective action
- To issue GLP compliance certificate to the applying test facilities
- To maintain the records and documents regarding the inspections, audits, as well as compliance status of the test facilities
- To take appropriate actions for the serious deviations observed during inspection/audit
- To create awareness about GLP in the country by conducting educational and training programs

GLP-compliance certification is voluntary, not mandatory. The compliance certification is valid for 3°years. The NGCMA will conduct a surveillance inspection once a year. National GLP program covers the application of GLP principles in non-clinical safety testing of Industrial Chemicals, Pharmaceuticals, Veterinary drugs, Pesticides, Cosmetic products, Food additives, and Feed additives. The test facilities undertaking following non-clinical studies can apply for GLP certification to NGCMA:

- Physical-Chemical testing
- Toxicity studies
- Analytical and clinical chemistry testing
- Mutagenicity studies
- Environmental toxicity studies on aquatic and terrestrial organisms
- Residue studies
- Studies on behavior in water, soil, and air bio-accumulation
- Studies on effects on mesocosms (studying natural environment under controlled conditions) and natural ecosystems

Bibliography

NGCMA. Department of Science & Technology. n.d.. https://dst.gov.in/ngcma (accessed February 24, 2020).

OECD Series on principles of Good Laboratory Practice (GLP) and compliance monitoring – OECD. n.d.. http://www.oecd.org/chemicalsafety/testing/oecdseriesonprinciplesofgoodlaboratorypracticeglpandcompliancemonitoring.htm (accessed February 24, 2020).

Part II

Biochemical Pharmacology

Basic Principles and Applications of Simple Analytical Methods

51

Abialbon Paul

Abstract

Analytical techniques are the methods used in the isolation, identification and quantification of chemical substances. They are also referred to as drug assays and are useful from drug development to clinical use of the drugs. They can be divided into qualitative and quantitative techniques. While sophisticated instrumentation makes advanced quantitative techniques possible, qualitative techniques are still used to detect the presence of drugs in various specimens. The following chapter describes some of the commonly used quantitative and qualitative assays for commonly used drugs. A brief description for sample preparation is also provided as the accuracy is heavily dependent on the proper sample preparation.

Keywords

Drug assay · Qualitative assay · Quantitative assay · Analytical techniques

51.1 Introduction

Analytical techniques refer to the methods used in the isolation, identification, and quantification of chemical substances in various solutions. Analytical techniques in the pharmaceutical analysis are being used in the laboratory investigation of raw drugs, drugs in drug formulations, drug impurities, and the presence of the drug and its metabolites in various biological samples. These analytical methods are sometimes referred to as drug assays.

A. Paul (✉)
Department of Pharmacology & Clinical Skills, Medical University of the Americas, Nevis, West Indies
e-mail: a.paul@mua.edu

M. Lakshmanan et al. (eds.), *Introduction to Basics of Pharmacology and Toxicology*, https://doi.org/10.1007/978-981-19-5343-9_51

Analytical techniques are useful from drug development to the clinical use of the drugs. Analytical techniques can be divided into qualitative and quantitative techniques. While sophisticated instrumentation makes advanced quantitative techniques possible, qualitative techniques are still used to detect the presence of drugs in various specimens.

51.2 Qualitative Drug Tests

Qualitative drug tests are often employed in medical school teaching for students' practice and basic laboratories as they are simple to perform, cheaper, and less time-consuming. Few commonly used qualitative tests are enlisted below (Table 51.1).

51.3 Analytical Techniques

The analytical techniques can be done in various ways and are enlisted below:

- Titrimetric techniques
- Spectroscopic techniques
- Chromatographic techniques
- Electrochemical techniques
- Kinetic techniques
- Electrophoretic techniques
- Flow injection and sequential injection techniques
- Hyphenated techniques

51.3.1 Titrimetric Techniques

Basic principle: Titration (also called volumetric analysis) is a method of quantitative analysis wherein a standard solution of known concentration and volume (titrant/titrator) is combined with the solution containing the analyte (titrant) until a measurable endpoint is reached. The volume of the titrate required for the reaction is called the titration volume and it is used to determine the concentration of the analyte.

Types:

Acid-base titration: The titration is a neutralization reaction between an acid and a base and the endpoint is measured by the color change of the indicator or a pH meter.

Redox titration: The titration is a reaction between an oxidizing agent and a reducing agent. A redox indicator or a potentiometer is used to assess the endpoint of the titration.

Table 51.1 Commonly practiced qualitative drug tests

Drug and solubility	Reagents required	Procedure
Aspirin	Sodium hydroxide solution (20% w/v) Dilute sulphuric acid (5.7 ml of conc.H_2SO_4 in 100 ml of H_20) Ferric chloride test solution (5%w/v) 5 M Acetic acid	Boil a portion of the sample with 10 ml of NaOH solution for 3 min. Cool and add 10 ml of dilute H_2SO_4. Filter the solution and collect the precipitate. Dissolve the precipitate in 1-2 ml of distilled water and add a ferric chloride test solution. The presence of aspirin is confirmed by the development of deep purple color. The deep purple color remains on the addition of 0.1 ml of 5 M acetic acid.
Carbamazepine	Conc. nitric acid (70% w/w HNO_3) Conc. sulphuric acid (95%w/w H_2S0_4) Formaldehyde solution	Add a portion of the sample to test tube containing 2 ml of conc. HNO_3 and heat in a water bath for 3 min. Orange-red color is formed. Add a portion of the sample to a test tube and add 2 ml of conc. H_2SO_4 and 0.1 ml of formaldehyde. On mixing, a yellow color is formed which turns orange on standing.
Diazepam	Dilute H_2SO_4	To a portion of the sample, add 2 ml of dilute sulphuric acid and heat for 3 min. A yellow solution is formed.
Digoxin	Glacial acetic acid Ferric chloride solution (2% w/v) Conc. H_2SO_4 Dichloromethane, Dinitrobenzoic acid	Add 2 ml of glacial acetic acid to a portion of the sample and mix. To the mixture, add 2 ml of ferric chloride and add concentrated H_2SO_4 and keep it standing for a while. A brownish-red color is formed with a bluish-green layer on the underside. Dissolve few mg of substance in dichloromethane then acidify with few drops of the dilute acid solution and add 4 to 5 drops of dinitrobenzoic acid, an intense color forms
Paracetamol	Conc. HCl 0.1 N Potassium dichromate	Dilute the sample with distilled water and add 0.05 ml of ferric chloride, a violet-blue color is produced. To a portion of the sample, add 1 ml of conc. HCl and boil for 3 min. Add distilled water and cool the mixture. Add 0.05 ml of 0.1 N potassium dichromate solution. Violet color is formed which does not become red on standing.
Phenytoin	Dilute HCl Copper sulfate with pyridine solution (dissolve 4 g of $CuSO_4$ in 90 ml of water and add 30 ml of pyridine) Pyridine solution (10%w/v)	Add a few drops of dilute HCl to the sample and a white precipitate is formed. Add 1 ml of pyridine solution and 1 ml of copper sulfate with pyridine to a portion of the sample and allow it to stand for 10 min. A blue precipitate is formed.
Phenobarbitone	Dilute HCl, Conc. sulphuric acid Sodium nitrite powder	Add 100 mg of the sample and 2 ml of concentrated sulphuric acid and a few milligrams of sodium nitrite powder. Golden yellow color is produced.

Gas-phase titration: The titration is between two gases and the reactivity of these gases is used to measure the concentration of the analyte by measuring the formation of the end products.

Complexometric titration: This titration involves the formation of a complex between the titrant and the titrant.

Zeta potential titration: The titration involves measurement of the zeta potential and is often used for colloids.

Advantages:

- It can be performed in a very short duration of time.
- Low cost of instrumentation and reagents
- Less skill required

Disadvantages:

- Lack of selectivity

Drugs tested by titrations:
Stavudine, Nordiazepam, Hydroxyzine, etc.

51.3.2 Spectrophotometry and Beer Lambert's Law

Basic principle: Spectroscopy studies the interaction between matter and electromagnetic radiation. Spectrophotometry is a type of spectroscopic analysis wherein the interaction between the analyte and visible light is used to determine the concentration of the analyte. As visible light passes through the specimen containing the analyte, light is absorbed as it passes through in accordance with Beer Lambert's law.

Beer Lambert's law: The measured absorbance of light passing through a sample is directly proportional to the path length and the concentration of the analyte.

$$A \propto C.L.$$

where A is the absorbance measured, C is the concentration of the analyte in the medium, and L is the path length of the light through the medium.

$$A = .C.L.$$

In the equation above, the epsilon is called as the 'molar absorptivity' or the 'absorption coefficient'. It is the absorbance measured for the unit concentration of the solution and unit path length.

The absorbance can also be expressed in terms of the ratio of the intensities as given below:

$$A = \log_{10}(I_0/I) = .C.L.$$

Here, I_0 is the intensity measured with the sample while 'I' is the intensity measured with the reference.

Limitations of Beer Lambert's law:

There are deviations from the law and they can be categorized as the following:

- Real deviations
- Chemical deviations
- Instrument deviations

Real deviations are due to fundamental deviations in the law itself.

- At higher concentrations and very low concentrations, the linearity relationship is lost due to various molecular interactions.
- Higher concentrations can also alter the refractive index of the solution which can alter the absorbance which is not accounted for in the equation.

Chemical deviations are specific molecular interactions between the solvent and the analyte resulting in changes to the absorbance measured.
Instrument deviations are due to the use of polychromatic radiation, the presence of stray radiation, and due to mismatched cells or cuvettes. This can be studied by multiple references without the presence of the analyte.

51.4 Sample Preparation for Drug Analyses

Sample preparation is the process by which the analyte in a sample is concentrated by removing the interfering substances from the matrix. It is an extremely important step for the accurate and reproducible quantification of the analyte. The choice of the sample preparation technique also determines the type and amount of sample collected, the time taken for the total analysis, and the cost of the procedure. The following are the main objectives of sample preparation:

- Removal of unwanted matrix components
- Concentration of the analyte
- Exchange of the solvent or solution in which the analyte is dissolved if it is incompatible with the testing method
- Removal of certain components that might reduce the accuracy of the procedure
- Removal of substances that could damage the instruments used
- Stabilization of the analyte so that it does not decompose during the procedure.

***Removal of Proteins*:**

- The first method is called salting out. Ammonium sulfate is often the preferred salt used because of its high solubility and high ionic strength. Addition of this salt results in protein precipitation.
- The second method is solvent precipitation. The addition of an organic solvent and/or addition of a buffer will often result in precipitation of proteins in the biological matrix. This is followed by centrifugation using membrane filters and the sediment is discarded.
- A minimum of 1 ml of the sample is required for successful protein removal.
- Other techniques for protein removal include special stationary phase which can directly remove the proteins when chromatography is run on the sample. Further, two columns can be used where the first acts as a pretreatment column, and the second column separates the analyte of interest and helps in measurement.

Lyophilization:

- Lyophilization is used to prepare large amounts of biological samples for storage and subsequent analysis
- The sample is frozen in dry-ice acetone bath or liquid nitrogen and moisture is removed by the process of vacuum sublimation. This is stored in cold temperatures. When needed, it is dissolved in a suitable medium for the assay.

Hydrolysis of Conjugates:

- If the sample contains the glucuronides, acetates, or sulfate esters of the drug, they need to be hydrolyzed so they can be extracted.
- This is done by enzymatic or acid hydrolysis. After hydrolysis, the unconjugated metabolites are less hydrophilic and can be extracted easily.

Homogenization:

- For samples containing insoluble proteins homogenization is required to make a liquid that can be further processed.
- This process is carried out either with 1 N HCl in a homogenizer or sonicator. Methanol can be used for some samples.

Liquid-Liquid Extraction (LPE):

- Liquid-liquid extraction is the traditional method for the extraction of drugs from biological samples. It uses the principle of partition of substances between an organic phase and an aqueous phase. The pH of the aqueous phase is adjusted in such a way that the drug molecules get partitioned in the organic phase in a neutral form.
- A combination of non-polar organic solvent and an aqueous solvent in the sample is added and prepared. The pH of the aqueous phase is adjusted depending on the nature of the drug being isolated. For acidic drugs, the pH is kept acidic while for basic drugs, the pH of the aqueous phase is alkalized.
- Useful when a sample containing multiple drugs require pretreatment to measure all the drugs.

Disadvantages:

- Large amounts of organic solvents are required
- Time-consuming
- Certain interferences are not effectively removed
- Chance of emulsion formation

Solid-Phase Extraction (SPE):

- Advantages of SPE over LPE
  1. Less organic solvent requirement
  2. Less time consuming
  3. Less complex processing steps leading to less chances of interference.
- Solid-phase extraction has replaced LLE to a large extent because of its superior efficacy. The mechanism involved separation by intermolecular interactions between the analyte and the functional groups present in the solid phase.
- Examples of the solid phase used are XAD-2 resin, silica, alumina, charcoal, and aluminum silicate.

Bibliography

Coskun O. Separation techniques: chromatography. North Clin Istanb. 2016;3(2):156–60.

India, Indian Pharmacopoeia Commission, editor. Indian pharmacopeia 2018. 8th ed. Ghaziabad: Indian Pharmacopoeia Commission; 2018.

Siddiqui MR, AlOthman ZA, Rahman N. Analytical techniques in pharmaceutical analysis: a review. Arab J Chem. 2017 Feb;10:S1409–21.

Swinehart DF. The Beer-Lambert law. J Chem Educ. 1962 Jul;39(7):333.

Principles of Quantitative Estimation of Drugs, Endogenous Compounds, and Poisons—1

52

Aarthi Manoharan

Abstract

Chromatography, mass spectrometry (MS), and enzyme-linked immunosorbent assay (ELISA) are three widely used quantitative analytical techniques. Chromatography analyzes a substance's components by distributing them between a stationary (either a solid or liquid) and a mobile phase (either liquid or gas). Retention time t_R, the time taken for the solute to travel from the sample injector and elute from the chromatographic column, is used to determine the analytes. MS directly detects and determines the amount of a given analyte by ionizing a given substance in the gas phase. The molecular ions undergo fragmentation and are subsequently separated based on their mass-to-charge (*m/z*) ratios in an electromagnetic field to measure the relative abundance of each ion species present. More recently, chromatography and mass spectrometry are combined to accrue the benefits of both techniques. ELISA is a sensitive and rapid technique to quantify an analyte using an enzyme-labeled antibody, and it depends on the ability of an antibody to bind specifically to an antigen. ELISA uses a spectrophotometer to quantify the end product

Keywords

ELISA · Chromatography · Mass spectrometry · Retention time

A. Manoharan (✉)
Multi-Disciplinary Center for Biomedical Research, Vinayaka Missions Research Foundation, Aarupadai Veedu Medical College and Hospital, Puducherry, India
e-mail: aarthi.manoharan@avmc.edu.in

M. Lakshmanan et al. (eds.), *Introduction to Basics of Pharmacology and Toxicology*, https://doi.org/10.1007/978-981-19-5343-9_52

52.1 Chromatography

The definition of chromatography, according to the International Union of Pure and Applied Chemistry (IUPAC), is "A physical method of separation in which the components to be separated are distributed between two phases, one of which is stationary (stationary phase) while the other (mobile phase) moves in a definite direction."

Solutes (different components in the sample) interact with phases differently and travel at different rates through the stationary phase, referred to as differential migration. Differential migration of the components in a sample forms the basis of separation in chromatography.

Chromatography can be preparative (purification) or analytical (identification and quantification).

52.2 Types of Chromatography

- Chromatography can be broadly classified as either planar and column chromatography based on the structure of the stationary phase or as gas and liquid chromatography based on the mobile phase's physical state.
- Paper chromatography and thin-layer chromatography are the two type's of planar chromatography. The two methods use a planar stationary phase.
- In column chromatography, the stationary phase is a solid or liquid either packed into a column or coated onto its wall.
- Gas chromatography uses gas for the mobile phase, while liquid chromatographic techniques use a liquid mobile phase. Paper and thin layer chromatographic techniques are also examples of liquid chromatography.

52.3 Paper Chromatography

Paper chromatography is the earliest and the simplest of chromatographic techniques. The Russian botanist M.S. Tswett first described chromatography in 1903 when he studied colored plant pigments using paper chromatography. In fact, the word chromatography is derived from the Greek words *chroma*-color and *graphein*-write.

52.3.1 How It Works

- The sample is placed as a small dot on a paper strip placed in a container with a solvent's shallow layer.
- The solvent rises through the paper by capillary action.
- Different components in the sample migrate at different rates.

- The solvent is allowed to advance (development) till it reaches a predetermined distance.

52.4 Thin-Layer Chromatography (TLC)

- Thin-layer chromatography (TLC) uses a thin layer of an adsorbent such as silica gel on a glass plate's inert flat surface.
- Commercially available plates come pre-coated for higher performance with a thickness ranging from 0.1 to 2 mm.
- A binder such as starch or salt of polyacrylic acid is added to provide greater mechanical strength, durability, and abrasion resistance.
- It is a rapid and straightforward method to check the purity of the samples.
- It is an effective method for low-cost analysis of samples. Traditionally, TLC finds wide application in forensic toxicology to identify drugs of abuse and toxic substances in biological fluids.

52.4.1 How It Works

- As in paper chromatography, a small spot or band of the sample is applied at the plate's bottom edge and placed in a chamber with the solvent.
- The solvent advances through the dried adsorbent layer by capillary action. The decrease in free energy as the solvent migrates allows it to pass through the porous structure of the layer.
- The different analyte components separate out in the direction in which the mobile phase travels up to a predetermined distance.
- After the development is complete, the mobile phase evaporates, and the separation becomes immobilized on the thin layer in the plate. This is followed by optical detection.

52.5 Gas Chromatography (GC)

Gas chromatography was introduced in 1952 by James and Martin.

52.5.1 Working Principle

- GC volatilizes the analyte using a heated sample injector and then uses a *gaseous mobile phase* to carry the analyte from the injector through the column (stationary phase) and into the detector.

52.5.2 Mobile Phase

- The mobile phase is an inert carrier gas, usually helium. Sometimes hydrogen and nitrogen may also be used.

52.5.3 Stationary Phase

- The column is coated with a stationary phase. The stationary phase can be liquid (gas-liquid chromatography, GLC) containing liquid silicone-based material or solid (gas-solid chromatography, GSC).
- The column contains chemically inert materials. The column is either packed or capillary.
- Packed columns: A non-volatile material is coated onto an inert solid material which is packed into the column.
- Capillary columns: In the stationary phase, a non-volatile material is coated as a thin layer on the capillary wall.

52.5.4 Samples for GC Analysis

- Chromatography separation is based on the partition equilibrium of the analyte between the two phases. Only samples that are thermally stable and withstand temperatures above 400 °C (upper limit) required for vaporization are suitable for GC analysis.
- Volatile compounds such as amino acids, amines, amides, saccharides, steroids need to be derivatized before GC analysis.

52.5.5 Instrumentation

- GC requires a regulated flow of the carrier gas, an inlet system that vaporizes the sample mixed with the gas, a thermostat oven that controls the vaporization temperature.
- The column (stationary phase) is placed in the oven. Separation of analyte components takes place in the heated column.
- Detectors: There are different kinds of detectors used for GC analysis.
  - Flame ionization detector: The analyte components that elute out of the column pass through a flame. The ions and electrons thus formed cause a current to flow, which is used as the signal for the detector.
  - Flame photometry detector: The Detector also uses flame similar to the flame ionization detector. However, it uses optical filters to select wavelengths specific to sulfur and phosphorous. A photomultiplier tube is employed to detect these compounds selectively.

– Electron capture detector: The carrier gas is ionized as it migrates towards an anode in the detector to produce a steady current. The analyte components capture the electrons thus produced by this ionization. The changes in the steady current as read as the signals.

52.5.6 Gas Chromatography: Mass Spectrometry (GC-MS)

- Gas chromatography was combined with the other powerful analytical technique, mass spectrometry (MS), in the 1950s. See Sect. 52.7 for MS
- While GC separates the components in the sample mixture, MS provides the structural information and identification of those components. The GC detector is replaced by MS.
- The eluent from GC must be carried over to the MS for which an interface is required. The interface does not allow the eluent to get condensed or decomposed during transit.
- The interface must therefore be heated up to a boiling point that surpasses the highest boiling point of any of the analyte components.
- An enrichment device called a jet separator is commonly used in GC-MS. The jet separator contains a vacuum region sandwiched between two capillary columns. The atoms of the carrier gas are lighter than the analyte molecules and are diverted from the linear path to the MS. Only the analyte molecules are channelized into the MS.

52.6 Liquid Chromatography (LC)

Liquid chromatography (LC) includes almost all chromatographic techniques that use a liquid mobile phase, from paper chromatography and thin-layer chromatography to more advanced and modern techniques such as high-performance liquid chromatography, affinity, adsorption, size exclusion, and ion-exchange chromatography. This chapter focuses on high-performance liquid chromatography.

52.6.1 Modern Liquid Chromatography

- The classical LC methods required the columns to be packed anew for every separation, skillful sample application, and manual analysis of individual fractions of the analyte. They were tedious and time-consuming.
- The modern LC offers greater ease of use with reusable columns, higher performance, greater accuracy, and reproducibility.
- Liquid chromatography is well suited for separating most macromolecules that are thermally unstable, which is not possible with gas chromatography.

- Proteins, polysaccharides, nucleic acids, drugs and metabolites, and naturally labile compounds can be effectively separated by LC.
- Based on the polarity of the two phases, liquid chromatography can be normal or reversed-phase. If the mobile phase is more polar than the stationary phase, it is called reversed-phase HPLC. Reversed-phase HPLC is more popular and commonly used for faster elution of more polar analytes.

52.6.2 Working Principle

- In high-pressure liquid chromatography, also called high-performance liquid chromatography (HPLC), the mobile phase solvents are forced through the column under high pressure.
- Analytes are solubilized in the mobile phase.
- Differential migration and retention characteristics of an analyte's components are determined by their interaction with mobile and stationary phases.
- The composition of the mobile phase is an essential factor that affects the migration of the components.
- Differential migration is mirrored by the differences in the retention time. Retention time t_R is the time taken for the solute to travel from the sample injector and elute from the chromatographic column.

52.6.3 Instrumentation

52.6.3.1 Solvent Delivery System

- The solvent delivery system comprises solvent reservoirs, pumps, and valves.
- The system delivers the mobile phase constantly to the column's head under high pressure (up to 5000 psi) and precisely sets flow rates.
- The high pressure overcomes the resistance of the particles used in the column packing. The flow rates range between 10 μl/min and 2 ml/min.
- Degassing is a critical step before the solvents are ready to be pumped into the column.
- Degassing is the application of vacuum combined with continuous stirring and warming to remove any bubbles that are formed when solvents are mixed.
- The presence of dissolved oxygen decreases the sensitivity during detection. It also prevents undesired reactions with the different phases.

A block diagram of liquid chromatography is shown in Fig. 52.1.

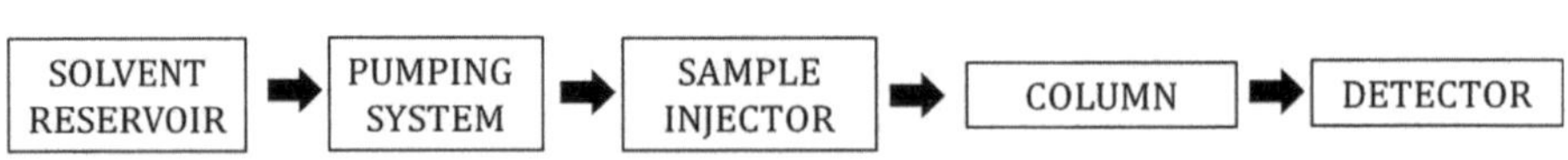

Fig. 52.1 Block diagram of liquid chromatography

LC can be operated with two kinds of elution methods, isocratic and gradient elusions. In isocratic elution, the composition of the mobile phase remains unchanged throughout the elution. In gradient elution, the composition of the mobile phase changes during the process. Gradient elution is commonly used. It employs an admixture of two or more solvents which uses an initial weak solvent followed by stronger solvents.

52.6.3.2 Sample Injectors

Samples are introduced using either syringe injectors or valves.

52.6.3.3 Columns

The columns are made of stainless steel tubing or heavy wall glass carrying materials that withstand high pressures.

52.6.3.4 Detectors

- Several detectors are used with LC, such as UV-Visible absorption, fluorescent and chemiluminescent detectors.
- UV detectors are most commonly used. A deuterium source lamp is used for UV range measurements, and a tungsten source lamp for visible range.
- Samples in the flow cell of the detector absorb wavelengths in the UV or visible region.
- The amount of sample in the flow cell in the detector is proportional to the light transmitted through the flow cell, i.e., A*bsorbance* is directly proportional to the sample concentration (Beer's law).

52.6.4 Chromatogram Output

The height of the peak in a chromatogram is defined by the component's retention time in the analyte. Quantification is based on comparing the peak height/area of a component of known concentration with that of the component in the analyte (see Fig. 52.2).

52.6.5 LC-MS

- Chromatographic techniques are limited to the separation of components in a mixture and their quantification. They cannot discriminate between different components having the same retention characteristics. This inherent limitation requires spectroscopy to augment the chromatographic outputs.
- Mass spectrometry provides absolute identification, structural information, and molecular weight of the analyte,
- Unlike GC, coupling LC with MS was faced with many challenges. The samples amenable to analysis with GC were thermally stable and volatile, which was

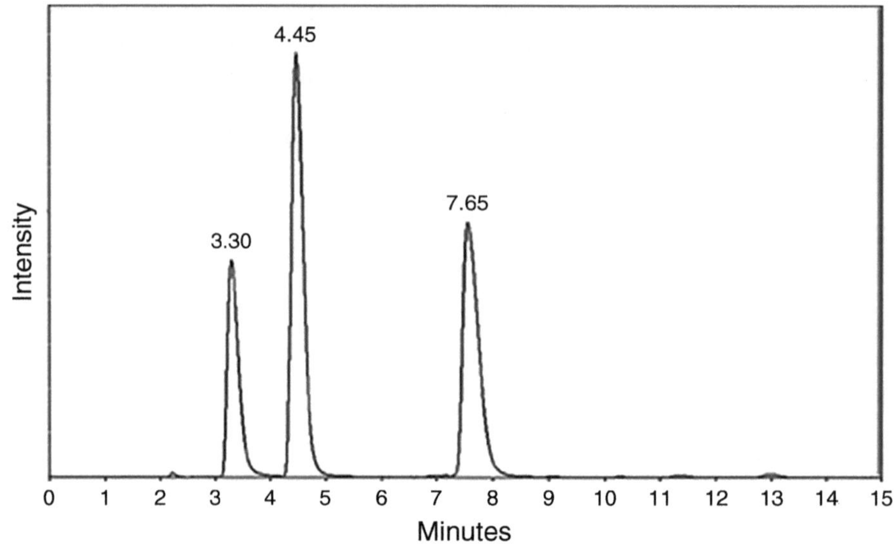

Fig. 52.2 Example of an HPLC chromatogram of three different substances

highly compatible with the requirements of MS that uses electron or chemical ionization to analyze compounds.

- LC, on the other hand, analyses compounds that are thermally unstable, non-volatile, and labile. In HPLC, the mobile phase solvents are pumped at a typical flow rate of 1 ml/min, while MS functions at a 10-6 torr pressure. Thus the eluate from LC requires an interface before getting introduced into MS.
- The interface is the MS inlet that links LC with MS. It removes the mobile phase enriching the sample that enters the MS with only the analyte components.
- The MS component coupled to LC only uses 'soft ionization techniques' that produce molecular ions with limited fragmentation, making structural elucidation difficult. Thus, LC is often combined with tandem MS (see Sect. 52.8.4).
- LC-MS/MS is a superior analytical technique compared to HPLC or immunoassays for low-molecular-weight analytes and offers better throughput than GC-MS.

52.7 Mass Spectrometry

Mass spectrometry (MS) is an old yet powerful analytical technique to detect and determine the amount of a given analyte directly. The basic principle was first described by Nobel laureate, Sir Joseph John. Mass spectrometers ionize a given substance in the gas phase. The molecular ions undergo fragmentation and are subsequently separated based on their mass-to-charge (*m/z*) ratios in an electromagnetic field to measure the relative abundance of each ion species present. The output is called the mass spectra. MS can also be used to determine the composition and some structural aspects of a molecule.

52.8 Working Principle and Instrumentation

52.8.1 Sample Introduction

- Samples are required to be introduced into the high vacuum of the MS.
- Samples are either directly placed on a probe and inserted into the ionization chamber through a vacuum interlock.
- The vacuum interlock is critical to maintain the high vacuum in the MS
- Samples can also be directly introduced into the ionization source using a capillary column. This method can efficiently inject minute quantities of the sample without affecting the vacuum.
- An ion path enables the transition of the electrons from the near-atmospheric pressure of the source into the high vacuum of the mass analyzer.

52.8.2 Ionization

- An electrically neutral atom gains or loses electrons; it is said to be ionized.
- The minimum amount of energy required to produce ions is called ionization potential
- MS ionizes the analyte in the gas phase either by laser desorption or direct heating.
- The different ionization sources include Matrix-Assisted Laser Desorption / Ionization (MALDI) (see Fig. 52.3), Electrospray Ionization (ESI) (see Fig. 52.4), Atmospheric Pressure Chemical Ionization (APCI) and Fast Atom/

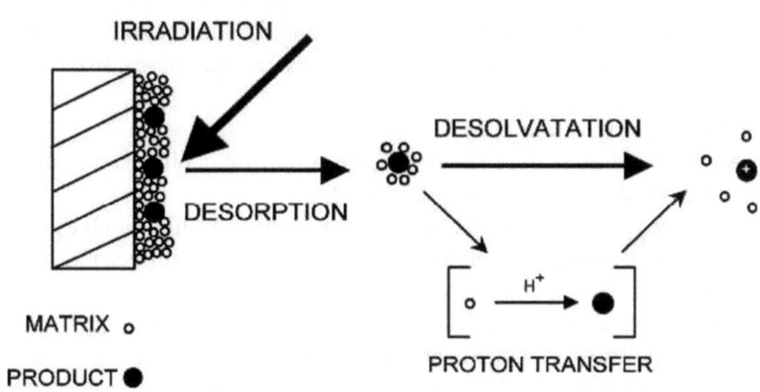

Fig. 52.3 Diagram of the principle of MALDI (Image courtesy Hoffmann E de, Stroobant V. Mass spectrometry: principles and applications. 3rd ed. Chichester: John Wiley & Sons; 2007)

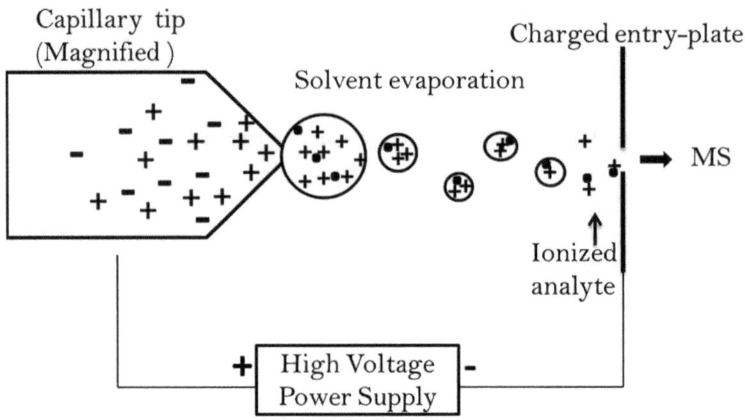

Fig. 52.4 Diagram of the principle of ESI

ion Bombardment (FAB), Electron Ionization (EI) and Chemical Ionization (CI), each method working differently.

- The most popular ionizations are briefly described below
  - Matrix-Assisted Laser Desorption/Ionization (MALDI):

    MALDI produced intact gas-phase ions from a wide range of large, non-volatile, and thermo-labile compounds from proteins, oligonucleotides to synthetic polymers and large inorganic substances. It is a highly sensitive and more commonly used method.

    The analyte is doped into and embedded throughout the matrix. A high intense laser irradiates the matrix ablating the matrix crystals.

    This allows the matrix to expand into the gas phase, tagging along with the analyte into the expanding matrix plume.

    Ionization usually occurs by gas-phase photoionization or desorption of preformed ions.

    As the matrix molecules outnumber the analyte molecules, MALDI minimizes the damage brought about by the incident energy on the analyte molecules and efficiently transfers the energy from the laser to the analyte.
  - Electrospray Ionization (ESI)

    As the solvent-analyte flows through a narrow capillary, a strong electric field is applied at the capillary end.

    This nebulizes the solvent-analyte molecules into highly charged droplets that pass through a curtain of heated inert gas (nitrogen) to remove any solvent.

    The analyte then disintegrates in a Coulomb explosion.
  - Atmospheric Pressure Chemical Ionization (APCI)

    Similar to ESI, APCI ionizes samples at atmospheric pressure before transferring the ions into a mass analyzer.

Table 52.1 Different ionization methods for different kinds of analytes

Ionization method	Description	Nature of sample
Protonation	Addition of a proton to produce a net positive charge of +1	Peptides
Deprotonation	Removal of a proton to produce a net negative charge of −1	Phenols, carboxylic acids, sulfonic acids
Cationization	Non-covalent addition of a cationic adduct to a neutral molecule to create a charged complex	Carbohydrates
Electron ejection	Produces a net positive charge of +1	Non-polar compounds
Electron capture/ absorption	Produces a net negative charge of −1	Molecules with a high electron affinity

Here, a heated nebulizer gas vaporizes the analyte in the solvent.

The polar components of the solvents are ionized by the high current discharge of the corona needle.

The ionized solvent molecules transfer their charges to the analyte molecules.

- The molecules are ionized by various methods, as described in Table 52.1

Most of the methods can be used to generate both positively and negatively charged ions. MALDI and ESI are the commonly used ionization sources to produce a wide mass range and increased sensitivity.

52.8.3 Mass Analyzers

- Mass analyzers sort the ions based on their mass-to-charge (*m/z*) ratios in an electromagnetic field to obtain a mass spectrum.
- Ionization of the analyte affects the movement of the ions in the electric field, and the ions are sorted based on their characteristic behavior in the electromagnetic field
- The ions pass through a vacuum in the mass analyzer. The vacuum ensures that ions do not collide during the sorting process.
- The collision of ions may result in an altered trajectory and failure to reach the detector. It can decrease the resolution and sensitivity of the system.
- The different mass analyzers include quadrupoles, quadrupole ion trap (QIT), and time-of-flight (TOF).
- Quadrupoles employ oscillating electric fields to selectively destabilize or stabilize the paths of the ions passing through a radio frequency quadrupole field. QIT uses four opposing electric fields where ions are trapped and sequentially ejected.

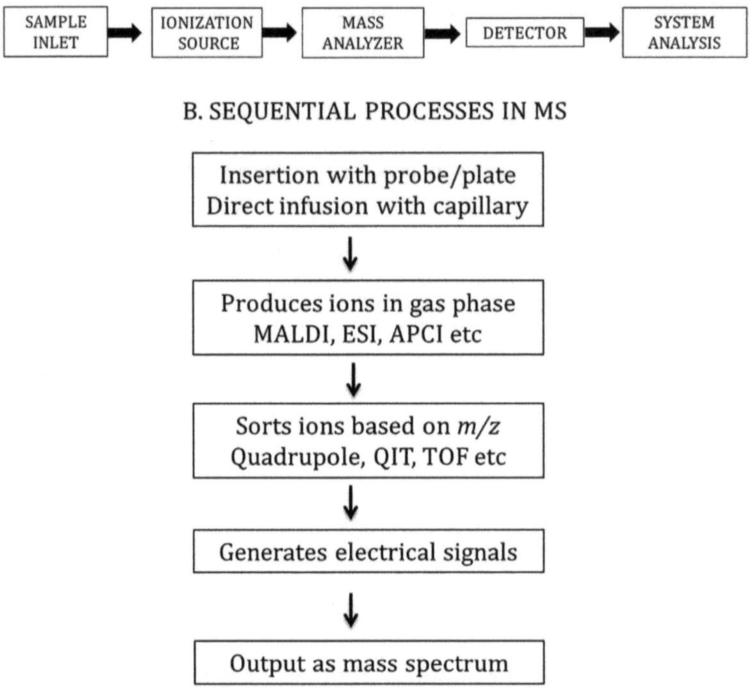

Fig. 52.5 Basic instrumentation of mass spectrometry. A. Schematic representation of MS components of MS. B. Stepwise events in MS analyses

- TOF uses an electric field of uniform potential to accelerate the movement of the ions. TOF thus measures the time taken for an ion to travel from the ionization source to the detector. Figure 52.5 shows the basic instrumentation and workflow of mass spectrometry

52.8.4 Tandem MS

- Mass spectrometers use a single ion source and two mass analyzers in tandem separated by a reaction region (MS/MS).
- Tandem MS can fragment ions from the parent ions and measure the mass of these ions to decipher the structure of the parent ions of a molecule.
- In tandem MS, selected ions are allowed to collide with inert gas molecules such as argon or helium.
- The triple quadrupole is a tandem MS that consists of a linear combination of three quadrupoles. While the first and third quadrupoles serve to scan the ions, the

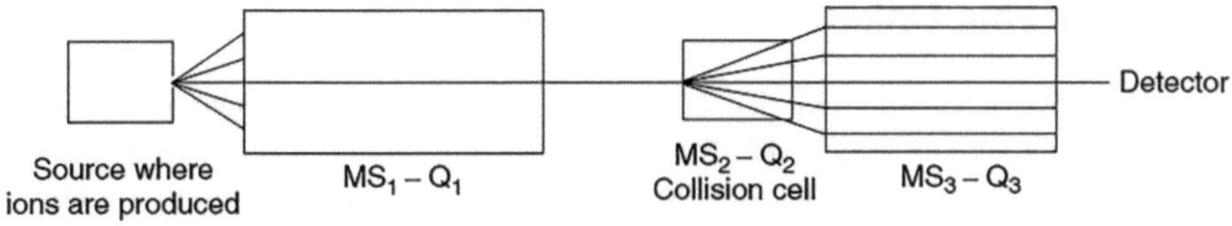

Fig. 52.6 Schematic representation of triple quadrupole mass spectrometer. MS_1 separates ions based on *m/z*, collision cell fragments the selected ions, and MS_2 analyzes the fragments (Image courtesy: Liquid Chromatography-Mass Spectrometry: An Introduction. Robert E. Ardrey, 2003)

second quadrupole functions as the collision cell (fragmentation of selected ions). A schematic representation of triple quadrupole is shown in Fig. 52.6.

- Hybrid quadrupole time-of-flight (QTOF) is another tandem MS that offers a full product ion scan.
- Tandem MS is thus used in the analysis of the sequence of peptides and the structural characterization of oligomers, lipids, and carbohydrates.

52.8.5 Detector

- When the ions reach the detector, the detector converts the ion flux proportionally into an electrical signal. The ions are detected in proportion to their abundance to produce a mass spectrum.
- A data processing system linked with the detector converts the electrical signal's magnitude into the mass spectrum.
- The mass spectrum is a graph of relative abundance (number of ions) plotted against *m/z*. The most intense or the tallest peak is referred to as the BASE PEAK and is arbitrarily assigned a relative abundance value of 100%.
- The number of ions/abundance at the other observed peaks are given as their proportional values (percentages of the base peak).
- The detector Faraday cup measures the direct charge as the ion hits a surface. In other kinds of detectors, such as electron multipliers and electro-optical ion detectors, the ions collide with the surface, and the kinetic energy transfer generates secondary electrons, which are further amplified to produce an electric current.
- Only a small number of ions enter the detector at a given point of time. It is thus necessary to significantly amplify the signal to make it more readable. Except for the faraday cup, other detectors amplify the signals by cascade effect.

52.9 Enzyme-Linked Immunosorbent Assays

Immunoassays detect and quantify an analyte using antibodies as reagents. Enzyme-linked immunosorbent assays (ELISA) utilize enzymes attached to the antibodies to quantify analytes following the development of color on the addition of a suitable substrate. The technique exploits the extreme specificity with which an antibody binds an antigen to form a stable complex. The site on the antigen to which an antibody binds specifically is called an epitope. The availability of the epitope and the antigen-binding selectivity determine the formation of an exquisite and stable antigen-antibody complex. There are numerous applications of ELISA that include identification of disease biomarkers in diagnostic and research laboratories and quality control in various industries.

52.9.1 Working Principle

- ELISA is traditionally performed in a 96- or 364- well polystyrene microtitre plate which is the solid-phase.
- Antigen or antibody is coated onto the surface of the solid phase. The enzyme-linked antibody is added to the solid-phase.
- Following incubation and washing the unbound (free) reagent, a substrate is added to either produce a color change or a luminescent signal in measurable units that correspond directly to the amount of analyte present in the sample is quantified using a spectrophotometer.

52.9.2 Types of ELISA

ELISA can be performed in four different ways: direct, indirect, sandwich, and competitive ELISA

52.9.2.1 Direct ELISA

- Antigen added to the solid phase gets immobilized onto the surface by passive adsorption on incubation.
- Any unbound antigen is washed away. The solid phase is thus 'coated' with the capture antigen immobilized on its surface.
- Enzyme-labeled (conjugate) antibody specific for the coated antigen is added and allowed to incubate.
- Any unbound conjugate is washed away, following which a suitable substrate of the enzyme is added. Figure 52.7 shows the schematic diagram of the principle of direct ELISA.

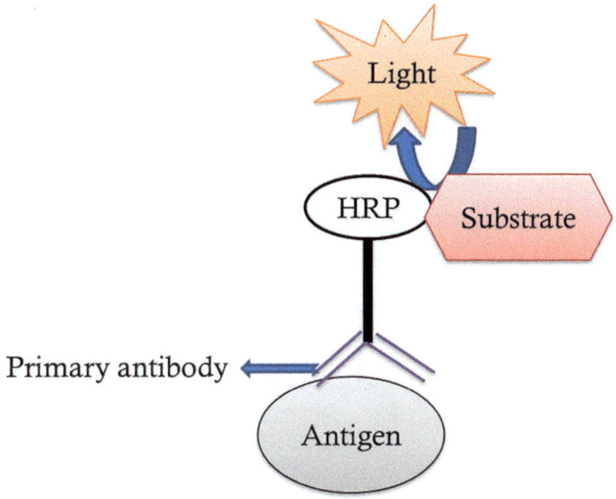

Fig. 52.7 Schematic diagram of the principle of direct ELISA. The light signal is generated from antigen immobilized on the plate

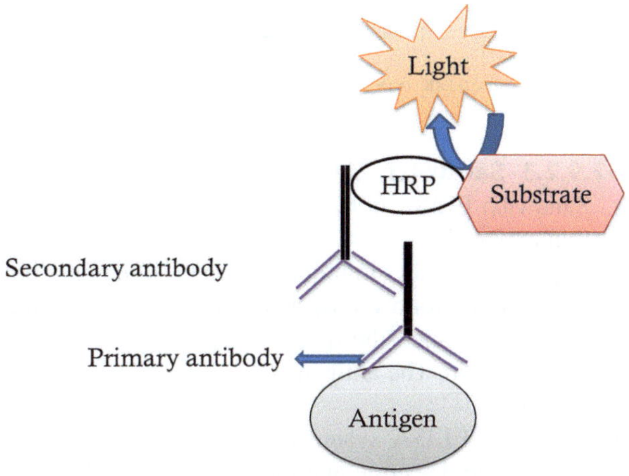

Fig. 52.8 Schematic diagram of the principle of indirect ELISA. The light signal is generated from antigen immobilized on the plate

52.9.2.2 Indirect ELISA

- As in direct ELISA, the solid phase is coated with the capture antigen.
- The detecting antibodies are, however, not conjugated with enzymes. These are called primary antibodies.
- The primary antibodies are detected by the addition of enzyme-conjugated secondary anti-antibodies.
- These anti-antibodies are raised against the immunoglobulins of the species in which primary antibodies are produced and are called anti-species conjugates.
- Indirect ELISA is more sensitive than direct ELISA as the secondary antibodies amplify the primary antibodies' signal. Figure 52.8 shows the schematic diagram of the principle of indirect ELISA.

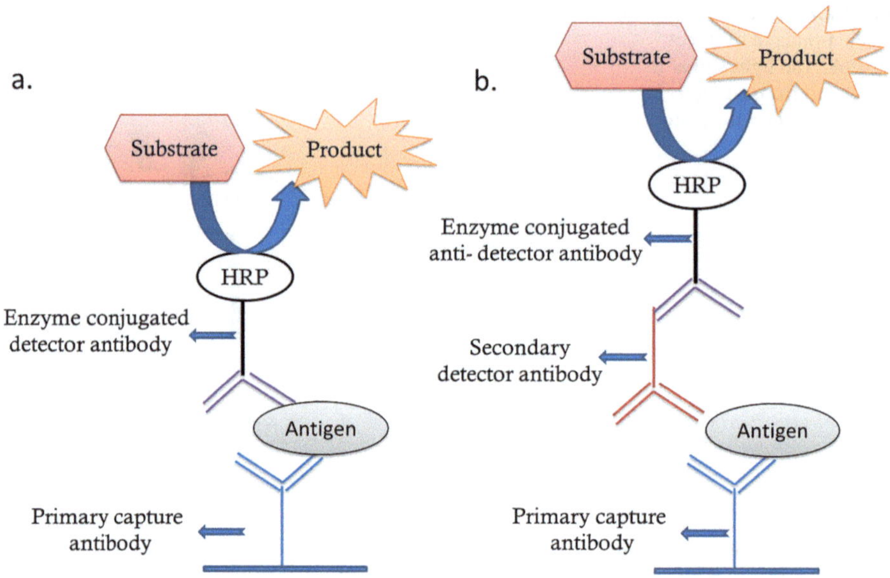

Fig. 52.9 Schematic representation of direct and indirect sandwich ELISA

52.9.2.3 Sandwich ELISA

- Sandwich ELISA utilizes two types of antibodies: Capture antibodies and detection antibodies.
- Capture antibodies immobilized on to solid phase bind the antigens in the solution (test sample) followed by detection with a different enzyme-conjugated antibody.
- The latter, called the detection antibody, recognizes a different epitope on the same antigen. The antigen is thus sandwiched between the capture and detection antibodies.
- Sandwich ELISA systems can be either direct or indirect. In direct sandwich ELISA, the detection antibody is directly linked with an enzyme conjugate.
- In indirect sandwich ELISA, a third antibody which is an anti-species enzyme conjugate, binds the detection antibody.
- Sandwich ELISA is the most popular method as it offers the highest sensitivity and greater confidence to detect even low levels of analyte in a complex sample. Figure 52.9 shows the schematic representation of direct and indirect sandwich ELISA.

52.9.2.4 Competitive/Inhibition ELISA

- In competitive ELISA, two different antigens compete for binding the antibody.
- The test sample is pre-incubated with a primary unlabeled antibody to allow antigen-antibody interaction.
- The sample antigen-antibody complexes and unbound antibodies are added to the solid phase coated with the reference antigen.

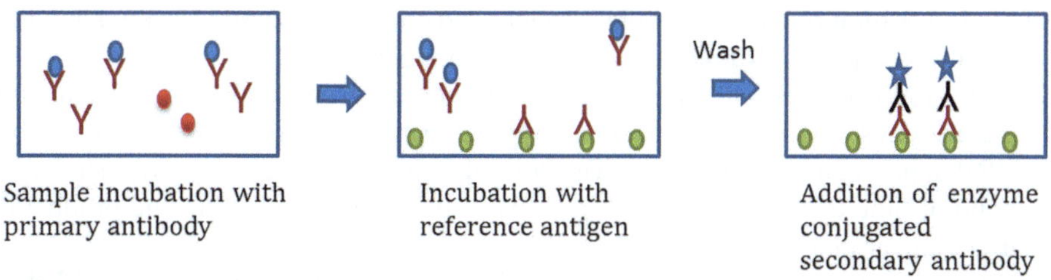

Fig. 52.10 Schematic diagram of competitive ELISA

- Only unbound antibodies will be available for binding with the reference antigen, hence the competition. Incubation is followed by washing.
- Enzyme-conjugated secondary antibodies specific to the primary antibodies are added. On addition of substrate, if more primary antibodies were available for binding the reference antigen, a strong signal would be generated. On the other hand, the more the antigen present in the sample, the lesser the number of free primary antibodies, and hence only a weak signal would be produced. Figure 52.10 shows a schematic diagram of competitive ELISA

52.9.3 Enzymes Commonly Used in ELISA

- Antibodies are covalently linked to enzymes such as horseradish peroxidase (HRP), alkaline phosphatase, and β-galactosidase.
- HRP is a glycoprotein whose polysaccharide chains are used for cross-linking the enzyme to an antibody.
- The relatively smaller size of HRP allows easy access to epitopes on the antigen molecules. HRP is also preferred for its lower detection limit.
- On addition of hydrogen peroxide, HRP catalyzes the reaction with an electron-donating substrate to produce color or luminescent product upon oxidation.
- Alternatively, streptavidin-HRP bioconjugates can be used for enhanced chemiluminescence, providing a highly sensitive reporter to detect the analyte. This method employs a primary antibody coupled to biotin and an enzyme-conjugated streptavidin secondary antibody.

52.9.4 Different Steps Performed in ELISA

STEP 1: Adsorption of either an antigen/antibody to the solid phase microtitre plate.

STEP 2: Addition of test sample and reagents for antigen-antibody interaction followed by incubation.

STEP 3: Separation of the bound and unbound free reagents by washing. This step involves flooding and emptying of the wells with a buffer, usually phosphate-buffered saline (PBS) [0.1 M, pH 7.4].

STEP 4: Addition of enzyme-labeled reagent and enzyme detection system allowing color development or producing luminescent signals.

STEP 5: Addition of STOP reagents. Reagents such as strong acids or bases denature the enzymes, and enzyme-specific stopping agents such as sodium azide inhibit HRP to halt color development. The color development is usually terminated at a point of time when the relationship between the enzyme-substrate and product is in the linear phase.

STEP 6: Measurement of the colored product using a spectrophotometer (ELISA reader) by transmitting light of a specific wavelength through the product.

Bibliography

Ardrey R. Liquid chromatography:mass spectrometry: an introduction. Chichester: John Wiley & Sons;2003.

Crowther JR. The ELISA guidebook, methods molecular biology. 2nd ed. Austria: Humana Press Springer; 2009.

Douglas DJ, Frank AJ, Mao D. Linear ion traps in mass spectrometry. Mass Spectrom Rev. 2005;24 (1):1–29.

Grebe SK, Singh RJ. LC-MS/MS in the clinical laboratory–where to from here? Clin Biochem Rev. 2011;32(1):5.

Hnasko R. ELISA: methods and protocols, methods in molecular biology. New York: Humana Press Springer; 2015.

Makarov A, Denisov E, Kholomeev A, Balschun W, Lange O, Strupat K, Horning S. Performance evaluation of a hybrid linear ion trap/orbitrap mass spectrometer. Anal Chem. 2006;78(7): 2113–20.

Pitt JJ. Principles and applications of liquid chromatography-mass spectrometry in clinical biochemistry. Clin Biochem Rev. 2009;30(1):19–34.

Hoffmann E de, Stroobant V. Mass spectrometry: principles and applications. 3rd ed. Chichester: John Wiley & Sons; 2007.

Synder LR, Kirkland JJ. Introduction to modern liquid chromatography. 2nd ed. New York: John Wiley & Sons; 1979.

Watson JT, Sparkman OD. Introduction of mass spectrometry: instrumentation, applications, and strategies for data interpretation. 4th ed. Chichester: John Wiley & Sons; 2007.

Wild DG. The immunoassay handbook, theory, and applications of ligand binding, ELISA and related techniques. 4th ed. London: Elsevier; 2013.

Principles of Quantitative Estimation of Drugs, Endogenous Compounds, and Poisons—2

53

Charukesi Rajulu, Arun Kumar Tharkeshwar, Sivaranjani Arumugam, and R. Padmapriya

Abstract

Several analytical instruments and molecular techniques are used in pharmacology to study the mechanism of action, therapeutic applications, pharmacokinetics, and adverse effects of drugs. They are used to analyze biological samples from humans, animals, plants, and other species. The availability of different instruments with unique features, accuracy, and precision is indispensable in the pharmaceutical industry to advance research, development and ensure quality assurance. Several techniques are available for determining the concentration or structure of the drug and biomolecules of interest, including colorimetry, spectrophotometry, photometry, flame photometry, fluorimetry, radioimmunoassay (RIA), and nuclear magnetic resonance (NMR). Each instrument has its specific working mechanism, principle, and applications; for example, a colorimeter is used to evaluate the concentration of colored substances in a solution based on the intensity of color. A spectrophotometer is an optical instrument that determines the concentration of a substance based on the absorption of light falling in the electromagnetic spectrum covering UV, visible and infrared regions. A photometer determines the concentration of substances by measuring the intensity of the visible region of the electromagnetic spectrum. In addition to analyzing the

C. Rajulu
Department of Biotechnology & Bioinformatics, JSS Academy of Higher Education & Research, Mysuru, Karnataka, India

A. K. Tharkeshwar
Laboratory of Neurobiology, VIB-Center for Brain & Disease Research, Leuven, Belgium

S. Arumugam
Department of Micro-Nano Mechanical Science and Engineering, Graduate School of Engineering, Nagoya University, Nagoya, Japan

R. Padmapriya (✉)
Department of Pharmacology, JIPMER, Puducherry, India

M. Lakshmanan et al. (eds.), *Introduction to Basics of Pharmacology and Toxicology*, https://doi.org/10.1007/978-981-19-5343-9_53

analyte's concentration, a photometer also measures photometric quantities. A flame photometer determines the concentration of ions or inorganic chemicals by measuring the intensity of light emitted by the substance when introduced into the flame. A Fluorimeter measures the amount of the substance based on the fluorescence emitted at a specific wavelength. RIA, an *in vitro* molecular assay, is used to determine the concentration of an antigen in a biological sample using an antigen-specific antibody. NMR, an advanced technique, is used to find the functional groups, chemical structure, and purity of an unknown compound by assessing the interaction of nuclear spins of the compound in a magnetic field.

Keywords

Colorimetry · Spectrophotometry · Photometry · Flame photometry · Fluorimetry · Radioimmunoassay (RIA) · Nuclear magnetic resonance (NMR)

53.1 Colorimetry

53.1.1 Introduction

Colorimetry is an analytical technique used to measure the concentration of colored substances in a solution. Colorimetry means "measurement of color." It is based on the property of visible light absorption by colored compounds present in a solution (Fig. 53.1). The concentration of a colorless solution can also be determined, in which case, a locating agent (a chemical that helps in detecting another substance by making the latter visible) is used that reacts with the compound present in the solution and produces the colored product. For example, ninhydrin reagent reacts with colorless amino acids and gives a purple/brown colored product whose absorbance can be measured to determine the concentration of the protein present in the solution.

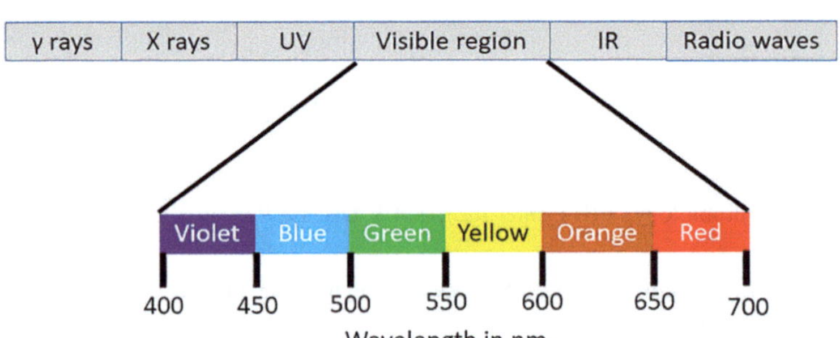

Fig. 53.1 Colorimetry measures absorbance of light in the visible region (400-700 nm) of the electromagnetic spectrum

Colorimetry is done using a light-sensitive instrument known as a Colorimeter. A colorimeter determines the concentration of the analyte based on the quantity of light absorbed. The colorimeter was invented by Louis Jules Duboscq in 1854.

53.1.2 Principle

When the light of a particular wavelength falls on a solution, based on the nature of the substance present, light can be reflected, absorbed, or transmitted (Fig. 53.2).

The light that falls on a subject is called the incident light of intensity I_o. The part of the incident light that is reflected is I_r, absorbed light (I_a), and the transmitted light is I_t. i.e. $I_o = i_r + i_a + i_t$.

The light that is reflected gives color to the substance. The amount of light absorbed (I_a) and transmitted (the light that passes through the sample - I_t) are quantified for colorimetric measurement.

By applying the principle of Lambert-Beer law (also referred to as Beer's law), light absorption is measured. Lambert-Beer law is the combination of two fundamental laws of photometry, namely, Beer's law and Lambert's law.

Beer's Law

As per Beer's law, proposed by August Beer, the quantity of light absorbed by a substance is a direct measure of its concentration. The higher the concentration of a substance, the higher the light absorbance.

The absorbance of a solution is defined as the logarithm (to the base10) of the ratio of incident light (I_0) to that of the transmitted light (I_t). i.e. ($\text{Log}_{10}\, I_0/I_t = A$).

Beer's Law equation is represented as

Fig. 53.2 Paths of light when it hits an object

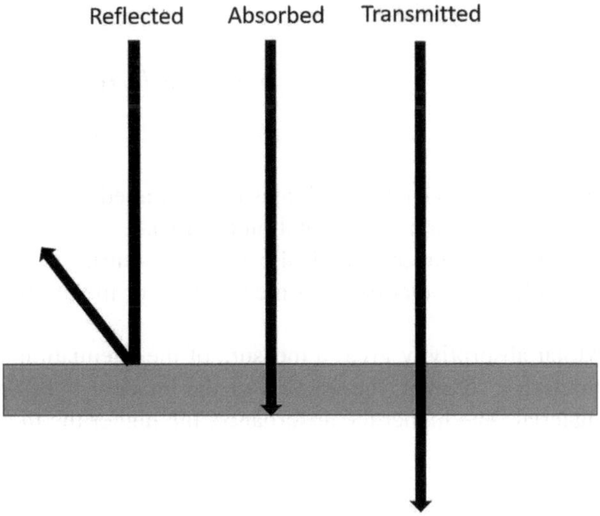

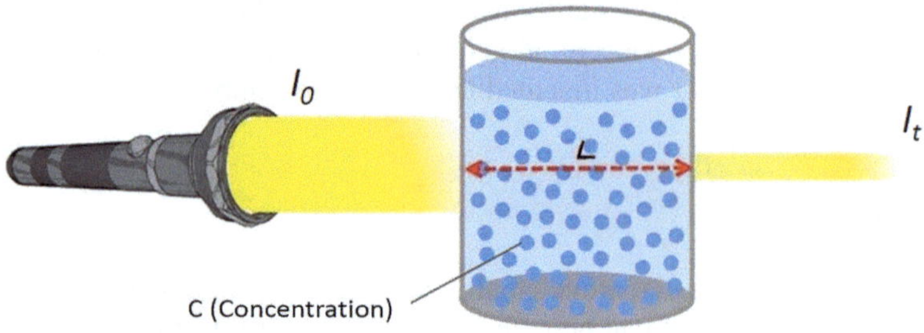

Fig. 53.3 Beer-Lambert law A = ε cl
Image adapted from: Pisano A. (2017) Light, Air Pollution, and Pulse Oximetry: The Beer-Lambert Law. In: Physics for Anesthesiologists. Springer, Cham. https://doi.org/10.1007/978-3-319-57330-413

$$A\ (\mathrm{Log}_{10}\ I_0/I_t) \propto c.$$

where, A = Absorbance; c = Concentration of the solution; I_0 is the incident light; I_t is the transmitted light.

Lambert's Law

According to Lambert's law (formulated by Bouguer), the amount of light absorbed is proportionate to the distance it travels through the absorbing medium.

$$A \propto l.$$

where A = absorbance of light; l = optical distance (distance traveled by the light in a solution).

The combination of the Beer's and Lambert laws (Fig. 53.3) that forms the basis of colorimetry is as follows:

$$A = \mathrm{Log}_{10}\ I_0/I_t = \varepsilon\ cl$$

Where,

A = Absorbance or optical density (expressed as absorbance units).
c = concentration of the solution (mol/L).
l = optical distance or optical path length (cm).
ε (epsilon) = molar extinction coefficient or molar absorptivity ($M^{-1}\ cm^{-1}$).

Molar absorptivity gives a measure of the attenuation of light intensity as it passes through a solution; the fraction of the incident light absorbed per thickness of the material. The higher the absorbance, the higher the molar absorptivity.

53.1.3 Instrumentation

The components of a colorimeter include the following (Fig. 53.4):

(a) Light source – the most commonly used light source in the colorimeter is the tungsten lamp.
(b) Slit and lens – allow the light to pass to the lens, which focuses the polychromatic light on the filter.
(c) Monochromatic filter – allows only monochromatic light (single color) to pass through the sample. Hence colorimeter is also known as a filter photometer. A color filter is chosen to select the wavelength in which the solute will have maximum absorption. Different filters with specific wavelength ranges are given in Table 53.1.
(d) Sample holder—is a cuvette (commonly made of glass or quartz) to hold the sample.

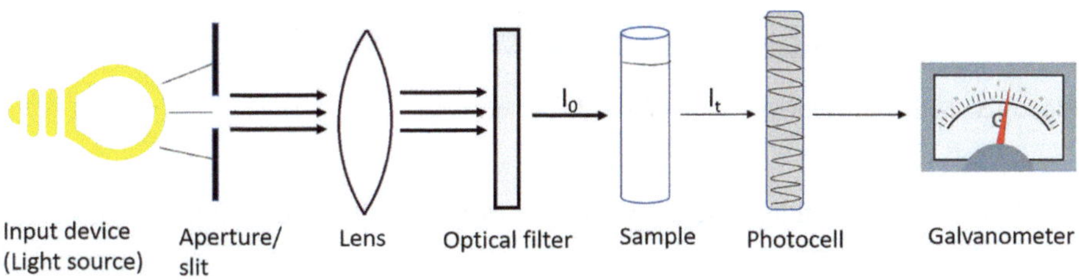

Fig. 53.4 Parts of a colorimeter
Image adapted from: Kumar V., Gill K.D. (2018) Photometry: Colorimeter and Spectrophotometer. In: Basic Concepts in Clinical Biochemistry: A Practical Guide. Springer, Singapore. https://doi.org/10.1007/978-981-10-8186-6_5

Table 53.1 Different wavelength filters used in colorimetry

Wavelength (nm)	Colour absorbed	Colour transmitted
400-420	Violet	Green-Yellow
420-500	Blue	Yellow
500-570	Green	Red
570-600	Yellow	Blue
600-630	Orange	Green-Blue
630-700	Red	Green

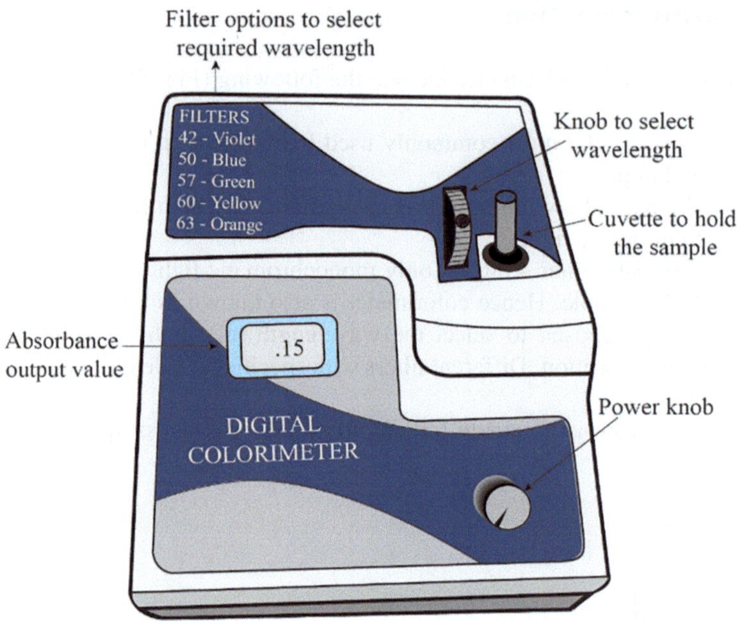

Fig. 53.5 Schematic representation of a digital colorimeter

(e) Photodetector—a light-sensitive device (a photocell) to measure the transmitted light.
(f) Measuring device—a meter that gives the measure of light intensity to a readable output (either in absorbance units or as % transmittance).

Light supplied by the input light source is navigated through the slit to the lens, which focuses the polychromatic light on the filter. The selected filter directs only monochromatic light (of a single wavelength) towards the cuvette containing the analyte. Some portion of the incident monochromatic light is absorbed by the sample, and some of it is transmitted, which is detected by the photocell. The photodetector converts the transmitted light into electrical signals, which are in turn measured by the Galvanometer that measures the electrical signals and gives a digital output as absorbance. Colorimeters are commercially available either as portable (hand-held), useful for on-site analysis, or benchtop instruments used for laboratory analysis (Fig. 53.5).

53.1.4 Methodology

1. Set filter: Select the filter or wavelength which will give maximum absorbance (λ max) for the test sample.

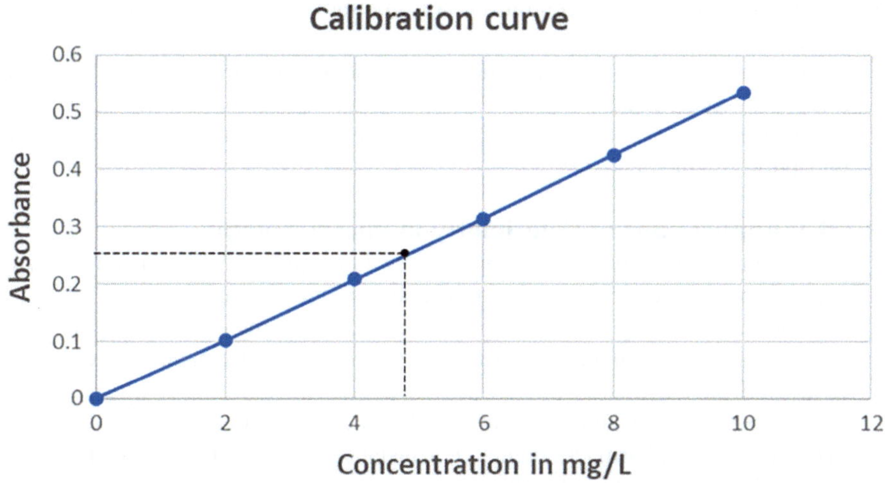

Fig. 53.6 Determination of the concentration of the unknown solution using Beer's law calibration curve. *Note*: X-axis: concentration in mg/L. Y-axis: Absorbance

2. Calibration: The first step involved in performing a colorimetric measurement involves calibration. Calibration is done using the solvent alone (without solute), known as the 'blank.' Measuring the blank absorbance is necessary to remove background interferences.
3. Calibration curve: Prepare a set of standard solutions with known solute concentrations and measure its absorbance using a colorimeter. Make a calibration curve by graphically plotting concentration and absorbance on the X and Y-axis, respectively.
4. Determine the concentration of the unknown sample: Note down the absorbance of the test sample and, using the calibration curve, determine the concentration of the unknown solution (an example is shown in Fig. 53.6).

53.1.5 Applications

Colorimetry is widely used in many disciplines to quantify the concentration of unknown substances. Some of the commonly used applications include the following:

(a) Urine analysis—testing levels of urine parameters (glucose, creatine) by portable colorimetry.
(b) Colorimetry is used to calculate the rate of the reaction; the rate of formation of colored products and the disappearance of the substrate over time can be monitored using colorimetry.
(c) Blood analysis—determining the concentration of hemoglobin (Hb) in blood.

(d) Determination of concentration of biomolecules like carbohydrates, proteins, etc.
(e) Water quality analysis—by measuring the content of chloride, fluoride, cyanide, dissolved oxygen, etc.
(f) Determines the levels of plant nutrients (for example, nitrate, ammonia, and phosphorous) in soil.

53.1.5.1 Applications of Colorimetry in Pharmacology

(a) Colorimetry is widely used for drug analysis to quantify the concentration of drugs in pharmaceutical dosage forms and biological samples.
(b) Tramadol, a centrally acting analgesic, is used in the management of moderate to severe pain. Colorimetry is used for the estimation of tramadol hydrochloride in tablet dosage forms.
(c) Colorimetric analysis of biological samples like blood and urine is very useful to understand drug pharmacokinetics. For instance, during tuberculosis treatment, urine colorimetry is done for therapeutic drug monitoring of pyrazinamide.

53.1.5.2 Advantages of Colorimetry

- Inexpensive method.
- Sensitive, specific, and easy to perform.
- Available in portable hand-held (easy for on-site analysis) and largescale bench-top models (laboratory analysis).

53.1.6 Limitations of Colorimetry

- Determining the concentration of colorless compounds is possible only with the usage of a locating agent.
- It is limited to measuring wavelengths falling in visible spectrum regions. The colorimeter has been replaced by a spectrophotometer that covers wavelengths of ultraviolet-visible-infrared regions of the spectrum.

53.2 Spectrophotometry

53.2.1 Introduction

Spectroscopy is the study of the interaction of light (more precisely electromagnetic radiation) with the matter. By examining the interaction between light and matter, it is possible to deduce the wavelength of the interacting light (λ) and/or quantify the amount of light (frequency (f)) being emitted, reflected, scattered, or absorbed at that particular wavelength. The term spectroscopy is a Latin-Greek hybrid derived from the Latin word *spectron,* meaning "image," and the Greek word *skopein* meaning, "to see." Historically, spectroscopy originated from studies involving the dispersion of visible light by a glass prism, and later the concept was expanded to other portions of the electromagnetic spectrum.

Spectrophotometry is a branch of spectroscopy that measures (quantifies) the absorbance, transmission, or reflection properties of a substance or a material as a function of wavelength. A spectrophotometer is an instrument used to determine the intensity of various wavelengths of light. For instance, every chemical compound absorbs, transmits, or reflects light at a particular (or over an assured range of) wavelength. A spectrophotometer measures the intensity of light absorbed by the chemical compound, by which the amount (concentration) of a known chemical constituent in the solution can be deduced. The science of spectrophotometers emerged in 1940, but it was not until 1941 when Arnold J. Beckman and his colleagues at National Technologies Laboratories developed the first improved spectrophotometer.

Today, multiple types of spectrophotometers exist, and each is used to detect different wavelength ranges of the electromagnetic spectrum. Further, spectrophotometers are also available in both single beam and double beam configurations. In laboratories, the most commonly used spectrophotometers are: (i) UV-visible spectrophotometer: uses light over the ultraviolet range (185–400 nm) and visible range (400–700 nm) and (ii) IR spectrophotometer: uses light over the infrared range (700–1500 nm) of the electromagnetic radiation spectrum. In addition to these, a fluorescence spectrophotometer, which can analyze a broad light spectrum depending on the fluorophore or fluorescent molecule employed, is also routinely used in laboratories.

53.2.2 Principles of Spectrophotometry

Spectrophotometers are based on photometric law, which states that when a beam of incident light (I_0) passes through a solution, a part of the incident light is reflected (I_r), a part is absorbed (I_a), and the rest of the light is transmitted (I_t) (Fig. 53.7).

Thus, $I_0 = I_r + I_a + I_t$

Fig. 53.7 Paths of light

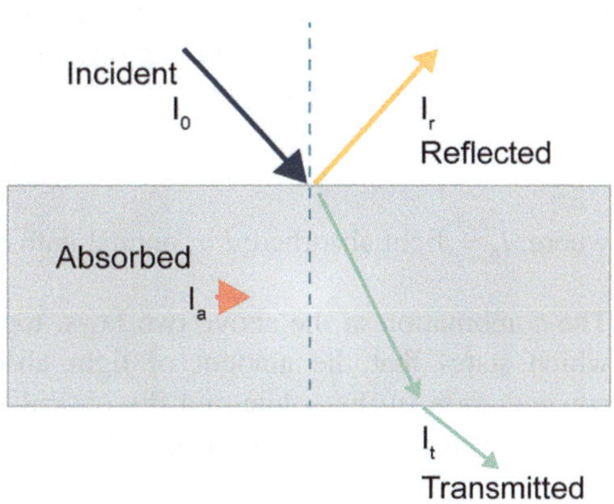

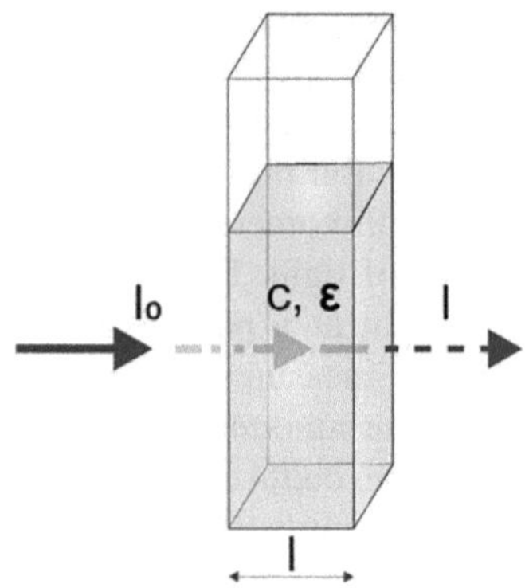

Fig. 53.8 Beer-Lambert law**:** The intensity of light decreases from I_0 to I as it passes through a solution of concentration (c) and optical path length (l)
Image adapted from: Jidraph Njuguna, O. Arda Vanli, Richard Liang, "A Review of Spectral Methods for Dispersion Characterization of Carbon Nanotubes in Aqueous Suspensions", Journal of Spectroscopy, vol. 2015, ArticleID 463,156, 11 pages, 2015. https://doi.org/10.1155/2015/463156

In a spectrophotometer, I_r is kept constant by using cells (cuvette) that have identical properties, and the amount of light absorbed (I_a) is calculated by measuring I_t and I_0. The relationship between the amount of light absorbed and the concentration of a constituent can be deduced using the following two laws:

(i) Beer's Law: Formulated by German mathematician and chemist August Beer, states that the absorptive capacity of a dissolved substance is directly proportional to its concentration in a solution:

$$I_a\ (\log_{10} I_0/I_t) \propto \mathrm{c}.$$

where, I_a = light absorbed; c = concentration of a dissolved substance (solute) in the solution.

(ii) Lambert's Law: Formulated by French scientist Pierre Bouguer, states that the loss of light intensity when it propagates in a medium is directly proportional to intensity and path length:

$$I_a \propto l.$$

where, I_a = light absorbed; l = optical path length.

The combination of the above two laws, together referred to as Beer-Lambert law which states that the amount of light absorbed is directly proportional to the concentration of the solute and the optical path length of the solution, forms the basis of a spectrophotometer (Fig. 53.8).

$$I_a = \varepsilon\, cl$$

where, I_a = light absorbed (optical density (O.D.) expressed in absorbance units); ε = molar absorption coefficient (M^{-1} cm^{-1}); c = molar concentration (M); l = optical path length (cm).

53.2.3 Instrumentation

The spectrophotometer combines the functions of a spectrometer, a photometer and consists of the following parts:

(i) Light source: An ideal electromagnetic radiation (light) source of a spectrophotometer must be able to emit radiations with a constant intensity over the whole wavelength range and should be long-lasting. Further, the light source must also be remarkably stable, as measurements are based on comparing samples with standards measured at different times. However, a light source with all such features does not exist. Therefore, different sources of light are used to produce light of different wavelengths. The most common source for the visible spectrum is a tungsten lamp. For the ultraviolet (UV) spectrum, a hydrogen or deuterium lamp is used, while a Nernst filament or globar is used as a source for the infrared (IR) spectrum.

(ii) A monochromator (Discriminator): Monochromator is an optical device used for selecting a specific wavelength from a range of frequencies emitted by the light source. They are of two types: slit and dispersive element. Spectrophotometers usually use a diffraction grating for wavelength discrimination, and in modern devices, holographic gratings are used for this purpose. These are glasses that are patterned with extremely thin lines by a holographic optical process. The lines and the wavelength of light they disperse are of the same order of magnitude.

(iii) Sample holder: Generally, cuvettes are used to hold the sample/test solutions. In a spectrophotometric analysis, the absorption of light is measured, and therefore the sample holder used must not absorb the light, or the absorption must be minimum. Hence, for measuring different regions of the spectrum, different materials are used (for example, quartz cuvettes are used for UV range, glass cuvettes for visible range, NaCl/KBr cuvettes for IR range, etc.).

(iv) Beam splitter: It is present only in a double beam spectrophotometer. As the name goes, the beam splitter splits the light beam into two and directs one beam through the sample cuvette and the other through the reference or blank cuvette.

(v) Photodetector system: The light initiated from the source passes through the sample holder, discriminator, and finally must be detected and quantified. In the early days, photographic films were used as detectors. Modern spectrophotometers use electro-optical detectors to detect, amplify, and digitize the signal. Ideally, a detector should be linear in the widest range possible and present high sensitivity and low noise. The detectors that are used today can be classified into three types: (a) photovoltaic cell (barrier layer cell), (b) phototubes (photoemissive tube), and (c) photomultiplier tubes. The photomultiplier tubes convert the signal and amplify it to such an extent that a single photomultiplier can grant a high sensitivity in the whole UV–vis range. Nowadays, photodiodes are more commonly used as they are robust, do not need a high voltage to work, present a wider range of detection, are smaller, and are economical compared to photomultiplier tubes. A photodiode is made up of semiconductor materials, and they generate "photocurrent," which is proportional to the intensity of the incident light. However, different detector materials are required for different wavelengths. For instance, photodiodes made of silicon are used to detect ultraviolet and visible light, while indium gallium arsenide is used for near-infrared and mercury cadmium telluride is used to detect infrared.

(vi) Measuring device: The current from the detector is fed to the measuring device (galvanometer). The reading shown on the meter is directly proportional to the intensity of light.

53.2.4 Types of Spectrophotometers

Different types of spectrophotometers presenting different configurations, advantages, and disadvantages are commercially available. They are broadly classified into two types:

(i) Single beam spectrophotometer
A single-beam spectrophotometer measures relative light intensity before and after test samples are inserted. In such devices, polychromatic light emitted by the source (one or more) is concentrated onto the entrance slit of a monochromator that selects and transmits a very narrow band of light which is then focused on the sample. After passing through the sample, the light reaches the detector (Fig. 53.9). The absorbance of the test samples is calculated by first measuring a blank (reference) followed by the test sample(s). Single-beam instruments have a compact design, are traditionally built with few moving parts, low cost, and offer large dynamic ranges. They are available solely in the

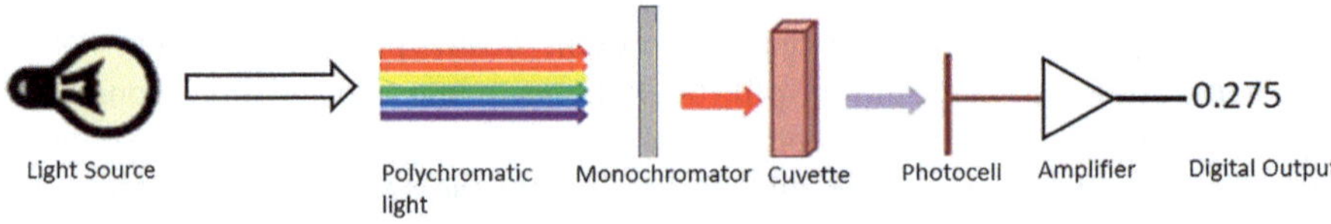

Fig. 53.9 Single beam spectrophotometer
Image adapted from: Sanjay M. Nilapwar, Maria Nardelli, Hans V. Westerhoff, Malkhey Verma, Chap. 4- Absorption Spectroscopy, Editor(s): Daniel Jameson, Malkhey Verma, Hans V. Westerhoff, Methods in Enzymology, Academic Press, Volume 500, 2011, Pages 59–75, ISSN 0076–6879

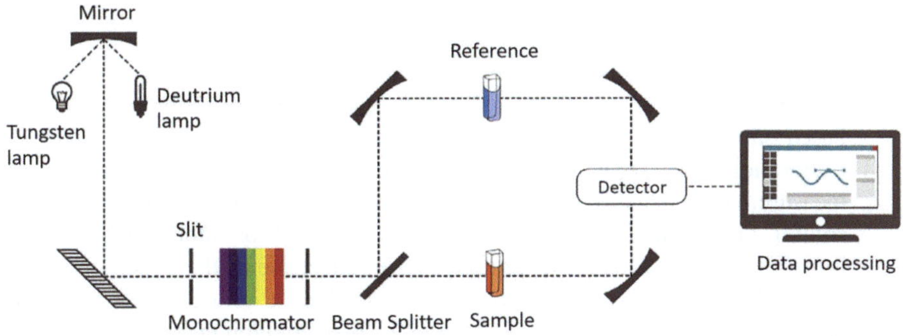

Fig. 53.10 Double beam spectrophotometer
Image source: P. Senthil Kumar, K. Grace Pavithra, Mu. Naushad, Chap. 4 - Characterization techniques for nanomaterials, Editor(s): Sabu Thomas, El Hadji Mamour Sakho, Nandakumar Kalarikkal, Samuel Oluwatobi Oluwafemi, Jihuai Wu, Nanomaterials for Solar Cell Applications, Elsevier, 2019, Pages 97–124, ISBN 9780128133378. https://doi.org/10.1016/B978-0-12-813337-8.00004-7

visible range of 340–750 nm or as a combination of visible and UV ranges, usually around 190–1100 nm. Single-beam spectrophotometers are routinely used in laboratories, for instance, to determine the concentration of an analyte in solution.

(ii) Double beam spectrophotometer
As explained above, blank (reference) and test samples are measured one after the other in a single beam spectrophotometer causing slight delays (a few seconds to minutes) between each measurement. The risk of variation in light intensities, current fluctuations, or other problems during such unavoidable delays might give rise to errors in measurements. On the other hand, a double beam spectrophotometer allows real-time referencing using a separate reference position in the spectrophotometer (Fig. 53.10). Herein, the modulator alternates the path of the incoming light from the source once to pass through the blank and then through the sample. Its speed is so high that it is possible to record many alternate measurements per second. Alternately, beam splitters are also used to send light along the blank and sample paths simultaneously to two separate, identical detectors, thereby measuring the light intensities at the same time. Applications that require high speed, stability, and flexibility are better suited for such dual-beam configuration. Further, the measurements obtained are more reproducible, and therefore dual beam spectrophotometers are preferred in industrial and quality control laboratories.

53.2.5 Sample Measurement

(i) Set Filter: The first step while using a spectrophotometer is to select an appropriate filter or wavelength that will give maximum absorbance (λ max) based on the test sample(s).
(ii) Calibration: Spectrophotometers are to be calibrated first before taking any measurements. This is done by first measuring the solvent alone (without solute), known as the blank or reference. The blank or reference absorbance is necessary to remove background interference. Next, the absorbance of a standard solution(s) with a known concentration of solute(s) that has to be determined in the test solution is measured. A calibration curve is prepared by graphically plotting concentrations and absorbance of the standard solution (s) on the X and Y-axis, respectively.
(iii) Measuring test sample: The absorbance value of the test sample is then measured. If the absorption of the solution is higher, more light is absorbed, and if the absorption of the solution is lower, more lights are transmitted through the solution, which, in turn, affects the reading observed on the galvanometer. By substituting the values in the below formula, it is possible to determine the concentration of the solution.

$$A = \varepsilon\, cl.$$

(ε = Constant; l = Constant (when same cuvette is used); c = concentration of the solution)

For two solutions (i.e. Test and Standard),

(i) $A_T = C_T$
(ii) $A_S = C_S$

From (i) & (ii):

$A_T \times C_S = A_S \times C_T$

$$C_T = (A_T/A_S) \times C_S.$$

where, C_T = Concentration of test solution; C_S = Concentration of standard solution; A_T = absorbance or optical density of test solution; As = absorbance of the standard solution.

53.2.6 Applications of Spectrophotometer

Spectrophotometers have been around for more than 75 years, and their applications are limitless. They are widely used by scientific laboratories across various disciplines. A few major applications in which spectrophotometers are widely used are:

(i) A spectrophotometer is commonly used to determine the concentration of colored and colorless compounds by measuring their optical density or absorbance. A perfect example of its day-to-day application is the determination of nucleic acid and protein concentrations in a molecular biology laboratory.
(ii) It is used to determine the course of a reaction by measuring the rate of formation and disappearance of the light-absorbing compound.
(iii) It is useful in qualitative analysis, especially when identifying classes of compounds in both biological and pure states.

53.2.7 Drawbacks or Disadvantages of Spectrophotometer

A few drawbacks or disadvantages of a spectrophotometer are:

(i) The samples should be prepared as homogeneous solution(s). The optical density of the solution must not be too low or too high.
(ii) Some compounds can be dissolved only in nonpolar solvents and therefore cannot be used with plastic cuvettes. On the other hand, glass cuvettes cannot be used while measuring at the UV range for such compounds.
(iii) Single beam spectrophotometer: Time-consuming—It takes a significant amount of time between measurements (reference followed by sample), giving rise to errors due to drifts. However, when using modern instruments, this problem can be kept to a minimum, but it does come with a heavy cost associated with the purchase of such instruments.
(iv) Double beam spectrophotometer: High cost, more complexity, and low sensitivity due to poor light throughput.

53.2.8 Safety Considerations

Before working with chemicals and solvents, always consult the material safety data sheets (MSDS) provided by the manufacturer and wear appropriate personal protective equipment (PPE). When working with lamps that emit UV radiation, it is recommended to wear UV-resistant safety goggles. Always use shutters or block the path of light from the source after disconnecting from a fiber optic. Although modern instruments do have such built-in safety shutters, beware of such occurrences.

53.3 Photometry

53.3.1 Introduction

Photometry is the branch of optical science that deals with the measurement of light, perceivable by the human eye. Specifically, it refers to the measurement of light falling within the visible region of the electromagnetic spectrum covering wavelengths from to 700 nm (refer to Fig. 53.1). While radiometry measures the complete spectrum of electromagnetic radiation, photometry is a branch of radiometry restricted to the visible spectrum.

Using photometry, the concentration of the light-absorbing analytes present in the solution can be determined. Like colorimetry, this assay is dependent on the property of light absorption by molecules; that is, the solutions that follow Beer-lambert's law can be assessed. In simple terms, colorimetry measures color, whereas photometry measures light. While colorimetry measures the transmission of light, photometry measures, in addition to analyte's absorbance, photometric quantities like luminance, luminous intensity, luminous flux, illuminance, CRI (color rendering index), and color temperature.

53.3.1.1 Key Terms in Photometry

Radiant power or Radiant flux (P): measured in watts (W or J/s), refers to the total amount of energy emitted or radiated by the light source per second. It includes light and thermal radiation.

Luminous flux (Ø): total light energy radiated or given off by the light source. This quantity is smaller than the radiant flux. It is measured in lumens (lm).

Luminous intensity (l_v): measured in candela (cd), refers to the quantity of light energy radiated by a light source per unit solid angle. It is defined as lumen/steradian (lm/sr). *Note: steradian is the SI unit of solid angle.*

Luminance: refers to the quantity of light emitted, reflected, or transmitted per unit area per solid angle. It is expressed as $lm/m_2.sr$ or cd/m^2.

Illuminance (E_v): measured in lux refers to the quantity of light falling on a given surface area.

Luminous efficiency (η, eta): is the ratio of luminous flux to that of radiant flux, i.e. (Ø/P).

The quantity of light (Q_v): is calculated by multiplying luminous flux with time.

CRI: Color rendering index (expressed in terms of %) gives a measure of the light source's ability to precisely reproduce the color of the light radiated by the objects.

All the above described photometric units are measured using a photometer.

53.3.2 Principle

The brightness of visible light perceived by the human eye exhibits differences in sensitivity to different wavelengths. A photometer closely resembles the human

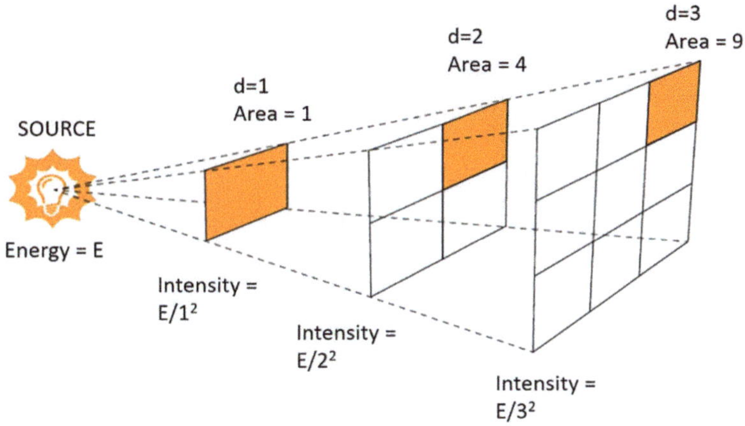

Fig. 53.11 Inverse-square law
Image source: Sara Simmons et al., 2017, Ultraviolet Light in Human Health, Diseases and Environment, Advances in Experimental Medicine and Biology 996. Springer International Publishing

eye's spectral response and accurately measures photometric quantities. The photometry principle is governed by two laws - Inverse-square law and Cosine law.

As per the Inverse-square law, a decrease in light intensity is inversely proportional to the increase in the distance between the illumination source and the object being illuminated (Fig. 53.11). If the distance between the light source and the subject is doubled, then the surface area illuminated is quadrupled; that is, the same amount of light energy is distributed to a larger surface. This, in turn, will proportionately decrease the light intensity. Therefore, the intensity of light is inversely proportional to the square of the distance between the subject and the light source. The photometer applies the Inverse-square law to quantify the luminous intensity of a luminaire.

Inverse-square law is given by the formula:

$$\mathrm{I} \propto 1/\mathrm{d}^2.$$

where,

I refer to the intensity of light and d implies the distance between the light source and the subject.

According to Lambert's cosine law, the illuminance (E_v) produced by a point light source is directly proportional to the cosine of the angle (θ) of incidence (Fig. 53.12). Lambert cosine law is applied when using point light sources.

$$E_v = \frac{\mathrm{I} \times \cos\theta}{d^2}$$

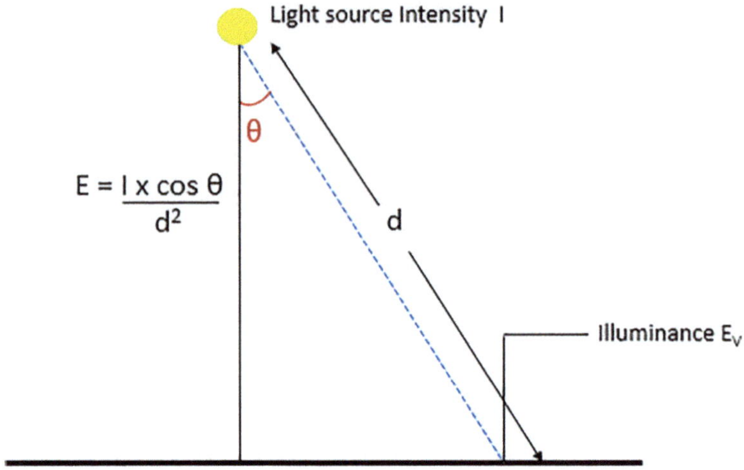

Fig. 53.12 Lambert cosine law
Image adapted from: McMullan R. (1983) Principles of Light. In: Environmental Science in Building. Macmillan Building and Surveying Series. Palgrave, London

where, E_v = Illuminance; I = luminous intensity; θ = angle of incidence; d = distance between the light source and surface.

A photometer can also determine the concentration of the analytes present in the test sample based on the principle of Lambert-Beer law.

Lambert-Beer law is represented as $A = \varepsilon\, cl$.

Where,

A = Absorbance or optical density (expressed in absorbance units).
c = concentration of the solution (mol/l).
l = optical distance or path length (cm).
ε (epsilon) = molar extinction coefficient ($M^{-1}\ cm^{-1}$).

53.3.3 Instrumentation

Different types of photometers available include filter photometer (also known as colorimeter), flame photometer, reflectance photometer (measures the reflected light), fluorophotometer (measures the fluorescence emitted from the interior of the eye subsequent to intravenous injection of fluorescein), fiber photometer (measures the fluorescence or light scattering from the tissue via implantation of a fiber optic cannulas), etc. Based on the experimental objectives, the type of photometer shall be selected for analysis.

Fundamental components of a photometer include the following (Fig. 53.13):

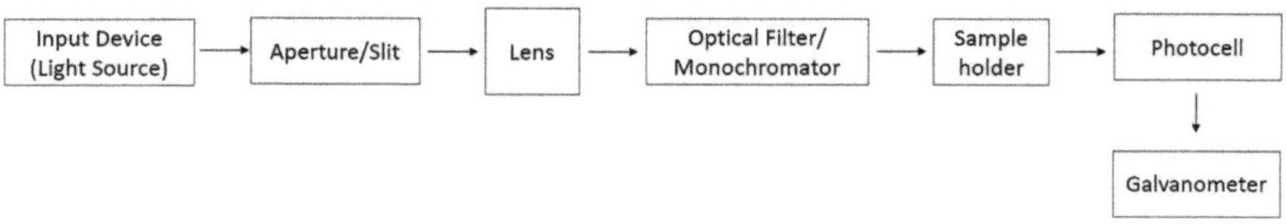

Fig. 53.13 Basic components of a photometer

(a) Input unit: Based on the type of the photometer, the input device either captures the incoming radiation from a light source or possesses a built-in light source. The light is passed through an aperture or slid and focussed through a lens onto the filter. The light energy is focussed onto a filter via an aperture or slit.
(b) Filter: The choice of a monochromator or an optical filter depends on the photometer model. An optical filter provides comparatively greater spectral sensitivity and precise wavelength separation. A monochromator requires a series of optical elements like narrow light apertures, prism, or diffracting grating and mirrors to separate wavelengths and focus the light on the sample. An optical filter, in general, transmits only selected wavelengths towards the sample. Broadly categorized into two types, an optical filter is either absorptive or dichroic. Absorptive filters, generally made of glass, will absorb specific wavelengths and transmit others. Dichroic filters (or interference filters) reflect specific wavelengths and transmit the remaining wavelengths of light.
(c) Sample holder: Provision to place the cuvette having the sample.
(d) Photodetector: Quantifies the radiation of the wavelength transmitted from the sample and proportionately converts it into an electrical current using a photomultiplier, photodiode, or photoresistor.
Output device: It is a galvanometer that measures the electrical current and displays the output readings.

53.3.4 Methodology

Please refer to the instruction manual of the photometer used in your laboratory for a detailed description.

Common steps involved in taking photometric measurements are given below:

1. Turn on and calibrate the instrument as per the instructions given in the photometer model opted. Calibration is done to 0% transmission (measure darkness by placing on the sensor cap) and 100% transmission using a reference solution.
2. Following calibration, select the required program or parameter using settings options.
3. Set the blank: Blank has the solvent components used to prepare the sample mixture. Always use the cuvette allotted for blank measurements. Fill the blank cuvette and set the readings to zero.

4. Test sample readings: Load the analyte to the cuvette allotted for the test sample and take the readings. Follow on with the remaining test samples.
   Note: Blank must be set every time after switching on the photometer and between different sample mixtures.
   In the case of using a photometer for determining the concentration of the sample, select the wavelength that will exhibit maximum absorption.
5. Save and export data: Based on the photometer model used, save your readings and export the data in the desired format.

Tips for accurate results and maintenance

- Wash and dry cuvette tubes before and after use.
- Before placing the cuvette in the sample holder, wipe the surfaces of the cuvette with a lens cloth or tissue paper.
- Always maintain the photometer instrument clean, devoid of spills or dust particles.
- Keep the instrument in a dry place away from reagents or chemicals for proper maintenance.

Calculations.
The calculation shown in (i) is used to determine the quantity of the unknown compound in the test sample, and (ii) is used to calculate the illuminance.

(i) Determination of molar absorptivity (ε)
Deriving from Beer-Lambert's law, $\varepsilon = A/cl$,
In this regard, A = absorbance of the standard solution (of known concentration), C = concentration of the standard solution, l = optical path length.
Calculate the concentration of the unknown using the formula:

$$C = A/\varepsilon l.$$

(ii) In case of calculating light parameters like illuminance, apply lambert cosine law as given below:

$$E_v = \frac{I \times \cos\theta}{d^2}$$

where, E_v = Illuminance; I = luminous intensity; θ = angle of incidence; d = distance between the light source and surface.

53.3.5 Applications of Photometry

(1) They are widely used in quantifying the flux (amount of energy) radiated by astronomical objects.

(2) To check the quality of the lightings used in buildings, light-emitting diode (LED), display devices like Television, automobile dashboards, etc.

Applications in pharmacology

(1) Filter photometry is widely used in the pharmaceutical industry to determine drug concentration.
(2) Fluorophotometry is used as a non-invasive diagnostic tool for detecting dry eye disease. Patients having dry eye disease exhibit approximately a five-fold increased corneal uptake of fluorescein compared to normal individuals.
(3) Flame photometry is used to determine the concentration of sodium and potassium ions in saline and infusions.

53.3.6 Advantages

- Easy to operate, versatile, and compact instrument. Photometers can be used in both the laboratory and onsite.
- Less expensive compared to the spectrophotometer.
- Comes in handy during conditions of limited resources.

53.3.7 Limitations

- Photometry does not cover the entire electromagnetic spectrum.
- Lacks precision compared to spectrophotometry.

53.4 Flame Photometry

53.4.1 Introduction

Flame photometry, also known as Flame Atomic Emission Spectrometry (FAES), is emission spectroscopy used for qualitative and quantitative analysis of elements present in the test sample. In FAES, thermal energy provided by flame is used to excite the atoms, and the intensity of light emitted at wavelengths characteristic of each atom is measured. Flame photometry uses the property of elements that emit light in the visible spectrum when exposed to flame. The invention of the flame photometry concept dates back to the mid-fifteenth century when German metallurgist - Georgius Agricola made a qualitative analysis of the color of the fumes that originated from different types of ores. Gustav Kirchhoff and Robert Bunsen (Fig. 53.14) jointly made a breakthrough contribution to the field of analytical chemistry by identifying trace elements like indium and gallium by flame emission spectroscopy.

Fig. 53.14 Gustav Robert Kirchhoff (left) and Robert Wilhelm Eberhard Bunsen (right)
Image courtesy: Edgar Fahs Smith Collection, Kislak Center for Special Collections, Rare Books and Manuscripts, University of Pennsylvania. https://www.sciencehistory.org/historical-profile/robert-bunsen-and-gustav-kirchhoff

Barnes B, Richardson D, Berry J, and Hood RL developed flame photometry for the rapid quantification of alkali metals (primarily sodium and potassium) in an aqueous solution. Since then, flame photometry has been widely used for determining the concentration of Group 1 and 2 elements of the periodic table (alkali and alkaline earth metals, respectively) such as sodium, potassium, lithium, calcium, magnesium, etc. Every alkali and alkaline earth metal emits light at a characteristic wavelength when passed through a flame and its emission intensity is directly proportional to the quantity of the element present. Alkali and alkaline earth metals are sensitive to flame photometry techniques because these metals have lower excitation temperatures (1500–2000 °C) compared to other elements, and their characteristic wavelengths can be easily isolated as they display wider emission spectral separation.

53.4.2 Principle

Alkali and alkaline-earth metals dissociate into atoms when exposed to flame. By absorbing the thermal energy provided by the flame source, some of the atoms get elevated to higher energy levels. These high-energy atoms, while returning from excited to the ground state, emit light at a specific wavelength. The light intensity emitted is equivalent to the number of atoms returning to the lowest energy state, which is proportionate to the concentration of the chemical species present. Every atom exhibits a unique emission spectrum that forms the basis of qualitative elemental analysis. i.e., the color of the radiation emitted depends on the characteristic element present in the sample. For example, sodium exhibits emission at 589 nm, giving a yellow color. Similarly, every metal produces a specific flame color, as shown in Table 53.2.

The intensity of light emitted is calculated using the Scheibe-Lomakin equation:

$$I = K \times C^m$$

Where, I = intensity of the light emitted; K = Proportionality constant; C = concentration of the chemical species; m = exponent that accounts for the deviation from the proportionality, primarily caused by self-absorption. Based on this equation, the light intensity is directly proportional to the sample concentration.

53.4.3 Instrumentation

Vital components of a flame photometer include the following (Fig. 53.15):

(a) Flame source: provided by a burner; the flame is maintained at a constant temperature. Flame temperature is dependent on the fuel-oxidant mixture selected (Table 53.3).
(b) Nebuliser and mixing chamber: In the mixing chamber, the fuel and oxidant mixture along with the sample is combined well to produce a homogenous mixture. The nebulizer aspirates the sample towards the flame at a steady rate.
(c) Optical system: consists of a convex mirror and convex lens. Convex mirror passes the light emitted by the atoms to the lens, which focuses the light on the

Table 53.2 Emission wavelength and flame colors of certain alkali and alkaline earth metals

Element	Emission wavelength (nm)	Flame Color
Barium (Ba)	550	Lime Green
Lithium (Li)	670	Carmine red
Sodium (Na)	589	Yellow
Potassium (K)	766	Violet
Strontium (Sr)	461	Scarlet
Calcium (Ca) CaOH	423 622	Orange

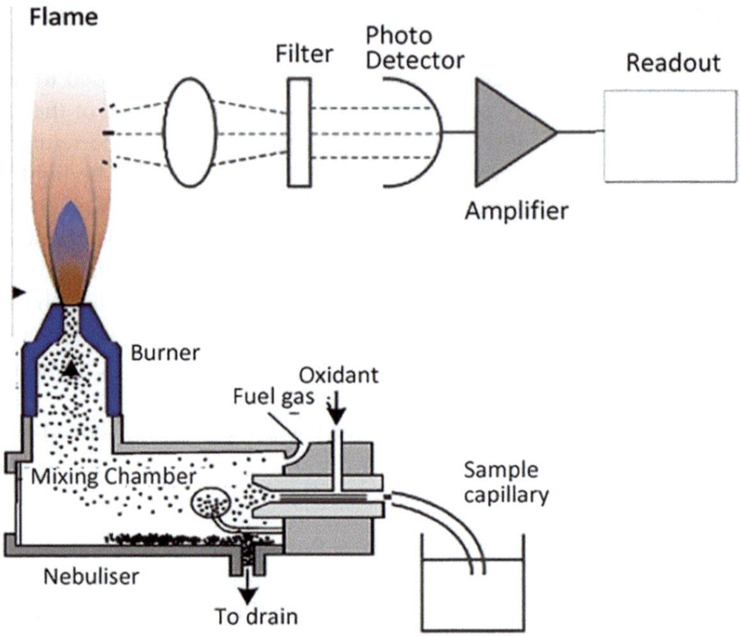

Fig. 53.15 Schematic representation of a flame photometer
Image adapted from: Kumar V., Gill K.D. (2018) To Estimate Sodium and Potassium in Serum by Using Flame Photometer. In: Basic Concepts in Clinical Biochemistry: A Practical Guide. Springer, Singapore. https://doi.org/10.1007/978-981-10-8186-6_36

Table 53.3 Flame temperatures of various fuel-oxidant mixtures used in flame photometry

Fuel-oxidant mixture	Temperature
Butane-air	1300–1900
Natural gas-air	1700–1800
Propane-air	1900–2000
Hydrogen-air	2000–2100
Acetylene-air	2100–2400
Hydrogen-Oxygen	2500–2700
Natural gas-oxygen	2700–2800
Acetylene-nitrous oxide	2900–3000
Acetylene-oxygen	3100–3200

slit. The light reflected from the mirror is passed through the slit towards the color filter.

(d) Monochromatic filter: Two types of filters are available (i) absorption filters used for separating wavelengths that fall in the visible spectrum and (ii) interference filters used for UV, visible, and IR regions of the spectrum. As the same suggests, a monochromatic filter transmits the wavelength emission spectra of only a single element while absorbing emissions of other wavelengths. Depending on the atom that is being examined, the corresponding filter is selected for use in flame photometry.

(e) Photodetector: Detects and converts the intensity of radiation released by the flame to an electrical signal which is amplified and given as output in a readable format. The electrical signal generated by the photodetector is directly proportional to the intensity of light emitted.

53.4.4 Working of Flame Photometry

In a flame emission spectrometry, the sample in aqueous form is aspirated into the flame by the nebulizer (or aspirator). Flame exposure results in desolvation (solvent evaporates) followed by atomization (elements dissociate to atoms) of the sample. Atoms get excited by absorbing thermal energy from the flame and, while returning to the ground level, emit radiation at the wavelength characteristic of the atom present in the flame. The emitted radiation gives both qualitative and quantitative measurements of the element present. The light emitted is passed through a monochromator which separates the wavelengths and transmits radiation of wavelength characteristic of a single element to the photodetector. An amplified version of the electrical signal proportionate to the emitted light is generated by the photodetector, and a readable output either in analog or digital format is displayed (Fig. 53.16).

53.4.5 Methodology

(a) Standard solution preparation:

Usually, primary standard solutions are prepared at higher concentrations (typically 1000 parts per million) and diluted to prepare required working standards.

The calculation for primary standard preparation.

1 ppm (parts per million) implies one part in a million. It is a dimensionless quantity (e.g. 1 μl/L or 1 mg/g etc.).

ppm in terms of g/L is represented as follows:

1 g/L = 1000 ppm (1 g in 1000 ml is 1000 ppm)

i.e. 1 mg/L = 1 ppm (one-thousandth of a gram in 1000 ml is 1 ppm).

One-gram pure element dissolved in one litre = 1000 ppm.

In the case of preparing 1000 ppm stock solution of Na metal from NaCl salt,

Molecular weight of NaCl = 58.44 g/mol.

The atomic weight of Na = 23u.

Amount equivalent to 1 g Na in NaCl salt = 58.44/23 = 2.54 g.

Thus, 2.54 g NaCl dissolved in 1000 ml solvent will give 1000 ppm Na.

The calculation for preparation of working standards:

Dilutions from stock solutions shall be prepared by applying the following equation.

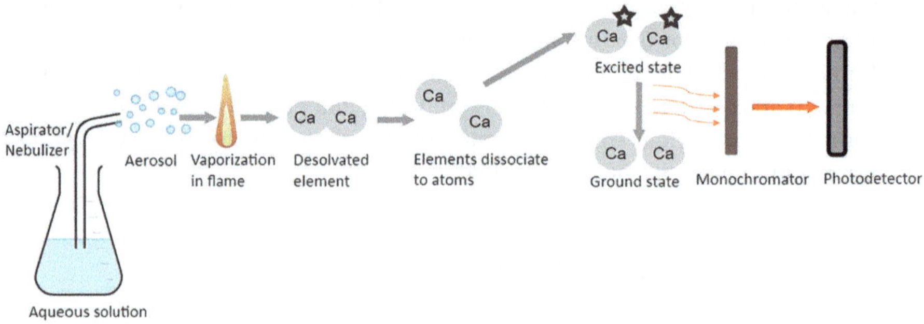

Fig. 53.16 Steps involved in flame photometry

$$C_1V_1 = C_2V_2.$$

where,

C_1 = initial concentration.
C_2 = final concentration.
V_1 = initial volume.
V_2 = final Volume.

For instance, to prepare 100 ml of 10 ppm working standard from stock 1000 ppm,

$$1000\text{ ppm} \times V_1 = 10\text{ ppm} \times 100\text{ ml}.$$

$$V_1 = 10 \times 100/1000 = 1\text{ ml}.$$

1 ml from the primary standard of 1000 ppm diluted to 100 ml with solvent will give a 10 ppm working standard. Similarly, desired working standard solutions can be prepared from the primary standard. Ideally, using five standard solutions would be optimum to construct a good calibration curve.

(b) Blank: Should have all the components of the standard solutions except the sample being analyzed.

(c) Sample preparation:
The sample must be in aqueous form to be introduced into the flame. Many extraction methods are used for removing solid particles based on the nature of the sample. For example, centrifugation or filtration is commonly used to remove solid debris from the sample. In the case of organic samples, solid particles are extracted using ashing followed by acid treatment for dissolving oxides. Irrespective of the extractive method used, the resultant sample should be devoid of any particulate matter to prevent clogging of the nebulizer (the capillary tube that aspirates the sample to the flame).

(d) Calibration curve:
A calibration curve is constructed using a minimum of 4 or 5 known concentrations of working standards of the element to be analyzed. Based on the expected concentration of the sample, the concentration range for preparing

the calibration curve will be selected to ensure sample concentration falls around the center of the calibration curve. As the emission spectrum of each element is unique, separate calibration curves should be generated.

Points to consider

- Store the solutions away from direct sunlight, ideally at temperatures not greater than 25 °C.
- Never handle samples with bare hands as it will lead to contamination.
- Use deionized water of the highest purity for achieving accuracy in analysis.
- Avoid long exposure of standards and sample solution to the atmosphere as it will evaporate the solvent leading to increased solute concentrations.

(e) Take measurements

1. Initiate the instrument operation based on the instructions given in the manual specific to the instrument model.
2. Blank control: Aspirate blank and set the readings to 00.
3. Top standard-setting: Aspirate the top standard (working standard with the highest concentration) and adjust the reading to the appropriate value. For example, for Na measurements, the top standard set is 10 ppm. Recheck blank readings.
4. Aspirate the remaining working standards in increasing concentrations and make a note of the readings to construct the calibration curve.
5. Plot a calibration curve with concentrations in ppm on X-axis and readout on Y-axis (as given in Fig. 53.17).
6. Once calibrated, aspirate the unknown sample and note down the readout value.
7. Determine the concentration of the unknown sample by plotting its readout value on the calibration curve.

Note: Calibrations need to be repeated by aspirating blank and standard solutions for every ten samples.

53.4.6 Applications of Flame Photometry

They are used effectively for quantifying alkali and alkaline earth metals' concentration in clinical, biological, and environmental samples.

- Water hardness can be determined using flame photometry by measuring the concentrations of Ca and Mg.
- Determination of levels of sodium, potassium, and calcium to set permissible limits in beverages like fruit juices, vegetable juices, and soft drinks.
- Soil quality analysis and determination of potassium and sodium concentrations in fertilizers.

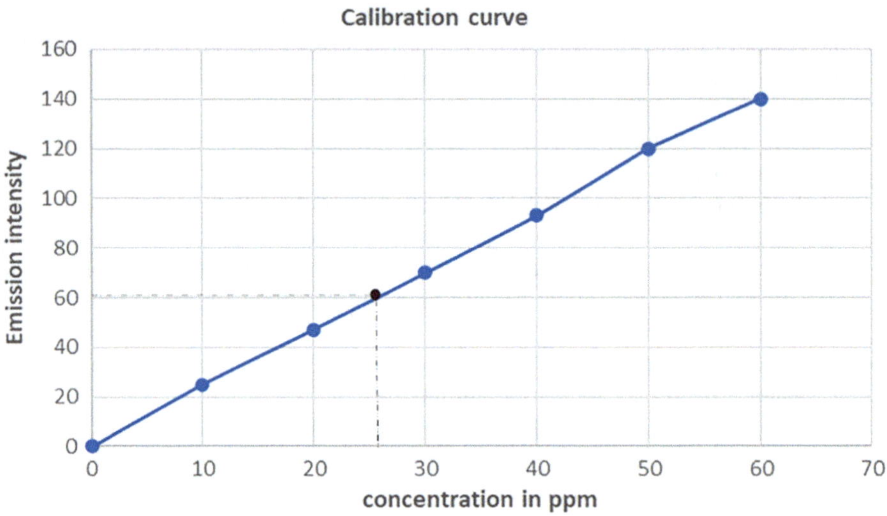

Fig. 53.17 Typical calibration curve measuring ppm of the unknown element present

Applications in pharmacology

- Flame photometry is used to measure the quantity of ions (e.g., potassium and sodium) in infusions like saline, Ringer's solution, etc.
- To check the presence of impurities and levels of ions in pharmaceutical drugs.
- Metal complexes are used widely both for the diagnosis and treatment of diseases. For example, lithium is used in the treatment of mania and prophylaxis of recurrences. Serum lithium concentrations are continuously monitored by flame photometry during the treatment course to make sure that the permissible limit is not exceeded.

53.4.7 Advantages

- Flame photometry is a simple, cost-effective, and highly sensitive technique.
- Enables both qualitative and quantitative analysis.
- It can be used for the analysis of trace elements as well.

53.4.8 Disadvantages

- Elements that lack the property of emitting light like hydrogen, halides, and inert gases cannot be analyzed.
- This technique is limited to liquid samples.
- The analysis does not provide structural information of the element present in the sample.

53.5 Fluorimetry

53.5.1 Introduction

Fluorimetry (or fluorometry) is the measurement of fluorescence. Fluorescence is a photoluminescence event (photo = light; luminescence = the emission of light), and it is a property of certain classes of molecules, chemical compounds that are referred to as fluorophores (described below). These fluorophores absorb light at a specific wavelength and subsequently emit (fluoresce) light of a different wavelength, usually a longer wavelength. In 1845, Sir John Frederick William Herschel reported the first observation of the fluorescence of a quinine solution in sunlight. All compounds (samples or objects) that fluoresce have their unique fingerprint, meaning it excites and emits predictable types of light. In other words, no two compounds have the same fluorescence signature. It is this property that makes fluorimetry a highly specific analytical technique.

A fluorometer (also referred to as fluorimeter, fluorescence spectrophotometer, or fluorospectrometer) is a spectrometer designed to measure the phenomenon of fluorescence. It generates a chosen wavelength of light required to excite a sample of interest, and it selectively transmits and measures the wavelength of light emitted by the sample. The intensity and wavelength of the light emitted can be used to identify the presence and the amount of specific molecules in the sample. Fluorescence analysis can be orders of magnitude more sensitive than other techniques. Therefore, they are used routinely in a variety of medical, industrial, environmental, forensics, and biotechnology applications.

53.5.2 Principles of Fluorimetry

A fluorophore is a chemical compound that can re-emit light upon excitation. They typically contain several aromatic rings (tyrosine, tryptophan, fluorescein, etc.) that make a fluorophore compound to fluoresce. A Jablonski diagram (Fig. 53.18) is typically used to illustrate the physics of fluorescence that involves the different electronic/energy (indicated by bold horizontal lines) and vibrational/rotational states (indicated by thin horizontal lines) in which a fluorophore can exist.

Electrons (e.g., fluorophores) are normally at the lowest energy state (indicated by S_0; Fig. 53.18). When a photon (indicated by the blue line; Fig. 53.18) at a particular wavelength interacts with a fluorophore molecule, the photon may be absorbed, causing the fluorophore to jump to an excited state (indicated by S_1 or S_2; Fig. 53.18). This transition process happens very fast (about 10^{-15} s). As soon as the energy input from the photon (excitation) stops, the fluorophore loses its energy and falls rapidly (about 10^{-12} s) to the lowest level of the first (S_1) excited state. The fluorophore remains in this state for some time (10^{-9}–10^{-8} s; also referred to as fluorescence lifetime) before returning to the electronic ground state. The return to the ground state is accompanied by the release of energy, known as fluorescence emission.

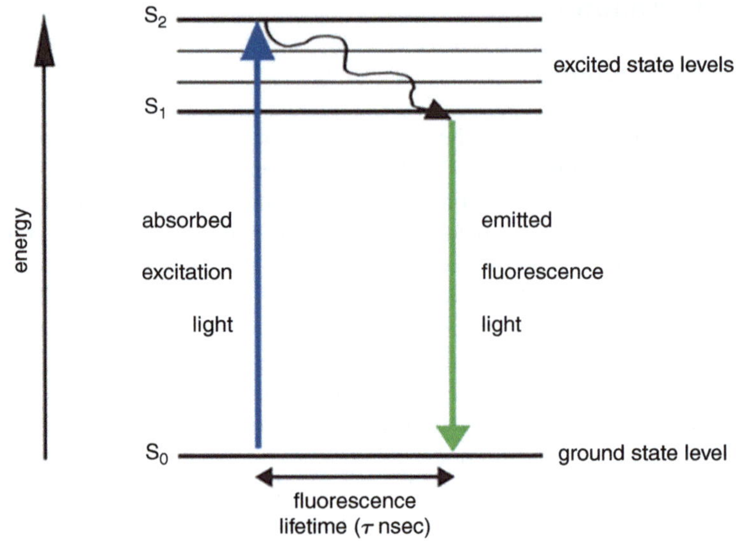

Fig. 53.18 Jablonski diagram for fluorescence
Image source: Llères D, Swift S, Lamond AI. Detecting protein-protein interactions in vivo with FRET using multiphoton fluorescence lifetime imaging microscopy (FLIM). Curr Protoc Cytom. 2007 Oct; Chap. 12:Unit12.10. doi: 10.1002/0471142956.cy1210s42. PMID: 18770849

One can deduce the fluorescence property of a molecule by measuring its excitation and emission spectra. The excitation spectrum would thus define the efficiency of a photon (light) at various wavelengths to energize or stimulate electrons in a molecule, while the emission spectrum would define the distribution pattern of the photon (light) released from the molecule. By calculating the number of photons emitted relative to the number of photons absorbed, one can derive the quantum yield of the molecule under study. Fluorophores with a large quantum yield (e.g., rhodamine) will display a bright emission as they would always emit light of a longer wavelength (lower energy) than the one that excites them.

53.5.3 Instrumentation

A fluorometer measures fluorescence parameters such as intensity and wavelength of the emission spectrum. These parameters help to identify the presence and/or the amount of specific molecules in a sample. In general, fluorometers can be classified into two types: (A) Filter fluorometer and (B) Spectrofluorometer.

(a) Filter fluorometer: Filter fluorometers have fixed excitation and emission wavelengths. They provide relative fluorescence measurements by exciting a sample and measuring its absorption and emission spectrum at a particular (fixed) wavelength. However, filter fluorometers require calibration (with a known standard concentration) before measuring test samples. A filter fluorometer is easy to handle and is a good choice when sensitive, quantitative measurements are desired for specific compounds. The essential components of a filter fluorometer are similar to that of a photometer, and it consists of an excitation source (such as a lamp), a primary filter (excitation filter), a sample chamber, a secondary filter (emission filter), and a fluorescence detection system (Fig. 53.19).

Filter fluorometers mostly have a single beam configuration with source intensity control (to reduce the effects of fluctuations and drift) and detector response. The light source used is often dependent on the type of sample being tested. A mercury lamp is the most commonly used light source as it provides many excitation wavelengths. However, when a continuous source of light (energy) is needed, a xenon arc lamp can be used. Both these sources are capable of inducing chemiluminescence.

The filters only permit the light of certain wavelengths. Typically, the primary filter permits light of short wavelengths needed for excitation, while the secondary filter permits long wavelengths of light associated with emission. Both these filters also do serve to eliminate residual light from scattering.

The fluorescence detection system consists of photomultiplier tubes (PMT) which amplifies the photon being emitted, record, and display the signal electronically.

Working: The light source sends out a light that passes through an excitation filter, which transmits wavelengths specific to the excitation spectrum of the sample and excites the sample. The sample, in turn, emits light of a particular wavelength, which then passes through the emission filter. The detector measures this emitted light, and the fluorescence value is displayed on the instrument (Fig. 53.19).

(b) Spectrofluorometer: A spectrofluorometer (also called fluorescence spectrophotometer) is a more advanced version of a filter fluorometer in that it can detect

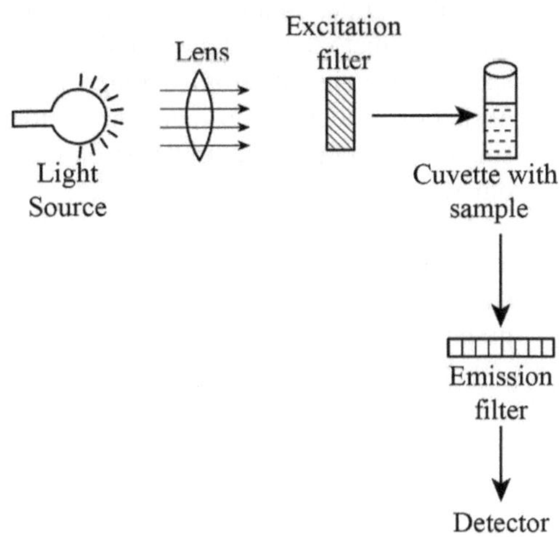

Fig. 53.19 Schematic diagram of a filter fluorometer

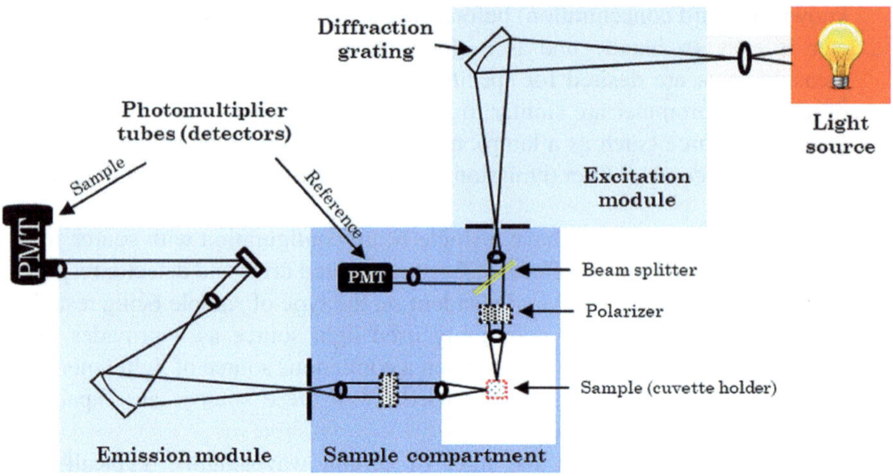

Fig. 53.20 Schematic diagram of a spectrofluorometer
Image source: Moro A.J., Lima J.C. (2019) Fluorescence Spectroscopy. In: Pereira A., Tavares P., Limão-Vieira P. (eds) Radiation in Bioanalysis. Bioanalysis (Advanced Materials, Methods, and Devices), vol 8. Springer, Cham. https://doi.org/10.1007/978-3-030-28247-9_3

fluorescence with higher precision and extraordinary sensitivity. The spectrofluorometer is superior in terms of wavelength selectivity, flexibility, and convenience. A spectrofluorometer is often equipped with a high-pressure xenon arc lamp (excitation source), monochromators, a sample chamber, and a fluorescence detection system (Fig. 53.20).
Most spectrofluorometers are equipped with high-pressure xenon arc lamps (excitation source) that act as a continuous radiation (energy) source and at the same time cover a wide spectral range (from ultraviolet to infrared spectrum).

Unlike filter fluorometers, instead of filters, spectrofluorometers have monochromators (excitation and emission) which allow the selection of a specific wavelength from a range of frequencies. Since they have separate excitation and emission monochromators, it is possible to keep one fixed at a single wavelength and monitor or record different wavelengths through the other.

Just like the filter fluorometer, spectrofluorometers use PMT for emission amplification. Electronic systems that are attached are then used to quantify the signal and display the output, usually in the form of a digital recording or as a graph.

53.5.4 Fluorescence Measurement

Fluorescence measurements can be classified into:

(i) Steady-state measurements
In steady-state measurements, the sample is excited continuously with a beam of light at a fixed wavelength while the intensity or wavelength(s) of fluorescent light emitted is recorded.
(ii) Time-resolved measurements
In time-resolved measurements, the sample is excited by a short pulse of the light, following which the fluorescence lifetime (also referred to as the intensity decay) is measured.

53.5.5 Applications of fluorometry

It is hard to imagine the chemical and biological sciences without fluorometers (both filter fluorometers and spectrofluorometers). Therefore, only a few selected examples are listed below:

(i) In biosciences, one of the most frequent applications of fluorescence spectroscopy (fluorometers) is the high precision quantification of DNA and RNA.
(ii) It is used in several industrial settings as a fast, non-invasive technique in the assessment of contamination.
(iii) In agriculture, fluorometer-based methods are widely used. For instance, tea manufacturers use total luminescence spectroscopy to discriminate between similar types of tea.
(iv) In environmental monitoring, the technique has a wide application. High-resolution fluorescence spectroscopy is used to characterize dissolved organic matter in landfills and optimize treatment processes.
(v) Spectrofluorometric techniques are also used in the pharmaceutical field to analyze and deduce the composition of various drugs.

53.6 Radioimmunoassay (RIA)

53.6.1 Background

The immune system is the body's defense mechanism elicited in response to any infection. The foreign body/toxin (antigen) released by the infectious agent induces the B cells of the host's immune system to produce specific antibodies. Antigen's activity is neutralized by the antibody to curb the manifestation of the disease. Detecting the antigen in the patient's sample is the key to the early disease diagnosis. One of the widely used immunoassay techniques for this purpose is radioimmunoassay (RIA).

53.6.2 Introduction

Radioimmunoassay (RIA) is an analytical *in vitro* technique used to detect and measure the concentration of an analyte antigen in the test sample using an antigen-specific antibody. Since it employs the use of an antigen-specific antibody, RIA is highly selective and sensitive; picomolar concentrations of an analyte can be measured. RIA was developed by Rosalyn Yalow and Solomon Berson (Fig. 53.21) in 1959 to assess the concentration of insulin hormone in human plasma. Yalow and Berson attached the insulin molecule with radioactive iodine isotope ^{125}I and administered minute concentrations of it to volunteers, including themselves. They analyzed blood samples at regular intervals to track the fate of injected ^{125}I tagged insulin. They found that Type-II diabetic patients were unable to process the insulin hormone, a significant discovery that laid the foundation for the treatment of diabetes. Yalow's discovery revolutionized the applications of radioisotopes in medicine. Biological substances like hormones, vitamins, enzymes, drugs, viruses, proteins, etc., can be measured using RIA. Yalow and Berson, in 1977, received the Noble Prize in Medicine or Physiology for the development of the RIA technique.

53.6.3 Principle

RIA is based on the competitive binding between radiolabeled and unlabeled versions of the antigen/hormone for binding to its complementary antibody (specific for the antigen). The known concentration of an antigen is tagged with a radioactive isotope and mixed with limited and defined concentrations of its complementary antibody. Test samples or standards having the unlabeled version of the antigen are mixed with the radiolabeled antigen-antibody mixture. Unlabeled antigens (also called cold antigens/analyte antigen) present in the test sample competes with radiolabeled antigens (hot antigens/tracer) to bind to the antibody (Fig. 53.22). Competitive binding exerted by the unlabeled antigens displaces the radiolabeled counterpart and binds to the antibody. The quantity of radioactive antigens displaced is equivalent to the ratio of labeled to the unlabeled antigen. As the concentration of unlabeled antigens increases, the amount of radio-labeled variants able to bind to the antibody decreases. So, the higher the quantity of unlabeled antigen in the test sample, the higher the displacement of the radiolabeled variants from the antibody. Hence, higher will be the radioactivity from the displaced free radioactive antigens. Using known concentrations of unlabeled antigens in the standard reaction mixtures, a standard curve is constructed from which the concentration of the analyte antigen can be computed.

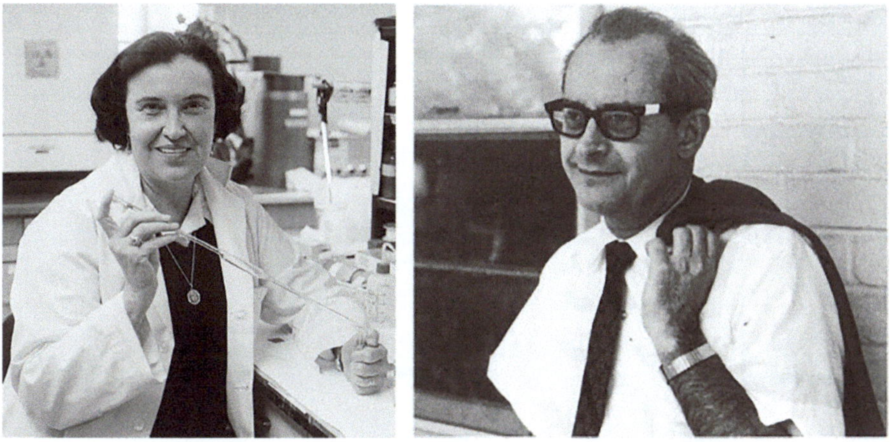

Fig. 53.21 Rosalyn Yalow (Left) at Bronx Veterans Administration Hospital, 1977 and Dr. Solomon Berson (right)
Image source: Glick, S. Rosalyn Sussman Yalow (1921–2011). Nature' **474,** 580 (2011) and William S. Middleton, Marc J. Musser, Jesse Roth, Irving Graef, Louis J. Soffer, Rolf Luft, Josepoh E. Rall, Rosalyn Yalow, Solomon A. Berson memorial service, Metabolism, Volume 22, Issue 8, 1973, Pages 963–971, ISSN 0026–0495

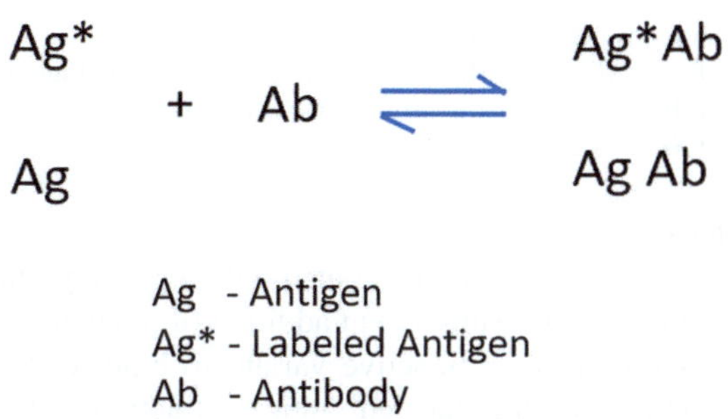

Fig. 53.22 competitive binding between radiolabeled and un-labeled antigens in RIA
Image adapted from: Sharma, A., Pillai, M. R. A., Gautam, S., & Hajare, S. N. (2014). MYCOTOXINS: Immunological Techniques for Detection and Analysis. Encyclopedia of Food Microbiology, 869–879. doi:10.1016/b978-0-12-384,730-0.00233-0

53.6.4 Types of RIA

There are two types of RIA

(i) Double-antibody RIA
In this subtype, a secondary antibody is used to precipitate the bound primary antibody. Once the antigen-bound primary-secondary antibody complex precipitates, the unbound antigen fraction is removed, and radioactivity is measured in the pellet using a gamma counter (Fig. 53.23).

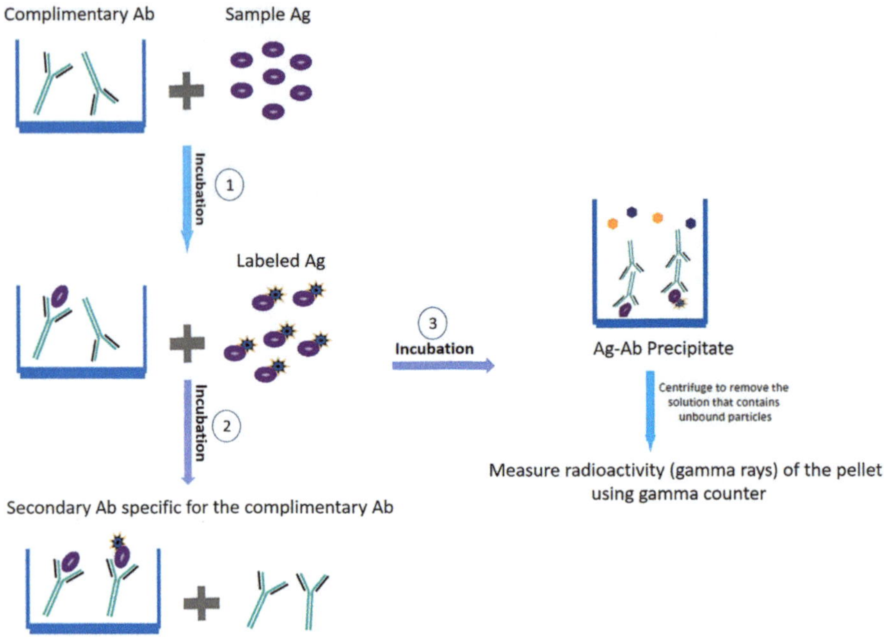

Fig. 53.23 Double-antibody RIA method
Image adapted from: R. Y. Alhabbab, Basic Serological Testing, Techniques in Life Science and Biomedicine for the Un-Expert, Springer International Publishing AG, part of Springer Nature 2018 77. https://doi.org/10.1007/978-3-319-77694-1_11

(ii) Coated-tube RIA
In this type, the reaction tube is pre-coated with a primary antibody and mixed with radiolabeled antigens. Subsequent addition of unlabeled antigen (from test sample) competes with its radioactive variant to bind to the coated antibody (Fig. 53.24). The unbound labeled antigen fraction can be collected by decanting the supernatant.

Both the supernatant and pellet fractions can be analyzed for radioactivity using a Gamma counter.

The radioactivity measured in the pellet having the antibody bound with unlabeled antigen is inversely proportional to the concentration of the analyte antigen.

53.6.5 Materials Required

1. Radiolabeled antigen/Tracer
   Although ^{14}C and ^{3}H radioisotopes (beta emitters) are available, ^{125}I is the preferred isotope for antigen labeling because it is a gamma emitter, possesses a convenient half-life period (60 days), and has a high specific activity. The ^{125}I

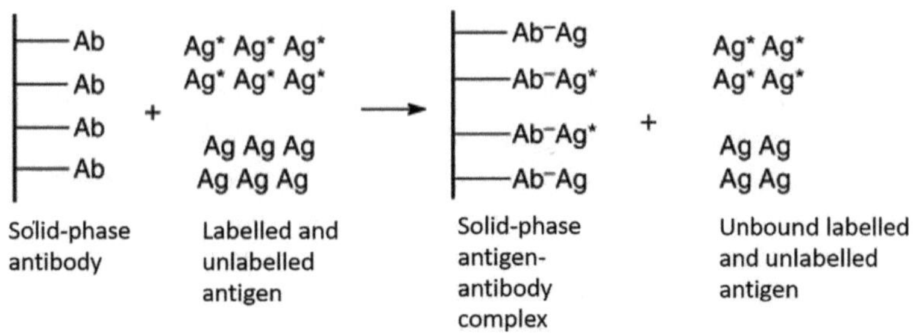

Fig. 53.24 Schematic representation of coated-tube RIA
Image adapted from: Sharma, A., Pillai, M. R. A., Gautam, S., & Hajare, S. N. (2014). MYCOTOXINS: Immunological Techniques for Detection and Analysis. Encyclopedia of Food Microbiology, 869–879. doi:10.1016/b978-0-12-384,730-0.00233-0

radiolabeled antigen is prepared by iodination of the antigen on its tyrosine residues using chloramine T or peroxidases followed by gel filtration or High-Performance Liquid Chromatography (HPLC) separation.

Specific Activity: refers to the amount of radioactivity present per quantity of radioisotope. One Becquerel is defined as the disintegration per second of the radionuclide. Expressed in units of Ci/mmol (curie/millimole of a radioisotope) or Ci/g or Bq/g (SI unit), specific activity value (usually provided in the assay kit) is important in knowing the mass of the tracer present per unit assay volume.

Radioactive concentration/Activity concentration: expressed as Bq/L, is the radionuclide activity per unit volume of diluent. This gives a measure of the number of counts added per assay. Gamma counter used to measure the gamma radiation emitted by ^{125}I gives a measure of cpm/μl (counts per minute per microlitre).

Note: Radioactive concentration is usually pre-optimized in assay kits. Follow the instructions of the RIA kit and appropriately dilute the Tracer with assay buffer.

2. Test sample (having the unlabeled antigen).
3. Antibody (specific to antigen/hormone).
4. Secondary antibody (if using the double-RIA method).
5. Antibody pre-coated microtiter plates (if using the coated-tube method).
6. Assay buffer (supplied by RIA kit).
7. Geiger counters to measure radioactivity.

53.6.6 Methodology

Note: For appropriate dilutions and final assay volume, refer to the manual of the assay kit used in the laboratory.

1. Reconstitute radiolabeled antigen, antibody vials, and positive control with the recommended volume of assay buffer.
2. Standard curve: Based on the instructions of the assay kit, prepare a minimum of five standard solutions with known concentrations of unlabeled antigen.
   Note: An ideal concentration range of standards will enable the concentration of the unknown sample to fall in the center of the range.
3. Add RIA buffer, positive control, test samples/standards, and antibody to respective tubes as shown in the table below. Prepare each tube in duplicates. Vortex the tube contents and incubate overnight (16–24 h) at 4 °C (Table 53.4).
4. Tracer preparation: Dilute the tracer with the appropriate volume of assay buffer as recommended by the kit. This is the primary Tracer stock. From primary stock, prepare a working stock solution to obtain the desired radioactive concentration. Check the concentration using the Gamma counter to achieve the desired tracer concentration in cpm/μl.
5. After the overnight incubation (step3), add the radiolabeled antigen (tracer) to all the reaction tubes and incubate for 16–24 h at 4 °C.
6. Add secondary antibody or precipitating reagent to all the tubes except the Total Counts tube. Mix well the contents and incubate at 4 °C for the duration specified by the kit.
7. Centrifuge all the tubes (except Total Counts) at 3000 rpm at 4 °C for 30 min.
8. Carefully aspirate the supernatant without disturbing the pellet in all tubes except the Total counts tube (measuring radioactivity in the supernatant will give a measure of the displaced free radiolabeled antigen).
9. Using a gamma counter, measure the radioactivity remaining in the reaction tubes (including the total count tube). Gamma counter gives the measure in terms of CPM (counts per minute). Measuring the radioactivity in the pellet will give a measure of the unlabeled antigen (from the test sample) bound to the antibody.
   Calculations
10. Find the average of the counts for each set of replicates.

Table 53.4 Tube contents of a typical RIA experimental setup

Label	Contents	Test sample/ standard *	Antibody	RIA buffer	Overnight incubation at 4 °C	Tracer
TC	Total counts	–	–	–		+
NSB	Non-specific binding	–	–	+		+
TB	Total binding	–	+	+		+
Std 1–5	Standard 1–5	+	+	–		+
PC	Positive control	+	+	–		+
TS	Test sample	+	+	–		+

[a] Dilutions of test samples and standards are done using RIA buffer

Table 53.5 Calculation model using hypothetical values

Label	Contents	CPM	Avg CPM	Net Avg CPM (B)	% B/ B_0
TC1	Total count	10900	11200		
TC2		11500			
NSB1	Non-specific Binding	313	327		
NSB2		340			
TB1	Total binding	8000	8212	7885	100
TB2		8423			
Std 1a	5 pg	6700	6712	6385	81
Std 1b		6723			
Std 2a	10 pg	5800	5817	5491	70
Std 2b		5834			
"	"	"	"	"	"
Std 5a	100 pg	2090	1991	1665	21
Std 5b		1892			
TS 1a	Test sample	4232	4117	3791	48
TS 1b		4002			
PC 1a	Positive control	2050	2027	1700	22
PC 1b		2003			

11. Calculate B_0 (Net total binding or maximum binding) by subtracting the average NSB count (nonspecific binding) from average TB (total binding) counts.

$$B_0 = TB - NSB$$

12. Calculate the net average CPM of all standards and test samples by subtracting the average NSB count from their average CPM. Example Net Avg CPM of standard solution 1 = Avg CPM of standard 1 – NSB.
13. Calculate the percentage of actual binding of standard or unknown sample antigens using the formula:

$$\%B/B_0 = \frac{\text{Net Avg CPM of standard or sample}}{B_0} \times 100$$

14. Determine the concentration of the test sample by plotting its % B/B_0 on the y-axis of the standard curve. In case of any sample dilutions, multiply the dilution factor to calculate the original sample concentration.

Table 53.5 and Fig. 53.25 are model examples showing calculations and standard curves, respectively.

Precautions to be followed.

Wear personal protective equipment – Face shield, gloves, laboratory coat, and close-toed shoes while working with radioactive materials.

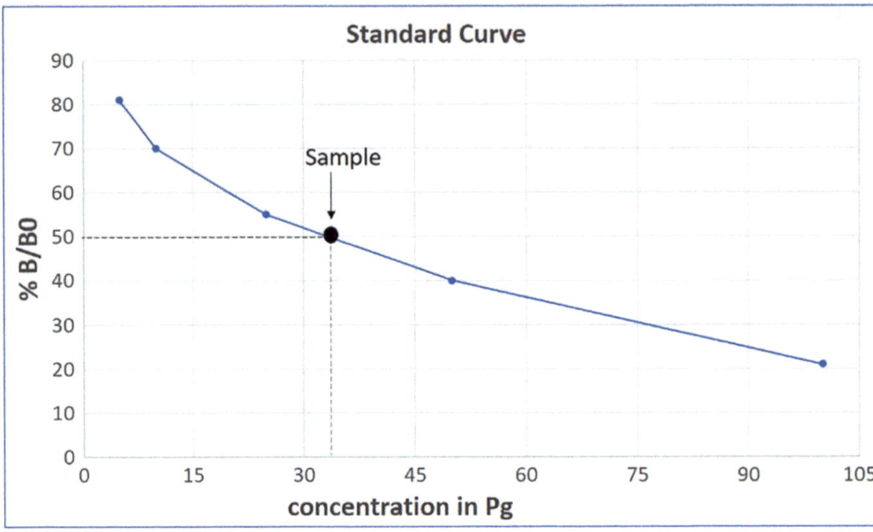

Fig. 53.25 Standard curve model using imaginary values

Always handle radioactive materials only at the designated area and check all areas and equipment with the Geiger counter before and after the procedure to avoid exposure to the radioactive isotope.

Safe handling and proper disposal of radioactive reagents and wastes are mandatory.

53.6.7 Applications

RIAs find wide applications in the analysis of clinical, biological, and environmental samples. Any molecule for which an antibody is available can be analyzed using RIA. Some of the common applications are listed below:

1. RIA aids in the early diagnosis of diseases. e.g., early diagnosis and treatment of cancer and peptic ulcers.
2. RIA is commonly used to measure the levels of hormones and vitamins.
3. Screening and tracking of blood banks for viruses like hepatitis or diseases like leukemia.

Applications in pharmacology.

RIA is one of the most specific and sensitive techniques to detect the level of drugs in biological fluids. In 1977, RIA was used to detect drugs of abuse in body fluids like saliva, urine, and blood. The assay involving ^{125}I was used to detect abused drugs - morphine, barbiturates, methadone, etc., in the urine. Dried paper discs that had been pre-immersed in urine were used in place of the liquid sample.

Some of the drugs analyzed in plasma samples by RIA include Digoxin (to treat heart failure), Isepamicin (antibiotic), Acyclovir (antiviral drug), etc.

53.6.8 Advantages

RIA offers the following advantages:

1. Highly sensitive (picograms of analyte/ml sample can be determined by using the high-affinity antibody).
2. Simple, rapid, accurate, and highly specific.
3. Allows quantification of both smaller (peptides) and larger molecules (proteins) in biological samples.
4. RIA is both qualitative and quantitative.

53.6.9 Limitations

1. Usage restricted to licensed laboratories: As it involves working with radioactive isotopes, its usage is restricted to licensed laboratories. Only trained and licensed professionals are permitted to work with radioisotopes.
2. This technique cannot be used for on-site analysis like agricultural fields, patients' homes, etc.
3. Safe disposal of radioactive waste is still a challenging issue and environmental concern.
4. Radioactive isotopes possess shorter shelf life (for example, ^{125}I has a half-life of approximately 60 days) due to radioactive decay. Therefore, non-isotopic immunoassays like EIA (enzyme immunoassay) such as ELISA (enzyme-linked immunosorbent assay) have largely replaced RIAs.

53.7 Nuclear Magnetic Resonance (NMR) Spectroscopy

53.7.1 Introduction

Over the past five decades, nuclear magnetic resonance (NMR) has been considered a well-known technique to determine the structure of carbon compounds ranging from simple organic molecules to highly complicated pharmaceutical drugs. It can also be used (i) to find an unknown compound, (ii) to verify the purity, and (iii) to determine the molecular structure of the compound. In the medical field, magnetic resonance imaging (MRI) applies nuclear magnetic resonance to identify the hydrogen nuclei (especially in water and lipids) in the body. NMR spectroscopy is used to detect the local magnetic field around the atomic nuclei. The term "nuclear" used in the NMR technique is mainly focused on the properties of the nucleus of an atom. Generally, the nucleus is made up of protons and neutrons.

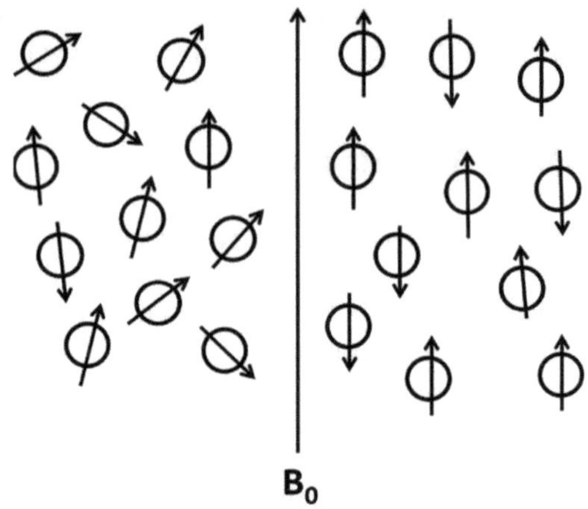

Fig. 53.26 The nuclear field in the absence (left) and presence (right) of the external magnetic field
Image adapted from: Tanzi MC, Farè S, Candiani G. Chap. 7 - Techniques of Analysis. Foundations of Biomaterials Engineering, 2019, 393–469

Moreover, the proton numbers of the nucleus are used to identify the specific element from the periodic table. The term "magnetic" represents the nuclei that act like a tiny magnet. A nucleus with an odd number of protons can have the ability to create a magnetic field. For example, proton (^{1}H) NMR is the most widely used NMR technique due to the odd number of protons. If a sample with these protons is located in a magnetic field (B_0), they will be aligned in the same (α spin state) or opposite direction (β spin state) of the external magnetic field (Fig. 53.26). The energy of the β spin is greater than the α spin. The term "resonance" represents that when using external radiofrequency energy, the proton with a low energy state of α spin is converted to a higher energy state of β spin. If it falls from β to α spin, it will also emit radiofrequency energy.

Moreover, the electrons which are surrounded by the nucleus can create diamagnetic shielding to the nucleus from the effect of the magnetic field applied by the NMR instrument. Therefore, different quantities of radiofrequency energies are required for different nuclei to change their spin state from α to β. When all the nuclei flip between the two states, they are said to be in resonance.

A spinning nucleus can create a magnetic field like a bar magnet. These nuclei are oriented randomly and pointed in different directions. Once kept in a strong external magnetic field, nuclei are aligned with the external magnetic field.

53.7.2 NMR Principle

The principle behind the NMR is that nuclei are electrically charged, and most of them have spin. When applying an external magnetic field, the energy is transferred from a lower level to a higher level. The energy transfer occurs at a specific wavelength that coincides with radiofrequency. The emitted energy will be at the same frequency when the spin reaches its lower energy level. Therefore, the NMR spectrum of the relevant nucleus can be obtained by measuring the signal that matches the transfer.

53.7.3 NMR: Sample handling and working

Usually, a deuterated solvent is used as a common solvent for NMR analysis, which (i) dissolves many compounds, (ii) is stable, and (iii) does not absorb moisture. Specifically, deuterated chloroform ($CDCl_3$) solvent is most often used for NMR measurements. Approximately 5–25 mg of sample is required to analyze ^1H NMR. To get a sharp NMR spectrum, the sample must be uniformly dissolved with NMR solvent. Therefore, the solubility of the sample should already be tested with an appropriate deuterated solvent. The sample should be located in a magnetic field. The NMR signals are formed by the excitation of the sample's nuclei with radio waves, and the obtained NMR signals are detected by radio receivers.

53.7.4 Instrumentation

The major components of the NMR instrument are shown in Fig. 53.27.

(i) Sample holder: usually, a glass tube with a length of 8.5 cm and a diameter of 0.3 mm is used as a sample holder. It should be pre-washed with acetone to ensure

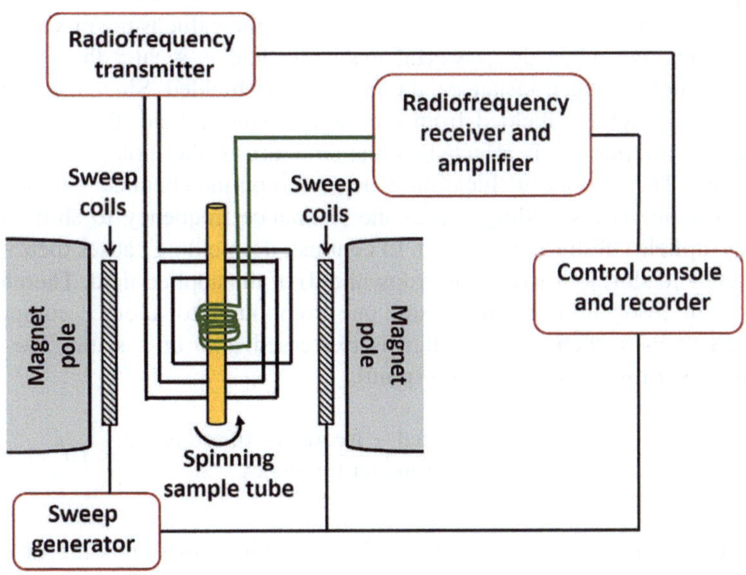

Fig. 53.27 Major components of the NMR instrument
Image adapted from: Tanzi MC, Farè S, Candiani G. Chap. 7 - Techniques of Analysis. Foundations of Biomaterials Engineering, 2019, 393–469

cleanliness. (ii) Sample probe: sample probe is used to hold the sample holder in a magnetic field and rotates the sample holder along its axis. Based on the instrument type, the sample probe might be a single coil or a system of coils. (iii) Permanent magnet: the permanent magnet can provide a uniform magnetic field (60–100 MHz). (iv) Magnetic coil: the magnetic field is induced when current flows through the magnetic coil. (v) Sweep generator: the sweep generator of the instrument is used to create an equivalent quantity of magnetic field to pass through the sample. (vi) Radiofrequency transmitter: the radio frequency transmitter coil is used to produce a short and powerful pulse of radio waves. (vii) Radiofrequency receiver: it detects the emitted radio frequency, and the signal is amplified for easy visualization. (viii) Readout systems: the computer is used to analyze the signals coming from the detector and convert them into the NMR spectrum.

53.7.5 NMR Interpretation

- *Resonant Frequency*
  The resonant frequency is referred to as the frequency at which resonance occurs in the protons of the sample. The resonant frequency is related to the strength of the magnetic field. Since all the nuclei require different energies from radio frequency to switch to the β spin state, this will cause different peaks on the spectrum.
- *Chemical Shift*
  The powerful magnetic field received by the nucleus is lesser than the applied magnetic field, which is known as shielding, and the value is positive. This value can be negative when the powerful magnetic field is greater than the applied magnetic field. In such instances, the proton is deshielded. Since the proton of a molecule is slightly shielded from the large magnetic field, the resonance frequency of the protons is affected by the structure of the molecule (aliphatic or aromatic). The amount of shielding is depended on the chemical composition of the molecule. The shielding causes the resonance frequency to shift to higher values (upfield) of the applied field. In contrast, deshielding causes the resonance frequency to shift to lower values (downfield) of the applied field. Therefore, the change in resonant frequency from one compound to another compound is referred to as a chemical shift that is expressed in δ or τ scale. The unit of chemical shift is parts per million (ppm).

$$\delta = \frac{\text{frequency of signal} - \text{frequency of reference}}{\text{spectrometer frequency}} \times 10^6$$

$$\tau = 10 - \delta$$

The values of the applied magnetic field of organic compounds are determined precisely in the presence of certain reference compounds. The most common reference compound used in NMR is tetramethylsilane (TMS). TMS is specifically used in proton magnetic resonance due to its twelve equivalent protons (Fig. 53.28).

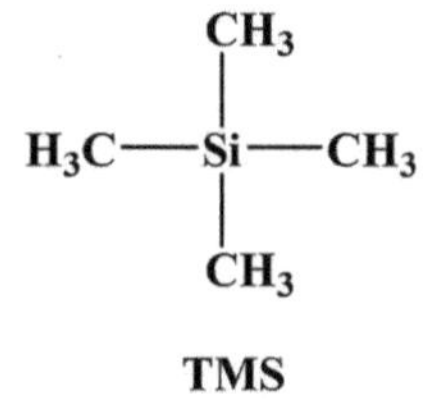

Fig. 53.28 Chemical structure of TMS

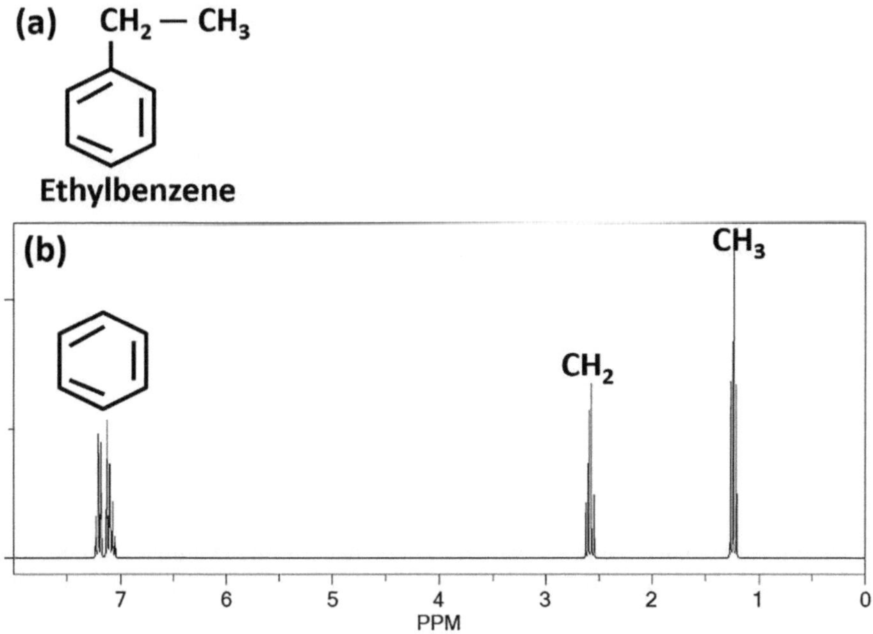

Fig. 53.29 (**a**) Chemical structure of ethylbenzene and (**b**) ^{1}H NMR spectrum of ethylbenzene

^{1}H NMR and ^{13}C NMR are most widely used to identify a specific compound based on the type and number of hydrogen and carbon atoms. In the ^{1}H NMR spectrum derived from ethylbenzene (Fig. 53.29), the methyl group (CH_3) is highly electronegative (electron-withdrawing); hence it resonates at the lowermost chemical shift. The phenyl group (aromatic ring) is the highest electron-donating group; hence it resonates at the highest chemical shift, and the methylene (CH_2) group is located in the middle. Moreover, the increased ppm of the phenyl group is due to the delocalized ring current (Fig. 53.30).

- *Spin-spin coupling*
  The relation between the spins of the adjacent nuclei of a molecule may split the NMR and which is known as spin-spin coupling. It corresponds to the number of feasible combinations of the spin alignments of adjacent protons. The splitting

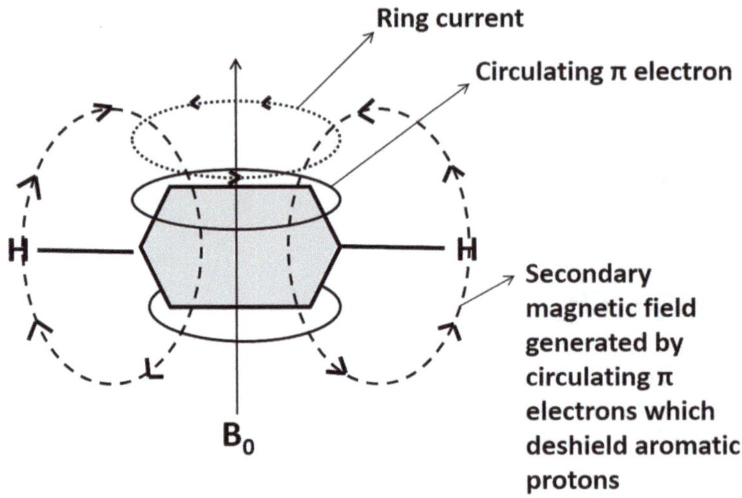

Fig. 53.30 Schematic illustration of the deshielding of aromatic protons by the effect of the ring current
Image adapted from: Poranne RG and Stanger A. Magnetic criteria of aromaticity. Chem. Soc. Rev., 2015, 44, 6597

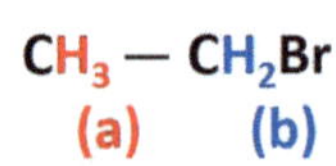

Fig. 53.31 Chemical structure of ethyl bromide

pattern is explained using the Pascal triangle pattern by the number of neighboring hydrogen atoms. The "n" equivalent neighboring hydrogen atoms can split a proton signal into n + 1.

For a clear understanding, ethyl bromide is taken as an example of the splitting of signals. The chemical structure of ethyl bromide (CH_3-CH_2Br) contains two different types of protons (Fig. 53.31), and the NMR spectrum of the molecule consists of two signals composed of multiple peaks (i.e., triplet for CH_3 and quartet for CH_2) (Fig. 53.32). The reason for the triplet is that the spin of -CH_3 can be coupled with the adjacent CH_2 group in three different ways in relation to the external magnetic field. The quartet for –CH_2 is the three protons of the adjacent -CH_3 group coupled with the –CH_2 group in four distinct ways in relation to the external magnetic field.

53.7.6 Applications of NMR Spectroscopy

- NMR is a high-precision analytical technique used to determine the purity, functional groups, and molecular structure of the sample. Moreover, NMR can be quantitatively used to analyze a mixture containing unknown compounds.

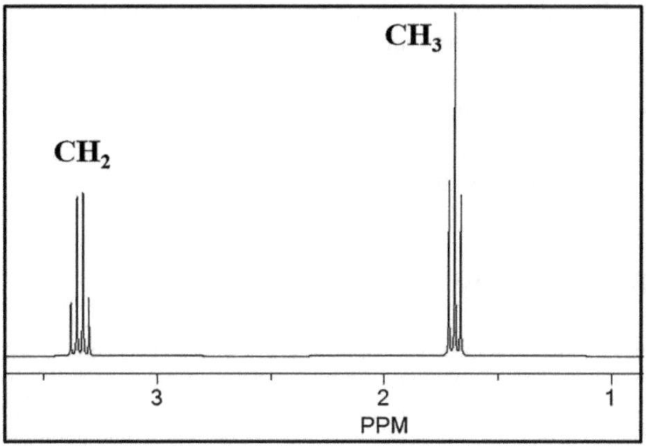

Fig. 53.32 ^1H NMR spectrum obtained from ethyl bromide

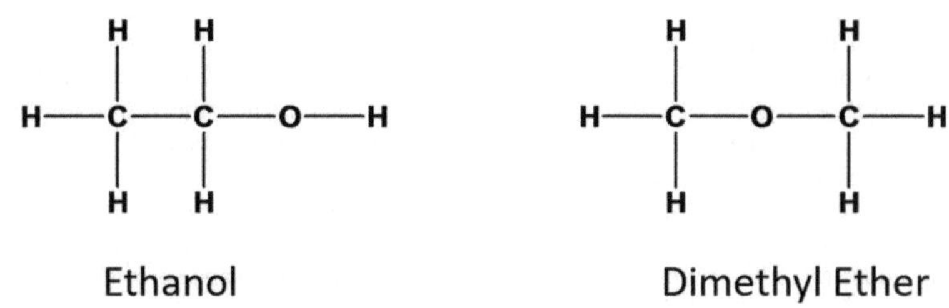

Fig. 53.33 Chemical structure of ethanol and dimethyl ether

- It is a sensitive and powerful technique that provides detailed information on the molecular structure and functional groups of organic molecules and small globular proteins.
- Sometimes two different molecules can have the same molecular formula, but the molecular structure and properties can differ based on their bonds and orientation. For example, ethanol and dimethyl ether have the same molecular formula (C_2H_6O), but the molecular structure and properties of these molecules are completely different (Fig. 53.33), i.e., ethanol is in liquid form, and dimethyl ether is a gas. Therefore, it is very important to identify the right structure to understand the properties. Under this circumstance, the NMR spectroscopic technique is precisely used to identify the molecular structure of the compound.
- The one-dimensional technique is used to study the simple chemical structure, and a two-dimensional study is used to study more complicated molecular structures.
- Complex formation viz., ligand binding to the enzyme, agonist to a receptor, the antigen to antibody, and a drug to DNA can be examined by NMR study.

- The alterations in metabolic reactions can be examined over time. Also, membrane transport can be studied by the NMR technique. It is used to study the transport of alanine and lactate in human erythrocytes.
- Compartmentation and biological structures can be studied by the NMR technique. In addition, quantitative, thermodynamic, and intact organ studies can be performed by using the NMR technique.

Bibliography

Albani JR. Principles and applications of fluorescence spectroscopy. Wiley-Blackwell; 2007.

Annesley TM. It's about the journey, not the destination: the birth of radioimmunoassay. 1960. Clin Chem. 2010;56:671–2.

Barnes RB, Richardson D, Berry JW, Hood RL. Flame photometry: a rapid analytical procedure. Ind Eng Chem Anal Ed. 1945;17(10):605–11.

Black AP, Whittle GP. New methods for the colorimetric determination of halogen residuals. Part I. Iodine, iodide, and iodate. J Am Water Works Assn. 1967;59:471–90.

Boumans PWJM. Variables in spectrochemistry and spectral-line intensity. Theory of Spectrochemical Excitation. 1966:1–29. https://doi.org/10.1007/978-1-4684-8428-1

Bursey MM. 'A brief history of spectroscopy'. Access science. McGraw-Hill Education; 2017.

Choudhury AKR. Principles of colour appearance and measurement: object appearance, colour perception and instrumental measurement. Woodhead Publishing Limited; 2014.

Church R. Clinical biochemistry and Pathology. Edtech press; 2018.

Clayden J, Greeves N, Warren S. Organic chemistry. Oxford University Press; 2012.

Cleeland R, Christenson J, Usategui-Gomez M, Heveran J, Davis R, Grunberg E. Detection of drugs of abuse by radioimmunoassay: a summary of published data and some new information. Clin Chem. 1976;22(6):712–25.

Darwish IA. Immunoassay methods and their applications in pharmaceutical analysis: basic methodology and recent advances. Int J Biomed Sci. 2006;2(3):217–35.

Fahim MM, Haji S, Koonapareddy CV, Fan VC, Asbell PA. Fluorophotometry as a diagnostic tool for the evaluation of dry eye disease. BMC Ophthalmol. 2006;6:20. https://doi.org/10.1186/1471-2415-6-20.

Lakowicz JR. Principles of fluorescence spectroscopy. 3rd ed. Springer US; 2006.

Jain and Jain. Engineering chemistry. Dhanpat Rai Publishing Company; 2017.

Katoch R. Analytical techniques in biochemistry and molecular biology. New York, NY: Springer; 2011. https://doi.org/10.1007/978-1-4419-9785-2.

Margoshes M, Vallee BL. Flame photometry and spectrometry; principles and applications. Methods Biochem Anal. 1956;3:353–407.

McMullan R. (1983) Environmental science in building. Macmillan Building and Surveying Series. Palgrave, London. doi:https://doi.org/10.1007/978-1-349-06279-9_6.

Middleton WS, Musser MJ, Roth J, Graef I, Soffer LJ, Luft R, Rall JE, Yalow R. Solomon A. Berson memorial service. Metabolism. 1973;22:963–71.

Ohta N, Robertson. AR. 2005. Colorimetry: fundamentals and applications. John Wiley & Sons, Ltd.

Phipps RJ, Smith JJ, Darwin WD, Cone EJ. Chapter 2 current methods for the separation and analysis of cocaine analytes. Forensic Sci. 2008:73–125. https://doi.org/10.1016/s1567-7192(06)06002-5.

Risby TH. Flame atomic emission spectrometry. Methods Enzymol. 1988;158:180–90. https://doi.org/10.1016/0076-6879(88)58055-2.

Salvi V, Cat Berro A, Bechon E, Bogetto F, Maina G. Litio e anticonvulsivanti nel trattamento della mania e nella profilassi delle ricorrenze [Lithium and anticonvulsants in the treatment of mania and in the prophylaxis of recurrences]. Riv Psichiatr. 2011;46(3):172–81. https://doi.org/10.1708/889.9807.

Schreuder D. Outdoor lighting: physics, vision and perception. Dordrecht: Springer; 2008. https://doi.org/10.1007/978-1-4020-8602-1_3.

Sharma A, Pillai MRA, Gautam S, Hajare SN. Mycotoxins: immunological techniques for detection and analysis. Encyclopedia of food microbiology. 2nd ed. Oxford: Academic; 2014. p. 869–79. https://doi.org/10.1016/b978-0-12-384730-0.00233-0.

Simmons S, Dale C, Holt J, Velasquez K, Stibich M. Role of ultraviolet disinfection in the prevention of surgical site infections. Adv Exp Med Biol. 2017;996:255–66. https://doi.org/10.1007/978-3-319-56017-5_21.

Thomas SP, Sankar HK. A simple colorimetric method for estimation of tramadol hydrochloride in pure and tablet dosage forms. Indian J Pharmacol. 2016;48(5):550–4. https://doi.org/10.4103/0253-7613.190746.

Twyman RM, Atomic emission spectrometry. Principles and instrumentation, Editor(s): Paul Worsfold, Alan Townshend, Colin Poole, Encyclopedia of analytical science 2nd ed., Elsevier, 2005, pp. 190–198. , https://doi.org/10.1016/B0-12-369397-7/00029–7.

Upadhyay A, Upadhyay K, Nath N. Biophysical chemistry principles and techniques. Mumbai: Himalaya Publishing House; 2009.

Ye J, Li N, Lu Y, Cheng J, Xu Y. A portable urine analyzer based on colorimetric detection. Anal Methods. 2017;9(16):2464–71. https://doi.org/10.1039/c7ay00780a.

Yoon JY. Introduction to biosensors: from electric circuits to immunosensors. 2nd ed. New York: Springer; 2016.

Wajchenberg BL, Pinto H, de Toledo T, e Souza I, Lerário AC, Ribeiro Pieroni R. Preparation of iodine-125-labeled insulin for radioimmunoassay: comparison of lactoperoxidase and chloramine-T iodination. J Nucl Med. 1978;19(8):900–5.

Zentner I, Modongo C, Zetola NM, Pasipanodya JG, Srivastava S, Heysell SK, Mpagama S, Schlect HP, Gumbo T, Bisson GP, Vinnard C. Urine colorimetry for therapeutic drug monitoring of pyrazinamide during tuberculosis treatment. Int J Tuberc Lung Dis. 2018;68:18–23.

Plant Extraction Methods

54

Mageshwaran Lakshmanan

Abstract

Extraction methods (sample preparation methods) determine the outcome and result for the study of any medicinal plants. The concerned bioactive phytochemicals are bound with other compounds in the plants most of the time. Hence, various crucial factors of extraction methods like the proper selection of plant parts, solvent, pressure, temperature, and time of process can affect the method's success. The classical techniques of plant extraction like maceration, percolation, Soxhlet extraction are used in 'small-scale' manufacture or research settings. Novel methods like microwave extraction, supercritical fluid extraction, accelerated solvent extraction, sonication, and enzyme-assisted extraction have made the extraction process of bioactive compounds relatively easier. This chapter will describe in detail the various methods, procedural steps, requirements, uses, and limitations available to extract bioactive phytochemicals from the plants.

Keywords

Extraction methods · Sample preparation methods · Percolation · Maceration · Soxhlet · Microwave extraction · Supercritical fluid extraction

54.1 Introduction

- For the study of medicinal plants, extraction methods (also called 'sample preparation methods') determine the outcome and result. Most of the time, the concerned bioactive phytochemicals are bound with other compounds in the

M. Lakshmanan (✉)
Department of Pharmacology, Thanjavur Medical College, Thanjavur, Tamil Nadu, India

M. Lakshmanan et al. (eds.), *Introduction to Basics of Pharmacology and Toxicology*, https://doi.org/10.1007/978-981-19-5343-9_54

plants. Therefore, various crucial factors of extraction methods like the proper selection of plant parts, solvent, pressure, temperature, and time of process can affect the success of the method.

- With the invention of modern chromatographic/spectrometric techniques, the process of extraction of bioactive compounds has become relatively more comfortable. However, the classical techniques of plant extraction like maceration, percolation, Soxhlet extraction are being used in 'small-scale' manufacture or research settings.
- An outline about the basic principles, choice of solvent selection, and pre-extraction procedures has already been covered in this book series (Refer section: 'Discovery of drugs from plants' in Chap. 18 of Volume 1).
- This chapter will describe the various methods, procedural steps, requirements, uses, and limitations available to extract bioactive phytochemicals from the plants.

54.2 Classification

- The plant extraction methods can be broadly classified into two types
  - Traditional methods (Conventional or classical methods):
    - Maceration, re-maceration, and sequential maceration
    - Soxhlet extraction
    - Reflux extraction
    - Decoction
    - Infusion
    - Percolation and Re-percolation
  - Novel (Non-conventional) methods:
    - Microwave extraction
    - Ultrasound-assisted extraction (Sonication extraction)
    - Supercritical fluid extraction
    - Pressurized liquid extraction (Accelerated solvent extraction)
    - Pulsed electric field extraction
    - Vibrocavitation homogenizer extraction
    - Enzyme assisted extraction
    - Hydro/Steam distillation
    - Two-phase extraction

54.3 Traditional Extraction Methods

54.3.1 Maceration

- Maceration is one of the oldest used methods worldwide. Since no heat is usually applied in this method, the term 'cold-maceration' can also be used. Traditionally, for many centuries, it is used for wine preparation.
- The finely powdered or coarsely powdered plant parts are soaked in the appropriate solvent (also called menstruum) for 3–7 days inside the tightly sealed container. The finely powdered material yields better in maceration than coarsely powdered by increasing the contact area between the solvent and plant parts.
- The mixture is repeatedly agitated to ensure uniform soaking and increase solvent penetration
- The solvent slowly digests the plant cell wall and releases the phytochemicals during the storage time. After 3 to 7 days, the initial liquid part is strained off from the container. The residual solid part (also called marc) is then pressed to release the remaining occluded solutions.
- The solutions contain phytochemicals mixed with a solvent. It is then subjected to further purification by filtration methods.
- *Re-maceration:* It is also called fractional maceration. The solvent used is first divided into three or four parts and added one by one after extracting the first part. Since new solvents are added periodically over the whole process, it results in extractant liquid production with different concentrations due to a diffusion rate difference. The extractant liquid can be later combined or used for analysis separately.
- *Sequential maceration:* This method is also slightly different from the traditional maceration method. Here, instead of one solvent, different solvents are used after completing the entire procedure with the first solvent. The 'marc' obtained after the first solvent is dried and reused with different solvents using the same maceration principle. The extractant of each step should be mixed with others, unlike in remaceration.
- *Advantages:*
  - Simple technique and suits best for small scale extraction
  - Thermolabile components can be extracted using this technique
  - A suitable method of choice for extraction of phenolic and polyphenolic compounds.
- *Disadvantages:*
  - Time for extraction is significantly longer
  - Extraction efficiency for other phytochemicals besides phenols like flavonoids is very low

54.3.2 Soxhlet Extraction

- It is also called as 'hot-continuous extraction' method.
- A specific apparatus called the 'soxhlet' apparatus is required for this method. The setup consists of a thimble, siphon arm, extraction chamber, condenser, and an outlet for a round-bottomed flask to collect the extraction solvent.
- A thimble is simply a porous bag made of either cellulose or filter paper that fits into the 'well' present in the apparatus's siphon arm. A heating mantle is required to heat the round-bottomed flask.
- First, the round bottom flask is filled with an adequate amount of appropriate solvent and kept inside the heating mantle. The round bottom flask is then fixed with the supporting stand.
- The siphon arm is now then fixed with the supporting stand such that the outlet is fitted to the mouth of the round-bottomed flask and sealed tightly. Next, the finely powdered plant parts are then placed inside the thimble and placed into the thimble chamber (well) present inside the siphon arm
- The condenser is then connected to the top of the siphon arm and fixed with the supporting stand. The condenser contains one 'water inlet port' at the bottom and one 'water out-let port' at the top. Both these ports are connected to the water source such that the continuous flow of cold water is ensured.
- When the solvent is heated continuously by heating the mantle inside the round-bottomed flask, the solvent's vapors reach the condenser via arms present in the siphon.
- Once the vapors are condensed, it drips back into the 'thimble' present inside the siphon, gets mixed with the plant parts, and produces 'extractant.'
- Once the maximum volume of extractant is attained inside the siphon via another small arm present in the siphon, the extractant gets emptied down into the round-bottomed flask (called one extraction cycle).
- The process keeps repeating until the extractant is concentrated inside the round-bottom flask.
- *Advantages:*
  - Required only a small quantity of solvent as the solvent is recycled using a condenser.
  - An excellent method to extract non-polar compounds
- *Disadvantages:*
  - Increased risk of fire accidents
  - Continuous exposure to flammable and hazardous solvents
  - The requirement of solvent purity is very high, leading to the high cost
  - Phytochemicals that are thermolabile could not be extracted by this method. For e.g., anthocyanin undergoes oxidation and degradation when extracted using Soxhlet apparatus.

54.3.3 Reflux Extraction

- The principle of reflux extraction similar to the Soxhlet extraction method.
- However, it differs from the Soxhlet method in terms of set-up
  - It does not require a siphon and thimble.
  - The round-bottomed flask is directly connected to the condenser.
  - Plant parts are added directly into the round-bottomed flask instead of the thimble in the Soxhlet apparatus.
  - Upon heating, the solvent evaporates and is recycled back to the round-bottomed flask after condensation and provides extractants.
- Similar to Soxhlet extraction, this is method cannot be used for thermolabile phytochemicals

54.3.4 Decoction

- This method is very suitable for extracting hard plant parts like roots and barks.
- The plant parts are boiled with water for a few hours until the plant cell wall is destroyed by the heat leading to the release of phytochemicals.
- Decoction, compared with maceration, increases the dissolution of compounds like baicalin, wogonoside, etc. This is because decoction utilizes a high temperature that prevents the beta-glucuronidase enzyme-based transformation of baicalin and wogonin to baicalein and wogonin, respectively.
- This method is often used in various traditional medicines like Ayurveda, Sidha, and Chinese traditional medicine to prepare herbal mixtures and paste.
- This method also utilizes heat and cannot be used for thermolabile compound extractions.

54.3.5 Infusion

- This is the simplest method of extraction in which the plant parts as a whole or cut minimally are suspended in the solvent (usually water or alcohol).
- It differs from the maceration method as follows
  - Infusion involves a shorter time
  - Plant parts are not powdered or crushed
- Both cold and hot methods can be used in the infusion.

54.3.6 Percolation and Re-percolation

- This method involves a constant passage of solvent into the plant parts at a predetermined flow rate. This method mainly works by gravitational force. The coffee maker is a simple example of a percolator in day-to-day life.

- It requires specialized equipment called 'percolator.' Percolator is a simple conical instrument with an outlet at the bottom. The conical space is filled with layers of different materials like plant parts, solvent, sand, filter paper, etc., depending upon the type of compound extracted.
- Percolation is done in three steps as follows
  - *Wetting (Imbibition):* Half amount of solvent is used in this step. The plant part swells and becomes easier for penetration of solvent in later steps. This step is conducted outside the percolator. It usually requires 4–5 h.
  - *Soaking:* The swollen raw material is now placed inside the percolator, filled up with the remaining volume of solvent, and left undisturbed. This step lasts for 1–2 days, and extraction efficiency is similar to maceration.
  - *Percolation:* Solvent is passed over the raw material at a constant rate, and the extract is collected until the raw material ran-out. The drop rate is usually fixed at a moderate value (6–7 drops/min).
- *Re-percolation:* Here, percolation is done in repeated steps by dividing plant parts into quotas. This is followed by the extraction of consecutive parts with the extractant from the former one. This increases the yield of bioactive phytochemicals.
- *Advantages:*
  - Simple technique and easy to operate
  - Alteration in the procedure can be done quickly to increase the extractant yield.
  - It can be used for both thermolabile and thermostable compounds.
- *Disadvantages:*
  - Requires large volume of solvent
  - Exposure to toxic and hazardous solvent for a long duration of time
  - Organic waste generation is high and requires a proper disposal method.

54.4 Novel/Modern (Non-conventional) Plant Extraction Methods

54.4.1 Microwave Extraction

- *Principle:* Analyte from the sample can be partitioned into the solvent by using microwave energy. The microwave produces heat by interacting with the dipoles of polarizable materials. This heat is then transferred by conduction. Rotation of dipole molecule under electromagnetic field by microwave results in the breaking of hydrogen bonds in the sample, enhancing solvent penetration.
- *Advantages:*
  - It can also be used for sample preparation techniques like drying besides extraction methods
  - The best method to rapidly evaporate the solvent in the extracted solution.

  - Shorter time for extraction (15–30 min) with less solvent consumption
  - Temperature and time can be fine-tuned when compared to conventional methods for a better quality of yield.
  - More number of the analyte can be isolated from the sample using this method
- *Disadvantages:*
  - Standard guidelines for temperature range and time are not available for various type of raw materials for extracting
  - Thermolabile compounds cannot be extracted by this method. Few thermostable compounds like isoflavones, gallic acid, and trans-resveratrol can undergo oxidative damage when the temperature crosses 100°C or subjected to microwave for more than 20 min.
  - It cannot be used to extract the non-polar compounds

54.4.2 Ultrasound-Assisted Extraction (Sonication Extraction)

- *Principle:* Ultrasound (20 kHz to 100 MHz) creates a mechanical force called 'cavitation' via sound energy. Cavitation is nothing but the formation, growth, and collapse of a bubble inside the medium formed when microwave travels and produces rapid compression and expansion in the medium. The cavitation results in the disruption of the plant cell wall and releases the phytochemicals by rinsing the cellular contents. Due to this mechanical effect by ultrasonic waves, the surface contact between the sample and solvent is also increased, resulting in enhanced solvent permeability.
- When used in a liquid-liquid-containing solid medium, ultrasound can produce a temperature up to 5000 k with 1000 atm pressure by forming a cavitation effect. This leads to leaching out of various organic and inorganic compounds from the plant parts rapidly.
- *Advantages:*
  - Drastic reduction in the extraction time and quantity of solvent used.
  - Simple equipment, simple procedure, and relatively low-cost technology
  - It can be adopted in both large and small-scale manufacturing.
- *Disadvantages:*
- Free radical generation may occur when more than 20 KHz is used results in an alteration in the phytochemical properties
- Efficacy is decreased when used for extraction in thick woody compact materials.

54.4.3 Supercritical Fluid Extraction

- This method is extensively used in the cosmetic, pharmaceutical, and food sectors to purify and isolate the active ingredients from raw materials.

- *Principle:* Supercritical fluid (called dense-gas) is a substance with the physical property of gas and liquid at a given critical temperature and pressure, leading to a dramatic change in its solving property.
- The CO_2 can act as a supercritical fluid above 7380 kPa and 31.1°C. At this condition, the solving property of CO_2 changes dramatically and can be used as a supercritical extractant (S-CO_2). The S-CO_2 is a commonly used solvent due to its lower critical temperature, inertness, non-toxicity, and low cost.
- *Advantages:*
  - By varying the pressure of the supercritical fluid, fine-tuning of extraction can be done by this method, i.e., the selectivity of extraction by this method is very high. E.g., Volatile terpenes in the plants are extracted while carotenoid pigments and waxes can be stopped from being dissolved by changing the pressure of the critical fluid. Hence impurities like waxes and carotenoid pigment will not appear in the final extract.
  - An excellent method for extracting non-polar compounds
  - Usually operated at room temperature and hence thermolabile compounds can be extracted by this method
  - Easy modification of procedure so that even the spectrum of extraction can be increased. E.g., Though S-CO_2 has low extraction efficacy for polar compounds, it can be increased significantly by adding ethanol and methanol in a small amount. As another example, dicarboxylic acids are insoluble in S-CO_2 and cannot be extracted. However, adding tertiary amines along with S-CO_2 will shield carboxylic acid and increases its solubility in S-CO_2
  - Utilizes a small amount of solvent (green-type- extractor), environment-friendly, and less pollutant
  - Supercritical fluid can be reused, and hence minimal waster generation
  - The solvent can be easily separated from the solute just in a matter of seconds by decompressing (liquid form of S-CO_2 will turn into gas-CO_2 instantly)
- *Disadvantage:*
  - Very high cost and can be used only in large-scale manufacturing units.

54.4.4 Pressurized Liquid/Solvent Extraction

- It is also called as enhanced fluid/solvent extraction, accelerated fluid/solvent extraction, high-pressure liquid/solvent extraction.
- *Principle:* Application of high pressure maintains the solvent in the liquid phase resulting in an increase of solubility and diffusion rate of solutes into the solvent. This results in high penetration of solvent in the plant parts and yields better extractants.
- *Advantages:*
  - A suitable method for extraction of saponins, essential oils, and flavonoids.
  - Decreases the extraction time significantly

  - Less use of solvent and eco-friendly
  - Better repeatability
- *Disadvantage:*
  - Inconsistent results for thermolabile compounds though few researchers successfully extracted using this technique

54.4.5 Pulsed Electric Field Extraction

- *Principle:* An electric potential is passed in the suspension containing the living cells of plant parts. The generated electric field will disrupt the dipole nature of the cellular membrane molecules as per their charge. When the electric potential exceeds a critical value (usually >1 V), pores are formed in the cell membrane due to repulsion between the charged molecules. This results in increased permeability and leaching out of phytochemicals from the cell into the solvent
- Usually, plant cell membranes are damaged when the electric field is generated as 500–1000 V/cm and applied in a pulsed manner (0.01–0.0001 s duration range).
- *Advantages:*
  - This method will not increase the temperature drastically and can be used to extract thermolabile compounds
  - It can also be used as a pretreatment method before extraction by traditional methods.
- *Disadvantages:*
  - Various factors like field strength, pulse number, energy input, and temperature play a role, and hence consistency of outcome is not uniform.

54.4.6 Vibrocavitation Homogenizer Extraction

- *Principle:* Mechanical vigorous mixing can increase the interfacial area of components in the mixer and concentrate energy in small volumes, leading to mixing compounds with different viscosity. This enables the excellent dissolution of solutes in the solvent.
- This method runs as a multi-step process.
- Preliminary mixing is the first step, followed by dispersion/homogenization to the particles of <5 μm.
- Later, the mixer is fed to the multi-stage homogenizers, which further disperse and homogenize the mixer by vigorous mixing. Later the generated emulsion is fed to the separation device, which extracts the phytochemicals from the raw material.
- Advantage:
  - It can be used as a pretreatment method for conventional extraction methods
  - Liquids with a broad range of viscosities can be extracted.

54.4.7 Enzyme Assisted Extraction

- *Principle:* Instead of heat-induced destruction, which causes the formation of micelles, denaturation, and coagulation of proteins, the cell wall and other macromolecules can be destroyed with enzymes. Cellulase, pectinase, and alpha-amylase are added to the extract, which destroys the cell wall and macromolecules and releases phytochemicals.
- The polysaccharide-lignin matrix formed by hydrogen bonds is inaccessible by the solvent in other methods. However, enzymatic pre-treatment breaks these bond and release the phytochemicals better.
- Two variants exist in this method:
  - Enzyme-assisted aqueous extraction—enzymes are used for the extraction of various seed oils.
  - Enzyme-assisted cold pressing—the enzyme is used to dissolve the cell wall of the seed by hydrolysis.
- Factors like particle size, enzyme concentration, water-solid ratio, and time to hydrolysis determine the outcome of extraction
- *Advantages:*
  - Water is used as a solvent and hence eco-friendly method
  - Non-toxic and non-inflammable method
  - One of the best method for extraction of thermolabile compounds
- *Disadvantage:*
  - Raw material with high moisture content cannot be processed effectively by this method as moisture interferes with enzymatic activity.

54.4.8 Hydro/Steam Distillation

- *Principle:* Water, either directly or in the form of steam, is used as a solvent for extraction. This method is commonly used for extracting the volatile oils
- Few natural compounds may undergo decomposition by this technique
- It takes a long time for extraction, and water-insoluble compounds cannot be extracted by this method

54.4.9 Two Phase Extraction

- This method is under development and standardization and not used widely.
- Most of the time, while extraction, the remaining pulp still contains useful biologically active compounds. This is because the solvent used is not appropriate for those compounds.
- Hence instead of using a single solvent, a two-phase solvent can completely extract the biologically active phytochemicals from the plant parts. E.g., using two-phase vegetable oil- aqueous EtOH as a solvent, both lipophilic and hydrophilic compounds can be extracted.

- Alternatively, surfactants (has both lipophilic and hydrophilic property) are mixed in the plant part, and by controlling the ratio of surfactants in the different phases, a controlled extraction can be done.
- Two-phase extraction is still in the standardizing phase, and the outcome is based on various factors like ratio of oil to plant part, type of oil used, and polar-phase composition, etc.,

54.5 Conclusion

The choice of extraction techniques depends upon various factors like the type of scale of production, area of research, and nature of phytochemicals being studied. Hence no extraction method is ideal for all types of plant parts, and the researcher should put maximum effort into choosing a particular method such that the final yield should be repeatable and of good quality in nature.

Bibliography

Bouras M, Grimi N, Bals O, Vorobiev E. Impact of pulsed electric fields on polyphenols extraction from Norway spruce bark. Ind Crop Prod. 2016;80:50–8.

Chemat F, Cravotto G. Microwave-assisted extraction for bioactive compounds. Boston: Springer; 2013.

Chen H, Zhou X, Zhang J. Optimization of enzyme assisted extraction of polysaccharides from Astragalus membranaceus. Carbohydr Polym. 2014;111:567–75.

Delazar A, Nahar L, Hamedeyazdan S, Sarker SD. Microwave-assisted extraction in natural products isolation. Methods Mol Biol. 2012;864:89–115.

Jones WP, Kinghorn AD. Extraction of plant secondary metabolites. Methods Mol Biol. 2012;864: 341–66.

Santos-Buelga C, Gonzalez-Manzano S, Dueñas M, Gonzalez-Paramas AM. Extraction and isolation of phenolic compounds. Methods Mol Biol. 2012;864:427–64.

Seidel V. Initial and bulk extraction of natural products isolation. Methods Mol Biol. 2012;864:27–41.

Strati IF, Gogou E, Oreopoulou V. Enzyme and high pressure assisted extraction of carotenoids from tomato waste. Food Bioprod Process. 2015;94:668–74.

Vinatoru M, Mason TJ, Calinescu I. Ultrasonically assisted extraction (UAE) and microwave assisted extraction (MAE) of functional compounds from plant materials. TrAC Trends Anal Chem. 2017;97:159–78.

Wei Q, Yang GW, Wang XJ, Hu XX, Chen L. The study on optimization of Soxhlet extraction process for ursolic acid from Cynomorium. Shipin Yanjiu Yu Kaifa. 2013;34(7):85–8.

Wrona O, Rafińska K, Możeński C, Buszewski B. Supercritical fluid extraction of bioactive compounds from plant materials. J AOAC Int. 2017;100(6):1624–35.

Yahya A, Yunus RM. Infuence of sample preparation and extraction time on chemical composition of steam distillation derived patchouli oil. Process Eng. 2013;53:1–6.

Zhang H, Wang W, Fu ZM, Han CC, Song Y. Study on comparison of extracting fucoxanthin from Undaria pinnatifid with percolation extraction and refuxing methods. Zhongguo Shipin Tianjiaji. 2014;9:91–5.

Part III

Research Methodology

Literature Search

55

Mourouguessine Vimal

Abstract

Litcraturc scarch is onc of thc most daunting tasks, especially for the novice researchers. However, it is one of the most essential steps in the research process and every researcher should know the various literature search techniques in order to develop the skill of performing a quick and effective literature search. Literature search refers to gathering information in a systematic manner from various published and non-published literature sources. This chapter discusses the different types of medical literature like primary, secondary and tertiary literature. It also discusses about the two broad types of literature search like physical literature exploration and electronic literature search. Also, the various steps involved in performing a literature search including the various techniques to improve literature search like MeSH term search, phrase search, field search, use of Boolean operators, wildcard search and truncation search have also been elucidated in this chapter.

Keywords

Literature review · Keyword search · Boolean operators · MeSH search

55.1 Introduction

- "Literature search is a systematic and well-organized search from the already published (e.g. journals, databases, and textbooks) and non-published (e.g. dissertation, thesis, conference reports, and preprints) data to identify a breadth of good quality references on a specific topic".

M. Vimal (✉)
Department of Pathology, Sri Manakula Vinayagar Medical College and Hospital, Puducherry, India

M. Lakshmanan et al. (eds.), *Introduction to Basics of Pharmacology and Toxicology*, https://doi.org/10.1007/978-981-19-5343-9_55

- Various reasons for conducting a literature search are given in Box 55.1
- Although literature search is a laborious and time-consuming process, good knowledge of literature search techniques and simple search tips will aid in simplifying the process

Box 55.1. Various Reasons for Conducting a Literature Search

- Improving the knowledge about a particular topic
- Understanding which aspects of a topic were researched and which were not
- Identification of research problem(s)
- Formulating research question(s)
- Devising/refining a research protocol (identification of literature gaps, trends in literature, the justification for research, appropriate study design, methodology, sample size and sampling techniques, statistical analysis, etc.)
- Identification of weaknesses/limitations in other similar studies to rectify the same in the present work
- Writing original research articles (especially the introduction and discussion section)
- Writing systematic reviews or meta-analysis
- Formulating evidence-based guidelines
- Identification of appropriate journals to publish the research findings
- To identify the seminal research works published in the area of interest
- To identify experts working in the same research area

55.2 Types of Medical Literature

There are three types of medical literature (Fig. 55.1), namely:

- *Primary literature:* it includes unpublished research (dissertation, thesis, conference reports, preprints) and original articles published in reputed peer-reviewed journals including ideas/opinions, case reports, case series, observational and interventional studies. The unpublished literature is also called as "gray literature".
- *Secondary literature:* it includes indexing and abstracting systems that organize and provide easy retrieval of primary literature. It also includes systematic reviews and meta-analyses, where the information from the primary literature are analyzed and interpreted
- *Tertiary literature:* it includes reference books, treatment guidelines, drug formularies, etc., which compile information from primary and secondary literature

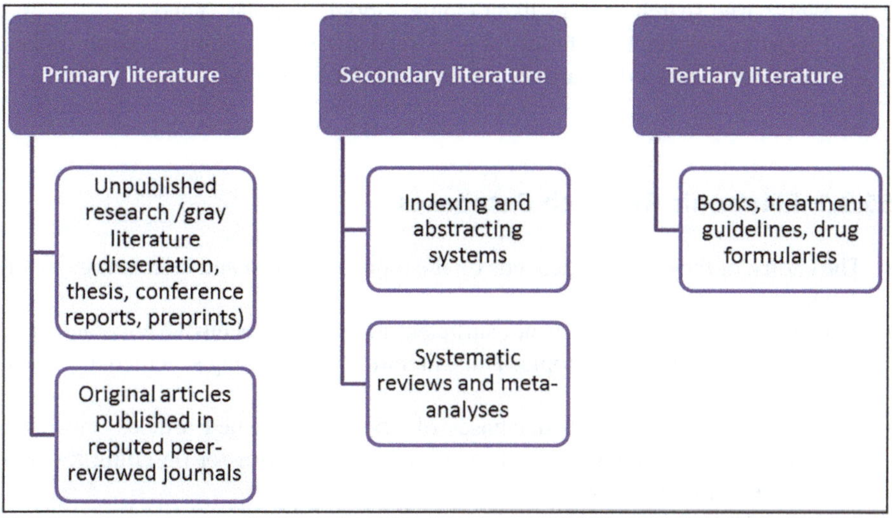

Fig. 55.1 Types of medical literature

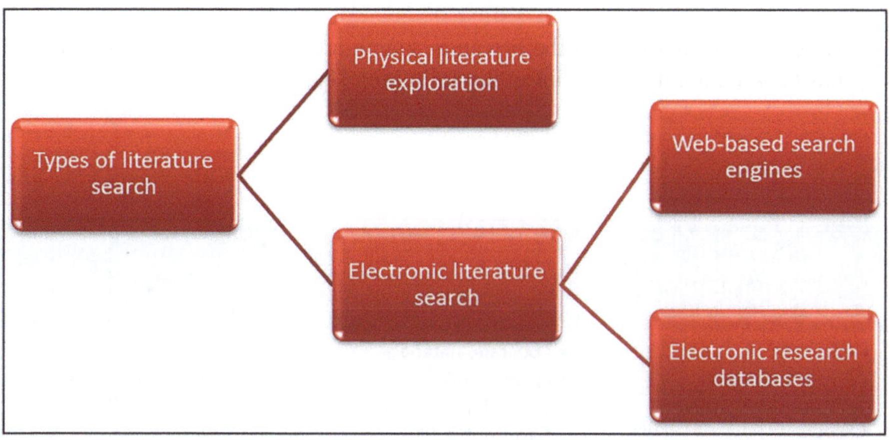

Fig. 55.2 Types of literature search

55.3 Types of Literature Search

Types of literature search can be broadly classified (Fig. 55.2) into the following:

- *Physical literature exploration:* searching the libraries manually for books, journals, etc.
- *Electronic literature search:* this can be done using the following methods:

- Web-based search engines like Google, Google Scholar, Yahoo
- Electronic research databases for published articles, on-going research studies, and evidence-based databases for integrated information

55.3.1 Electronic Research Databases

- The choice of the database depends on the topic of interest and its coverage by the databases
- Searching a single database is not sufficient to get comprehensive coverage of all information pertaining to a topic. Thus, literature search using several databases is required
- Various electronic research databases of original published articles, evidence-based databases for integrated information, and databases for on-going research studies are enlisted in Fig. 55.3.

55.4 Methodology of Literature Search

The various steps involved in the literature search are shown in Fig. 55.4 and each step has been elaborated in detail in the following sections.

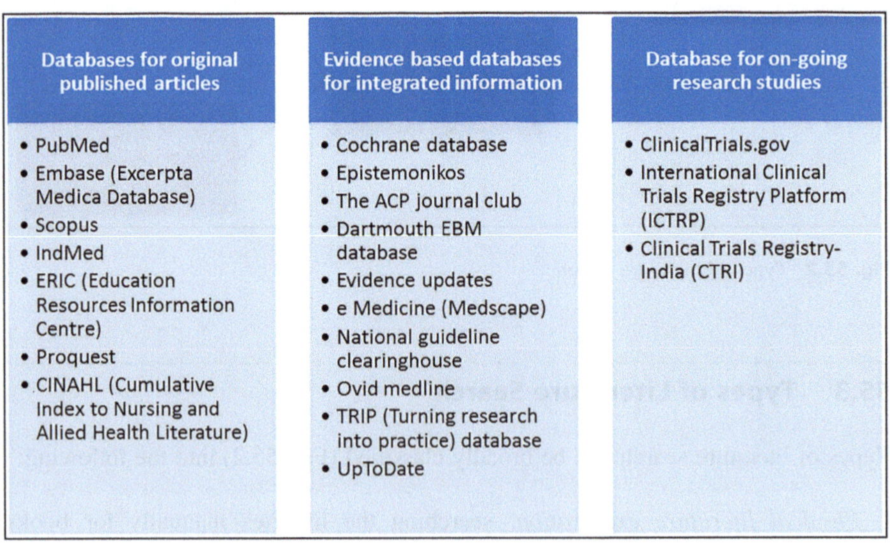

Fig. 55.3 Electronic research databases of original published articles, evidence-based databases for integrated information, and databases for on-going research studies

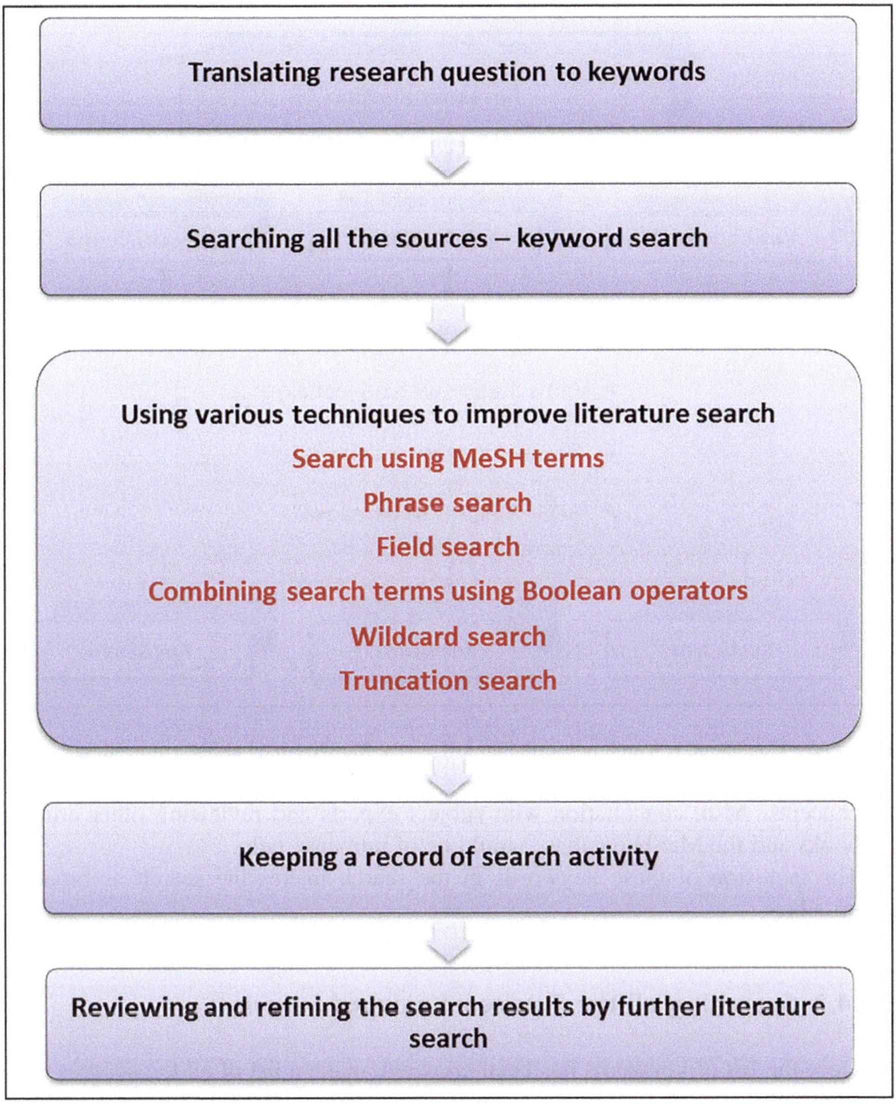

Fig. 55.4 Various steps involved in the literature search

55.4.1 Translating Research Question to Keywords (Refer Fig. 55.5)

- We can begin with the identification of key concepts/ideas in the research question
- Choose keywords for each concept/idea—this can be done by brainstorming to find synonyms/alternate terms that could be used by various authors for the key

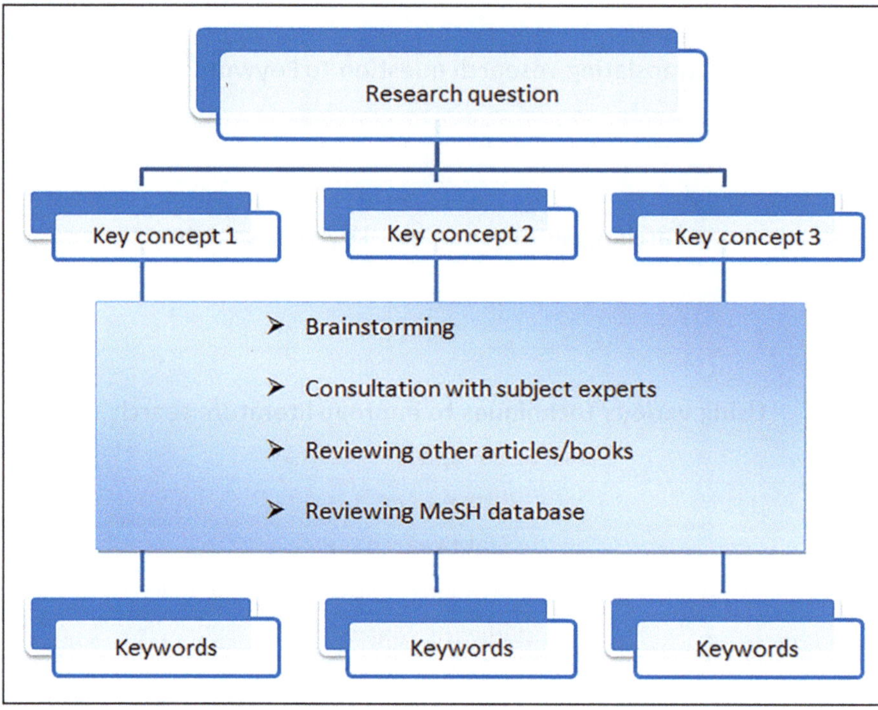

Fig. 55.5 Translating research question into keywords. MeSH: Medical Subject headings

concepts. Also, consultation with subject experts and reviewing other articles/ books and the MeSH database would be of immense help.

- The inclusion of these keywords in the search makes the search as broad as possible

55.4.2 Searching all the Sources: Keyword Search

- Once the list of keywords has been prepared, make a list of all the sources to be searched
- Both physical literature exploration and electronic literature search can be carried out
- Electronic literature search can be done with the help of web-based search engines like Google, Google Scholar, Yahoo or by exploring the electronic research databases
- Preparing a list of the electronic research databases to be searched for before initiation of the search process would be of immense benefit (refer Fig. 55.3)
- While searching the electronic literature, it requires exertion of caution to prevent the acquisition of information from unreliable sites. Thus, the search should be carried out only in authentic sites or research databases

55.4.3 Techniques to Improve Literature Search

- *Search using MeSH terms:*
  - The MeSH term refers to Medical Subject headings which are the "National Library of Medicine's (NLM's) controlled hierarchical vocabulary thesaurus that is used for indexing articles in PubMed, with more specific terms organized underneath more general terms"
  - Each article in PubMed is tagged with various MeSH terms. The MeSH term (s) with the asterisk (*) symbol (e.g. Vitamin D Deficiency*) indicate that they are the major topic(s) of the article, usually derived from the title and/or statement of purpose. The non-asterisked MeSH terms are non-major topics or additional topics discussed within the article
  - Identifying the right MeSH term from the MeSH database and searching using MeSH terms helps us in getting more relevant search results. E.g. the MeSH term for "pregnancy induced diabetes" is "gestational diabetes"; and that for "heart attack" is "myocardial infarction"
  - The MeSH term can be identified from the MeSH browser, which can be accessed from https://www.ncbi.nlm.nih.gov/mesh/ or by clicking on the MeSH Database on the PubMed homepage (https://www.ncbi.nlm.nih.gov/pubmed/)
  - Once the MeSH terms are identified, they can be searched in PubMed by typing the MeSH term followed by [mh]. E.g. the search "angina [mh]" helps to retrieve articles which are tagged with angina as a MeSH term.
  - Also, the MeSH database homepage contains brief tutorials on how to use MeSH terms for the literature search
- *Phrase search:*
  - Phrase search refers to "searching for two or more words together in a phrase"
  - However, the phrase should be mentioned within closed quotes ("") while performing the literature search. E.g. "adverse drug reaction", "medication error", "diabetes mellitus" etc.
  - This type of search decreases the number of search results by increasing their relevance to the particular topic
  - Although the majority of the databases permit phrase search (e.g. PubMed), the database guide needs to be referred, to know whether it offers such a facility or not
- *Field search:*
  - Some databases, like PubMed, permit searching specific fields of the PubMed record using certain search field tags
  - A list of PubMed search field tags and their descriptions are available in the PubMed help page (available from https://www.ncbi.nlm.nih.gov/books/NBK3827/) (refer Table 55.1 for a list of PubMed search field tags and their description with some examples)

Table 55.1 List of some PubMed search field tags and their descriptions

Search field tag	Description	Example
[ad]	Affiliation	Jipmer[ad]
[ti]	Title	Asthma[ti]
[tiab]	Title/abstract	Hypertension[tiab]
[ir]	Investigator	Vimal M[ir]
[au]	Author	Vimal M[au]
[mh]	MeSH terms	Glaucoma[mh]
[majr]	MeSH major topic	Neoplasm[majr]
[pt]	Publication type	Review[pt]
[la]	Language	Chinese[la]
[dp]	Publication year	2010:2020[dp]orlast 10 years[dp]

- Field search can also be done by using search filters which aids in focused searching
- This can be done by applying limits or filters to the search. The following filters can be used while searching in PubMed:
  Article types—case reports, clinical study, clinical trial, comment, editorial, guideline, meta-analysis, observational study, practice guideline, systematic reviews, review, or validation study, etc.
  Text availability—Abstract, free full text, or full text
  Publication dates—last 5 or 10 years. Even a custom date range can be entered to identify articles published within that period
  Species – Humans, or other animals
  Additional search filters like languages (e.g. English, Spanish, etc.); sex (male, female); subjects (e.g. AIDS, cancer, etc.); journal categories (Core clinical journals, dental journals MEDLINE, etc.); ages (newborn, infant, adult, etc.); and search fields (author, journal, title, etc.) are also available

• *Combining search terms using Boolean operators:*
  - The search terms can be linked together for searching, with the help of three terms, namely AND, OR, NOT; called as the Boolean operators
  - The Boolean operators are always written in capital letters/upper case
  - Use of Boolean operators may aid in either narrowing or widening the search results
  - The various Boolean operators and their influence on the search results are as follows (refer Fig. 55.6):
    AND—every result item will have every search term (helps in **narrowing** the search results). E.g. the search "vitamin D AND asthma" will retrieve only those articles containing both the search terms vitamin D and asthma.
    OR—every result item will have at least one search term (helps in **widening** the search results). E.g. the search "vitamin D OR asthma" will retrieve

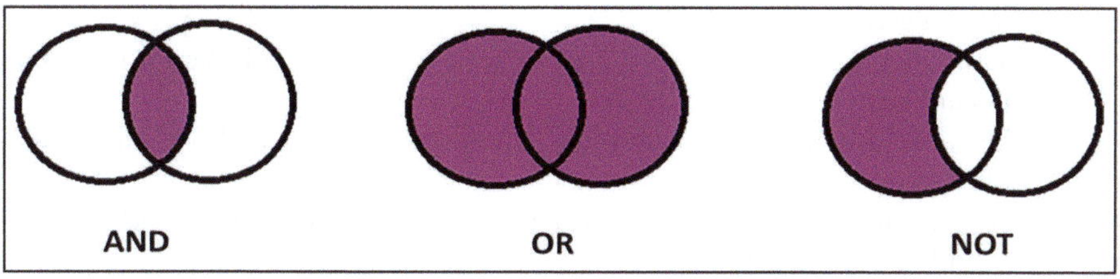

Fig. 55.6 Combining search terms using Boolean operators and their influence on the search results. The shaded area in violet indicates the results that would be obtained by the use of each type of Boolean operator

Table 55.2 Literature search by wild card and truncation search

Search technique	Symbol	Description	Examples: Search term	Examples: Words captured
Wild card search	?	Used for single character searching, if it exists (zero or one character), especially searching for alternate spellings	colo?r	color and colour
			an?esthesia	anesthesia and anaesthesia
	#	Used for single character searching (one character), especially capturing spellings with "s" or "z".	Organi#ation	organisation and organization
			Nebuli#er	nebuliser and nebulizer
Truncation (also a type of wild card search)	*	Captures the search term with variant endings (zero, one or more characters)	Bacteri*	bacterium, bacteria, bacterial, etc
			Gene*	gene, genes, genetic, etc.

all the articles containing either of the search terms namely, vitamin D or asthma.

NOT—every result item contains the first search term and not the remaining ones (helps in **narrowing** the search results). E.g. the search "vitamin D NOT asthma" retrieves all articles on vitamin D but not those containing the word asthma.

– If using more than one Boolean operator in a search, use brackets or parentheses to specify the order in which they are interpreted. The search terms within the parentheses are read first, followed by the search term (s) outside the parentheses E.g. (glaucoma OR cataract) AND complications

- *Wildcard search:*
  - When letters inside a keyword are replaced with a symbol, then it is called a wildcard search (Table 55.2)
  - Wildcards are used to search for alternate spellings and variations in a search term
  - Various symbols can be used for wildcard search like a question mark (?), hash (#), and an asterisk (*)
  - However, it is prudent to consult the database guide to know whether a particular database allows for wild card search and if so, what is the symbol it permits for wildcard searching
  - The question mark (?), and hash (#) symbols are used for wildcard searching in the Ovid databases. However, these two symbols are not permitted for wildcard searching in PubMed.
  - A question mark (?) is used for single character searching if it exists (zero or one character). It can be used in the part of the word which helps in capturing articles using various alternate spellings. Some examples are given below:
    E.g: Inserting a question mark instead of alphabet "u" in the word "colour", namely "colo?r" will help in capturing articles with both the American and British spellings "color" and "colour"
    E.g: an?esthesia searches for both the words "anesthesia" and "anaesthesia" (American and British spellings)
  - Using the hash (#) symbol in the part of the word also helps in finding a variable single character (one character), especially capturing spellings with "s" or "z". Some examples are as follows:
    E.g.: the search "nebuli#er" retrieves articles with the words "nebuliser" and "nebulizer"
    E.g.: the search "organi#ation" retrieves articles with the words "organisation" and "organization"
  - The asterisk symbol (*) is commonly used for truncation search, which is a type of wildcard search. (It is discussed in detail below in the sub-section on truncation search)
- *Truncation search:*
  - Truncation search is also a type of wildcard search in which the last part of the search term is replaced by a symbol
  - The majority of the databases (including Ovid, PubMed, and PROQUEST databases) use an asterisk (*) as their truncation symbol
  - The use of the symbol * at the end of the word captures the search term with variant endings. It helps to retrieve every word that begins with the letters entered. In other words, the symbol* aids in multiple character searching (zero, one or more characters). Some examples are as follows:
    E.g. the search "bacteri*" produces search results with the terms bacterium, bacteria, bacterial, etc.
    E.g. the search "gene*" produces search results with the terms gene, genes, genetic, etc.

- In PubMed, truncation search retrieves the first 600 possible variations of the truncated term
- A truncation search is performed when the search term has several endings, but all of the variations signify nearly the same idea. Thus, it is not required to manually type in and search for each variation of the search term, thus helping to save time.

55.4.4 Keeping a Record of Search Activity

- A search diary or document can be used to record the search activities
- A record of the following details can be maintained
  - Key concepts identified
  - Keywords used
  - Search strategies employed (e.g. search techniques used like truncation/wild card search and how the search terms were combined using the Boolean operators)
  - Databases searched
  - Number of search results from each source and each strategy employed

55.4.5 Reviewing and Refining the Search Results by Further Literature Search

- The literature review is a continuous process that should be undertaken before, during, and after completion of the research study, and even during the process of drafting the manuscript.
- This will aid in keeping abreast with the recent developments in the topic of research and incorporating the same during the research process or manuscript drafting

Bibliography

Anandabaskar N. Drug information. In: Raj GM, Raveendran R, editors. Introduction to basics of pharmacology and toxicology. Volume 1: general and molecular pharmacology: Principles of drug action. 1st ed. Singapore: Springer; 2019. p. 223–34.

Charles Sturt University Library. 2020. Literature review. [Internet]. [cited 2020 Mar 5]. Available from: https://libguides.csu.edu.au/review.

Ecker ED, Skelly AC. Conducting a winning literature search. Evid Based Spine Care J. 2010 May;1(1):9–14.

Grewal A, Kataria H, Dhawan I. Literature search for research planning and identification of research problem. Indian J Anaesth. 2016;60(9):635–9.

Literature Search. In: Raveendran R, Gitanjali B, Manikandan S, editors. A practical approach to PG dissertation. 2nd ed. Hyderabad: PharmaMed Press; 2012. p. 21–9.

Mansoor I. Tips on searching the internet for medical information. J Family Community Med. 2002 Jan;9(1):57–61.

Study Designs

56

Avinash Arivazhahan and Jhananey Senthilvelan

Abstract

The crux of any study proposal lies in the way the study is designed, which determines the outcomes and successful conduct of the study, be it primary research or secondary research. While there exist several forms of study designs, one cannot be claimed to be better than another, since each study design has its pros and cons, thus making individualization of study design selection extremely crucial. This choice depends on various factors, including the research question, study population, resources at hand, etc. Broadly, the study designs can be descriptive or analytical in their nature. The common study designs that are used in our practice are randomized controlled trials, case-control studies, cohort studies, cross-sectional studies, case reports, case series, etc. This chapter gives an exhaustive overview of the commonly used study designs, while also touching upon the less commonly used ones in brief.

Keywords

Randomized controlled trials · Cohort studies · Case control studies · Cross-sectional studies

A. Arivazhahan (✉)
Sanofi India Ltd., Chennai, Tamil Nadu, India

J. Senthilvelan
Department of Internal Medicine, Chengalpattu Govt. Medical College, Chengalpattu, Tamil Nadu, India

M. Lakshmanan et al. (eds.), *Introduction to Basics of Pharmacology and Toxicology*, https://doi.org/10.1007/978-981-19-5343-9_56

56.1 Introduction

There existed an era when medical decision-making was based purely on expert opinions. However, as is obvious, this can be associated with personal and professional bias and favoritism. In today's scientific era, we have evolved to a state where we practice 'evidence-based medicine' (EBM). This concept has greatly aided in debunking several myths and unraveling several unknown domains. Further, the so-called 'grey areas' of science are also being converted to white or black, with the help of this 'building wall of evidence'.

56.2 Evidence-Based medicine

The origin of the concept of EBM is still debatable, but the earliest recorded literature on the lack of concrete evidence to support several 'perceived' or 'hypothesized' medical judgments came from Archibald Cochrane in the early 1970s. Today, EBM has become the order of the day, being defined as 'meticulous, judicious and explicit application of the best evidence available today in clinical decision-making to ensure the best patient outcomes'. EBM can be said to have three distinct components, as below:

- Understanding of patient's expectations
- Best available evidence
- Clinical expertise

Ultimately, the goal of EBM is to ensure the best possible improvement in patient outcomes.

56.3 Concept of Variables

Any measurable or assessable attribute in a research study is called a variable. This is different in different studies. Variables that are commonly found across studies include gender, age, height, weight, disease status, health status, etc.

Variables can be of two types: exposure and outcome. Exposure variables refer to the risk factors that cause a particular effect. These are also called as intervention or independent variables. On the other hand, outcome variables refer to the consequence that happens as a result of the intervention variable. These are also known as dependent variables.

56.4 Types of Research

Medical research, as such, can be of two broad categories: primary and secondary. Primary research comprises those evidence that are generated freshly through new studies or reports, whereas secondary research includes results obtained from the analysis of existing studies or reports. The key differences are highlighted in Table 56.1.

56.5 Importance of Study Designs

Knowledge of study designs is essential for any potential investigator. Designing a study is more important than analyzing the results of a study. This is because once the study design is decided, all resources are employed, and it is humanly impossible to come back to the designing step without significantly wasting resources. On the other hand, if needed, if results are analyzed incorrectly, we have the option of analyzing the same again with a minimal escalation of resource utilization. Study designs are the first step in determining the credibility and integrity of the data generated.

56.6 Types of Study Designs

There exist several types of clinical study designs, the aptest of which has to be picked based on the type of study, population studied, parameters in consideration, etc. Common classifications of study designs can be done in several ways:

(a) Based on the purpose of the investigation
  - Descriptive or analytical research

(b) Based on the applicability of results
  - Basic or applied research

(c) Based on the directionality of studies

Table 56.1 Key differences between primary and secondary research

Primary research	Secondary research
Conducted by self	Conducted by others in the past
Necessary to answer specific research questions designed by the investigator(s)	Necessary to understand background data about a particular research question
Requires extensive investment of time, money and other resources	Requires minimal time and money
Usually follows some form of secondary research	Usually helps in forming the structure of the primary research

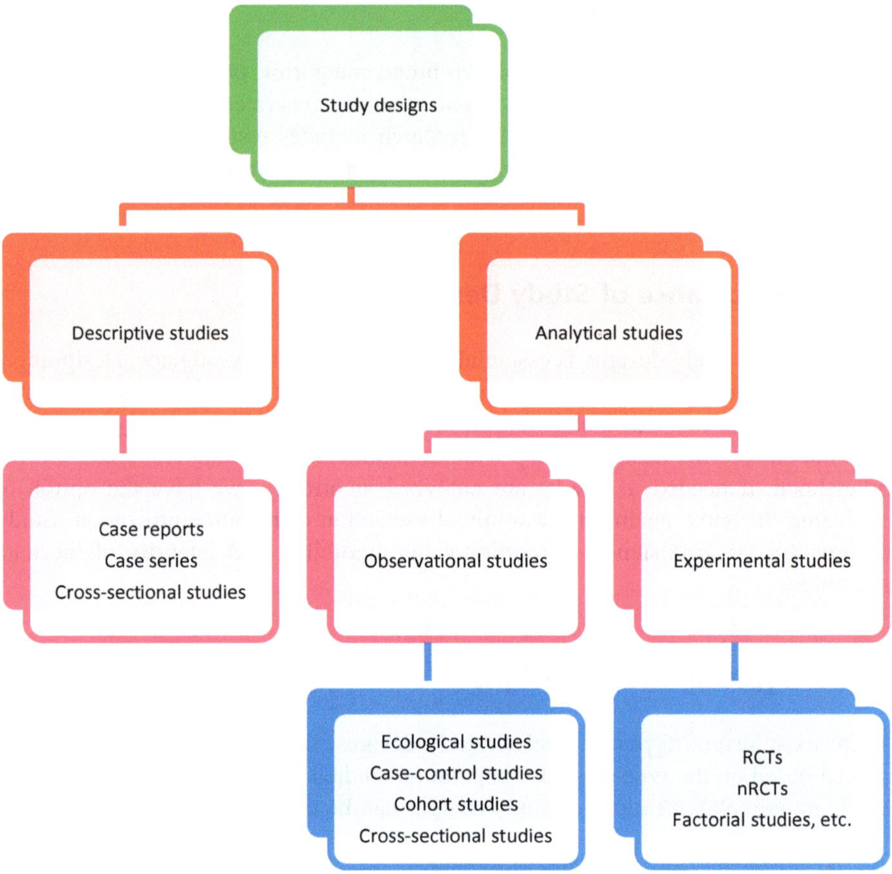

Fig. 56.1 Overview of study design types

- Retrospective or prospective research

(d) Based on the outcome measures
- Prophylactic, diagnostic, therapeutic or theranostic research

(e) Based on the investigators' role
- Observational or interventional research

The most commonly described broad categories of clinical studies include descriptive and analytical studies. The classification is detailed in Fig. 56.1.

56.6.1 Descriptive Studies

Descriptive studies, as the name implies, are those that do not consider causality assessment, and are only designed to describe variables and their distribution in the study population. There is no active intervention on the part of the investigators. In simple words, descriptive studies are designed to give an overview of what is happening in a particular population (in terms of incidence, prevalence, etc. for example), and does not attempt to prove or ascertain any kind of relationship among variables. These studies are performed to study patterns or trends in a particular disease area, and not to study the cause and effect of the variables involved in the disease area.

There are several forms of descriptive studies, the most common ones being case studies (in the form of reports or series), cross-sectional studies, ecological studies, etc.

(a) Case reports

Case reports are simplistic descriptions of clinical scenarios citing one single patient as an example. These can be index cases of a new disease or rare presentations of common diseases. There is no control patient or group involved in this type. Most case reports are purely anecdotal and hence may hold very limited clinical importance in advancing medicine. However, there have been instances in the past where case reports have guided newer advances in scientific understanding.

A typical example is the identification of HIV/AIDS through a case report about a homosexual man with features of disseminated sarcoma of Kaposi.

Case reports are simple and easy to perform and publish with minimal investment of time and resources. However, the downside is that the credibility and evidence level is the least with case reports. Also, consent is required from the participating patient, particularly if their photographs are to be published.

(b) Case series

Case series are very similar to case reports, however with a set or series of patients with similar clinical presentations. There is no fixed minimal and a maximal number of patients that can be included in a case series but most case series have only a small number of patients included in them. Again, there are no control groups involved in these studies as well. An example is a series of patients who present with hair growth as an adverse effect on minoxidil therapy initiated for hypertension.

As with case reports, case series studies are also easy and simple to perform, again with a low level of evidence.

(c) Descriptive cross-sectional studies

Cross-sectional studies are like a snapshot or a photograph that gives us an overview of the presence and distribution of variables at a particular point in time. These studies are designed to collect data on variables in a population and how they exist in that time frame. These cross-sectional studies are purely descriptive since they do not attempt to study or establish any relationship among the variables (analytical cross-sectional studies, described in the next section of this chapter, also analyze this relationship). The most common outcome of these descriptive cross-sectional studies is to estimate the prevalence of a particular disease (or even a risk factor) during a particular time.

Table 56.2 Merits and demerits of descriptive studies

Merits of descriptive studies	Demerits of descriptive studies
Easy and simple to perform	Low level of evidence generated
Limited resources needed	May not be a true representation of the general population
Ethical clearance is usually quick and easy to obtain	Impact of bias is high
Helps in the estimation of the prevalence of diseases or risk factors	
Temporal trends can be obtained	

An example is a cross-sectional study on the prevalence of diabetes mellitus in the Indian subcontinent during the year 2019.

A key advantage with the descriptive cross-sectional studies is that a similar study performed again after a certain point in time helps us in obtaining the temporal trend of progression of the particular prevalence. For example, the study on diabetes mellitus in 2019 can be repeated with similar parameters in 2029 to visualize the trend of progression of diabetes mellitus over these 10 years (from 2019 to 2029). However, it has to be ensured that the same (or similar) population is being studied again.

Descriptive cross-sectional studies are very easy to perform with minimal resources and expertise. Also, ethical clearance is easily granted since no active intervention occurs in these studies. However, these studies are associated with selection or measurement bias, which can skew the results of the study, and end up being non-representative of the general population.

An example is a study conducted to assess for the prevalence of short sight (myopia) in school-going children. While we can get a good estimate of the same with a descriptive cross-functional study, the results may not be a true reflection of the actual numbers since only school-going children are included here. The actual scenario might be that the children with very high myopia could have dropped out of school due to the condition. On the whole, descriptive studies come with a certain set of merits and demerits, as mentioned in Table 56.2.

56.6.2 Analytical Studies

Analytical studies are those studies that 'analyze' the relationship between/among variables in a population. The most common objective of analytical studies is to estimate whether a variable is a risk factor in the development of a disease or not or to evaluate whether a particular therapeutic intervention helps in prevention or therapy of disease. Analytical studies, as mentioned earlier, are of two forms: observational and experimental (or interventional). The observational analytical studies have a minimal intervention on the part of the investigator(s), in the sense that risk exposure and the outcome are naturally determined. On the other hand, in the observational experimental studies, the investigator(s) takes on the responsibility of delegating each of his study participants to be exposed to the risk or not. Either of these studies have their own merits, demerits, and limitations, and hence, one cannot be claimed to be better than the other in the true sense.

The vector (or directionality) of these studies is another important feature. The studies may be directed forward (prospective) or backward (retrospective). Again, both these designs have their own advantages and disadvantages, which are elucidated below.

The observational studies are of different subtypes, a few of which are described in detail here.

(a) Ecological studies

Ecological studies are also referred to as correlational studies. The objective of these descriptive studies is to describe the relationship between/among variables in a wide population, but not at an individual level.

The most popular example for ecological studies is the cholera study by John Snow in London, wherein a pump on Broad Street was identified as the source of the cholera epidemic. A correlation was established between the incidence of cholera and the use of the water pump on Broad Street by John Snow (Robert Koch later identified the bacterium responsible for cholera).

Ecological studies are simple to perform since the background data or information is already available or collected, and only descriptions of correlation are to be made by the investigator(s). These studies are highly suitable when inter-individual variations are expected to be minimal within the population studied. For example, if an ecological study is planned to be conducted on the impact of diet on Alzheimer's disease, then it is ideal to choose a population whose eating habits are likely to be similar (e.g., South Indian elderly males) rather than taking a wider population (e.g., Asian elderly males). If this selection is not done adequately, this may lead to the phenomenon of "ecological fallacy", wherein the results may be applicable to the wide population but not at an individual level. Also, as with other studies evaluating correlation, the impact of 'confounding factors' is to be considered before generalizing the results.

(b) Case-control studies

Case-control studies are conventionally referred to as "retrospective studies" owing to the backward nature of their design. As the name suggests, the investigator(s) enroll both cases (with disease outcomes) and controls (normal or without disease outcomes) into the studies. The cases are first identified and included in the study by looking to registers and databases. Then, controls are selected in a way that they resemble the cases in every manner possible except in the disease exhibition. This is done to ensure that the control group at large is representative of the population from which the cases are selected, to ensure homogeneity. Since these studies are designed to look from the outcome to the risk or exposure, they are retrospective in nature (Fig. 56.2).

A classic example that is quoted in every literature on this topic is that of the association between cigarette smoking and the incidence of lung carcinoma. In

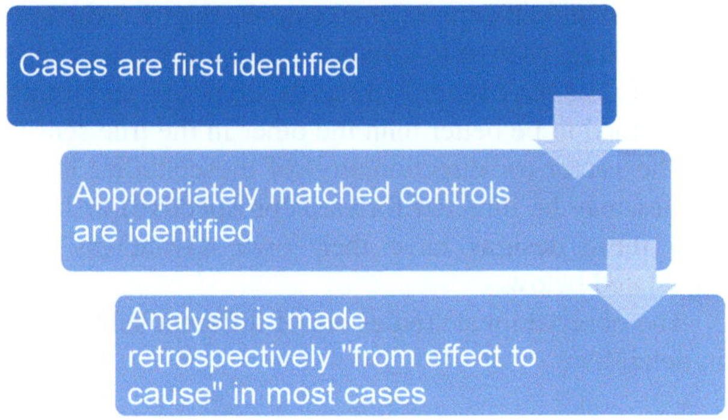

Fig. 56.2 Case-control studies

this instance, for every 'case' of lung carcinoma, a 'control' who is ideally matched for factors like age, gender, ethnicity, etc. is chosen. It is not mandatory that the ratio of cases to controls needs to be 1:1. This ratio can be altered according to the needs (2 controls per case, 3 controls per case, etc.).

Nested case-control studies are a specialized form of this study design, which has a component of cohort study design incorporated. Here, the cases (who develop the studied outcome) and controls (who do not develop the studied outcome) are selected from within a selected cohort. This is more suitable for acute diseases rather than for chronic diseases.

The key advantages of conducting case-control studies that they are relatively inexpensive and time-consuming than the cohort studies. Also, multiple variables can be studied simultaneously once the cases and controls are adequately matched. Case-control studies are particularly useful when rare diseases are being studied. This is because it is simply easier to collect data from existing cases rather than prospectively following up with subjects for these rare diseases to occur.

However, in reality, it is very difficult to establish whether the cause occurred first or the risk occurred first. Bias is always a factor in the selection of cases and controls (selection bias). While selection bias is a big hurdle in conducting these studies, others like recall bias and misinformation bias are also highly likely since the crux of the study relies on data from records or documentation. Confounding factors might interfere and give us wrong and inaccurate interpretations. A major downside that is seen by epidemiologists is that these studies cannot be used to determine the relative risk ratio as these are retrospective in nature (instead, the "odds ratio" is calculated).

(c) Cohort studies

The term "cohort" denotes a group of individuals who have a common feature or characteristic. Although the classical cohort studies are longitudinal and prospective, they can also be conducted retrospectively. In the prospective type, the cohorts are followed up to evaluate if the studied outcome is attained or not (prospective because the outcome has not yet happened at the beginning of the

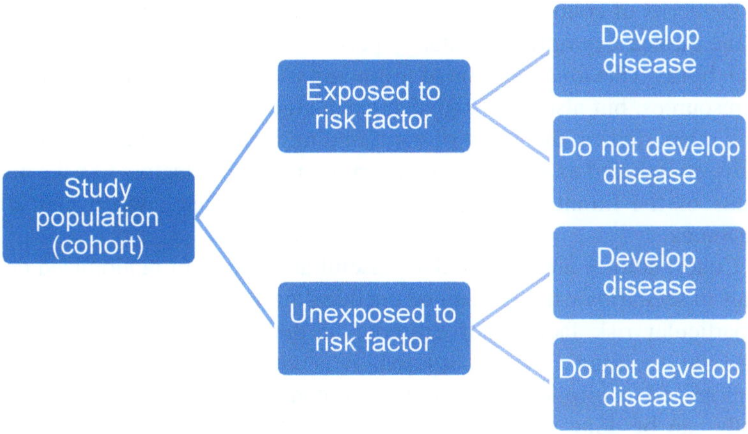

Fig. 56.3 A typical prospective cohort study design

study), as shown in Fig. 56.3. On the other hand, in the retrospective cohort studies, data from the past are extracted for these cohorts.

An example of (prospective) cohort studies is a study to assess the relationship between maternal hemoglobin levels and milestone achievement in infancy. Here, a group of mothers is chosen, and their hemoglobin levels are estimated. After delivery, the achievement of milestones by the infants is assessed on a prospective basis. The infants would be grouped into two categories: infants of mothers with normal hemoglobin, and infants of mothers with low hemoglobin levels.

An example of the retrospective type of cohort studies is when the investigator tries to contact alcoholic subjects based on a hospital database from 15 years ago, and checks if they have developed any long-term complications related to alcoholism. This is considered to be retrospective because the complication has already had its onset.

The cohort study design has a few enticing advantages. It is the best observational study design to establish temporal causation of disease since at baseline, there is no disease present. Hence, it is sometimes referred to as the "gold standard" of observational studies. These studies can help us in the estimation of incidence (incidence rate and cumulative incidence). Relative risk, risk ratio, and hazard ratio can also be estimated from these cohort studies. Also, multiple risk factors (or exposures) can be assessed and studied simultaneously, thus saving time and other resources for the investigator(s). Similarly, multiple outcomes can also be studied together. For example, in the scenario stated above, the investigator can add additional exposures to maternal hemoglobin levels like maternal thyroxine levels and mode of delivery, etc. Also, outcomes other developmental milestones like birth weight, APGAR scores, etc. can be added to the same study.

Demerits of cohort studies are long follow-up periods associated with them. In the quoted example, the infancy period has to be fully covered (12 months from birth). This long duration of follow-up is not just associated with utilization of resources, but also with logistical challenges including subject dropouts. As with other studies, the effect of confounding factors is to be considered. In the above-mentioned example, there may be confounding factors like the nutritional status of the infant, developmental manipulation, etc. which can affect the outcomes.

Cohort studies are a particularly useful alternative to randomized controlled trials in scenarios where it may be unethical or impractical to expose subjects to a particular risk factor. For example, if the causation of lung cancer with cigarette smoking is being studied, it is unethical to ask participants of a randomized controlled trial to start or continue smoking. Instead, a cohort study can be done in smokers and non-smokers, and the relative risk can be arrived at.

56.6.3 Interventional Studies

Interventional or experimental studies are those studies wherein the investigator's intervention is on the higher side, usually by allocating the subjects to different groups of therapeutic options. Other than this allocation, the investigator may also go on to modify the natural course by introducing newer drugs or other therapies in all or a part of the subject population. Interventional studies are always prospective (however, different from prospective cohort studies described earlier in this chapter).

The basic difference between a prospective cohort study and an interventional study is explained in the following example. Let us consider that the investigator wants to assess the effect of insulin glargine on cardiovascular protection. In an interventional study, the investigator would select diabetic patients, start them on insulin glargine (with a control group on placebo or other therapeutic options), and follow these patients over some time to check for the occurrence of cardiovascular events. In a prospective cohort study, the investigator would simply choose patients who are on insulin glargine and follow them up for the occurrence of cardiovascular events.

Although the most common and most widely described type of interventional studies is the randomized controlled trial, there do exist other types of interventional designs. A few of these types are described here, along with the relevant concepts involved.

(a) Randomized controlled trials

As mentioned earlier, the randomized controlled trials (RCTs) are the most common form of interventional research. It is also considered to be the strongest form of primary research and sits on top of the evidence level pyramid. Here, the trial is conducted in a "randomized" and "controlled" manner. In RCTs, "randomized" means that the trial participants are allocated to the different

Fig. 56.4 Overview of a typical randomized controlled trial

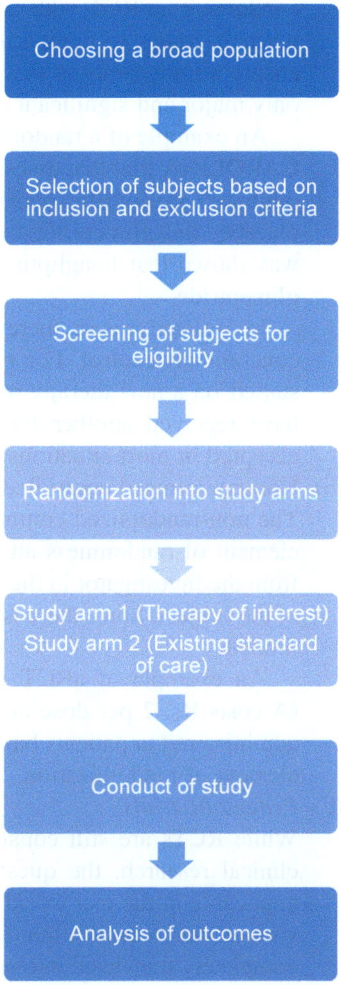

arms of the study in a random fashion, with each participant having an equal chance or probability of being assigned to each group. The concept of randomization is further explained in the next section of this chapter. "Controlled" trial reflects that there is a comparator group that is available to compare the effects of the active treatment arm. This comparator group can be a placebo arm or another active treatment (usually, the currently accepted standard of care treatment). In recent times, there have been several recommendations and regulations that it is unethical to include a placebo arm if there is a standard of care therapy that is already approved and available. A broad overview is shown in Fig. 56.4.

RCTs are the order of the day when it comes to new drug approvals, and cardiovascular safety outcome trials (CVOTs). An important prerequisite for conducting an RCT with good results is that the baselines have to be matched when it comes to comparing the groups among themselves. This is done to ensure homogeneity in the population, thus making the active intervention the only major and significant difference among the groups.

An example of a randomized active-controlled trial is the recently published CAROLINA trial that assessed the cardiovascular safety of linagliptin (a DPP-4 inhibitor) in diabetic patients. The comparator used in this trial was not a placebo, but glimepiride (a modern sulfonylurea). At the end of the trial, it was shown that linagliptin's cardiovascular safety was non-inferior to that of glimepiride.

In isolated cases, there may be a "historical control" used instead of a concomitant control. For example, a set of patients who are currently being started on a new therapy may be compared with a past group of patients who have received another form of therapy. However, these trials are not well accepted in most situations.

(b) Non-randomized controlled trials

The non-randomized controlled trials (nRCTs) are just like RCTs but with no element of randomness attached to them. Here again, there is an intervention from the investigator in the form of introducing the active treatment in a section of the subject. Since there is no randomization, the acceptance of these studies is limited.

An example of nRCT would be a study comparing two drugs A and B (A costs Rs. 2 per dose and B costs Rs. 8 per dose). Here, the drugs may be administered to patients based on their affordability quotient. Hence, there is no element of randomization.

(c) *Pragmatic trials*

While RCTs are still considered the gold standard of evidence generation in clinical research, the question of whether the results from RCTs would be applicable in the real world setting always does exist. It has been shown across various data searches that in all reality, the results from these RCTs are never completely translated into real-world practice. This "evidence gap" that exists is mainly because of the highly controlled environment in which RCTs are conducted, whereas in the real world, we do not have control over several factors (sometimes over the patient himself). This has led to the introduction and acceptance of the "pragmatic trials".

Although the earliest available literature on pragmatic trials dates back to the late 1960s, it is only in the last decade that they have gained popularity and acceptability. Pragmatic trials are designed to ensure that the results generated from these studies are applicable to the wide general population that is encountered in common practice. In short, pragmatic trials provide a more practical view of clinical research. The key differences between proper RCTs and pragmatic trials are summarized in Table 56.3.

(d) *Adaptive trials*

Adaptive trials function as an addendum to existing clinical trial protocols. RCTs are conducted based on the protocol that is prepared at the very beginning.

Table 56.3 Key differences between randomized controlled trials and pragmatic trials

Randomized controlled trials	Pragmatic trials
Small sample size	Large sample size
Strict inclusion and exclusion criteria	No/minimal inclusion and exclusion criteria
Complex trial design	Simple trial design
Internal validity is high	External validity is high
High costs involved	Lower costs involved
E.g., most of phase II and phase III studies related to new drugs	E.g., several of the phase IV studies done in the post-marketing period

However, it is highly likely that the actual scenario might require minor or major tweaking of the protocol to get the best results. These tweaks may include (but not limited to) the following:
- Sample size modification
- Inclusion/exclusion criteria modification
- Change or replacement of subjects
- Addition of deletion of study arms for various reasons
- Changes in drug dosing pattern, size, etc.

In short, adaptive trial designs facilitate adequate utilization of resources, and in most scenarios, help in shortening the duration of the study itself.

(e) Pre-post studies

Pre-post studies also called as before-after studies are studies that only have a single treatment arm. There is no control arm available, and all comparisons are drawn against the same group at two different points in time.

An example of this pre-post study design is when the investigator studies the effect of a daily dose of salbutamol in a chronic asthmatic patient. For the initial period of the study, the patients would not be on salbutamol (unless required as rescue medication), and their clinical status would be evaluated. In the latter half, they will be initiated on a daily dose of salbutamol, and the same parameters would be evaluated again. Comparisons are drawn between the two sets of readings, and results are published.

As evident from this design, this is not a very accurate depiction of the effects of salbutamol. In the above-quoted example, there is a time-lapse that happens between the two parts of the study. Various other factors could have impacted the results in the meantime, including factors like worsening of the disease, change of climate, or just simple aging. Hence, these studies are less accepted. Also, few experts believe that these studies should not be grouped under "interventional" study designs, although technically, the investigator does "intervene" in the clinical course of the subjects. However, this type of study design might be useful to identify intra-individual variations that can happen within the same individual and to avoid the effect of inter-individual variations. A few of the concepts involved in these interventional trials are listed here for a better understanding of the applicability of study designs.

Table 56.4 Common types of randomization techniques

Type of randomization	Feature
Simple randomization	• Based on a single sequence • The simplest and the weakest form of randomization
Block randomization	• Based on a certain fixed group-based algorithm • Ensures equal samples in all arms • Covariates may not be matches across arms
Stratified randomization	• Combination of the block and simple randomization techniques • Blocks of participant groups are first created, within which simple randomization is performed • Considers matching of covariates across various arms in the study
Adaptive randomization	• Considers both covariate matching and the sample size during randomization • Ensures adequate samples in all arms with covariate matching

(a) Concept of randomization

As explained earlier, randomization is a vital feature of any RCT, which eliminates the important aspect of selection bias (explained later in the chapter). Randomization allows for the study participants to have equal chances of being allotted to any of the treatment arms in the study design. There are several types of randomization, a few of which are described in Table 56.4.

Today, various software applications are available that can help in performing randomization. The adoption of technology also eliminates human error bias.

(b) Concept of cross-over

Cross-over is a concept that is frequently encountered in the realm of RCTs (although parallel study design is the most common design). Here, there are two or more arms in the study with multiple phases. Typically, the study starts with a run-in phase (for participants to adapt to the environment), followed by phase 1 (where arm A receives drug 1 and arm B receives drug 2). Following this, there is a phase 2 (where arm A receives drug 2 and arm B receives drug 1). In between phase 1 and phase 2, there is a recommended wash-out period (not mandatory) to eliminate any carry-over effects from the first treatment. The design is shown in Fig. 56.5. The same design can be extrapolated to trials with more than two treatment arms as well.

The major applicability of these cross-over trials are in treatment switch trials (for example, comparing the effects of two statins in patients with hyperlipidemia). The biggest advantage is that the confounding factor of inter-individual variation is eliminated completely since both treatment regimens are administered to both sets of patients (enabling pre- and post- comparisons). Also, lesser subjects are required when compared to the parallel (conventional) group study design. However, this design is applicable only for chronic disorders, and not for acute conditions.

(c) Concept of factorial designs

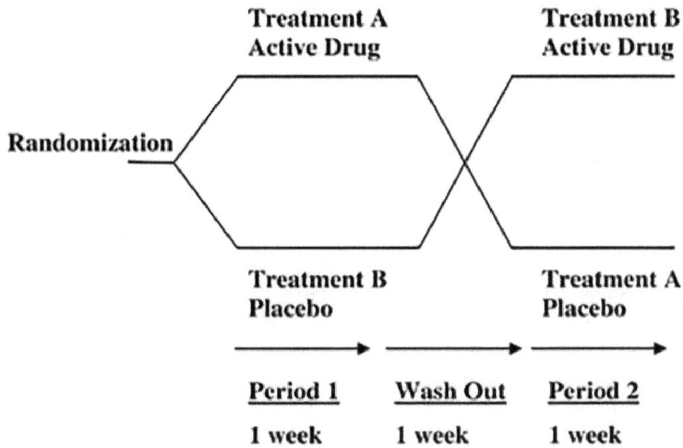

Fig. 56.5 Schematic representation of a two-way cross-over design

Often in medicine, there are multiple drugs available for a single pathological condition, and two or more of these drugs might be used in combination in practice. This is where the factorial design serves useful. Here, the drugs are compared alone versus in combination to draw results on efficacy and safety (and other parameters, if needed). The simplest form of a factorial design is a 2×2 factorial design trial.

For example, let us consider that there are two drugs (X and Y) available for the treatment of systemic hypertension, and we are not sure if combining both drugs is better than giving a single drug. A 2×2 factorial design would help us in solving this research question. Here, there can be four arms (2×2): arm 1 with drug X, arm 2 with drug Y, arm 3 with the combination of X and Y, and arm 4 with neither X nor Y (serves as the control group).

(d) Concept of bias
Bias is said to occur in research whenever there is a systematic error introduced into the research, thus making the researcher select and tend towards one outcome over the other in an erroneous manner. Simply put, bias is defined as any deviation away from the actual fact and truth, resulting. In false outcomes and interpretations. Bias can occur at any point in the research chain, right from designing the study to the conduct of the study to analysis and publication of the study results. It is nearly impossible to have a study that is completely free of bias. Hence, some amount of bias is allowed, but the investigator has to minimize bias as much as possible. The various common types of biases and the methods to avoid or minimize them are elucidated in Table 56.5.

(e) Concept of blinding
Blinding is one of the most common procedures followed in conducting RCTs. Blinding simply denotes that one or more of the individuals involved in the research are unaware of the treatment arm allocation that has happened. Also known as masking, blinding is part of several clinical trials, which are then

Table 56.5 Various types of bias and methods to minimize or avoid bias

Type of bias	Method to minimize or avoid
Bias occurring before the study	
Confirmation bias (investigator designs the study methodology to positively confirm or prove his hypothesis)	An external reviewer can be included (if not the ethical clearance committee) to review the study design and methodology
Sampling bias (the sample chosen for the research is not truly representative of the larger population)	Pre-work and prior research to be performed based on available data to select the correct samples
Selection bias (incorrect selection of study subjects on several bases)	Proper selection criteria to be framed at the beginning of the research
Admission bias (also known as Berkson's bias; denotes the over-expression and influence of admission rates on the outcomes)	Selection of properly matched controls
Bias occurring during the study	
Interviewer bias (investigator's awareness of the treatment arm of each subject leading to deviations in interviewing or reporting techniques)	Standardized questionnaire to be followed
Recall bias (differences and deviations from actual incidents and narrated events based on the memory of the incidents)	Prefer prospective studies over retrospective memory-based studies
Hawthorne bias (when study subjects voluntarily change their behavior or performance once they know that they are being monitored or observed)	Indirect observation techniques to be incorporated
Attrition bias (when study subjects drop out from the study, usually in an uneven erratic pattern)	Proper care and communications to the study subjects; inclusion of an ITT (intention-to-treat) analysis
Bias occurring after the study	
Publication bias (when the researcher strives for positive results in his study to get the study published)	Education of researchers (several journals exist with the core objective of reporting negative results from studies; negative results are sometimes equally or more important than positive results)
Funding bias (results when the researcher thinks he has to be faithful to the study sponsor by biasing the results in their favor)	Improving the integrity of the researchers

labeled as “blinded RCTs”. If blinding does not happen, then they are “open-label RCTs”.

Blinding can happen at multiple levels:

- Single blinding: Here, only one individual (usually the trial subject) is blinded
- Double blinding: In this case, two individuals (trial subject and investigator) are blinded. This is a highly recommended blinding technique and is followed across several RCTs.

- Triple blinding: Here, in addition to the trial subject and the investigator, the data analyst (or the sponsor) is also additionally kept in the dark. This can be called as the ideal form of blinding but is practically not feasible in several cases.

In most blinded RCTs, the placebo (or the active comparator) is made to appear identical in its physical properties to the treatment of interest. This is possible if both placebo (or the active comparator) and the treatment of interest have the same route of administration. However, if the routes of administration are different, then a "double-dummy" technique is followed. For example, consider that the treatment of interest is an intravenous drug while the active comparator is an oral agent. In this case, dummies are created in both arms (a dummy oral placebo for the active arm and a dummy intravenous placebo for the control arm).

56.7 Miscellaneous Study Designs

(a) Meta-analysis: An analysis that attempts to interpret the results of several scientific studies (usually RCTs). For example, there is a drug X that has been subjected to four RCTs in terms of its effectiveness in malaria. A meta-analysis would analyze all these four RCTs on a common platform, and come up with a conclusion. This increases the power of the study since subjects from all four RCTs are being included here.

(b) Systematic reviews: These are reviews of the literature on a specific clinical issue. They may be descriptive or quantitative in nature. When they are quantitative in nature, then they are referred to as meta-analyses.

(c) Mechanistic studies: As the name implies, these studies are performed to unravel the mechanism behind a particular effect or action. For example, a study conducted to explain the mechanism of the anti-inflammatory effect of drug X is a typical mechanistic study.

(d) Expert opinion statements: These are performed by bringing together a group of experts in a particular domain, and brainstorming together before coming to a consensus. They may be international, national, or local in nature. Examples are the ADA-EASD consensus statements for management of diabetes, BE-SMART international expert opinion on basal insulin use, etc.

56.8 Levels of Evidence

Now that we have gone through the basics of the different types of study designs, it can be comprehensively said that not one design is better than the other in the true sense. The choice of study design depends on several factors. However, the acceptability of each of these study types differs. The levels of evidence are depicted in Fig. 56.6.

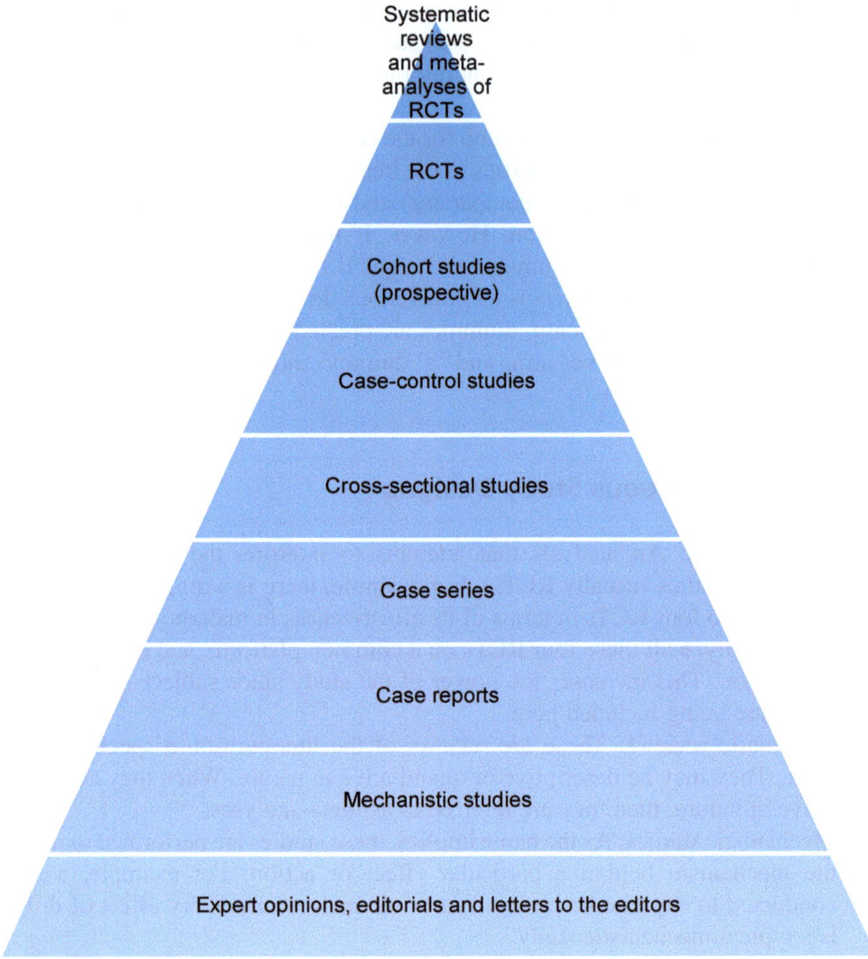

Fig. 56.6 Pyramid depicting the levels of evidence (higher the position, greater is the level of evidence)

In this comprehensive depiction (Fig. 56.6), the higher the position of the study design, the greater is the quality of its evidence. Also, as the level increases, the risk of bias keeps decreasing.

Bibliography

Aggarwal R, Ranganathan P. Study designs: part 4 – interventional studies. Perspect Clin Res. 2019a;10(3):137–9.

Aggarwal R, Ranganathan P. Study designs: part 2–descriptive studies. Perspect Clin Res. 2019b;10(1):34–6.

Parab S, Bhalerao S. Study designs. Int J Ayurveda Res. 2010;1(2):128–31.
Ranganathan P, Aggarwal R. Study designs: part 1–an overview and classification. Perspect Clin Res. 2018;9(4):184–6.
Ranganathan P, Aggarwal R. Study designs: part 3 – analytical observational studies. Perspect Clin Res. 2019;10(2):91–4.
Thiese MS. Observational and interventional study design types: an overview. Biochem Med. 2014;24(2):199–210.

Ethical Issues Related to Medical Research on Human Participants

57

Gerard Marshall Raj

Abstract

The 'practice of ethics' takes the upper hand over all other obligations including legal enforcements and moral prejudices—even so during medical research. Ethics in medical practice is a composite term including the exercise of ethical principles both during clinical practice (as a healthcare provider) and research activities (as a biomedical researcher). *The Belmont Report* had put forth the four basic tenets of medical ethics such as the *principle of respect for autonomy*, the *principle of nonmaleficence*, the *principle of beneficence*, and the *principle of justice*. The many guidelines published on the ethical issues related to research in human population are built upon a common principle that the "well-being of the research participants" would never be compromised. The Chapter includes discussion on the various internationally acclaimed guiding regulations, principles, and policies on biomedical ethics, various components of medical ethics—including publication ethics, evolution and current status of medical research ethics.

Keywords

Medical ethics · Biomedical ethics · Research ethics · Autonomy · Nonmaleficence and beneficence · Justice and equality · ICMR Ethical Guidelines · The Belmont Report · The Nuremberg Code · The CIOMS Guidelines

G. M. Raj (✉)
Department of Pharmacology, All India Institute of Medical Sciences (AIIMS) Bibinagar, Hyderabad, Telangana, India
e-mail: gerard.pharma@aiimsbibinagar.edu.in

M. Lakshmanan et al. (eds.), *Introduction to Basics of Pharmacology and Toxicology*, https://doi.org/10.1007/978-981-19-5343-9_57

57.1 Introduction

Though medical research is pursued in the premise to delineate the safety (toxicity), efficacy (under controlled settings), effectiveness (in real-world settings), efficiency (cost-effectiveness), accessibility, and also quality of the interventions, the robustness of such studies is largely determined by the ethical standards upheld throughout the research process. The rights and wellbeing of the individual research participants are always the priority over the other studied parameters.

57.1.1 Definitions

Broadly, *ethics* can be defined as "a set of principles of right conduct, such as those governing the actions of an individual or of a professional group, or the philosophy underlying such principles". And *medical ethics (healthcare ethics)* deals with "the study of a moral ideals, rules and codes of conduct that govern behavior of medical professionals". The term medical ethics is a hybrid term which requires both technical knowledge and ethical wisdom on part of the medical professionals. Medical ethics underpins the commitments of a physician towards the patients (physician-patient relationship) as well as towards fellow healthcare providers (physician-physician relationship). As medical professionals are also expected to be involved in biomedical research, the term medical ethics encompasses both medical practice- and medical research-related ethics.

Bioethics is the term used "to differentiate conventional medical ethics from ethical issues that arise from current advances in biology and medicine".

Research ethics is applied "to those who conduct research involving human (or animal) participants, to assure the integrity of the research process and to protect the interests of participants"

Further, there is term known as *health policy ethics* (*public health ethics*) which is "concerned with ethical issues relevant to the organizing, financing and delivering of health care services". As such ethics functions as a connecting bridge between health policy and human values.

The four basic tenets of medical ethics (like, any other *applied ethics*) are the *principle of respect for autonomy*, the *principle of nonmaleficence*, the *principle of beneficence*, and the *principle of justice*. Sometimes, particularly in medical research parlance, the principles of nonmaleficence and beneficence are clubbed to a single entity—though they are not always interdependent.

The above guiding ethical principles [except for the principle of nonmaleficence] were part of the *Belmont Report* (Ethical Principles and Guidelines for the Protection of Human Subjects of Research) submitted by the National Commission for the Protection of Human Subjects of Biomedical and Behavioral Research under the aegis of the U.S. Department of Health, Education and Welfare.

57.1.2 Respect for Autonomy

Autonomy (*auto* means *self* and *nomos* means *rule*) is the provision to exert voluntary decisions (for themselves) to either to participate or not in medical research (freedom of choice) after obtaining complete information (full-disclosure) regarding the project. Hence, the participants are expected to possess the *decision-making capacity* (DMC) after thorough understanding of the settings and possible sequelae of such decisions. Informed consent (sometimes, referred to as informed-understood consent) (Box 57.1) forms the basis for implementation of autonomy. The informed consent process encompasses three components, namely, information, comprehension, and voluntariness. The purpose of autonomy is to act so as to enhance participants' self-determination (coercion-free) and support self-governance (self-reliance) and not interfering with the individual's choice or will. However, it is sometimes felt that the burden of knowledge can creep into research ethics also like that of medical ethics—wherein the participant is put in a catch-22 situation to make a decision.

Box 57.1. Essential Components of Informed Consent Document

- Disclosing statement that it is research (and not part of therapy/diagnosis)
- Need for the study (objectives) and plan of execution (method of experimentation)
- Duration of participation (treatment schedule), frequency of contact, and data collection methods (also invasive methods, if any)
- Probability for random assignment to each treatment arm (for randomized trials) and no therapeutic effect in placebo-administered subjects (for placebo-controlled trials); and also about blinding, if any
- Availability of specific appropriate alternative procedures or therapies
- Potential benefits to the participants (direct), community, or others (indirect) [If there are no direct benefits to the participants, the same should be divulged]
- Possible risks (hazards), discomfort or inconvenience
- Confidentiality of case records (source documents) and privacy of participants. Direct or secondary research use of the medical records and biological specimens, if any.
- Monetary or non-monetary compensation (reimbursement) for participation (anticipated prorated payment)
- Plan for medical management or compensation for research-related injury
- Responsibilities of the participant during the trial
- Right to refuse to participate, freedom to continue or withdraw from the study at any time without adversely compromising on the possible present or future benefits
- Sources of funding and relevant conflicts of interest of all parties
- Outcomes particularly that of life-saving information and data of immediate clinical utility will be disclosed during or after study completion (right to access the data)

(continued)

Box 57.1 (continued)

- Complete contact details (with affiliations) of the principal investigators and also that of the Ethics Committee members (for study-related queries and other study-related events)

A point to be emphasized is obtaining consent for participation in research study is not a one-off event and does not end with the participant signing the consent. Rather, the investigator is expected to apprise the participants about the various developments (expected and unexpected) of the study and finally, the results need to be communicated (Post-trial access and benefit sharing). Additionally, the participants can withdraw at any point of time based on the feedback received (interim trial results, if made available) again with complete knowledge of the consequences of their withdrawal thereafter.

57.1.3 Nonmaleficence

As Hippocrates stated about treating physicians, *primum non nocere* which means *first do no harm*, the same holds good in research ethics. Nonmaleficence (*mal* means *bad*) refers to the act of avoiding harm—either through acts of commission or omission. It entails balancing risks and seriousness of potential harms against potential benefits.

57.1.4 Beneficence

Beneficence (*bene* means *good*) refers to the act of benefiting (doing good to) others and balancing good results over potential harms. Researchers should act in the best possible interests of their patients (acts of kindness).

In both nonmaleficence and beneficence, the future consequences are assessed with regards to potential goods and harms with a pinch uncertainty. In summary, it should be ensured that by participating in the study they are not harmed and may derive some benefit.

57.1.5 Justice

Justice (fairness) refers to the act of distributing the benefits and burdens of society fairly irrespective of the gender, religion, caste, creed, race (ethnicity), societal status or financial background. This principle of justice (and equality) can be achieved by equitable selection (distribution) of study subjects. Participants with similar background have the right to access to similar interventional exposures and rendering what an individual participant deserves or needs is essential. This principle gains more relevance during the conduct of research in resource-limited settings.

All the above four principles are binding and mutually exclusive. Sometimes, there can be conflicts among these principles; wherein the onus lies with the research team to decide based on the specific governing regulations and/or case-by-case basis (*judgmental ethics*).

Though appear to be interconnected and over-lapping, there are intricate differences between the terms, *ethical*, *legal*, and *moral*. 'Ethical principles' are set of guidelines prescribed by a professional body (code of professional conduct) regulating a specified group of professionals (*professional ethics*)—for example, medical ethics put forth by medical organizations. 'Legal rules and regulations' are laws of the land and if violated, penalties can be levied; laws are required to uphold social decorum and to resolve disputes fairly. 'Moral virtues' reflect individual behavior (beliefs) that are relatively stable and often have emotional component (conscience of the individual). Moral virtues aids in deciding what one *should* do, whereas, law dictates what one *must* do. In short, ethical principles vary from profession to profession; legal norms vary across regions, states, or nations (have territorial restrictions); and moral values vary across individuals, societies, or communities.

57.2 History of Medical Research Ethics: *How It Started?*

As stated above, even in the fourth century, the physicians had awareness regarding ethical issues in medical practice—as it is evident from the words of Hippocrates. Subsequently, the codes of ethics—both in medical practice and research sectors—have evolved through the years, particularly in the middle of the twentieth century with the advent of the Nuremberg Code (1947) and the Belmont Report (1979).

The Nuremberg Code (Nürnberg Principles) was put into force following the rulings (10 points defining legitimate medical research) passed by the Nuremberg Military Tribunal (Doctors' Trial) against those who conducted inhuman experimentation in the Nazi concentration camp during the Second World War. The U.S. military experiments, Sino-Japanese war (experiments on Chinese inmates in Japanese camps, Unit 731), the Tuskegee syphilis studies (1932 to 1972) in collaboration with the U.S. Public Health Service [which led to the creation of the Belmont Report], the U.S. Public Health Service Sexually Transmitted Diseases Inoculation Study (in Guatemala), the Milgram social psychological experiments at Yale University and the Johns Hopkins Lead Abatement Study are some of the infamous and inhumane examples of human experimentation.

There are many international regulations (guidance documents) that discuss about ethical conduct of biomedical research. Though the underlying principles are the same, the overall presentation is not harmonized and necessitate careful interpretation of the guidelines in accordance with the needs. The bottom line of all these guidelines is to place the well-being of the research participants as the most crucial factor in planning and executing studies—and it is the responsibility of the investigator to uphold the same (Table 57.1).

Table 57.1 List of guidelines on ethical issues in biomedical research

Title	Organization, Country	Year of publication[a]	Weblink[b]
Declaration of Helsinki (DoH)	World Medical Association (WMA) at the 64th WMA General Assembly, Brazil	2013	https://www.wma.net/policies-post/wma-declaration-of-helsinki-ethical-principles-for-medical-research-involving-human-subjects/
International Ethical Guidelines for Health-related Research Involving Humans [Council for International Organizations of Medical Sciences (CIOMS) Guidelines]	Council for International Organizations of Medical Sciences (CIOMS), Geneva, Switzerland	2016	https://cioms.ch/wp-content/uploads/2017/01/WEB-CIOMS-EthicalGuidelines.pdf
ICH harmonised guideline integrated addendum to ICH E6 (R1): Guideline for Good Clinical Practice ICH E6(R2) ICH Consensus Guideline	International Council for Harmonisation (ICH)	2016	https://database.ich.org/sites/default/files/E6_R2_Addendum.pdf
WHO Standards and Operational Guidance for Ethics Review of Health-Related Research with Human Participants	World Health Organization (WHO)	2011	https://www.who.int/publications/i/item/9789241502948
WHO Handbook for Good Clinical Research Practice (GCP): Guidance for Implementation	World Health Organization (WHO)	2005	https://apps.who.int/iris/bitstream/handle/10665/43392/924159392X_eng.pdf?sequence=1&isAllowed=y
WHO Handbook: Quality Practices in Basic Biomedical Research	Special Programme for Research and Training in Tropical Diseases (TDR), World Health Organization (WHO)	2006	https://www.who.int/tdr/publications/documents/quality_practices.pdf?ua=1
National Ethical Guidelines for Biomedical and Health Research Involving Human Participants	Indian Council of Medical Research (ICMR), India	2017	https://main.icmr.nic.in/sites/default/files/guidelines/ICMR_Ethical_Guidelines_2017.pdf
National Ethical Guidelines for Biomedical Research Involving Children	Indian Council of Medical Research (ICMR), India	2017	https://main.icmr.nic.in/sites/default/files/guidelines/ICMR_Ethical_Guidelines_2017.pdf

(continued)

Table 57.1 (continued)

Title	Organization, Country	Year of publication[a]	Weblink[b]
National Guidelines for Stem Cell Research	Indian Council of Medical Research (ICMR) and Dept. of Biotechnology (DBT), India	2017	https://main.icmr.nic.in/sites/default/files/guidelines/Guidelines_for_stem_cell_research_2017.pdf
Handbook on National Ethical Guidelines For Biomedical And Health Research Involving Human Participants	Indian Council of Medical Research (ICMR), India	2018	https://ethics.ncdirindia.org/asset/pdf/Handbook_on_ICMR_Ethical_Guidelines.pdf
The New Drugs and Clinical Trials Rules	Ministry of Health and Family Welfare, Government of India	2019	https://cdsco.gov.in/opencms/export/sites/CDSCO_WEB/Pdf-documents/NewDrugs_CTRules_2019.pdf
National Guidelines for Gene Therapy Product Development and Clinical Trials	Central Drug Standards Control Organisation (CDSCO) and Indian Council of Medical Research (ICMR), India	2019	https://main.icmr.nic.in/sites/default/files/guidelines/guidelines_GTP.pdf
National Guidelines for Ethics Committees Reviewing Biomedical & Health Research During COVID-19 Pandemic	Indian Council of Medical Research (ICMR), India	2020	https://main.icmr.nic.in/sites/default/files/guidelines/EC_Guidance_COVID19_06_05_2020.pdf
The Common Rule (The Final Rule)	Office for Human Research Protections (OHRP), U.S. Department of Health and Human Services (HHS), United States	2017	https://www.hhs.gov/ohrp/regulations-and-policy/regulations/45-cfr-46/index.html#46.116
Guidelines by Nuffield Council on Bioethics	Nuffield Foundation, the Medical Research Council and Wellcome United Kingdom	–	https://www.nuffieldbioethics.org/
Ethical and Policy Issues in Research Involving Human Participants: Volume I and II	National Bioethics Advisory Commission (NBAC) (1996–2001) President's Council on Bioethics (2001–2009), United States	2001	https://repository.library.georgetown.edu/bitstream/handle/10822/559360/nbac_human_part.pdf?sequence=1&isAllowed=y https://repository.library.georgetown.edu/bitstream/handle/10822/559361/nbac_human_part_vol2.pdf?sequence=1&isAllowed=y

(continued)

Table 57.1 (continued)

Title	Organization, Country	Year of publication[a]	Weblink[b]
Core practices for all those who are involved in publishing scholarly literature: editors and their journals, publishers, and institutions	Committee on Publication Ethics (COPE), United Kingdom	2017	https://publicationethics.org/core-practices
Recommendations for the Conduct, Reporting, Editing, and Publication of Scholarly Work in Medical Journals [ICMJE Recommendations ("The Uniform Requirements")]	International Committee of Medical Journal Editors (ICMJE), United States	2019	http://www.icmje.org/icmje-recommendations.pdf
Handbook for Applicants and Reviewers of Clinical Trials of New Drugs in India	Indian Council of Medical Research (ICMR) and Central Drug Standards Control Organisation (CDSCO), India	2017	https://main.icmr.nic.in/sites/default/files/reports/Handbook%20for%20Applicants%20and%20Reviewers%20of%20Clinical%20Trials.pdf
Policy on Research Integrity and Publication Ethics	Indian Council of Medical Research (ICMR), India	2019	https://ethics.ncdirindia.org//asset/pdf/ICMR_PRIPE2019.pdf

[a]Only the latest version is included
[b]Last accessed on 6th August, 2021

57.2.1 Declaration of Helsinki

The Declaration of Helsinki (DoH) was developed and adopted by the World Medical Association (WMA) at the WMA General Assembly, Helsinki, Finland in 1964 and underwent multiple revisions since then with the recent version (7th) being released in 2013 (64th WMA General Assembly, Fortaleza, Brazil). The DoH is a statement of ethical principles and provides guidance for conducting medical research involving human subjects. At the outset, it enlists certain *general principles*, followed by discussion on *risks, burdens and benefits, vulnerable groups and individuals, scientific requirements and research protocols, research ethics committees, privacy and confidentiality, informed consent, use of placebo, post-trial provisions, research registration and publication and dissemination of results* and finally, *unproven interventions in clinical practice*—in total there are 37 points (also referred to as articles or paragraphs).

Owing to lesser adherence to the earlier Nuremberg Code, this Declaration came into vogue and has become a vital document in international deliberations of research ethics. The informed consent process and the ethical review by independent committees are the two main pillars of the DoH—both were introduced in the 2nd revision (1975).

Considered as a living document, the DoH is bound to address many evolving and novel ethical issues in biomedical research.

57.2.2 Council for International Organizations of Medical Sciences (CIOMS) Guidelines

The *International Ethical Guidelines for Biomedical Research Involving Human Subjects* was published in 1993 by the CIOMS (an international, non-governmental, non-profit organization)—shortly known as the CIOMS Guidelines. Previously, the first draft (1982) of the CIOMS Guidelines was titled *Proposed International Ethical Guidelines for Biomedical Research Involving Human Subjects*. The CIOMS was formerly known as the *Council for Coordination of International Medical Congresses* and was constituted by the United Nations Educational, Scientific and Cultural Organization (UNESCO) in collaboration with the WHO in 1949.

CIOMS also publishes the ethical guidance for the conduct of public health research in the name of *International Guidelines for Ethical Review of Epidemiological Studies* published in 1991 and updated in 2009.

Presently, the *International Ethical Guidelines for Health-related Research Involving Humans*, published in 2016, functions as a unifying document from CIOMS and replaces both the erstwhile guidelines on biomedical and epidemiological research domains separately.

The purpose of the CIOMS guidelines, as stated in the first draft, was "to indicate how the ethical principles that should guide the conduct of biomedical research involving human subjects, as set forth in the DoH, could be applied, particularly in developing countries, given their socioeconomic circumstances, laws and regulations, and executive and administrative arrangements." The same holds true even now and the guidance document exhibits particular courtesy towards research conducted in low- and middle-income countries (LMICs). The 2016 version contains the following 25 guideline heads:

- Scientific and social value and respect for rights
- Research conducted in low-resource settings
- Equitable distribution of benefits and burdens in the selection of individuals and groups of participants in research
- Potential individual benefits and risks of research
- Choice of control in clinical trials
- Caring for participants' health needs
- Community engagement
- Collaborative partnership and capacity-building for research and research review
- Individuals capable of giving informed consent
- Modifications and waivers of informed consent

- Collection, storage and use of biological materials and related data
- Collection, storage and use of data in health-related research
- Reimbursement and compensation for research participants
- Treatment and compensation for research-related harms
- Research involving vulnerable persons and groups
- Research involving adults incapable of giving informed consent
- Research involving children and adolescents
- Women as research participants
- Pregnant and breastfeeding women as research participants
- Research in disasters and disease outbreaks
- Cluster randomized trials
- Use of data obtained from the online environment and digital tools in health-related research trials
- Requirements for establishing research ethics committees and for their review of protocols
- Public accountability for health-related research
- Conflicts of interest

57.2.3 The Common Rule (The Final Rule)

The Common Rule refers to the federal rules regulating conduct of clinical research. This federal policy for the protection of human subjects was adopted in 1991. Among the 15 federal departments and agencies that have codified these rules, the U.S. Department of Health and Human Services (HHS) regulations [45 Code of Federal Regulations (CFR), Part 46] comprise five subparts, namely, 'Subpart A', for regulation (protection) of human subjects participation in research (also referred to as the Federal Policy or the *Common Rule*); 'Subpart B', additional protections for pregnant women, human fetuses, and neonates involved in research; 'Subpart C', additional protections pertaining to biomedical and behavioral research involving prisoners as subjects; 'Subpart D', additional protections for children involved as subjects in research; and 'Subpart E', registration of institutional review boards. The Office for Human Research Protections (OHRP) (part of the Office of the Assistant Secretary for Health in the Office of the Secretary of HHS) facilitates in the protection of the rights, welfare, and wellbeing of human participants by multiple ways—including the provision of decision charts (graphic aids) for ethical review of clinical research. The Common Rule was updated in 2017 and came to be known as *The Final Rule*. The purpose of this revision is for modernizing, strengthening and enhancing the effectiveness of the Common Rule. This Final Rule has implemented novel steps for better protection of human participants involved in research, while facilitating valuable research and concurrently decreasing unforeseen liabilities, unnecessary delays, and uncertainty for investigators.

Further, the NIH (National Institutes of Health) of the HHS abides by the following seven main principles for guiding the conduct of ethical clinical research. They are *social and clinical value*, *scientific validity*, *fair subject selection*, *favorable risk-benefit ratio*, *independent review*, *informed consent*, and *respect for potential and enrolled subjects*.

Apart from these regulations, the ICH (International Council for Harmonisation) Consensus Guideline for Good Clinical Practice (E6, 2016), the Nuffield Council on Bioethics (Nuffield Foundation, UK), the WHO Standards and Operational Guidance for Ethics Review of Health-Related Research with Human Participants (2011), the President's Council on Bioethics, US [previously, National Bioethics Advisory Commission (NBAC)], the WHO Tropical Diseases Research Programme, and the UNESCO Universal Declaration on Bioethics and Human Rights (2005) are some of the other internationally acclaimed guidelines on biomedical ethics.

57.2.4 Indian Context

- Indian Council of Medical Research (ICMR)—Biomedical Research:
  - Indian Council of Medical Research (ICMR), under aegis of the Department of Health Research (DHR), Ministry of Health and Family Welfare, Government of India, is the apex body for the formulation, coordination, and promotion of biomedical research in India.
  - The fourth and the latest update of Indian ethical guidance document was titled as *National Ethical Guidelines for Biomedical and Health Research Involving Human Participants, 2017*—often referred to as the ICMR Ethical Guidelines.
  - The ICMR Ethical Guidelines is structured under 12 sections. The first 6 sections contains basic discussion on all topics of biomedical and health research. The next 6 sections are more topic-specific. The sections are as follows:
    - (i) Statement of general principles
    - (ii) General ethical issues
    - (iii) Responsible conduct of research*
    - (iv) Ethical review procedures
    - (v) Informed consent process
    - (vi) Vulnerability
    - (vii) Clinical trials of drugs and other interventions
    - (viii) Public health research
    - (ix) Social and behavioural sciences research for health*
    - (x) Human genetics testing and research
    - (xi) Biological materials, biobanking and datasets
    - (xii) Research during humanitarian emergencies and disasters

      (*Newer sections)
  - The scope of these guidelines is categorically stated as "these guidelines are applicable to all biomedical, social, and behavioral science research for health conducted in India involving human participants, their biological material and data"
  - The section on general principles comprises the following twelve principles:

(i) Principle of essentiality
(ii) Principle of voluntariness
(iii) Principle of non-exploitation
(iv) Principle of social responsibility*
(v) Principle of ensuring privacy and confidentiality
(vi) Principle of risk minimization
(vii) Principle of professional competence
(viii) Principle of maximization of benefit
(ix) Principle of institutional arrangements
(x) Principle of transparency and accountability
(xi) Principle of totality of responsibility
(xii) Principle of environmental protection*
(*Newer principles)

- Coincidentally, the *National Guidelines for Stem Cell Research* (third revision) and the *National Ethical Guidelines for Biomedical Research Involving Children* (first edition) were released in 2017. In 2018, ICMR also released a concise user friendly *Handbook on National Ethical Guidelines* based on the primary source document.

• *CDSCO—CTR rules:*
  - *The New Drugs and Clinical Trials Rules, 2019* came into force after due publication in the Gazette of India on 25th March 2019 by the MoHFW, GoI. These rules supersede the erstwhile *Schedule Y* under the Drugs and Cosmetics Rules, 1945.
  - There are a total of 107 rules prescribed under 13 different Chapters as follows:
    • Chapter I—Definitions
    • Chapter II—Authorities And Officers
    • Chapter III—Ethics Committee for Clinical Trial, Bioavailability and Bioequivalence Study
    • Chapter IV—Ethics Committee for Biomedical and Health Research
    • Chapter V—Clinical Trial (Part A), Bioavailability and Bioequivalence Study of New Drugs and Investigational New Drugs (Part B)
    • Chapter VI—Compensation
    • Chapter VII—Bioavailability and Bioequivalence Study Centre
    • Chapter VIII—Manufacture of New Drugs or Investigational New Drugs for Clinical Trial, Bioavailability or Bioequivalence Study or for Examination, Test and Analysis
    • Chapter IX—Import of New Drugs and Investigational New Drugs for Clinical Trial, Bioavailability or Bioequivalence Study or for Examination, Test and Analysis
    • Chapter X—Import or Manufacture of New Drug for Sale or for Distribution
    • Chapter XI—Import or Manufacture of Unapproved New Drug for Treatment of Patients in Government Hospital and Government Medical Institution

    - Chapter XII—Amendments of Drugs and Cosmetics Rules, 1945
    - Chapter XIII—Miscellaneous
  - Further there are 8 Schedules to account for the 107 rules
    - First Schedule—General Principles and Practices for Clinical Trial
    - Second Schedule—Requirements and Guidelines for Permission to Import or Manufacture of New Drug for Sale or to Undertake Clinical Trial
    - Third Schedule—Conduct of Clinical Trial
    - Fourth Schedule—Requirements and Guidelines for Conduct of Bioavailability and Bioequivalence Study of New Drugs or Investigational New Drugs
    - Fifth Schedule—Post Market Assessment
    - Sixth Schedule—Fee Payable for Licence, Permission and Registration Certificate
    - Seventh Schedule—Formulae to Determine the Quantum of Compensation In the Cases of Clinical Trial Related Injury or Death
    - Eighth Schedule—Forms CT-01 to CT-27

57.3 Ethics Review Committees

The ethics review committees (also referred to as 'research ethics committees' or 'institute review boards') play a vital role in maintaining the scientific and ethical standards of medical research. It should be noted that ascertaining the scientific merit of study is the first step, followed by which ethical judgment of the same can be made. Independent scientific review committees are also in place for the aforementioned activity. Irrespective of the names, these research oversight systems are to be appropriately constituted and competent enough.

Ethical review and approval is mandatory before commencing any biomedical study and the same is a prerequisite while approaching research funding bodies or during the time of publication.

The ethics review committees scrutinize the submitted research proposals, conduct review meetings, make quorum-decisions, and also continuously monitor the approved proposals. It is necessary that these committees are adequately represented with multidisciplinary members and function as per well-structured standard operating procedures (SOPs). Apart from ethical appraisal, continuous quality assurance of research activities is also the responsibility of the ethics review committees.

Ultimately, the review committees should be the *torchbearers* for legitimate conduct of research and should act as the *troubleshooters* whenever necessary but never the *troublemakers* for rightful research.

57.4 Ethics in Different Contexts

57.4.1 Underprivileged Populations

Ethical regulations for conduct of research in the minority group of populations (like, racially underrepresented), the vulnerable (underprivileged or disadvantaged) populations and also in the LMICs require further consideration. Field trials in such socially and economically deprived group of populations are to be initiated only when there is a perceptible uncertainty with regards to potential benefit or harm caused by the intervention. Mere exploitation of minority populations is dissuaded and appropriate justification for inclusion of such groups is to be given. Coercion and deception of any type—even under the premise of overall benefit for the participants—are to be avoided particularly among these populations. And in a community-based trial, the decision by individual participants (individual autonomy) overturns the general group (broad) consent obtained from the community leaders (communal consent) or local health authorities. Double standards of any kind should be avoided—like administration of placebo to a vulnerable group (in the poor world) and standard care (in the rich world) to less vulnerable group. The amount of monetary compensation should be meticulously computed so as to not to be an undue inducement over and above the daily wages lost or the usual travel expenses.

57.4.2 Publication Ethics

Publication misconduct include data fabrication and falsification, manipulation of images, plagiarism (including self-plagiarism), simultaneous submissions, duplicate/redundant publications (or 'salami' publications), improper author contribution or attribution (gift/ghost authorship), conflicts of interest (personal, financial, scientific, or political), excessive inappropriate self-citations.

COPE (Committee on Publication Ethics), over the past 20 years, has supported the editors, publishers, and other stakeholders involved in academic publishing and strives to maintain the integrity and genuineness of scholarly publishing. COPE releases best practice guidelines (guidance documents) for attaining and sustaining highest standards in publication ethics. Likewise, the ICMJE (International Committee of Medical Journal Editors), a closed working group of general medical journal editors, also gives "Recommendations for the Conduct, Reporting, Editing and Publication of Scholarly Work in Medical Journals" (The Uniform Requirements). And the global nonprofit voluntary association of editors of peer-reviewed medical journals WAME (World Association of Medical Editors) also strives to uphold publication integrity through its Professional Code of Conduct.

Suppression of negative or equivocal study reports is also considered as an ethical misconduct and dissemination of the same is highly encouraged along with recording of clinical trial protocols (and their results) in publically accessible databases, like the Clinical Trials Registry of India (http://ctri.nic.in/Clinicaltrials/login.php) and the US ClinicalTrials.gov (https://www.clinicaltrials.gov/).

57.5 Medical Research Ethics: *What is the Current Status?*

- Keeping in pace with burgeoning growth of science of medical research, the intricacies of the ethical component of medical research is evolving continuously.
- Though as discussed earlier, obtaining informed consent is a prerequisite prior to participation, it is felt that act of signing (thumb printing) the informed consent document sometimes translates into undue anxiety and fear—particularly among the vulnerable populations. Hence, the stress should be more on the (consent) *process*—i.e. the complete knowledge of participation and the thorough explanation of research—rather on the (consent) *signatures*.
  - Difficulties in comprehension, degree of voluntariness, and the capacity to decide are some of the major challenges in obtaining informed consent.
- The current place of placebos in clinical research—when and where to utilize them (Box 57.2).
- Apart from the physical risk (and the subsequent harm), psychological, economic, social (cultural) and legal risks should also be looked out for while undertaking a benefit-risk assessment. Benefit-risk assessment incorporates benefit-risk analysis followed by benefit-risk evaluation.
- As more and more lawsuits are being accumulated particularly in the case of pharmaceutical industry-sponsored clinical trials, ethics committees are keen on producing insurance certificate or atleast plan for the same during the review process.
- Practical implications of multicentre and multinational (North-South) collaborative research projects. To take cognizance of 'ethics dumping'—which can either be intentional or unintentional
  - Intentional ethics dumping occurs when researchers hailing from high-income countries exploit on the compromised and less-regularized research setting existing in low- and middle-income countries (LMICs).
  - On the other hand, ethics dumping can be unintentional when a researcher is trying to conduct biomedical/clinical research in an ethically and culturally inappropriate manner—i.e., not in line with the regional communal practices—as the researcher is not aware of the prevailing norms.
- Ethical issues are relevant even in the case of Phase IV (post-licensure or label extension) studies—where special informed consents are sought for performing additional interventions over and above the administration of the approved drug.
- Continuous monitoring of ongoing research by ethics committees—passive 'off' site (ongoing or continuing) and active 'on' site (routine/risk based or for cause).
- Emphasis on registration and also accreditation of ethics committees with authorized overseeing bodies.
  - National Ethics Committee Registry for Biomedical and Health Research (NECRBHR), Department of Health Research, MoHFW [Naitik portal] → For registration of ethics committees handling biomedical and health research involving human participants.
  - Central Licencing Authority/Central Drugs Standard Control Organization (CDSCO), Directorate General of Health Services, MoHFW [SUGAM portal]

Box 57.2. Placebos in Clinical Research

Guidelines	Use of placebo
WMA Declaration of Helsinki—Ethical Principles for Medical Research Involving Human Subjects (2018)	• Where no proven intervention exists, the use of placebo, or no intervention, is acceptable; or • Where for compelling and scientifically sound methodological reasons the use of any intervention less effective than the best proven one, the use of placebo, or no intervention is necessary to determine the efficacy or safety of an intervention • And the patients who receive any intervention less effective than the best proven one, placebo, or no intervention will not be subject to additional risks of serious or irreversible harm as a result of not receiving the best proven intervention. • Extreme care must be taken to avoid abuse of this option.
ICMR – National Ethical Guidelines for Biomedical and Health Research involving Human Participants (2017)	• When there is no established effective therapy available; • When withholding an established effective therapy would not expose participants to serious harm, but may cause temporary discomfort or delay in relief of symptoms; • If the disease is self-limited; or • When the use of an established effective therapy as a comparator would not yield scientifically reliable results and the use of placebo would not add any additional risk of serious or irreversible harm to the participants.
CIOMS—International Ethical Guidelines for Health-related Research Involving Humans (2016)	• When there is no established effective intervention for the condition under study, or when placebo is added on to an established effective intervention. • When there is an established effective intervention, placebo may be used as a comparator without providing the established effective intervention to participants only if: – there are compelling scientific reasons for using placebo; and – delaying or withholding the established effective intervention will result in no more than a minor increase above minimal risk to the participant and risks are minimized, including through the use of effective mitigation procedures.

→ For registration of ethics committees handling clinical trials, bioavailability and bioequivalence studies.

- Some of the accreditation bodies for IEC are Forum for Ethics Review Committees in India (FERCI), National Accreditation Board for Hospitals & Healthcare Providers (NABH), and Association for the Accreditation of Human Research Protection Programs (AAHRPP).
  - FERCI is the Indian Chapter of Forum for Ethical Review Committees in the Asian and Western Pacific Region (FERCAP). FERCAP is a regional conglomeration under the Strategic Initiative for Developing Capacity in Ethical Review (SIDCER) as part of the World Health Organization (WHO) Special Training and Research Programme in Tropical Diseases (TDR).
  - NABH is a constituent board of Quality Council of India. NABH is a member of the International Society for Quality in Health Care (lSQua) and is also accredited to the International Society for Quality in Health Care External Evaluation Association (ISQua EEA).
  - AAHRPP is a world-renowned, not-for profit, independent, accrediting body that helps in upholding human research protection programs (HRPPs).

- Introduction of me-too drugs (copy of an existing drug) to hold the market share or obtaining patent is again ethically debatable—as testing of such drugs in human participants may not be more beneficial than the current therapies.
- Payment by cash (or by any kind) for participation in clinical research should be reviewed meticulously—so that the same is neither exorbitantly inducive nor paltry.
- Further, there is a transformation from overcautious protection (underrepresentation) of the vulnerable populations to a more participatory-yet-regulated conduct of trials in these disadvantaged groups (with optimal representation).
- Compensation and insurance coverage of the research participants in clinical trials is emphasized. In investigator-initiated academic clinical trials, the investigator plays the additional role of sponsor. However, the onus lies with the host institution to provide necessary compensation and/or for insurance coverage for research related injury or harm and also appropriate financial support for the conduct of study.
- Privacy with regards to participants enrolled (even during AV informed consent) and confidentiality with regards to data collected (even as part of registries)—both are equally important. However, the right to life supersedes the right to privacy.
- The importance of anonymization and retention of identifiability in biobanking and genetic research.
- Complexities of post-trial responsibilities—so as to:
  - What should be provided: Drug (investigational product), medical care, or moral support?
  - Who should be the provider: Sponsor, researcher, or government?

  - Who should be the beneficiary: Trial participants, general public, or risk groups?
  - What should be the pricing: Free, low-cost, or unsubsidized?
- The outcome of research should be made *available*, *accessible*, *acceptable*, and *affordable* to the benefit of the larger community—the same should be emphasized during the review process of the research proposal.
- Conflicts of Interest (COI) at all levels of research—starting from the researchers (individual), sponsors, institution, and IEC members—should be addressed diligently. COI in academic and professional matters is inevitable and hence, it would be more pertinent to manage COI rather than to eradicate COI.
- Recently, an umbrella-term of Responsible Conduct of Research (RCR) is often used. RCR incorporates "components such as planning and conducting research, reviewing and reporting research, responsible authorship and publication of the research work".
- Hence, it is the responsibility of the EC members to get trained adequately and acquiring requisite skills in evaluating the research proposals (including that of clinical trials)—by actively participating in training conducted by the National Institutes of Health (NIH), the Collaborative Institutional Training Initiative (CITI program), and other recognized bodies. Programs organized by national, regional, or institutional bodies are also equally valid.

57.6 Conclusion

In summary, the application of bioethics guidelines is relevant to all stakeholders, starting from the researchers, participants, review committees, trial sponsors, institute administrators, policy makers, program managers, funding agencies to regulatory bodies, among others. Hence, it is a prerequisite for all these patrons of clinical research to work cohesively and translate these ethical guidelines into action. And medical researchers, at the heart of biomedical and clinical research, have to exert utmost care in diligent execution of human research—being an 'ethical' researcher is equally important as being a 'good' physician!

Bibliography

CIOMS. International ethical guidelines for health-related research involving humans [Internet]. 2016 [cited 2021 Apr 4]. Available from: https://cioms.ch/wp-content/uploads/2017/01/WEB-CIOMS-EthicalGuidelines.pdf

COPE. Promoting integrity in scholarly research and its publication [Internet]. 2017 [cited 2020 Dec 12]. Available from: https://publicationethics.org/core-practices

Ethical Considerations. In: Smith PG, Morrow RH, Ross DA, editors. Field trials of health interventions: a toolbox. Oxford University Press; 2015. ISBN: 9780198732860.

Horner J. Morality, ethics, and law: introductory concepts. Semin Speech Lang. 2003;24:263–74. https://doi.org/10.1055/s-2004-815580.

ICMJE. Recommendations for the conduct, reporting, editing, and publication of scholarly work in medical journals [Internet]. 2019 [cited 2020 Dec 12]. Available from: http://www.icmje.org/recommendations/

ICMR. National ethical guidelines for biomedical and health research involving human participants [Internet]. 2017 [cited 2021 Apr 4]. Available from: https://main.icmr.nic.in/sites/default/files/guidelines/ICMR_Ethical_Guidelines_2017.pdf

ICMR. Handbook on national ethical guidelines for biomedical and health research involving human participants [Internet]. 2018 [cited 2021 Apr 4]. Available from: https://ethics.ncdirindia.org/asset/pdf/Handbook_on_ICMR_Ethical_Guidelines.pdf

ICMR. ICMR policy on research integrity and publication ethics, 2019 [Internet]. 2019 [cited 2020 Dec 12]. Available from: https://sts.icmr.org.in/Result/ICMR_Policy_On_RIPE_30.7.19.pdf

New Drugs and Clinical Trial Rules. 2019. Ministry of health and family welfare. Govt of India. Notification. The gazette of India: extraordinary, Part II, Section 3, subsection (i), New Delhi: 2019 [cited 2021 Apr 4] Available from: https://cdsco.gov.in/opencms/export/sites/CDSCO_WEB/Pdf-documents/NewDrugs_CTRules_2019.pdf

NIH. Patient recruitment. Ethics in clinical research. Ethical guidelines [Internet]. 2020 [cited 2020 Dec 12]. Available from: https://clinicalcenter.nih.gov/recruit/ethics.html

Sengupta S, Honavar SG. Publication ethics. Indian J Ophthalmol. 2017 Jun;65(6):429–32. https://doi.org/10.4103/ijo.IJO_483_17.

Taylor RM. Ethical principles and concepts in medicine. Handb Clin Neurol. 2013;118:1–9. https://doi.org/10.1016/B978-0-444-53501-6.00001-9.

U.S. Department of Health and Human Services (HHS). 45 CFR 46, The Common Rule [Internet]. 2017 [cited 2021 Apr 4]. Available from: https://www.hhs.gov/ohrp/regulations-and-policy/regulations/45-cfr-46/index.html#46.116

UW Medicine. Department of bioethics and humanities. McCormick TR. principles of bioethics [Internet]. 2018. Available from: https://depts.washington.edu/bhdept/ethics-medicine/bioethics-topics/articles/principles-bioethics

WAME. About Us [Internet]. 2020 [cited 2020 Dec 12]. Available from: http://www.wame.org/about

White MG. Why human subjects research protection is important. Ochsner J. 2020;20(1):16–33. https://doi.org/10.31486/toj.20.5012.

WHO, Bankowski Z. Ethics and health. World Health 1989;April:2–6. Available from: https://apps.who.int/iris/handle/10665/53516

World Medical Association. WMA declaration of helsinki – ethical principles for medical research involving human subjects [Internet]. 2018 [cited 2021 Apr 4]. Available from: https://www.wma.net/policies-post/wma-declaration-of-helsinki-ethical-principles-for-medical-research-involving-human-subjects/

Part IV

Biostatistics

Basic Principles and Application of Statistics in Drug Research

58

Abialbon Paul

Abstract

Statistics is the branch of mathematics that deals with the collection, organization and analysis of data collected from research on a study sample and drawing conclusions about the whole population. Statistics help make decisions which are less biased, establish causality and evaluate interference from other confounding factors. Statistics helps in multiple scenarios including determining the dose response relationship, quantifying the drug's action and differentiating it from confounding factors, to establish a superior treatment from other candidates, to establish the safety of the drug, etc. The most important component of good statistics is good data. Data can be either qualitative or quantitative. The statistical tests depend on the type of data selected measuring the research objective. Differentiating causality and association is discussed in the following chapter along with common terms and concepts used commonly in research.

Keywords

Statistics · Data · Drug research · Experimentation

58.1 Why to Use Statistics?

Before a drug is marketed and made available for clinical use, it's efficacy and safety must be rigorously established with preclinical studies and clinical trials. Statistics is the branch of mathematics that deals with the collection, organization, and analysis of data collected from research on a study sample and drawing conclusions about the

A. Paul (✉)
Department of Pharmacology & Clinical Skills, Medical University of the Americas, Nevis, West Indies
e-mail: a.paul@mua.edu

M. Lakshmanan et al. (eds.), *Introduction to Basics of Pharmacology and Toxicology*, https://doi.org/10.1007/978-981-19-5343-9_58

whole population. Statistics help to make less biased decisions, establish causality, and evaluate interference from other confounding factors.

58.2 Uses of Statistics in Drug Research

Statistics help us in the following ways:

- To evaluate the activity of the drug and establish its dose-response relationship
- To quantify the drug's action and differentiate it from other confounding variables
- To compare the action of two or more drugs and identify the superior mode of treatment
- To find an association between adverse drug reactions and drug use.
- To find the appropriate subgroup of patients in whom the drug will be more efficacious and safe

58.3 Some Important Terms

Population: The complete collection of all elements that need to be studied. For example, for the study of diabetes, all patients with diabetes in the world constitute the population.

Sample: A subgroup of the population that is representative of the population. The findings from the sample can be extrapolated to the entire population. This makes clinical trials easier to perform as it would be tedious and impossible to study all participants in a population.

Parameter: A parameter is a numerical measurement describing some characteristic of a population.

Statistic: A statistic is a numerical measurement describing the characteristics of a sample.

Variable: Variable is a characteristic that varies from one subject to another in a study population. Variables help us measure quantitative and qualitative information.

Quantitative data can either be continuous or discrete. Discrete data generally takes integers as values while continuous data take any numerical value. For example, the number of cancer patients dying in 5 years is discrete data while the intracranial pressure measurement is a continuous measure.

Qualitative data can either be categorical, ordinal, or interval and ratios.

Categorical data are unordered and based on their names. It is also referred to as nominal data. Example: Male and female gender. If only two categories are present (example Yes/No) they are referred to as dichotomous data or binary data.

Ordinal data is arranged in some order. Examples are stages of particular cancer.

Interval variables are similar to ordinal data where the intervals are equidistant from each other. Example: Temperature measurements on the Fahrenheit scale

Ratios are interval scales with a true zero point. Example: Milliliters.

Variables are also divided into *dependent variables* and *independent variables.* The dependent variable is the parameter that is studied. Various factors can influence the dependent variable. The independent variables are other factors that differ among study groups and are independent of the effects that are being studied.

Placebo: A substance that does not have any effect on the dependent variable that is being measured and studied.

Control group: A control group is a group receiving a placebo or sham treatment and acts as a baseline for comparison of the effect of the drug or other treatment procedure.

Causality:

Many factors may be associated with the outcome but may not be causative of it directly. Improper use of statistics could result in wrong conclusions of the causative factor. The mere association does not mean causation. The following criteria help in elucidating the causation from the association.

Box 58.1

Bradford Hill Criteria	Elwood's criteria
Plausibility There must be a reasonable explanation of the mechanism, a pathway to the outcome **Consistency** The same results should be observed if the experiments are repeated many times in different times at different places by different people **Temporality** The outcome studied must occur after the exposure **Strength** There should be a reasonable dose-response relationship **Specificity** The causal factor should have specificity to the outcome **Dechallenge** The outcome should stop once the exposure is removed	Descriptive evidence must exist Non-causal explanation should be ruled out E.g. change, bias and confounding Should contain positive features: Time Strength Dose-response relationship Consistency Specificity Should be generalizable to a larger population than the study group Comparison with other evidence should show consistency, specificity, plausibility, and coherence

Bias:

Bias is a preference for a particular value and avoids certain other values. Bias is a factor that can lead to incorrect conclusions about association or causation. Some of the types of biases that could exist in scientific research are given below:

Selection bias: Selection bias occurs when only a particular subtype of study participants are chosen which might influence the final outcome.

Sampling bias: Sampling bias occurs due to very small or inadequate sample size. The findings of the study will not be generalizable to the entire population.

Observation bias: Observation bias results from an improper recall of information by the study participants resulting in inadequate or incorrect data collection (Recall bias). Observation bias might also occur due to interindividual differences between various observers in the study.

Attrition bias: Attrition bias occurs due to people leaving the study prematurely.

Confirmation bias: Influence of previous beliefs and ideas resulting in the alteration of observations or judgments by the researcher is called confirmation bias. This happens when the researcher strongly believes nor disbelieves in a particular idea.

Channeling bias: Channeling bias occurs when the patient's degree of illness or prognosis is more influential in the study group they are enrolled in.

Priming bias: This is seen where a questionnaire is used to collect data from the study participants. The effect a question has on the subsequent questions in the questionnaire is called priming bias.

Interviewer bias: Interviewer bias is seen when the researcher knows the study subject's health status before questioning. They tend to ask more probing questions to patients who have a particular condition and hence influencing the data collected by the study.

Publication bias: Publication bias often influences the results from systematic analysis and meta-analysis. Positive studies tend to be published more than the negative studies resulting in influencing the results from a large meta-analysis.

Strategies to control bias:

Bias cannot be completely removed from any scientific study but measures can be taken to understand the presence of bias and systematically reduce it. Here are some of the strategies that are commonly employed:

- Use of multiple control groups
- Standardization of observation protocols, questionnaires, instruments, etc.
- Use of dummy variables with the known association so the effect of bias can be evaluated to some extent.
- Selection of patients with rigorous criteria
- Use of objective data sources to reduce recall bias whenever possible.
- Define a plan for lost-to-follow-up patients before the study begins.
- Register the trial with a recognized registry.

Confounding:

Confounding variables are those variables that are associated with the outcome but do not play any role in the causation. Confounding can result when a researcher makes a wrong conclusion about a confounding variable. Example: Subjects who gamble tend to develop liver disease and cancer. Gambling is a confounding variable because people who gamble often consume alcohol and smoke cigarettes.

Various strategies Are Employed to Reduce Confounding in Clinical Trials

- Use of randomization. (This helps in a random distribution of the confounding variable among the various groups in the study). Various types of randomization techniques are employed.
  - Simple randomization
  - Block randomization (Blocks of participants are randomized instead of single entities resulting in comparable sample size among the groups)
  - Stratified randomization (This method specifically measures the levels of confounding variables among different groups and tries to equalize them)
  - Adaptive randomization (The randomization technique adapts to the incoming participants in a study)
- Use of restricted selection criteria. (Patients with the confounding variable may be excluded from the study)
- Use of matched groups. (Study subjects can be matched for confounding variables in various study groups)
- Use of stratified group in the analysis. (Subgroup analysis can be done so the effect of confounding variable is removed)
- Use of multivariate analysis (Confounding variables should be known before the study starts)

58.4 Important Principles for the Application of Statistics in Clinical Research

- Statistical significance and clinical significance are not the same. While statistical tests might say the difference is statistically significant, the difference has to be evaluated for clinical significance.
- The way the observations are scaled can influence the final outcome of the trial. For example measurement of pain as present or absent versus a pain scale.
- Sampling time is very important and might be very sensitive to influence the outcome of various drug trials. Since the blood levels of the drug keep changing, very accurate timing should be maintained for pharmacokinetic studies.
- Carry-over effects are likely to happen when a crossover study design is used and adequate time should be given for proper washout of drugs from the system and return of parameters to the baseline.
- Before using the various statistical tests of significance, care should be taken to study and understand the various assumptions that need to be met to make correct use of the tests.
- Statistical analysis can be easily misleading. It is important to make sure we do not use statistical analysis to prove what we think is right but instead use statistics to understand the processes in nature.
- The p-value is an important measure of statistical significance but not the only important measure.
- Power calculation must be carried out to understand how much we can rely on the conclusions of the study

Bibliography

Barkan H. Statistics in clinical research: important considerations. Ann Card Anaesth. 2015;18(1):74.

Bhandari M, Chan S, Jonsson A. Planning a clinical research study. Indian J Orthop. 2007;41(1):16.

Pannucci CJ, Wilkins EG. Identifying and avoiding bias in research. Plast Reconstr Surg. 2010;126(2):619–25.

Calculation of Basic Statistical Parameters 59

Gomathi Ramaswamy

Abstract

Statistics is a discipline that involves the collection, analysis, and interpretation of data. Statistics provides information from the collected data, which aids in acquiring knowledge about the situation. The primary role of statistics is to understand the population level details through representative sample data. Statistics used in the biomedical field is also called biostatistics. Biostatistics operates under the principle of statistics. Incorrect understanding of statistics results in misinterpretation of scientific information and flawed conclusions, which the scientific theories cannot explain. This chapter focus on the basic statistical parameters commonly used in the field of biomedical research. The chapter will enlighten the reader on the variables, descriptive statistics, measures of central tendency, and measures of dispersion. The reader will also get familiarized with the concepts and calculations of these parameters. It also gives a brief overview of inferential statistics.

Keywords

Statistics · Measures of central tendency · Measures of dispersion · Descriptive statistics · Inferential statistics

59.1 Introduction

Statistics is the science that deals with collecting, analyzing, interpreting, understanding, summarizing, and presenting data. Epidemiologists, clinicians, microbiologists, pharmacologists, pathologists, and almost everyone in the field of

G. Ramaswamy (✉)
Department of Community Medicine & Family Medicine, All India Institute of Medical Sciences (AIIMS) Bibinagar, Hyderabad, Telangana, India

M. Lakshmanan et al. (eds.), *Introduction to Basics of Pharmacology and Toxicology*, https://doi.org/10.1007/978-981-19-5343-9_59

medicine and allied sciences use statistics in their research, clinical practice, reporting of the patient census, etc. This chapter focuses on the details and calculation of basic statistical parameters to assist the post-graduate students to understand, choose the correct statistical methods, and interpret their research findings accurately. Also, this chapter enables the reader to critically analyze data and related information provided in the scientific manuscript. Understanding the basic statistical measures is essential for conducting sound research, especially from the initial phase of one's career. Biostatistics is a domain under statistics and functions fundamentally according to the statistical propositions.

Statistics can be broadly classified as descriptive statistics and inferential statistics based on the summary measures and interpretation of the data. Descriptive statistics primarily describes and summarizes the data from the sample population which is being studied. Inferential statistics infers the sample data findings to the larger population from which the sample is collected. Inferential statistics use margins of error to extrapolate the sample findings to the whole population. This chapter focuses on types of variables and the appropriate descriptive statistics to handle specific variables. We will discuss measures of central tendency, measures of dispersion, rate, ratio, and proportions in detail. We will also shed light on the inferential statistics such as 95% confidence interval and standard error.

59.2 Variable

What is a variable? A Variable is an attribute that takes different values. In simple words, a variable is something that varies across your observation or data points. For example, gender can change and take the attributes such as male, female or transgender. Such attributes describe the characteristics of the variable for a given observation. Few other examples of variables are height, weight, blood sugar, blood pressure, etc. In research, the list of variables has to be selected based on the aims and objectives of the study. It is always a good practice to identify the variables, their possible attributes, approach for data collection, plan for analyzing and summarizing during the protocol development stage itself.

59.2.1 Types of Variables

Variables are classified based on the nature of the attributes. The variables can be categorized into quantitative and qualitative variables. Quantitative variables can take numerical details ranging from zero to infinity. Quantitative variables with the whole numbers are called discrete variables, and those with decimal points are called continuous variables.

59.2.1.1 Quantitative Variable (Numeric Variable)

Discrete Variable

A discrete variable is always a whole number and, in reality, cannot have a decimal value. For example, the number of children in a family, the number of admissions in hospital per day, and the number of tablets prescribed to the patient are a few examples of discrete variables. (Fig. 59.1).

Continuous Variable

A continuous variable can have a decimal value. For example, height, weight, body mass index, the dose of a drug, hemoglobin level are a few examples of a continuous variable, and it is agreeable to have decimal points in these examples.

59.2.1.2 Qualitative Variable (Categorical Variable)

Nominal Variable

As the name implies, a nominal variable has the name as categories or attributes. A nominal variable with only two categories or attributes is called a binary categorical variable.

Ex. Name of the individual, name of the drug injected to the rat, route of administration drug.

Binary variable: Diabetes: Present/ Absent, adverse drug reaction to drug: present/ absent.

Ordinal Variable

An ordinal variable is a variable that has an order in its attributes.

Ex. The severity of pain captured using pain scale (no pain, mild, moderate, severe, and very severe pain), types of anemia (no anemia, mild, moderate and severe anemia), stages of diabetes (no diabetes, pre-diabetes and diabetes).

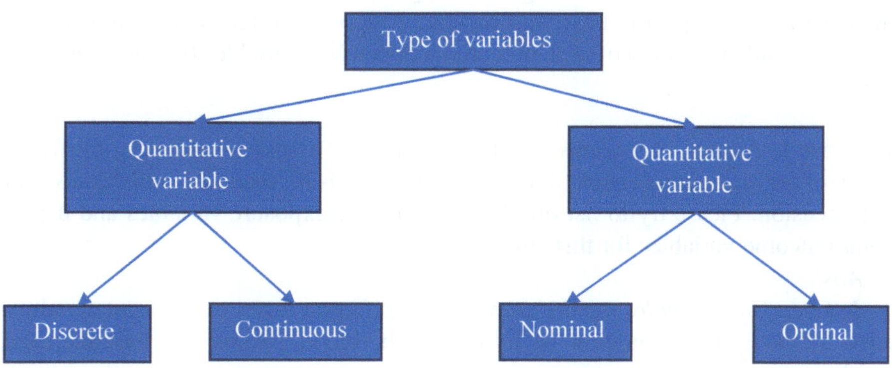

Fig. 59.1 Types of variable

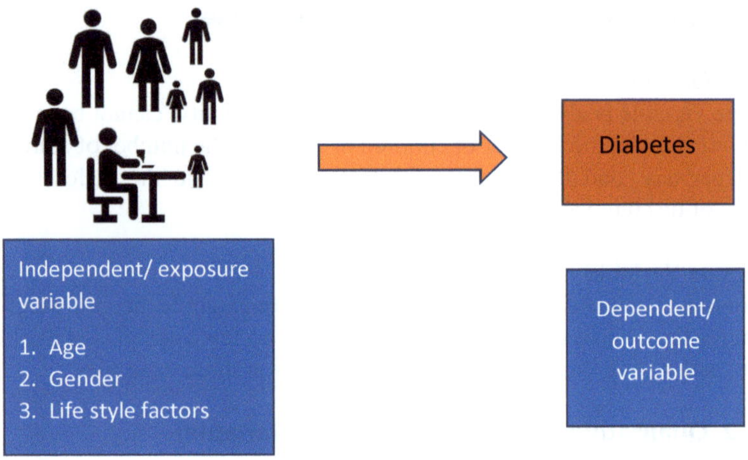

Fig. 59.2 Example for independent and dependent variables

Variables can also be classified as *independent/ dependent variables* or *exposure/ outcome variables*. These terms are more frequently used in research than the quantitative and qualitative terminologies. Independent or exposure variables are expected to cause or influence the dependent/ outcome variable. The outcome or dependent variables will occur or be affected by an independent/ exposure variable.

Ex. Age, gender, and lifestyle factors are few independent/ exposure variables for diabetes, which is an outcome or dependent variable. (Fig. 59.2) The dependent variable, diabetes, can be influenced by many other independent variables. Therefore, it is the responsibility of the researcher to identify the list of independent variables potential to influence the dependent variable (either directly or indirectly) and include it in the research. Strictly speaking, it is not always feasible to accommodate all independent / exposure variables in the study. However, the researcher should brainstorm and identify the crucial list of independent variables and include them in their research. Conducting a thorough review of literature, discussion with the experts in the specific field, and review of reports and relevant documents will assist in the identification of such potential independent variables for inclusion in the study.

Exercise 1 A researcher wants to conduct a study to assess the efficacy of different doses of telmisartan to control blood pressure in 40–70 years old individuals with hypertension. Please try to list out the independent/ exposure variables and dependent/ outcome variables for this study.

Ans:

Independent variables for the above study: different doses of telmisartan, adherence to medication, duration of hypertension, blood pressure at baseline, age, gender, etc.

Dependent variable: Control of blood pressure – Yes or no (If the researcher wants to consider the outcome as categorical variable, he will check whether the study participant has the desired blood pressure at the end of the study. Another option is that the researcher can consider the outcome variable as a continuous variable. They can report the mm/Hg reduction in blood pressure compared to the baseline.

59.3 Measures of Central Tendency

Using suitable summary measures for the given data is essential for appropriate interpretation and arriving at or proving a hypothesis. In the previous section, we learned about types of variables (quantitative and qualitative). The quantitative variables can be summarized using (i) measures of central tendency, which indicates the central values of the variables, and (ii) measures of dispersion which indicates the distribution of the individual data across the whole data set. Therefore, it is always good practice to mention the appropriate measures of central tendency and dispersion together to provide meaningful insight into the data.

The measures of central tendency summarize the data into a single value. The single summary value is usually the central value of the data set. However, it depends on the distribution of the data. Measures of central tendency include mean, median, and mode. Let us see about these in detail now.

59.3.1 Mean

Mean is the average value of the data. For example, if there are n number of recorded values for a variable X, the mean is calculated by adding the total recorded values of the variable X and divided by the number n.

The arithmetic mean is the average of the given quantitative or continuous numeric data.

For example, if the dataset has *n1, n2, n3, n4, n5, n6, n7, n8, n9*, and *n10*, the mean can be calculated by the following method.

$$\text{Mean}\left(\overline{X}\right) = \frac{n1 + n2 + n3 + n4 + n5 + n6 + n7 + n8 + n9 + n10}{10}$$
$$= \frac{\text{Sum of the numeric data}}{\text{Total number of the values}}$$

Let us see an example:

A researcher has measured the blood sugar values of eight individuals following an anti-diabetic drug administration.

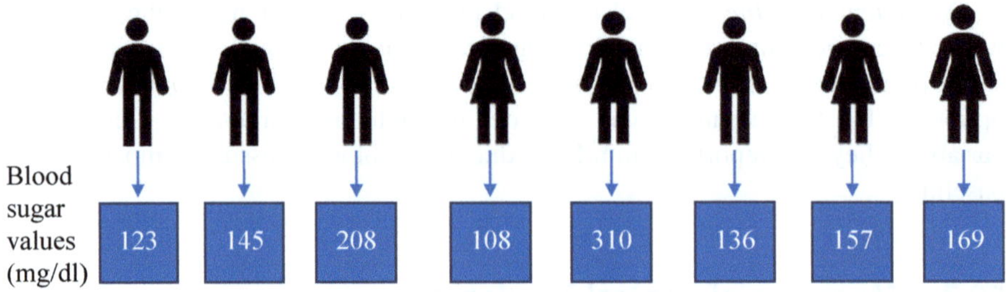

$$\text{Mean}\ \left(\overline{X}\right) = \frac{123 + 145 + 208 + 108 + 310 + 136 + 157 + 169}{8}$$

Mean blood sugar value = 136 mg/dl.

Advantages of using mean:

Represent the whole data values.

Easy to interpret and do further statistical analysis.

Relatively less variation while doing repeat sampling.

Disadvantages of mean:

Mean is affected by large outlier values. When the dataset has a large value which is an outlier, it will pull the mean towards the higher value resulting in skewed results. For instance, in the above example, if the eighth person's blood sugar value is 569 instead of 169, the calculated mean will be 219.5 instead of 136. It is clear from this example that if there is a large value (one or a few values as outliers), the mean will be skewed towards the large value, and other values will be minimally represented.

It is challenging to locate the exact position of the mean in the given set of data values.

There are three different types of mean, namely, *geometric mean*, *arithmetic mean*, and *harmonic mean*.

The calculation mentioned above yields us an arithmetic mean, which is the average value of the given data set. Arithmetic mean is most widely used in bio-statistics. Thus, wherever mean is mentioned in biomedical research, arithmetic mean is reported in the majority of the cases. On the other hand, the geometric mean (used when data has a multiplicative relationship between values) and harmonic mean (used in rate, speed, etc.) are not commonly used in the biostatistical calculation, hence not eloborated here.

59.3.2 Median

Another measure of central tendency is the median. Median indicates the central value of the given data set.

The values in the data set have to be arranged in ascending or descending order, and the middle values are the median values.

Ex. Body weight of 11 individuals are as follows: 45, 67, 90, 65, 32, 14, 69, 89, 59, 94, 98.

What is the median height of these individuals?

Step 1: Arrange the values in ascending or descending order:

14, 32, 45, 59, 65, 67, 69, 89, 90, 94, 98.

Step 2: Find out the middle value:

14, 32, 45, 59, 65, ***67,*** 69, 89, 90, 94, 98.

Median = 67.

Ex. Find out the median hemoglobin values from the following data: 9.6, 5.8, 8.7, 6.9, 7.5, 8.3, 9.2, 7.0.

Step 1: Step 1: Arrange the values in ascending or descending order:

9.6, 9.2, 8.7, 8.3, 7.5, 7.0, 6.9, 5.8.

Step 2: Find out the middle value:

9.6, 9.2, 8.7, ***8.3, 7.5***, 7.0, 6.9, 5.8

$$\text{Median} = \frac{8.3 + 7.5}{2} = 7.9$$

When the total number of values is odd, the middle value is the median after arranging the data in ascending/ descending order. At the same time, in the case of an even number of values, the data to be arranged in ascending/ descending order and the average/ arithmetic mean of the middle two numbers has to be calculated to derive the median value.

The position of median in the given "n" of observations can be located by the given formula:

$$\text{Location of median number} = \frac{n+1}{2}$$

Advantages of median:

1. Median is not affected by the extreme/outlier values.
2. Median is not affected by skewed data.
3. Easy to compute.

59.3.3 Mode

Mode represents the most frequent number in the data.

For example, in the given data, find out the mode.

56, 79, 53, 54, 56, 45, 34, 98, 89, 56.

The most frequent value in the above data is 56. Here only one mode value is there, and this kind of data is unimodal.

Let's see another example:

8, 9, 8, 4, 5, 3, 2, 8, 1, 6, 8, 5,

Table 59.1 Frequency table for the above data

Value	Frequency
8	**4**
9	1
4	1
5	**2**
3	1
2	1
1	1
6	1

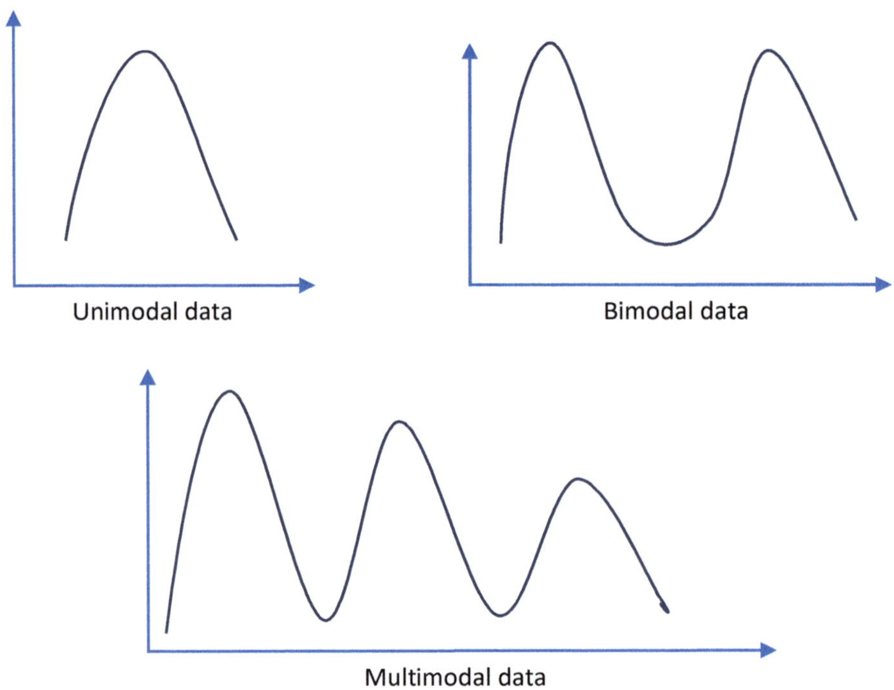

Fig. 59.3 Unimodal, Bimodal and multimodal data

The maximum repeated frequency in the above data is 8 and 5. Hence, this data is bimodal. We can also use a frequency table to find out the mode. (Table 59.1).

Some data might have multiple repeated frequencies and will be multimodal in nature. (Fig. 59.3).

Advantages of mode:

1. Easy to obtain.

Disadvantages:

1. Mode doesn't represent the complete data
2. Difficult to interpret multiple modes, however, it may be a useful information.

When to use mean, median and mode?

- The data has to be quantitative
- Use mean when the data is normally distributed. (We will learn about normal distribution later in the sections)
- Use median/mode when the data is not normally distributed or skewed.

59.4 Measures of Dispersion

So far, we have learned about the measures of central tendency such as mean, median, and mode. Measures of central tendency aid the researcher in understanding the average, or middle, or most frequent numbers in the data. However, these central values do not give a complete knowledge of the data since we don't know how the data points are distributed throughout the data. The measures of dispersion provide a solution for such an issue. Measures of dispersion give an idea of how far the data is distributed or spread from the center. Standard deviation, range, and interquartile range are the commonly used measures of dispersion in medical research. In this section, we will learn about the measures of dispersion.

59.4.1 Standard Deviation

Standard deviation (SD) is one of the commonly used measures of dispersion for quantitative variables. Standard deviation is the positive square root of the mean of squared deviations from the mean. The following steps are to be followed for the calculation of standard deviation.

Step 1: Calculate the mean for the given data.
Step 2: Calculate each values deviation from the mean value
Step 3: Calculate the squares of the deviation values
Step 4: Calculate the positive square root of the sum of squared deviation values.

The formula for calculation of standard deviation:

$$\mathrm{SD} = \sqrt{\frac{\sum (x - \overline{X})^2}{n - 1}}$$

In the above formula, SD = standard deviation, x = individual data value, $\overline{X}$ = mean of the given data, n = total number of values.

If the total number of values is less than 30, the "$n-1$" should be used as the denominator, and if the total number of values is more than 30, "n" has to be considered as the denominator.

Ex. For the given value calculate the standard deviation:

67, 54, 78, 89, 67, 90, 43, 84, 63, 35.

Step 1: Calculate the mean:

$$= \frac{67 + 54 + 78 + 89 + 67 + 90 + 43 + 84 + 63 + 35}{10}$$

$$= 670/10 = 67$$

Step 2: Calculate the deviation of each value from the mean:

Values (x)	Mean ($\overline{X}$)	Deviation from mean ($x-\overline{X}$)
67	67	0
54	67	−13
78	67	11
89	67	22
67	67	0
90	67	23
43	67	−24
84	67	17
63	67	−4
35	67	−32

Step 3: Calculate the squares of the deviation values.

Values (x)	Mean ($\overline{X}$)	Deviation from mean ($x-\overline{X}$)	Squares of deviation $(x-\overline{X})^2$
67	67	0	0
54	67	−13	169
78	67	11	121
89	67	22	484
67	67	0	0
90	67	23	529
43	67	−24	576
84	67	17	289
63	67	−4	8
35	67	−32	1024
	Total	0	3200

The sum of the deviation of the mean will always be zero since we are calculating the deviations from the mean, which consists of positive and negative values.

Step 4: Calculate square root:

$$\begin{aligned}\mathrm{SD} &= \sqrt{\frac{\sum (x-\overline{X})^2}{n-1}} \\ &= \sqrt{\frac{3200}{10-1}} \\ &= \sqrt{355.5} \\ \mathrm{SD} &= 18.8\end{aligned}$$

59.4.1.1 Interpretation of Standard Deviation

Standard deviation is a measure of dispersion used consistently with the mean, which is a measure of central tendency. Mean, and the standard deviation is reported in scientific journals as 67 (18.8) or 67 ± 18.8 along with the unit of measurement. Standard deviation indicates the spread of data from the mean. For example, one standard deviation indicates that 68% of data values are spread on either side of the mean. Similarly, 95% of data values will be spread across both sides of the mean, that is, two standard deviations. (Fig. 59.4) We will learn about this in the section on the normal distribution.

In the above example, the value 355.5 is the variance. Variance is calculated by the below formula.

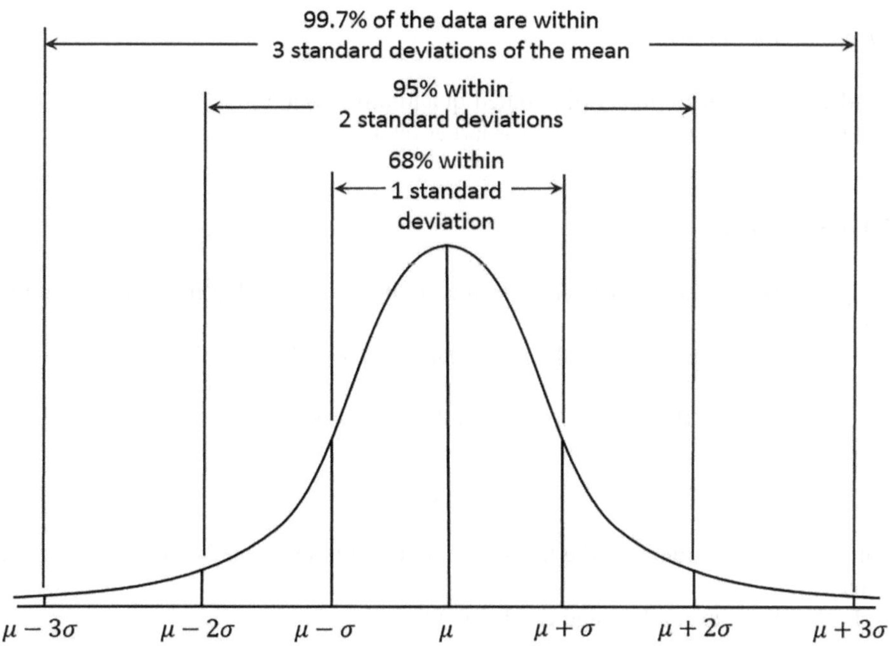

Fig. 59.4 Relationship between mean and standard deviation

$$\text{Variance } \sigma^2 = \frac{\sum (x - \overline{X})^2}{n - 1}$$

$$\text{Standard deviation} = \sqrt{\sigma^2}$$

59.4.2 Variance

Variance is the average squared deviation from the mean and one of the measures of dispersion. Variance uses the whole data set in its calculation. However, it is affected by extreme values as it uses original units. Thus, for large datasets, it is challenging to estimate variance manually.

When to calculate standard deviation:

1. Quantitative data.
2. Data is normally distributed without any skewness.

Advantages of SD:

1. Uses whole data set for estimation
2. Better representation of the data, including the tail values.
3. SD can be used for further statistical analysis and inferential statistics.

Disadvantages of SD:

1. Ideal to use for normally distributed quantitative data
2. Difficult to calculate manually for larger data

59.4.3 Range

Range is the difference between the maximum/ largest value and the minimum/ smallest value.

Ex. Calculate the range for the given data. 1, 6,3, 6, 4, 7, 12, 7, 9, 16, 34, 23, 37.

$$\begin{aligned} \text{Range} &= \text{maximum value} - \text{minimum value.} \\ &= 37 - 1 \\ &= 36 \end{aligned}$$

Some authors just mention the maximum and minimum values while reporting the range.

In the above example range can be mentioned as 1, 37.

Advantages of range:

1. Easy and quick to calculate
2. Easy to interpret

Disadvantages of range:

1. Affected by extreme values
2. Range does not give completed representation of the dispersed data
3. Range cannot be calculated for open ended frequencies
4. Difficult to use in further statistical calculations

59.4.4 Interquartile Range

As the name implies, quartiles are values that divide the whole data into four equal fragments. Hence there will be three quartile values, viz., 25%, 50%, and 75% (positioned at 25th, 50th, and 95th percentiles of the data) (Fig. 59.5).

Let's see an example for the calculation of interquartile range (IQR). The number in blue color in the above image indicates minutes after drug administration and the occurrence of the desired response among individuals.

The IQR is always reported along with the median, which is a measure of central tendency. Similar to the median calculation, the data has to be arranged in ascending/ descending order to calculate IQR. The 25th(Q1) quartile and 75th (Q3) quartile values are the interquartile range values, and 50th (Q2) quartile value is the median value. If the data is arranged in ascending order, the IQR can be calculated directly.

In the example mentioned above, the data is already arranged in ascending order. IQR can be reported can 68, 278 or 278–68 = 210.

The following formula can also determine the Q1 value:

$$=\frac{N+1}{4}$$

In this data, Q1 = (19 + 1)/ 4 = 5, fifth value that is 68 is the Q1 value.

Q3 value can also be determined by the following formula:

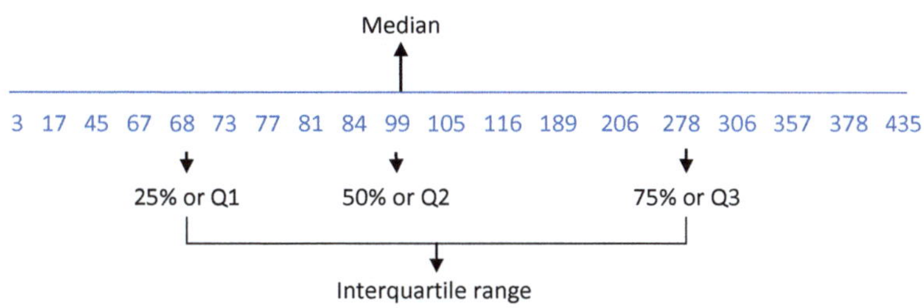

Fig. 59.5 Interquartile range

$$= 3 \times \frac{N + 1}{4}$$

In this dataset, Q3 = 3 × 5 = 15th value, i.e. 278 is the Q3 value.

59.4.4.1 Interpretation of IQR

IQR and median values divide the data into four equal quarters. Thus, when the data is arranged in ascending order, the 25th (Q1) quartile value indicates that 25% of values are below the 25th(Q1) quartile value and 75% of values are above this value. Similarly, 75th(Q3) quartile value indicates that 75% of the values in the data are below this value and 25% of the values are above the 75th (Q3) quartile value.

Advantages of IQR:

1. IQR is not affected by extreme values.
2. It can be used in data that is not normally distributed or skewed data
3. Provides better representative of data

Disadvantages of IQR:

1. IQR ignores 50% of the data (25% data above Q3 and 25% data below Q1.
2. Doesn't provide any information on outliers
3. Difficult to interpret

59.4.5 Normal Distribution Curve

For continuous or quantitative data, the mean is the measure of central tendency, and the standard deviation is the measure of dispersion. While plotting such quantitative data in a histogram, if the data is distributed equally on either side of the midline, we can say that the data is normally distributed. (Fig. 59.6) A few statistical tests, such

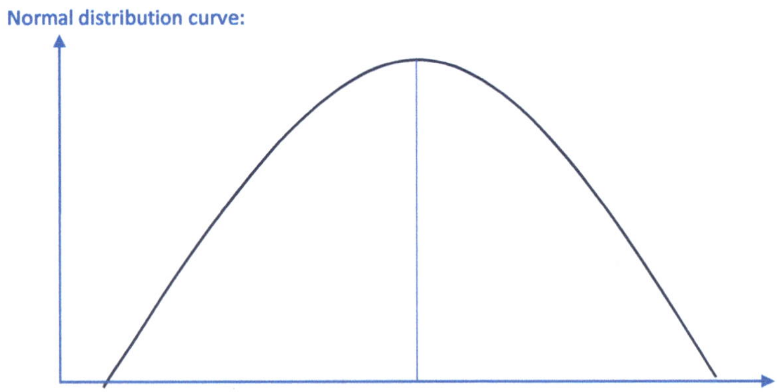

Fig. 59.6 Normal distribution curve

as the Kolmogorov-Smirnov test, Shapiro-Wilk test, etc., are available to determine the normality of the data. These tests require either software for statistical analysis or manual mathematical calculation to ascertain the normality of the data. Another way to determine normality is by following the rule of thumb. According to the rule of thumb, the standard deviation should be less than half of the mean to have data with normal distribution. If the standard deviation is more the half of the mean, then we can say that the data is not normally distributed.

When the data is normally distributed, we can use mean as the measure of central tendency and SD as the measure of dispersion. For non-normal data, the median should be used as the measure of central tendency, and the IQR is the measure of dispersion. However, the normal distribution of the data largely depends on the sample size. If the sample size is small, there is a high chance that the data follows non-normal distribution compared to data with a large sample size.

59.4.5.1 Few Characteristics of the Normal Distribution Curve

A normal distribution curve is a smooth bell-shaped curve distributed equally on either side of the middle line. (Fig. 59.7) The mean, median, and mode coincide in the normally distributed data. When the mean is zero and variance is 1, then the normal distribution is called a standard normal distribution.

The empirical rule for the distribution of data in a normal distribution curve is as follows:

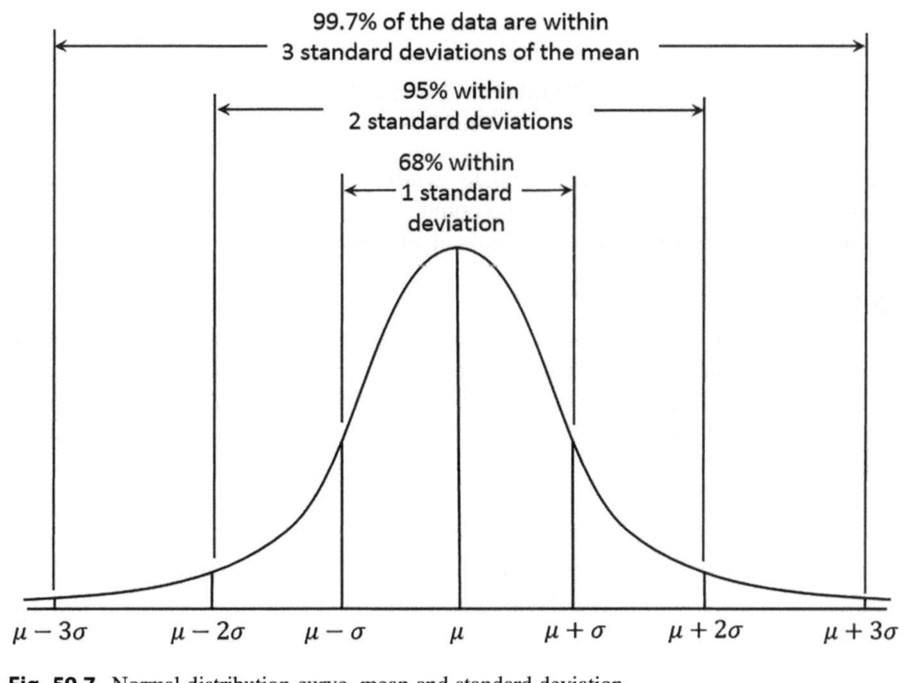

Fig. 59.7 Normal distribution curve, mean and standard deviation

In normally distributed data, 68.3% of the values fall within mean ± 1SD values. Similarly, 95.4 and 99.7% of the data will be distributed within mean ± 2SD values and ± 3SD values, respectively.

59.4.6 Summary Measures for Qualitative or Categorical Data

So far, we have read about the measures of central tendency and dispersion for quantitative data. Let us learn about qualitative data and its summary measures in this session. Qualitative variables are categorical variables that have percentages or frequency standards. It is not appropriate to summarize them using mean (SD) or median (IQR). Instead, the qualitative and categorical variables are summarized using rate, ratio, or proportions, which are frequency measures.

59.4.6.1 Ratio

The ratio is the division of two different numbers or magnitudes or quantities. The two values (numerator and denominator) need not be related to each other. The numerator need not be part of the denominator. Ratios generally do not have units of measurement attached to them while reporting.

Ex:

$$= \frac{\text{Number of rats in the laboratory}}{\text{Number of frogs in the laboratory}}$$

Ex: Gender ratio:

$$= \frac{\text{Number of girl children}}{1000 \text{ boys}}$$

The above formula is used to calculate gender or sex ratio, which the officials measure for reporting gender ratio in a country.

The ratios can be reported as 987 girls: 1000 boys.

59.4.6.2 Rate

Rate is a measure of the occurrence of events in a specified period and specified population. The numerator is part of denominator. Unlike ratios, rates have units attached to them. Thus, rates can be used for the occurrence of disease in a population in a specified time, occurrence of adverse effects in a group of people taking specific medication, etc.

$$\text{Rate} = \frac{\text{Number of individuals with the events in a specified population in a specified time}}{\text{Total number of population in the same time}} \times 1000$$

Ex. In a study involving 5000 population in a village, 50 developed adverse events for a new drug in 2020.

$$\text{Rate} = \frac{\text{Number of individuals with the adverse events in a specified population in 2020}}{\text{Total number of population in 2020}} \times 1000$$
$$= (50/5000) \times 1000$$

Rate of adverse events = 10 adverse events per 1000 population in the year 2020.

If the events are newly occurring events, the rate is called as *incidence rate*. If the numerator consists of both newly arising and already existing events, such a rate is called *prevalence rates. Attack rate* consists of the numerator who develops disease during an outbreak, and the denominator includes the population at risk or susceptible to the disease during the outbreak.

59.4.6.3 Proportion

Proportion is also similar to rate, which measures the occurrence of events or numbers compared to a specified population. However, the proportion does not have the time component like the rate. Proportions can be expressed as percentages or fractions.

$$\text{Proportion} = \frac{\text{Number of individuals with the events in a specified population}}{\text{Total number of population}} \times 100$$

Example. In an outpatient clinic with 1000 patient attendance, 450 individuals were identified to have diabetes. Calculate the proportion.

$$\text{Proportion} = \frac{\text{Number of individuals with the diabetes}}{\text{Total number of patients attending OPD}} \times 100$$
$$= (450/1000) \times 100 = 45\%$$

Here there was no time component. Proportions are also used when the population is not from a defined population. For example, in a clinic, people from many places can attend. It is difficult to define the catchment area of the population. In such cases, instead of rate, proportions are a better measure of summary.

59.5 Inferential Statistics

We usually select a small population from a large population as our sample (or study participants) for conducting research. When we describe or summarize the findings from our study participants, it is called descriptive statistics. However, if we extrapolate the sample findings to the large population from where the sample population has been derived, it is called inferential statistics. The interpretation in inferential statistical happens beyond the number of people that are included in the study. The primary role of inferential statistics is testing hypothesis and making effective decisions. There are two types of hypothesis, viz., the null hypothesis (Ho) and alternative hypothesis (H1). Ho indicates there is no difference between the characteristics of the population or groups. H1 suggests that there is a difference between the characteristics of the population or groups.

You must have heard of the term *p*-value or statistical significance (α). If the *p*-value is less than 0.05, then it is considered statistically significant. Lesser the *p* value higher the statistical significance. *P*-value of less than 0.05 indicates that the null hypothesis is wrong or rejected, and there is a significant difference between the population characteristics. In other words, we can say that the alternative hypothesis is true.

59.5.1 95% Confidence Interval

Confidence interval (CI) is also the measure for inferential statistics and is used widely along with *p*-value. 95% confidence interval (95%CI) has an upper value and a lower value symbolizing the true population-level values. The sample parameters are extrapolated to the population to derive the 95% CI. 95%CI is defined as when the sample is repeated "*n*" number of times, and each time 95% CI is calculated, 95% times these will represent true population parameters. A simple definition of 95% CI is that one can be 95% confident that the true population parameters fall between the two values of 95% CI derived from the study.

For example, the prevalence of hypertension in a city is 15% (95%CI: 10–22%) following a representative sample survey. We can explain that if the survey is repeated *n* number times, 95% times the true population hypertension prevalence will fall between 10 and 22%.

Bibliography

Douglas G Altman, J Martin Bland. Variables and parameters. BMJ [Internet] 1999 [cited 2021 Jul 8];318:1667. Available from: https://www.ncbi.nlm.nih.gov/pmc/articles/PMC1116021/pdf/1667.pdf.

Dunn OJ, Clark VA. Basic statistics: a primer for the biomedical sciences. 4th ed. New Jersey: John Wiley & Sons; 2009.

Kaur S. Variables in research. Indian J Res Reports Med Sci. [Internet]. [cited 2021 Apr 29]. 2013;3(4):36–38. Available from: https://sites.usp.br/rnp/wp-content/uploads/sites/830/2020/10/complemento-aula-1.pdf.

Manikandan S. Measures of dispersion. J Pharmacol Pharmacother. [cited 2021 May 3]. 2011;2(4):315–316.

L MSykes, F Gani, Z Vally. Statistical terms part 1: the meaning of the MEAN, and other statistical terms commonly used in medical research. South African Dent J [Internet] 2016 [cited 2021 Jul 8];71. Available from: http://www.scielo.org.za/scielo.php?script=sci_arttext & pid = S0011-85162016000600009.

Statistical Parameters. Measures of central tendency and variation.

Watkins JC. An Introduction to the science of statistics: from theory to implementation. Preliminar.

Hypothesis Testing

60

Lakshmi Balasundaram

Abstract

Hypothesis testing forms an important domain of inferential statistics, it also forms a stepping stone to evidence-based medicine (EBM). A narrow research question leads to the formulation of a research hypothesis which is put into test, in order to draw a statistical inference. A researcher believes in the alternative hypothesis which states that there is a treatment difference between the two groups, in contrast to null hypothesis which believes in no difference. The hypothesis is tested on a sample population after setting the significance level. The researcher uses the right statistical procedure to arrive at a statistical decision which helps is rejecting the null hypothesis or otherwise. A *p*-value with confidence interval helps in interpreting the results of the hypothesis testing better. In case where multiple hypotheses need to be tested simultaneously, the alpha error rate must be controlled, which can be done by various methods. A newer method of hypothesis testing is by using Bayesian methods which uses Bayes factor in assessing the plausibility of the hypotheses.

Keywords

Null hypothesis · Alternative hypothesis · *p*-value · Confidence interval

60.1 Introduction

A practising physician is faced by a constant dilemma of choosing the right drug amongst so many drugs from the same class or a different class, for his patient. To add to the confusion, there are new drugs entering the market, each being marketed

The original version of this chapter was revised. Revised figure updated this chapter. A correction to this chapter can be found at https://doi.org/10.1007/978-981-19-5343-9_64

L. Balasundaram (✉)
Department of Pharmacology, Jawaharlal Institute of Postgraduate Medical Education & Research (JIPMER), Puducherry, India

M. Lakshmanan et al. (eds.), *Introduction to Basics of Pharmacology and Toxicology*, https://doi.org/10.1007/978-981-19-5343-9_60

with a promise of better efficacy and good safety profile. To ease these woes, evidence-based medicine (EBM), acts as a guide to physicians in their clinical practice. In this chapter, we shall see how hypothesis testing forms a stepping stone to EBM.

60.2 Inferential Statistics and Its Domains

- Descriptive statistics describes the data of a sample such as by measures of central tendency. There is no element of uncertainty in descriptive statistics as we have obtained the data from a sample.
  Example: *In a village X in India, 80 out of the 150 women in reproductive age group were found to have haemoglobin levels more than 12.5 g/dL.*
- Whereas inferential statistics helps at making predictions/extrapolations using the data of a sample, the element of uncertainty remains as we have relied only on a sample.
  We can use the data of the above study to make a prediction for the entire population such as by this hypothetical example—*Only 56% of women in the reproductive age group in India have normal haemoglobin levels.*
- Inferential statistics is used for determining point estimate, interval estimate and for hypothesis testing.
  - Point estimate: The data of a sample is used to estimate a population parameter such as population mean.
  - Interval estimate: A range of values (interval) is calculated within which the population parameter lies. Example is the 95% confidence interval, which is the range of values that have 95% chance of containing the population mean.
  - Hypothesis testing: Statistical hypothesis refers to an assumption made regarding a population parameter, for example, a researcher would like to know if there a difference in the means between two populations. Hypothesis testing helps in arriving at conclusions.

60.3 Formulating a Research Question and a Hypothesis

- During clinical practice or during medical research, there may be a discovery of a lacuna that leads to formulation of a research question. For example, a physician may wonder which class amongst DPP-4 inhibitors and sulfonylureas, provides a better blood glucose lowering effect?
- A broad research question such as '*Which class, amongst DPP-4 inhibitors and sulfonylureas is better?*' needs to be narrowed down to '*What is the percentage of HbA1c reduction achieved by sitagliptin in comparison with glimepiride, when used as add-on therapy to metformin, over a period of 3 months in Type 2 diabetes mellitus patients?*' A good research question must remain focussed.

- A research question must ideally follow the PICO(T) format where, population or problem of interest (P), intervention (I), comparison (C), outcome (O) and time frame (T) should be included.
- The investigator then formulates a hypothesis based on the research question. One should remember that, to have a hypothesis, the investigator must take some position based on relevant literature or evidence that is available and the hypothesis should be testable, for example:
  - ***Research question**: What is the cholesterol lowering effect of daily administration of drug A in comparison with drug B at the end of 8 weeks in south Indian patients with primary hypercholesterolemia?*
  - ***Hypothesis**: Drug A has a better cholesterol lowering effect than drug B when administered daily for 8 weeks in south Indian patients with primary hypercholesterolemia.* (Here, the researcher does not take a position blindly towards drug A, rather, a guess is made based on information available on both drugs A and B).
- In this hypothesis, we are specifying a direction that drug A is better, here we do a one-tailed hypothesis test but when the research hypothesis states that '*There **is a difference** between drug A and drug B in lowering LDL-c*' where drug A is a new drug, it could be possible that drug A is better that drug B or it could be vice versa. In such a scenario, we must use a two-tailed hypothesis, which can test the superiority on both sides.
- Once a hypothesis is postulated, the investigator requires a representative sample of the population which provides necessary data to test the hypothesis.
- Hypothesis testing is a statistical means of testing the validity and meaningfulness of the results of a study.
- To draw a statistical inference, the researcher must apply an appropriate statistical test which is based on the variables used in the study (numerical or categorical) and the type of distribution the data follows (parametric or non parametric).

60.4 Null and Alternative Hypothesis

- Biostatistics takes a neutral stance when a statistical inference is being made, that is, when a hypothesis is formed, it is assumed that there is no difference in treatment effect between the experimental and control groups (drug X = drug Y) or there is no association between study variables and this assumption is called the null hypothesis. It is represented by the notation H_0. The mathematical signs commonly used in stating a null hypothesis include '$=$', '$\leq$' or '$\geq$'. Just as a person facing trial is presumed to be innocent until proven guilty, statistics believes that there is no difference between the groups (null hypothesis) unless proven otherwise. The concept of null hypothesis is to make a universal statement such as:
  - The sun rises in the east.
  - Vitamin D deficiency causes rickets in children.

- The investigator usually believes in the alternative hypothesis (H_1 or H_a) that is, drug X is more effective than drug Y (drug X > drug Y). Alternative hypothesis is also referred to as the research hypothesis. The mathematical signs used in alternative hypothesis are '≠', '<' or '>'. It can be observed that these signs are just opposite to those used in null hypothesis.
- Let us understand how the null and alternative hypothesis work: Say for example, the average birthweight of Indian babies is 2.8 kg, this is the null hypothesis. A few years later, a researcher may feel that because of improving healthcare and other factors, the birthweight now must be higher, so he challenges the null hypothesis with the alternative hypothesis, that is
  - H_0: $\mu = 2.8$ (null hypothesis)
  - H_a: $\mu > 2.8$ (alternative hypothesis)
- To prove his point, he cannot measure the birthweight of all babies born in India, instead he chooses to measure the birthweight from a sample population and then extrapolates. The mean birth weight if found to be higher than 2.8 kg means that the null hypothesis is rejected. Hence hypothesis testing can help in understanding and updating current scenario better.
- Hypothesis testing is focussed only upon acceptance or rejection of null hypothesis. It does not verify if the alternative hypothesis is true or not.

60.5 Type I and Type II Errors

During statistical inference, the following untoward errors may creep in:

1. The null hypothesis may be falsely rejected, that is, although drug X and drug Y have similar treatment effects, our study can show that drug X is better than drug Y (false positive). This error probability is called a Type I error or an alpha (α) error (Table 60.1).
2. The null hypothesis may be falsely accepted, that is, even though drug X is better than drug Y, our study shows that the treatment difference between the two is zero (false negative). This error is termed as Type II error or beta error(β).
   Considering the impact of the errors, one can observe the following:

Table 60.1 Type I and Type II errors with respect to hypothesis testing

Outcome of the study	Null hypothesis is actually true (drug X = drug Y)	Null hypothesis is actually false (drug X > drug Y)
Difference in outcome between treatment groups (drug X > drug Y)	False positive (Type I error)	True positive
No difference in outcome between treatment groups (drug X = drug Y)	True positive	False negative (Type II error)

- Falsely rejecting a true null hypothesis (α error) may have a serious impact, a drug with no treatment benefit could be falsely claiming to show benefit, which can impact treatment outcome or rather, can have an adverse outcome.
- Whereas, a beta error (claiming that a drug has no more benefit than a standard drug, when in fact, it is better) is less a serious error when compared to alpha error, in the sense that we may be missing a better drug, but it will not have any adverse treatment outcome.
- In order to have zero error in a study, the sample must be infinite which is not possible, hence to have a good representative sample, we must determine a significance level (alpha) which is the probability of rejecting the null hypothesis when it is actually true. This level is kept at 5% (0.05) meaning that there is only a 5% probability of erroneously rejecting the null hypothesis.
- The power (P) of a study is the ability of the study to correctly reject the null hypothesis when it is false (that is to find out a difference or association, when actually there is one) It is calculated as $P = 1\text{-}\beta$ (β = type II error). Power of 90% and 80% correspond to β value of 0.1 and 0.2 respectively. Power in simple terms, is the probability of not making a beta error. As the sample size increases, the power of a study also increases.
- The targets of statistical decision theory are to ensure that Type I error is at 5% and the power is maximum.

60.6 Steps in Hypothesis Testing

The basic steps involved in hypothesis testing is represented in a flowchart as following:

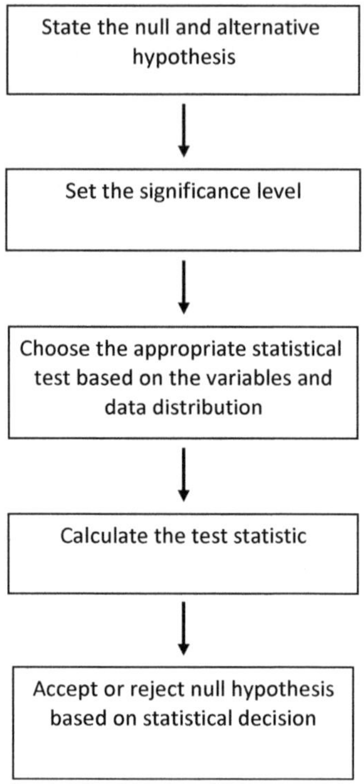

60.6.1 Example of Hypothesis Testing

- We shall work out an example using one-sample *Z* test to do hypothesis testing.

A medical officer claims that in a particular village, women of reproductive age group have higher haemoglobin levels, a random sample of 100 women showed a mean haemoglobin level of 16.5 g/dL. The population mean is 12.0 g/dL with a standard deviation of 3. We need to draw a statistical inference if the mean haemoglobin of the 100 women in that village is indeed higher than the population mean or not.

Step 1: Stating the null hypothesis. The mean Hb g/dl of the women in the reproductive age group is 12.0 that is H_0: $\mu = 12$.

Step 2: Stating the alternative hypothesis. The women have a haemoglobin level higher than that of population mean, that is, H1: $\mu > 12$, we will be choosing one-tailed test as we are assuming that the mean Hb levels are higher than the population mean.

Step 3: Set the significance (alpha) level. Usually set at 0.05 or 5%. In case of a two-tailed test, where in we are considering difference in either side, the alpha level is calculated as $0.05/2 = 0.025$.

Table 60.2 Common hypothesis tests with the corresponding test statistic

S. No.	Hypothesis test	Test statistic used
1	*Z* test • One sample • Two samples	*Z*-statistic
2	T test • One sample • Two samples – Paired – Independent	T- statistic
3	Chi-square	Chi-square statistic
4	ANOVA	F- statistic

Step 4: Use the Z-table to determine the rejection region based on alpha value. Alpha level of 0.05 corresponds to a *Z* score of 1.645. In case of a two-tail test, *Z* score of 0.75(1–0.025) is 1.96.

Step 5: Choose the test statistic. Common hypothesis tests with the appropriate test statistic is tabulated in Table 60.2. In this example we shall be using the formula

$$Z = \frac{\overline{X} - \mu_0}{\sigma/\sqrt{n}}$$

$\overline{X}$ = sample mean.μ_0 = population mean.σ = standard deviation.n = sample size.

Using the data in our example, $Z = (16.5–12)/ (3/ \sqrt{100}) = 15$.

Step 6: Accepting/rejecting null hypothesis. If our test statistic result is greater than the *Z* score (here $15 > 1.645$) we can reject the null hypothesis.

60.7 Interpreting the *p*-Value

- Sir Ronald A. Fischer (considered to be the Father of modern statistics) provided an insight to the concept of *p*-value.
- A clinical trial reports that drug A is better that drug B in reducing systolic blood pressure, thereby rejecting the null hypothesis. A question may arise, was the difference between the two drugs in this trial, a result of a mere chance or was it a true difference?
- The *p*-value tells us by how much probability this much or greater difference could have occurred by chance. A *p*-value less than the significance level rejects the null hypothesis and implies that the test is repeatable. The cut-off value of *p* is 0.05. In simple words, *p*-value of 0.02 implies that there is only 2% probability that the treatment difference occurred by chance and we are 98% confident that the treatment difference between the two drugs was not due to chance.
- One should clearly understand the difference between significance level α and *p*-value. Significance level is something that is set before the study while *p*-value is computed at the end of the study.

60.7.1 Problems in Interpretation of *p*-Value

- A cut off *p*-value of 0.05, determines if the result is statistically significant or not, resulting in dichotomous results. One can be put in a tough situation if the *p*-value stands at 0.045 or 0.051.
- A fall in BP by 1 mm Hg by a new drug when compared with a placebo, may show statistical significance ($p < 0.05$), but is it relevant clinically? So, *p*-value does not depict the effect size.
- A smaller *p*-value such as 0.001 in comparison to 0.02 does not indicate that the hypothesis is stronger, rather it indicates that the quality of evidence generated by the study to support the hypothesis was good.
- A study when replicated can result in different *p*-values, significant in one and non-significant in another, this is due to random variations that occurs in studies.
- In order to interpret the *p*-value properly it is imperative to consider various other aspects such as sample size, quality of data and study design.

60.8 Confidence Interval

- Due to the problems that can be encountered by just depending on *p*-value, researchers have come up with testing the hypothesis by bringing in the confidence interval (CI).
- Hypothesis testing by *p*-value gives a dichotomous decision, just a quick glance at the *p*-value can help us decide whether to accept the null hypothesis or not whereas a confidence interval gives an estimate of the effect size with a certain degree of confidence.
- Confidence interval is calculated by the formula

$$\mathrm{CI} = \overline{x} \pm z\left(\mathrm{s}/\sqrt{n}\right)$$

Where $\overline{x}$ = sample mean, z = confidence value, s = standard deviation, n = sample size.

- Let us see an example to understand the concept better, a new drug was found to reduce the systolic blood pressure by 10 mm Hg in a study, when compared to a standard drug. The study was repeated five times under similar conditions on different population, the mean fall in SBP was found to 6, 8, 7.5, 8 and 12 mm Hg. We can see that the effect size ranges from 6 to 12 mm Hg. When we say 95% confidence interval [6,12] we mean that our calculated confidence interval contains the population mean with 95% probability.
- Although 95% CI is most commonly presented, confidence intervals can also be presented as 90% and 99% (Table 60.3).
- As the sample size increases, confidence interval becomes narrower.
- Whilst *p*-value, helps in accepting or rejecting the null hypothesis, confidence interval provides a direction to the effect size.

Table 60.3 Confidence interval and corresponding *z* value

Confidence interval (CI)	*z* value
90	1.645
95	1.960
99	2.576

- The study has no statistical significance if the confidence interval contains the null value which is zero in case of difference between groups and one in case of ratios such as relative risk or odds ratio, for example 95% CI [−5, 5] includes 0 which means there is no difference between the treatment groups. Likewise, in a study considering odds ratio or relative risk, a confidence interval that contains 1.0 means that the odds or relative risk is similar with both groups, for example: 95% CI [0.8, 1.2].

60.9 Multiple Hypothesis Testing

- Sometimes, there arises a need to test multiple null hypothesis at the same time. For example, we may want to test two hypotheses, for example:
  - Reducing dietary sugar reduces blood glucose and
  - A 30-minute walk every day reduces blood glucose.
- If we test these hypotheses simultaneously at a significance level of 0.05, the probability of finding a significant result by chance can be measured by the following formula:

$$\begin{aligned}\text{Probability of 1 significant result} &= 1 - (\text{Probability of no significant result}) \\ &= 1 - (1 - 0.05)^2 \\ &= 1 - (0.95)^2 \\ &= 0.10\end{aligned}$$

  Which means there is 10% probability that a significant result was due to chance. Let us assume we have 25 hypotheses to test (say to find out if 25 genes are related to a disease), the probability to declare a significant result by chance increases to 73% [1-$(0.95)^{25}$].
- In multiple hypothesis testing, there are two important error types that must be understood:
  - *Family wise error rate (FWER)* which is the probability of making at least one type I error and is represented by FWER = $P(V \geq 1)$, where V = Type I error.
  - *False discovery rate (FDR)* which is the expected proportion of rejection of null hypothesis out of all rejections.
- To avoid such errors, it is important to adjust the value of α, and one of the methods is Bonferroni correction wherein the significance level is calculated by α/n, where n = number of hypothesis. For example, if we have 25 hypotheses to be tested, the significance level is set at 0.05/25 = 0.002. Bonferroni correction therefore controls the FWER.

- Benjamini-Hochberg procedure controls the false discovery rate. Here the *p*-values are arranged in ascending order of ranks and assigned a value (i). Benjamini-Hochberg critical value is calculated by the formula $(i/m)\ *Q$ where, $i =$ rank assigned to *p*-value, $m =$ number of tests, $Q =$ false discovery rate which is set by the researcher. The largest *p*-value less than this critical value is determined and all the *p*-values smaller than this value are considered to be significant.
- There are several other multiple testing correction methods which are beyond the scope of this chapter.

60.10 Bayesian Hypothesis Testing

- Of late, Bayesian approach of hypothesis testing is gaining popularity, as it follows a pragmatic approach. Some terms that need to be understood before we see the Baye's theorem are as following:
  - *Prior probability:* probability of an event occurring before data is collected
  - *Posterior probability:* probability of an event after incorporating data. In statistics, posterior probability P(A/B) is of event A occurring when event B has occurred.
- Bayes theorem relates the conditional probabilities of events A and B and is given by the formula:

$$P(A/B) = \frac{P(B/A) * P(A)}{P(B)}$$

P(A) is the probability of A without taking into account event B, that is, prior probability.

P(A/B) is the conditional probability of event A given B.

P(B/A) is the conditional probability of event B given A.

- Bayes theorem can also be expressed in terms of likelihood as:

$$\text{Posterior} = \frac{\text{likelihood} * \text{prior}}{\text{Normalising constant}}$$

Or as

$$\text{Posterior} = \text{Normalised likelihood} * \text{prior}$$

- Bayesian theorem can aid in assessing the plausibility of two hypotheses H_0 (effect size = 0) and H_1(effect size$\neq$0)
- Here posterior odds = Prior* Bayes factor (where Bayes factor is the likelihood ratio)
- The researcher begins by assigning prior odds to H_0 and H_1

– $P(H_0) = p_0$, $P(H_1) = p_1$, where $p_0 + p_1 = 1$

At the end of the study with the data available, if the posterior odds of H_0 over H_1 was found to be increased by 20, we can conclude that our belief on the null hypothesis has increased by 20 times.

60.11 Conclusion

The utility of hypothesis testing has been criticised for decades. It has been considered to result in misinterpretation of data and is said to be devoid of any practical purpose, in the sense that it provides no information on the effect size. In spite of these criticisms, reporting of hypothesis testing with *p*-value is still widely used, the usage of 95% confidence interval enhances the interpretation of results. One should always bear in mind that ultimately it is the 'clinical significance' that is weighed upon more and aids in decision making than the statistical significance.

Bibliography

Biau DJ, Jolles BM, Porcher R. P value and the theory of hypothesis testing: an explanation for new researchers. Clin Orthop Relat Res. 2010 Mar;468(3):885–92.

Faulkenberry TJ. A simple method for teaching Bayesian hypothesis testing in the brain and behavioral sciences. J Undergrad Neurosci Educ. 2018 Jun 15;16(2):A126–30.

Guyatt G, Jaeschke R, Heddle N, Cook D, Shannon H, Walter S. Basic statistics for clinicians: 1 Hypothesis testing. CMAJ. 1995;152(1):27–32.

Hazra A, Gogtay N. Biostatistics series module 2: overview of hypothesis testing. Indian J Dermatol. 2016;61(2):137–45.

Sacha V, Panagiotakos DB. Insights in hypothesis testing and making decisions in biomedical research. Open Cardiovasc Med J. 2016 Sep 30;10:196–200.

Shreffler J, Huecker MR. Hypothesis testing, P Values, confidence intervals, and significance. In: StatPearls [Internet]. Treasure Island (FL): StatPearls Publishing; 2020. [cited 2021 Feb 1]. Available from: http://www.ncbi.nlm.nih.gov/books/NBK557421/.

Parametric Tests

61

Mourouguessine Vimal, Vinayagamoorthy Venugopal, and Nishanthi Anandabaskar

Abstract

Inferential statistics help researchers by two major means. Firstly, they are used to extrapolate the result obtained from studying sample to a greater population. Secondly, they are used for hypothesis testing. They are otherwise called as statistical test of significance. These tests are broadly classified into parametric and non-parametric tests of significance. Parametric tests are those that make assumptions about the parameters of the population. The general assumption is that the population data are normally distributed. Hence, the sample data that is collected from population also to be normally distributed in order to apply the parametric test of significance to test hypothesis. Apart from that, there are various other assumptions to be fulfilled by the sample data to use parametric test of significance. If any of these assumptions is violated, then its equivalent non-parametric tests have to be applied. The appropriate choice of selection of parametric test depends on, type of dependent and independent variables, number of groups to be compared and the relatedness of data. Various parametric tests commonly used are, student's *t* test, paired *t* test, one way and two ways ANOVA, one way and two way repeated measures ANOVA, Pearson's correlation and linear regression.

Mourouguessine Vimal and Vinayagamoorthy Venugopal contributed equally to this work.

M. Vimal
Department of Pathology, Sri Manakula Vinayagar Medical College and Hospital, Puducherry, India

V. Venugopal
Department of Community and Family Medicine, All India Institute of Medical Sciences, Deoghar, Jharkhand, India

N. Anandabaskar (✉)
Department of Pharmacology, Sri Manakula Vinayagar Medical College and Hospital, Puducherry, India

M. Lakshmanan et al. (eds.), *Introduction to Basics of Pharmacology and Toxicology*, https://doi.org/10.1007/978-981-19-5343-9_61

Keywords

Parametric test · *T*-test · ANOVA · Correlation · Regression

61.1 Introduction

Statistics help researchers to arrange, analyze and make meaningful interpretations out of much information collected from the study participants. It also ensures accuracy in interpretation. Thus, the term statistics refers to a set of mathematical procedures used for organizing, summarizing, analyzing, and interpreting data. Bio-statistics deals with data obtained mainly from the biological experiment. Medical bio-statistics applies the principles of statistics to manage uncertainties in the field of medicine and health-related issues and events.

Research in the field of health typically begins with a general question about a specific group (or groups) of individuals. The researcher is usually interested in studying the effect of exposure (risk factors or causative agents) on the outcome (disease/health-related event) among a specific group of population. For example, it may be studying the effect of exclusive breastfeeding (exposure) on the occurrence of respiratory illness (outcome) among infants. In statistics, the population is the term used to represent the entire group of people that a researcher wishes to study. In practice, populations are typically very large. In the above-mentioned example, the infants of the entire district/country for whom the study results are intended to be applicable. Because populations tend to be very large, it is usually impossible for a researcher to examine every individual in the population of interest. Therefore, researchers typically select a smaller, more manageable group from the population and limit their studies to the selected group. They are usually intended to represent the population in a research study and are called a sample.

A characteristic that describes a population, for example, the mean birth weight of neonates in the total population, is called a parameter. Thus, a parameter is a value, usually, a numerical value that describes a population. A parameter is usually derived from measurements of the individuals in the entire population. A characteristic that describes a sample is called a statistic. Thus, the average birth weight of neonates selected from the population (the study sample) for research purposes is a statistic. It is inferential statistics that helps the researcher to use collected information from a sample to make intelligent and accurate interpretations about a greater population. The other purpose of inferential statistics is hypothesis testing. It is the statistical procedures that use the sample data to evaluate the hypothesis generated about a population (Fig. 61.1). They are otherwise called the statistical test of significance. These tests are broadly classified into parametric and non-parametric tests of statistical significance.

Parametric tests are those that make assumptions about the parameters of the population in terms of the distribution of parameters under study from which the

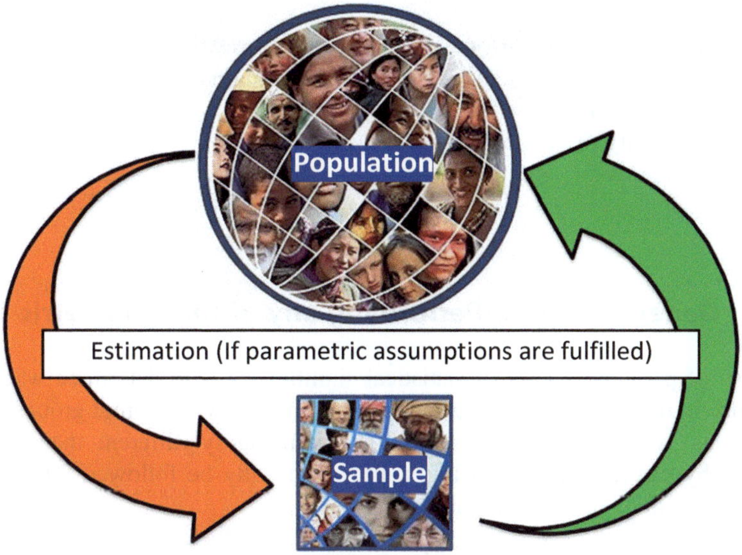

Fig. 61.1 Illustration of the concept of inferential statistics

sample is drawn. The general assumption is that the population data are normally distributed. Hence, the sample data collected from the population is also normally distributed to apply the parametric test of significance to test the hypothesis. Apart from that, there are various other assumptions to be fulfilled by the sample data to use parametric tests of significance that are mentioned below. On the other hand, non-parametric tests are applied when the basic assumption is that the sample data is not normally distributed.

61.2 The Basic Assumptions Underlying Parametric Tests are as Follows (Box. 61.1)

- The scale of measurement of the data: Data must be measured on a continuous numerical scale (e.g., blood pressure, serum cholesterol, body weight, etc.)
- Population distribution – The observations must be drawn from a normally distributed population, and the sample data (the variable of interest) should be normally distributed.
- Homogeneity of variances – variances of the group to be compared should be equal.
- Independence of samples - The samples are randomly drawn from the population, and the observations within a group are independent of each other

Box 61.1. Basic Assumptions Underlying Parametric Tests

- The scale of measurement of the data – continuous numerical
- Population distribution - normally distributed
- Homogeneity of variances
- Independence of samples

61.3 General Steps in Performing any of the Parametric Tests

The goal of parametric tests of statistical significance is to determine whether the difference in mean between the groups of the sample (two different groups of study participants) and the stated hypothesis is significantly different than would be expected by chance. The following steps will usually be followed for any of the parametric tests of statistical significance:

- Setting up of the null hypothesis and alternate hypothesis
- Setting up the significance level (normally set at 5% or 0.05)
- Verifying the fulfillment of various assumptions for the use of parametric test
- Selecting the appropriate statistical test for analysis
- Selecting and calculating the appropriate test statistic
- Calculating the *p*-value
- Rejecting the null hypothesis (or accepting the alternate hypothesis) if the *p*-value is less than the chosen level of significance (usually 0.05) or accepting the null hypothesis (or rejecting the alternate hypothesis) if the *p*-value is greater than the chosen level of significance (usually 0.05)

61.4 Selection of Appropriate Parametric Test

There exist many parametric tests of significance and their equivalent non-parametric tests. The choice of selection of appropriate parametric tests depends on the following features (Fig. 61.2).

- Type of dependent and independent variables (categorical, continuous)
- Number of groups to be compared (two or more than two)
- The relatedness of data (paired or unpaired data)

61.5 Parametric Testing Methods

The various parametric testing methods are listed below

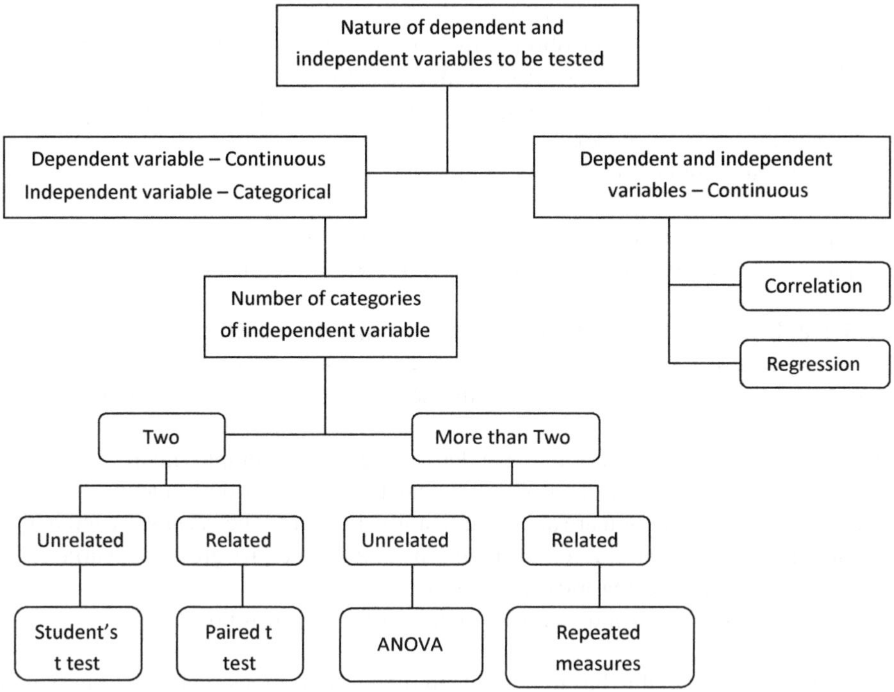

Fig. 61.2 Algorithm to choose the appropriate parametric test

- The '*t*' test
  - One sample *t*-test
  - Independent sample *t*-test/unpaired *t*-test/Student's *t*-test
  - Paired *t*-test
- Analysis of variance (ANOVA)
  - One way ANOVA
  - Two way ANOVA
  - Repeated Measures ANOVA
- Pearson's correlation coefficient (*r*)
- Linear regression

61.5.1 *T*-Test

61.5.1.1 One Sample *t*-Test

The one-sample *t*-test is used when the researcher wants to determine whether the sample mean obtained from study participants is statistically different from a known value, or the hypothesized population mean.

Data requirements for one sample *t*-test are as follows:

- Test variable is of continuous type (e.g., age, weight. Height, etc.)
- The individual participant observations are to be independent (i.e., there is no relationship between values between individuals)
- The study sample needs to be selected randomly from the population
- The data needs to be normally distributed
- There should not be any outliers in the data

Example: A researcher wanted to know whether the serum chloride level of the study sample is different from the normal values reported in the medical textbooks.

61.5.1.2 Independent Sample *t*-Test/unpaired *t*-Test/Student's *t* Test

The independent samples *t*-test is used when the researcher is interested in identifying the significant difference in means of any outcome variable in two independent sample groups (samples could be divided into two with respect to any of the independent variables like gender, age category, employment status, test result, etc.). The aim is to determine whether the means of that outcome variable in the associated population are also significantly different. The variables used in this test are (i) outcome variable (continuous in nature) and (ii) grouping variable or independent variable (categorical in nature).

Data requirements for independent samples *t* test are as follows:

- The dependent variable has to be measured on a continuous scale (i.e., interval or ratio level) (e.g., age, blood pressure, etc.)
- The independent variable should be a dichotomous categorical variable which is used to categorize the study participant into two groups
- The individual participant observations are independent (i.e., there is no relationship between values between individuals. This means that the participants in the first group cannot also be in the second group
- The study participants need to be selected randomly from the population
- The outcome variable in both groups have to be normally distributed
- There should not be any outliers in the data
- Homogeneity of variances to be followed (i.e., variances have to be approximately equal between two groups). When this assumption is violated, the Welch *t*-test statistic, which is based on the assumption that the variance of two groups is not equal could be used. The Welch *t*-Test is also known as an Unequal Variance *t*-Test or Separate Variances *t*-Test.
  Examples:
- A researcher wanted to know is there any difference in the systolic blood pressure of smokers and non-smokers
- In experimental design, this test is used to find out the difference in the outcome variable like blood sugar between those who were treated by two different drugs.

Note: When the listed assumptions for independent samples *t*-test are not met, then one may have to run its nonparametric equivalent that is the Mann Whitney U test.

61.5.1.3 The Paired-Samples *t*-Test

The paired *t*-test is used when the researcher wants to test if the population means estimated by two dependent samples differ significantly. The term two dependent sample means that the variables are measured from the same individual but collected from two different time points. The variables used in this test are (i) outcome variable (continuous in nature that was measured twice in different two pint or conditions or instruments) and (ii) grouping variable or independent variable (categorical in nature).

The two paired observations can fall under any of the below-mentioned conditions:

1. Any continuous outcome variable measured at two different time points among the same study participants (e.g., comparison of the difference in the baseline and end-line systolic blood pressure following any intervention delivered between the two-time points)
2. Any variable (continuous) measured by two different instruments or methods among the same study participants at the same point of time (e.g., performing a blood sugar test using two different instruments at the same time point)
3. Any measurements taken from the left and right sides of a subject (e.g., measuring intraocular pressure in a subject's left and right eyes).

Data requirements for paired sample *t* test are as follows:

- The dependent variable has to be measured on a continuous scale (i.e., interval or ratio level) (e.g., age, blood pressure, etc.)
- The study participants need to be selected randomly from the population
- The outcome variable in both groups have to be normally distributed
- There should not be any outliers in the data
- The data measured are related to each other, which means they belong to related samples or groups (i.e., dependent observations). The study participants in the first group are also there in the second group.

Example:

- To find out the difference in blood pressure before and after physical activity sessions
- To find out the difference in the size of the right and left kidney

Note: When the normality or outlier's assumptions for paired sample *t*-test are not met, then one may have to run its nonparametric equivalent, that is, Wilcoxon signed-rank test.

61.5.2 Analysis of Variance (ANOVA)

61.5.2.1 One Way ANOVA Test

One-Way ANOVA (The term ANOVA stands for 'analysis of variance'). It compares the means of more than two independent groups to determine whether there is statistical evidence that the associated population means are significantly different. The two types of variances studied in ANOVA are between-group variability and within-group variability. The between-group variability (or effect variance) results from any intervention (exposure) or any difference concerning the grouping independent variable between groups. The within-group variability (error variance) is based on random differences present in the study samples, which cannot be accounted for in the study design. An F-test is used to compare the above-mentioned two estimates of variances, and F statistic is calculated in the ANOVA test to look for the significance between groups. Based on the F statistic, if the null hypothesis is rejected, that means there is a significant difference in the outcome variable measured between at least one pair of means. In that scenario, a post-hoc test can be used to identify between which pair(s) the significant difference exists. The various post-hoc tests used are Tukey, Dunnett, and Bonferroni. The selection of the test is based on the homogeneity of variance between groups. The variables used in this test are (i) outcome variable (continuous in nature) and (ii) one grouping variable or independent variable (categorical in nature, more than two categories).

Data requirements for one way ANOVA are as follows:

- All the data requirements that need to be fulfilled for the independent sample *t*-test holds good for one-way ANOVA as well. The only difference is that there are more than two groups in ANOVA.
- Homogeneity of variances to be followed (i.e., variances have to be approximately equal between groups). When there is a difference in variance between groups, then any test (Browne-Forsythe or Welch statistics) that is not based on this assumption could be used. The choice of post hoc tests also depends on the variance status between groups. If it is different across the groups, then a post hoc test like Dunnett's C could be used.

  Examples:
- A researcher wanted to know is there any difference in the systolic blood pressure of current smokers, past smokers, and non-smokers
- In experimental design, this test is used to find out the difference in the outcome variable like blood sugar between three groups of patients who were treated with three different drugs.

Note: When one or more of the assumptions mentioned above for One-Way ANOVA are not met, one may have to run the nonparametric equivalent of one-way ANOVA, which is the Kruskal-Wallis test.

61.5.2.2 Two Way ANOVA Test

When a researcher wants to compare the continuous outcome variable between the groups that are split based on two independent variables. Here, the independent variables can have more two or more than two categories. It is mainly used to know the interaction between two independent variables on a continuous outcome variable. The assumptions and data requirement for running the two-way ANOVA is same as that of one-way ANOVA.

Example: When a researcher wants to determine the interaction between physical activity level (mild/moderate/severe) and smoking status (smoker/non-smoker) on blood cholesterol. Here, the outcome variable is blood cholesterol, and there are two categorical independent variables, namely physical activity level, and smoking status.

61.5.2.3 Repeated Measures ANOVA

It is a combination of ANOVA and paired sample *t* test. Similar to the ANOVA test, depending on the number of independent variables, it can be of two types: repeated measure ANOVA and two-way repeated-measures ANOVA. One-way repeated measure ANOVA is equivalent to one-way ANOVA, but the difference is that the outcome variable is measured at more than two-time points. The same participant is being measured at more than two-time points on the same outcome variable measured on a continuous scale. Two-way repeated-measures ANOVA is applied when the continuous outcome variable is measured at more than two-time points, and there seems to be more than one independent variable to be associated with it.

The data requirements of one-way ANOVA have to be fulfilled by the repeated measures ANOVA as well. The repeated measures equivalent of homogeneity of variance is a statistical concept called the sphericity in repeated measures ANOVA.

Example: When a researcher wanted to study the effect of the physical activity training program on the systolic blood pressure of participants, and the systolic blood pressure was measured at three different points of time two, four, and 6 months after the initiation of the program then the repeated measures ANOVA is used.

61.5.3 Pearson's Correlation

The bivariate correlation is a statistical procedure that is used to identify the linear relationship between two continuous variables. Based on the analysis of sample data, the linear relationship is brought out by correlation coefficient, which is denoted as "*r*". It conveys both the strength and direction of linear relationships between the two continuous study variables. When both the continuous variables are normally distributed, it fulfills the criteria for running the parametric correlation test that is named Pearson's correlation. When any of the variables are not normally distributed or when they are measured in discrete scale, then its non-parametric equivalent, that is, Spearman's correlation, has to be used.

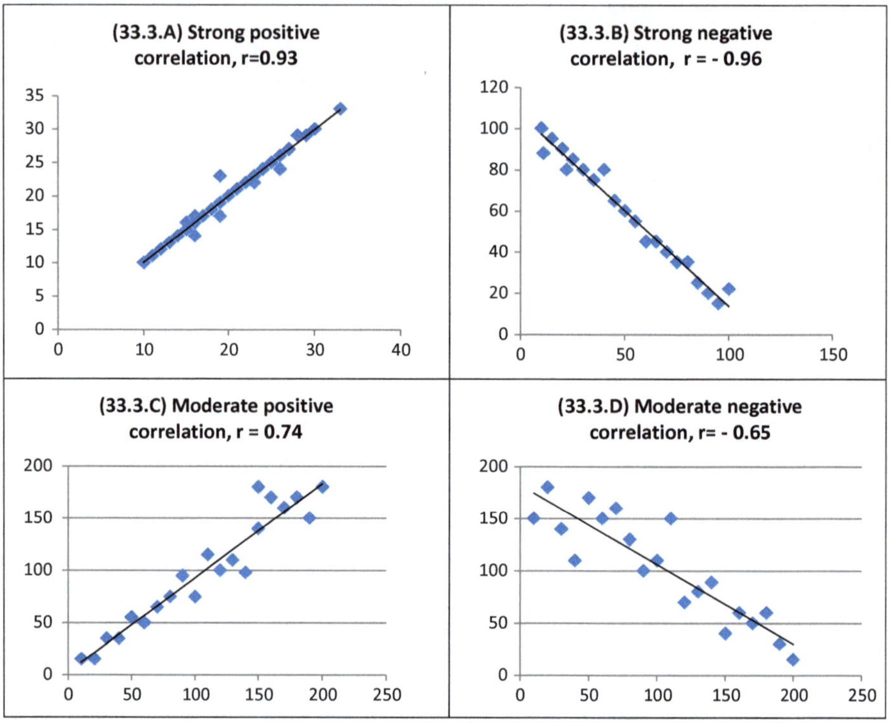

Fig. 61.3 (**a** to **e**): Scatter plots depicting the relationship between two continuous variables. The letter "*r*" stands for the correlation coefficient. Note that the higher the *r*-value, the stronger the relationship. The direction is depicted in these plots

Table 61.1 Strength of correlation based on the value of '*r*' the correlation coefficient (Cohen et al.)

Correlation coefficient value	Strength
< 0.3	Weak
0.3–0.49	Moderate
0.5–0.89	Strong
0.9–1.0	Very strong

The bivariate Pearson's correlation test result indicates three of the following:

1. Based on the *p*-value, it implies that is there a statistically significant correlation that is linear relationship exists between two variables (if the *p*-value is <0.05, then statistically significant)
   Based on the value of the correlation coefficient "*r*," it conveys the strength and direction of the association. The *r*-value can range between +1 to −1. Graphically they are represented by a scatter plot (Fig. 61.3).
2. The strength of the relationship
   The strength of the relationship between the two continuous that is how strong the relationship, is graded by Cohen (depicted in Table 61.1).

3. The direction of association between them can be positive or negative. A positive relationship means that when the value of one variable increases, the other also increases. For example, as age increases, blood pressure also increases. A negative relationship means that when the value of one variable increases, the other decreases. For example, as the BMI increases, the vital capacity decreases.

Note: The Pearson correlation cannot be applicable if there is a non-linear relationship exists between variables. The bivariate Pearson correlation only assesses the relationship between two continuous variables. If any of the variables are categorical in nature, then one needs to use any other tests of association.

61.5.4 Regression

Regression is a statistical procedure that links the two correlated variables (x, y) in the form of a linear equation ($y = a + bx$). This helps the researcher to predict the value of one variable (y) based on the other variable (x). In order to make a prediction, there should be a good correlation between them which is indicated by the correlation coefficient. The method of least squares is used to obtain the equation of the regression line. The variable that is getting predicted is called as outcome variable or dependent variable. Those variable(s) that determine the prediction or occurrence of the dependent variable is/are called independent variables. When the outcome variable is measured on a continuous scale and is normally distributed, then the regression used is termed linear regression, which is a parametric test. The independent variable can be measured in any type of scale (nominal/ordinal/continuous/discrete), and it can be of any number. When there is a single independent variable, it is termed simple linear regression, and in case of many, it is called multiple linear regressions.

Data requirement /assumptions for linear regression:

1. The outcome variable is measured on a continuous scale (ratio or interval scale).
2. The dependent variable has to be normally distributed.
3. A linear relationship between the dependent and independent variables should be present. This can be tested by running the scatter plot.
4. There should not be any significant outlier
5. There should be independence of observation. This can be tested by the Durbin-Watson statistic.
6. The data needs to show homoscedasticity, which is where the variances along the line of best fit remain similar as one moves along the line.
7. The residuals or errors of the regression line have to be normally distributed. This can be checked using a histogram or Normal P-P plot.

Example: when a researcher wants to predict the systolic blood pressure of an individual based on his age. Then, simple linear regression can be applied. Blood pressure is the dependent variable, and age is the independent variable. If the researcher includes blood cholesterol value as well in the equation, then the blood pressure can be predicted by multiple linear regression as there is more than one independent variable (that is, age, blood cholesterol) involved.

61.6 Advantages of Parametric Test Over Non-parametric Test

- Statistical power is more with parametric tests compared to non-parametric tests. Thus, parametric tests are more robust in detecting a significant difference between groups when it really exists.
- When the dispersion of the groups is different, the parametric test produces more valid results compared to the non-parametric test.
- Interpretation of the results drawn from parametric tests is easier and more meaningful compared to non-parametric tests.
- Apart from testing the hypothesis, parametric tests help us to estimate the confidence interval of the estimated parameter (outcome variable) studied.
- Performing flexible modeling and identifying confounders, and doing regression is possible only with parametric tests

61.7 Circumstances Where Parametric Tests Cannot Be Used

- The parametric test cannot be used to analyze ordinal data or nominal data
- The parametric test cannot be used if the data is skewed (non-normally distributed) or has outliers that cannot be removed
- The parametric test cannot be used if the sample size is small since it is not possible to assess the normality of the data (the normality tests lack sufficient power to provide meaningful results if the sample size is less)
- Parametric tests cannot be used when one or more assumptions of the parametric tests are not fulfilled

Bibliography

Ali Z, Bhaskar SB. Basic statistical tools in research and data analysis. Indian J Anaesth. 2016;60: 662–9.

Altman DG, Bland JM. Parametric vs non-parametric methods for data analysis. BMJ. 2009;338: a3167. https://doi.org/10.1136/bmj.a3167.

Campbell MJ, Swinscow TDV. Statistics at square one. 11th ed. Wiley-Blackwell: BMJ Books; 2009.

Chan YH. Biostatistics 102: quantitative data – parametric & non-parametric tests. Singap Med J. 2003;44(8):391–6.

Cohen J. Statistical power analysis for the behavioral sciences. 2nd ed. Hillsdale, NJ: Erlbaum; 1988.

Gravetter JF, Wallnau BL. Statistics for the behavioral sciences. 9th ed. Jon-David Hague; 2013.

Rana RK, Singhal R, Dua P. Deciphering the dilemma of parametric and nonparametric tests. J Pract Cardiovasc Sci. 2016;2:95–8.

Non-parametric Tests

62

Amod Tilak and Avinash Arivazhahan

Abstract

Statistical tests are of two major types: parametric and non-parametric. This chapter details on the non-parametric tests, also known as distribution-free statistical tests. These tests are generally based on the concept that an assumption of normal distribution of data is not made. While the parametric tests are most preferred for statistical analysis due to higher reliability, the non-parametric counterparts are used wherever parametric tests cannot be applied. The commonly used non-parametric statistical tests are detailed here in the chapter, including sign tests, Mann-Whitney test, Friedman test, etc. The merits and limitations of non-parametric tests are also included towards the end of this chapter.

Keywords

Statistics · Sign tests · Mann-Whitney · Friedman · Kruskal-Wallis

62.1 Introduction to Non-parametric Statistics

Statistical tests based on fewer statistical assumptions are the non-parametric tests also known as the distribution-free tests. As discussed earlier, specific probability distributions (e.g., normal distribution) are considered in the parametric tests and the estimation of the key parameters in such a distribution (e.g., the mean or difference in means) from the sample data is part of the statistical evaluation. However, the

A. Tilak
Pfizer India Ltd., Mumbai, Maharashtra, India

A. Arivazhahan (✉)
Sanofi India Ltd., Chennai, Tamil Nadu, India

M. Lakshmanan et al. (eds.), *Introduction to Basics of Pharmacology and Toxicology*, https://doi.org/10.1007/978-981-19-5343-9_62

non-parametric tests are generally less powerful as compared to their respective counterparts owing to the fewer assumptions.

Appropriate use of the parametric or non-parametric test for the evaluation of a continuous outcome may be difficult as it is essential to assess whether the outcome follows a normal distribution. The goodness-of-fit tests like Kolmogorov-Smirnov test, Anderson-Darling test, and Shapiro-Wilk test help assess the normal distribution of data under consideration.

62.2 Scope of Non-parametric Tests

Parametric tests are the preferred statistical tests for most datasets. Non-parametric tests can be preferred over parametric tests in the following scenarios:

- When there is no normal distribution of given data.
- When data are represented in nominal or ordinal scales.
- Violation of the assumption/s of a parametric test.
- A small size doesn't allow a parametric test for statistical evaluation.
- A dataset with outliers that cannot be adjusted for.
- When the median analysis would be more useful than the mean.

62.3 Steps in Performing Non-parametric Tests

1. Set up the hypothesis and level of significance
2. Selection of appropriate test statistic—ranks in case of non-parametric tests
3. Decision rule—Decision on when and under what circumstances the null hypothesis is accepted or rejected
4. Computation
5. Conclusion— Comparing the test statistic using the decision rule

62.4 Non-parametric Testing Methods

1. *1-sample sign test:* This test allows a comparison of a population median with a reference value or target value.
2. *1-sample Wilcoxon-signed-rank test:* This test too allows comparison of population median with target/reference value with the assumption that the data is taken from a symmetric distribution (like the Cauchy distribution or uniform distribution).
3. *Friedman test:* This test evaluates differences between groups when looking at ordinal dependent variables. This test also evaluates continuous data sets where violation of one or more assumptions does not allow one-way ANOVA with repeated measures.

4. *Goodman Kruskal's Gamma:* Ranked variables can be evaluated for the association.
5. *Kruskal-Wallis test:* With the use of ranks for data points in the calculations, this is used to test if two or more medians are different instead of using the one-way ANOVA.
6. Trends in time-series data can be tested using the Mann-Kendall Trend Test.
7. *Mann-Whitney test:* This test helps compare the differences between two independent groups with either ordinal or continuous dependent variables.
8. *Mood's Median test:* This is used with two independent samples in place of a sign-test.
9. *Macnemar chi-squared test:* To test whether the observed frequency of events is significant or not

Brief descriptions of these tests are given below.

62.4.1 Sign Test

- This test is used to evaluate ordinal data. It measures the relative orders for different categories of the variables. It is the simplest non-parametric evaluation for matched or paired data i.e. data looking at observations before and after an intervention in the same sample.
- The evaluation is done based on the signs allotted to each observation i.e. '+' sign for improvement after intervention and '–'sign for no improvement or deterioration. If we accept the null hypothesis, then we expect approximately half of the values to demonstrate a positive outcome and the other half to demonstrate negative outcomes.
- Lower of the positive or negative signs from the observations is the test statistic for the sign test. After deciding the critical value based on the sample size and pre-decided level of significance, the sign test is evaluated based on whether the test statistic is greater or smaller as against the critical value. The null hypothesis is rejected when the statistic derived is smaller than the critical value but we hold the null hypothesis if the test statistic returns a value larger than the critical value based on the desired level of significance.
- Implementation of the sign test may alter when the difference scores of zero are recorded with one or more participants (i.e., their paired measurements are identical). In case of a single such observation, investigators may find it reasonable to drop the data point and reduce the sample size by 1 for deducing the final calculations ($n-1$).
- However, in case of more than one such observations, the following alternatives can be utilized.
  - An even number of zeroes allows the distribution of the data points in either a positive or negative value and sign.

– An odd number of observations combine the above two alternatives. One observation is randomly dropped, the remaining observations are randomly assigned positive or negative signs.

62.4.2 Wilcoxon Signed-Rank Test

- It is also called as the Wilcoxon matched-pairs test.
- It is the non-parametric statistical equivalent of the paired t-test.
- More powerful as compared to the sign test since it considers the magnitude of the difference unlike the sign test
- The evaluation is done in similar fashion to the sign test
- The differences in the paired samples are noted and ordered from lowest to highest ranks.
- The application of the signs for improvement converts these ranks into signed ranks.
- Test statistic— the lower of the summations of positive ranks (W+) and negative ranks (W-).
- The null hypothesis is accepted if there should be a similar number of positive and negative ranks, both lower as well as higher ranks. If there are higher and positive ranks in the observation, the research hypothesis is accepted indicating positive change due to the intervention rather than just due to chance.
- The significance is calculated using the critical values for the predetermined level of significance for the desired sample size.
- Test statistics below the critical value concludes the Wilcoxon signed-rank test saying there is a significant difference after the intervention.
- The sign test may fail to demonstrate these results due to only the signs being taken into account to draw conclusions rather than the entire information obtained from the data set.

62.4.3 Friedman Test

- Comparison between three or more matched groups
- Also known as Friedman's two-way ANOVA
- Ideal statistic to be used repeated measures type of experiment to determine if a particular factor has an effect.
- The test involves the ranking of variables across multiple points or factors of evaluation enabling a comparison of the same variable across time points in the same subject
- A Friedman statistic (Fr) is determined, significance decided using the table on critical values for the Friedman two-way variance analysis by ranks statistic.
- Dunn's test is used to compare different treatments after Friedman's analysis.

- The results are reported indicating the number of subjects in the test, the Fr, the p-value and the degree of freedom followed by the comparison of multiple interventions using the Dunn's test

62.4.4 Goodman Kruskal's Gamma Test

- Evaluation of the correlation of ranks
- Estimates the association of ordinal variables from cross-tabulated data
- +1(perfect agreement) to −1(perfect inversion) indicates the range of values.
- No association if the value is 0
- A special case of this is the Yule's Q or the Yule coefficient of association specific for 2x2 matrices

62.4.5 Kruskal-Wallis Test

- Outcomes in more than two independent groups can be compared using this non-parametric test
- To compare medians among more than 2 comparison groups
- ANOVA evaluating ranks of data available
- Equivalent to the analysis of variance for parametric tests
- This test helps determine if a set of independent samples are random samples from the same population or different populations
- Null hypothesis indicates no difference between the samples
- Research hypothesis indicates that at least one of the samples if different from the set of samples
- Being a non-parametric test, it does not make predictions about the means of the different populations but compares the medians
- It is not mandatory to have the same sample size across the compared groups but should not differ much and can be used to evaluate small sample sizes
- Results are reported using the H score(test statistic), the degrees of freedom, and the p-value of the test used. Also, the test used to differentiate between the groups in the subsequent procedure must be mentioned. Eg. Dunn's test or the Mann Whitney U test.

62.4.6 Mann-Kendall Trend Test

- Non-parametric test to analyze increasing or decreasing trends in data collected over time
- The parametric equivalent of the linear regression model
- Data should not demonstrate the serial correlation
- Minimum number of recommended data points is 8–10 as fewer data points may not indicate an existing trend

- Null hypothesis—there exists no monotonic trend
- Alternate hypothesis—positive, negative or non-null trend exists
- The test compares sign differences in the subsequent data points
- A trend indicates subsequently increasing or decreasing signs in subsequent data points
- Every value is compared to the preceding value in the data set. Eg. If we have 10 data points, the total number of comparable data pairs would be 10(10–1)/2 i.e. 45.
- Thus even a small data set allows comparison across a large number of pairs to derive a possible trend in the data set.

62.4.7 Mann Whitney *U*-Test

- Also known as the Wilcoxon two-sample test, Wilcoxon test, Wilcoxon Mann Whitney test, Wilcoxon rank-sum test
- It helps to compare two independent groups irrespective of the sample size in each
- Equivalent of the parametric unpaired *t*-test
- More sensitive than the median test
- The ranks of data from one sample/group are added and the sums are compared with a similar sum from some other group
- The test evaluates whether the scores are randomly mixed or if one of the statistic clusters towards one end of the distribution
- Variable in a comparison between the two groups is a continuous variable with approximately equal variance
- All observations are ranked in ascending order, then split into groups and the U statistic computed
- The test looks at the shape and spread of scores in the data sets
- Thus even if the medians between two groups are similar, the test identifies the spread of scores and enables more power to the results deduced
- Test report indicates the U statistic, the number of observations per group and the significance values observed

62.4.8 Mood's Median Test

- Simple statistical test to evaluate the difference in any between two independent samples.
- Like sign test, the number of observations in a category is marked irrespective of the magnitude of difference between data points
- Then, contingency tables are set up based on the median of observations
- Chi-Square test is applied and significance determined to comment on the medians of the two independent sample under observation

62.4.9 Macnemar's Chi-Squared Test

- To test whether the observed frequency of events is significant or not
- Used to evaluate categorical variables
- The goodness of fit test evaluates if an observed distribution fits the predicted distribution
- Results are expressed in terms of the X^2 value, the level of freedom(number of categories—1) as well as the p-value for significance
- The chi-square table is used to derive the levels of significance
- The test of independence is also used to evaluate the independence of samples.
- This can be used for the evaluation of a particular intervention for a particular disease and to check whether the intervention affects the outcomes in a positive or negative test
- The degrees of freedom are determined from the number of rows and columns
- In case the values in any particular cell of a contingency table are less than 5, Fischer's exact test is used to carry out the statistic, expressed in a similar way as the chi-squared test.

62.5 Advantages of Non-parametric Tests

- Can be used with nominal, ordinal, interval, or ratio data
- Are not restrictive about assumptions concerning distribution and variance
- These tests are not affected by extreme outliers in the data
- Can sometimes detect differences between groups that parametric statistics do not
- Can be used with very small sample sizes
- Can be used even when the data are skewed
- Easy to calculate
- Easy to understand

62.6 Disadvantages of Non-parametric Tests

- Less powerful than parametric statistics
- Can be used only with simple experimental designs
- Not all information from a sample is used and conveyed
- Analyses rank rather than actual experimental values
- Less specific hypothesis testing, thus general conclusions rather than specific

Bibliography

Altman DG, Bland JM. Parametric v non-parametric methods for data analysis. Br Med J. 2009;338:a3167.

Bagdonavičius V, Kruopis J, Nikulin MS. Non-parametric tests for complete data. ISTE/ Wiley; 2011.

Conover WJ, Conover WJ. Practical nonparametric statistics.

Corder GW, Foreman DI. Nonparametric statistics: a step-by-step approach. John Wiley & Sons; 2014.

Fay MP, Proschan MA. Wilcoxon-Mann-Whitney or t-test? On assumptions for hypothesis tests and multiple interpretations of decision rules. Stat Surv. 2010;4:1.

Hollander M, Wolfe DA, Chicken E. Nonparametric statistical methods. John Wiley & Sons; 2013.

Common Errors in Using Statistical Tools and Data Presentation

63

Mageshwaran Lakshmanan

Abstract

Errors in using the statistical tool can occur at multiple stages of the study. It can occur in sample size calculation, defining the parameter, assigning appropriate statistical tests for them, checking all the assumptions of a particular statistical tool, and finally in proper tabulation and graphical representation of data. Manipulation of alpha and beta values to achieve low sample size, extrapolating conclusions for an objective from the sample size estimated for different objectives, and missing drop out cases are common errors in sample size calculation. In data handling, unnecessary conversion of continuous to categorical data, setting improper cutoff value while categorization, and unnecessary log transformation can lead to loss of data validity. Similarly, choosing the wrong statistical test, not checking the 'normality distribution' of data before analysis, and not meeting the test assumptions are often encountered while doing statistical analysis. This chapter summarizes common statistical errors encountered in various biomedical researches and their implications and steps to avoid them.

Keywords

Sample size · Data representation · Statistical test · Statistical analysis · Normality checking · Data transformation · Error

M. Lakshmanan (✉)
Department of Pharmacology, Thanjavur Medical College, Thanjavur, Tamilnadu, India

M. Lakshmanan et al. (eds.), *Introduction to Basics of Pharmacology and Toxicology*, https://doi.org/10.1007/978-981-19-5343-9_63

63.1 Introduction

Errors in using the statistical tool can occur at multiple stages of the study. It can occur in sample size calculation, defining the parameter, assigning appropriate statistical tests for them, checking all the assumptions of a particular statistical tool, and finally, in proper tabulation and graphical representation of data. This chapter summarizes common statistical errors that are being encountered in various biomedical researches.

63.2 Errors Related to Sample Size

63.2.1 Manipulation of Confidence Interval (1-α) and Power (1-β) Values to Achieve Low Sample Size

- Sample size calculation is a vital part of research as it determines the impact and significance of the outcome. It also enables the researchers to obtain a robust decision from the limited information and generalize it to the larger population.
- While calculating the sample size, as a standard norm, the α and β error values are kept as 0.05 and 0.2, respectively, and hence the CI value will be 0.95, and power will be 0.8. It is well known that a study with a smaller sample size requires less money, less effort, and minimal time. Thus, the manipulation of α and β error values can yield a smaller sample size that is beneficial for the researcher. However, one should understand the following pitfalls.
  - The α error value indicates the probability of finding a statistically significant difference when there no difference actually (i.e., false-positive error). Thus when the α error value is increased (thereby decreasing the CI value as CI = 1-α) for the sake of achieving a smaller sample size, then the investigator is allowing the higher false positive in his study results that directly affect the validity of the study conclusion.
  - The β error value indicates the probability of not finding a difference between the groups when the difference actually exists (i.e., false-negative error). When β error value is increased, the power of the study (indicates the ability to avoid false-negative error) is decreased (as the power of study = 1 – β). Thus, by increasing the β value, the investigator allows higher false-negative errors in his study results at the expense of the validity of the conclusion obtained from the results.
- The effect of manipulation of α and β error values by keeping other factors constant with respect to sample size is illustrated in Fig. 63.1. It is evident from the figure that the sample size obtained by keeping standard norms (CI as 0.95; β error as 0.2) is roughly three times (126 vs. 46) higher than the lower standards (CI as 0.85; β error as 0.4). Thus, without proper justification, manipulation of α and β error values for the sake of achieving a smaller sample size should be avoided.

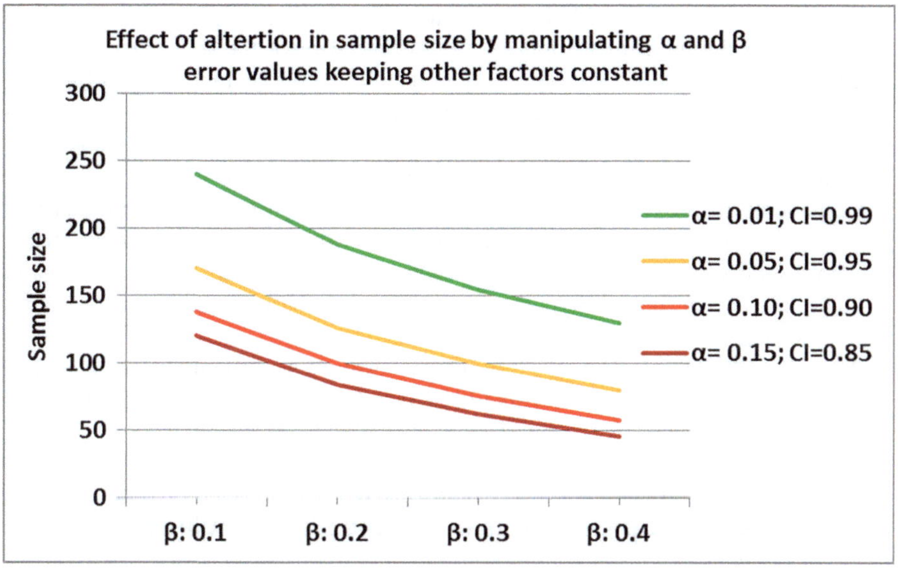

Fig. 63.1 Effect of alteration in sample size by manipulating α and β error values keeping other factors constant

63.2.2 Extrapolating Conclusions for an Objective From the Sample Size Estimated for Different Objectives

- Researchers often adopt multiple secondary objectives besides their primary objective in the study and estimate the sample size for the primary objective. However, occasionally, the investigator can conclude the result for secondary objectives based on the sample size estimated for primary objectives.
- This scenario is not entirely wrong, provided the sample size for the primary objective is larger than that of the secondary objective. However, the pitfall is that the investigators often miss estimating sample size for the secondary objective and extrapolate the results obtained based on the sample size for the primary objective.
- Consider the following hypothetical example in which an investigator wants to study the effect of iron preparation-A on improving hemoglobin level (primary objective) when compared to Iron preparation-B. Based on previous studies, the mean difference (δ) obtained was 2 g/dl, and setting standard error values (CI as 0.95; β error as 0.2), the sample estimated was 32 (16/group). However, the investigator has also set a secondary objective of improvement of serum ferritin by iron preparation-A when compared to iron preparation-B. From previous studies, the iron preparation-A and B have ferritin values of 101 ± 17 ng/ml and 90 ± 16 ng/ml, respectively. Setting standard error values (CI as 0.95; β as 0.2), the sample size for the secondary objective is 72 (36/group).

- Ideally, the sample size 72 should be taken as a study sample size in the above scenario. If the investigator ignores the sample size for the secondary objective, chooses 32 (sample size of a primary objective), and concludes results based on that for the secondary objective, then there is a risk for occurrence of type II error. This is because 32 patients are insufficient to prove 'no differences between the change in ferritin level between the two iron preparations,' as it requires 72 patients to prove it actually. Clearly, the variation of ferritin value is high (as observed in previous study results) and requires a higher sample size to identify the significant difference between them. Hence, using 32 patients is not enough to accept the null hypothesis (no difference between the groups), leading to type II error.
- Thus, if multiple objectives are involved, the sample size should be calculated for each objective, and the results should be concluded based on the objective with larger sample size.

63.2.3 Missing Drop-Out Cases in Sample Size Calculation

- Often, the final sample size calculation incorporates the 'drop-out' rates of study participants. Though the dropout rates vary highly based on study duration and methodology, the investigators should always try to select appropriate drop-out rate while estimating the final sample size
- However, to minimize the sample size, the investigators often select a non-practical dropout rate. For example: selecting a dropout rate of 5% for a 10-year follow-up study. Once again, due to low dropout rate estimation or ignoring it totally while calculating the final sample size will result in the accumulation of type II errors.

63.3 Errors Related to Data Handling

63.3.1 Unnecessary Conversion of Continuous Data to Categorical Data

- Continuous data like exact-age, height, and weight, etc., are often converted to categorical data for the ease of summarization.
- Nevertheless, the continuous data, when converted to categorical data, will lose its credibility in the following ways
  - Categorization reduces statistical efficiency to estimate the relationship between variable and outcome
  - Categorization decreases the contribution of individual data for analysis (equivalent to data discarding).
  - Categorization increases the risk of type I error
  - While categorizing, the data that are similar and very close are considered different and results in a wrong conclusion. Example: Consider the data of

blood glucose level: 86, 91, 99, 102, 108, 109.1, 109.4, 109.7, 109.9, 110.2, 110.4, 113, 117, 119, and 126. When it is categorized into normal and abnormal at the cut-off value of 110, then the values 109.7, 109.9, 110.2, and 110.4 falls under different categories, though the difference between them is meaningless.
  - Categorization masks the presence of any non-linearity in the relationship between the variable and outcome.
- Thus, the investigators should avoid the unnecessary conversion of continuous data. If it is converted to a category, then a proper justification should be provided.

63.3.2 Setting a Groundless Cut-off Value During Categorization

- Defining a cut-off value while converting a continuous variable to a categorical value is a challenging one. The investigator must take proper care in selecting the cut-off values for category based on clinical usefulness, easy interpretation, etc.,
- Consider the following hypothetical table (Tables 63.1 and 63.2) horizontal bar diagrams (Figs. 63.2 and 63.3) in which the incidence of the adverse effect of drug X with respect to the age of the participant is shown.

Table 63.1 Hypothetical data of the incidence of the adverse event of drug X with respect to age

S. No	Age group in years	n	Adverse event present n(%)	Adverse event absent n(%)	Chi-square value	df	*p*-value
1	≤10	700	100 (14.3)	600 (85.7)	168.5	6	<0.0001*
2	11 to 20	1800	300 (16.7)	1500 (83.3)			
3	21 to 30	2400	300 (12.5)	2100 (87.5)			
4	31 to 40	2600	400 (15.4)	2200 (84.6)			
5	41 to 50	2000	700 (35)	1300 (65)			
6	51 to 60	1700	800 (47.1)	900 (52.9)			
7	>60	400	200 (50)	200 (50)			

A Chi-square test was used to compare the frequency between the age categories
*indicates $p < 0.05$ and considered statistically significant

Table 63.2 Hypothetical data of the incidence of the adverse event of drug X with respect to age

S. No	Age group in years	n	Adverse event present n(%)	Adverse event absent n(%)	Chi-square value	df	*p*-value
1	<=10	700	100 (14.3)	600 (85.7)	244.4	3	<0.0001*
2	11–20	1800	300 (16.7)	1500 (83.3)			
3	21–60	2400	300 (12.5)	2100 (87.5)			
4	>60	400	200 (50)	200 (50)			

A Chi-square test was used to compare the frequency between the age categories
*indicates $p < 0.05$ and considered statistically significant

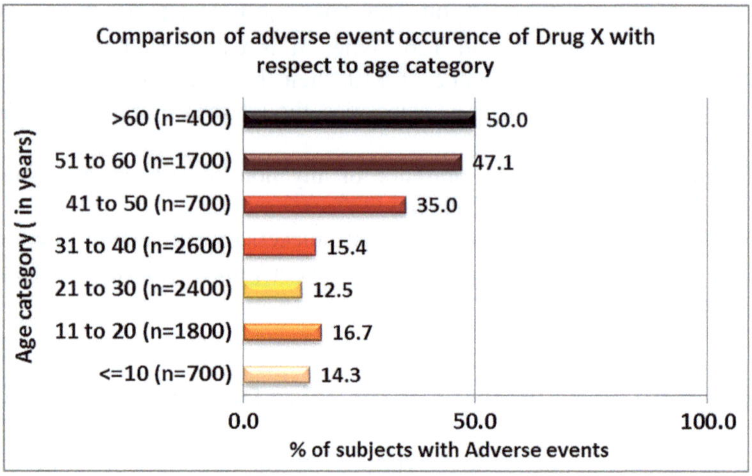

Fig. 63.2 Comparison of adverse event occurrence of drug X with respect to age category (uniformly divided)

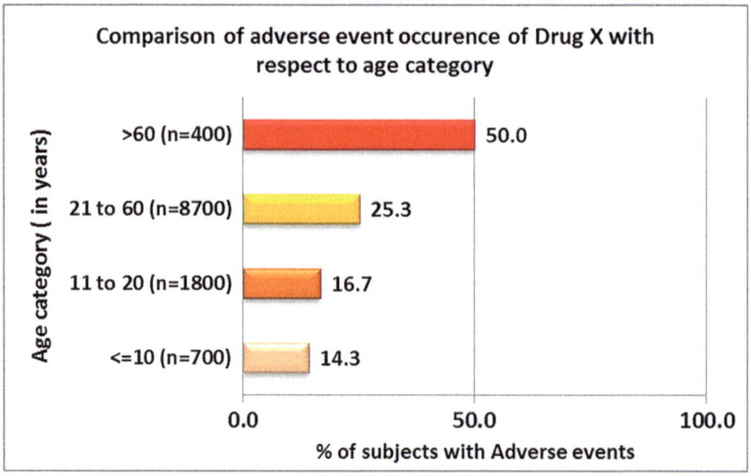

Fig. 63.3 Comparison of adverse event occurrence of drug X with respect to age category (unequally divided)

- In Table 63.1 and Fig. 63.2, the investigator has reasonably divided the age into seven categories with 10 years interval and observed the frequency of adverse events for each category. On the other hand, in Table 63.2 and Fig. 63.3, the investigator categorized age into a bizarre range (clubbing 21–60 years as one group). When the outcome is compared with respect to the age category, both the presentations show a significant difference between the age groups. However, in Table 63.2 and Fig. 63.3, the trend of increase of adverse events with an increase of age could not be identified due to unequal categorization. Additionally, the adverse events are significant from 40 years of age itself when interpreted by using Table 63.1 and Fig. 63.2. Nevertheless, from Table 63.2 and Fig. 63.3, the significant adverse event is noted only above 60 years of age. Hence masking of the contribution of individual data groups to the whole picture can occur due to illogical categorization.

- Thus, as a good practice, uniform categorization of continuous data should be adopted. Even if categorization is done in other ways, then a proper justification should be offered by the researcher in detail.

63.3.3 Unnecessary Data Transformation

- It is often said that 'if you torture the data too much, it will yield.' This indicates that researchers often involve in data transformation to obtain significant results to support their hypothesis. Log transformation is one of the most common methods to convert non-normal to the normal distribution of data.
- Though log-transformation reduces/removes the skewness of original data, the researchers often fail to recheck the skewness of derived data. Often the log-transformed data is also skewed. Using a parametric test for such data results in wrong conclusions.
- Additionally, log-transformed data makes the estimation of certain model parameters (e.g., linear regression) values difficult and inaccurate.
- For parameters with negative values, it is often practiced to add a small constant value uniformly to all data to make them positive and then subjected to log-transformation. Though it seems to be harmless, the addition of 'constant' can significantly influence the magnitude of the *p*-value and result in type I error.
- Non-normally distributed data should be analyzed using the non-parametric test, as they are robust. Log transformation of non-normally distributed data to the normal distribution and using a parametric test should be avoided.

63.4 Errors Related to Statistical Tests

63.4.1 Choosing Correct Statistical Tests

- The interpretation of the *p*-value is proper when the correct statistical test is applied to check the differences between the groups.
- Often, researchers fail to select an appropriate statistical test. E.g., For comparing the means of three independent normally distributed data groups One-way ANOVA with posthoc test should be used. However, the researcher can use the student '*t*' test to compare the two groups in a combination of three times (A vs. B, B vs. C, and A vs. C). This leads to α error inflation because we are testing three times using the student '*t*' test. Each time we test, we set an alpha value of 0.05 limit, and hence the limit increases to 0.15 for testing three times. This decreases the validity of the *p*-value interpretation.

- Similarly, for data with non-normal distribution, appropriate non-parametric tests should be used. Many researchers often fail to test data distribution or unaware of the testing and blindly run the parametric tests.
- Likewise, paired and unpaired data types also require their own statistical tools. This is because the variation between the samples is absent in paired data. Not choosing an appropriate test for paired data can lead to a wrong interpretation of the outcome. E.g., Repeated measures ANOVA should be chosen over One-way ANOVA to compare certain biochemical parameters obtained from the same patient at multiple time points. In addition to this appropriate posthoc test should also be selected for a particular type of ANOVA for obtaining a valid *p*-value.
- Hence selection of appropriate statistical tools should be ensured, mentioned in the protocol, and scrutinized by the peer group before the initiation of the study to avoid errors.

63.4.2 Violating the Assumptions of Statistical Test

- All the statistical tests work based on certain assumptions. The investigator must be aware of those assumptions and check whether all those assumptions are met by the data before subjecting it to the analysis.
- The following are the common violations that often encountered in biomedical research
  - Missing to check for normality distribution of data (most common) while using parametric tests.
  - Unaware about testing for the homogeneity of variance of the data
  - Using the Chi-square test when more than 20% of cells have an expected value/numerical value of <5
  - Not using Yates continuity correction while using Chi-square for a small sample size
  - Using student '*t*' test for unevenly matched group data
  - Not checking for linearity of the data, significant outliers, and homoscedasticity while using Pearson's correlation.
- Hence, to avoid these errors, the researchers must aware and ensure that all the assumptions of the particular statistical test are met before interpreting the *p*-value.

63.5 Errors in Data Representation

63.5.1 Errors in Data Tabulation

- 'All tables and figures should be self-explanatory' is a rule of thumb. Thus all the tables should contain a fitting title, parameters with units, and appropriate footnote.

- Often the name of the statistical test used to calculate the *p*-value is not mentioned in the table's footnote. The table should also contain the test's statistic value, degree of freedom, and sample size for all the rows and columns.
- In addition to this, the data expressed in table form should bring clarity but not confusion. For example, while expressing the age in the table, it is sufficient to express it to one precision unit (62.1 years, 40 years, etc.). Expressing age as 62.45 years, 40.05 years often brings confusion, errors and reduces clarity.
- Similarly, large units like 1,250,000, 2458,00, etc., should be converted to and summarized shortly in thousands or millions.
- The investigator should avoid using too much '*labeling of data'* inside the table with symbols like '$,' '@,' '^,' '#' etc. Often this is encountered in biomedical research to keep up the rule- 'table should be self-explanatory.' Nevertheless, non-judicial use of data labeling and extensive footnote incorporating many explanations for symbols often reduces the clarity. One approach for solving the above issue is by splitting the table.

63.5.2 Errors in Graphs

- Most of the time, graphs are used to achieve easy comprehension of data summarization, the trend of changes, retention of concepts, etc. However, graphs are often misused in research publications.
- The following are the few errors often encountered regarding graphical data representation
  - *Data repetition:* Same data is presented in the graph beside the table in the same article, which is redundant and unscientific.
  - *Unnecessary dimensional effect:* using 3D graphs to express data with two axes is unnecessary and creates confusion. The 3D graphs should be used when data changes in three axes. (Figs. 63.4 and 63.5)

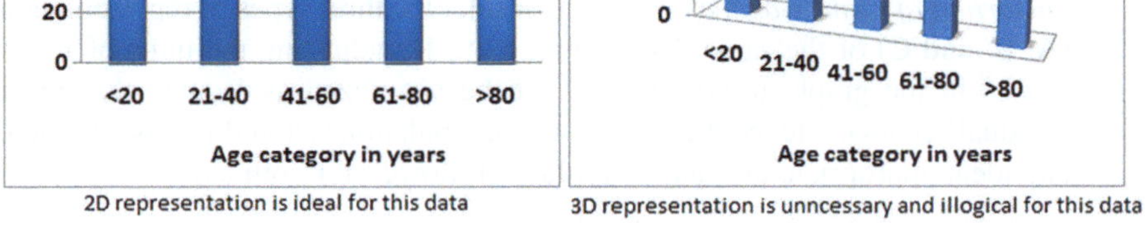

Fig. 63.4 Proper and improper use of 3D graphs in research studies

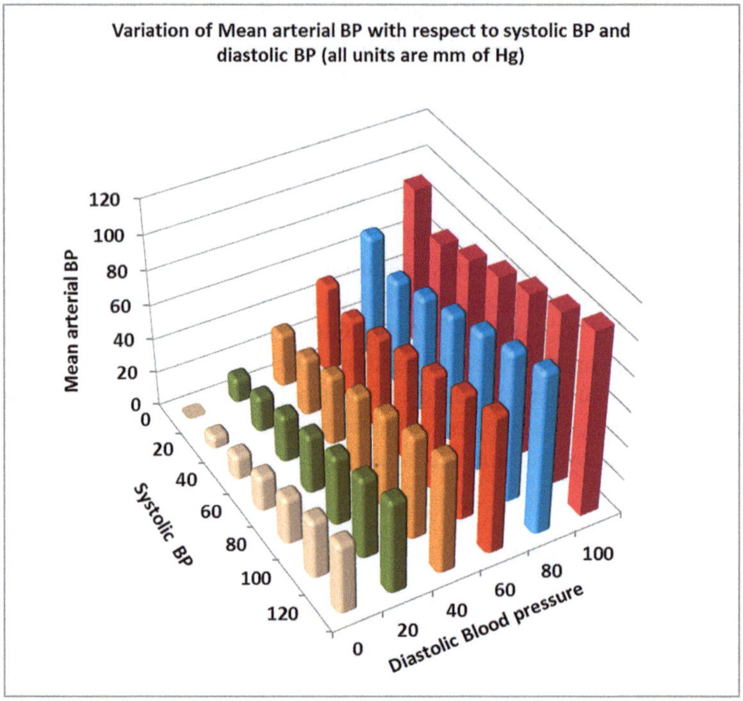

Fig. 63.5 Proper use of a 3D graph in biomedical research. Variation of mean arterial pressure (z-axis-depth) is influenced by the change in systolic blood pressure (x-axis) and change in diastolic blood pressure (y-axis)

- *Histogram and vertical bar diagram-confusion:* Histogram is used for continuous data distribution (no gap between the bars), and the vertical bar diagram is used for categorical data distribution (equal space between the bars). Often these two graphs are used interchangeably, which is a gross error from a statistical point of view (Figs. 63.6 and 63.7)
- *Lost-zero problem:* Often, the test drug's adverse effect is compared with the standard drug by using graphs. However, to make the illusion that test drugs produce fewer ADRs than standard, the baseline value is adjusted (i.e., baseline does not start from zero). In Fig. 63.8, the ADR of the test drug appears to less than half of the standard drug, but after adjusting the baseline to zero, the illusion disappears.
- *Distortion of x and y-axis in graphs:* Consider the three types of representation (A, B, and C) of the same data in Fig. 63.9. By adjusting the units of x and y-axes in the graph, trend changes can be visualized as 'acute' change or 'gradual' change. Hence these types of manipulations should be discouraged, and ideal/appropriate unit gaps should be followed for both axis.

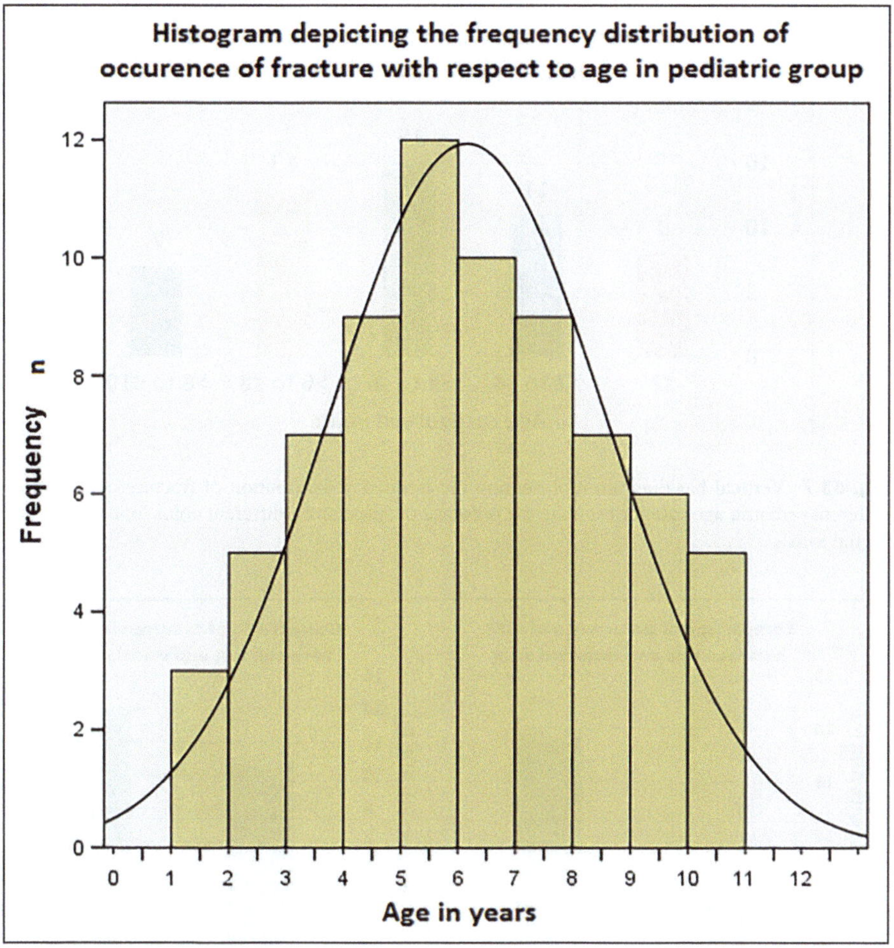

Fig. 63.6 Histogram depicting the frequency distribution of occurrence of fracture with respect to age in the pediatric group. Note the absence of spacing in the X-axis as age in this context is continuous data

- Similar to the table, all the graphs should also be self-explanatory. Hence the graphs should contain the proper title, labeling of axes, units of the parameter, sample size, group name, and proper legend.

63.5.3 Errors in Descriptive Data Summarization

- Data are summarized and described using 'measures of central tendency' like mean and median, and 'measures of dispersion' using standard deviation and interquartile range.

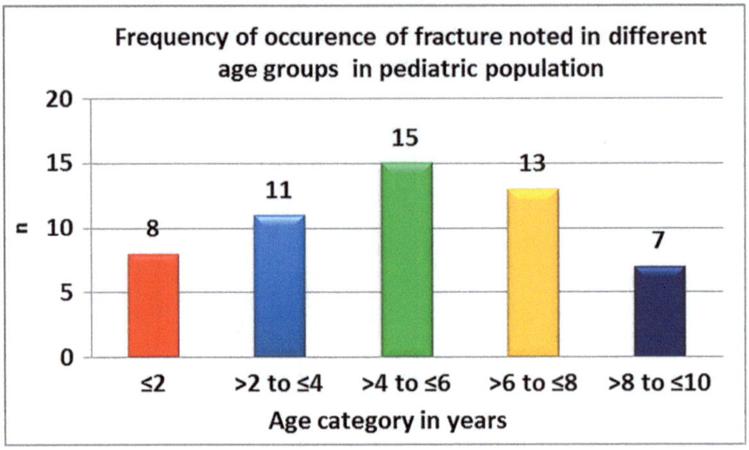

Fig. 63.7 Vertical bar diagram representing the frequency distribution of fracture occurrence in different pediatric age categories. Note the presence of space and a different color for each category on the x-axis

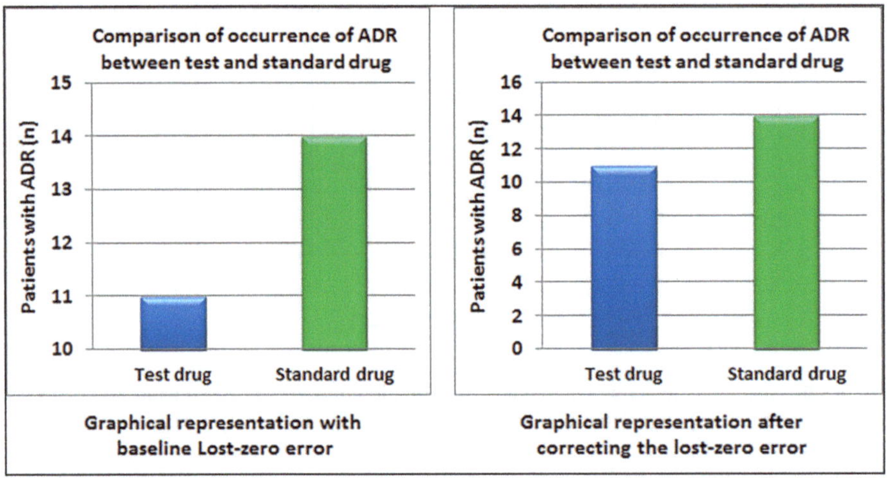

Fig. 63.8 Vertical bar diagram representing the occurrence of ADR in test drug in comparison with standard drug. After correcting the lost-zero error in the baseline, the illusion of 'test drug has significantly lesser ADR than standard drug' is abolished

- The normally distributed data should be summarized using mean with standard deviation. Because the data is not skewed, the mean is the ideal measure of central tendency for normal distribution.
- Due to the presence of outliers, data distribution may be skewed and becomes a non-normal distribution. In such cases, the mean will be affected by 'outliers.' Hence median with interquartile range is used for data summarization.

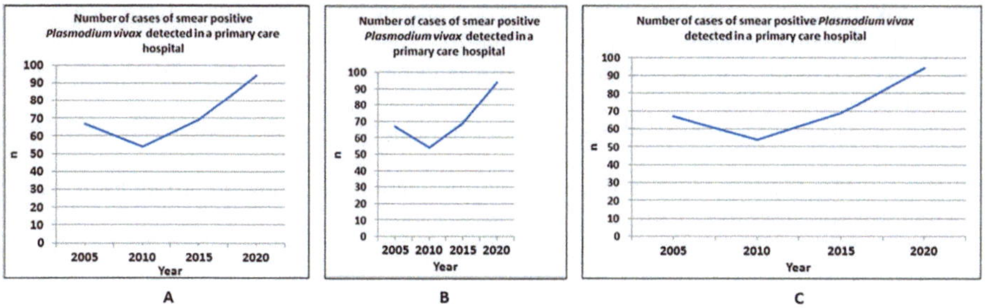

Fig. 63.9 Part '(**a**)' represents the trendline of the number of cases of smear-positive P.vivax detected over 20 years. Manipulation of x and y axes can distort the graph leading to the creation of illusion as an acute rise of incidence (**b**) or a gradual increase of incidence (**c**)

- However, various biomedical research articles did not check the normality distribution and use mean with SD for non-parametric distribution.

63.5.3.1 Errors in *P*-Value Expression

- Often *p* values are expressed simply as '*p* value<0.05 and considered statistically significant' in the result section of many biomedical research papers. Ideally, this statement is not reflecting the magnitude of the '*p*.' In the above statement, the *p*-value could be 0.0495, which is very close to the cut-off range of 0.05 and gives problems in arriving at conclusions.
- Recent statistical guidelines of many journals demand the author to express the absolute *p*-value number (e.g., $p = 0.0024$, etc.) to avoid such confusion. However, when the *p*-value is 0.000025, it is sufficient to express '$p < 0.001$'. Universally, it is accepted that $p < 0.001$ is the least *p*-value that may be reported. For values above this, absolute magnitude should be mentioned.
- In addition to this, the 95% confidence interval of the particular parameter should also be expressed along with the *p*-value. Mentioning the confidence interval of mean change brings clarity regarding its clinical significance besides statistical significance. E.g., the *p*-value statement can be expressed as 'the mean change in the systolic blood pressure observed between drug A and drug B was 16 mm of Hg with 95% CI of 4–41 mm of Hg; $p = 0.0014$* which is statistically significant.'
- Recently, researchers are asked to mention the degree of freedom and corresponding statistic value of tests (*F* value for ANOVA, Chi-square value for Chi-square test, *t* value for student '*t*' test, etc.) along with confidence interval and *p*-value.

Bibliography

Bajwa SJS. Basics, common errors, and essentials of statistical tools and techniques in anesthesiology research. J Anaesthesiol Clin Pharmacol. 2015;31(4):547–53.

Brown AW, Kaiser KA, Allison DB. Issues with data and analyses: errors, underlying themes, and potential solutions. PNAS. 2018;115(11):2563–70.
Lang T. Twenty statistical errors even you can find in biomedical research articles. Croat Med J. 2004;45(4):361–70.
Young J. statistical error in medical research – a chronic disease? Swiss Med Wkly. 2007;13(3): 41–3.
Zinsmeister AR, Connor JT. Ten common statistical errors and how to avoid them. Am J Gastroenterol. 2008;103(2):262–6.

Correction to: Hypothesis Testing

Lakshmi Balasundaram

Correction to:
Chapter 60 in: M. Lakshmanan et al. (eds.),
Introduction to Basics of Pharmacologyand Toxicology,
https://doi.org/10.1007/978-981-19-5343-9_60

The original version of this chapter was inadvertently published with an error in Fig. 60.1, which has been updated with the correct figure as follows:

The updated original version for this chapter can be found at
https://doi.org/10.1007/978-981-19-5343-9_60

M. Lakshmanan et al. (eds.), *Introduction to Basics of Pharmacology and Toxicology*, https://doi.org/10.1007/978-981-19-5343-9_64

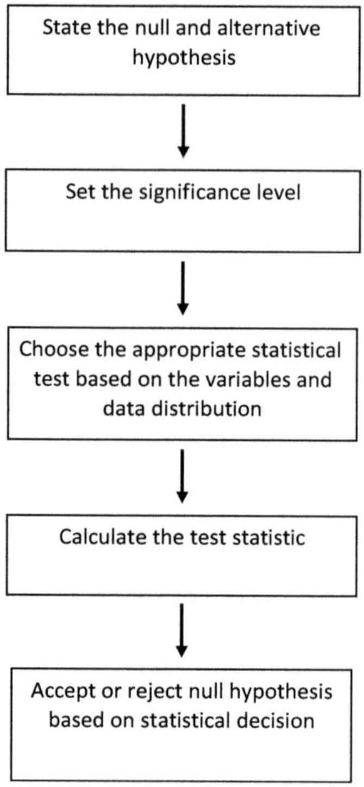
State the null and alternative hypothesis
Set the significance level
Choose the appropriate statistical test based on the variables and data distribution
Calculate the test statistic
Accept or reject null hypothesis based on statistical decision